Early Vertebrates

PHILIPPE JANVIER

Centre National de la Recherche Scientifique
Muséum National d'Histoire Naturelle, Paris

CLARENDON PRESS ● OXFORD

Oxford University Press, Great Clarendon Street, Oxford OX2 6DP

Oxford   New York
Athens   Auckland   Bangkok   Bogota   Bombay
Buenos Aires   Calcutta   Cape Town   Dar es Salaam   Delhi
Florence   Hong Kong   Istanbul   Karachi
Kuala Lumpur   Madras   Madrid   Melbourne
Mexico City   Nairobi   Paris   Singapore
Taipei   Tokyo   Toronto   Warsaw
and associated companies in
Berlin   Ibadan

Oxford is a trade mark of Oxford University Press

Published in the United States
by Oxford University Press Inc., New York

© Philippe Janvier, 1996
Reprinted 1998

A catalogue record for this book is available from the British Library

Library of Congress Cataloging in Publication Data
Janvier, Philippe.
Early vertebrates / Philippe Janvier.
(Oxford monographs on geology and geophysics; 33)
Includes bibliographical references and indexes.
1. Fishes, Fossil.   2. Vertebrates, Fossil.   I. Title.
II. Series. Oxford mongraphs on geology and geophysics; no. 33.
QE851.J36   1996   567–dc20   95–26847

ISBN 0 19 854047 7

Printed in Great Britain by
The Bath Press, Bath

Preface

'*Early vertebrates*', the title of this book, admittedly sounds vague and uninformative by comparison with '*Palaeozoic fishes*', the title of a now-famous book by J. A. Moy-Thomas (1939, revised by R. S. Miles 1971), but I deliberately chose these words to avoid linking groups of animals with particular geological periods, even though most of the species dealt with here are in fact 'fishes' and are of Palaeozoic age. Since the mid-nineteenth century, to most people, 'early' has meant primitive, and my chosen title may be expected to tell something of origins—origins of the vertebrates, of jawed vertebrates, of bony jawed vertebrates, or bony, four-legged jawed vertebrates, etc. I shall indeed allude to questions about origins of structures where the fossils permit, but this book is mostly about relationships: relationships between characters (homologies) and between taxa (that is, characterized and named groups of organisms). Looking for relationships is the only way of depicting diversity with some logic or order, and orderliness is just what is needed to save time when there are many things to learn.

In writing this book, my concern was to do something different from the few other works on the same subject published during the last thirty years or so, in particular Moy-Thomas and Miles's (1971) *Palaeozoic fishes* and Jarvik's (1980–1) *Basic structure and evolution of vertebrates*. The former had a considerable success among students because of its simplicity, and the second was much appreciated by professional biologists and palaeontologists because of the large amount of anatomical data it contains. Both books also deal with relationships among higher taxa, but with different methods and different results. Moy-Thomas and Miles's *Palaeozoic fishes* appeared at the dawn of a major change in comparative biology, the rise of Hennig's (1950) phylogenetic systematics (now known as 'cladistics'), and was the first textbook on early vertebrates in which this method of phylogenic reconstruction is foreshadowed. The power of cladistics is to propose a hierarchy of the homologies, from the most general to the most particular, which defines nested taxa and brings theories of relationships within the reach of refutation. Although agreeing with the cladistic principle of rejecting shared primitive characters in phylogeny reconstruction, Jarvik (1968, 1980, 1981*a,b*) discards the widespread consensus on vertebrate interrelationships. This is because of his transcendental approach to vertebrate structure, i.e. his view is that any character can appear independently from a basic, ideal vertebrate pattern (see Schultze and Trueb 1981). The result is that much argument was needed to refute congruent character distributions that were otherwise clear and obvious to most anatomists. In sum, the choice is between oversimplistic and overcomplicated approaches. I felt that even if simplicity may be the source of mistakes, e.g. by overlooking details that suggest homoplasy, excessive complexity often makes a theory inaccessible to refutation and thus of little use in a scientific debate. I therefore opt for simplicity, even at the cost of reducing the description of anatomical details that I do not regard as crucial to the understanding of the phylogenetic pattern.

In order to make this book of more general use, I have included some classical considerations of 'transformations', i.e. evolutionary morphology, which can be superimposed on patterns of relationships, yet involving a large measure of speculation. Moreover, I have ventured into the realm of 'life history', which is even more speculative.

All in all, I have tried to provide the reader with the essential data for a general survey of early vertebrates. Some illustrations may seem to be redundant (i.e., the same feature may appear twice in different figures), but this is because I have tried to make most of the figures self-contained. For the same reason, some data are repeated in different chapters.

The amount of new information and, above all, the complexity of the discussions on large numbers of characters in modern phylogeny reconstruction have become so great that they are now difficult to account for in detail. Therefore, in most of the cladograms that are illustrated here, only selections of synapomorphies (shared derived character states or unique characters) are cited at each node, and wherever possible I have tried to select the synapomorphies that are illustrated in the figures or are familiar to students of comparative anatomy. The defining characters of the major taxa under consideration are always cited first in the paragraphs on systematics, and a classification (not necessarily the consensus) is given for each major taxon.

For the sake of simplicity, I have avoided mentioning species names, since many of these early vertebrate genera are monospecific. Locality names or geographical distributions of taxa are mentioned only where relevant (endemic taxa, biogeographical comparisons). Citation of authors in the text is kept to the minimum in order to make the book easier to read. Classical works, monographs, and general or controversial papers on the subject are cited as 'further reading' at the end of each section or chapter.

The stratigraphical scale and ages used in this book are taken from Harland *et al.* (1990). The appendix provides references and short comments on important papers that have appeared while the book was in the press.

Paris   P. J.
August 1995

Acknowledgements

I would like to thank all the colleagues who have allowed me to use and redraw their illustrations, and in particular those who provided me with original material (M. Arsenault, W. Bemis, P. Y. Gagnier, L. Grande, J. Long, R. Lund, D. Martill, T. Rowe, and A. Tamarin). I am also indebted to those who have reviewed and edited my manuscript, as well as to colleagues who helped me in finding my way in piles of writings on unfamiliar taxa, in particular Per Ahlberg (London), Alain Blieck (Lille), Mee-Mann Chang (Beijing), Jennifer Clack (Cambridge), Richard Cloutier (Lille), Michael Coates (Cambridge), Peter Forey (London), Pierre-Yves Gagnier (Paris and Edmonton), Daniel Goujet (Paris), Brian Gardiner (London), Daniel Heyler (Paris), Oleg Lebedev (Moscow), Hervé Lelièvre (Paris), John Long (Perth), Richard Lund (New York), John Maisey (New York), Elga Mark-Kurik (Tallinn), Colin Patterson (London), Cécile Poplin (Paris), A. Raynaud (Toulouse), Armand de Ricqlès (Paris), Bobb Schaeffer (Rochester), Hans-Peter Schultze (Berlin), Moya Smith (London), Susan Turner (Brisbane), Sylvie Wenz (Paris), and Min Zhu (Beijing). Although they would certainly disagree with many statements in this book, I want to express my gratitude to my masters in Stockholm and Paris, who introduced me to early vertebrate palaeontology, in particular Erik Jarvik, Hans Bjerring, and the Late Jean-Pierre Lehman, Erik Stensiö, and Tor Ørvig. Equal gratitude goes to those who introduced me to the logic of systematics, namely Daniel Goujet, Gareth Nelson (New York), Colin Patterson, and Pascal Tassy (Paris). I would like to thank Richard Hazlewood for stylistic improvements, Brigitte Chopinet for the labelling, S. Barta, for bibliographic research, and Denis Serrette and Lionel Merlette for the photographs. I would also like to thank the copy-editor, Bruce Wilcock, who put in an enormous amount of work on the typescript.

Figure acknowledgements

Academic Press (Inc.) London Ltd. Figs 3.16A–D (redrawn from Jarvik, 1980, Figs 41, 53); 3.18A (redrawn from Jarvik, 1980, Fig. 238); 3. 22A, B (redrawn from Jarvik, 1980, Figs 236A, 337); 4.75B2 (redrawn from Jarvik, 1980, Fig. 218B); 4.77A2, B1 (redrawn from Jarvik, 1980, Figs 186A, 187A); 4.83D (redrawn from Jarvik, 1980, Fig. 326B); 4.88C1 (redrawn from Jarvik, 1980, Figs 75); 4.90A, C, D, F1 (redrawn from Jarvik, 1980, Figs 86A. 97, 109 112A); 4.91A, (redrawn from Jarvik, 1980, Figs 100, 106A); 4.95B–E, D4 (redrawn from Jarvik, 1980, Figs 156, 170, 171, 174B). Copyright 1980 Academic Press London Ltd.

Alcheringa Figs 4.2A, H1; 4.40D; 4.92B, C; 7.3D; 7.7.

Almqvist & Wiksell, Stockholm Figs 3.2A, B; 3.24D2, 3; 4.59C1; 4.83B2.

The American Association for the Advancement of Science Figs 4.1B2 (redrawn with permission from D. Bardack, *Science*, 254, 701–3, Fig. 4, Copyright 1991 American Association of the Advancement of Science); 4.3A2 (redrawn with permission from D. Elliot, *Science,* 237, 190–2, Fig. 3. Copyright 1987 American Association of the Advancement of Science).

American Journal of Science Figs 4.17C; 7.11B.

The American Museum of Natural History Figs 3.3C2; 3.13D; 3.14F2–J2; 3.15D; 3.20A–C; 3.23D; 4.11B; 4.27A; 4.33E1; 4.34C, D; 4.35C1, D; 4.40B; 4.77D; 4.88C4; 4.89A2; 6.8B2–4.

American Zoologist Figs 4.30D3; 4.33B, C; 4.34B1, 2.

Association of Australasian Palaeontologists Figs 4.64C; 4.65B; 4.92D.

Athlone Press Figs 4.11C; 7.11A2.

The Australian Museum Figs 4.53D; 4.54C; 4.86C; 4.87A; 4.88D, E; 4.92A.

Blackwell, Melbourne Fig. 4.82B.

Cambridge University Press Figs 4.80A, B; 8.1D.

Carnegie Museum of Natural History Fig. 4.32A1.

Cornell University Press Figs 4.94A–C; 4.103D; 6.5C.

Dansk Polar Center Figs 4.31C; 4.52A, B; 4.54H; 4.77B1, C; 4.78A–E; 4.88F; 4,96A, B; 6.5B; 7.10C.

Department of Energy, Mines and Resources, Ottawa Fig. 4.6A.

Editions du CNRS, Paris Figs 2.5A; 3.21D, G–J; 4.4A2; 4.5F, I; 4.6C, D; 4.7A, C; 4.8F; 4.30D1; 4.31B2; 4.41; 4.42A–E; 4.61A2; 4.66E; 4.68B; 4.90B; 7.4.

Estonian Academy of Science Figs 1.17; 4.24B, C; 4.53F1, G; 4.69A; 7.10B; 7.14B2.

E. Schweitzerbart'sche Verlagsbuchhandlung Figs 4.9E1; 4.11D2; 4.30B, C; 4.33A; 4.44C; 4.45A; 4.51A1; 4.52F; 4.56B; 4.62B; 4.63A, B3, C3; 4.65C1; F2; 4.67H; 4.69C; 4.74B; 4.75A, B1; 4.78D; 4.84A; 4.103G; 6.5E; 7.11D; 7.12A; 8.3.

Field Museum of Natural History Figs 4.1A1; 4.3B1; 4.9D; 4.13C1; 4.31D; 4.32A1; 4.38A, B1.

Geobios. Figs 4.13A2, 3, D; 4.33F; 4.76B–D.

Geological Museum (Copenhagen University) Fig. 4.96A, B.

Geological Publishing House, Beijing Fig. 4.20A, B.

Geological Society of Denmark Figs 3.13I; 4.36G2.

Geological Society of Glasgow (for *Scottish Journal of Geology*) Fig. 4.12A2.

GOTAB, Stockholm University Fig. 4.84B1.

Gustav Fischer Verlag Figs 4.30D2; 4.31A, B1, C, F; 4.32A2, B1; 4.33D; 4.36F; 4.40C, D3; 4.64A, D–J; 4.101D.

Ichthyological Society of Japan Fig. 4.36B–D.

John Wiley & Sons Ltd Figs 7.5C; 7.6B; 7.8B. (from Figs 3.1, 2.4, and 2.6 *In* J. A. Long, ed., *Palaeozoic Vertebrate Biostratigraphy and Biogeography*. Copyright 1993. Redrawn by permission of John Wiley & Sons, Ltd.

Journal of Experimental Zoology Fig. 3.15F.

Journal of Morphology Figs 3.6D (redrawn from Hilliard *et al.* 1983, Copyright 1983 Alan Liss, Inc., by permission of John Wiley & Sons, Inc.); 4.81B3, 4.83C (redrawn from Campbell and Barwick 1987, Copyright 1987 Alan Liss, Inc., by permission of John Wiley & Sons, Inc.); 4.83B1 (redrawn from Schultze and Campbell 1987, Copyright 1987 Alan Liss, Inc., by permission of John Wiley & Sons, Inc.); 6.4A2, C2–E2 (redrawn from Schultze 1987, Copyright 1987 Alan Liss, Inc., by permission of John Wiley & Sons, Inc.).

Journal of Paleontology Figs 4.36A; 4.103A; 7.10D; 7.12C.

Kungliga Svenska Vetenskapsakademien, Stockholm (for *Kungliga Svenka VetenskapsAkademiens Handlingar, Acta Zoologica,* and *Arkiv for Zoologi*) Figs 3.6M; 3.16F, H; 4.17A; 4.28; 4.43A; 4.47C; 4.48C; 4.60C; 4.74A, F; 4.75B3; 4.77A5; 4.80F; 4.88A1, 3, B; 4.89A1; 6.3C; 6.7B.

Lethaia Figs 4.65D, E1, 3; 4.67A, B; 7.1B1; 7.11C.

The Linnean Society of London Figs 3.17G; 3.18C-E; 4.27C1; 4.14A; 4.46; 4.47D; 4.48A; 4.51, A3, 4, B; 4.58B; 4.59C2, D; 4.61; 4.63B1; 4.65A, F; 4.74E; 4.79A; 4.80C; 4.81A, B1, 2; 4.90E; 6.4A1–E1; 6.10B.

The Linnean Society of New South Wales Figs 4.19B; 4.44A; 4.45B; 4.56A; 4.83A.

Macmillan Magazines Ltd Figs 4.24E, F (redrawn with the permission from *Nature*, 361, 442–4, Fig. 2A, C, Copyright 1993 Macmillan Magazines Ltd.); 4.95F; 4.96D (redrawn with the permission from *Nature*, 347, 66–9, Fig. 2, Copyright 1990 Macmillan Magazines Ltd.); 4.104 (redrawn with the permission from *Nature*, 365, 346–50, Fig. 2a, Copyright 1990 Macmillan Magazines Ltd.); 7.14B (redrawn with the permission from *Nature*, 238, 469–70, Fig. 2, Copyright 1972 Macmillan Magazines Ltd.).

Macmillan Publishing and Co. Figs 3.8C, D; 3.12A; 3.13E; 3.15G; 3.22D–F; 3.23B; 3.25A, B.

Masson, Paris Figs 3.8A, J; 3.13B, C, G; 3.22G; 3.24E; 4.2B, C, F; 4.11A2; 4.14C.

Meckler Publishing Corporation Fig. 3.10D–F.

Metropolitan Museum of Art Fig. 8.5A.

Moray District Council, Museums Division Fig. 8.1B.

Mosklas, Vilnius Fig. 4.26A, B, D–F.

Museum of Comparative Zoology (Harvard) Figs 3.9B; 3.14N2; 4.102.

Muséum National d'Histoire Naturelle, Paris Figs 4.45E; 4.53A2, B; 4.78F; 4.103C.

National Geographic Society Fig. 4.51C; 4.89B.

New Mexico Bureau of Mines and Mineral Resources Fig. 4.59B.

Norsk Polar Institutt Fig. 4.17B.

Norsk Videnskapsakademi, Oslo Fig. 4.11A1, A3.

Palaeontographica italica Figs 4.74C; 4.86D; 4.88C4; 4.98.

The Palaeontographical Society, London Figs 4.53E; 4.54D, F.

The Palaeontological Association Figs 1.6; 4.7B; 4.8C, H, I; 4.9A; 4.10B, C, D; 4.12A1, D7; 4.34A2, A3; 4.35B; 4.40G; 4.52 C–E; 4.91B; 4.101A–C; 4.103F; 7.9E.

Palaeovertebrata Fig. 4.15A, B, D; 4.16A, B.

Plenum Publishing Corp. Figs 3.4I; 3.5; 3.9E; 4.95A; 6.3A; 6.9A; 6.13; 6.15; 6.16.

Royal Ontario Museum Fig. 4.6B.

The Royal Society of Edinburgh Figs 4.45C; 4.47A; 4.65A; 4.67G; 4.69B, D; 4.86A; 4.87B, C; 4.90F2. (Redrawn by permission of the Royal Society of Edinburgh and S. M. Andrews, R. S. Miles, and D. M. Pearson, from *Transactions of the Royal Society of Edinburgh*: Earth Sciences, 65, 1952, 59–77; 67, 1968, 373–476; 68, 1970, 391–489; 70, 1979, 337–399; 76, 1985, 67–95.)

The Royal Society of London Figs 4.36E; 4.45D; 4.60A, B; 4.96E; 4.99B; 4.100; 7.1B2; 7.3C.

The Royal Society of Victoria Fig. 4.58A.

The Russian Academy of Science (*Nauka, Paleontologicheskii Zhurnal, Priroda*) Figs 1.11, 1; 4.4C1; 4.5H; 4.30A; 4.94D, F; 4.97; 4.103E; 4.97; 7.9B.

Science Press, Beijing Figs 4.12B1, C; 4.20A, B; 4.21E, F; 4.22B; 4.23; 4.24D; 4.26C, E; 4.50B; 4.79B.

Senckenbergische Naturforschende Gesellschaft Fig. 4.75C.

Société Royale Zoologique de Belgique Fig. 6.8A1–A3.

Society of Vertebrate Paleontology Figs 1.18; 3.9F; 3.18E; 4.10A; 4.13A1; 4.28C2; 4.30E; 4.32C; 4.33E; 4.58C; 4.65D3; 4.81C; 4.86B; 4.87D; 4.88C2; 6.1A, B, D; 7.11F; 7.12D; 7.13A; 8.5B.

Springer Verlag Fig. 4.96C.

Swedish Museum of Natural History Figs 3.13H; 3.21A–C; 8.4A.

Trustees of the Natural History Museum, London Figs 2.6; 4.27D1; 4.59C3; 4.61, 4.65C2; 4.66A–D; 4.67D–F; 4.68A; 7.1A; 7.2B, D, F, G.

The University of Chicago Press Fig. 3.2C. (Copyright 1993, The Unversity of Chicago Press)

Vertebrata PalAsiatica, Beijing, Figs 4.19C–E; 4.22C, E–G; 4.53A1; 4.84B2.

Znatne, Riga Fig. 4.53F2.

Elsevier Science Ltd Figs 4.51D (redrawn from Ørvig, 1980, *Zoologica Scripta*, 9, 235–40, Fig. 4); 6.1C (adapted from Bjerring, 1977, *Zoologica Scripta*, 6, 127–83, Fig. 11A); 6.2B (adapted from Bjerring, 1979, *Zoologica Scripta*, 8, 235–40, Fig. 5C), with kind permission from Elsevier Science Ltd, The boulevard, Langford Lane, Kidlington 0X5 1GR, UK.

Zoological Society of London Fig. 3.10A–C; 4.16D; 4.35A1; 6.3B; 7.11E.

Every effort has been made to trace the copyright holders of unadapted material but if any have been inadvertently overlooked the publishers will be pleased to make the necessary arrangement at the first opportunity.

Contents

1 What, where, and when? Early vertebrates as we imagine them   **1**
1.1   The earliest known vertebrates   1
1.2   Silurian vertebrates: the rise of diversity   2
1.3   Devonian vertebrates: the shaping of the major extant groups   5
1.4   Carboniferous vertebrates and the great change   13
1.5   The shaping of the modern vertebrate faunas   15

2 From rocks to theories: techniques of preparation and methods of analysis   **26**
2.1   The search for fossils   26
2.2   Travelling   26
2.3   Collecting   27
2.4   Types of rock and methods of preparation   27
2.5   Tools   30
2.6   Chemicals   31
2.7   Special techniques   32
2.8   Photographs and drawings   34
2.9   Characters   34
2.10   Comparisons   34
2.11   Character analysis and phylogeny reconstruction   36
2.12   Beyond the cladogram   40
2.13   Beyond phylogeny   41

3 A survey of extant vertebrates   **42**
3.1   Vertebrates versus craniates   42
3.2   Hyperotreti or hagfishes   44
3.3   Hyperoartia or lampreys   47
3.4   Gnathostomata, or jawed vertebrates   50
    3.4.1   General features   50
    3.4.2   The two skeletons   51
    3.4.3   Morphology of the endoskeleton   55
    3.4.5   Soft anatomy   58
3.5   Diversity and relationships of Recent gnathostomes   59
    3.5.1   Chondrichthyes   59

3.5.2   Osteichthyes   65
3.6   A cladogram of extant craniates   81

4 Early vertebrates and their extant relatives   **83**
4.1   The earliest known vertebrate remains   83
4.2   Fossil hagfishes (Hyperotreti)   85
4.3   Arandaspida   85
4.4   Astraspida   87
4.5   Heterostraci   89
    4.5.1   Exoskeleton and internal anatomy   90
    4.5.2   Histology   95
    4.5.3   Diversity and phylogeny   95
    4.5.4   Stratigraphical and geographical distribution, habitat   99
4.6   Anaspida   100
    4.6.1   Exoskeleton and external morphology   100
    4.6.2   Internal anatomy   101
    4.6.3   Histology   103
    4.6.4   Diversity and phylogeny   103
    4.6.5   Stratigraphical and geographical distribution, habitat   104
4.7   Fossil lampreys (Hyperoartia, Petromyzontiformes)   104
4.8   Osteostraci   104
    4.8.1   Exoskeleton and external morphology   106
    4.8.2   Internal anatomy   107
    4.8.3   Histology   112
    4.8.4   Diversity and phylogeny   113
    4.8.5   Stratigraphical and geographical distribution, habitat   115
4.9   Galeaspida   115
    4.9.1   Exoskeleton and external morphology   116
    4.9.2   Internal anatomy   117
    4.9.3   Histology   120
    4.9.4   Diversity and phylogeny   120
    4.9.5   Stratigraphical and geographical distribution, habitat   121

4.10 Pituriaspida   122
4.11 Thelodonti   123
        4.11.1 Exoskeleton and external
                morphology   123
        4.11.2 Internal anatomy   124
        4.11.3 Histology   126
        4.11.4 Diversity and phylogeny   126
        4.11.6 Stratigraphical and geographical
                distribution, habitat   127
4.12 Gnathostomata: the characters and
        morphotype   127
        4.12.1 The neurocranial fissures   128
        4.12.2 Jaw suspension and teeth   129
        4.12.3 Gill arch structure   130
        4.12.4 Skeletal histology and structure   131
        4.12.5 Paired fins and girdles   133
        4.12.6 Unpaired fins   134
        4.12.7 Vertebrae and ribs   134
        4.12.8 Sensory-line pattern   134
4.13 Chondrichthyes   135
        4.13.1 Cladoselachidae   137
        4.13.2 Eugeneodontida   137
        4.13.3 Petalodontida   137
        4.13.4 Symmoriida: Symmoriidae and
                Stethacanthidae   139
        4.13.5 Elasmobranchii   141
        4.13.6 Holocephali   144
        4.13.7 Iniopterygia   145
        4.13.8 Phylogeny and classification   147
        4.13.9 Stratigraphical and geographical
                distribution, habitat   148
4.14 Placodermi   150
        4.14.1 General characters   151
        4.14.2 Arthrodira   155
        4.14.3 Petalichthyida   161
        4.14.4 Ptyctodontida   163
        4.14.5 Antiarcha   164
        4.14.6 Rhenanida   169
        4.14.7 Acanthothoraci   170
        4.14.8 Stensioellida and
                Pseudopetalichthyida   171
        4.14.9 Placoderm phylogeny   172
        4.14.10 Stratigraphy, distribution,
                habitat   172
4.15 Acanthodii   173
        4.15.1 Exoskeleton and external
                morphology   174
        4.15.2 Endoskeleton   176
        4.15.3 Histology   177
        4.15.4 Diversity and phylogeny   177

        4.15.6 Stratigraphical and
                geographical distribution,
                habitat   181
4.16 Osteichthyes: fossils and characters   183
4.17 Actinopterygii   183
        4.17.1 General morphology and
                exoskeleton   184
        4.17.2 Endoskeleton   185
        4.17.3 Histology   187
        4.17.4 Diversity   188
        4.17.5 Phylogeny   188
        4.17.6 Stratigraphical distribution and
                habitat   193
4.18 Sarcopterygii   194
        4.18.1 Onychodontiformes   198
        4.18.2 Actinistia   199
        4.18.3 Dipnomorpha   200
        4.18.4 Rhizodontiformes   214
        4.18.5 Osteolepiformes   216
        4.18.6 Panderichthyida   224
        4.18.7 Tetrapoda   225

5   **Interrelationships of the major craniate
    taxa: current phylogenetic theories and
    controversies**   237
5.1   Interrelationships of the Craniata   237
        5.1.1 Monophyly versus paraphyly
               of the Cephalaspidomorphi   239
        5.1.2 The Pteraspidomorphi   240
        5.1.3 A cladogram of the Craniata   241
5.2   Interrelationships of the Gnathostomata   241
        5.2.1 Relationships of the Placodermi   241
        5.2.2 Relationships of the Acanthodii   245
5.3   Interrelationships of the Sarcopterygii   246
        5.3.1 Sarcopterygian monophyly   247
        5.3.2 Rhipidistian monophyly   247
        5.3.3 Choanate monophyly   248
5.4   Craniate classification   250

6   **Anatomical philosophy: homologies,
    transformations, and character
    phylogenies**   252
6.1   The segmented vertebrate   252
        6.1.1 The three historical periods of
               the segmental theory   252
        6.1.2 How many head segments?   254
        6.1.3 A segmented snout?   254
        6.1.4 The anatomy of a segmentalist
               theory   255

6.2 Fossils and segmentalism 257
   6.2.1 The earliest heads 257
   6.2.2 Visceral contributions to the braincase 257
   6.2.3 The origin of jaws 258
   6.2.4 The intracranial articulation 258
6.3 The paths to dermal bone homologies 261
6.4 Monorhiny versus diplorhiny 263
6.5 Nostrils, tear ducts, and choanae 264
6.6 The origin of the tetrapod limbs 266
6.7 Homologies in the labyrinth 270
6.8 The origin of the middle ear 272
6.9 Trends in brain morphology 273
6.10 Evolution of the hard tissues: polarities and transformations in histological structures 274
   6.10.1 The endoskeleton 274
   6.10.2 The exoskeleton 275
6.11 Organization of the exoskeleton 278

7 Evolution, palaeobiogeography and life history 281
7.1 Origins 281
   7.1.1 Origin of the Craniata 281
   7.1.2 Origin of the Gnathostomata 284
   7.1.3 Origin of the Tetrapoda 285
7.2 Early vertebrates and evolutionary processes 285
7.3 Selection among higher groups 287
7.4 Biodiversity and mass extinction 288
7.5 Environmental constraints and evolutionary radiations 291
7.6 Survival and living fossils 292
7.7 Marine versus fresh water 293
7.8 Historical palaeobiogeography 294
7.9 Diet 297
7.10 Organic matter in early vertebrates 300
7.11 Biomechanics or functional morphology 300
7.12 Individual variation 303
7.13 Growth 303
7.14 Sexual dimorphism and behaviour 305
7.15 Trace fossils 306

8 A short history of research on early vertebrates 308
8.1 Early vertebrates as organisms 308

8.1.1 Early vertebrates and Scotsmen 308
8.1.2 Louis Agassiz 311
8.1.3 Early vertebrates, the Russians, and the Baltics 313
8.1.4 Thomas H. Huxley and the time of evolution 314
8.1.5 Erik Stensiö and the time of the endoskeleton 315
8.1.6 Gareth Nelson and the time of cladistics 317
8.1.7 Present-day centres and perspectives in early vertebrate studies 318
8.2 Explorations, geology, and early vertebrates 319
   8.2.1 The Arctic 320
   8.2.2 Russia and Siberia 321
   8.2.3 China and Indochina 321
   8.2.4 North America 322
   8.2.5 Africa 323
   8.2.6 Australia and Antarctica 323
   8.2.7 The Middle East, India 323
   8.2.8 South America 325
   8.2.9 New discoveries in Europe 325

9 Epilogue 326
9.1 On the large amount of data 326
9.2 On success in phylogenetics 327
9.3 On 'odd phylogenies' 327
9.4 On new data, new techniques, new methods 330
9.5 On relationships with other disciplines 332

References 333

Appendix 361

Author index 365

Index of genera 371

Subject index 375

1

What, where, and when?
Early vertebrates as we imagine them

Most fossils of what I refer to here as 'early vertebrates' are about 470 to 250 million years old. They are unattractive, often fragmentary or crushed, and are preserved in very hard rocks. Amateur palaeontologists therefore prefer them when reconstructed, as museum displays, drawings, or paintings, and scientists like to see them in textbooks as line drawings illustrating transformation series. Such reconstructions are the ultimate result of a lengthy process of collection, preparation, comparison, and interpretation. One of the purposes of this book is to show how this work was—and still—is done, and how scientists have struggled with these miserable remains over more than a century and a half to integrate them into the classification of the vertebrates and into the evolutionary history of life.

I suspect that if I had begun by showing the usual 'rough material' of early vertebrate palaeontology—crushed carapaces or squashed slug-like imprints—the reader would have gone no further. I therefore, prefer to begin by depicting—if in a rather impressionistic way—some aspects of early vertebrate life at different times in the Palaeozoic and in different environments (see Fig. 1.23). The reconstructions of these animals (essentially fishes) are of course subject to future modifications, but most of those shown here are now fairly well known and should not change radically. I am conscious that such reconstructions are naïve and rooted in a nineteenth-century tradition, which in turn derives from biblical views of the creation, as pointed out by Rudwick (1992), but at least they show the state of knowledge, or of the collective scientific imagination, at a given time. The risk is that this practice may ossify our way of looking at the past, but the great propensity of present-day students to contradict the opinions of their peers should lead to these views being fundamentally modified in the future.

Some of the fossil-bearing localities mentioned in this chapter are famous; others are less spectacular. I have chosen them either because they are unique or because I have studied them and their environment is familiar to me.

1.1 THE EARLIEST KNOWN VERTEBRATES

The earliest known vertebrates appear at the beginning of the Ordovician period, about 470 million years ago (470 Ma). Some of them are abundant in the Ordovician shales and sandstones of the Precordilleran area of Bolivia. They are found in association with marine invertebrates, such as trilobites and, in particular, lingulid brachiopods (lampshells), which are the usual companions of early vertebrate fossils. These Ordovician vertebrates are very simple in shape, with a fusiform head covered with large bony plates and a tail covered with rod-shaped scales (Fig. 1.1). They are devoid of jaws and their mouth is armed only with thin rows of bony platelets, which formed a scoop for scraping the muddy bottom of the sea. The eyes were situated at the tip of the head, flanking a pair of nostrils. On the top of the head were two pineal 'eyes' and, laterally, there was a long series of gill openings. These animals had no paired fins and no other median fin than a caudal fin (for details see Chapter 4, p. 85).

The form reconstructed here (Fig. 1.1) in its environment is called *Sacabambaspis*, after the village of Sacabamba, near Cochabamba in Bolivia, where the first specimens were discovered by Gagnier in 1986. *Sacabambaspis* is very similar to another animal, *Arandaspis*, discovered earlier in Ordovician sandstones in Australia. Both genera are therefore now included in the taxon Arandaspida. *Sacabambaspis* lived in the sea, probably in shallow waters near the coast, where algae are most abundant. When found in the field, specimens of *Sacabambaspis* are accumulated in a way that suggests sudden death: they are almost complete, lying side by side with nearly the same orientation (Fig. 1.2). It is noteworthy that lampshells, which are constantly associated with them in large numbers, are known to tolerate a wide range of variation in salinity and are among the last invertebrates to survive when a sudden flow of fresh water is poured into the sea. The fact that lampshells died in such large numbers suggests that *Sacabambaspis*, too, was killed by a sudden change in salinity.

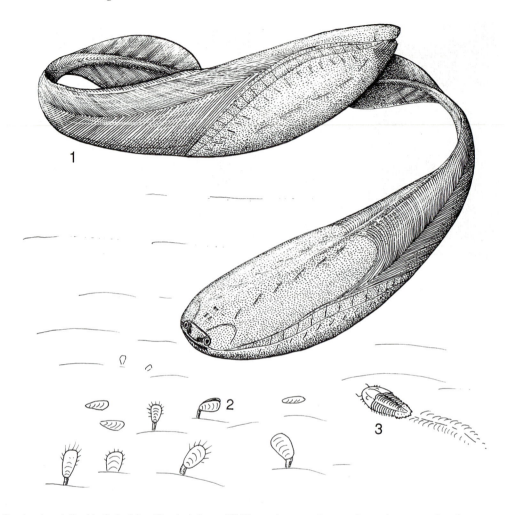

Fig. 1.1. (1) *Sacabambaspis* lived in Ordovician (Caradoc) times, 450 Ma ago in a near-shore marine environment, where it occurs in association with lingulids (2) and trilobites (3). (Reconstruction based on data from Gagnier (1993*a*).)

Ordovician vertebrate localities are very rare. Arandaspids occur in Bolivia and Australia; somewhat different forms, the Astraspida, occur in North America (see Chapter 4, p. 87). They are all jawless, fish-like animals, and are devoid of paired fins. The carapace is made up of a kind of bone that lacks cells. These Ordovician vertebrates thus display a very low diversity in overall aspect. However, jawed vertebrates may already have existed by this time, though nothing has been recorded with certainty.

1.2 SILURIAN VERTEBRATES: THE RISE OF VERTEBRATE DIVERSITY

Near the end of the Ordovician period, a glaciation occurred, though it was probably less severe than the Pleistocene one. An ice cap with glaciers covered most of Africa (which at that time was near the south pole), and the retention of a large quantity of water on the continents in the form of ice caused a fall in sea level. The effect of the cold climate on the marine faunas is not clear, but a

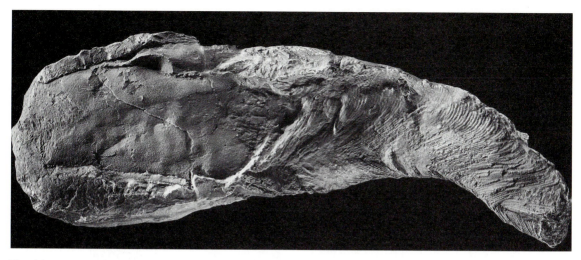

Fig. 1.2. One of the earliest known vertebrates, *Sacabambaspis*, from the Ordovician of Bolivia. This articulated head armour and tail is the most complete Ordovician vertebrate specimen known to date (×0.7). (From Gagnier (1993*a*).)

reduction in species diversity is apparent in the coastal areas, possibly as a consequence of eustatic sea-level changes (see Chapter 7, p. 290). After this glaciation, in early Silurian times (about 435 Ma), vertebrate life suddenly boomed in the sea. Many groups of jawless and jawed vertebrates appeared that were unknown earlier, such as heterostracans (often known as 'pteraspids'), osteostracans (also known as 'cephalaspids'), galeaspids, thelodonts, acanthodians, shark relatives (chondrichthyans), placoderms (armoured jawed fishes), and perhaps also bony fishes (osteichthyans).

It is difficult to find a single outcrop of Early Silurian rocks that displays a wide variety of these vertebrate groups. However, we may take as an example the Silurian dolomites of Saaremaa Island, in Estonia (Fig. 1.3). Saaremaa, a low-lying island covered with heath, is made up exclusively of Silurian dolomites and limestones, which have been quarried and used since the nineteenth century for building stone throughout the Baltic.

The environment in which these rocks were deposited was that of a shallow coast, below the tidal zone, in a warm sea. Coral reefs grew not far away, and a rich invertebrate life developed on both muddy and hard substrates. Brachiopods, ostracods, large crustaceans such as ceratiocarids (now extinct), and eurypterids ('sea scorpions') were among the most common animals here. The vertebrates are represented by four types of jawless fishes: the heterostracans (Fig. 1.3, 2), which are relatives of the Ordovician arandaspids and astraspids; the osteostracans (Fig. 1.2, 4–9); the anaspids (Fig. 1.3, 3, possibly close

relatives of modern lampreys); the thelodonts (Fig. 1.3, 1); and two types of jawed vertebrates, the acanthodians (Fig. 1.3, 10) and actinopterygians (Fig. 1.3, 11). The last-named are the earliest known representatives of this large group of jawed vertebrates, which includes the modern teleost fishes, such as herring and cod. The acanthodians are of debated affinity, though possibly related to bony fishes. They became extinct in Early Permian times (about 260 Ma ago). Heterostracans and Osteostracans lived near the shore, probably in lagoons protected from the waves, whereas anaspids, thelodonts, and jawed fishes were pelagic animals, and were able to live in the open sea.

The osteostracans of Saaremaa are small. They all belong to a particular group, the Thyestiida. These were formerly believed to be primitive osteostracans, because of their geological age, but are now known to be highly specialized forms within this group. The shape of the head ranges from that of a horseshoe to that of an olive. A genus in the former category, *Procephalaspis* (Fig. 1.3, 4), is a primitive member of this local group and had paired fins. Genera with olive-shaped heads, such as *Tremataspis* (Fig. 1.3, 8) or *Oeselaspis* (Fig. 1.3, 9) are the most derived forms and have lost the paired fins, possibly as an adaptation to a burrowing mode of life. But all of them share the characteristics of the osteostracans: the peculiar 'fields' on the dorsal surface of the head-shield (possibly electric or dynamosensory organs), and the keyhole-shaped median and dorsal 'nostril' (the naso-hypophysial opening). The mouth and gill openings of

Fig. 1.3. A typical Silurian (Wenlock–Ludlow) vertebrate fauna from Saaremaa Island, Estonia. This 420-million-year-old fauna is dominated by jawless fishes, such as thelodonts (1, *Phlebolepis*), heterostracans (2, *Tolypelepis*), anaspids (3, *Rhyncholepis*), and osteostracans (4, *Procephalaspis*; 5, *Witaaspis*; 6, *Thyestes*; 7, *Dartmuthia*; 8, *Tremataspis*; 9, *Oeselaspis*). Jawed fishes are also present, in particular acanthodians (10, *Nostolepis*) and possibly also the earliest known osteichthyans (11, *Andreolepis*, known only from isolated scales and teeth). All these fishes lived in a marine environment, but osteostracans were more confined to lagoonal, shallow-water facies, whereas acanthodians and thelodonts were probably epipelagic. (Based on Janvier (1985*b*) and Märss (1982).)

osteostracans are situated ventrally, and their tail is upwardly directed, as in jawed vertebrates.

Heterostracans were rare in this type of environment, and were represented by rather generalized members of the group, such as *Archaegonaspis* or *Tolypelepis* (Fig. 1.3, 2), the carapace of which looks as if it were made up of fused scales. The carapace consists of separate plates, including large median dorsal and ventral shields, and is pierced laterally by a single branchial opening on each side (the character of the group).

Anaspids, such as *Rhyncholepis* (Fig. 1.3, 3) were also rare, but were not armoured like the osteostracans and heterostracans. Their heads were covered with minute scales and a few larger plates, and their bodies with elongated scales arranged in chevrons. Their tails were tilted downward, with a large dorsal fin web. With their presumed nasohypophysial opening in a dorsal position, and a row of gill openings arranged in a slanting line, anaspids looked more or less like stout-bodied and scaly lampreys.

Thelodonts, such as *Phlebolepis* (Fig. 1.3, 1) were entirely covered with minute scales resembling superficially the placoid scales of modern sharks. They had paired flaps overlying the gill openings, and, like anaspids, a posteriorly tilted caudal fin.

Acanthodians looked like small, spiny sharks. They had a long bony spine in front of each fin except the caudal one. Two groups of acanthodians were common in the Silurian of Saaremaa: the climatiids, like *Nostolepis* (Fig. 1.3, 10), with broad-based spines and small teeth, and the ischnacanthids, like *Gomphonchus*, with slender spines and a series of teeth attached to a massive jawbone.

The bony fishes *Lophosteus* and *Andreolepis* (Fig. 1.3, 11) doubtfully referred to actinopterygians (ray-finned fishes), are still known only from isolated scales, teeth, and bone fragments, and cannot be reconstructed in detail. In this fauna, they are probably the only fishes that belong to a group still living. A few million years later, in Early Devonian times, the number of forms belonging to major extant groups increased rapidly in fossil faunas. The shaping of our major modern vertebrate taxa seems to have taken place by the end of Silurian times.

1.3 DEVONIAN VERTEBRATES: THE SHAPING OF THE MAJOR EXTANT GROUPS

In mid-Silurian times, about 420 Ma, plate movement led to the rise of mountain ranges in several places, such as Scandinavia. During the subsequent twenty million years, the erosion of these ranges resulted in the deposition of large amounts of sand on the margins of the continents. Because of warm climatic conditions, this sand often gained a red colour and became consolidated into sandstone, known in northern Europe as the 'Old Red Sandstone'. The vast flood-plains, deltas, lagoons, and tidal flats that extended over these sandy areas seem to have been favourable to the development of vertebrate life, as is indicated by the numerous vertebrate localities in the 'Old Red Sandstone' of Britain, Spitsbergen, Greenland, North America, and other places where similar rocks occur, such as China, Australia, Russia, and the Middle East.

An example of this type of vertebrate assemblage is provided by the fauna of the Old Red Sandstone of Spitsbergen, more precisely the Wood Bay Formation, dated as Pragian (about 400 Ma). In the north of Spitsbergen there are extensive outcrops of unmetamorphosed red sandstone that were deposited at the margins of an almost barren continent. A variety of fishes are now found as fossils throughout the entire sandstone series, with some particularly dense accumulations at given levels, in association with lampshells (Fig. 1.4). Many of these fishes belong to the same groups as those in Saaremaa: heterostracans, osteostracans, thelodonts, and acanthodians. They are in many instances larger and more diverse, but there are also new groups such as the sarcopterygians and the placoderms (Fig. 1.5).

Among heterostracans, large forms, like *Gigantaspis* or *Zascinaspis* (Fig. 1.5, 5), lived together with strange variants like *Doryaspis* (Fig. 1.5, 6), all with the same basic structure and plate arrangement. The osteostracans also included large forms, such as *Parameteoraspis* (Fig. 1.5, 11), as well as minute ones, like *Norselaspis* (Fig. 1.5, 7), *Gustavaspis* (Fig. 1.5, 8), *Boreaspis* (Fig. 1.5, 10), and *Belonaspis* (Fig. 1.5, 9), with rather different aspects. Thelodonts (Fig. 1.5, 13) and acanthodians (Fig. 1.5, 3) were very much the same as in Saaremaa, but anaspids had disappeared from the scene. Placoderms (Fig. 1.5, 1, 2) were then the most diverse group. They were jawed vertebrates, possibly related to

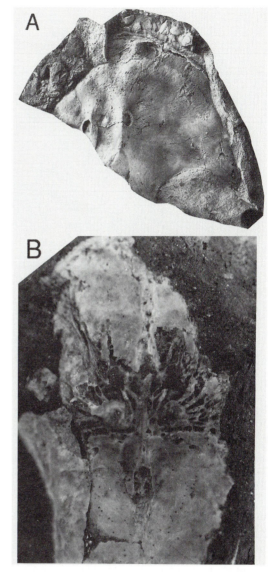

Fig. 1.4. Heads of two osteostracans from the Early Devonian red sandstones of Spitsbergen. *Parameteoraspis* (A, ×0.3) is one of the largest forms, but details of the internal anatomy are provided by the smallest forms, such as this acid-prepared specimen of *Norselaspis* (B, ×5), with the brain cavity, internal ear, and cranial nerves exposed in ventral view. (From Janvier (1981*b*, 1985*a*).)

sharks, and armoured with two sets of bony plates covering the head and the body respectively. Instead of true teeth, their jaws were armed with cutting or crushing

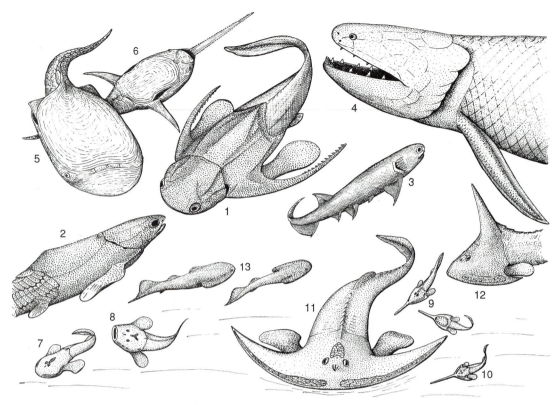

Fig. 1.5. Early Devonian fishes from the Old Red Sandstone of Spitsbergen (Wood Bay Formation, Pragian, about 400 Ma). Only 20 million years have passed since the scene illustrated in Fig. 1.3, but this fauna displays a remarkable number of jawed fishes, such as the placoderms (1, *Dicksonosteus*; 2, *Sigaspis*), acanthodians (3, *Mesacanthus*), and close relatives of extant lungfishes, the porolepiforms (4, *Porolepis*). Jawless fishes nevertheless remain fairly abundant and diverse, with the same major taxa as in the Silurian, i.e. heterostracans (5, *Zascinaspis*; 6, *Doryaspis*), osteostracans (7, *Norselaspis*; 8, *Gustavaspis*; 9, *Belonaspis*; 10, *Boreaspis*; 11, *Parameteroraspis*; 12, *Machairaspis*), and thelodonts (13, *Turinia*). As a whole, this fauna differs from most Silurian ones by the large size of some species (1, 4, 11) which could reach about a metre in length. Although they have long been regarded as freshwater, these fishes associated with the Old Red Sandstone facies are now believed to have lived in either marine, near-shore environments or deltas. (Reconstructions based on Blieck and Goujet (1983); Goujet (1973, 1984*a*), Heintz (1968), Janvier (1985*a*), and Jarvik (1980).)

bony plates. Most placoderms from Spitsbergen belong to the arthrodires, such as *Arctolepis* or *Dicksonosteus* (Fig. 1.5, 1), which were of moderate size. It is probable that when they first appeared in the sandstone series of Spitsbergen placoderms already had a long history of development elsewhere, for we know that they were already present, though rare, in the Early Silurian of China.

Among bony fishes, typical actinopterygians, or ray-finned fishes, are still scarce, but another group, the sarcopterygians, is represented in Spitsbergen by such large predatory forms as *Porolepis* (Fig. 1.5, 4). Sarcopterygians, or lobe-finned vertebrates, are 'our'

group, since they include the tetrapods (all four-limbed vertebrates). In Early Devonian times, they were mostly represented by primitive lungfishes or relatives of that group. These were rather large fishes, reaching a metre or so in length, with slender pectoral fins and scales covered with a shiny enamel layer. Their teeth were pointed and, in section, display a strongly infolded structure, which later persists also in the teeth of the early land vertebrates.

Let us now turn to a different fauna, or community, of the same age as the Spitsbergen one but in a different part of the world. In the north of Vietnam, between Hanoi and Langson, sugar-loaf-shaped mountains of Carboniferous and Late Devonian limestone rest on Early Devonian red

Fig. 1.6. Head-shield of *Polybranchiaspis* (×1), a galeaspid from the Early Devonian of Tong Vai, northern Vietnam. (From Tong-Dzuy *et al.* (1995).)

sandstone and shales that contain a rich vertebrate fauna (Fig. 1.7). The composition of this fauna, which also occurs widely in southern China, differs markedly from those of Europe in the same period. Among jawless vertebrates, heterostracans and osteostracans are lacking, but instead there is a variety of species of galeaspids, a group that is unknown outside China and Vietnam (Fig. 1.6). Among placoderms arthrodires are comparatively rare, but there are many different types of antiarchs, a group of placoderms that does not occur elsewhere until much later. In contrast, sarcopterygians are represented by primitive lungfishes and *Porolepis*-like forms, and the acanthodians are not much different from the European forms.

Galeaspids such as *Polybranchiaspis* (Fig. 1.7, 8) or *Bannhuanaspis* (Fig. 1.7, 9) have a flattened head-shield vaguely resembling that of osteostracans, but they lack the dorsal 'fields', have widely separated eyes, and in the place of the nasohypophysial opening there is a large median hole, which served both for olfaction and the inhalation of respiratory water. As in osteostracans, the mouth and gill openings (often very numerous) were on the ventral side of the shield.

The antiarchs are peculiar placoderms whose eyes and nostrils are situated on the top of the head, and whose pectoral fins are transformed into articulated appendages covered with bony platelets (Fig. 1.7, 4–7). These characters are now believed to be an adaptation to burrowing habits. The Early Devonian antiarchs from China and Vietnam are the earliest known examples. They display a variety of types of articulation of their pectoral fins, which correspond to successive stages in the evolution of this specialized feature.

Among sarcopterygians, the early lungfishes, like *Diabolepis* (Fig. 1.7, 3), have the typical tooth plates of the later members of the group, an adaptation to a durophagous diet. *Youngolepis* (Fig. 1.7, 1), another sarcopterygian in this fauna, resembles *Porolepis* from Spitsbergen, although it seems more closely related to lungfishes.

Why is this Asian fauna so peculiar, or so different from that of Europe in the same period? There are several explanations, including endemism due to isolation of the Chinese continental blocks, but more likely is a scenario that involves the presence of an almost closed sea or gulf over most of this area during Silurian and Devonian times (see Chapter 7, p. 296). This 'Chinese' fish fauna would thus have been a relict of a once more widespread vertebrate fauna, 'trapped' in a confined area.

Later, by the beginning of Middle Devonian times, shallow seas began to invade the continents all over the world, bringing with them a rich marine life with many fish groups. The jawless vertebrates, which were essentially bound to deltas, lagoons, and sandy tidal flats, then tend to disappear. A good example of these new vertebrate faunas dominated by placoderms and osteichthyans is the locality of Lethen Bar, near Nairn in Scotland, which has been famous since L. Agassiz's (1844–5) early works on the fishes of the Old Red Sandstone. Lethen Bar, an outcrop of marls and sandstones in a quarry, is placed in the Middle Old Red Sandstone, at the Eifelian–Givetian boundary (380 Ma). Geologically, it belongs to the Orcadian basin, which has long been believed to have been occupied in Devonian times by a large lake covering most of northern Britain. However, recent reconsiderations of its Devonian faunas, which include many widespread taxa, suggest that the 'Orcadian Lake' was in fact connected to the open sea to the south. The large concretions found at Lethen Bar have yielded a rich fish fauna comprising placoderms, actinopterygians, and numerous sarcopterygians (Fig. 1.8). Among placoderms, the largest forms are the arthrodires, such as *Coccosteus* (Fig. 1.8, 2), but the antiarch *Pterichthyodes* (Fig. 1.8, 1), and the ptyctodont *Rhamphodopsis* are also present. Acanthodians are still abundant, with *Cheiracanthus* and *Diplacanthus* (Fig. 1.8, 3). Actinopterygians are represented by a large form, *Cheirolepis* (Fig. 1.8, 4), and sarcopterygians by the lungfish *Dipterus* (Fig. 1.8, 5) and various members of extinct groups that are thought to be related either to lungfishes (*Glyptolepis*) or to the land vertebrates (*Osteolepis*, Fig. 1.8, 6, *Thursius*, *Gyroptychius*, and *Tristichopterus*).

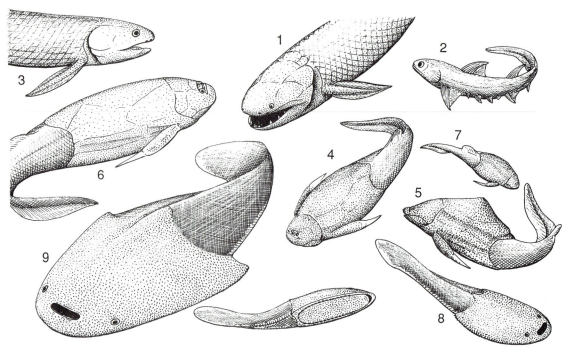

Fig. 1.7. An Early Devonian fish fauna in the 'Chinese realm': the 400-million year-old fauna of the Bac Bun Formation of Vietnam (Lochkovian–Pragian). This fauna is contemporaneous with that of Spitsbergen, but most of its components are different. Only the dipnomorph *Youngolepis* (1) and the acanthodian *Nostolepis* (2) closely resemble the porolepiforms and acanthodians in the Spitsbergen fauna (Fig. 1.5). In contrast, there are true lungfishes (3) and a number of antiarchs (4, 5, *Yunnanolepis*; 6, *Chuchinolepis*; 7, *Vanchienolepis*), which are unknown in Europe at the same period. In addition, there is a group of jawless fishes, the galeaspids, which is unique to this part of Asia (8, *Polybranchiaspis*; 9, *Bannhuanaspis*), whereas the classical European, Siberian, and North American taxa (heterostracans and osteostracans) are lacking. These fishes lived in coastal lagoons or shallow marine waters, in association with various marine invertebrates (e.g. lampshells). (Reconstructions based on Janvier *et al.* (1993) and Tong Dzuy and Janvier (1990).)

At Miguasha, a village in the Chaleur Bay, in the east of Quebec, cliffs of sandstone and thin-bedded siltstone, the Escuminac Formation, extend along the shore (Fig. 1.9), yielding a great variety of Late Devonian fishes dated as Frasnian (370 Ma). The rocks of the Escuminac Formation have long been considered as having been deposited in a lake, but as for the Scottish localities, faunal comparisons and geochemical analyses have shown that they were formed in an estuary discharging into a shallow marine gulf. The environment of the Escuminac Formation was admittedly not truly marine, as is shown by the abundance of plant remains and the scarcity of invertebrates, which are mostly represented by scorpions and related eurypterids and by some crustaceans. In contrast, the vertebrates display a great diversity (Fig. 1.10), which may well be due to the excellent conditions of preservation. Most of the fishes are jawed fishes: acantho-

dians, placoderms, and bony fishes, but sharks are lacking (again an indication of distance from the open sea, since sharks are abundant in typically marine sediments of the same period). There are still some survivors of the armoured jawless vertebrates in the form of osteostracans (*Escuminaspis*, Fig. 1.10, 10). Other jawless vertebrates (*Endeiolepis*, *Euphanerops*: Fig. 1.10, 11, 12) look like anaspids, though devoid of scales, and may be close relatives of lampreys.

The placoderms of Miguasha are represented by extremely abundant antiarchs (*Bothriolepis*, Fig. 1.10, 7) and an arthrodire (*Plourdosteus*, Fig. 1.10, 8). There are many acanthodians (Fig. 1.10, 9), which grew to a large size, one actinopterygian (*Cheirolepis*, Fig. 1.10, 6), and many sarcopterygians. The last-named include lungfishes (*Scaumenacia*, Fig. 1.10, 3) and their porolepiform relatives (*Holoptychius*, Fig. 1.10, 2), a coelacanth

Fig. 1.8. Fishes of the Middle Devonian locality of Lethen Bar, in Scotland (Givetian, about 377 Ma). They show the predominance of the jawed vertebrates, in particular the placoderms (1, *Pterichthyodes*; 2, *Coccosteus*) and acanthodians (3, *Diplacanthus*) with an increasing number of representatives of extant major taxa, such as the ray-finned fishes (4, *Cheirolepis*), lungfishes (5, *Dipterus*) and a group of lobe-finned fishes that is regarded as closely related to land vertebrates, the osteolepiformes (6, *Osteolepis*). (Reconstructions based on Hemmings (1978), Jarvik (1948*a*), Miles and Westoll (1968), Moy-Thomas and Miles (1971), and Pearson and Westoll (1979).)

(*Miguashaia*, Fig. 1.10, 1), and two fishes that are believed to be close relatives of the tetrapods, the osteo-lepiform *Eusthenopteron* (Fig. 1.10, 4) and the pan-derichthyid *Elpistostege* (Fig. 1.10, 5). *Eusthenopteron* is historically interesting, because it is the first fossil fish that was used to build up a theory of the origin of tetrapods. It has therefore been studied in great detail, in particular by the Swedish palaeontologist Erik Jarvik (Fig. 8.4). Its position as a putative tetrapod ancestor or close relative has, however, been questioned after the re-evaluation of the panderichthyids, which are in fact amazingly similar to the earliest known tetrapods.

As the deep gulf of Miguasha was filled with mud, vast amounts of sand and clay were deposited in rivers and deltas over what are now the Baltic States (Estonia, Lithuania, and Latvia). Some close relatives of the Miguasha fishes lived here: the placoderms *Asterolepis* (Fig. 1.11, 6) and *Plourdosteus* (Fig. 1.11, 3), many acan-thodians (Fig. 1.11, 4), the sarcopterygians *Laccognathus* (Fig. 1.11, 2) and *Panderichthys* (Fig. 1.11, 1), as well as the lungfish *Dipterus* (Fig. 1.11, 5). There are also fishes that are unknown at Miguasha, such as the huge arthrodire *Livosteus*, and the last heterostracans, the psammosteids (Fig. 1.11, 7). The psammosteids were

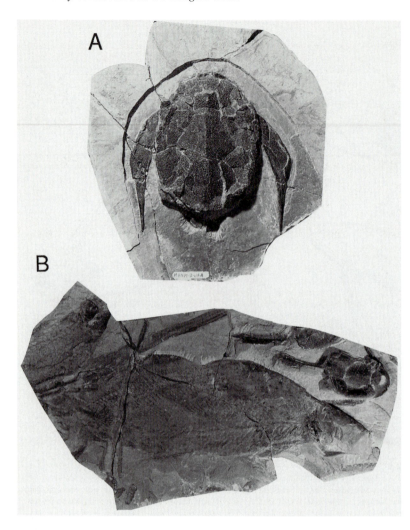

Fig. 1.9. The Late Devonian fish locality of Miguasha, on the coast of the Chaleur Bay in Quebec, was discovered by Gesner in 1842. The cliffs cut through the Escuminac Formation, which consists of sandstones and siltstones, and have yielded an impressive variety of vertebrates in an outstanding state of preservation, such as this articulated armour of the antiarch *Bothriolepis* (A, ×0.5), and this complete specimen of the lungfish *Scaumenacia* (B, ×0.5), associated with *Bothriolepis* armours. (Courtesy R. Cloutier, Lille, and M. Arsenault, Miguasha Museum, Quebec.)

rather large, with branchial expanded plates, and part of their head armour was made up of small bony platelets.

In Frasnian times the coral reefs of Gogo in north-western Australia (Fig. 1.12) supported one of the most diverse fish faunas known from this period. Its preservation is so outstanding that the anatomy of some of the Gogo fishes is now better known than that of many extant fishes. The Gogo fauna was a world of lungfishes and arthrodires that was also inhabited by some actinopterygians (*Mimia*, *Moythomasia*, Fig. 1.13, 12), osteolepids (*Gogonasus*, Fig. 1.13, 11), onychodontids (*Onychodus*), and a few antiarchs (*Bothriolepis*, Fig. 1.13, 7) and ptyctodonts (*Ctenurella*, *Campbellodus*, Fig. 1.13, 8). There were some large arthrodires such as *Eastmanosteus* (Fig. 1.13, 1), *Holonema*, *Kendrickichthys*, *Kimberley-ichthys*, and *Harrytoombsia* (Fig. 1.13, 5) and many

smaller ones, like *Incisoscutum* (Fig. 1.13, 4), *Simosteus*, and *Bruntonichthys*, some of them having a peculiar long pointed snout (*Tubonasus*, Fig. 1.13, 3). Lungfishes were duck-billed (*Griphognathus*, Fig. 1.13, 10) or blunt-snouted (*Chirodipterus*, *Holodipterus*, Fig. 1.12, 9) and, unlike their extant relatives, lived in the sea and breathed mostly—if not exclusively—with their gills.

If the Gogo fauna is impressive, this is mainly due to the quality of its preservation, which permits a thorough survey of its diversity. It is, however, probable that all comparable environments at comparable latitudes in Frasnian times contained equally rich fish faunas, of which we now find only scattered scales and teeth.

Let us now jump a few million years and reach the Famennian age, at the end of the Devonian (365 Ma). On what is now the east coast of Greenland, we would then

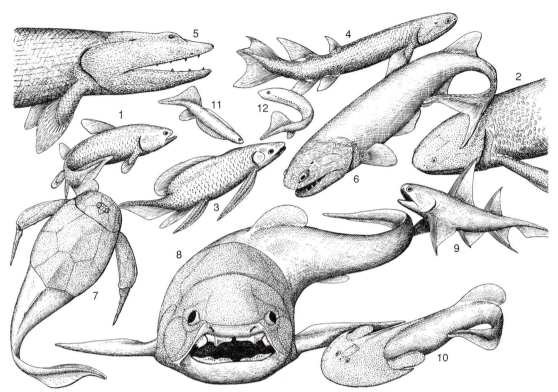

Fig. 1.10. The Miguasha fish fauna is regarded as Late Devonian (Frasnian) in age, i.e. approximately 370 million years old. It is remarkable in its diversity of taxa, although chondrichthyans (sharks), which are largely represented in other localities of similar age, are totally lacking here. In contrast, there is a variety of lobe-finned fishes, or sarcopterygians, belonging to five major taxa: the coelacanths (1, *Miguashaia*), porolepiforms (2, *Holoptychius*), lungfishes (3, *Scaumenacia*), osteolepiforms (4, *Eusthenopteron*), and panderichthyids (5, *Elpistostege*), the last-named being regarded as the closest relative of the tetrapods. Ray-finned fishes, or actinopterygians, are rare, with only one large form (6, *Cheirolepis*). In addition, there are survivors of earlier faunas, such as the antiarchan and arthrodiran placoderms (7, *Bothriolepis*, 8, *Plourdosteus*), acanthodians (9, *Diplacanthus*), and even the youngest known osteostracan jawless fishes (10, *Escuminaspis*). Two anaspid-like naked jawless fishes (11, *Endeiolepis*, 12, *Euphanerops*) may be close relatives of the extant lampreys. The Miguasha fishes are now believed to have lived in an estuary surrounded by fern forests. (Reconstructions based on Arsenault and Janvier (1991), Jarvik (1980), Ørvig (1957), Pearson and Westoll (1979), Schultze (1973), Schultze and Arsenault (1985), Stensiö (1948), and Vézina (1990).)

still have found vast tidal flats and lagoons filled with red sand and clay, but the hilly background would be covered with bushes and small trees (mostly ferns, lycopods, and horsetails, in fact). Along the sea-shore, in the deltas and lagoons, lived the last antiarchs (Fig. 1.13, 3), together with long-snouted lungfishes and huge sarcopterygians (*Eusthenodon*, Fig. 1.13, 4). But a new group now enters the scene: the four-legged vertebrates or tetrapods. These early tetrapods lived more like fishes than land animals. Their paired fins had, however, lost their web and each of their internal fin rays was segmented and enclosed in a fleshy lobe of its own: a digit. *Acanthostega* (Fig. 1.13, 2), perhaps the most fish-like of all known tetrapods, still had fin rays in its caudal fin. It was partly covered with

scales, and had a functional gill apparatus. Its paired appendages had eight digits. *Ichthyostega* (Fig. 1.13, 1) was more massively built and perhaps more advanced in having only six or seven digits. It is probable that these early tetrapods could not walk properly on land, as their limbs looked more like flippers than like true legs. At the most they could crawl on a beach like seals, but they certainly fed in the water, on fish trapped in pools by low tide.

At this time, Morocco and eastern North America were joined (the Atlantic Ocean did not exist) and this area was covered by a warm sea inhabited by many invertebrate groups and large pelagic fishes. In southern Morocco vast exposures of limestone deposited in this sea have yielded

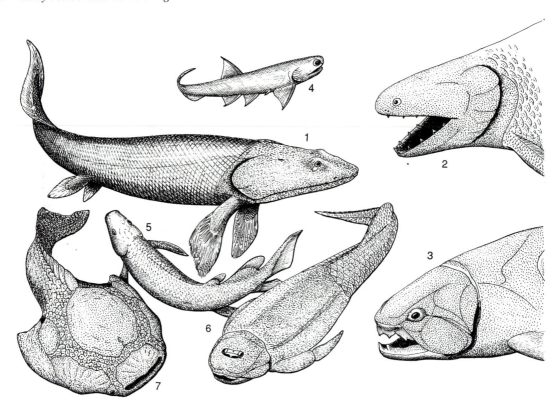

Fig. 1.11. The vast Late Devonian sandy tidal flats and deltas of Estonia and Latvia were formed at almost the same time as the Miguasha deposits (Fig. 1.10). There are thus several closely similar types of fish in common, such as the panderichthyids (1, *Panderichthys*), porolepiforms (2, *Laccognathus*), and placoderms (3, *Plourdosteus*). The acanthodid acanthodians (4) and lungfishes (5) also recall those from Miguasha. In addition there are somewhat different antiarchs (6, *Asterolepis*) and huge armoured jawless fishes (7, *Psammolepis*) which are the youngest known heterostracans. (Reconstructions based on Lyarskaya (1981), Obruchev (1964), Vorobyeva and Schultze (1991), and Vorobyeva, (1993); 1, after a reconstruction by N. Panteleyev, Moscow.)

the armour of huge arthrodiran placoderms, such as *Dunkleosteus* (Fig. 1.15, 1), which also occurs in the Cleveland Shale of North America. The carapace of these arthrodires grew to two metres and their total body length is estimated at six or seven metres. The bony plates of their jaws developed beaks and cutting edges, which suggests that they were the largest predators of their time. *Titanichthys* (Fig. 1.14; Fig. 1.15, 2) was another large arthrodire, but probably of benthic habit. Their companions in the open sea were primitive sharks (*Ctenacanthus*, Fig. 1.15, 4), with large spines in front of the unpaired fins, small actinopterygian fishes (Fig. 1.15, 5), and some sarcopterygians such as lungfishes, coelacanths (Fig. 1.15, 6), and close relatives of *Eusthenopteron* (Fig. 1.15, 7).

To complete this series of snapshots of the Devonian period, we now turn to the arid cliffs of Cucurka, in south-eastern Turkey, near the Iraqi border. Here, thin-bedded limestones were deposited in a shallow sea during the transition between Late Devonian and Early Carboniferous times (360 Ma). They have yielded a mix of Devonian and Carboniferous vertebrate faunas (Fig. 1.16). Although no articulated fish has been found in this series, a number of forms can be identified from isolated bones. The latest known arthrodire placoderm, *Groenlandaspis* (Fig. 1.16, 1), which is characterized by a peculiar crest on its back, and the latest known lungfishes with shiny, enamel-covered scales (*Rhinodipterus*, *Chirodipterus*, Fig. 1.16, 2) belong to the typically Devonian members of the fauna, whereas some deep-

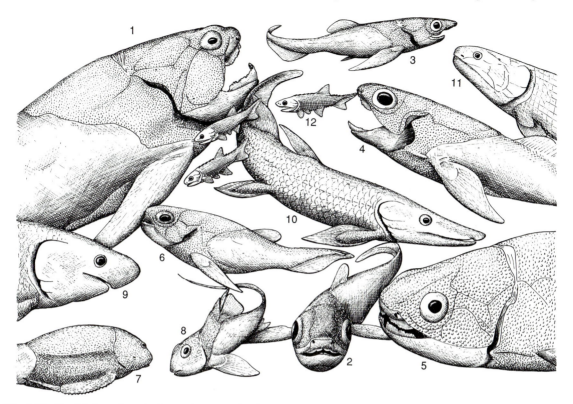

Fig. 1.12. The fish fauna from the famous Late Devonian (Frasnian) locality of Gogo, north-western Australia, displays an amazing diversity of placoderms, most of them belonging to the arthrodires (1, *Eastmanosteus*; 2, *Latocamurus*; 3, *Tubonasus*; 4, *Incisoscutum*; 5, *Harrytoombsia*; 6, *Torosteus*). Other placoderm groups are represented by the antiarchs (7, *Bothriolepis*) and ptyctodonts (8, *Campbellodus*). The lobe-finned fishes are essentially lungfishes (9, *Holodipterus*; 10, *Griphognathus*) and osteolepiforms (11, *Gogonasus*). Small ray-finned fishes (12, *Mimia*) were also fairly abundant. (Reconstructions based on Dennis-Bryan (1987), Dennis and Miles (1979, 1981), Gardiner (1984*a*), Gardiner and Miles (1990), Long (1985*c*, 1988*a*, 1988*c*), Miles (1977), Miles and Dennis (1979), and Young (1984).)

bodied actinopterygian fishes (Platysomidae, Fig. 1.16, 5), large rhizodontiform sarcopterygians (*Strepsodus*, Fig. 1.16, 3), and peculiar cartilaginous fishes related to extant chimaeras, the bradyodonts, are all classical components of the Carboniferous fish faunas.

Well-dated occurrences of the latest Devonian fishes are relatively rare. Another famous locality is Andeyevka, near Tula, in Russia, which has also yielded remains of a tetrapod, *Tulerpeton* (Fig. 1.17, 7). This differs from *Acanthostega* and *Ichthyostega* in its very slender digits and in some of its skull bones, which suggest affinities to the later anthracosaurs. It also lived on the sea shore, together with bradyodonts (Fig. 1.17, 8), antiarchs (*Asterolepis*, *Remigolepis*, Fig. 1.17, 1, 2), lungfishes (Fig. 1.17, 3), and other large sarcopterygians such as

Chrysolepis (Fig. 1.17, 4), *Eusthenodon* (Fig. 1.17, 5), and *Onychodus* (Fig. 1.17, 6).

1.4 CARBONIFEROUS VERTEBRATES AND THE GREAT CHANGE

The Carboniferous period (about 360 to 290 Ma) began with a rise in sea-level—a marine transgression—that favoured the development of invertebrate life, in particular corals, brachiopods, sea-lilies, and various molluscs. The vertebrate faunas of the continental margins had been impoverished by the extinction of the placoderms, the armoured jawless fishes, and a large number of the sarcopterygians, in particular the cosmine-covered lungfishes (Fig. 1.23). As a consequence of this, two

Fig. 1.13. By the end of the Devonian period, in Famennian times, 365 Ma ago, the flood-plains and deltas of the 'Old Red Sandstone continent' (North America and Europe) were inhabited by the earliest land vertebrates, or tetrapods. In East Greenland, two of these animals (1, *Ichthyostega*; 2, *Acanthostega*) spent most of their time in the water, where they preyed on placoderms (3), lungfishes, and possibly also on larger sarcopterygians (4). They could certainly crawl on land with the help of their limbs, which, however, were not as flexible as those of later tetrapods. These early tetrapods retained many piscine features, such as a complete gill apparatus, scales, and endoskeletal and dermal fin rays in the tail. (Reconstructions based on Clack and Coates (1993) and Jarvik (1980).)

groups greatly diversified: the chondrichthyans, or cartilaginous fishes, and the actinopterygians, or ray-finned fishes. In both groups a number of species developed thick, crushing teeth (bradyodonts among the former, and platysomids among the latter) in order to eat shelly invertebrates. They then occupied the niches left empty by the extinction of the placoderms.

Over a large part of the continents, a dense vegetation of ferns, seed ferns, giant horsetails, and club-mosses developed around bays and lakes. Here also lived the first truly freshwater vertebrate faunas, with lungfishes (which were less 'bony' than their Devonian ancestors), ray-finned fishes, some large sarcopterygians, the rhizodontiforms, and a variety of tetrapods, including amphibians, and also early amniotes.

I have chosen as two examples of the Carboniferous vertebrate life the marine fish fauna of Bear Gulch in Montana (USA) and the freshwater fauna of Dora (Cowdenbeath), north of Edinburgh in Scotland.

The Bear Gulch locality consists of thick series of calacareous flagstones deposited in a shallow and confined bay in Late Namurian times (about 315 Ma). Thanks to these quiet conditions of deposition, fragile fishes such as cartilaginous fishes or lampreys, and also ray-finned fishes and coelacanths, are perfectly preserved, even with imprints

of their soft parts (Fig. 1.18). The most striking feature of this fauna is the great diversity of the chondrichthyans (Fig. 1.19). Some of these are vaguely shark-like, such as *Stethacanthus* (Fig. 19, 2), *Damocles* or *Falcatus* (Fig. 1.19, 1), though different from modern sharks. Others are chimaera-like (*Echinochimaera*, (Fig. 1.19, 6), and some do not resemble anything we know in our modern seas. These are the durophagous bradyodonts, such as *Harpagofututor* (Fig. 1.19, 3) and various cochliodontids (Fig. 1.19, 4, 5), the petalodontids (*Belantsea*, Fig. 1.19, 7), and the edestids. Bradyodonts are related to chimaeras, whereas the petalodontids and edestids still have uncertain affinities within the chondrichthyans. The coelacanths display a variety of body-shape, ranging from quite generalized ones, such as *Caridosuctor* (Fig. 1.19, 9), which is not too different in overall aspect from the modern coelacanth, to deep-bodied forms (*Allenypterus*, Fig. 1.19, 10), whose shape is suggestive of some modern reef-fishes. Ray-finned fishes are also quite diverse, with some peculiar eel-shaped species (*Partarrassius*, Fig. 1.19, 8). One of the earliest known lampreys, *Hardistiella* (Fig. 1.19, 11), lived in this environment and was closely similar to a small modern lamprey.

The deposits at Dora, in Scotland, are nearly of the same age (Namurian, about 325 Ma) as those of Bear Gulch,

Fig. 1.14. While tetrapods were beginning the conquest of land, vertebrate life in the sea was almost unchanged, with a variety of placoderms, some of which could reach six or seven metres in length. This almost complete armour found in the Tafilalt (southern Morocco) belongs to one of these large placoderms, the arthrodire *Titanichthys* (×0.08). (From Lehman (1956).)

but represent a quite different environment, most probably a lake. Unlike the Bear Gulch fauna, an important part of the fauna at Dora consists of tetrapods (Fig. 1.20). Among them are the strange aquatic form *Crassigyrinus* (Fig. 1.20, 1), the loxommatids *Spathicephalus* (Fig. 1.20, 2) and *Doragnathus*, and also anthracosaurs, which are relatives of the amniotes (*Proterogyrinus*, *Eoherpeton*, Fig. 1.20, 3). These tetrapods probably preyed on small actinopterygians (Fig. 1.20, 5) or lungfishes (Fig. 1.20, 4). Their competitors were large predatory fishes, such as the rhizodontiform sarcopterygians, freshwater sharks, and large acanthodians (*Gyracanthus*, Fig. 1.20, 6).

After the Carboniferous, the survivors of the Silurian and Devonian early vertebrate groups became more and more scarce as more advanced ray-finned fishes, sharks, and—on land—tetrapods developed and diversified.

1.5 THE SHAPING OF THE MODERN VERTEBRATE FAUNAS

During the following geological period, the Permian (about 290–245 Ma), the warm seas that extended over most of the northern continents were still inhabited by large, shark-like chondrichthyans such as the edestids (Fig. 1.21, 1), which had a pointed snout and a sharp median tooth row on the lower jaw, or large petalodontids with median teeth like parrot bills (Fig. 1.21, 2). These lived together with hybodont sharks and even perhaps with the first sharks of modern type (neoselachians), as well as many ray-finned fishes (Fig. 1.21, 4). Such marine, pelagic Permian fish faunas can be exemplified by that of the Early Permian limestone of the Copacabana Formation of Bolivia (Fig. 1.21), which is of nearly the same type as is found in the Permian limestones of Russia, eastern Asia, North America, and Greenland. By the end of the Early Permian (260 Ma) the last two survivors of the Devonian vertebrate faunas became extinct: the acanthodians and the osteolepiforms. The rich Permian vertebrate fauna of the Wichita Group in Texas still yields abundant remains of two of these relict groups: acanthodians (*Acanthodes*, Fig. 1.22, 2) and osteolepiforms (*Ectosteorhachis*, Fig. 1.22, 1), in association with other, more 'modern' fishes, such as advanced lungfishes (Fig. 1.22, 5), hybodont sharks (Fig. 1.22, 4), and actinopterygians (Fig. 1.22, 6), and pelycosaurian amniotes (Fig. 1.22, 7), which are relatives of the mammals.

The 'world of early vertebrates' may be regarded as having ended by the beginning of Mesozoic times, in the Early Triassic (about 245 Ma), although this statement is admittedly subjective, as shown in Fig. 1.23. Most Palaeozoic chondrichthyan groups had disappeared by the end of the Permian, leaving only the true rays and sharks and some chimaeras. The durophagous 'bradyodonts' (stem-group holocephalans) were replaced ecologically, first, by some durophagous hybodont sharks and later, more successfully, by skates and rays. Amphibians remain relatively abundant throughout the Triassic. They comprise some piscivorous forms probably adapted to marine life (trematosaurs). True frogs appear in the Early Triassic. All modern amphibians (lissamphibians) were probably already established by Late Triassic times, while the other temnospondyl amphibians (the 'stegocephalans') became extremely rare.

The 'modern' vertebrate world, with advanced ray-finned fishes (teleosts), sharks and rays (neoselachians), modern amphibians (frogs, salamanders, and apods), lizards, crocodile, turtles, birds and mammals was shaped between the end of the Permian and the end of the Jurassic. A few groups nevertheless, survived almost

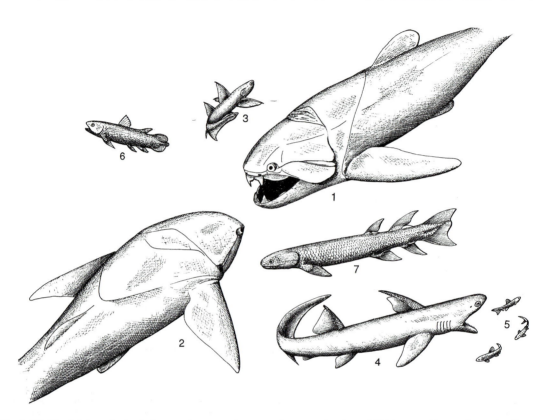

Fig. 1.15. The fish fauna in the shallow sea that covered Morocco in Famennian times, 365 Ma ago, with two giant arthrodiran placoderms (1, *Dunkleostus*; 2, *Titanichthys*), acanthodians (3), large sharks (4, *Ctenacanthus*), actinopterygians (5), and sarcopterygians (6, a coelacanth; 7, an osteolepiform close to *Eusthenopteron*). (Reconstructions based on Lehman (1956) and Lelièvre *et al.* (1993).)

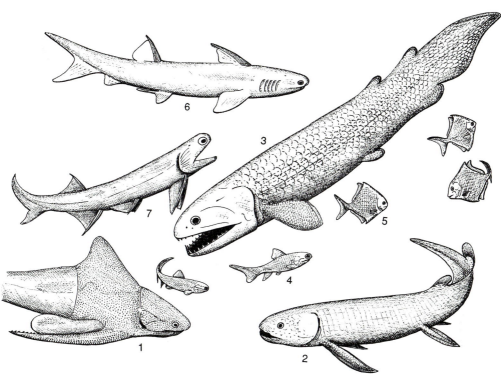

Fig. 1.16. By the very end of the Devonian period, in Late Famennian times, 362 Ma ago, some marine faunas displayed a peculiar assemblage of typically Devonian taxa with others that are more and more common in the Carboniferous. A fish-bearing locality discovered in the uppermost Devonian of Cukurca, in south-eastern Turkey, illustrates these transitional ecosystems. It still contains placoderms (1, *Groenlandaspis*) and dipterid lungfishes (2, *Chirodipterus*), but also yields large rhizodontiforms (3, *Strepsodus*) as well as amblypterid (4) and platysomid (5) ray-finned fishes, which occur essentially in the fish faunas of the Carboniferous. Sharks (6, *Ctenacanthus*) and acanthodid acanthodians (7, *Acanthodes*) are also frequent components of the Carboniferous fish faunas. (Reconstructions based on Andrews (1985), Janvier *et al.* (1984), and Ritchie (1971).)

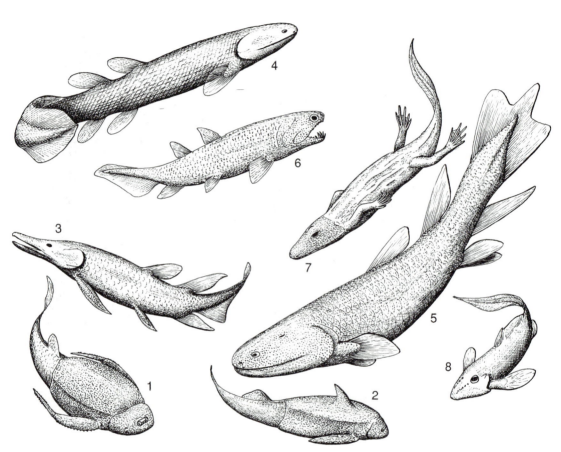

Fig. 1.17.   The Late Famennian locality of Andreyevka, near Tula, in Russia was deposited at the same time as that of Cukurca in Turkey (Fig. 1.16), i.e. just before the Devonian–Carboniferous boundary. As at Cukurca there are placoderms, in particular antiarchs (1, *Asterolepis*; 2, *Remigolepis*) and lungfishes (3, *Andreyevichthys*), as well as very large osteolepiforms (4, *Chrysolepis*; 5, *Eusthenodon*) and onychodonts (6). This fauna also includes one tetrapod (7, *Tulerpeton*), which bears some resemblance to the Carboniferous anthracosaurs, although it retained six digits in the hand. Another 'Carboniferous' element of this fauna is represented by some sharks and relatives of modern chimeras, the bradyodonts (8). The Andreyevka fauna was marine, and it shows that the earliest tetrapods were essentially aquatic, and even marine animals. (Reconstructions based on Lebedev (1985, 1990).)

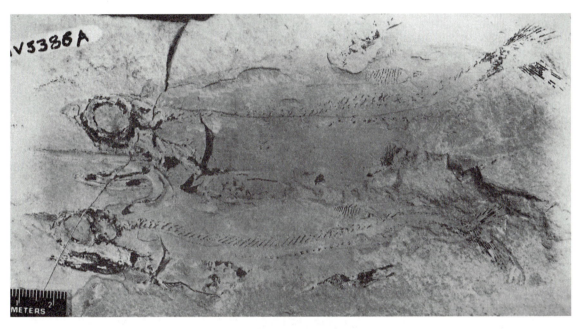

Fig. 1.18. The Carboniferous locality of Bear Gulch in Montana (USA) has yielded an amazingly diverse fish fauna which consists essentially of chondrichthyans (relatives of sharks and chimaeras), but also includes a number of ray-finned fishes and coelacanths. Thanks to the outstanding preservation of articulated specimens such as these, of a male and female stethacanthid shark *Falcatus* (which possibly died during courtship or mating, ×1), the excavations carried out at Bear Gulch by Richard Lund in the 1980s have considerably increased our knowledge of the Carboniferous chondrichthyan diversity. (Courtesy of Dr R. Lund, Adelphi University.)

Fig. 1.19. By the end of the Early Carboniferous, in Namurian times (315 Ma), the composition of the marine fish faunas was radically different from those of the Late Devonian, in particular through the extinction of the placoderms and the considerable reduction of most sarcopterygian groups, which survived only in lagoonal or freshwater ecosystems. The Bear Gulch fauna, partly reconstructed here, is composed essentially of chondrichthyans. Some of them are shark-like, though not necessarily close relatives of modern sharks, such as the stethacanthids (1, *Falcatus* [a, male, b, female]; 2, *Stethacanthus* [male]). The bradyodonts, were more closely related to extant chimaeras (3, *Harpagofututor* [a, male, b, female]; 4, 5, cochliodonts; 6, *Echinochimaera* [male]). Some of these chondrichthyans are of uncertain affinities, such as the stout-bodied petalodontids (7, *Belantsea*). In addition, the Bear Gulch fauna includes some ray-finned fishes, some of which are eel-shaped (8, *Paratarrassius*), as well as several coelacanths (9, *Caridosuctor*; 10, *Allenypterus*). The earliest known true lamprey (11, *Hardistiella*) also occurs there. (Reconstructions based on Janvier and Lund (1983, 1985), Lund (1982, 1985b, 1986b,c, 1989), and Lund and Lund (1984).)

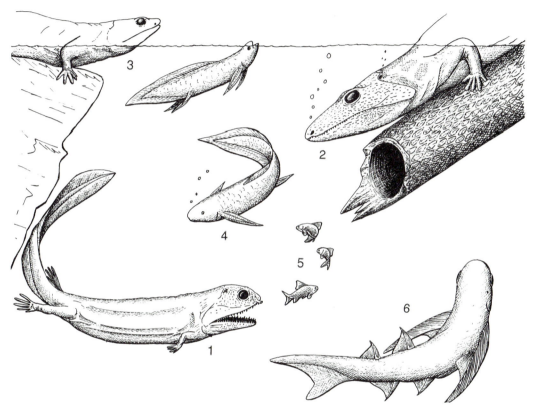

Fig. 1.20. The Early Carboniferous (Namurian, 325 Ma) vertebrate locality at Dora in Scotland is almost of the same age as that of Bear Gulch, but it was deposited in a shallow lake. The most striking difference between them is the abundance of the tetrapods at Dora. Some of these were obviously aquatic, whereas others were washed out into the lake. Of the Dora tetrapods, two are supposedly primitive members of the group (1, *Crassigyrinus*; 2, *Spathicephalus*); others are typical anthracosaurs, i.e. relatives of the amniotes (3, *Proterogyrinus* and *Eoherpeton*). The fauna also includes the classical Carboniferous fish assemblages, i.e. chondrichthyans, large acanthodians (6, *Gyracanthus*), ray-finned fishes (5), and lungfishes (4). (Based on Milner *et al.* (1986).)

Fig. 1.21. A marine fauna of early Permian times, 260 Ma ago: the fishes from the Copacabana Formation of Bolivia. It still includes some of the peculiar chondrichthyans of the Carboniferous period, such as huge eugeneodontids (1, *Parahelicoprion*) and petalodontids (2, *Megactenopetalus*), accompanied by primitive sharks (3) and a variety of ray-finned fishes (4, platysomids). (From Janvier (1991).)

Fig. 1.22. In the Early Permian, the last refuge for some of the vertebrate taxa that were widespread in Devonian times was the marginal, possibly brackish or freshwater, environment, exemplified here by the red beds of the Wichita Group of Texas. In this environment survived the youngest osteolepiform (1, *Ectosteorhachis*) and acanthodians (2, *Acanthodes*), in association with xenacanthiform (3) and hybodontiform sharks (4), as well as with coelacanths, lungfishes (5), and various ray-finned fishes (6). Remains of various tetrapods (amphibians and mammal-related amniotes, 7) also occur together with this fish fauna. (Reconstructions based on information in Romer (1958) and Milner (1993a).)

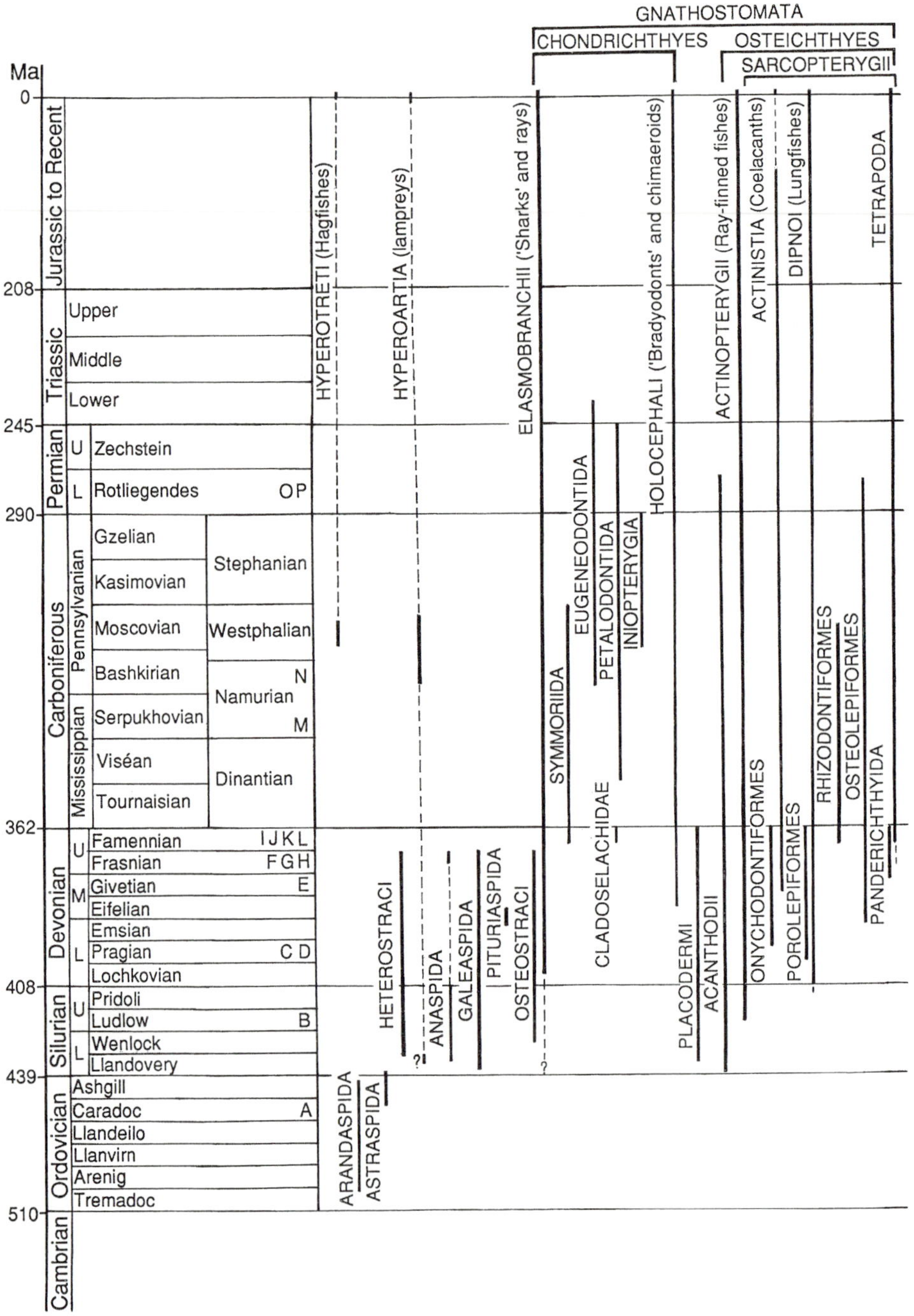

Fig. 1.23. Distribution of the major vertebrate taxa in time (stratigraphical scale from Harland *et al.* (1991), distribution of the taxa based mainly on Benton (1992).) The temporal position of the localities and faunas cited in Chapter 1 is indicated by letters to the right of the stratigraphical column: A, Sacabamba, Bolivia; B, Saaremaa, Estonia; C, Spitsbergen, Wood Bay Formation; D, Bac Thai, Vietnam; E, Lethen Bar, Scotland; F, Miguasha, Quebec; G, Lode, Latvia; H, Gogo, Australia; I, Mt Celsius, East Greenland; J, South Morocco; K, Cukurca, Turkey; L, Andreyevka, Russia; M, Dora, Scotland; N, Bear Gulch, Montana; O, Copacabana Formation, Bolivia; P, Wichita Group, Texas. Abbreviations on the left side of the column: L, lower; M, Middle; U, Upper.

unchanged from the dawn of the history of the vertebrates or, at least, the Late Palaeozoic: lampreys and hagfishes among jawless vertebrates, chimaeras among chondrichthyans, some ray-finned fishes, and two groups of sarcopterygian fishes: the coelacanths and lungfishes.

This brief survey of the first 250 million years of the recorded history of the vertebrates shows how the composition of the modern vertebrate faunas, in particular the fish faunas, developed stepwise. Some of the 'early vertebrate' groups have almost unmodified extant survivors; very few underwent a sudden extinction; and some were suddenly impoverished but survived for many millions of years. However different a Silurian fish fauna may look from a Recent one, the search for relationships, which is the basic task of the palaeontologist, shows that we are dealing simply with relatives, and not with a different 'world'.

2

From rocks to theories:
techniques of preparation and methods of analysis

The series of reconstructions reviewed in Chapter 1 is the result of a long process of study that begins with the discovery of a fossil and may end with the elucidation of its relationships with extant animals. Additional information, on how, where, and when an animal lived may be obtained from its anatomy or from the rock in which it is preserved. Other fossils, associated with it, or the nature of the sediment can provide such information. Conversely, the fossils can generate theories about the rocks, that is, hypotheses on the age of the beds or the palaeogeography. The purpose of this chapter is to illustrate this process of discovery as applied to early vertebrates. The methods and techniques are generally the same as for younger fossil vertebrate groups or even for any fossil, but some are particular to material of this type.

2.1 THE SEARCH FOR FOSSILS

Fossil vertebrate localities are often discovered by field geologists, but early vertebrate remains may look so unprepossessing that they are rarely noticed by people who do mapping work, for example. When they eventually pay attention to them, they collect only a few scraps, since vertebrates—and particularly Palaeozoic ones—have the reputation of being of little stratigraphical interest. Early vertebrate remains are also discovered by micropalaeontologists who look for conodonts, which are small tooth-like or comb-like fossils of great stratigraphical significance that occur widely in Palaeozoic marine sediments. Small fish remains, such are teeth, scales, or bone fragments, are made up of calcium phosphate, like conodonts, and are not digested by diluted weak acids (acetic or formic acid). During the processing of calcareous rocks with these weak acids, conodonts and fish remains are unaffected and can easily be extracted. Such fish remains, often referred to as 'ichthyoliths' have long been neglected by stratigraphers. They have, however, now been shown to be useful for dating geological strata. In any case, they provide an indication that larger fish remains may be found in the same rock and may trigger further field investigation by vertebrate specialists.

I have many times experienced the excitement of finding early vertebrate localities on the basis of such tenuous indications provided either by field geologists or by micropalaeontologists, and in nearly every instance I have discovered, if not articulated specimens, at least substantial numbers of isolated but complete bones. Close collaboration with geologists in the field may thus help in showing them which types of rock or facies are best for finding such remains.

It is commonly said that certain palaeontologists are good 'fossil finders' or are 'sharp-scented' for fossils. In fact, this ability rests on wide field experience and great energy. After field campaigns on different types of rocks, one learns that, for example, Devonian fishes commonly occur in calcareous sandstone in association with lingulids. Also, spending a lot of time in breaking a large quantity of rock (given that there are some indications of fish remains) is usually rewarding: one cannot expect natural exposures to yield the best specimens immediately.

2.2 TRAVELLING

For more than a century, most advances in early vertebrate palaeontology have been made on the basis of rich material collected in 'comfortable' countries, such as western North America and northern Europe. Some expeditions in the Arctic (Spitsbergen, Greenland) have also considerably enriched our knowledge of these Silurian and Devonian faunas (see Chapter 8, p. 320), but they too were associated with a large palaeogeographical unit, the 'Old Red Sandstone continent', which comprises what is now north-eastern North America, Greenland, Spitsbergen, and northern Europe. For a long time in the history of this subject, the discoveries therefore concerned essentially the same higher taxa. Some palaeontologists were ironical about those who tried to find early vertebrate material elsewhere in the world, especially in remote and very difficult countries. They thought that wherever one went the same fishes that were found in Scotland or Canada would turn up. This may be partly true for some periods, such as the Late Devonian and

Carboniferous, but it is certainly not the case for earlier periods such as the Early Devonian, Silurian, or Ordovician. In the 1960s, the discovery of the Early Devonian fish faunas from China, for example, was a great surprise to palaeontologists; and some new groups, such as the galeaspids or yunnanolepiform antiarchs, were totally unexpected. The discovery of the Ordovician arandaspids in Australia and Bolivia, or that of the Devonian pituriaspids in Australia, caused similar astonishment. Discovering new localities anywhere in the world increases the probability of finding not only new species or new taxa, but also remarkable instances of preservation. The Late Devonian fish fauna from Gogo, Australia, for example, is not much different from what was previously recorded from Late Devonian localities in Germany, but the preservation of the Gogo fishes, in limestone concretions, is so remarkable that it has increased considerably our knowledge of the internal anatomy of Devonian lungfishes and actinopterygians. Similarly, the Carboniferous fauna of Bear Gulch, Montana, has yielded a large amount of information on chondrichthyans that were previously known only from isolated teeth, scales, and fin spines.

Exploring in detail reputedly well-known areas may also provide some surprisingly rich localities. A Carboniferous fish and amphibian locality comparable to Bear Gulch in quality of the preservation was found by the fossil hunter Stanley Wood in 1984 at East Kirkton, Scotland, in a quarry that had been carefully investigated by palaeontologists since the early nineteenth century.

Finally, there is yet another way of looking for early vertebrate localities: by examining carefully the samples left aside as 'undetermined fish remains' in the collections of museums and geological surveys. This is how the first arandaspid remains from Australia were discovered by Alex Ritchie in the South Australian Museum (Ritchie and Gilbert-Tomlinson 1977). This is also how Jenny Clack and Per Ahlberg discovered new Devonian tetrapod material from Greenland and Scotland respectively (Clack 1988; Ahlberg 1991*b*).

2.3 COLLECTING

Collecting early vertebrates in the field, in particular in remote places or difficult countries, may sometimes be a question of choice. Material of this kind is usually of small size (except some large Late Devonian arthrodires, for example), but embedded in hard rock. A large quantity of useless sediment has thus to be carried back to the lab, sometimes at great expense if transported by air. Unless transport facilities are available, one must thus choose the most informative specimens, such as skulls, articulated specimens (if any), or dermal bony plates with complete natural margins. A good impression of a bone with natural margins is often more useful than a real bone without any natural margins preserved. Isolated scales and teeth or plate fragments are of secondary interest, unless they are the only elements in the fauna. There are only a few places in the world (Estonia, Latvia), where early vertebrate remains occur in loose sands or clay, thereby requiring them to be encased in a plaster jacket before transport.

2.4 TYPES OF ROCK AND METHODS OF PREPARATION

Early vertebrates occur in sandstones, shales, limestones, and, more rarely, dolomites. According to the hardness of the sandstone, the vertebrate remains can be prepared either by removing the rock (grain by grain with a needle, under a binocular microscope, if the anatomical structures are delicate), or by removing the bone (mechanically or with hydrochloric acid) if the sandstone is compact and strongly siliceous. In the latter case, a rubber or silicone cast of the natural mould can then be made (Fig. 2.1A). A calcareous matrix is ideal for acid preparation of early vertebrate remains, but delicate endoskeletal structures (e.g. perichondral bone) may be microfractured and fall apart during the digestion of the matrix. Careful examination of the specimen is therefore recommended before processing it. The ideal concentration for both acetic and formic acid preparation is 10 per cent. It may be useful to buffer the solution with tricalcium phosphate to avoid etching of the bone or rendering it fragile. In any case, acid preparation must be done stepwise and interrupted by washing, drying, and consolidation of the specimen by glue. At the beginning of the process, these interruptions should be more frequent, in order to check how the specimen is reacting. The detail of this technique is well explained in Rixon (1976) and Crowther and Collins (1987). The best preservation of Palaeozoic (and later) fishes is found in calcareous concretions, such as in the Devonian of Gogo, Australia (Fig. 2.2). These concretions are believed to have formed rather rapidly, as a consequence of the metabolism of bacteria around the decaying fish. The independent bony elements therefore generally remain in their natural position or are only very slightly displaced. In the famous Late Jurassic and Early Cretaceous fish localities of Sierra de Varas (Chile) and Araripe (Brazil), calcium phosphate or carbonate precipitation (probably due either to bacteria or to calcinosis of the fish) has preserved soft tissues such as muscles, blood vessels, blood cells, and even intracellular organelles, such as mitochondria (Schultze 1989; Martill 1988; Fig. 2.3). Such remarkable preservation has never been found in Palaeozoic concretions. A few traces of muscle blocks and kidneys have been recorded in the

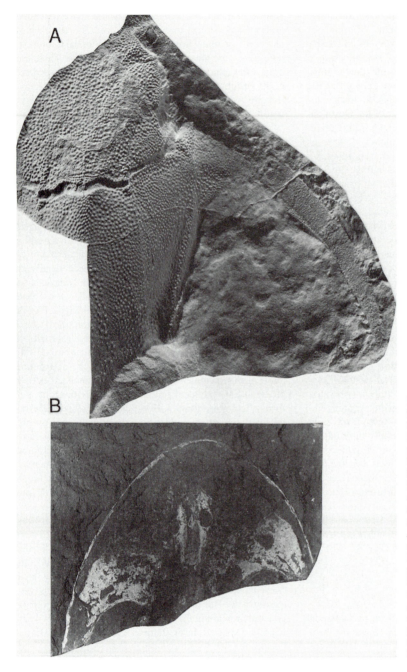

Fig. 2.1. When fossils are preserved as impressions, either because the bone has been dissolved away, or because it has been removed chemically during preparation, a latex or elastomer cast can be made. This provides an accurate image of the external ornamentation. To increase the contrast in photographs, a dark-coloured cast can be whitened with ammonium chloride or magnesium (A, elastomer cast of the armour of *Arctolepis*, an Early Devonian arthrodire of Spitsbergen, in dorsal view). Photographs in ultraviolet light may reveal thin traces of bone that would have gone unnoticed otherwise (B, impression of the head-shield of *Parameteoraspis*, an Early Devonian osteostracan from Spitsbergen, photographed in ultraviolet light). (A, courtesy of D. Goujet, Muséum National d'Histoire Naturelle, Paris.)

Devonian shark *Cladoselache* in the Cleveland Shales, and part of the soft tissues of the head of a similar shark have been recorded in a silicified concretion from the Devonian of Tennessee (Maisey 1989*a*, Fig. 4.30). There are exceptional cases of preservation of immature fishes

and even eggs (Schultze 1972; Grande 1989; Figs 2.4, 4.75C).

Fossil fishes preserved in shales are generally flattened and often distorted. Black or dark shales were generally deposited in an anoxic environment which has prevented

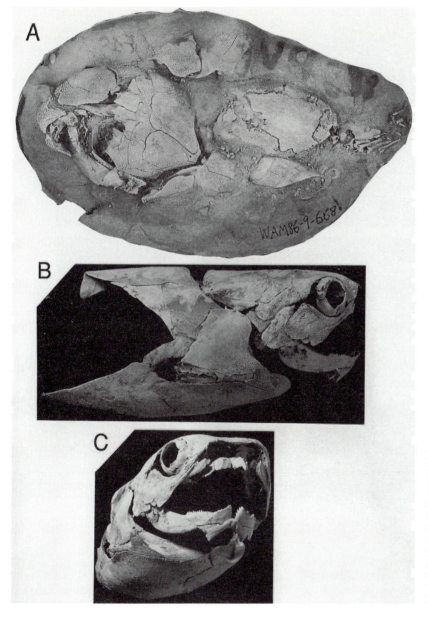

Fig. 2.2. Acetic or formic acid preparation is now widely applied to early vertebrate material. The most spectacular results have been obtained on the Late Devonian fishes (here an arthrodire) from Gogo, Australia, which are preserved in limestone concretions (A) and can subsequently be mounted like skeletons of extant fishes (B, C). (Courtesy of J. Long, Western Australian Museum, Perth.)

the rapid decay of some of the non-mineralized structures such as cartilage, eyes, liver, the digestive tract, and sometimes muscle fibres. These are then preserved in the form of tarry imprints. Preparation of such fossils is often unnecessary since they are preserved in only two dimensions. It is even harmful because needle marks made on the specimen become conspicuous when photographed in immersion. Good examples of such preservations are the Early Devonian Hunsrück shales of Germany or the Carboniferous Mecca Shale fossils of Indiana. A particular case is the Bear Gulch locality of Montana, where fishes are perfectly preserved in two dimensions, with some soft parts, but in thin limestone beds.

The chemical preparation of fossils preserved in shales and clay can be effected by chemical disruption, using detergents or hydrogen peroxide. Other chemical methods, such as the use of chelating agents, may harm the fossils.

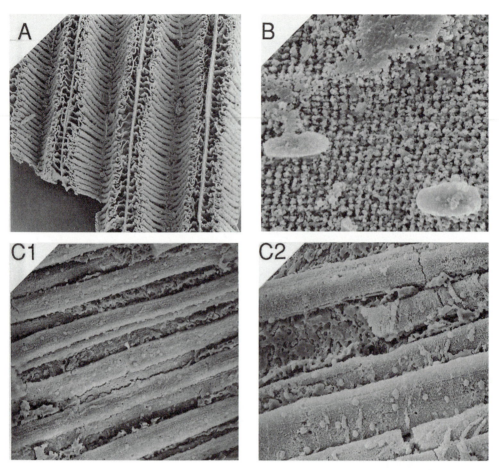

Fig. 2.3. Exceptional preservation of soft tissues may occur in fossil fishes, though rarely in Palaeozoic forms. Shown here are examples of phosphatized soft tissues in *Notelops*, one of the Cretaceous fishes of the Santana Formation in Brazil, prepared with dilute formic acid. Scanning electron microscope examination reveals the gill filaments with erect secondary lamellae (A). A close-up of muscular tissues (B) reveals the banding of the striated muscles and oblong cell nuclei on their surface. Muscle fibres (C) also display mitochondria and T-tubules (C2). A, ×100; B, ×3000; C1, ×350; C3, ×600. (Courtesy of D. Martill, University of Portsmouth.)

Preservation in dolomite is the worst case because of the impossibility of processing the sample in acetic or formic acid preparation (dolomite does not dissolve in weak acid) or of preparing a natural mould with hydrochloric acid (the rock and the fossil would both be destroyed). Even mechanical preparation is difficult because the matrix is firmly attached to the bone, which is largely recrystallized and breaks into minute prisms. Preservation in phosphatic concretions, even if it rules out chemical preparation, can, however, be as good as in calcareous concretions. The phosphatic concretions from the Carboniferous of Kansas or the Triassic of Madagascar and Spitsbergen are good examples of this type of preservation, which is also due to the activity of bacteria.

Such concretions can be prepared mechanically, or serially ground and cut (see below).

Rare types of matrix, for early vertebrates, are sand and clay (e.g. Late Devonian of Estonia and Latvia), or ferruginous concretions, the latter being treated by chemical preparation with thioglycolic acid (difficult and dangerous) or, more fruitfully with a percussive tool.

2.5 TOOLS

Early vertebrates can be prepared with the usual tools of the palaeontologist (chisel, needle, pneumatic engraving hammer), as well as the personal 'tricks' that any of us

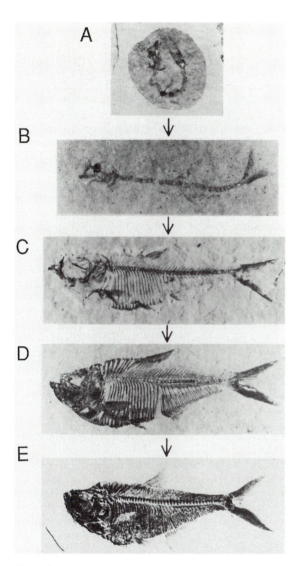

Fig. 2.4. Exceptional preservation of fossil fishes can provide growth series that start from the egg. This example shows the teleost fish *Diplomystus*, from the Eocene Green River Formation of the USA. Embryos can be found in eggs (A), and growth series range from 24 mm (B) to 480 mm (E). (From Grande 1989; courtesy of L. Grande, Field Museum of Natural History, Chicago.)

may have. The obligatory tools are a very sharp steel or tungsten carbide needle and a grindstone to sharpen it regularly and, of course, a good binocular microscope with a *camera lucida*. A porcupine spine or the thinnest entomological needle are the ideal tools for the preparation of delicate structures, in particular during the course of acid preparation. Pneumatic hammers and saws are also very useful, in particular for the gross removal of the matrix before operating close to the fossil. One of the most useful tools is the air abrader, which projects particles on to the specimen and removes the matrix, provided that it is softer than the fossil. The air abrader works best with specimens in soft limestone or dolomite, or specimens that have already been treated with acid or hydrogen peroxide to soften the matrix. The danger is that the fossil itself may become sculptured or abraded. It is therefore desirable to control abrasion under a binocular microscope. The best abrasive to use for early vertebrate fossils is dolomite powder.

The scanning electron microscope is now widely used to study fish scales, histological structures, and even anatomical structures of minute, acid-prepared, fish skulls. To my knowledge, it was first used by Gross (1968c) for the study of early vertebrate histology and confirmed details that had been previously observed only in thin section (namely the thin perforated lamellae in the pore-canal system of tremataspid osteostracans; Chapter 4, p. 112). This technique has since evolved considerably. Most interesting for palaeontologists is the possibility of inserting large specimens in to the microscope without damaging them. Scanning electron microscopy has not, however, replaced the classical thin section method employing a light transmission microscope. Using both means of observation is always recommended by palaeohistologists.

2.6 CHEMICALS

Acid preparation of early vertebrates requires acetic, formic, or hydrochloric acid. Observation of specimens during or after preparation can be improved if the fossil is wettened with alcohol. The contrast between a fossil and a black shale is accentuated by immersion in alcohol or aniseed oil (avoid xylene, which is extremely toxic and will dissolve tarry imprints). If the bone is white and difficult to distinguish from the rock, it can be coloured red or pink with alizarin disolved in alcohol.

A wide variety of glues and resins can be used to consolidate the fossil, in particular during acid preparation. Polyvinyl butyryl resin or polymethylmethacrylate disolved in ethyl acetate are widely used, but the latter tends to shrink on drying. Cynoacrylate glues, disolved in acetone, are also good hardeners but their long-term stability is uncertain, and many prefer to use polybutylmethacrylate for this purpose.

When the fossil is preserved as a natural mould, latex or silicone rubber can be used to make a cast. Rubber (latex) can be used directly, without intervening

anti-adhesive matter. It can be made very fluid to reproduce extremely small details (of ornament, for example),
but it shrinks (and more so on large surfaces) and decays
after 10 to 15 years. Silicone or elastomer neither shrinks
nor decays, but it must be used with an anti-adhesive
(soap diluted in water, for example—but the specimen
must be perfectly dry before being covered by elastomer).
The anti-adhesive may, however, mask some fine detail of
the natural mould. Elastomer casts can be enlarged by immersion in kerosene, which causes the cast to swell by up
to 10 per cent, without any major distortion. This enlarged cast can then, in turn, be moulded, and a new elastomer cast can be made and enlarged in the same way. If
a rubber or silicone cast is to be photographed, it is better
coloured dark grey (not black!) with a few drops of Indian
ink, and then superficially whitened with ammonium
chloride or magnesium (Fig. 2.1A).

2.7 SPECIAL TECHNIQUES

Sollas's method of grinding sections has been widely
used by Erik Stensiö (see Chapter 8, p. 316) and his disciples on early vertebrates (in particular on Triassic
actinopterygians and Devonian osteostracans, placoderms, and coelacanths). The method consists of grinding
sections of the fossil at very close spacing (from a few
microns to one or two millimetres) and making enlarged
photographs or *camera lucida* drawings of each section.
By this means, complex internal structures that could
hardly be observed or reconstructed by any other method
have been thoroughly described. The reconstruction of the
three-dimensional internal anatomy or structure from a
series of photographs or drawings can be effected in two
ways. One is to make a graphic reconstruction on squared
paper by linear projection of the details observed in the
section. The other is to transfer the drawing of each
section to a plate of wax mixed with about 10 per cent
paraffin. The thickness of the plates is calculated according to the intervals between two sections in the fossils,
multiplied by the enlargement of the drawing. The wax
plates are then cut up (following the outline of either the
external or the internal limits), assembled, and glued together by means of a warm knife blade. Some points of
reference on the fossil and the sections serve as guides to
ensure an accurate assembly. This technique for constructing wax plate models has been used since the nineteenth century, mainly by embryologists. Although it may
seem outdated, it has the advantage of being cheap and
easy, if time-consuming. (See Stensiö 1927 or Jarvik
1980 for illustrations of wax models.) Wax plates can be
replaced by polystyrene plates, but these cannot be
thinner than 2 mm (Jefferies and Lewis 1978).

Sollas's grinding section method destroys the fossil,
and can thus be used only on specimens that are available
in large quantity. For this reason, its use on early vertebrates has been strongly criticized outside the Stockholm
'school' (see Chapter 8, p. 315). Nevertheless, it permits
observations that no other method would allow, and I
doubt whether the internal anatomy of osteostracans, for
example, would be as well known as it is now without the
use of this technique by Stensiö in the 1920s. To avoid
the shortcomings of Sollas's method, and to keep at least
a physical trace of each section before it is destroyed, one
can use the 'peel' technique, which is widely used by
coral specialists. This consists of covering the fresh
section with a peel of transparent acetate, and dissolving
it in a few drops of acetone. When the peel is dry it is
removed. Minute particles of bone remain attached to it,
which reproduce the pattern of the sectioned structure.
The peels do not, however, reproduce the finest details of
the structures, and some minute elements may be missed.
A much more reliable method has been described by
Poplin and de Ricqlès (1970). They use a special type of
microtome (Jung microtome, model K with a tungsten
carbide razor), which can cut very hard materials such as
metals (it is used for the study of crystalline structures in
steel). The specimen, preferably in a calcareous or phosphatic matrix, is first impregnated with a very fluid resin
and is sliced with the microtome at intervals ranging from
15 to 30 microns. The thin sections are then fixed on a
glass slide, like a normal histological preparation, and observed under the microscope or photographed (Fig. 2.5A).
This technique, however, has its own shortcomings: the
knife of the microtome is easily damaged, especially if
sand grains are present in the matrix, and it therefore has
to be sharpened regularly by specialist firms (which is
expensive); also, the size of the specimen to be sectioned
is limited by the size of the razor, i.e. to two to three
centimetres.

Reconstruction by means of computed images obtained
with a high-resolution computed tomographic (CT)
scanner is possible but is still rather expensive. This
method is, however, nondestructive and takes only a few
hours, whereas grinding sections take months, if not
years. Digitized tomographic sections made in transverse,
horizontal, and parasagittal planes make it possible to reconstruct the image of the entire fossil, to 'cut' it in any
plane, and to rotate on a computer screen. This method
will no doubt develop to such an extent in the future that
physical sectioning methods (grinding or slicing) will
virtually disappear. Rowe *et al.* (1992) have produced a
CT atlas of the skull of the therapsid *Thrinaxodon*
(Fig. 2.5B). There are, however, still physical impedi-
ments to computed tomographic scanning, for it depends
on the nature of the rock and the density of the bone. A

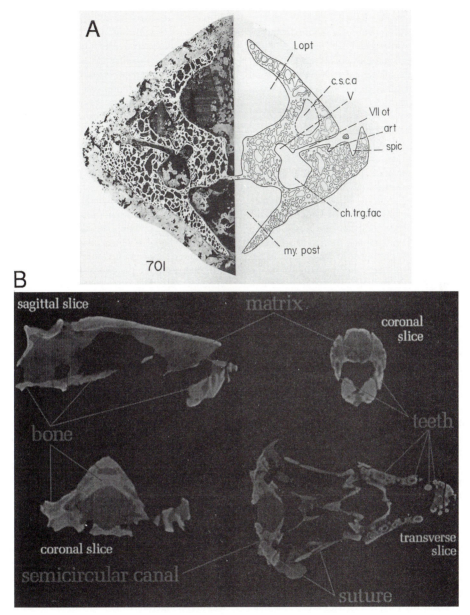

Fig. 2.5. The classical Sollas method of grinding sections and the preparation of enlarged reconstructions in the form of wax models have been greatly improved by Poplin and de Ricqlès's slicing technique, which makes it possible to keep a record of each section. A, thin transverse section through the left half of the braincase of the Carboniferous actinopterygian *Kansasiella*, with its interpretation on the right. The development of computerized tomographic methods and image analysis will no doubt result in considerable progress. B, various sections through the skull of the therapsid *Thrinaxodon*, as they appear on the computer screen. (A, from Poplin (1974); B, courtesy of T. Rowe, University of Texas at Austin, Texas.)

heterogenous matrix, or hematite- or pyrite-bearing matrices prevent X-rays from penetrating the rock. Problems of another kind may arise from the nature of the skeleton. The very thin perichondral bone coating of the osteostracan, galeaspid, or placoderm endoskeleton, for example, may not be detected by this method.

Cathodoluminescence has been used in palaeohistology to unravel growth lines in the dentine of teeth and dermal denticles (Smith *et al.* 1984, Derycke 1990). Although rather destructive for the thin sections, this technique may be interesting to study incremental zones that may be caused by changes in environmental conditions.

2.8 PHOTOGRAPHS AND DRAWINGS

The methods of photography and radiography for early vertebrate fossils are the same as those for both other fossils, invertebrate and vertebrate. Whitening with ammonium chloride or magnesium is frequently used, since many of these fossils are quite flat (dermal bones with ornamentation, impressions, etc.). Specimens preserved in a dark matrix are better photographed immersed in water, alcohol, or aniseed oil. Slight traces of white bone in red sandstone become clear when photographed in ultraviolet light. Retouching photographs was a common practice in the beginning of this century, but should be avoided. When important details are hardly visible in a photograph, an explanatory sketch, drawn from the specimen at the same scale, can be published beside the photograph. Radiographs are sometimes used to study articulated specimens that are preserved in shales, hence difficult to prepare.

Drawings are important in the illustration of early vertebrates, because some of the fossils are difficult to photograph and details appear in direct observation because they are in a different colour, and are hardly distinguishable in black-and-white photographs. (Most scientific journals accept only black-and-white photographs.) The best drawing technique for this purpose is the use of a *camera lucida*, sometimes combined with a photograph. One should not expect to make a good *camera lucida* drawing the first time. The sheet of paper must be slightly more brightly lit than the fossil, and one must beware of distortion near the limits of the field of the microscope. The *camera lucida* must be used to draw only the outline and proportions of the structures to be illustrated. Normal vision is then used to complete the drawing.

The rules for drawing early vertebrates are the same as for other fossils, with the light coming from the upper left corner of the sheet. Fishes are generally drawn with the head pointing toward the left in lateral view, and upwards when in ventral or dorsal view.

2.9 CHARACTERS

What kind of information does an early vertebrate fossil give us? There are two kinds of information, extrinsic and

intrinsic, that is, information provided by the fossil itself, which is chiefly relevant to its relationships and function, and information provided by the fossil *and* the rock, which is relevant to adaptation, environment, or age.

Intrinsic information is provided by characters, which can be taken from the anatomy, histology, or measurements (biometry). A character is an attribute and all its subsequent states of transformation, such as a pectoral fin of a lungfish, a tetrapod fore leg, or a bird wing. Characters define taxa. Ensembles of organisms that cannot be defined by at least a single character are not taxa, although they are sometimes termed paraphyletic taxa.

Examples of morphological characters are the pattern of dermal bone, ornamentation, the course of the sensory-line canals and pit lines, the position of teeth, the position of the openings for nerves and blood vessels in the braincase, the number and position of elements in gill arches, the structure of the paired fin skeleton, the structure of the vertebrae, etc. (Fig. 2.6). Histological characters concern the structure of the hard tissues, cartilage, bone, or teeth. They bear on the presence and arrangement of calcification (in cartilage), the presence or absence of bone cells, lineation, orientation of Sharpey's fibres, and the pattern of branching of the dentine tubules. Most of these characters can be observed in thin sections or with a scanning electron microscope. In addition, ultrastructural characters, better observed with a scanning electron microscope, are the arrangement of the apatite crystals in the hypermineralized tissues, such as dentine, acrodine, or enamel. Finally, quantitative characters can be the number of serially arranged elements (vertebrae or arcualia, mesomeres in paired fins, supports in unpaired fins, teeth, scale rows, etc.). They can also concern measurements or proportions, which are often quite difficult to compare from taxon to taxon and, thus, to assess as homologies.

Extrinsic information relates to things that are other than morphological or systematic. It is obtained from the fossil and the matrix and concerns chiefly the relationships between the organism and its environment, as documented by the rock. The rock can provide information on salinity, temperature, age, environment, etc. Some information on environment can be provided by the fossil alone (e.g. isotope ratios; see Chapter 7, p. 293).

2.10 COMPARISONS

The work of interpreting structures and comparing them with those of other living or fossil organisms can begin before full completion of the technical work, but sometimes ideas on the affinities of the fossil may emerge in

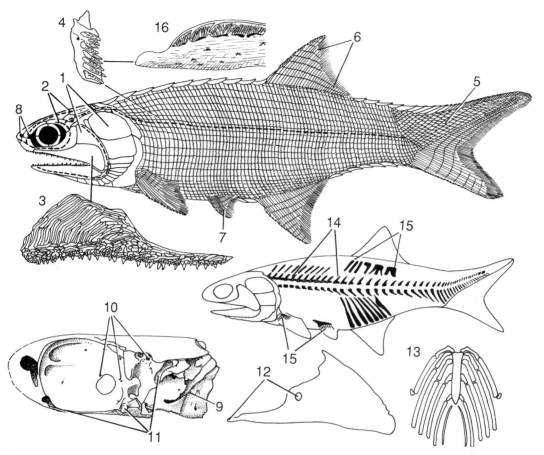

Fig. 2.6. The kinds of characters that are generally used in reconstructions of early vertebrate comparative anatomy and phylogeny are exemplified here by the Late Devonian actinopterygian *Mimia*. They are essentially taken from the morphology: 1, dermal bone pattern; 2, sensory-line pattern; 3, dermal bone ornamentation and tooth morphology; 4, scale morphology; 5, tail morphology; 6, number of dermal fin rays; 7, relative position of the fins; 8, position of the nostrils; 9, cranial fissures; 10, position of foramina for the cranial nerves; 11, 12, areas of attachment for the visceral skeleton (in particular the palatoquadrate); 13, structure of the branchial arches; 14, axial skeleton; 15, paired and unpaired endoskeletal fin supports. Characters are also taken from the histological structure (16, scale or tooth histology) as well as from quantitative or metric data (number of dermal fin rays, radials, vertebrae, scale rows, and proportions of the entire body). (From Gardiner (1984*a*).)

the field when the specimen is discovered. One must, however, be very careful with this kind of intuitive 'at first glance' conclusion, which may influence further observation and reasoning. One should show one's specimen to many colleagues in order to make sure that they see the same details. Ideas also come or change as preparation progresses, but one should be very careful if the reconstruction of a comparable or closely related fossil or extant form is used, as a guideline for a preparation. There are tales of palaeontologists who have made a hole in a fossil because the reconstruction they used during preparation showed a foramen in the same position!

The comparison, character by character, of the newly prepared specimen with other organisms, fossil or extant, is generally made on the basis of published illustrations, unless one has access to large collections. In most cases, before or during preparation, one gets an idea of the major group to which the specimen belongs (placoderms, lungfish, etc.) and the comparison often remains within the limits of a monophyletic order, family, or genus. In some cases, however, and particularly when dealing with a fossil referred to a presumably paraphyletic group, the comparison must be extended to a wide range of taxa.

How to start a comparative analysis cannot be presented in the form of a rule, for each of us has his own propensity to follow his intuitions or to look first for particular characters. One can never be fully objective during this part of the work, even though one *must* be as objective as possible. The greatest sign of objectivity is to doubt every single conclusion one comes to, and to try and refute it by all means. This is not easy, for when most palaeontologists start to examine a fossil they look for characters that may be of interest because they are linked in one way or another with a previous theory, and they sometimes overlook other characters that may be of significance. There was a tradition in early vertebrate anatomy that canals which housed nerves were more 'important' or 'interesting' than those housing blood vessels, and that, among the latter, those housing arteries were more 'important' than those housing veins. Moreover, endoskeletal structures were regarded as more important, theoretically, than exoskeletal ones (similar traditions occur in embryology; Presley 1993). There were certainly some bases for this tradition (individual variation, bearings on a '*bauplan*', etc.). Current methods of character analysis (see below) tend, however, to be as free as possible from such prejudices.

Again, the first steps in character analysis are always more fruitful when not made in isolation. Discussions with colleagues often trigger the discovery of new details, new elements of comparison that one would not even have thought of. A good basic knowledge of extant taxa is always useful, if only to avoid over-interpreting fossil structures.

The basis of comparative biology is said to be the search of homologous characters among a variety of organisms. But how to recognize homologies has long been—and still is—debated. In practice, a hypothesis of homology is based on mere resemblance in either position or structure, or both, but this hypothesis can be corroborated only after the analysis of many other presumed homologies in the same organisms. Congruence in the distribution of several characters that are similar in aspect or position is certainly the best indication that these characters are homologous. The eyes of a vertebrate and an octopus, though somewhat similar, are considered non-homologous because there are no other homologies shared only by these two forms, whereas there are many that are shared by the octopus and a clam, which has no such eye. This is called the congruence test of homology. There are other possible tests, such as the ontogenetic test (characters arising from the same embryonic segment, germ layer, or presumptive area are regarded as homologous) and the conjunction test (two supposedly homologous characters that occur together in the same organism cannot be homologous) but, as pointed out by

Patterson (1982*b*), most of these tests fail in some instances. Many of the problems of relationships among major vertebrate taxa discussed in this book are due to the fact that these taxa share only a few particular homologies with other taxa, and that the number of congruent homologies supporting one theory of relationship is not large enough by comparison with those supporting another theory.

The cavities and canals inside the head of a Devonian osteostracan, for example, housed the brain, eye, otic capsules, nerves, and blood vessels, and they reproduce quite accurately the shapes of the organs they contained (see Fig. 4.16). There is no question about the homology between the labyrinth of an osteostracan and that of any other craniate, or between its two vertical semicircular canals and those of, say, a shark (Fig. 4.16, 7, Fig. 4.34, 13). This conclusion is based on resemblance, which is regarded as beyond the bounds of chance. But if we turn now to the nasal capsule of the same osteostracan, we notice that it opens to the exterior by a keyhole-shaped opening on the top of the head and is associated with a median recess that is comparable with the naso-hypophysial complex known only in lampreys among extant vertebrates (Fig. 4.16, 3, 25, Fig. 3.6B). This anatomical resemblance and uniqueness leads to a hypothesis of homology and seems to give strong support to close relationships between the two groups. The hypothesis is, however, refuted by only a few other homologies shared exclusively by osteostracans and jawed vertebrates (see Chapter 5, p. 239). This example shows that characters are not homologous by essence, nor by authority, but that a provisional decision on homology is the result of a feedback process between resemblance at the level of one character and congruence with many other characters.

The ultimate goal of comparison between homologous characters is to discover a hierarchy in which homologies range from the most particular (restricted to two species, or even two individuals) to the most general. This hierarchy will then be the guideline for constructing a phylogeny, that is a theory of genealogical relationships between species and taxa, assuming that each node of branching equates with the presence of a common ancestral species, which possessed at least one homology of all its descendants.

2.11 CHARACTER ANALYSIS AND PHYLOGENY RECONSTRUCTION

From the early days of evolutionary theory until the 1960s, the analytical work of comparative biologists, including palaeontologists, was based on the search for overall resemblances between species. Groupings, or classifications were then, made on the basis of these resemblances, and evolutionary relationships between these groups were proposed, essentially in the frame of an ancestor–descendant model, with an increasing weight on the geological criterion that ancient groups are ancestral to more recent ones, given that they share resemblances.

Ordering the data in this way resulted in the construction of evolutionary trees, based on a classification that reflects overall resemblance. The method is therefore called *evolutionary systematics*. Evolutionary systematics has two goals: to evaluate the degree of divergence, or difference between organisms or taxa, and to suggest ancestor–descendant relationships between species or groups of species.

Evolutionary systematics tends to select characters according to various criteria, which depend largely on the idiosyncrasies of the systematicist as well as on the availability of characters in fossils. Skeletal parts are therefore preferred to other characters. Statements about 'adaptive' *versus* 'non-adaptive' characters, i.e., whether a character is influenced by the environment, are widely used to select or discard particular characters.

This principle was later modified by considering a very large number of characters, taken at random, but groupings were still based on the overall resemblance of these characters. This method, called *numerical taxonomy*, or *phenetics*, is thus based on the probability that the greater the number of character the closer to the natural order. The goal of pheneticists was in fact to obtain a stable classification, and the greater the number of characters, the more stable the classification.

Both evolutionary systematics and phenetics have a major flaw: they could define groups of organisms on the basis of shared absence of characters ('privative' characters), fishes being thus 'non-tetrapod' vertebrates, amphibians 'non-amniotes', and reptiles 'non-birds and non-mammals'. Such 'privative' groupings had already been criticized by Aristotle 2500 years ago, because 'privative' characters cannot be further subdivided. Aristotle did not, however, hesitate to erect such groups as the anema (animals without blood). By doing this, both methods can create groups defined by the absence of characters, which thus match the definition of many other groups. Fishes, defined by the absence of limbs, for example, have the same definition as amoebas, which also have no limbs.

In the 1960s a new method of evaluating the relationships between organisms appeared, called *phylogenetic systematics*, and now known as *cladistics*. This method is foreshadowed in the works of various systematicists of the nineteenth century and the early twentieth century. It was completed and formalized by the German entomologist Willi Hennig in 1950, but its application became widespread only in the late 1960s when Hennig's book was translated into English (Hennig 1966).

The novelty in Hennig's principle is that resemblance between organisms does not only mean either close relationship or convergence (characters that have appeared independently in two groups), as was the tradition in evolutionary systematics, but also common ascent with no particularly close relationships. To take a classical example, the presence of five toes in a turtle and in man does not mean that man is more closely related to the turtle than to the horse, which has only one toe. In fact, a considerable number of characters are unique to man and horses (mammalian characters). Given that the five toes of the turtle and of man have not appeared independently, the character thus means all three animals share a common ancestor which had five toes, and that the number of toes has been reduced in horses.

Cladists therefore classify resemblances into *synapomorphies*, shared by an ancestor and *all* its descendants; *homoplasies*, shared by two taxa but not their common ancestor; and *symplesiomorphies*, shared by an ancestor and only *some* of its descendants. A *reversion* is a particular case in which a character disappears or returns to a more generalized state, becoming thus an apparent symplesiomorphy (imagine a modern horse gaining five toes). These four types of resemblance may concern a character or a the state of a character. The notions of synapomorphy and symplesiomorphy are relative, since a symplesiomorphy at one level of the phylogeny was once a synapomorphy. For example, the five toes were once the synapomorphy of the early five-toed land vertebrates, but are a symplesiomorphy for mammals or amniotes, since many of them have less than five toes, by reduction.

Another principle of Hennig's method is that resolved relationships between species or taxa in general are expressed by a pairwise grouping. Two taxa or species that share a synapomorphy are *sister-taxa* or *sister-species*. This means that they are more closely related to each other than to any other species.

Grouping according to these various types of resemblance produces *monophyletic*, *polyphyletic*, or *paraphyletic* groups respectively. Two sister-species or sister-taxa form a monophyletic taxon. To Hennig, only monophyletic groups or *clades* are to be considered in a classification, because others are 'non-groups'. In other words, *only clades are taxa*. Clades are defined by at least one character or character state that occurs nowhere else. Polyphyletic and paraphyletic groups are not taxa.

It is thus clear that in this method the first task of the comparative biologist is to find a hierarchy of synapomorphies, from the most general to the most particular.

This hierarchy is expressed in the form of a branching diagram called a *cladogram*, where the branching points, or nodes, are defined by at least one synapomorphy. Character distributions that are inconsistent with the cladogram are provisionally regarded as homoplasies, reversions, or symplesiomorphies, as long as they remain less numerous than the synapomorphies (principle of parsimony).

Character analysis is not limited to the distribution of characters among organisms, but also includes the definition of *character states*. Character states are 'sub-characters', or states of transformation of a character, and may be arranged in order of decreasing generality, that is, from the *plesiomorphous* to the *apomorphous* state. In other cases, the apomorphous states are not ordered (for example, the flippers of cetaceans, the wings of birds and bats). There are three main ways of evaluating the state of a character: *out-group comparison, ontogeny*, and *palaeontology*.

Out-group comparison consists of assessing a character state as plesiomorphous within a given group if the same state occurs outside this group, preferably in its closest relative, or sister-group. To take the previous example, mammals generally possess from one to five functional toes, but five is regarded as the plesiomorphous state, because it occurs in most other tetrapod groups, in particular the sauropsids (turtles, lizards, crocodiles, and birds), the supposed sister-group of mammals (Fig. 3.26).

During embryonic development, or ontogeny, characters change from one state to another, the most generalized embryonic state appearing first, and the most particular last. Again, in most embryonic tetrapods, including those with a reduced number of toes, the number of toes is five in the early stages of limb development.

Palaeontology shows empirically that the earlier a character state occurs in the history of a group the more generalized it is. In fact, all early tetrapods have five, or even more, toes. But this is not a rule, for the distribution of fossils in time is largely due to the likelihood of preservation, and there are many examples where the oldest known representatives of a group do not display the most generalized, or plesiomorphous, character states suggested by out-group comparison and ontogeny. The most plesiomorphous and apomorphous character states frequently occur together in the earliest representatives of a group.

Most of the cladograms presented in this book will concern major extinct clades (e.g. heterostracans, osteostracans, placoderms, acanthodians, and osteolepiforms), the sister-group of which is often undecided. The ontogeny is unknown and, in some instances, only limited data on growth are available. In many instances, the order of appearance in the fossil record is of little help. The

cladistic analysis of characters is thus made on the basis of out-group comparisons with all other vertebrates, or even chordates, thereby leading to a simple hierarchy of unique characters. If one considers, for example, the various states of the caudal fin in Recent and fossil craniates (Fig. 2.7B), the only possible out-group comparison is with the cephalochordates, which bear only a vague resemblance to craniates in tail morphology. Growth has sometimes been used to polarize character states, as in the case of the 'tessellate' structure of the head armour of psammosteid heterostracans (Fig. 2.7A). This is absent in juvenile individuals, as in all other heterostracans. It is therefore assumed that the partly 'tessellate' pattern of the adult psammosteid armour is an apomorphous state, which has only a superficial resemblance to the primitive tessellate armour of various other vertebrates.

Stratigraphical position is often misleading when used to assign a polarity to character states in early vertebrates. An example of this is the earlier view that tremataspid osteostracans (Fig. 2.7C3), which are devoid of paired fins, are ancestral to other osteostracans because they occur slightly earlier than those with paired fins (Fig. 2.71C1, 2). On the contrary, a cladistic analysis of several other characters in osteostracans clearly shows the reverse polarity, i.e. that tremataspids have lost the paired fins.

In contrast to evolutionary systematics, where fossils are supposed to provide a key to the ancestry, cladistics considers fossils in the same way as living organisms, though they are less rich in information since they are incomplete. Their age is thus not a property or an attribute, but an item of information that can be taken into consideration in the construction of a phylogenetic tree (but not of a cladogram). Cladogram reconstruction should therefore be first based on extant organisms (if any), which convey the largest amount of information (i.e. morphological, ethological, or molecular data). Fossils can later be inserted into the cladogram on the basis of the unique characters they share with particular extant taxa. Many fossils, which lack such characters linking them with other particular extant or fossil taxa (or whose characters shared with other taxa are contradictory), are *incertae sedis*. The hierarchy of a cladistic classification reflects the branchings of the cladogram. There are different ways of expressing this hierarchy (e.g. the use of the term *plesion* for fossil clades, or the phyletic sequencing; see Nelson (1974), Patterson and Rosen (1977).). The term *sedis mutabilis* added to the taxa in some of the classifications in this book indicate a polytomous branching in the cladogram (unresolved relationships). *Sedis mutabilis* taxa have thus equal rank.

Since the 1960s, cladistics has evolved towards greater simplicity, becoming as free as possible from evolution-

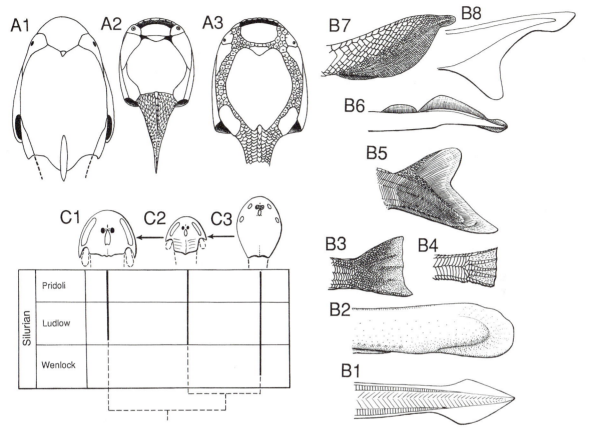

Fig. 2.7. The assessment of the character states or of the degree of generality of characters in early vertebrates is often rendered difficult by uncertainties concerning the out-group. A, growth stages may be used as ontogenetic arguments in some cases; here, the partly 'tessellate' pattern of the dermal armour of the psammosteid heterostracans (A3) is regarded as apomorphous (derived) because young psammosteids (A2) lack the fields of polygonal platelets, like most other pteraspids (A1). B, when dealing with the interrelationships of major extant and fossil craniate taxa, out-group comparison is uncertain; here, the apomorphous states of the various tail morphologies in craniates are assessed by comparison with the cephalochordates (B1), their presumed sister-group (B2, hagfishes; B3, 4, heterostracans; B5, anaspids; B6, lampreys; B7, osteostracans; B8, gnathostomes); considered in this way, the hypocercal (B5, B6) and epicercal (B7, B8) tails are apomorphous, whereas the pad-shaped diphycercal tail of hagfishes (B2) is plesiomorphous. C, the use of the stratigraphical criterion to assess character states often leads to mistaken polarities; for example, some of the Silurian thyestidian osteostracans (C1, *Procephalaspis*; C2, *Thyestes*; C3, *Tremataspis*) are devoid of paired fins (C3); classical evolutionary sequences based on stratigraphy suggested that they have *gained* paired fins (arrows), whereas cladistic analyses of many other characters imply that *Tremataspis* has *lost* the paired fins, despite its greater geological age. (A1, from Blieck (1984); A2, A3, and B3 based on Gross (1963a); B1, B2, and B6 from Janvier (1981a); B4 from Dineley (1976), B5 from Ritchie (1980), B7 from Heintz (1967), and B8 from Heyler (1969).)

ary theories or preconceived ideas on how characters may have changed. Some cladists therefore now tend to equate characters and homologies (non-homologies not being characters) and to consider that the distinction between monophyletic and non-monophyletic (paraphyletic and polyphyletic) taxa is useless, since non-monophyletic taxa are not taxa in a cladistic classification. In order to avoid assumptions on character evolution, that is, how one char-

acter may have transformed into another, they also prefer to use the term 'general' instead of 'plesiomorphous', and 'particular' or 'unique' instead of 'apomorphous'. In this way, they avoid ambiguous names for homologies. For example, the distinction between fins and limbs, which are one single homology (of extant gnathostomes) obliges us to say that fins are plesiomorphous relatively to limbs. In contrast, it sounds less ambiguous to say that paired extremities are general for gnathostomes, and limbs particular to tetrapods. Using the term 'fins' without additional assumptions about how fins may have transformed into limbs can lead us to consider that fins are one homology, unique to fishes, and that fishes are taxon.

The elaboration of cladistic computer program based on parsimony analysis, such as PAUP (Swofford 1990), HENNIG 86 (Farris 1988) or MACCLADE (Maddison and Maddison 1987) has enlarged considerably the scope of cladistic analysis by making it possible to handle a large number of characters and taxa. Above all, these programs provide all the equally most parsimonious cladograms, that is, cladograms in which homoplasies and reversions are reduced to the minimum. They also provide the consensus trees (Adams and Nelson, or strict, consensus trees) of all the most parsimonious cladograms. These can then serve as a basis for the elaboration of phylogenetic trees or scenarios. In some instances the injection of data other than characters (time, geography, or whatever else) in the process of selecting the cladograms may lead to the choice of solutions that are not really the most parsimonious; this is a matter of personal evaluation that has to be argued. Cladistics is parsimony first, but what you then do with parsimony is your problem, as long as you say what you are doing. Parsimony is a means of discovering relationships, not the goal of evolutionary biology.

Cladistics is not the subject of this book, and this brief introduction to the method is aimed only at helping the reader to understand terms that are used in the following chapters, since I shall use only results obtained by means of cladistics. There are numerous papers and books on the subject and the reader is referred to them for further information (e.g. Hennig 1966; Wiley 1981; Nelson and Platnick 1981; Schoch 1986; Wiley *et al.* 1991; Forey *et al.* 1992*b*; Darlu and Tassy 1993).

I shall not discuss here the merits or pitfalls of the respective methods in systematics. Cladistics may be only a little better than other methods, but it *is* better in the sense of being more heuristic, i.e. it makes relationships and classifications easier to argue and refute. The fact that many cladograms have turned out to be 'wrong', i.e. refuted, does not mean that the method has failed but, on the contrary, that its results are easier to criticize; in this sense, it represents progress. Returning to evolutionary systematics would thus be justified only by an appeal to romanticism, ambiguity, or authority, and returning to phenetics would provide only an illusion of stability.

For historical reasons, early vertebrate systematics was one of the earliest fields in which cladistics was practised (see Chapter 8, p. 317). Strangely, the use of this method in palaeontology varies according to the taxa under consideration and the respective authorities. (The use of

cladistics to study the interrelationships of tetrapods, amniotes, or mammals was far from widespread until the late 1980s.) As to fishes in general, cladistic analyses have been carried out to such a large extent that I can hardly imagine how a student could enter this field without working within the same methodological framework.

2.12 BEYOND THE CLADOGRAM

Elucidating the pattern of relationships of living beings, which is the result of the history of life, can be regarded as a goal *per se*. One may, however, consider that there is something to discover beyond the pattern of relationships between organisms. This 'something' is generally called 'evolution'. It comprises a body of assumptions based on empirical data taken from extant organisms, such as population genetics, embryology, and historical biogeography. It is often regarded as the flesh on the bones of a cladogram that subsequently becomes a *phylogenetic tree* (with a time-scale and actual ancestors) or a *scenario* (including all possible assumptions on processes, in particular biogeography).

Whatever may be the evolutionary process that can be envisaged, it must fit the pattern of relationships implied by character distribution, that is, the cladogram. This step beyond the cladogram may be also regarded as one step away from the truth and toward further speculation.

There is, however, one step beyond the cladogram that may be more fruitful than others: cladistic biogeography. Here, endemic taxa are replaced in the cladogram by the areas in which they are endemic. Relationships between these areas, which may or may not be the results of geological history, are then revealed. This is called *vicariance biogeography*. It rests on the postulate that when two species appear by division of the area occupied by their ancestral species (vicariance), their relationships are relevant to the history of their respective areas, and, conversely, that random dispersal is uninformative. The method is also based on the search for common patterns of relationships displayed by the areas of endemism of several different groups, such as plants, butterflies, mammals, and spiders. The discovery of a common pattern is a strong argument in favour of a common cause for the diversification of the groups, essentially a geological cause (continental drift, eustatic movements, etc.).

Vicariance biogeography, which is directly homothetic with cladistic analysis (apomorphy = endemism, plesiomorphy = widespread taxa, homoplasy = random dispersal), has given fruitful results, but vicariance palaeobiogeography, in particular for remote periods such as the Palaeozoic, is extremely difficult in practice (see

p. 294). The main problem is the scarcity of the data and the theoretical impossibility of testing endemism for fossils. There is, for example, a strong probability that galeaspids (see p. 294) are endemic to China and Vietnam, but no one can be sure that they did not exist in other areas where Silurian and Devonian rocks are no longer exposed. However, when we can define at least three apparently endemic monophyletic taxa, a theory can be proposed about their relationships, and consequently about the relationships of the areas to which they are endemic.

Whichever method one may use, one must always remember that even if fossils are incomplete and often disappointing, they are *unique* in three respects: they provide the minimum age for a character, and hence for a taxon; they may reveal associations of characters that no longer exist today; and they may show a geographical distribution that is different from that of extant members of a given group.

2.13. BEYOND PHYLOGENY

Problems of relationships and evolution are not the only things of interest in connection with early vertebrates. Various aspects of their life, such as biomechanics, growth, diet, mode of life, environment, individual variation, and sexual dimorphism can be examined or inferred from the fossils or from their matrix (see Chapter 7). Information of this kind is often cherished by those who dare not or do not wish to enter debates on phylogeny. A phylogenetic framework is nevertheless always useful in this kind of study, if only to link functions or habits with particular character states or particular taxa (Bonde 1985).

Further reading

Exceptional preservations: Briggs and Crowther (1990); Grande (1989); Lyarskaya and Mark-Kurik (1972); Maisey (1989*a*); Martill (1988): Schultze (1989); Zangerl and Case (1976). *General techniques of preparation, illustration and curation*: Briggs and Crowther (1990); Crowther and Collins (1987); Rixon (1976); Toombs (1948); Whybrow and Lindsay (1990). *Special techniques*: Jarvik (1980); Jefferies and Lewis (1978); Poplin and de Ricqles (1970); Rowe *et al.* (1992); Sollas (1904); Stensiö (1927). *Homology*: De Pinna (1991); Hall (1994); G. Nelson (1994); Patterson (1982*b*); Rieppel (1988). *Cladistics*: Darlu and Tassy (1993); Forey *et al.* (1992); Hennig (1950, 1966); Nelson and Platnick (1981); Schoch (1986); Wiley (1981); Wiley *et al.* (1991). *Cladistic computer programs*: HENNIG 86, version 1.5 (Farris 1988); MAC-CLADE, version 2.1 (Maddison and Maddison 1987); PAUP, version 3.0 (Swofford 1990); TAX (Nelson and Ladiges 1991).

3

A survey of extant vertebrates

Most extant relatives of the early vertebrates, such as teleosts or tetrapods, have evolved in their own way and are now quite different from their Silurian and Devonian relatives. Recent fishes and amphibians are often referred to as 'lower vertebrates', a term inherited from gradistic representation of the history of life. In fact, these 'lower vertebrates' branch off in the 'lower' part of the cladogram of the vertebrates, but most of them display unique characters that may be extremely complex and 'advanced' in evolutionary terms. The 'tongue' of lampreys, for example (see p. 49), is certainly more complex than the jaw mechanism of sharks, and many extant teleosts are perhaps much more derived, in terms of number of transformation steps, in relation to the earliest actinopterygians than we are in relation to the earliest amniotes.

This chapter is not concerned with featuring the detailed anatomy of extant vertebrates but with providing a simple framework of extant vertebrate characters and relationships that may help in comparing or interpreting the early fossil taxa. Most of the basic features depicted here are therefore selected from those that are relevant to group definitions or theories of relationships. How vertebrates develop in ontogeny, or what is their basic structure can be found in other books. Among them, I recommend the still-unequalled *Structure and development of vertebrates* by E. S. Goodrich (1930) and, of course, the monument of German comparative anatomy, the *Handbuch der vergleichenden Anatomie der Wirbeltiere*, edited by Bolk *et al.* (1931–9). De Beer (1937), Jollie (1962), Romer (1970), Waterman (1971), Wake (1979), Jarvik (1980, 1981*a*), and Hanken and Hall (1993) also provide a large amount of information on Recent and some fossil vertebrate anatomy.

The question of the relationship of the vertebrates (or craniates, see below) among metazoans will be discussed in Chapter 7 in connection with the theories on their origin. Suffice it to say here that the most widely accepted theory is that the deuterostomes (i.e. the metazoans in which the anus opens first in ontogeny, and the mouth second) are monophyletic. The deuterostomes comprise five major taxa, which are generally regarded as monophyletic: the Echinodermata (sea urchins, starfishes, sea lilies, and sea cucumbers; Fig. 3.1E), the Hemichordata (Fig. 3.1A), the Tunicata (Fig. 3.1B), the Cephalochordata (lancelets; Fig. 3.1D), and the Vertebrata (or Craniata; Fig. 3.1C). The last three taxa are put into the clade Chordata, on the basis of the presence of an axial support, the notochord. The cladogram in Fig. 3.1 illustrates the two current views on the relationships of these taxa, chiefly on the basis of developmental, morphological, and physiological characteristics. The major controversy is about the three-taxon problem concerning the tunicates, cephalochordates, and vertebrates. Most zoologists now consider that the cephalochordates are the closest relatives of the vertebrates, but the palaeontologist R. P. S. Jefferies (1986) gives this position to the tunicates on the basis of still-debated palaeontological arguments (see Chapter 7). Molecular sequence data generally support the monophyly of the chordates, but scarcely resolve the question of vertebrate affinities of either the tunicates or cephalochordates (Lake 1990). When they do, they generally agree with the vertebrate–cephalochordate relationship.

3.1 VERTEBRATES VERSUS CRANIATES

Extant vertebrates are distributed into three major groups; the Hyperotreti (Fig. 3.3A1), or hagfishes (also known as myxinoids), the Hyperoartia (Fig. 3.3A2), or lampreys (also known as petromyzontids), and the Gnathostomata (Fig. 3.3A3), or jawed vertebrates. The terms 'Hyperotreti' and 'Hyperoartia', which should be used by priority sound so similar that I shall use thereafter the vernacular names 'hagfishes' and 'lampreys' respectively. That these three taxa form a monophyletic group is now widely agreed. The only dissent comes from Bjerring (1984), who regards the jawless vertebrates (hagfishes and lampreys) as more closely related to the cephalochordates than to the gnathostomes. These three taxa share many unique characters, but the most peculiar of them is the neural crest (Fig. 3.2, 2). This is an ectodermal thickening of the margins of the dorsal groove that, in the early

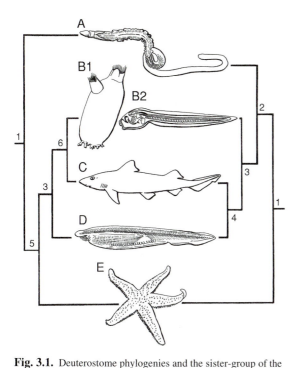

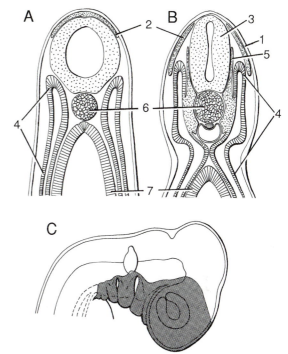

Fig. 3.1. Deuterostome phylogenies and the sister-group of the Craniata (cladogram on the right-hand side from Schaeffer (1987) and on the left-hand side from Jefferies (1986).)
Terminal taxa: A, Hemichordata; B, Tunicata (B1, adult; B2, larval); C, Craniata; D, Cephalochordata; E, Echinodermata. *Nested taxa and synapomorphies*: 1, Deuterostomes (blastopore becomes larval anus, enterocoelous development of coelom and mesoderm, which is derived from vegetal cells); 2, Pharyngotremata (ciliated pharyngeal slits with septal rods and synapticulae); 3, Chordata (equatorial areas of potential mesoderm in embryos, notochord, and neurectoderm giving rise to nervous system, mesodermal induction involving vegetal–animal axis signal inferred from fate map, notochord with lateral muscle bands, neural induction by the roof of the embryonic gut, endostyle); 4, Myomerozoa (segmented paraxial mesoderm, or somites, dorsal, hollow nerve chord with metameric spinal nerves); 5, Dexiothetes (gain of asymmetry, calcitic skeleton); 6, tripartition of hind tail (inferred from the 'calcichordates', see p. 281 and Fig. 7.2).

Fig. 3.2. The neural crest. A, B, transverse sections through the trunk of a shark (*Squalus*) embryo at two successive stages (A, 3.75 mm, B, 5.5 mm), showing the organization of the germ layers and neural crest; C, final ventral position of the neural crest tissues (shaded) in a chick embryo. 1, ectoderm; 2, neural crest cells; 3, spinal cord; 4, mesoderm (somites and lateral plate); 5, sclerotomic (skeletogenic) part of mesoderm; 6, notochord; 7, endoderm (embryonic gut). (A, B, from Bjerring (1968); C, from Langille and Hall (1993), after D. M. Noden.)

embryo, gives rise to the central nervous system. The cells from the neural crest migrate ventrally toward the mesoderm, forming the ectomesoderm, or ectomes-enchyme (i.e., they are ectodermal in origin and mesodermal in position) and subsequently contribute to the formation of, e.g. the gill arches, pigment cells, and dermal skeleton (Fig. 3.2C). The neural crest is often said to generate the dermal skeleton in the head only, but this is due to the fact that it has mainly been studied in tetrapods, which have no dermal skeleton over the body.

In fishes, however, the neural crest cells are now known to contribute to the formation of scales and dermal fin rays (Smith and Hall 1990, 1993). The neural crest is well known in the lampreys and gnathostomes, but it has never been clearly observed in hagfishes, because of the lack of early embryos.

How these three groups are related in the first controversy that is met with in vertebrate phylogeny, and it has bearings on the contents of the taxon Vertebrata. Linnaeus, in the tenth edition of his *Systema naturae* (1758), named the vertebrates as 'Vertebrata-Craniata', that is, animals with vertebrae and a skull. He put in this group all jawed vertebrates and lampreys, which he considered as mere cartilaginous fishes, like sharks and rays. The peculiar jawless mouth structure of lampreys did not arouse attention until much later and was simply regarded as a degenerate state of the jaws. Linnaeus knew of hagfishes, but he placed them among intestinal worms ('*Vermes intestinalis*'), because they often enter decaying fishes and had long been believed to be endoparasites. Although Bloch recognized the piscine nature of hagfishes as early as 1797, it was only in 1806 that Duméril placed hagfishes with lampreys in the group Cyclostomi ('rounded mouth') (Fig. 3.3A, *2*). The basis for this inclusion of hagfishes in the cyclostomes was that they shared with lampreys a large notochord and horny teeth.

In this pre-evolutionary period, the cyclostomes were seen just as a group of vertebrates, but were in no way regarded as more primitive than other groups. Although Duméril (1807) regarded lampreys as a possible 'link' between worms and fishes, the cyclostomes were thought to be 'degenerate', and this was supported by their inferred parasitic mode of life (which is actually the case for most lampreys). The distinction between jawless and jawed vertebrates, seen as a major dichotomy of the vertebrates, was not formalized until the late nineteenth century, when Cope (1889) erected the names 'Agnatha' and 'Gnathostomata'. This distinction was further emphasized when the gills of lampreys and gnathostomes were supposedly shown to develop from the endoderm and ectoderm respectively during ontogeny (Goette 1901). Early evolutionary trees (e.g. Haeckel 1866) nevertheless show the cyclostomes in a 'basal' position of the vertebrate phylum (see also the classification of Milne Edwards 1844). The two major taxa, Agnatha (jawless) and Gnathostomata (jawed) were thus regarded as two sister-groups (with some reservations about certain fossil groups, such as the Heterostraci) until the late 1970s, when Løvtrup (1977) and, after him, a number of anatomists and palaeontologists showed that there were many characters common to lampreys and the gnathostomes that occurred neither in hagfishes nor in any of the closest relatives of the vertebrates, such as the cephalochordates and tunicates. These characters are, for example, the nervous control of the heart, a complex adenohypophysis, extrinsic eye muscles, vertebral elements, true lymphocytes, or a lateral-line system. The absence of these features in hagfishes had long been known before, but was interpreted as a loss in connection with the legendary parasitic mode of life and 'degeneracy', although the fact that hagfishes were not parasites had been well established since the 1950s. These lamprey–gnathostome synapomorphies appeared to be so numerous, by comparison with the few characters defining the cyclostomes, that a sister-group relationship between these two groups was proposed by Løvtrup (1977), hagfishes being, in turn, the sister-group of all other vertebrates (Fig. 3.3A, *3*). This new clade containing lampreys and gnathostomes, was named then the

Vertebrata, because both possessed elements of the vertebrate column, that are lacking in hagfishes. Hagfishes, however, share with the Vertebrata a skull, though a very simple one, and could thus be gathered with them in the clade Craniata. In sum, the taxon Vertebrata now returns to its original sense as given by Linnaeus, i.e. to include only lampreys and gnathostomes (Fig. 3.3, *4*). There still are anatomists who consider that cyclostome monophyly remains well supported by an impressive number of similarities in hagfishes and lampreys, in particular the structure of their 'tongue' (Yalden 1985). These, and other common features, may well be general craniate characters, lost in the gnathostomes. The closer relationships between lampreys and gnathostomes are admitted here only on the basis of parsimony, i.e. because the unique hagfish–lamprey characters are outnumbered by the unique lamprey–gnathostome characters.

3.2 HYPEROTRETI OR HAGFISHES

Hagfish are eel-shaped craniates which live essentially in cold marine waters (there are, however, a few tropical species). They are characterized by a long ventrolateral series of large slime glands (Fig. 3.4, *3*), a peculiar opening, the oesophagocutaneous duct (Fig. 3.4, *13*), on the left side only, behind the series of gill openings, and four pairs of tentacles on the snout (Fig. 3.4, *3*). There are four extant genera: *Myxine*, *Paramyxine*, *Eptatretus*, and *Neomyxine*.

Hagfish have no paired fins and their only unpaired fin is the caudal, which is strengthened by large, non-muscularized cartilaginous fin rays (Fig. 3.3A1). The branchial openings open ventrolaterally in the midpart of the body length (Fig. 3.3A1, *1*). Their number ranges from 15 (*Eptatretus*) to one single common opening (*Myxine*; Fig. 3.4, *12*). At the anterior end of the head is the mouth opening (Fig. 3.4, *1*) which is jawless and armed with retractile comb-shaped horny teeth (Fig. 3.4, *4*). Dorsally to it is the 'nostril' (Fig. 3.4A, *2*), which is in fact a median, rounded opening of a tube-like structure, the prenasal sinus (Fig. 3.3, *7*), leading to the olfactory organ (Fig. 3.4, *6*) and further posteriorly to the pharynx (nasopharyngeal duct; Fig. 3.4B, *2*).

The eyes are small, almost conical in shape, and are devoid of musculature as well as of a true lens (Fig. 3.4H). The labyrinth is enclosed in a cartilaginous capsule (Fig. 3.3, *10*) and apparently has only one vertical semicircular canal (Fig. 3.4D, *15*) and two ampullae (Fig. 3.4D, *14*). (Whether this single canal results from the fusion of two separate vertical canals or not is still debated; see Chapter 6). The acoustic nerve does not send off any lateral-line component toward the skin, and there

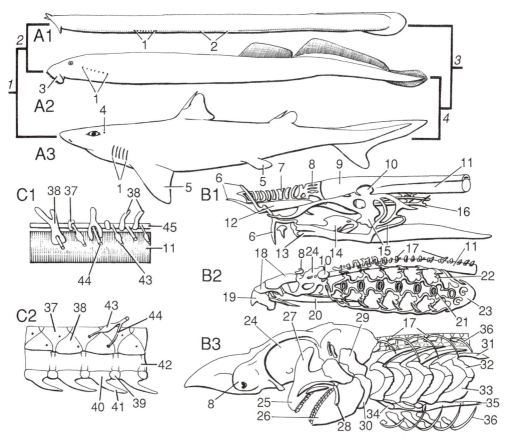

Fig. 3.3. A, morphology and interrelationships of the three major extant craniate taxa; *terminal taxa*: A1, Hyperotreti, the hagfish *Eptatretus* in lateral view (about ×0.5); A2, Hyperoartia, the lamprey *Petromyzon* in lateral view (about ×0.3); A3, Gnathostomata, the shark *Squalus* in lateral view (about ×0.1); *nested taxa and selected synapomorphies*: *1*, Vertebrata (skull, vertebral column, neural crest); *2*, Cyclostomi (horny teeth, lingual apparatus, flattened spinal cord, sperms shed in the coelom, pouch-shaped gills, gills formed from endoderm, gill arches lateral to gills, velum); *3*, Craniata (skull, neural crest); *4*, Vertebrata (arcualia of the vertebral column, extrinsic eye muscle, radial muscles in fins, innervated heart, two vertical semicircular canals in the labyrinth; sensory-line system, osmoregulation, spleen, concentrated pancreas, numerous physiological characters). B, skulls, in lateral view, of a Hyperotreti (*Myxine*, B1), Hyperoartia (*Lampetra*, B2), and Gnathostomata (*Squalus* B3). C, basic elements of the vertebral column and their relationships to the spinal nerve roots in a lamprey (C1, at the transition between the head and the trunk) and a gnathostome (C2, shark). 1, gill openings; 2, slime glands; 3, oral sucker; 4, spiracle; 5, paired fins; 6, tentacles; 7, prenasal sinus of nasopharyngeal duct; 8, nasal capsule; 9, fibrous braincase; 10, otic capsule; 11, notochord; 12, subnasal cartilage; 13, median basal plates of lingual apparatus; 14, dentigerous cartilage; 15, lateral head cartilages (planum viscerale); 16, velar cartilage; 17, arcualia; 18, tectal cartilages; 19, annular cartilage; 20, piston cartilage; 21, trematic rings; 22, branchial arches; 23, pericardic cartilage; 24, braincase; 25, palatoquadrate; 26, Meckelian cartilage; 27, orbital process of palatoquadrate; 28, labial cartilage; 29, epihyal; 30, ceratohyal; 31, pharyngobranchial; 32, epibranchial;33, ceratobranchial; 34, hypobranchial; 35, basibranchial copula; 36, extrabranchials; 37, interdorsal; 38, basidorsal; 39, interventral; 40, basiventral; 41, rib; 42, notochord or calcified centrum; 43, dorsal spinal root; 44, ventral spinal root; 45, spinal cord. (B, redrawn from Marinelli and Strenger (1954, 1956, 1959); C2, from Rosen *et al.* (1981).)

are no neuromasts of the lateral-line system. Peculiar grooves containing villosities have been described in the head of *Eptatretus* (Fig. 3.4I, 24) and regarded as lateral-line grooves (Fernholm 1985). There is, however, no evidence that these are innervated by *lateralis* fibres and they do not contain any neuromast-like structure.

The hagfish skull (Fig. 3.3B1) is a complex structure made up of cartilaginous bars that display no clear homology either to the lamprey or to the gnathostome skull. There is no braincase proper, the brain being surrounded by a cylindrical sheath of fibrous tissue (Fig. 3.3, 9), and there is no branchial skeleton either (apart from possible

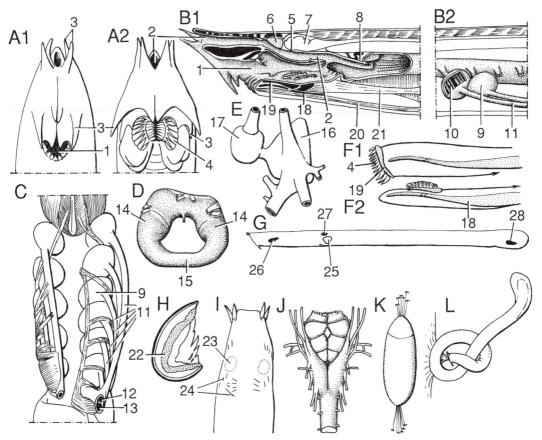

Fig. 3.4. Morphology and behaviour of hagfishes (Hyperotreti). A, head of *Myxine* in ventral view, with teeth retracted (A1) and protracted (A2); B, sagittal section through the head (B1) and lateral view of two gill pouches (B2) (sections of cartilages in black); C, gill pouches and external branchial ducts of *Myxine* in ventral view; D, labyrinth in dorsal view; E, heart in dorsal view; F, mechanism of the 'lingual' apparatus, during protraction (F1) and retraction (F2) of the teeth; G, *Eptatretus* in lateral view, showing the position of the main and accessory hearts; H, transverse section of the eye of *Eptatretus*; I, enigmatic 'sensory-line' grooves on the dorsal side of the head of *Eptatretus*; J, brain of *Myxine* in dorsal aspect; K, egg of *Myxine* (×1); L, the 'knotting' used by hagfishes to tear food. 1. mouth; 2, nasopharyngeal duct; 3, tentacles; 4, horny teeth; 5, hypophysis; 6, olfactory organ; 7, brain; 8, velum; 9, gill pouch; 10, gill lamellae; 11, external branchial duct; 12, external branchial opening; 13, oesophagocutaneous duct; 14, ampullae; 15, semicircular canal; 16, atrium; 17, ventricle; 18, basal plates; 19, dentigerous cartilage; 20, protractor muscle; 21, retractor muscle; 22, retina; 23, eye spot; 24, 'sensory-line' grooves; 25, main heart; 26, cephalic heart; 27, portal heart; 28, caudal heart. (A, B, C, D, E, and J, redrawn from Marinelli and Strenger (1956); H, redrawn from Fernholm and Holmberg (1975); I, from Fernholm (1985).)

cartilaginous rings around the gill openings). The most peculiar structure in the skull is the 'lingual apparatus' which serves to protract and retract the horny teeth. It consists of a series of large median cartilaginous bars or basal plates (Fig. 3.3, 13) on the anterior tip of which rotates vertically a plate bearing the teeth, the dentigerous cartilage (Fig. 3.3B, 14), by means of a complex protracting and retracting musculature (Fig. 3.4B, 20, 21).

The postcranial skeleton consists of a large notochord (Fig. 3.3B1, 11), which reaches anteriorly the level of the

midbrain, and of the skeleton of the caudal fin. The basal part of the rearmost caudal radials fuse into a single cartilaginous plate.

The gills are mere folds of the epithelium enclosed in fibrous pouches (Fig. 3.4, 9, 10). These form paired, but slightly asymmetrical series situated far behind the rest of the head. The respiratory water enters through the prenasal sinus and is conveyed toward the pharynx by means of a pumping apparatus, the velum (Fig. 3.4, 8), which is supported by a complex skeleton (Fig. 3.3B1, 16). The mouth plays no role in respiration.

The brain is anteroposteriorly compressed, devoid of a cerebellum and a true pineal organ (Fig. 3.4J). The neurohypophysis, on the ventral side of the diencephalon, lies against the wall of the nasopharyngeal duct, but the adenohypophysis (Fig. 3.4, 5) is poorly developed. The latter develops from an endodermal expansion of the embryonic gut (Fig. 3.5, 5, 10), and therefore differs from that of all other craniates, in which the adenohypophysis is ectodermal in origin and forms from the wall of the Rathke's pouch (Fig. 3.7, 2; Fig. 3.11, 1).

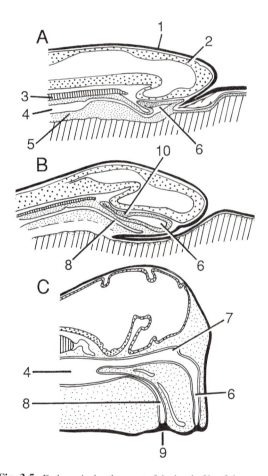

Fig. 3.5. Embryonic development of the head of hagfishes. A–C, longitudinal sections through the embryonic head at three successive stages. Ectoderm in black, endoderm and surrounding mesoderm in fine stipple, brain in coarse stipple, yolk obliquely hatched. 1, ectoderm; 2, brain; 3, notochord; 4, pharynx; 5, endoderm; 6, nasopharyngeal duct; 7, olfactory organ; 8, mouth cavity; 9, stomodeum; 10, position of developing adenohypophysis. (From Gorbman and Tamarin (1985).)

The cranial nerves of hagfishes are basically like those of other craniates, but there are no oculomotor, trochlear, or abducens nerves, which in other craniates innervate the extrinsic eye muscles. The trigeminus branches arise from two separate ganglia, one for the *ramus trigeminus profundus* and one for the *ramus maxillaris*. The *ramus mandibularis* is separated from the latter ganglion. There is no *lateralis* component of the acoustic nerve. Other cranial nerves are the facial (VII), acoustic (VIII), glossopharyngeus (IX), and vagus (X) nerves. There is an optic tract (II), but the olfactory nerve (I) is in the form of numerous olfactory nerve fibres.

The ventricle and atrium of the heart are well apart (Fig. 3.4E, 16, 17), not as in all other craniates, in which these heart chambers are closely set. Hagfishes also have accessory venous hearts which help in blood circulation. There is a cardinal, or cephalic, heart in the head (Fig. 3.4G, 26), a portal heart above the main heart (Fig. 3.4G, 27), and a caudal heart in the tail (Fig. 3.4G, 28). This condition is regarded as general for the craniates, and lost in lampreys and gnathostomes.

Hagfishes generally have nocturnal habits and are carnivorous. They feed on various small invertebrates or on decaying fishes. They have a special way of tearing off the flesh of fishes by 'knotting' themselves (Fig. 3.4L). The eggs of hagfishes are large, and are united into a chain by means of anchor filaments. Although eggs occur frequently on the bottom, the spawning behaviour is completely unknown, even in aquariums. Recently fertilized eggs are therefore seldom found, and information on early embryonic stages is scarce.

3.3 HYPEROARTIA OR LAMPREYS

Like the hagfishes, lampreys are also eel-shaped vertebrates which are generally anadromous, that is, they live in the sea but migrate to rivers to spawn, lay eggs, undergo larval development and metamorphosis, and then return to the sea. Some species however, spend all their life in fresh water (*Lampetra*). Most lampreys are haematophagous ectoparasites and attach themselves to other fishes or even dolphins by means of a large sucker armed with horny teeth, which surrounds the mouth (Fig. 3.3, 3; Fig. 3.6, 2, 3). Depression of the sucker is effected by a velum (Fig. 3.6, 10), and a complex lingual apparatus (Fig. 3.6F), which both produce a pumping effect. The same apparatus also serves in rasping the skin of their victims and sucking their blood (Fig. 3.6O). This sucker and the associated pumping device are the main characters that define this taxon. Lampreys are unique among extant vertebrates in having a single median nostril associated with a blind hypophysial tube (Fig. 3.6, 7), which both open on the

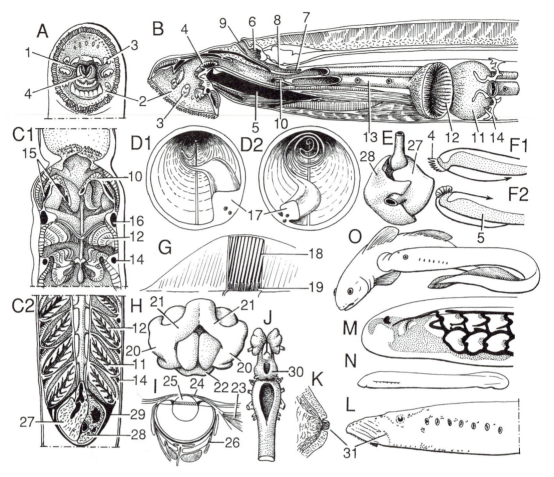

Fig. 3.6. Morphology and behaviour of lampreys (Hyperoartia). A, mouth, 'tongue', and sucker of *Lampetra* in ventral aspect; B, sagittal section through the head, with two gill pouches (one sectioned) exposed (sections of cartilages in black); C, horizontal section through the anterior part of the branchial apparatus and velum of a larval lamprey (C1; sections of cartilages in black), and through the posterior part of the branchial apparatus and heart of an adult lamprey (C2), in ventral view; D, section through intestine of a larval (D1) and adult (D2) lamprey, showing the typhlosole; E, adult heart in dorsal view; F, mechanism involved in the protraction (F1) and retraction (F2) of apical teeth of the 'tongue'; G, dorsal fin with radials and radial muscles exposed; H, adult labyrinth in dorsal aspect; I, horizontal section through the adult eye, showing the corneal muscle; J, adult brain in dorsal aspect; K, section through a neuromast of the lateral line; L, distribution of the lateral-line neuromasts in the adult head; M, larval head skeleton (cartilage in black, mucocartilage stippled); N, young larva ('ammocoete') in lateral aspect (×1); O, attack by an adult lamprey on a fish. 1, mouth; 2, sucker; 3, horny teeth of sucker; 4, apical horny teeth of 'tongue'; 5, piston cartilage; 6, olfactory organ; 7, hypophysial tube; 8, hypophysis; 9, nasohypophysial opening; 10, velum, 11, gill pouch; 12, gill lamellae; 13, pharyngobranchial duct (in adult only); 14, gill arch; 15, velar skeleton (mandibular arch); 16, hyoid arch; 17, typhlosole; 18, radials; 19, radial muscles; 20, ampulla; 21, vertical semicircular canals; 22, ciliated sacs; 23, corneal muscle; 24, cornea; 25, crystalline; 26, extrinsic eye muscles; 27, ventricle; 28, atrium; 29, pericardic cartilage; 30, optic tectum (optic lobes); 31, lateral-line neuromast. (A, B, C2, E, H, J, K, and L, redrawn from Marinelli and Strenger (1954); I, from Franz (1934); D, from Hilliard *et al.* (1983); M, from Johnels (1948).)

dorsal surface of the head through a common naso-hypophysial opening (Fig. 3.6, 9). The hypophysial tube, when pressed by the wall of the pharynx, causes water to enter and leave the olfactory organ (Fig. 3.6, 6).

Like hagfishes, lampreys have no paired fins, but have both dorsal and caudal fins supported by numerous, thin cartilaginous fin rays, or radials (Fig. 3.6, 18), each associated with a single radial muscle (Fig. 3.6, 19). This type of fin musculature enables the fin web to undulate.

The eyes are well developed (Fig. 3.6I), with a lens and extrinsic muscles arranged in very much the same way as in the gnathostomes (except for the superior oblique muscle). Since there is no intrinsic eye muscle to modify the shape of the lens, accommodation is effected by a corneal muscle (Fig. 3.6, 23), which spans the cornea (Fig. 3.6, 24), thereby pushing the lens (Fig. 3.6, 25) closer to the retina. This represents a unique mode of visual accommodation among extant vertebrates.

There are seven gill openings on each side (Fig. 3.3A2, 1), arranged in slanting rows behind the eyes. On the surface of the head are rows of neuromasts of the sensory-line system (Fig. 3.6, 31), innervated by the *lateralis* branch of the acoustic nerve. Between the eyes there is a median translucent area covering the photosensory pineal and parapineal organs.

The skull is more complex than in hagfishes, essentially because of the presence of a large branchial skeleton, the 'branchial basket', which extends outside the gill pouches and surrounds the heart (Fig. 3.3, 23; Fig. 3.6, 29). The braincase is also more developed with lateral walls and transverse bars (Fig. 3.3B2, 24). The structure of the 'lingual apparatus' is based on very much the same principle as that of hagfishes, i.e. a bar (e.g. the 'piston cartilage'; Fig. 3.3, 20; Fig. 3.6, 5) with a tooth-bearing element, the apical cartilage, which bears the horny apical teeth (Fig. 3.6, 4) and rotates on its apex (Fig. 3.6F). In lampreys, however, the entire apparatus can also be moved back and forth to bring the apical teeth into contact with the skin of the prey. Its musculature is thus more complex than in hagfishes. This 'lingual apparatus', which probably has nothing to do with the tongue of the gnathostomes, has long been claimed to be a either a well-established unique character of the cyclostomes, or the result of a convergence. It is, however, likely to be a general character of the craniates, lost by the gnathostomes when the jaws appeared. This basic in-and-out movement of the mouth skin and associated horny teeth may well also have been present in fossil jawless craniates.

The lamprey snout is strengthened by a series of cartilaginous plates (Fig. 3.3, 18), and there is a cartilaginous ring, the annular cartilage (Fig. 3.3, 19), around the mouth.

The olfactory organ in small, though paired, and enclosed in a cartilaginous capsule with a single common opening (Fig. 3.3B2, 8). The otic capsule (Fig. 3.3B2, 10) contains a labyrinth with two distinct vertical semicircular canals and their respective ampullae (Fig. 3.6, 20, 21), and a number of ventral ciliated sacs (Fig. 3.6, 22) that play a role in maintaining equilibrium.

As in hagfishes, there is a large notochord (Fig. 3.3B2, 11), but the overlying spinal cord is flanked with arcualia (Fig. 3.3B2, C1, 17) as in the gnathostomes. These arcualia occupy the same position as those of the gnathostomes, relatively to the spinal nerve-roots (Fig. 3.3C2, 43, 44), and segmental blood vessels, and are probably homologues of the basidorsals and interdorsals of the gnathostomes (Fig. 3.3C, 37, 38).

In the adult lamprey, the gills are enclosed in muscular sacs, each of which contains an anterior and posterior hemibranch (Fig. 3.6C2, 11, 12). In fact, larval anatomy shows that the posterior hemibranch of one pouch and the anterior hemibranch of the next posterior one form a single branchial unit (Fig. 3.6C1, 12), very much like in the gnathostomes (Fig. 3.10). The foremost functional gill is the posterior hemibranch borne by the hyoid arch. Although there is a spiracular slit between the mandibular and hyoid arches in the lamprey embryo, it disappears early in development, and the larva shows no spiracular opening. Instead, the two arches, mandibular and hyoid, are closely united (Fig. 3.6, 15, 16). In adult lampreys, the gill pouches open medially into a blind expansion of the gut, the pharyngobranchial duct (Fig. 3.6, 13), which develops during metamorphosis.

The brain is elongated in shape (Fig. 3.6J); a slight dorsal flexure appears in the adult. There is a cerebellum, in the form of thin fold between the optic lobes and the rhombencephalon. The pineal and parapineal organs are photosensory, and the habenular ganglion of the left side is larger than that of the right side.

The cranial nerve series comprises oculomotor, trochlear, and abducens nerves for the extrinsic eye muscles. In contrast to hagfishes, the *ramus mandibularis* of the trigeminus nerve is closely associated with the *ramus maxillaris*, as in the gnathostomes. The acoustic nerve develops a *lateralis* component that innervates the lateral-line neuromasts (Fig. 3.6, 31). The spinal cord is dorsoventrally flattened, yet less than in hagfishes (Fig. 3.8G), and the dorsal and ventral spinal roots are separate.

The heart differs from that of hagfishes in the close association of the ventricle and atrium (Fig. 3.6E, 27, 28) and the presence of a cardiac innervation. Its enclosure in a rigid cartilaginous capsule (Fig. 3.6C2, 29) produces a suction effect on the atrium during the ventricular systole. Unlike hagfishes, lampreys have no accessory heart.

The intestine, in particular in the adult, displays a slight spiral fold, the typhlosole (Fig 3.6D, 17), which is foreshadowed in the hagfish intestine, and may be regarded as a plesiomorphous state of the spiral valve of the gnathostomes (Fig. 3.9H).

Lampreys possess a spleen, and their endocrine pancreas is concentrated in one part of the digestive tract, whereas in hagfishes the pancreas is disseminate along the intestine.

Lampreys undergo a larval development, which may last for several years, and a rapid metamorphosis (Fig. 3.6N). The larva lives in fresh water and is a microphagous suspension feeder, which traps food particles with mucus produced in the pharynx. The larval skeleton is quite different from that of the adult. There is no lingual apparatus or sucker, and the velum, which pumps water from the mouth toward the gills, is shaped differently from that of the adult. Unlike that of the adult, which is a peculiar hand-shaped structure (Fig 3.6B, 10), the larval velum (Fig. 3.6C1, 10) is a valvule-shaped organ situated at the entrance of the pharynx, and is supported by a cartilaginous skeleton derived from the mandibular mesodermal segment in the embryo. In addition to the cartilaginous bars, the larval skeleton consists of a special tissue, the muco-cartilage (Fig. 3.6M, stippled), which breaks down during metamorphosis into an undifferentiated tissue from which musculature and cartilage seem to develop anew. The sucker, lingual skeleton, and many other structures of the head thus form only in metamorphosis.

The embryonic development of lampreys, in particular in the earliest stages, is now well known, and displays very much the same pattern as in the gnathostomes. The lamprey embryo differs, however, from that of the gnathostomes in that the ventral part of the mesoderm of the head (the parasomitic mesoderm) is segmented in the same way as the dorsal part (somites), whereas it forms a single 'lateral plate' in the gnathostomes (subsequently pierced by the gill slits). This condition is regarded as primitive because a similar complete mesodermal segmentation also occurs in the cephalochordates. As far as the skeleton is concerned, most of the braincase and the arcualia derive from the sclerotomic part of the somites, and the ectomesenchyme (neural crest cells) gives rise to the visceral skeleton (gill arches, skeleton of the velum, some snout plates, and the trabecles of the embryonic braincase).

A classical feature of the ontogeny of lampreys is the dorsal migration of the nasal placode (Fig. 37, 3) and associated hypophysial tube (Fig. 3.7, 2). In the early stages, both embryonic structures are ventral in position (Fig. 3.7, A–C), as in the gnathostome embryo; they then migrate dorsally as the 'oral hood' develops (Fig. 3.7, D, E). This supports the idea that the dorsal nasohypophysial opening in lampreys is a derived state.

3.4 GNATHOSTOMATA, OR JAWED VERTEBRATES

The gnathostomes, or jawed vertebrates, represent the majority of extant vertebrates. They are characterized chiefly by a vertically biting apparatus, the jaws, but also by a large number of other characteristics ranging from paired nostrils, myelinated nerve fibres, germ cells passing through excretory apparatus in males, paired fins, and bone to various biochemical characteristics. Extant gnathostomes fall into two major taxa, the Chondrichthyes, or cartilaginous fishes, and the Osteichthyes, or bony fishes, the latter including the four-legged land vertebrates.

3.4.1 General features

The general state for the overall external morphology of jawed vertebrates is an elongate body-shape, pectoral and pelvic fins, two dorsal fins, an anal fin, and an epicercal caudal fin, i.e. the vertebral column points posterodorsally. Secondary modifications, such as the loss of one or all dorsal fins, of the anal fin, or the gain of a diphycercal (symmetrical) caudal fin, have occurred more than once, and through different processes. As in lampreys, all the fins contain radials with associated radial muscles.

Extant gnathostomes differ from both groups of extant jawless craniates by the presence of paired fins, pectoral and pelvic (Fig. 3.8B2, 4, 5), which are modified into pectoral and pelvic limbs in land vertebrates (Fig. 3.8B1). There is, however, palaeontological evidence that paired fins, probably homologous to the pectorals, existed before jaws. Separate pelvic fins are known exclusively in the gnathostomes. Both pelvic and pectoral paired fin skeletons articulate on endoskeletal girdles, the pectoral girdle being generally more developed more than pelvic one. The mouth of the gnathostomes is armed with jaws, i.e. a biting apparatus that comprises an upper (fixed) and a lower, (mobile) component (Fig. 3.8A, 1, 2). The upper and lower jaws may bear bones, teeth, or epidermal phaners. They may be reduced (Fig. 3.8A2) but are never lacking.

The nasal sacs are separate and open directly toward the exterior on each side (Fig. 3.8A, 3). There are anterior, incurrent, and posterior, excurrent nostrils, which are completely separated only in osteichthyans.

The eyes do not differ much from those of lampreys but they have, in addition to the extrinsic muscles, an intrinsic musculature that serves to accommodate the lens (Fig. 3.8C, 6). In extant gnathostomes, the superior oblique extrinsic eye muscle (innervated by the trochlear nerve) is attached to the wall of the orbital cavity anteriorly, whereas it is attached posteriorly in lampreys. The last-named condition can be regarded as general for the gnathostomes, for it occurs in fossil jawless forms (osteostracans) as well as in one fossil gnathostome taxon, the placoderms.

The labyrinth, housed in the otic capsule, is basically similar to that of lampreys but has, in addition, a horizon-

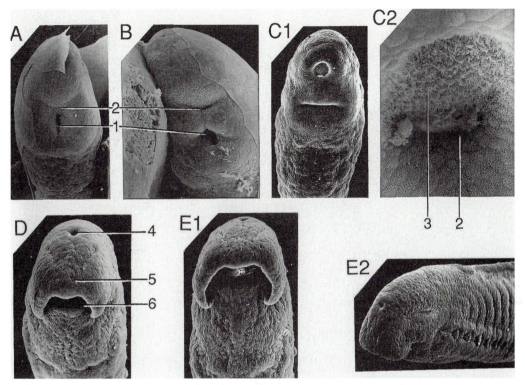

Fig. 3.7. Embryonic development of the head of a lamprey. A–E, scanning electron micrographs of the embryonic head of *Petromyzon* at five successive stages ('Piavis' stages), showing the dorsal migration of the nasohypohysial complex. A, stage 13, ventral view; B, stage 14, ventral view; C, stage 15, ventral view (C1) and detail of the nasohypophysial region (C2); D, stage 17, ventral view; E, stage 18, ventral (E1) and anterolateral (E2) view. 1, stomodeum; 2, hypophysial tube; 3, presumptive olfactory epithelium of the nasal placode; 4, nasohypophysial opening; 5, oral hood ('upper lip'); 6, mouth. (Courtesy of A. Tamarin, University of Washington, Seattle.)

tal semicircular canal (Fig. 3.8F, 7), and lacks the ciliated sacs. The labyrinth is also initially connected to the exterior by an endolymphatic duct (Fig. 3.8F, 8). This duct, which is always present in the embryo as well as in many early fossil taxa, is closed in most extant gnathostomes.

The lateral-line system is more developed than in lampreys, expanding over the entire body in the form of canals, opening of the exterior by pores, or pit lines, which both house neuromasts (Fig. 3.8D, 9–11). This sense organ, linked with an aquatic mode of life, is much reduced or lacking in most land vertebrates (Fig. 3.8E). They have a different sound-conducting device, the middle ear. This is made up by elements of the skull, and develops in various ways (Fig. 3.23E).

The respiratory apparatus in most originally aquatic gnathostomes consists of gills (Fig. 3.8J, 12). The gill chamber (Fig. 3.8J, 15), opens to the exterior either by separate gill slits or by a common gill opening covered either by separate (Fig. 3.8J1, 14), or by a single gill cover (Fig. 3.8J2, 16). Behind the eyes, there is sometimes a spiracle (Fig. 3.8J1, 17), for the intake of the respiratory water. In most Recent gnathostomes, however, the spiracle is lacking. Some osteichthyans possess diverticles of the pharynx that may serve, like lungs, to breathe air directly (Fig. 3.9G). Land vertebrates and some of their closest relatives, lungfishes, have either rudimentary gills or no gills at all, and use lungs exclusively for respiration.

3.4.2 The two skeletons

Most of the characters that are relevant to early vertebrate palaeontology belong to the skeleton. As we shall see in the next chapter, the cartilaginous skeleton of hagfishes and lampreys is of very little help in the interpretation of the strongly ossified and different structure of the fossil jawless forms. In contrast, a thorough knowledge of the

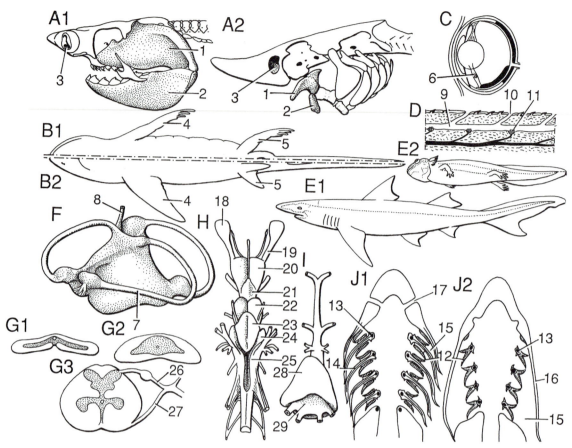

Fig. 3.8. Basic structure of the Gnathostomata. A, the mandibular arch (stippled) in two different states of either enlargement (A1, shark) or reduction (A2, sturgeon); B, the paired extremities in a piscine gnathostome (below) and a tetrapod (above); C, vertical section through the eye, showing the intrinsic musculature; D, longitudinal section through a sensory-line canal and neuromasts of a shark; E, distribution of the lateral-line canals in a shark (E1) and of the free neuromasts in a larval amphibian (E2); F, labyrinth of a shark in lateral view; G, transverse section through the spinal cord of hagfishes (G1), lampreys (G2), and gnathostomes (G3), to show the distribution of the grey matter (stippled); H, brain of a shark in dorsal view; I, heart of a shark in dorsal view; J, organization of the branchial apparatus in a shark (J1) and an osteichthyan (J2). 1, palatoquadrate; 2, Meckelian cartilage or bone; 3, external opening of nasal cavity; 4, pectoral fin or fore limb; 5, pelvic fin or hind limb; 6, intrinsic eye muscle; 7, horizontal semicircular canal; 8, endolymphatic duct; 9, lumen of sensory-line canal; 10, sensory-line pore; 11, sensory-line neuromast; 12, gill lamellae; 13, gill arch; 14, gill septum; 15, gill chamber and gill slit; 16, gill cover; 17, spiracle; 18, olfactory bulb; 19, olfactory tract; 20, telencephalon; 21, diencephalon; 22, mesencephalon (optic lobes); 23, metencephalon; 24, cerebellar auricles; 25, myelencephalon (medulla oblongata); 26, dorsal spinal nerve root; 27, ventral spinal nerve root; 28, ventricle; 29, atrium. (A, J, from Devillers (1958); C, from Waterman (1971); D, from Goodrich (1909); E, G, based on Bolk *et al.* (1934); F, from Retzius (1881); H, I, from Marinelli and Strenger (1959).)

skeletons of extant gnathostomes is an important pre-requisite for the study of fossil gnathostomes, which retain the same basic structure.

Among extant craniates, only the gnathostomes possess an ossified skeleton, whether external or internal, but bone was present in many fossil jawless vertebrates and thus has preceded jaws in time. The skeleton of the gnathostomes and of these fossil jawless vertebrates has a multiple embryonic origin. In fact, vertebrates have in effect two skeletons, which have separate histories: the *exoskeleton,* or *dermal skeleton* (including epidermal derivatives), and the *endoskeleton* (Fig. 3.9A, C). The history of the craniate or vertebrate skeleton will be discussed in Chapter 6, with a consideration of the large

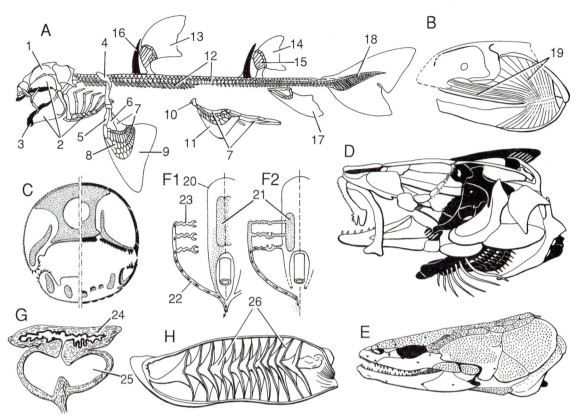

Fig. 3.9. Basic structure of the Gnathostomata. A, skeleton of a chondrichthyan (*Heterodontus*) in lateral view, showing the associated large dermal elements (teeth and fin spines in black); B, attempted reconstruction of the most generalized gnathostome jaw musculature in the Devonian chondrichthyan *Cladodus*; C, schematic transverse section through the head of a chondrichthyan (left-hand side) and an osteichthyan (right-hand side), showing the distribution of the exoskeleton (black) and endoskeleton (stippled); D, skull and shoulder girdle of a cod (*Gadus*) in lateral view, showing the distribution of the dermal (white) and endoskeletal (black) bones; E, skull and shoulder girdle of the cladistian actinopterygian *Polypterus* in lateral view, showing the dermal bones with ornamented parts (white), and the endoskeletal bones (black). F, schematic representation of the urinary and genital apparatus in male lampreys (F1) and gnathostomes (F2), gonads and sperms stippled; G, transverse section through the digestive tract and air bladder ('lung') of the cladistian actinopterygian *Erpetoichthys*; H, longitudinal section through the spiral intestine of a shark (*Squalus*). 1, braincase; 2, visceral skeleton (mandibular, hyoid, and branchial arches); 3, teeth; 4, scapulocoracoid; 5, propterygium; 6, mesopterygium; 7, metapterygium; 8, paired fin radials; 9, pectoral fin; 10, pelvic girdle; 11, pelvic fin; 12, vertebral column (arcualia and notochord) and ribs; 13, anterior dorsal fin; 14, posterior dorsal fin; 15, median fin radials; 16, fin spine; 17, anal fin; 18, chordal lobe of caudal fin; 19, adductor jaw muscle; 20, coelomic cavity; 21, gonads; 22, urinary duct; 23, kidney tubules; 24, digestive tract; 25, air bladder; 26, spiral valve. (A, from Dean (1895); B, from Lauder (1980*a*); E, from Bjerring (1985); F, from Janvier (1981*a*); G, from Marcus (1937); H, from Marinelli and Strenger (1959).)

amount of data provided by the fossil taxa, in particular the fossil jawless craniates.

3.4.2.1 The nature of the endoskeleton

The endoskeleton, initially made up of cartilage, is probably the earliest one to appear, for it occurs in all craniates and appears first in ontogeny. It is internal and forms during ontogeny from the mesoderm, mainly from cells derived from the sclerotomic part of the somites, as well as from the ectomesenchyme, that is, cells derived from the neural crest (Fig. 3.2). During ontogeny, the braincase (neurocranium), vertebrae, paired and unpaired fin skeleton, and girdles (Fig. 3.9A) derive chiefly from the somitic and lateral-plate mesoderm, whereas the visceral skeleton (splanchnocranium, i.e. jaws and gill arches) derives from the ectomesenchyme (Fig. 3.9, 2). The endoskeleton may turn into bone in different ways: either the superficial tissues lining the cartilage produces bone (perichondral bone; Fig. 3.16J), or the cartilage itself is invaded by bone (endochondral bone; Fig. 3.16J). Both processes can occur simultaneously, but the fact that only perichondral bone is known in such fossil jawless vertebrates as osteostracans or galeaspids suggests that it occurred earlier than endochondral bone (see Chapter 6, p. 274). In any case, bone arises by the production (by special cells, the osteocytes) of minute crystals of calcium phosphate (bio-apatite), which are arranged along fibres of collagen. Cartilage may also become calcified in the form of a crystalline (prismatic or spherulitic) deposit of calcium phosphate that is not guided by collagen fibres (Fig. 6.12A2, A3).

The cartilaginous endoskeleton of chondrichthyans (sharks, rays, and chimaeras), or 'cartilaginous fishes', has long been regarded as a primitive condition, because extant jawless vertebrates also have a cartilaginous endoskeleton. Both neontological and palaeontological evidence suggest, however, that endoskeletal bone (or at least perichondral bone) has been lost in these gnathostomes; an extremely thin layer of perichondral bone has been discovered lining the cartilage of some extant sharks (Fig. 3.12C). Instead of bone, however, the cartilaginous elements of the chondrichthyans are strengthened with prismatic calcified cartilage, that is, a layer of crystallized calcium phosphate arranged in the form of minute platelets (Fig. 3.12B) which cover many skeletal elements, such as the jaw elements, braincase, and girdles. In addition, the cartilage may contain globular calcified cartilage in the form of minute spheres of calcium phosphate.

The so-called 'membrane bone', which is formed by ossification of membranous connective tissue, is no longer regarded as identical with dermal bone; the term is now generally used for endoskeletal bones that are not preformed in cartilage, such as some bones of the braincase of Recent bony fishes.

3.4.2.2 The nature of the exoskeleton

The exoskeleton is initially superficial and formed in the skin. It consists of three different hard tissues: enamel or enameloid, dentine, and bone. Enamel is produced by the superficial epithelium derived from the embryonic ectoderm or endoderm (the epidermis of the skin or the epithelium of wall of the pharynx). Unlike bone and dentine, enamel contains no collagen fibres. However, enamel appears only when the epidermis is in contact with underlying ectomesenchyme. This phenomenon of induction between these two tissues has been crucial to the rise of the dermal skeleton in general. True enamel (i.e. enamel derived only from the epidermis) occurs only in osteichthyans; other fossil or extant craniates possess an enamel-like tissue, or enameloid, which involves both the ectomesenchyme and the epidermis.

Dentine (meso-, semi-, meta-, and orthodentine; Fig. 6.13) and dermal bone are formed by deeper skeletogenous tissues that all derive from the ectomesenchyme (i.e. neural crest cells) during ontogeny. Both tissues are formed by cells (odontoblasts, osteocytes), which eventually are included inside the matrix they produce, as for the endoskeleton. When this happens, they may either survive (cellular bone) or die (acellular bone).

In fishes, the scales, fin rays, fin spines, teeth, branchial denticles, and superficial head or shoulder girdle bones belong to the exoskeleton (Fig. 3.9A, D, E). Generally, dentine overlies dermal bone, forming a more-or-less continuous layer, but in some instances dentine alone can form rather deep inside the body (e.g. the fin spines of sharks; Fig. 3.9A). The dermal skeleton has long been said to have a tendency towards sinking deeper into the body, while new superficial layers of dermal skeleton form (Holmgren's principle of delamination; Jarvik 1959). In fact, many bones of the human skull, for example, are originally dermal, although lying deep below the skin and undistinguishable from originally endoskeletal bones. The same association of dermal and endoskeletal bones occurs in extant bony fishes, and it is impossible for an unaware student to distinguish the two bone types in the skull of a cod (Fig. 3.9D), for example. In extant sharks, rays, and chimaeras the two types of skeleton are clearly distinct. The exoskeleton consists almost exclusively of scales, teeth, and spines, and the endoskeleton of cartilage (Fig. 3.9A). In most fossil early vertebrates, such as placoderms or early bony fishes, as well as in some extant osteichthyans, this distinction is easier to make thanks to the presence of a dentinous ornament (tubercles, ridges) on the dermal bones (Fig. 3.9E). This 'delamination principle' is now viewed as a consequence of the tendency for ectomesenchymal tissues to pervade the embryo deeper and deeper before they start producing the skeleton (Maisey 1988). This would explain why dermal bones never pass through a cartilaginous state during ontogeny, except in birds and mammals (adventitious cartilage).

3.4.2.3 Diversity of exoskeleton morphology

The structure of the exoskeleton of the gnathostomes is more diverse than that of the endoskeleton. In sharks, rays, and chimaeras, it consists almost exclusively of minute scales, the 'placoid' scales (Fig. 3.13H), which possess a crown of dentine, a base of acellular bone, and a pulp cavity, and of independent teeth, tooth plates, and fin spines more-or-less firmly attached to the endoskeleton (Fig. 3.9A, 3, 16; Fig. 3.13I; Fig. 3.15, 10). In osteichthyans, including tetrapods, the exoskeleton consists of larger units, the dermal bones, which cover the head and girdles and support the teeth, and of larger scales covering the body and the fins (Fig. 3.16). The generalized condition for the gnathostome exoskeleton is regarded as microsquamose, i.e. having minute dermal elements and separate teeth attached to the skin, as in sharks (see p. 131). These small dermal components also cover the inside of the mouth and the wall of the pharynx. All these dermal units which are formed from a single papilla are called *odontodes* and are the basic components of the superficial layer of the exoskeleton. This is inferred from the belief that placoid scales are simple, and that teeth are derived from placoid scales. Palaeontology does not, however, clearly support this hypothesis. The closest fossil relatives of the gnathostomes among fossil jawless vertebrates are all covered with large dermal plates or scales, with the exception of the thelodonts (see Chapter 4, p. 123), and the earliest-known gnathostomes are either microsquamose (some chondrichthyans) or meso- to macrosquamose (i.e., with dermal units composed of more than one odontode, such as acanthodians, placoderms, and osteichthyans). The ontogeny of the dermal skeleton in osteichthyans nevertheless shows that odontodes appear first and then become fused to the bony basal plate, or serve as a core to accretion of other younger odontodes (Fig. 6.15C). There have been several theories of the history of the dermal skeleton, which will be briefly discussed in Chapter 6 (p. 275). There is a current consensus that the microsquamose condition is primitive, but, opinion varies as to whether small compound elements (micromery) or large ones (macromery) in macrosquamose exoskeletons are primitive, and how many times reversions can have occurred.

The structure of the exoskeleton in osteichthyans is one of the major sources of characters used in systematics, but its analysis embodies a large number of assumptions about homologies, fusions, or losses. This will be discussed below (see section on Osteichthyes, p. 67, and Chapter 6, p. 261). There is a close ontogenetic and topographic relationship between the dermal skeleton and the sensory-line neuromasts, and this still remains a major guideline to patterns of homology (Fig. 3.16H, I).

3.4.3 Morphology of the endoskeleton

Whether cartilaginous or bony, the endoskeleton of the gnathostomes displays basically the same components of the skull, axial skeleton, and appendicular skeleton. Its pattern is so profoundly different from that of lampreys and hagfishes that virtually all attempts at finding cranial homologies between the three groups have failed. Only the embryonic trabecles and the otic and nasal capsules may be regarded as homologous in these three groups. The dorsal arcualia of the axial skeleton are also most probably homologous in lampreys and the gnathostomes,

since they have the same relationships to the spinal nerves (Fig. 3.3C, 37, 38).

3.4.3.1 Neurocranium

The endoskeletal skull comprises the neurocranium, or braincase (Fig. 3.9A, 1), and the splanchnocranium, or visceral skeleton, which includes the jaws, hyoid, and branchial apparatus (Fig. 3.9A, 2). The braincase is chiefly formed from the tissues that originate from the sclerotomes (i.e. the wall of the embryonic somites lining the central nervous system and the notochord), but it also includes derivatives of the myotomes (which elsewhere form the body musculature) and dermatomes (which give rise to the dermis). Some parts of the braincase are known to include neural crest derivatives. For example, it has been demonstrated experimentally that the trabecles, which are the chassis of the embryonic braincase are, like the splanchnocranium, derived from neural crest cells. As we shall see in Chapter 6, there have been lengthy debates about the possible incorporation of other elements of the visceral skeleton into the braincase of the gnathostomes. Nevertheless, there is now good evidence that most of the anterior part of the braincase forms from neural crest cells. The braincase of the gnathostomes is generally closed, except in sharks, rays, and some actinopterygians, where it remains dorsally open by a variable number of fontanelles (Fig. 3.13, 3). It houses the brain, olfactory capsules (Fig. 3.3B3, 8), the eye and labyrinth, and major blood vessels as well as occasionally part of the notochord (Fig. 3.21, 11) and some blood-producing (haematopoietic) organs. It is fairly homogeneous in overall shape, and even more so when the most generalized morphologies for each taxon are considered on the basis of early fossil forms (see Chapter 4, p. 128, Fig. 4.27). Major differences rest on the breadth of the space between the eye (in fact, whether the brain extends between the eye or not), the presence or absence of transverse gaps (neurocranial fissures), and the posterior extension of the occipital region. The overall shape of the braincase is generally low and broad in chondrichthyans, and high and narrow in osteichthyans.

3.4.3.2 Visceral skeleton

The visceral skeleton consists of the endoskeletal support of the jaws, or mandibular arch (Fig. 3.3B3, 25, 26), and a subsequent series of arches, the hyoid (Fig. 3.3B3, 29, 30), and gill arches (Fig. 3.3B3, 31–5), which, unlike the unjointed arches of lampreys, are compounded of several elements. All these arches comprise at least two major elements arranged into backwardly directed chevrons: the epal (dorsal; Fig. 3.3B3, 29, 32) and cerata (ventral, Fig. 3.3B3, 30, 33) elements. There can be additional elements, the pharyngeal (infra- and supra-pharyngeal,

(Fig. 3.3B3, 31) elements, connecting the epal elements and the braincase in osteichthyans, and the hypobranchial and hypohyal elements (Fig. 3.3B3, 34; Fig. 3.13D, 9), connecting each arch to the median basibranchial series (Fig. 3.3B3, 35; Fig. 3.13D, 13). The diversity and modifications of these elements, in particular in the gill arches, are an important source of characteristics in the reconstruction of gnathostome interrelationships.

Mandibular arch   The mandibular arch comprises the palatoquadrate (dorsal, Fig. 3.3B3, 25) and the Meckelian cartilage (ventral, Fig. 3.3B3, 26), which articulate at the mandibular joint. The palatoquadrate is attached to the braincase by means of three possible articulations: ethmoid, postorbital, and basipterygoid (or basal), which do not always occur together. In chondrichthyans, the postorbital, and the ethmoid, articulation sometimes occur together, whereas in extant osteichthyans only the ethmoid and basipterygoid articulations always occur together. In some instances (chimaeras, lungfishes, some tetrapods), the palatoquadrate is reduced or fused with the braincase into a single unit (holostylic suspension). In the fossil placoderms and acanthodians, as well as in Recent osteichthyans, the palatoquadrate and Meckelian cartilage become ossified into a variable number of elements. In osteichthyans, the platatoquadrate generally comprises autopalatine and pterygoquadrate ossifications.

The suspension of the palatoquadrate is also effected by means of the following arch, the hyoid arch (hyostylic suspension). This is the case in sharks and many osteichthyans. There, a ligament, or a series of small bones, such as the symplectic, unite the ventral end of the epal element, or hyomandibula, to the posterior corner of the palatoquadrate. There is still controversy as to whether this type of jaw suspension is general or primitive for the gnathostomes, or whether it is the result of a secondary junction between the hyoid and mandibular arch.

Labial cartilages (Fig. 3.3B3, 28) are sometimes associated with the mandibular arch and are of debated origin. The old idea that they represent a remnant of the premandibular arch is now abandoned by most anatomists.

Hyoid arch   The epal element of the hyoid arch, or hyomandibula (Fig. 3.3B3, 29), articulates on the braincase anteriorly to the level of the otic capsule. Between the ceratal element, or ceratohyal (Fig. 3.3B3, 30), and the hyomandibula there are in osteichthyans, one or two small elements, the symplectic and interhyal; these are now regarded as derived, by segmentation, from the hyomandibula and ceratohyal respectively. The ceratohyal may be independent of the dorsal part of the arch when

the hyomandibula is modified. This is the case in four-legged land vertebrates, or tetrapods, where the hyomandibula is modified into a sound-conducting bone, the stapes. The hypohyal connects the ceratohyal to the median basihyal element. It has long been supposed that a complete gill slit once existed between the hyoid and mandibular arches, and that the spiracle of sharks or primitive bony fishes (Fig. 3.3A3, 4; Fig. 3.8J1, 17) is a rudiment of this gill slit. Out-group comparison with lampreys and even fossil jawless craniates does not, however, support this theory, for none of them possesses a spiracular gill slit (Maisey 1989*a*). Only the embryonic development of lampreys and the gnathostomes shows the presence of a cleft between the mandibular and hyoidean mesodermal segments, which subsequently disappears in lampreys and most gnathostomes. Nevertheless, the spiracle of gnathostomes may have a respiratory function when associated with a pseudobranch. The latter is not a true gill, and the water flow in it runs the opposite way to that in true gills.

Gill arches   The subsequent gill arches bear gills that are supported by cartilaginous rods, the gill rays. The number of gill arches in the gnathostomes varies from six to one, but most primitive members of each major gnathostome group have five gill arches, which thus seems to be the general condition (Maisey 1989*a*). In the air-breathing tetrapods, the gill arches are reduced.

The relationships between the gill lamellae and their supporting endoskeletal arches are different in the gnathostomes and jawless vertebrates (Fig. 3.10A–C). In both, each branchial unit consists of two hemibranchs, anterior and posterior (Fig. 3.10, 1), separated by a septum (Fig. 10, 7) and attached to a skeletal arch (Fig. 3.10, 2, 3). However, in the gnathostomes the skeletal arch (Fig. 3.10, 3) lies close to the pharynx, medially to the hemibranchs, afferent and efferent blood vessels, and branchial nerves (Fig. 3.10C, 4, 5, 6), whereas in a lamprey the arch is situated laterally to the gills, vessels, and nerves, inside the head wall (Fig. 3.10A, B, 4, 5, 6). This important difference has long been regarded as irreconcilable, and was further supported by the fact that the gills in lampreys are endodermal in origin and those in the gnathostomes are ectodermal in origin. Hence the idea that gills had evolved twice in the vertebrates, from a hypothetical form devoid of gills (see, e.g. Jarvik 1981*a*; Bjerring 1984). Recent studies on the structure and development of the gills in hagfishes, lampreys and gnathostomes have shown that they are similar at the cellular level, and that ecto- and endodermal contributions occur in both groups (Mallatt 1984*b*, Mallatt and Paulsen 1986). In contrast, the gill arches, although neural crest derivatives in both groups, are constructed in a totally different way, and lie quite dif-

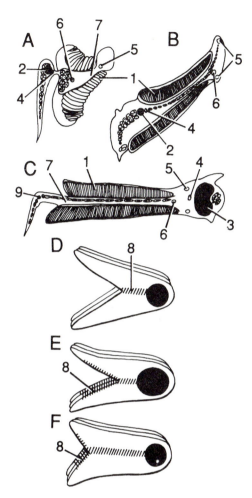

Fig 3.10. Gill structure in lampreys and gnathostomes. A–C, Horizontal section through a gill unit of a larval (A) and adult (B) lamprey and a shark (C); D–F, gill septum and gill rays (hatched) in osteichthyans: teleost actinopterygians (D), actinistians (E), and lungfishes (F). 1, gill lamellae; 2, external branchial arch; 3, internal branchial arch; 4, branchial nerve; 5, efferent branchial artery; 6, afferent branchial artery; 7, 8, interbranchial septum; 9, extrabranchial arch. (A–C, from Mallatt (1984*b*); D–F, from Maisey (1986).)

ferently, being jointed and medial in the gnathostomes, and unjointed and lateral in lampreys. It is now widely admitted that gills are homologous in all vertebrates, and even all craniates, but that the gnathostome gill arches are neomorphs, formed in the medial part of each interbranchial septum. Decisive support for this view is given by the fact that potentially skeletogenic neural-crest derived pigment cells are present in the medial part of the interbranchial septum of larval lampreys (Mallatt 1984*a*,*b*).

In elasmobranchs, there are endoskeletal elements, the extrabranchials (Fig. 3.3B3, 36; Fig. 3.10C, 9), which lie laterally to the gills in very much the same position as the gill arches of lampreys, and it has sometimes been suggested that these could be remnants of the primitive gill arches. No neural crest contribution to the formation of the extrabranchials has, however, ever been demonstrated.

In osteichthyans, the gill septum, or interbranchial septum, is not prolonged to the wall of the branchial chamber (Fig. 3.10D–F, 8). It is short in actinopterygians (Fig. 3.10D) and is divided into two blades along the gill lamellae in sarcopterygians (Fig. 3.10E, F), a condition that is regarded as derived.

3.4.3.3 Axial skeleton

The axial skeleton is, like most of the neurocranium, of somitic origin, i.e. formed from the sclerotomic part of the somites and the notochord. It consists chiefly of the notochord and a number of elements, the arcualia, surrounding it and the overlying spinal cord. The general condition for the gnathostomes, still found in Recent sharks, for example, is four elements for each segment: ventrally the basiventral (Fig. 3.3C2, 40) and interventral (Fig. 3.3C2, 39), and dorsally the basidorsal (Fig. 3.3C2, 38) and interdorsal (Fig. 3.3C2, 37). The basidorsal and interdorsal are probably the only elements that occur in lampreys (Fig. 3.3C1, 38, 37). All these elements may reduce or fuse in various ways, and their fate may be further complicated by a phenomenon of resegmentation which occurs during ontogeny in some groups (teleosts, tetrapods). In several gnathostome groups the notochordal sheath becomes calcified or ossified to form vertebral centra.

Ribs are present in several gnathostomes groups, but are of three distinct type that do not always coexist. There are dorsal, ventral (pleural), and myoseptal ribs. Myoseptal ribs (often referred to simply as 'myosteptal bones') are unique to teleost fishes. Ventral ribs are situated in the wall of the coelom. They grow centrifugally from the vertebral column, and probably represent the most generalized type, since they occur in sharks (Fig. 3.3C2, 41), actinopterygians, and lungfishes. Dorsal ribs are included in the horizontal septum of the musculature and develop centripetally from the body wall. They occur in actinopterygians and possibly also in tetrapods (although only urodeles have a horizontal septum). The identification of the rib type is nevertheless quite difficult, since all the criteria of recognition (in particular between pleural and dorsal ribs) may fail, according to the level of the vertebral column that is considered.

3.4.3.4 Median fins

The median, or unpaired, fins are strengthened by endoskeletal fin rays, or radials (Fig. 3.9A, 15), which articulate on basal elements (whether they are separate elements or a single cartilaginous plate) in the body wall. The basal elements connect radials to the neural and haemal spines of the vertebral column, sometimes with intervening supraneural elements in osteichthyans. The gnathostomes generally have two dorsal fins (Fig. 3.9A, 13, 14), an anal fin (Fig. 3.9A, 17), and a caudal fin (Fig. 3.9A, 18). The generalized condition for the caudal fin structure in the gnathostomes is believed to be epicercal, that is, with an upwardly directed axis containing the notochord and vertebral column, and lined ventrally by the caudal fin web (Fig 3.9A, 18). This is inferred from the fact that the group of fossil jawless vertebrates that shares the largest number of unique characters with the gnathostomes, the osteostraci, has a caudal fin of this type (Fig. 4.14), and also from the presence of an epicercal tail in most of the earliest-known gnathostomes (acanthodians, placoderms, most osteichthyans). Ontogeny also confirms the generalized state of the epicercal tail. An apparently diphycercal tail, that is with dorsal and ventral webs equal in size, occurs in many fish groups, but the endoskeletal support generally shows some trace of an epicercal condition (teleostean actinopterygians; Fig. 3.18E). Only extant sarcopterygians (the coelacanth, lungfishes, and tetrapods) possess a true diphycercal tail, with a horizontal lobe containing the notochord (Figs 3.21, 3.22).

3.4.3.5 *Paired fins*

The appendicular endoskeleton, i.e. that of the pectoral and pelvic girdles and fins (or limbs), develops from two different sources. The girdles derive from the lateral plate mesoderm, which migrates toward the body wall, whereas the radials (endoskeletal fin rays and limb bones) develop from a mesenchymatous filling of the limb and fin buds, which arises just behind the gill openings and in front of the anus.

The pectoral girdle is initially a single mass of cartilage or bone, the scapulocoracoid (Fig. 3.9A, 4), pierced by canals for nerves and blood vessels, but it may separate into two distinct units, the scapula and the coracoid respectively, in higher tetrapods (amniotes) and actinopterygians, presumably by convergence. The pelvic girdle (Fig. 3.9A, 10) is smaller than the pectoral one, and is divided into three components in amniotes. In actinopterygians, the pelvic girdle is supposed to be lacking, and its role is played by a basal fin element, the metapterygium, sunk into the body wall.

In extant chondrichthyans, the radials of the paired fins articulate on the girdles by means of intervening elements, the pro-, meso-, and metapterygium (Fig. 3.9A, 5–7), which are derived from the fusion of the proximal radial segments, and constitute the basipterygium. Fossils indicate, however, that unsegmented radials articulating directly on the girdles represent the most generalized condition for the gnathostomes (see Chapter 4, p. 133, Fig. 4.29B). In many extant gnathostomes, such as sharks, rays, chimaeras, and sturgeons, the posterior part of the paired fin skeleton consists of radials that articulate obliquely on a series of elements which seems to prolong the girdle posteriorly. This component of the fin is the metapterygium, which thus comprises a metapterygial axis, on which articulate metapterygial radials (Fig. 3.9A, 7). In one group of extant gnathostomes, the sarcopterygians, the paired fins are represented only by the metapterygium (Fig. 3.20). In chondrichthyans, the metapterygium of the pelvic fin in males is modified into a complex copulatory organ, the mixipterygium or pelvic clasper (Fig. 3.12A, 2, 3). In most piscine gnathostomes, the thinnest margins of the unpaired and paired fins are strengthened by flexible rods of collagen, the ceratotrichiae, which are of dermal origin.

3.4.5 Soft anatomy

The major basic morphological features of the soft anatomy of extant gnathostomes are to be found in the brain, heart and blood vessels, digestive tract, kidneys, and gonads. The musculature of a gnathostome does not differ greatly from that of a lamprey, except in the anterior part of the head, namely the musculature associated with the jaws, and in the paired fins, which are lacking in extant jawless craniates. In the gnathostomes, as in lampreys and hagfishes, the trunk musculature is arranged into myomeres, but these are divided horizontally by a horizontal septum (absent in the tetrapods, except in the urodeles). The jaw movements are effected by a powerful adductor mandibulae muscle (Fig. 3.9B, 19), which lies laterally to the palatoquadrate and Meckelian cartilage (or bone), and an abductor mandibulae muscle connected to the shoulder girdle. In the fins, each radial is associated with a radial muscle on each side, as in lampreys (Fig. 3.6, 19), but modifications of this arrangement occur in sarcopterygians in connection with the exclusively metapterygial skeleton of the paired fins.

The nervous system of extant gnathostomes differs from that of lampreys and hagfishes in, for example, the presence of myelinated nerve fibres. The brain also has a larger cerebellum with lateral and median divisions (Fig. 3.8H, 23, 24). There are two distinct olfactory tracts leading to respective olfactory bulbs (Fig. 3.8H, 18, 19). The three branches of the trigeminus nerve issue from a single trigeminus ganglion and the branchial nerves have a sensory pretrematic branch that passes anteriorly to the gill slits. The spinal cord is thicker than in hagfishes and lampreys, with typical 'horns' of grey matter in section

(Fig. 3.8G). The dorsal and ventral spinal nerve roots are united to form compound spinal nerves (Fig. 3.8G3, 26, 27). The *lateralis* component of the acoustic nerve runs far back in the body, to innervate the trunk and tail sensory lines (Fig. 3.8E). The eye of the gnathostomes has an intrinsic musculature (Fig. 3.8C, 6), which serves for accommodation. The labyrinth has a horizontal semicircular canal (Fig. 3.8F, 7). In some fishes, the endolymphatic duct (Fig. 3.8F, 8) is open to the exterior.

In the gnathostome heart, the atrium, or auricle, lies posterodorsally to the ventricle (Fig. 3.8I, 28, 29), instead of laterally as in hagfishes and lampreys (Figs 3.4E, 3.6E). The atrium is generally single-chambered, but a short septum dividing it into two auricles appears in lungfishes, and develops further in tetrapods. There are still questions about its homology in the two groups.

The digestive tracts of all primitive members of the various gnathostome groups (all chondrichthyans, primitive actinopterygians, coelacanth, and lungfishes) have a spiral valve in the intestine (Fig. 3.9H, 26). This is believed to be general for the gnathostomes, since a less developed spiral fold, or typhlosole, exists in lampreys (Fig. 3.6, 17). The structure of the spiral valve has, however, undergone separate specializations within the various groups in which it occurs.

Among the gnathostomes, only osteichthyans develop diverticles of the digestive tract, which may serve as either lungs or air bladder for buoyancy (Fig. 3.9G, 25). The homology of the lungs or air bladder in the sarcopterygians and actinopterygians, and even within actinopterygians, is still debated, on the ground that these organs arise from the gut differently and at different places. The diverticles have a respiratory function in cladistians, but are most efficient in lungfishes and tetrapods. Chondrichthyans have no buoyancy device other than a large amount of oil.

The kidneys in adult gnathostomes are formed only by the mesonephros and metanephros. In rare instances, owing to paedomorphosis, a functional pronephric component may persist. The gnathostomes differ from both hagfishes and lampreys in having the male gonads and kidneys linked by ducts that allow the sperms to pass through the excretory system (Fig. 3.9F2). In lampreys and hagfishes, sperms are shed directly in the coelomic cavity, a condition that is thought to be generalized for the craniates (Fig. 3.9F1).

The embryonic development of the gnathostomes does not differ considerably from that of lampreys, except in the anterior part of the head. There, the fate of the nasal capsules (Fig. 3.11, 2) and that of the ectodermal infolding, which is classically regarded as giving rise to the adenohypophysis (Rathke's pouch, Fig. 3.11, 1) are dissociated. The nasal capsules migrate towards a terminal position, whereas the Rathke's pouch remains in the roof of the developing oral cavity and meets the brain floor to contribute to the formation of the hypophysis. Consequently, no nasopharyngeal duct or nasohypophysial tube forms, and the two well-separated nasal cavities open directly to the exterior by means of nostrils. Since the inhalation of respiratory water is effected directly through the mouth, the water flow is led towards the nasal capsule by means of anterior (incurrent) and posterior (excurrent) nostrils. In extant gnathostomes, the buccohypophysial duct (the remnant of the Rathke's pouch), which connects the adenohypophysis to the oral cavity, remains only in the coelacanth, but it is a widespread character in all early gnathostomes.

3.5 DIVERSITY AND RELATIONSHIPS OF RECENT GNATHOSTOMES

Recent gnathostomes fall into two major clades, the chondrichthyans and osteichthyans. Their common ancestor is also the common ancestor to all known fossil gnathostomes, except perhaps the placoderms and a few Silurian fossils doubtfully referred to as chondrichthyans (see Chapter 5, p. 241).

3.5.1 Chondrichthyes

Chondrichthyans, or cartilaginous fishes, are characterized by a prismatic calcified cartilage layer covering the endoskeleton (Fig. 3.12, 5), and by a particular male copulatory organ, the pelvic clasper, or mixipterygium, which is a modified metapterygium (Fig. 3.12, 3, 2). There is a broad diversity in the structure of the pelvic clasper, hence the widespread belief that it is homoplasic in sharks and chimeras; but palaeontology rather supports homology. Another chondrichthyan character may be the large basal cartilaginous plates that support the radials in fins.

Recent chondrichthyans comprise the Elasmobranchii (sharks, skates, and rays) and the Holocephali (chimeras). A major difference between the two groups is the number of branchial openings: elasmobranchs have five to seven separate gill slits (Fig. 3.13, 2), whereas holocephalans have a single large gill opening (Fig. 3.15, 4), covered by an opercular flap, as in bony fishes. Since jawless vertebrates, except fossil heterostracans, have several separate gill openings, the elasmobranch condition may be regarded as general for the gnathostomes, and the operculate condition as particular to holocephalans and osteichthyans.

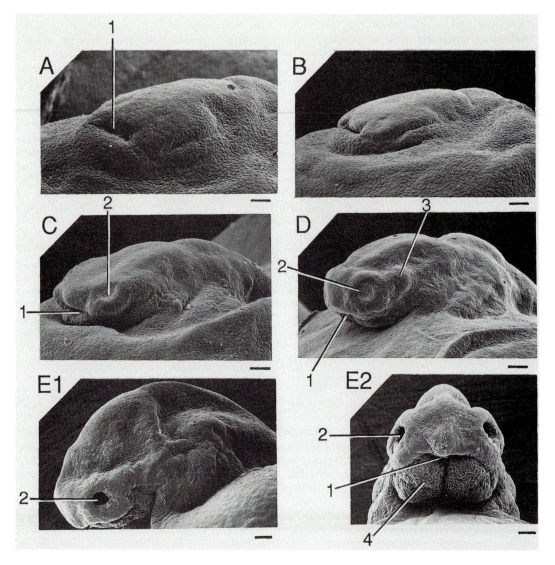

Fig. 3.11. Embryonic development of the gnathostome head. Scanning electronic microscope photographs of a series of embryos of the Actinopterygian *Polyodon* (A–E1, anterolateral view; E2, anteroventral view; scale bar: 100 μm), showing the independence of the olfactory placodes in relation to the Rathke's pouch (compare with Fig. 3.7). 1, Rathke's pouch; 2, olfactory placode; 3, optic vesicle; 4, special adhesive organ. (From Bemis and Grande (1992), by courtesy of the authors.)

3.5.1.1 Elasmobranchii

Extant elasmobranchs are characterized by basibranchial elements separated from the basihyal, posteriorly directed hypohyals and hypobranchials (Fig. 3.13D, 9), pectoral fins with three basal plates (pro-, meso-, and metapterygium), and a rather narrow articulation of the pectoral fin skeleton on the scapulocoracoid. Most elasmobranchs retain a spiracular opening behind the eye (Fig. 3.13A, 1)

and generally five (rarely six) gill slits (Fig. 3.13A, 2). The braincase is pierced anteriorly by a large fontanelle (Fig. 3.13B, C, 3), and the endolymphatic duct opens in a median dorsal endolymphatic fossa (Fig. 3.13C, 5). It also displays a more-or-less distinctly paired occipital condyle (Fig. 3.13C, 8). The nostrils are situated on the ventral side of the snout (Fig. 3.13F), and the eye is generally attached to the braincase by means of a cartilaginous

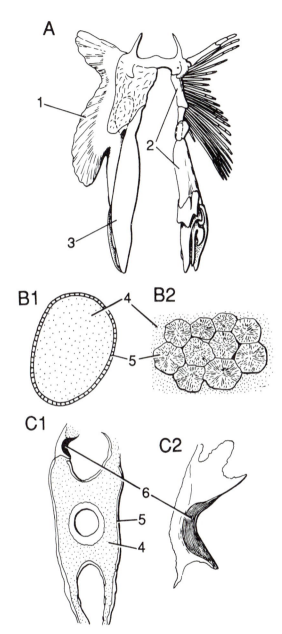

optic pedicle, or eye-stalk (Fig. 3.13G, 20). The vertebral column consists of arcualia that are inserted into strongly calcified centra (Fig. 3.13E, 14–16). The exoskeleton consists of minute, non-growing, placoid scales (Fig. 3.13H, 21–3), but some eleasmobranchs possess larger dermal elements in the form of dorsal fin spines (Fig. 3.9A, 16; Fig. 3.13A). The teeth of elasmobranchs are separate and form tooth 'families' in which the younger teeth form on the lingual side of the jaw, and the older ones are shed as they reach the labial side (Fig. 3.13I, 24, 25). The average 'life' of a shark tooth is about 12 days. The teeth of modern elasmobranchs are characteristically covered with shiny layer of enameloid tissue. In sharks, this enameloid is three-layered and each layer has a particular microstructure: 'shiny', 'parallel-fibred' and 'haphazardly fibred'. The haphazardly fibred layer also occurs in batomorphs.

Rays and their close relatives, or batomorphs (Fig. 3.14, 3), clearly form a clade and are characterized by enlarged pectoral fins, which are connected to the antorbital process by means of an antorbital cartilage (Fig. 4.14N2), as well as fused anterior vertebrae (synarcual). In contrast, sharks do not share any clearly unique character. The interrelationships of elasmobranchs are consequently a much-debated subject. Some consider sharks and batomorphs as forming two monophyletic sister-groups (Fig. 3.14, right-hand side), whereas others regard batomorphs as derived sharks, the squatinids (Fig. 3.14J) being anatomically intermediate between sharks and batomorphs (Fig. 3.14, left-hand side). There is, at the moment, no clue to a solution of this problem.

In current classifications, extant sharks comprise two major groups, the squalomorphs (Fig. 3.14, F–I) and galeomorphs (Fig. 3.14, A–E), the latter being poorly defined. *Squatina* (Fig. 3.14J), *Chlamydoselachus* (Fig. 3.14I), hexanchids (*Hexanchus, Heptranchias*; Fig. 3.13H), pristiophoriforms (Fig. 3.14G), and squaliforms (Fig. 3.14F) share a characteristic orbital process of the palatoquadrate which serves in attaching the latter to the braincase behind the optic nerve (Fig. 3.14 F2, I2, J2). This feature is regarded by some as a unique character of this ensemble of 'orbitosylic sharks' (Maisey 1980), since it does not occur in any of the earliest fossil chondrichthyans. In contrast, batomorphs have no orbital process, and this is one reason for rejecting their close relationship to *Squatina*. One must, however, be aware of the fact that the mandibular arch in batomorphs is strongly modified and has lost any contact with the braincase. It is thus possible that the orbital process has been lost secondarily in batomorphs.

Extant batomorphs comprise the torpedoes (torpedinids Fig. 3.14K), sawfish (pristioids, Fig. 3.14L), guitar-rays (rhinobatids, Fig. 3.14M), stingrays (dasyatids, myliobatids, Fig. 3.14O), and true rays (rajids; Fig. 3.14N).

Fig. 3.12. Characters of the Chondrichthyes. A, pelvic fins and girdle of a male ray, showing the pelvic clasper and its endoskeleton; B, section of a chondrichthyan radial, showing the hyaline and prismatic calcified cartilage (B1), and detail of the external surface of the prismatic calcified cartilage (B2, ×10); C, vertical section of a developing dogfish vertebra (C1), showing the thin lining of perichondral bone in the arcualia (C2, ×20). 1, pelvic fin; 2, metapterygium; 3, pelvic clasper; 4, hyaline cartilage; 5, prismatic calcified cartilage; 6, lamellar perichondral bone. (A, from Goodrich (1930); B2, from Goodrich (1909); D, redrawn from Peignoux-Deville *et al.* (1982).)

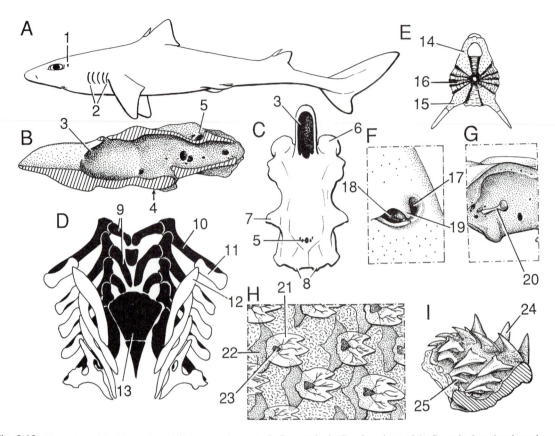

Fig. 3.13. Characters of the Elasmobranchii. A, general aspect of a Recent shark (*Squalus*; about ×0.1); B, sagittal section through the braincase (section of cartilage hatched); C, braincase in dorsal view; D, branchial skeleton in dorsal view (ventral elements in black); E, vertical section through a vertebra with calcified centrum (arcualia stippled); F, left nasal openings of *Squalus* in ventral view; G, right orbital cavity of *Squalus*, showing the eye-stalk; H, organization of the placoid scales in the skin of an elasmobranch, showing the pulp cavity in the scale crown by transparency (about ×20); I, portion of the lower jaw of a shark in labial view, showing the developing and functional teeth. 1, spiracle; 2, gill slits; 3, anterior fontanelle; 4, ventral flexure; 5, endolymphatic fossa; 6, nasal capsule; 7, postorbital process; 8, occipital condyles; 9, posteriorly directed hypobranchials; 10, ceratobranchial; 11, epibranchial; 12, pharyngobranchial; 13, basibranchial copula; 14, basidorsal; 15, basiventral; 16, centrum; 17, anterior nostril; 18, posterior nostril; 19, skin flap (ala nasalis); 20, eye-stalk; 21, scale crown; 22, scale root; 23, pulp cavity; 24, functional tooth; 25, developing tooth. (B, C, G, from Devillers (1958); D, modified from Rosen *et al.* (1981); E, from Goodrich (1930); F, from Marinelli and Strenger (1959); H, from Ørvig (1968*b*); I, from Bendix-Almgreen (1983).)

Fig. 3.14. Diversity and phylogenetic interrelationships of the Elasmobranchii. Sharks are paraphyletic in the cladogram on the left-hand side and monophyletic on the right-hand side. *Terminal taxa*: A, Heterodontiformes (Port Jackson shark); B, Carcharhinidae (blue shark); C, Scyliorhinidae (dogfish); D, Lamniformes (great white shark); E, Orectolobiformes (carpet shark); F, Squaliformes (spiny dogfish); G, Pristiophoriformes; H, Hexanchidae; I, Chlamydoselachidae (frilled shark); J, Squatiniformes (angel fish); K, Torpediniformes (torpedoes); L, Pristidae (sawfish); M, Rhinobatidae (guitar ray); N, Rajidae (rays); O, Myliobatidae (sting rays). *Nested taxa and selected synapomorphies*: 1, Neoselachii (all Recent Elasmobranchii; posteriorly directed basibranchials, paired occipital condyles, tooth enameloid with at least one layer of haphazardly fibred enameloid); 2, Galeomorphii (enlarged preorbital muscle inserted on the lateral surface of orbitonasal lamina); 3, Batomorphii (ventral gill slits, reduced palatoquadrate, synarcual, propterygium in contact with antorbital process); 4, Pleurotremata (sharks) (three-layered enameloid with 'shiny', parallel-fibred, and haphazardly fibred layers); 5, 'orbitostylic sharks' (orbital process of palatoquadrate); 6, Squalomorphii. F1, I1, J1, braincase and palatoquadrate of three orbitostylic sharks, showing the position of the orbital process of the palatoquadrate (arrow); N1, skull and pectoral fin of a ray, showing the contact between the antorbital cartilage (arrow) and the pectoral propterygium (F2, I2, J2, from Maisey (1980); N2, from Garman (1913); characters from Maisey (1984*b*) and Reif (1973).)

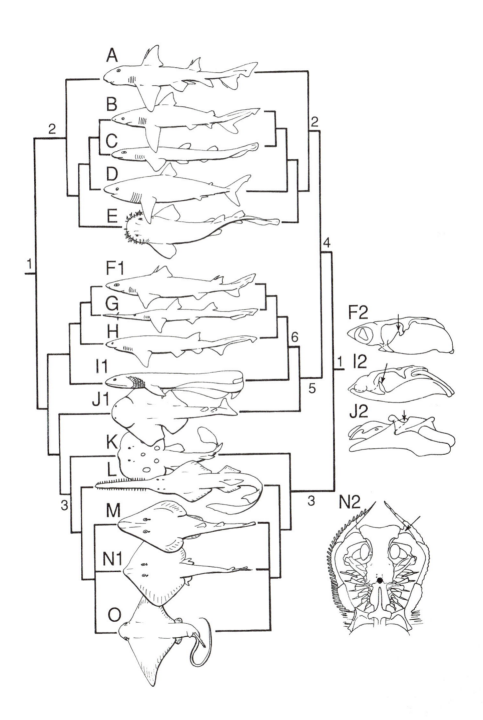

One of the major questions in elasmobranch phylogeny is whether the clade including all Recent taxa also includes major extinct Palaeozoic and Mesozoic elasmobranch taxa, or is restricted to the extant taxa and their fossil, Mesozoic and Cenozoic members. Some consider, for example, that the extant frilled shark *Chlamydoselachus* (Fig. 3.14, I) is more primitive than the extinct Hybodontiformes (see Chapter 4 p. 141; Fig. 4.35B, C), whereas others consider that all extant elasmobranchs form a clade, the Neoselachii, which excludes such Palaeozoic or Mesozoic fossil groups as the Hybodontiformes and Ctenacanthiformes (Fig. 4.35A). The neoselachians (Fig. 3.14, 1) would thus be characterized by a pair of occipital condyles (Fig. 3.13, 8), a slight ventral flexure of the braincase floor (Fig. 3.13, 4), calcified vertebral centra (Fig. 3.13, 16), an externally closed buccohypophyseal duct, and at least one layer of haphazardly fibred enameloid in teeth.

3.5.1.2 Holocephali

Holocephalans are represented by five extant genera only: *Chimaera* (Fig. 3.15A, H2), *Rhinochimaera* (Fig. 3.15H3), *Callorhinchus* (Fig. 3.15H1), *Hydrolagus*, and *Harriotta*. Among the numerous characters that are unique to holocephalans, one may select the tooth plates (Fig. 3.15, 10), the articulated spine of the dorsal fin (Fig. 3.15, 2), the frontal (Fig. 3.15, 1) and prepelvic claspers (Fig. 3.13, 5) in males, and the calcified rings surrounding the sensory-line canals (Fig. 3.15F). Unlike elasmobranchs, and like osteichthyans, they have a single branchial opening, with a fleshy gill-cover (Fig. 3.15, 4). Also, the hypohyals and anterior hypobranchials (Fig. 3.15, 14) are anteriorly directed.

Holocephalans are devoid of the small, independent teeth found in sharks and rays. Instead, their jaws bear powerful tooth plates made up of two particular types of dentine, tubate and pleromic dentine, which increase resistance to abrasion. These hard tissues enable holocephalans to eat shelly prey, and their durophagous diet is probably lined to the reduction and subsequent fusion of the palatoquadrate to the braincase (Fig. 3.15, 9). The tooth plates are formed on the lingual side of the jaw and grow towards the labial side. Thanks to a series of early fossil holocephalans (see Chapter 4, p. 144), it is now possible to understand how these tooth plates have developed from families of separate teeth with a low replacement rate (Fig. 4.36G). There are two pairs of tooth plates in the upper jaw and one pair in the lower jaw (Fig. 3.15, 10).

The large dorsal fin spine of holocephalans is articulated on a thick cartilaginous plate, the synarcual, formed, as in rays, by the fusion of the foremost vertebrae. Both the dorsal spine and the fin (Fig. 3.15, 2, 3) can be housed in a median slit in the back.

In addition to the pelvic claspers (Fig. 3.15, 6), which are common to all extant male chondrichthyans, male holocephalans also have a prepelvic clasper (Fig. 3.15, 5), which is retractible process of the pelvic girdle, and a frontal clasper, or tenaculum, on the snout (Fig. 3.15, 1).

The canals of the sensory-line system, which are widely open to the exterior by large pores (Fig. 3.15, 7), are strengthened by small calcified rings (Fig. 3.15F, 22) found in no other chondrichthyan group. As we shall see below, these minute rings, which appear early in holocephalan history, have been useful in assigning holocephalan relationships to peculiar fossil forms that are otherwise quite unlike extant chimeras. Apart from these calcified rings of the lateral-line system, the skin of holocephalans is essentially naked. Only a few hook-shaped placoid scales occur on the claspers and a few scutes on the back.

The nostrils of holocephalans are ventrally placed, the posterior one being inside the mouth, and the anterior one on the margin of the lip (Fig. 3.15E, 19, 20). This has been thought by some to relate them to lungfishes, which display a comparable condition (see Fig. 3.22, 10, 11). However, holocephalan nostrils remain united by a furrow covered with a skin flap, like those of elasmobranchs and thus differ from the well-separated nostrils of lungfishes and other osteichthyans.

The anatomy and development of holocephalans display puzzling features. For example, the maxillar branch of the trigeminus nerve, which in all other craniates is strictly sensory, contains some viscero-motor fibres. This feature has played a role in the rise of the premandibular arch theory (see Chapter 6, p. 254). Also, the hoid arch of holocephalans (Fig. 3.15C) has been said to possess a pharyngeal element (pharyngohyal), whereas this element is lacking in all other gnathostomes. The hyoid arch of holocephalans was therefore regarded as unmodified and as representing the most primitive condition among the gnathostomes. There is now, however, a consensus that this small 'pharyngohyal' is not a true pharyngeal element and is either part of the epihyal (Fig. 3.15, 11) or a neomorph (Maisey 1984a). Behind the epihyal and ceratohyal there is a large opercular cartilage (Fig. 3.15, 13) formed by the fusion of the epihyal rays.

One theory of relationships between the extant holocephalans shows the Callorhinchidae (*Callorhinchus*) as being the sister-group of the Chimaeridae and Rhinochimaeridae (Fig. 3.15H), both sharing a special mode of tooth-plate growth (Patterson 1992). This would, however, imply that the bifid pelvic clasper of callorhinchids and chimaerids is a general holocephalan character, and that the simple clasper of rhinochimaerids is particular to this group.

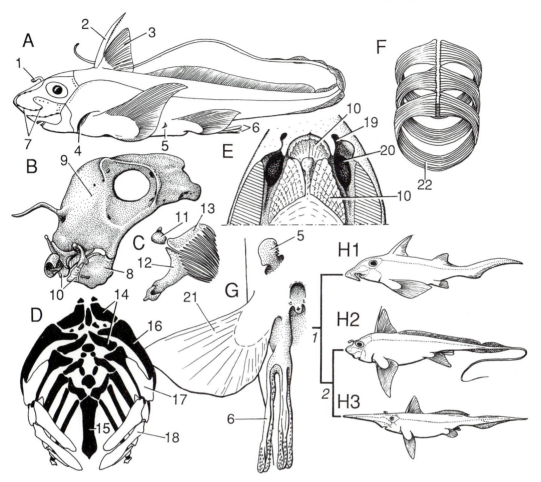

Fig. 3.15. Characters and interrelationships of the Holocephali. A, general aspect of a Recent Holocephalan (male *Chimaera*, ×0.3); B, braincase and lower jaw of *Hydrolagus* in lateral view; C, hyoid arch in lateral view; D, branchial skeleton in dorsal view (ventral elements in black); E, palate and snout of *Chimaera* in ventral view; F, calcified rings of the lateral-line canals; G, pelvic fin with prepelvic and pelvic claspers of *Chimaera* in ventral view; H, interrelationships of the major extant holocephalan taxa (H1, Callorhinchidae; H2, Chimaeridae; H3, Rhinochimaeridae); *nested taxa and selected synapomorphies*: *1*, Holocephali (Chimaeroidei; tenaculum, prepelvic claspers, tooth plates, calcified rings in lateral-line canals), *2*, Chimaeriformes (loss of descending lamina in the anterior upper tooth plates, elongated second dorsal fin). 1, frontal clasper, or tenaculum; 2, mobile spine; 3, dorsal fin; 4, gill slit; 5, prepelvic clasper; 6, pelvic clasper; 7, sensory-line pores; 8, Meckelian cartilage; 9, braincase; 10, tooth plates; 11, epihyal (and possible pharyngohyal); 12, ceratohyal; 13, opercular cartilage; 14, hypobranchials; 15, basibranchial copula; 16, ceratobranchial; 17, epibranchial; 18, pharyngobranchial; 19, anterior nostril; 20, posterior nostril; 21, pelvic fin; 22, calcified rings. (A, from Dean (1895); B, from Holmgren and Stensiö (1936); C, G, from Goodrich (1930); D, modified from Rosen *et al.* (1981); F, from Reese (1910).)

3.5.2 Osteichthyes

Osteichthyans, or bony fishes, are characterized by the presence of endochondral bone (Fig. 3.16J, 55) in the endoskeleton, i.e. the bones are not only ossified superficially (perichondrally), but also internally (endo-chondrally), in the form of what is usually called spongy bone. There are many other characters unique to extant osteichthyans: the head and shoulder girdle are covered with large dermal bones (Fig. 3.16A–F), some of them having a relatively constant shape and position; the teeth are attached to dermal bones (Fig. 3.16, 30, 32); the labyrinth contains large otoliths, or statoliths; the two nostrils (anterior and posterior) of each side are

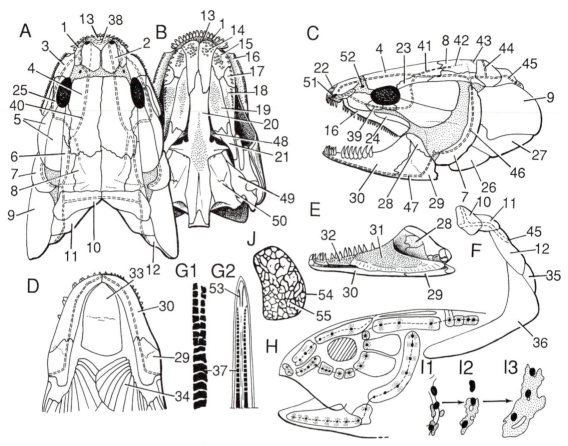

Fig. 3.16. Characters of the Osteichthyes. A–E, skull of the actinopterygian *Amia* in dorsal (A), ventral (B), lateral (C), and ventral (D) view, lower jaw in internal view (E). F, dermal shoulder girdle of *Amia* in lateral view (the clavicle and interclavicle are lost in neopterygian actinopterygians). G, series of lepidotrichiae of the caudal fin of *Amia* in lateral view (G1) and vertical section (G2). H, early development of the dermal bones of the head of *Amia*, to show their close association with the pattern of the sensory-line neuromasts (black dots). I, three successive stages of the development of the left frontal bone (stippled) in *Amia*, in association with the supraorbital sensory-line canal neuromasts (dots). J, section through an endoskeletal bone (basioccipital) of *Amia*, showing the structure of the endochondral bone. *Dermal bones*: 1, toothed antorbital; 2, nasal; 3, lachrymal; 4, parietal; 5, postorbitals; 6, supratemporal; 7, preopercular; 8, postparietal; 9, opercular; 10, extrascapular; 11, post-temporal; 12, supracleithrum; 13, premaxilla; 14, vomer; 15, anterior dermopalatine; 16, maxilla; 17, posterior dermoalatine; 18, ectopterygoid; 19, entopterygoid; 20, parasphenoid; 21, dermometapterygoid; 22, rostral; 23, dermosphenotic; 24, supramaxilla; 25, infraorbital; 26, subopercular; 27, interopercular; 28, surangular; 29, angular; 30, dentary;31, prearticular; 32, coronoids; 33, gular plate; 34, branchiostegals; 35, postcleithrum (or anocleithrum); 36, cleithrum; 37, lepidotrichiae; *sensory lines*: 38, ethmoid commissure; 39, infraorbital line; 40, supraorbital line; 41, 42, anterior pit line; 43, posterior pit line; 44, supratemporal commissure; 45, main lateral line; 46, propercular line; 47, mandibular line; *others*: 48, basal articulation of palatoquadrate; 49, hyomandibula; 50, braincase; 51, anterior nostril; 52, posterior nostril; 53, actinotrichia (ceratorichia); 54, perichondral bone; 55, endochondral bone. (A–E, from Jarvik (1980); F, from Jarvik (1944); H, I, from Pehrson (1940).)

completely separated (Fig. 3.16, 51, 52); and the fins are covered with a particular type of elongated scales, the lepidotrichiae (Fig. 3.16G, 37). There are primitively two dorsal fins (except in actinopterygians, which have only one, and in extant lungfishes and tetrapods), an anal fin, and the tail is primitively epicercal, as in sharks, but may undergo various modifications leading to a diphycercal aspect (apparent or real).

3.5.2.1 Pattern of the osteichthyan exoskeleton

The dermal bone patterns of the head and shoulder girdle of osteichthyans, whether Recent or fossil, display some constant features, that have been invoked in lengthy discussions on their homology, in particular on the relationships between piscine osteichthyans and tetrapods. This brief review is far from extensive, and the reader is referred to classical works (Westoll 1943*a*; Moy-Thomas and Miles 1971; Jarvik 1980; Borgen 1983; Panchen and Smithson 1987; Schultze 1993) for a deeper treatment. Like Jarvik (1980) and many other anatomists, I have chosen an actinopterygian, the North American bowfin, *Amia* (Fig. 3.16) as an example of a rather generalized osteichthyan head exoskeleton. In fact, this actinopterygian is the extant osteichthyan that retains the largest number of general osteichthyan features notwithstanding the fact that it displays a number of actinopterygian, halecostome, and halecomorph specializations.

One may distinguish four major and rather stable groups of dermal bones in the skull of a bony fish: the bones lining the mouth and covering the palate, the bones of the skull-roof and snout, the bones of the cheek and surrounding the eyes, and the bones forming the gill cover.

Jaw and mouth bones The bones lining the mouth, or jaw bones, are the premaxilla, maxilla, and dentary (Fig. 3.16, 13, 16, 30). In *Amia*, there is, in addition, a tooth-bearing antobital (Fig. 1.16, 1). The palate is covered by pairs of vomers, entopterygoids, dermopalatines, and ectopterygoids (Fig. 3.16, 14, 15, 17, 18, 19) and a median bone, the parasphenoid (Fig. 3.16, 20). The inside of the lower jaw is covered by a prearticular and coronoids (Fig. 3.16, 31, 32). Posteriorly, the surface of the pharynx may be covered by minute toothed platelets attached to the hyoid and branchial arches. In generalized fossil osteichthyans, the largest teeth are borne by the vomer, dermopalatine, and ectopterygoid in the upper jaw, and by the coronoids in the lower jaw. In addition many fossil forms possess small parasymphysial toothed plates flanking the symphysis of the lower jaw (Fig. 4.74, 6). The posterior part of the lower jaw is covered by other dermal bones, the angular and surangular (Fig. 3.16, 28, 29), but the jaw articulation is primitively ensured by two endoskeletal bones, the articular and the quadrate. The jaw articulation involves various dermal bones in some actinopterygians and tetrapods. Modifications of this general pattern of the dermal jaw-bones includes the loss of the parasymphysial toothed plates (all Recent osteichthyans), the presumed loss of the maxilla (actinistians and lungfishes), dermopalatine and ectopterygoids (lungfishes), and the development of infradentaries, or splenials (sarcopterygians).

Skull-roof and snout The bones of the skull-roof consist essentially of a paired series of large bones, the postparietals (Fig. 3.16, 8), parietals (Fig. 3.16, 4), and nasals (Fig. 3.16, 2), and a variable number of smaller bones which flank them laterally (e.g. dermosphenotic, dermopterotic, dermohyal, antorbital, supraorbital, prespiracular, postspiracular, intertemporal, supratemporal, post-temporal, and tabular, (Fig. 3.16, 6, 23). The nomenclature of these bones of the lateral skull-roof series fluctuates according to theories on their homology, fusion, or loss. Anteriorly, the skull-roof may be continued by a mosaic of small bones covering the snout and surrounding the nostrils or by larger nasals (Fig. 3.16, 2) and rostrals (Fig. 3.16, 22). Posteriorly, this skull-roof series ends with two, three, or four large modified scales, the extrascapulars (Fig. 3.16, 10).

Cheek The bones of the cheek comprise a number of small bones surrounding the eye, the cirumorbital series (suborbitals, postorbitals (Fig. 3.16, 5), and a variable number of larger bones, the preopercular (Fig. 3.16, 7), squamosal, jugal, and quadratojugal.

Operculo-gular series The operculo-gular series, covering the gill chamber, consists of the opercular (Fig. 3.16, 9), subopercular (Fig. 3.16, 26), interopercular (Fig. 3.16, 27), branchiostegal rays (Fig. 3.16, 34), and sometimes large, paired or median gular plates (Fib. 3.16, 33) covering the ventral side of the head.

Shoulder girdle The dermal shoulder girdle of osteichthyans comprises the cleithrum (Fig. 3.16, 36), clavicle, and interclavicle, plus a variable number of bones that either connect the cleithrum to the skull, such as, the post-temporal (Fig. 3.16, 11), supracleithrum (Fig. 3.16, 12), and anocleithrum, or lie behind the cleithrum, such as the postcleithrum (Fig. 3.16, 35).

Homologies Numerous papers have been published on homologies, fusions, fragmentations, and the evolutionary significance of the dermal bone pattern in osteichthyans. This is due to the fact that such a complex and variable pattern is a potential source of characters for tracing relationships, in comparison to the fairly constant structure of the endoskeleton. It is also due to the fact that the dermal head is often the only part of the skull that is preserved in fossil bony fishes, in particular the Palaeozoic actinopterygians. Palaeontologists therefore prefer to use this source of data in assessing relationships.

The methods of assessing homologies between dermal bones in different osteichthyan groups have been also vigorously debated. The canals of the lateral sensory-line system are known to pass through the centre of ossification of the dermal bones, and to induce their formation (Fig. 3.16H, I). The dermal head bones that bear such a canal or a pit line have therefore been widely used as points of reference for identifying other, adjacent bones. The relationships between the dermal bones and the underlying endoskeleton have also been claimed to be of great importance in assessing homologies. To some anatomists, for example, the bone that lies over the otic capsule has to be parietal, whatever group it belongs to. Others give much weight to 'intermediate' forms that enable them to reconstruct transformation series and thereby trace the fate of a dermal bone throughout a group.

Here we meet another problem, because many dermal bones in the fish skull have been given names defined in mammals, in particular man (nasals, frontals, parietals, etc.). Mammalian dermal bones can be traced with no great difficulty back to some early fossil tetrapods. But the morphological gap between tetrapods and fishes— extant or fossil—that were supposed to be their closest relatives has long impeded homologization. The discovery of the fossil panderichthyid sarcopterygians, which are the closest tetrapod relatives known at present, seems to have filled this gap and has led to a shift in bone names for the major skull-roof bones, not only in sarcopterygians, but also in actinopterygians (see Chapter 6, p. 261; Fig. 6.5). According to this pattern and nomenclature, which I use in this book, all osteichthyans initially possess two median pairs of skull-roofing bones that meet along the midline: the parietals and postparietals (Fig. 3.16, 4, 8). (This may, however, be misleading for some actinopterygian specialists who are accustomed to calling these bones 'frontals' and 'parietals' respectively.)

Scales   The scales of most extant osteichthyans, including some tetrapods, are thin and rounded in shape, sometimes ornamented superficially with ridges or tubercles of dentine. However, some actinopterygians possess thick, diamond-shaped scales (Fig. 3.17G), covered with thick layers of dentine and enamel. From fossil evidence we know that this latter scale type is probably general for all osteichthyans, and has subsequently been independently modified several times into the form of rounded scales (Schultze 1977). The dermal fin rays of all piscine osteichthyans consist of elongated scales, the lepidotrichiae (Fig. 3.16, 37), which are often sunk in the skin and replace most of the former dermal support of the fin web, i.e. the ceratotrichiae (Fig. 3.16, 53).

3.5.2.2 Osteichthyan diversity

Recent osteichthyans fall into two clades: the Actinopterygii (ray-finned fishes) and the Sarcopterygii (lobe-finned fishes and four-limbed vertebrates).

Actinopterygii

Actinopterygians are characterized by a transparent cap of hypermineralized hard tissue on the tip of the teeth: the acrodine (Fig. 3.17, 2), which differs in structure from the true enamel (collar enamel, Fig. 3.17, 4) lining the base of the teeth. (acrodine is bitypic—produced by both the ectoderm and ectomesenchymatous cells—whereas enamel is monotypic produced by ectoderm only.) They have only one dorsal fin (Fig. 3.17A–F) and the median fin rays are inserted directly into the body, with no intervening basal lobe. In many extant forms, in particular advanced teleosts, the dorsal fin can, however, be divided into two parts, giving the impression that there are two dorsal fins. The maxilla is primitively large and expands posteriorly. The scales are primitively diamond-shaped and are covered with ganoine (overlying layers of dentine and enamel), with a peg-and-socket type of articulation in which the peg (Fig. 3.17, 1) is pointed and clearly distinct from the dorsal margin of the scale. There are some other actinopterygian characteristics, such as the loss of the pelvic girdle and its replacement by part of the metapterygium, and numerous features of the soft anatomy (gill and jaw musculature, brain development; see Wiley 1979; Patterson 1982a; Lauder 1980a,b; Lauder and Liem 1983).

None of these characters occurs in all Recent actinopterygians. They have all been variously modified or lost in the history of the group, in particular in the largest actinopterygian group, the teleosts. The acrodine cap is probably the most widespread character, and is absent only in forms that have completely lost their teeth, or whose teeth are modified into crushing tooth plates. The maxilla is much reduced in most 'higher' actinopterygians (neopterygians) and the diamond-shaped 'ganoid' scales, though general in early actinopterygians, occur only in a few extant groups (cladistians and ginglymods, Fig. 3.17, A, D). In other actinopterygians, the scales are modified into a few large scutes (chondrosteans), or rounded (cycloid) scales. As the name suggests, 'actinopterygians' ('ray-finned') has long been restricted to fishes whose fin web, strengthened by numerous thin rays, arises immediately from the body wall, with no intervening meaty lobe. This feature occurs in the Actinopteri, i.e. teleosts, bow-fins, gars, and sturgeons (Fig. 3.17, 2). In cladistians the fin web of the paired fins lines a large scaly lobe (Fig. 3.17A).

For historical reasons explained in Chapter 8 (p. 327), actinopterygians are the first major vertebrate group whose interrelationships and systematics have been worked out in detail by means of the cladistic method (see p. 37). We can now therefore produce a fairly well-corroborated phylogeny of this group, at least for Recent taxa (Figs 3.17A–F, 3.19).

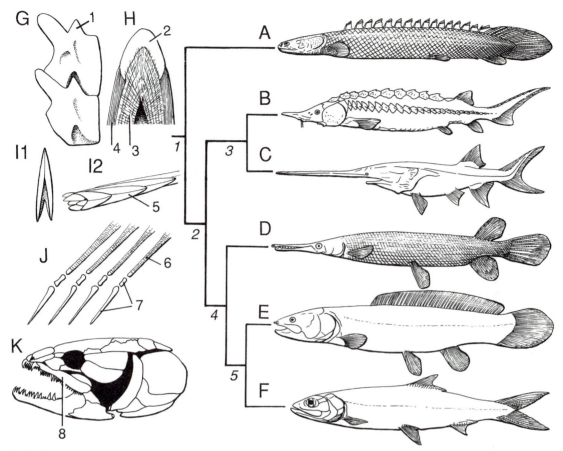

Fig. 3.17. Characters and interrelationships of the Actinopterygii (ray-finned fishes). A–F, interrelationships of
Recent actinopterygians; *terminal taxa*: A, Cladistia (*Polypterus*); B, Acipenseridae (*Acipenser*); C, Polyodontidae (*Polyodon*);
D, Ginglymodi (*Lepisosteus*); E, Halecomorphi (*Amia*); F, Teleostei (*Elops*) *nested taxa and selected synapomorphies*:
1, Actinopterygii (well-defined peg on diamond-shaped scales, ganoine, acrodine cap on teeth); *2*, Actinopteri (fulcral scales on
leading edge of all fins); *3*, Chondrostei (loss of posterior myodome, anterior symphysis of palatoquadrate); *4*, Neopterygii (dermal
fin rays, or lepidotrichiae, equal in number to their supports in dorsal and anal fin); *5*, Halecostomi (mobile maxilla, median neural
spines). G, scales of *Polypterus* in internal view; H, actinopterygian tooth of in vertical section (about ×10); I, fulcra of the pectoral fin
of *Lepisosteus* in anterior view (I1) and in their natural position (I2); J, median fin supports and lepidotrichs in *Amia*; K, skull of *Amia*
in lateral view. 1, dorsal peg; 2, acrodine cap; 3, dentine; 4, collar enamel; 5, fulcra; 6, lepidotrichiae; 7, radials and fin supports;
8, maxilla. (A–E, K, based on Lauder and Liem (1983); G, from Pearson (1981); J, based on Jarvik (1980).)

The Cladistia live in the fresh waters of Africa and
comprise two genera (*Polypterus*, Fig. 3.17A, and
Erpetoichthys). Both have the same shiny, ganoin-
covered rhombic scales, elongated dorsal fin (made up of
a series of spines bearing pinules), and lobed base in
paired fins. They also display many anatomical peculiarities that are responsible for their erratic systematic
position: e.g. they have like sharks, a spiracle and a spiral
intestine, which are now regarded as generalized gnathostome characters. Moreover, they have lungs that arise

from the ventral side of the digestive tract, like the lungs
of sarcopterygians.

When, as a participant in Napoleon's Egyptian cam-
paign in 1801, the French anatomist E. Geoffroy
St Hilaire discovered the polypter (or 'bichir', hence the
name *Polypterus bichir*), he was puzzled by the lobe-
shaped base of its pectoral fins, which he compared to
the fore limbs of the tetrapods. His paper on the pectoral
fins of this fish (Geoffroy St Hilaire 1807) was thus
the first attempt to find homologies between a fish fin and

a tetrapod limb. Much later, in 1861, T. Huxley included the polypter in a group, the Crossopterygidae, together with a number of fossil fishes having the same lobed paired fins. These fishes were subsequently regarded as tetrapod relatives or ancestors (e.g. *Osteolepis*). But the structure of the fin skeleton of *Polypterus* was soon shown to be quite different from that of other 'crossopterygians', and this fish became excluded from tetrapod ancestry. Yet, as it did not fit in with other actinopterygians, its earlier fame survived for a long time, and some of its vague resemblances to the coelacanth regularly reappear in the literature (lungs, structure of the basibranchial series, olfactory organ, and lower jaw; Jarvik 1980).

Although they share only a small number of characters with other actinopterygians, there is no better place for cladistians than as the sister-group of the Actinopteri (all other extant actinopterygians, Fig. 3.17, *2*), because these characters can be compared point-for-point and assessed as unique to the two taxa. This is also corroborated by comparisons of nucleotid sequences from molecular data (Lê *et al.* 1993).

Cladistians display an assemblage of general actinopterygian or osteichthyan features and unique apomorphous characters. Among the latter are the peculiar structure of the paired fin endoskeleton, in which two large basal elements articulate with a ball-shaped process of the scapulocoracoid. The tail also looks diphycercal, with the dorsal and ventral webs almost equal in size, separated by a rounded, fleshy lobe. The vertebral column in this lobe is, however, slightly epicercal, i.e. it displays the primitive gnathostome condition (Fig. 3.18A). Jarvik (1980) has listed eleven characters in which *Polypterus* differs from actinopterygians. Most of them are clearly unique to cladistians and do not suggest another relationship for this taxon; others are of undecided significance (i.e., they could either be general osteichthyan characters or homoplasies).

Fossil evidence for the Cladistia is scarce: it takes the form of isolated scales (in particular the typical pinnules of the dorsal fin) and dates back to the Cretaceous (Gayet and Meunier 1992), but according to current phylogenetic theories (see Fig. 4.70) the common ancestor to cladistians and actinopterans must have been at least Middle Devonian in age, i.e. as old as the earliest known Actinopteri.

All other actinopterygians are the Actinopteri (Fig. 3.17, *2*) and are characterized by special arrow-shaped scales, the fringing fulcra, on the leading edge of all fins (Fig. 3.17, *5*). Fringing fulcra are formed by the fusion, side-by-side, of two adjacent elongated, diamond-shaped scales. The paired fins are devoid of large basal lobes. Actinopterans include the Chondrostei (Fig. 3.17,

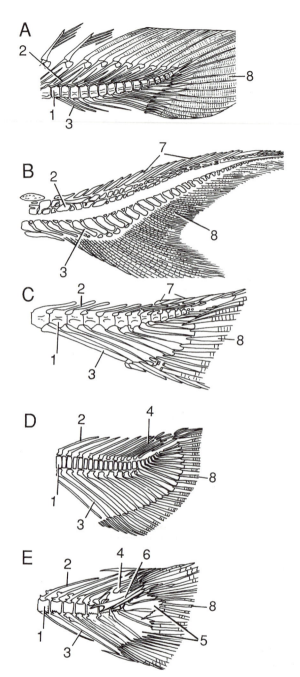

Fig. 3.18. The actinopterygian tail skeleton. A, Cladistia (*Polypterus*); B, Chondrostei (*Polyodon*); C, Ginglymodi (*Lepisosteus*); D, Halecomorphi (*Amia*); E, Teleostei (*Elops*). 1, vertebral centrum; 2, neural arch (basidorsal); 3, haemal arch (basiventral); 4, epurals; 5, hypurals; 6, uroneural; 7, fringing fulcra; 8, dermal fin ray (lepidotrichiae). (A, from Jarvik (1980); B, from Grande and Bemis (1991); C–E, from Patterson (1973).)

3), Ginglymodi (Fig. 3.17D), Halecomorphi (Fig. 3.17E), and Teleostei (Figs 3.17F, 3.19).

The chondrosteans include sturgeons, such as *Acipenser* (Fig. 3.17B) and *Scaphirhynchus* and paddle-fishes, such as *Polyodon* (Fig. 3.17C) and *Psephurus*. Both have a reduced body squamation, yet large scutes occur in sturgeons, and fringing fulcra in fins. The teeth are lost and the jaws are reduced in sturgeons (Fig. 3.8A2). The group is known with certainty from Early Jurassic times. There have been questions about the monophyly of this taxon, since it was largely defined by reductive characters, such as the loss of the posterior myodome (a large median recess of the orbital cavity, where recti eye muscles are attached in most other actinopterans; see Chapter 4, p. 189). However, other unique characters, such as the symphysis of the two pala-toquadrates anteriorly, support chondrostean monophyly.

Ginglymods (gars, Fig. 3.17D) are represented by a few species distributed between two extant North American genera, *Lepisosteus* and *Atractosteus*. Gars have an elongated snout in which the maxilla is much reduced and the teeth of the upper jaw are borne by a series of infraorbital bones. The caudal fin of gars is supported by a shorter, but still epicercal, portion of the vertebral column, and the ventral elements of the caudal vertebrae (haemal arches and spines) are enlarged into triangle-shaped hypurals (Fig. 3.18C, 3).

The halecomorphs (bow fins; Fig. 3.17E) consist of a single North American genus (*Amia*), although the group has many fossil representatives in the Mesozoic and Cenozoic. Halecomorphs are characterized by the fact that both the symplectic and the quadrate participate in the articulation of the jaw. In halecomorphs, the maxilla is mobile in relation to the other skull bones (Fig. 3.17, 8), and the haemal spines of the caudal skeleton are arranged in a fan-like manner as a result of the accentuated dorsal flexure of the rearmost vertebrae (Fig. 3.18D, 3).

Teleosts are represented by thousands of extant species and are characterized by the presence of uroneurals in the caudal skeleton (Fig. 3.18E, 6), a mobile premaxilla, unpaired basibranchial toothed plates, and by the fact that the foramen for the internal carotid arteries is enclosed in the parasphenoid, as well as by many characteristics of the soft anatomy. The uroneurals are derived from the rearmost neural spines and are paired, lying on either sides of the axis. The epurals are median and are also derived from neural spines (Fig. 3.18E, 4). The hypurals, derived from haemal spines, become considerably enlarged into a thick bony pad (Fig. 3.18E, 5), and the rearmost vertebral centra disappear. The structure of the caudal fin skeleton of teleosts displays a wide range of diversity and is a source of characteristics for working out

the phylogeny of the group. Teleosts are known only since the Late Triassic, and the phylogeny of this group therefore falls outside the scope of this book. However, the cladogram in Fig. 3.19 (combined from Lauder and Liem 1983, Nelson 1989, and Johnson and Patterson 1993) gives an idea of their diversity. Only a few selected characters supporting the major dichotomies are cited here.

The interrelationships of Recent actinopterans are now well corroborated. The ginglymods, halecomorphs, and teleosts form a clade, the Neopterygii (Fig. 3.17, *4*), characterized for example by the fact that the fin rays of the dorsal and anal fins are equal in number to their support inside the body (Fig. 3.17, 6, 7), and by the reduction or loss of the clavicle. Within neopterygians, halecomorphs and teleosts share several unique characters, e.g. the mobile maxilla (Fig. 7.17, 8) and the median neural spines. They thus form a clade, the Halecostomi (Fig. 3.17, 5). Olsen and McCune (1991) have recently argued, on the basis of fossils, that ginglymods were more closely related to teleosts than halecomorphs.

Sarcopterygii   Sarcopterygians are characterized by the structure of the endoskeleton of their paired fins, in which only the metapterygium remains (Fig. 3.20A, stippled), all other fin rays having disappeared (Fig. 3.20, 2). Thus, the entire fin skeleton articulates to the girdle by means of a single element (monobasal articulation, Fig. 4.20B–D, 1, 3). The radial muscles around this skeletal axis are thus arranged in a club-shaped mass covered with skin and scales; hence the name Sarcopterygii, which means 'fleshy fins'. The unpaired fins also generally display the same concentration of the skeleton and musculature within a fleshy lobe. Other sarcopterygian characters are the lack of any connection between the preopercular and supratemporal lateral-line canals, the presence of true prismatic enamel on the entire teeth (not only the collar, Fig. 3.21F, 18), and the sclerotic ring of the eye composed of more than five bony plates. As far as soft anatomy is concerned, two features are noteworthy: the presence of a vena cava posterior, which drains the blood from all the post-cardiac part of the body, in place of the posterior cardinal veins; and a lung with pulmonary veins and arteries (Fig. 3.21, 32; Fig. 3.22, 22, 24, 25).

Sarcopterygians comprise the Actinistia, Dipnoi, and Tetrapoda. The Actinistia are represented by a single extant species, *Latimeria chalumnae* (Fig. 3.21), discovered in 1938, which lives near the Comoro islands in relatively deep waters (about 400 m). Actinistians are also known as 'coelacanths' ('hollow spines'), a name given by palaeontologists on the basis of the partly hollow spines (in fact fused lepidotrichs) that support the anterior dorsal fin in most Mesozoic and Recent forms (Fig.

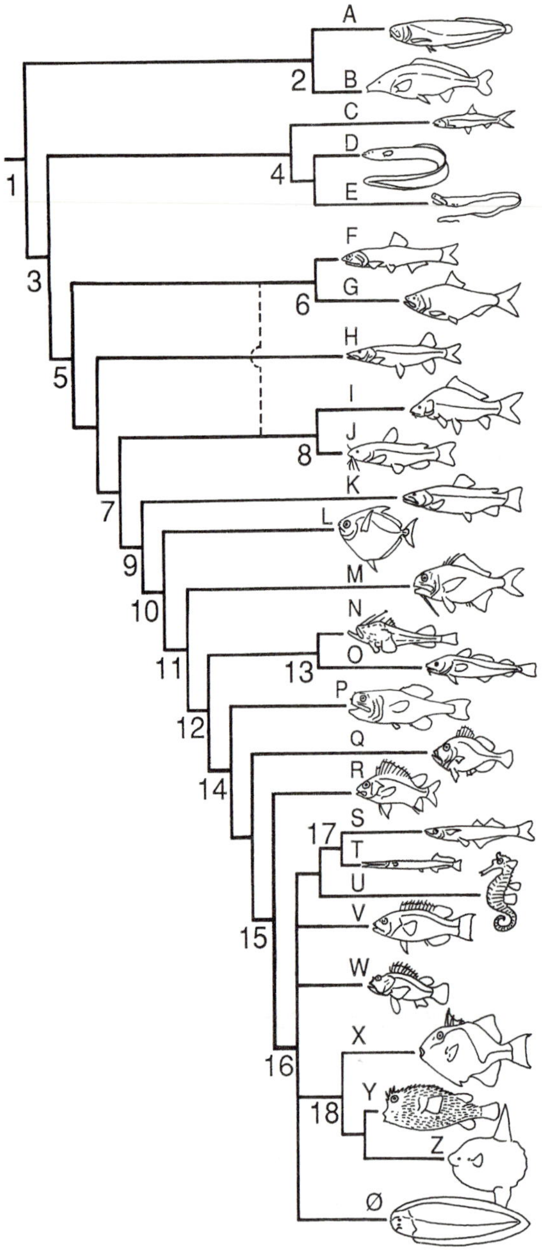

Fig. 3.19. Interrelationships of the Teleostei. *Selected terminal taxa*: A, Osteoglossidae (bonytongue); B, Mormyridae; C, *Elops* (tarpon); D, Angilloidea (eel); E, Saccopharyngoidea (pelican fish); F, Engraulidae (anchovies); G, Clupeidae (sardine, herring); H, Esocidae (pike); I, Cypriniformes (characin, carp); J, Siluriformes (catfish); K, Salmoniformes (salmon); L, Lampridiformes (moonfish); M, Polymyxiiformes; N, Lophiiformes (angler); O, Gadiformes (cod); P, Stephanoberyciformes; Q, Zeiformes (John Dory); R, Beryciformes (beryx); S, Atherinidae (silverside); T, Beloniformes (green bone); U, Gasterosteiformes (seahorse, stickleback); V, Perciformes (perch); W, Scorpaeniformes (scorpion fish); X, Balistidae (triggerfish); Y, Diodontidae (globe fish); Z, Molidae (sunfish); Ø, Pleuronectiformes (flatfishes). *Nested taxa and selected synapomorphies*: 1, Teleostei (endoskeletal basihyal, basibranchial and basihyal covered by median tooth palates, mobile premaxilla); 2, Osteoglossomorpha; 3, Elopocephala (two uroneurals); 4, Elopomorpha; 5, Clupeocephala (tooth plates fused with endoskeletal gill arch elements); 6, Clupeomorpha; 7, Euteleostei (adipose fin); 8, Ostariophysi (swimbladder divided into two unequal chambers, modification of the anterior vertebrae and pleural ribs in connection with a pressure-regulating device); 9, Neognathi (rostral cartilage between ethmoid and premaxillae); 10, Acanthomorpha; 11, Euacanthomorpha; 12, Holacanthopterygii; 13, Paracanthopterygii; 14, Acanthopterygii; 15, Euacanthopterygii (separate spinous and soft dorsal fins); 16, Percomorpha; 17, Atherinomorpha; 18, Tetraodontiformes. (Based on Lauder and Liem (1983) and Johnson and Patterson (1993), fishes after J. Nelson (1994); note a possible ostariophysan–clupeomorph sister-group relationship, which would contradict euteleost monophyly (Lecointre 1994).)

3.21, 1). In actinistians, this fin attaches directly to the back, and not to a basal lobe (unlike all other fins). The caudal fin of actinistians is also unique among extant fishes in being structurally diphycercal (or isocercal); that is, the dorsal and ventral webs are equal in size and are separated by a horizontal prolongation of the body, which contains the notochord and ends with a small rounded tuft (Fig. 3.21, 4). The scales (Fig. 3.21E) are large, rounded in shape, and ornamented with spine-shaped tubercles. Actinistians also possess a unique organ, the rostral organ (Fig. 3.21, 7), housed in the snout, which opens to the exterior by a series of pores (Fig. 3.21, 5). The rostral

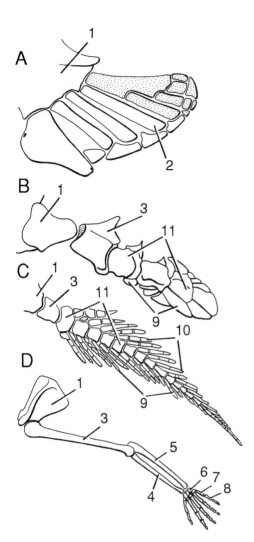

Fig. 3.20. Characters of the Sarcopterygii. A, endoskeleton of the pectoral fin of an actinopterygian (sturgeon) showing the metapterygium (stippled), which is presumed to be the homologue of the sarcopterygian monobasal paired fin skeleton. B–D, endoskeleton of the pectoral appendages (fin or limb) of sarcopterygians: Actinistia (B, *Latimeria*), Dipnoi (C, *Neoceratodus*), and Tetrapoda (D, *Homo*). 1, endoskeletal shoulder girdle (scapulocoracoid); 2, premetapterygial radial; 3, first mesomere, or humerus; 4, radius; 5, ulna; 6, carpals, 7, metacarpals; 8, phalanges; 9, preaxial radials; 10, postaxial radials; 11, metapterygial axis (mesomeres). (*Note*: the postaxial radials of lungfishes are supposed by Rosen *et al.* (1981) to be the preaxial radials, upturned into a postaxial position, owing to the torsion of the fin during the development; see Chapter 7, p. 267.) (A–C, from Rosen *et al.* (1981).)

organ is now believed to have an electrosensory function. There are also a number of features of the skull that are regarded as actinistian characteristics, but may occur elsewhere: a very large opercular (Fig. 3.21A), the lack (by loss) of the maxilla, the reduced dentary (Fig. 3.21, 15) and enlarged prearticular and infradentary (Fig. 3.21, 16), and main lateral-line canal passing along the sutures of the skull-roofing bones.

Latimeria is the only extant fish in which the braincase consists of two parts, anterior and posterior, articulated by means of an intracranial joint (Fig. 3.21, 12). The anterior half is referred to as the ethmosphenoid (Fig. 3.21, 8), and the posterior one as the otoccipital (Fig. 3.21, 9). As will be shown in Chapter 6 (p. 258), this remarkable feature is probably a general sarcopterygian character, subsequently lost in some groups, such as lungfishes and tetrapods.

The notochord (Fig. 3.21, 11) is a thick tube filled with a liquid, and extends forward into the otoccipital of the braincase. Anteriorly, it abuts against the ethmosphenoid, behind the hypophysial region (Fig. 3.21, 13), and probably plays a role in the movement of the entire skull as an antagonist to the powerful basicranial muscles (Fig. 3.21, 23). The vertebrae are represented by thin ventral and dorsal arcualia. Apart from that of the anterior dorsal, all unpaired fin skeletons consist of a series of small endoskeletal elements housed in the lobe, which articulate with a single basal plate.

The paired fin skeleton (Figs 3.20B, 3.21H) consists of a series of elements, or mesomeres (Fig. 3.21, 27), now regarded as the elements of the metapterygial axis, and adjoining cartilaginous nodules, which are modified metapterygial radials (Fig. 3.21, 28). The first, or proximal, mesomere (Fig. 3.20B, 3) articulates with the girdles by means of a ball-and-socket joint, in which the ball is borne by the girdle (Fig. 3.21, 26, 29).

For historical reasons, which date back to Huxley (1861; see p. 314), *Latimeria* was once regarded as a possible key to the understanding of the origin of land vertebrates. It has therefore been studied in great detail. Its soft anatomy has revealed a strange assemblage of primitive gnathostome features and more advanced characters that link it undoubtedly with lungfishes and tetrapods. Among the former are the elongated shape of the heart (Fig. 3.21I) and the structure of the pancreas, which resemble those of sharks. Among the latter is the presence of a vena cava posterior and a pulmonary vein related to a large fatty organ connected to the digestive tract, which is therefore likely to be a modified lung with a glottis (Fig. 3.21J, 31, 32).

Latimeria is ovoviviparous, and the female produces a few, very large, orange-shaped eggs. The young (about five) develop until a late stage inside the oviduct of the

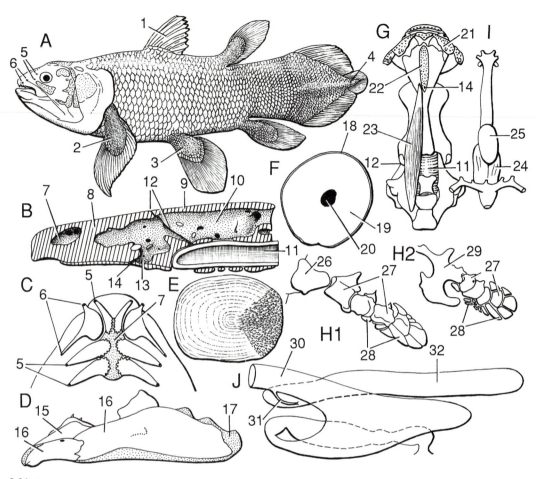

Fig. 3.21. Morphology of the Actinistia. A, general aspect of the extant coelacanth *Latimeria* (about ×0.1); B, sagittal section through the braincase of a fossil actinistian, showing the notochord; C, rostral orgal and nasal capsules in dorsal aspect; D, lower jaw in lateral aspect; E, scale in external aspect; F, horizontal section of a tooth; G, braincase and basicranial muscle in ventral view; H, endoskeleton of the pectoral (H1) and pelvic (H2) fins; I, heart in ventral view; J, 'lung' and digestive tract in lateral aspect. 1, anterior dorsal fin; 2, pectoral fin, 3, pelvic fin; 4, chordal lobe of caudal fin; 5, openings of the rostral organ; 6, anterior and posterior nostrils; 7, rostral organ; 8, ethmosphenoid (anterior half of braincase); 9, otoccipital (posterior half of braincase); 10 brain cavity; 11, notochord; 12, intracranial joint; 13, hypophyseal fossa; 14, buccohypophysial canal; 15, dentary; 16, infradentaries (splenials); 17, retroarticular process; 18, enamel; 19, dentine; 20, pulp cavity; 21, vomer; 22, parasphenoid; 23, basicranial muscle; 24, atrium; 25, ventricle; 26, scapulocoracoid; 27, mesomeres; 28, preaxial radials; 29, pelvic girdle; 30, digestive tract; 31, 'glottis'; 32, fatty 'lung'. (A, B, C, G, from Bjerring (1967, 1986*b*); D, E, H, from Millot and Anthony (1958); I, from Millot *et al.* (1978); F, from Castanet *et al.* (1975); J, from Robineau (1987).)

female. They are attached to a very large ventral yolk sac (Fig. 4.75D).

The life and growth of *Latimeria* is poorly known, although progress has been made recently thanks to direct observations from a submersible. The largest individuals reach 1.8 m in length and the coelacanth moves slowly near the bottom, with a peculiar alternating movement of its paired fins reminiscent of the way in which land ver-

tebrates walk. This movement is also seen in the paired fin movements of lungfishes. The coelacanth seems to be extremely sensitive to electric fields, which trouble its sense of orientation and make it perform strange headstands.

Another peculiarity of *Latimeria* is its endemic distribution, considering that, until Cretaceous times, actinistians had a world-wide distribution.

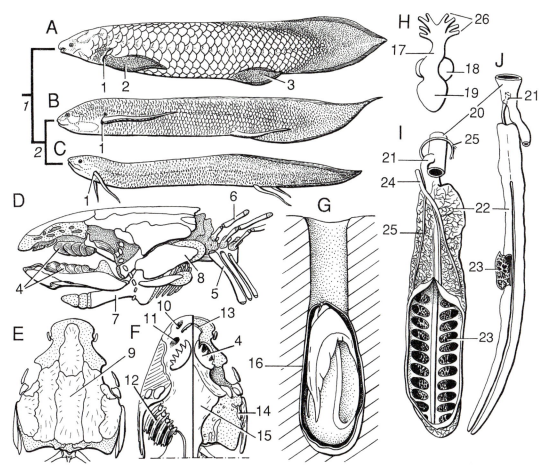

Fig. 3.22. Morphology, interrelationships, and behaviour of the Dipnoi (lungfishes). A–C, the three extant lungfish genera and their interrelationships; *terminal taxa*: A, *Neoceratodus*; B, *Protopterus*; C, *Lepidosiren*; *nested taxa and selected synapomorphies*: *1*, Dipnoi (tooth plates with radiating ridges); *2*, Lepidosirenidae (reduced paired fins, reduced tooth plates). D–F, skull of *Neoceratodus* in lateral (D), dorsal (E), and ventral (F) view (cartilage stippled; nostrils and gills shown on the left-hand side in F). G, aestivation burrow of *Protopterus*; H, heart of *Neoceratodus* in ventral view; I, J, lungs of *Neoceratodus* (I) and *Lepidosiren* (J) in ventral and dorsal view respectively. 1, gill slit; 2, pectoral fin; 3, pelvic fin; 4, entopterygoid (upper) and prearticular (lower) tooth plates; 5, cranial rib; 6, supraneural; 7, ceratohyal; 8, opercular; 9, median 'B' bone; 10, anterior nostril; 11, posterior nostril; 12, gills; 13, 'vomerine' tooth plates; 14, hyomandibula; 15, parasphenoid; 16 mucous 'cocoon'; 17, sigmoid arterial cone; 18, atrium; 19, ventricle; 20, digestive tract; 21, glottis; 22, lung; 23, alveolae; 24, pulmonary vein; 25, pulmonary artery, 26, afferent branchial arteries. (A, B, from Jarvik (1980); D–F, from Goodrich (1930); G, modified from Arambourg and Guibé (1958); H, from Janvier (1980*b*); I, modified from Günther (1871); J, modified from Marcus (1937).)

The Dipnoi, or lungfishes, are represented by three extant genera: *Neoceratodus* (Fig. 3.22A) in Australia, *Protopterus* (Fig. 3.22B) in Africa, and *Lepidosiren* (Fig. 3.22C) in South America. The last two belong to the same family, Lepidosirenidae (Fig. 3.22, *2*). Lungfishes are characterized by massive tooth plates (Fig. 3.22, 4), the surfaces of which are adorned with radiating ridges of dentine. These tooth plates rest on the palate and the

inside of the lower jaw. They correspond respectively to the entopterygoids and prearticulars of other osteichthyans. In addition, there are smaller paired upper tooth plates (Fig. 3.22, 13) and one unpaired (symphysial) lower tooth plate, the homology of which is debated. They may be derived from the vomers and dentary respectively. All lungfishes have well-developed lungs, one in *Neoceratodus* (Fig. 3.22I, 22) and two in lepidosirenids (Fig. 3.22J, 22), with alveolae (Fig. 3.22, 23), a glottis (Fig. 2.33, 21), and blood circulation (Fig. 3.22, 24, 25) closely similar to that of the tetrapods. The pulmonary artery (Fig. 3.22, 25) arises from the rearmost efferent branchial artery, and the pulmonary vein joins the atrium of the heart. The lepidosirenids breathe exclusively with their lungs; their gills are reduced and the gill slit is very small (Fig. 2.22, 1). In contrast, *Neoceratodus* uses essentially its gills. (Fig. 3.22, 12) Extant lungfishes gulp air by means of a complex pumping mechanism that involves the hyobranchial apparatus (Fig. 3.22, 7), the parasphenoid (Fig. 3.22, 15), and the pectoral girdle. The heart of lungfishes displays an S-shaped arterial cone (Fig. 3.22H, 17), a character that is shared uniquely with the tetrapods (Fig. 2.23F). The posterior nostrils (Fig. 3.22, 11) of lungfishes are situated well inside the mouth, and the anterior ones (Fig. 1.22, 10) at the margin of the upper lip. This condition has been compared to the choanae of the tetrapods, thereby raising considerable debate during the past 150 years (see reviews of the controversies in Rosen *et al.* 1981; Jarvik 1981*b*; Holmes 1985; Forey 1987*b*; Panchen and Smithson 1987; Schultze 1994).

The skeleton of extant lungfishes is poorly ossified and the endoskeleton consists largely of cartilage (Fig. 3.22D, D–F, stippled), but we know from fossil lungfishes that this is due to a process of secondary reduction that had started by the end of Devonian times, 360 Ma ago. Similarly, the unpaired fins, which in early lungfishes displayed the general gnathostome pattern (two dorsals, one anal, and an epicercal caudal), have been modified into a diphycercal tail with a single fin web (Fig. 3.22A–C). The paired fins, however, are characeristically elongated in shape (Fig. 3.22, 2, 3), and can move alternatively, like those of the coelacanths and the tetrapod limbs. Lepidosirenids, for example, use their pectoral fin as a support to raise the anterior part of the body and can even use them to help in conveying food into the mouth. This behaviour, which is widespread in tetrapods, is unknown in other fishes.

Lepidosirenids can stand drought and dessication by burying themselves in the mud and enclosing themselves in a protective cocoon of mucus (Fig. 3.22G, 16). The cocoon is pierced by an opening through which the fish can breathe for up to one and a half years. Lungfishes undergo a larval development and metamorphosis. The larvae have large external gills and vaguely resemble those of the caudates (salamanders and newts) among amphibians.

When extant lungfishes were first discovered in the early nineteenth century, their anatomical resemblance to land vertebrates soon led to animated debates about whether they were tetrapod-like fishes or fish-like tetrapods. Four features were particularly highlighted: the lungs, internal nostrils, heart, and paired fins. The lungs of lepidosirenids were regarded as particularly tetrapod-like in being paired and having alveolae (Fig. 3.22, 23). The position of the posterior nostrils inside the palate recalled the tetrapod choanae (Fig. 3.22, 11). The heart possesses an apparently two-chambered auricle (Fig. 3.22, 18), and a sinuous arterial cone (Fig. 3.22, 17), as in tetrapods; and the paired fin skeleton, which consists of a long series of mesomeres supporting two series of radials (preaxial and postaxial, Fig. 3.20C, 9, 10) was more closely comparable to the tetrapod limb skeleton than was that of the cladistian *Polypterus*. By the end of the nineteenth century, there was therefore a widespread consensus among evolutionary biologists that lungfishes were close relatives of, and possibly ancestors to, the tetrapods. As embryology and palaentology progressed, the lungfish–tetrapod resemblances have, however, been regarded as the result of convergences on the ground that the unique features shared by the two groups were not point-for-point similar. Lungfishes have consequently been left aside from debates on tetrapod origins, to the benefit of various Palaeozoic fossil sarcopterygians (see historical review in Patterson 1980 and Forey 1987*b*). As we shall see later (Chapter 5, p. 249), the debate has continued for over a century. In fact, if most zoologists accept that lungfishes are sarcopterygians, there is still a controversy as to whether they are the closest extant relatives of the tetrapods or whether the actinistians are to occupy this position (Fig. 3.26).

The Tetrapoda (Fig. 3.23) are sarcopterygians whose paired fins are modified into locomotory limbs (Fig. 3.20D). A number of their radials (generally five or less in extant tetrapods) are enclosed in separate lobes and are called metapods and phalanges (fingers or toes, Fig. 3.20D, 25). The rest of the limb skeleton consists mainly of a series of long bones (one proximal and two distal) which are regarded as homologous to the first two mesomeres and the first radial in other sarcopterygians (Fig. 3.20D, 3–5). The first radial and second mesomere (radius and ulna/tibia and fibula, Fig. 3.23, 27–28, 31–32) are almost equal in length. The major difference between the fore limb and the hind limb rests on the fact that the wrist (forelimb) and the knee (hindlimb) joints are hinge-like, whereas the ankle (hind limb) and the elbow (fore limb) are rotary. In extant tetrapods, the paired limbs are generally used for locomotion on land, unless secondarily lost or adapted to swimming or flying. Extant tetrapods also display a number of other anatomical features that are thought to be linked with life on land, such as the loss of the gills; the pulmonary respiration (Fig. 3.23G); a complete interauricular septum in the heart (Fig. 3.23F); the horny epidermal formations (phaners); sound

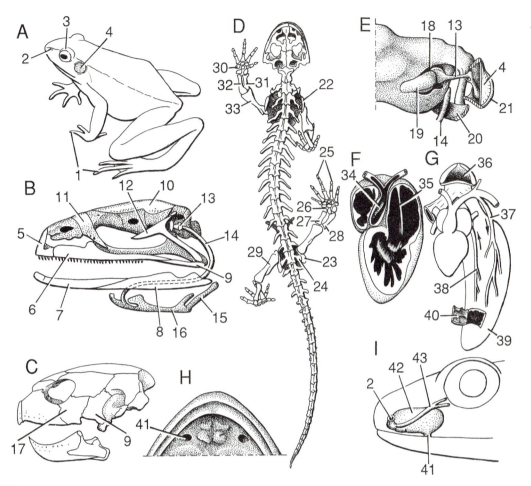

Fig. 3.23. Characters of the Tetrapoda. A, general morphology of a tetrapod (*Rana*); B, skull of a frog in lateral view; C, skull of a turtle in lateral view; D, skeleton of a salamander in dorsal view; E, middle ear and tympanic membrane of a frog in posterolateral view; F, vertical section through the heart of a frog; G, heart and left lung of a frog in ventral view (arrows indicate the direction of the blood circulation); H, palate and choanae of a frog in ventral view; I, olfactory organ and tear duct of a newt in lateral view. 1, digits; 2, external nostril; 3, eyelid; 4, tympanic membrane; 5, premaxilla; 6, maxilla; 7, dentary; 8, angular; 9, quadratojugal; 10, frontopartietal; 11, nasal; 12, squamosal; 13, stapes; 14, hyale (ceratohyal); 15, postero-medial process; 16, hyoid plate; 17, jugal; 18, fenestra ovalis; 19, opercular endoskeletal bone (unique to lissamphibians); 20, palatoquadrate; 21, tympanic ring; 22, scapular blade; 23, ilium; 24, sacral rib; 25, phalanges; 26, tarsals; 27, tibia; 28, fibula; 29, femur; 30, carpals; 31, radius; 32, ulna; 33, humerus; 34, arterial cone; 35, interauricular septum; 36, glottis; 37, pulmonary artery; 38, pulmonary vein; 39, lung; 40, pulmonary alveolae; 41, choana; 42, nasal sac; 43, nasolacrymal duct (tear duct). (B, F, based on Goodrich (1930); E, from Bolk *et al.* (1936); D, from Schaeffer (1941).)

conduction through bone (effected by modified hyomandibula, the stapes, which links the cheek to the labyrinth through a large opening in the braincase, the fenestra ovalis, (Fig. 3.23E); the tear duct (for keeping the olfactory organ wet, Fig. 3.23I); the eyelids (Fig. 3.23A); the internal nostril (choana, if homoplasic with that of lungfishes, Fig. 3.23, 41); the loss of unpaired fins and of the fin web on the limbs; a sacrum with at least one pair

of sacral ribs contacting the ilia on each side (Fig. 3.23, 24); a large scapular blade of the scapulocoracoid (Fig. 3.23, 22); the lack of dermal bones in the shoulder girdle dorsally to the supracleithrum; the broad sutural contact between the jugal and quadratojugal (Fig. 3.23, 9, 17); and a single pair of nasals (Fig. 3.23, 11).

During tetrapod ontogeny, the nasal sac opens at the margin of the stomodeum. When the anterior and posterior nostrils become separated, the anterior nostril lies outside and the posterior nostrils (choanae) inside the mouth (Fig. 3.23H), both being united by a nasolabial groove, which finally closes. In salamanders, however, the stomodeum is very shallow and the choanae form by direct contact between the anterior tip of the embryonic gut (endoderm) and the nasal sac.

Most extant tetrapods look quite different from any other extant sarcopterygian, hence the difficulty encountered in pre-palaeontological times (i.e. in the early nineteenth century) in trying to relate them to fishes. Only the Caudata (salamanders and newts; Fig. 3.24B) still retain an elongated body shape and a horizontal septum, which was compared to that of lepidosirenids when they were

first discovered. Extant tetrapods fall into two groups: the Amphibia (represented only by the Lissamphibia; Fig. 3.24, *1*) and the Amniota (Fig. 3.25; Fig. 3.26, *9*). Lissamphibians are characterized by fourteen unique features, among which are the pedicellate and bicuspid teeth (Fig. 3.24D). They comprise the Apoda (limbless amphibians, or the Gymnophiona when fossils are included, Fig. 3.24A); the Caudata (salamanders and newts, or the Urodela when fossils are included, Fig. 3.24B); and the Anura (frogs and toads, or the Salientia when fossils are included, Fig. 3.24C). Caudates and anurans are the most closely related of the three groups, and are gathered in the taxon Batrachia (Fig. 3.24, *2*). They share many unique characters, such as the

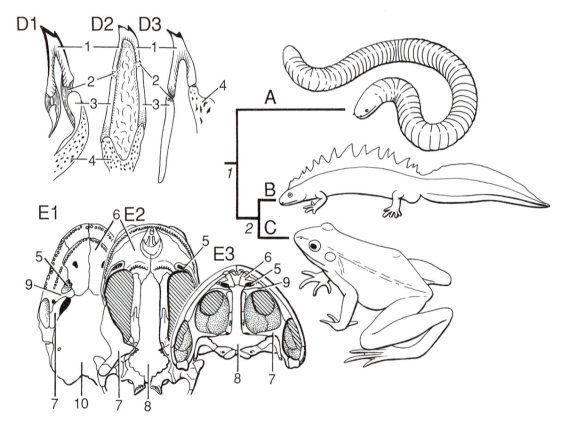

Fig. 3.24. Characters and interrelationships of the Lissamphibia. A–C, interrelationships of lissamphibians; *terminal taxa*: A, Apoda (limbless amphibians or apods); B, Caudata (salamanders and newts); C, Anura (frogs and toads); *nested taxa and selected synapomorphies*: *1*, Lissamphibia (pedicellate and bicuspidate teeth, papilla amphibiorum in ear, opercular endoskeletal bone associated with stapes); *2*, Batrachia (modification of the pronephros for sperm transport, loss of the papilla neglecta, loss of postfrontal and surangular bones, anterior ramus of entopterygoid not articulated with the dermopalatine). D, vertical section through the pedicellate and bicuspidate teeth of the Apoda (D1), Caudata (D2), and Anura (D3); E, skull in ventral aspect of the Apoda (E1), Caudata (E2), and Anura (E3). 1, articulated portion of bicuspidate crown; 2, ligament; 3, pedicel; 4, jaw-bone; 5, choana; 6, vomer; 7, entopterygoid; 8, parasphenoid; 9, palatine (fused to maxilla), 10 'os basale' (parasphenoid fused to braincase). (D1, from Casey and Lawson (1981); D2, D3, from Lehman (1968); E1, E2, redrawn from Laurent (1985); E3, from Goodrich (1930).)

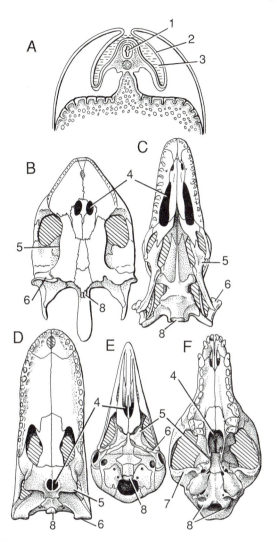

Fig. 3.25. Amniote characteristics. A, diagram of the extra-embryonic membranes of the chick, to show the amniotic cavity; B–F, ventral view of the skull of a turtle (B), lizard (C), crocodilian (D), bird (E) and mammal (F). 1, embryo; 2, amnion; 3, amniotic liquid; 4, choanae; 5, pterygoid flange; 6, quadrate; 7, glenoid fossa of the squamosal for the articulation of the dentary (lower jaw); 8, convex occipital condyle. (A, modified from Waterman (1971); B, from Goodrich (1930).)

reduction of the stomodeum in embryo (hence the partly endodermal derivation of the choanae); the loss of the postfrontal, surangular, splenial, and ectopterygoid, and of the scales; and the loss of the articulation between the anterior branch of the entopterygoid and the dermo-palatine (Fig. 3.24E2, E3, 7, 9). All extant amphibians undergo a larval development and a metamorphosis,

notwithstanding a few cases of secondary direct development. The larval development has sometimes been regarded as a general character of tetrapods and lungfishes. This may be true, but many features of amphibian larval development are in fact derived and may be the result of special adaptations to survival in temporary ponds.

The Amniota are characterized by the amniotic, or cleidoic, egg, i.e. they are provided with an additional, extra-embryonic membrane filled with water, in which the embryo can develop. This device, called the amnion (Fig. 3.25A, 2, 3), allows amniotes to lay eggs outside the water, in contrast to amphibians, whose eggs are generally laid in pools or, at least, on moist substrates. Amniotes can be oviparous, ovoviviparous, or viviparous, but they all retain an amnion.

There are also some anatomical features unique to the amniotes, such as the posterior lateral wing of the pterygoids or ptergoid flanges (Fig. 3.25B–F, 5), which is ventrolaterally projected and gives a bean-shaped outline to the space housing the jaw muscles in most early amniotes. Other osteological features are the separate scapular and coracoid ossifications, the convex occipital condyle (Fig. 3.25, 8), and the astragalus and calcaneum in the ankle. In some amniotes (turtles, crocodiles, and mammals), the choanae are displaced backward and more-or-less covered by expansions of the vomers, dermopalatine, and entopterygoids, which form a secondary palate (Fig. 3.25D, F, 4). Although homoplasic, a secondary palate is unknown outside the amniotes.

The amniotes comprise the Mammalia (mammals, Fig. 3.26, O), Aves (birds, Fig. 3.26, N), Crocodylia (crocodiles, Fig. 3.26, M), Lepidosauria (sphenodontids, lizards, and snakes, Fig. 3.26, J–L) and Chelonii (turtles, Fig. 3.26, I). Despite a thorough knowledge of their comparative anatomy, the interrelationships of the amniotes are still the subject of heated discussion (see review in Benton 1988). During the past ten years, several cladograms have been produced to depict amniote phylogeny. The major controversy bears on the position of the birds, which are generally believed to be the sister-group of the crocodiles, on account of their 'sauropsid' blood circulation and 'archosaurian' skeletal characteristics, but which also share a number of unique features with mammals (Goodrich 1930; Gauthier *et al.* 1988; Gauthier 1989; Gardiner 1982, 1993). Another question concerns the position of the turtles, which have sometimes been regarded as the sister-group of all other amniotes, or Eureptilia. This hypothesis is no longer favoured, and the Sauropsida, or the Reptilia (according to the position given to birds), i.e., turtles and diapsids, are now generally regarded as a clade.

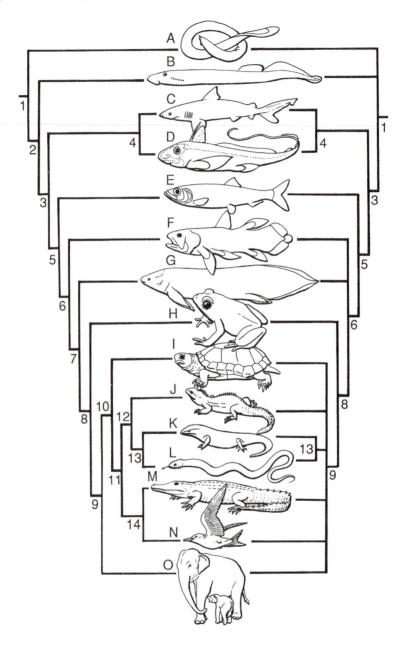

The relationships of extant amniotes above node 9 in Fig. 3.26 are shown according to the main current views (which are consistent with palaeontological inferences) on the left-hand side, and according to the consensus of the resolved cladograms (i.e. the relationships on which all authors agree) on the right-hand side. The currently admitted cladogram of the amniotes is largely based on the morphological data. In contrast, molecular data (on myoglobin, beta- and alpha-haemoglobin, cytochrome *c*, etc.) give conflicting results, perhaps because of methodological biases (choice of taxa, distance methods). Some of them, however, show a sister-group relationship between birds and mammals, which is rejected by most morphologists (except Gardiner 1982, 1993).

Fig. 3.26. Interrelationships of the extant Craniata. One of the fully resolved cladograms on the left-hand side and consensus of the current theories on the right-hand side. *Terminal taxa*: A, Hyperotreti (hagfishes); B, Hyperoartia (lampreys); C, Elasmobranchii (sharks and rays); D, Holocephali (chimeras); E, Actinopterygii (ray-finned fishes); F, Actinistia (coelacanths); G, Dipnoi (lungfishes); H, Lissamphibia (apods, salamanders, newts, and frogs); I, Chelonii (turtles); J, Sphenodontia (tuatara); K, Lacertilia (lizards); L, Serpentes (snakes); M, Crocodylia (crocodiles); N, Aves (birds); O, Mammalia (mammals). *Nested taxa and selected synapomorphies*: 1, craniata (neural crest, skull enclosing sensory capsules); 2, Vertebrata (vertebral elements, or arcualia, extrinsic eye muscles, cardiac innervation, adenohypophysis derived from ectoderm, two vertical semicircular canals, sensory-line system, spleen); 3, Gnathostomata (jaws, medial and segmented gill arches, gills essentially ectodermal in origin, intrinsic eye muscles, horizontal semicircular canal, sperms passing through urinary system, neuromasts of sensory lines enclosed in canals, calcified exo- and endoskeletons); 4, Chondrichthyes (prismatic calcified cartilage, pelvic claspers in males); 5, Osteichthyes (endochondral bone, large dermal bones in head, teeth attached to the premaxilla, maxilla, and dentary, lepidotrichiae, air-filled diverticles of the pharynx); 6, Sarcopterygii (monobasal paired fins, pulmonary vein, vena cava); 7, Rhipidistia (alveolae in lungs, partially divided and sigmoid arterial cone in heart, incipient atrial septum, numerous physiological and molecular characters); 8, Tetrapoda (locomotory limbs with digits, sacral ribs connected to pelvic girdle, choanae, epidermal phaners, tear duct, hyomandibula sound-conducting and inserted in the fenestra ovalis); 9, Amniota (amnion, downturned pterygoid flange, convex occipital condyles); 10, Sauropsida (production of ornithuric acid, nasal gland outside nasal capsule, iris composed of stritated muscles); 11, Diapsida (two temporal fenestrae); 12, Lepidosauria (lateral 'conch' of the quadrate supporting the tympanic membrane; bone extremities, or epiphyses, forming specialized joint surfaces); 13, Squamata (mobile quadrate); 14, Archosauria (antorbital fossa, mandibular fenestra, urinary bladder lost, transparent nictitating membrane in eye).

3.6 A CLADOGRAM OF EXTANT CRANIATES

The cladogram of the extant craniates in Fig. 3.26 shows on the right-hand side the strict consensus (i.e. the resolved relationships that occur in all cladograms) of the resolved cladograms proposed by various scientists over the past ten years. It therefore shows many unresolved branchings, which mean that the relationships are undecided. The cladogram on the left-hand side is completely resolved and represents the relationships that are accepted by the majority of biologists and palaeontologists, but is not necessarily closer to the truth.

The consensus cladogram shows three main problematical points of multiple branchings: the basal trichotomy (or trifurcation) of the hagfishes, lampreys, and gnathostomes (Fig. 3.26, 1); the trichotomy of the actinistians (coelacanths), lungfishes, and tetrapods (Fig. 3.26, 6); and the multiple branching of the amniotes (Fig. 3.26, 9). The first and third of these points have been briefly discussed above, but the second, i.e. the interrelationships of sarcopterygians, is perhaps one of the most fiercely debated. Many characters that are classically regarded as unique to lungfishes and tetrapods (e.g. those cited by Rosen *et al.* 1981) have been discarded on the basis of embroyological or palaeontological arguments (Jarvik 1981*b*; Holmes 1985; Panchen and Smithson 1987) but, nevertheless, some molecular data bring lungfishes and tetrapods together as sister-groups (see e.g. Hillis and Dixon 1989; Stock and Swofford 1991), whereas others treat the actinistians and tetrapods as sister-groups (Gorr *et al.* 1991; Meyer and Wilson 1992) or treat the actinistians and lungfishes as sister-groups (Forey 1991*a*). The majority of the characters that are advanced for treating the actinistians as sister-group of the tetrapods are general sarcopterygian or even osteichthyan characters (intracranial joint, dermal bone pattern) that are retained by early fossil tetrapods, but have been lost or deeply modified in all lungfishes (see, however, Fritzsch 1987; Schultze 1991, 1994). My reasons for adopting here the sister-group relationship between lungfishes and tetrapods are based on palaeontological arguments that, in turn, are not accepted by all palaeontologists (i.e. the theory that lungfishes are the sister-group of the fossil porolepiforms, which share a number of unique characters with tetrapods and some other fossil sarcopterygians, and none with actinistians; see Chapter 4, p. 200 and Chapter 5, p. 247). In order to avoid the name 'Choanata', which refers to a homoplasic structure, for the taxon including lungfishes and tetrapods, I prefer to use here, as did Ahlberg (1991*a*) the name 'Rhipidistia', originally erected for fossils (porolepiforms and osteolepiforms), to which lungfishes and tetrapods respectively are supposed to be closely related.

The brief review of the extant craniates, with particular emphasis on fishes, is aimed at giving an idea of the pattern of relationships of the group, and some of the arguments that support it, as if no fossil vertebrate were known. As we shall see in the following chapter, the numerous fossil early vertebrates (not to mention later forms) have been 'injected' into this pattern for more than a century; and the elements of uncertainty, interpretation, and ignorance that are inherent in the study of fossils have also caused distortions in interpreting the Recent data or have thrown doubt on relationships established for Recent groups. The widespread belief that 'ancient equals primitive' has also introduced biases because deductions about the primitive state of a character inferred from the analysis of Recent taxa were often disproved by the discovery of different patterns in fossils. Many palaeontologists are obsessed by the idea that many homologies of Recent taxa turn out to be homoplasies when the earliest fossils of one or the other taxon are studied (the 'basal taxa', Schultze 1987). This is true in some cases (e.g. the lungfish 'choana'), but following this principle *ad absurdum* would imply that neontological data are useless for reconstructing phylogenies, whereas they are the most abundant and unambiguous. Although I am a palaeontologist and I am convinced that fossils tell us many things, I still believe that it is wise to use neontological data first and then fossil data, even if we are aware of the risk of dealing with homoplasies.

Table 3.1. Phylogenetic classification of the extant Craniata

Craniata (craniates)
 Hyperotreti (hagfishes)
 Vertebrata (vertebrates)
  Hyperoartia (lampreys)
  Gnathostomata (jawed vertebrates)
   Chondrichthyes (cartilaginous fishes)
    Elasmobranchii ('sharks' and rays)
    Holocephali (chimeras)
   Osteichthyes (bony vertebrates)
    Actinopterygii (ray-finned fishes)
     Cladistia
     Actinopteri
      Chondrostei (sturgeons and paddlefishes)
      Neopterygii
       Ginglymodi (gars)
       Halecostomi
        Halecomorphi (bow-fin)
        Teleostei (teleosts)
    Sarcopterygii (lobe-finned or limbed vertebrates)
     Actinistia (coelacanths)
     Rhipidistia
      Dipnoi (lungfishes)
      Tetrapoda (four-legged vertebrates)
       Amphibia (amphibians)
        Apoda (limbless amphibians)
        Batrachia
         Caudata (salamanders and newts)
         Anura (frogs and toads)
       Amniota (amniotes)
        Mammalia (mammals)
        Sauropsida (= Reptilia)
         Chelonii (turtles)
         Diapsida (= Sauria)
          Lepidosauria
           Sphenodontia (tuatara)
           Squamata
            Lacertilia (lizards)
            Serpentes (snakes)
          Archosauria
           Crocodylia (crocodiles)
           Aves (birds)

The phylogenetic classification of the craniates in Table 3.1 reflects the fully resolved cladograms in Fig. 3.3, 3.17, 3.24, and 3.26.

Further reading (most of the works cited here contain, in turn, a large number of references on the subject):

Chordate morphology, relationships, and interrelationships: Fernholm *et al.* (1989); Hennig (1983); Jarvik (1988); Jefferies (1986); Lake (1990); Løvtrup (1977); Maisey (1986); Schaeffer (1987). *Craniate anatomy and development in general*: Bjerring (1977); Bolk *et al.* (1931–1939); De Beer (1937), Goodrich (1930); Hanken and Hall (1993); Holmgren and Stensiö (1936); Jarvik (1980, 1981a); Jollie (1962); Romer (1970); Wake (1979); Waterman (1971). *Craniate interrelationships*: Hardisty (1982), Hennig (1983), Løvtrup (1977), Janvier (1981a), Jarvik (1981a), Forey and Janvier (1994), Schaeffer and Thomson (1980), Yalden (1985). *Hagfish anatomy, development, and biology*: Brodal and Fänge (1963), Fänge (1973), Foreman et al. (1985), Gorbman and Tamarin (1985), Hardisty (1979), Holmgren (1946), Janvier (1993); Mallatt and Paulsen (1986); Marinelli and Strenger (1956). *Lamprey anatomy, development, and biology*: Damas (1944), Hardisty (1982), Hardisty and Potter (1974–1982); Janvier (1993); Johnels (1948), Mallatt (1981); Marinelli and Strenger (1954); Potter and Hilliard (1987). *Gnathostome anatomy, histology, development, and interrelationships*: Gardiner (1983); Greenwood *et al.* (1973); Goodrich (1930); Hanken and Hall (1993); Jarvik (1980); Maisey (1986, 1987, 1988, 1989a); Nelson (1969); Patterson (1977a); Rosen *et al.* (1981); Reif (1982a); Schaeffer (1977); Schultze (1993, 1994); Smith and Hall (1990, 1993). *Chondrichthyan anatomy and interrelationships*: Cappetta (1987); Compagno (1977); Daniel (1937); Garman (1904, 1913); Maisey (1984a, b); Marinelli and Strenger (1959); Stahl (1967). *Osteichthyan anatomy and interrelationships*: Gardiner (1973); Jarvik (1980); Rosen *et al.* (1981); Lauder and Liem (1983); Schultze (1993). *Actinopterygian anatomy and interrelationships*: Johnson and Anderson (1993); Lauder and Liem (1983); Marinelli and Strenger (1973); G. Nelson (1989); J. Nelson (1994); Patterson (1975, 1977b, 1982a). *Actinistian anatomy and biology*: Forey (1980, 1988); Fritzch (1987); Millot and Anthony (1958, 1965); Millot et al. (1978); Musick *et al.* (1991); Northcutt and Bemis (1993); Robineau (1987); Smith *et al.* (1975). *Lungfish anatomy and biology*: Bemis *et al.* (1987); Goodrich (1930); Günther (1871). *Tetrapod anatomy and interrelationships*: Benton (1988); Gardiner (1982, 1993); Gauthier *et al.* (1988); Hanken and Hall (1993); Løvtrup (1985); Schultze and Trueb (1991). *Sarcopterygian interrelationships*: Ahlberg (1991a); Forey (1984b, 1987b, 1991a); Forey *et al.* (1991); Holmes (1985); Jarvik (1980, 1981a); Kesteven (1950); Miles (1975); Panchen and Smithson (1987); Rosen *et al.* (1981); Schultze (1991, 1994).

4

Early vertebrates and their extant relatives

The phylogenetic and systematic framework of extant craniates given in the preceding chapter is based on characteristics of all kinds, although morphological ones are still by far the most reliable, in the absence of the methodological stability of molecular systematics. The source of information represented by these neontological characteristics is not unlimited but it is far from thoroughly explored. Although great weight is given now to molecular sequence data, the anatomy of many of the best-known extant craniates still reveals features that had gone unnoticed by zoologists for two centuries. Our interpretation of the phylogeny of extant craniates can be expected to become more stable, and better and better corroborated, as knowledge of their structure grows.

None the less, this framework is certainly the best basis we have at present for comparing and assessing the relationships of fossil craniates. Some of them, as is the case for most Cenozoic or Mesozoic taxa, are easy to place within one or the other of the extant taxa, but as one progresses into earlier epochs the picture is complicated by the presence of extinct taxa that have fewer and fewer anatomical characters in common with the Recent taxa. A glance at the reconstructions in Chapter 1 clearly shows this increasing 'weirdness' of the fishes, and gives the false impression of increasing 'disparity' in earlier ages. This goes so far that when one reaches Silurian and Devonian times—not to speak of Ordovician times—one is faced with fishes that we can relate to major modern groups only on the basis of very few characters. In most cases, we can tell what these organisms are not, but we can rarely say what they are, except that they are craniates or gnathostomes. The purpose of this chapter is to provide the essential data that are available on these fossil taxa and to discuss their possible interpretation and relationships with extant taxa.

4.1 THE EARLIEST KNOWN VERTEBRATE REMAINS

As we have seen in Chapter 1, fossil craniates, and presumably vertebrates in a strict sense, become abundant first in Early Silurian strata, more precisely in the Wenlock (424–28 Ma). Undoubted vertebrate remains occur earlier, in the Ordovician, and the earliest known articulated vertebrates are the Arandaspida (*Arandaspis*, *Porophoraspis*) from Australia, now dated as Early Llanvirn (470 Ma). Somewhat earlier (Tremadoc–Arenig, 485 Ma) dermal bone fragments resembling arandaspids have also been recorded from the Amadeus Basin in Australia. Our knowledge of these early vertebrates has been also greatly increased by the discovery of almost complete specimens of the Bolivian arandaspid *Sacabambaspis*, of Llanvirn or Caradoc age. The Ordovician vertebrate assemblages of North America, which have been known since the last century and have yielded, for example, the Astraspida, are generally regarded as Caradoc in age.

Ordovician vertebrate remains occur only in a few localities or geological formations, mainly the Harding Sandstone Formation and contemporaneous formations of North America, the Stairway Sandstone and some other older and younger formations of the Amadeus Basin in the Northern Territory of Australia, and the Anzaldo Formation of central Bolivia. In addition, there are a few local occurrences of astraspid-like forms on the Siberian Platform.

The Harding Sandstone Formation outcrops near Canyon City, Colorado, but extends into adjoining states. It consists of thick red sandstone series in which scattered vertebrate remains can be found in great abundance, in the form of scales and platelets ornamented with large rounded tubercles. These have been referred to as *Astraspis* and *Eriptychius* by Walcott (1892; see also Yochelson 1983). The discovery of these first Ordovician vertebrates was a great surprise at that time, because they were the first evidence of vertebrates older than the classical Old Red Sandstone fauna (Devonian and Late Silurian), which had long been regarded as the 'primordial' fish fauna. Only one partly articulated head-shield of *Astraspis* (Fig. 4.3.A) was found by Walcott. It was described much later by Eastman (1917) and Ørvig (1958), and it did not match any of the previously known Silurian or Devonian fishes. There was, however, a consensus on the affinities of these Ordovician forms with the Silurian and Devonian heterostracans (see section 4.5 below), which was based on the fact that their dermal skeletons were made up of acellular bone (aspidine). Since these early discoveries, *Astraspis* and *Eriptychius* have been found in the Ordovician of most of the USA and Canada. The conodonts of the Harding Sandstone Formation clearly indicate a Middle Caradoc age. In addition to these two genera, Ørvig (1958) described a third, *Pycnaspis*, from Wyoming, which differs from the other two in its histological structure, but has been regarded by Denison (1967) as a mere variant of *Astraspis*. The overall aspect of *Astraspis* is known only from two partly articulated specimens, the dorsal shield mentioned above, and part of another dorsal shield associated with a tail.

These three genera can be distinguished essentially by their histological structure, which is markedly different, in particular between *Astraspis* and *Eriptychius* (see section 4.4 below). It is, however, probable that the rich bone beds of the Harding Sandstone contain a more diverse vertebrate fauna. As early as 1902, Vaillant pointed out the presence of bone fragments with clear cell spaces in thin sections of Harding Sandstone samples, a discovery which was later confirmed by Ørvig (1951, 1965), Denison (1967), and M. M. Smith (1991) (Fig. 4.3.D, 12). Smith associated this cellular bone with tubercles of mesodentine, which are strongly indicative of the Silurian and Devonian osteostracans. Fragments of globular calcified cartilage from the Harding Sandstone are now known in association with *Eriptychius* (Fig. 4.3B).

The Harding Sandstone Formation was first regarded as a freshwater deposit, since it is made up of red sandstone, but it later proved to contain many marine invertebrates and is now regarded as a shallow-water marine marginal deposit (Spjaeldnaes 1979), despite Graffin's (1992) claim that the vertebrate remains are always associated with fluviatile channel facies in this formation and, consequently, may indicate a freshwater habitat.

The Ordovician rocks of the Amadeus Basin of Australia have yielded the earliest known vertebrate remains in the world (Shergold 1991), in particular those of the Paccota Sandstone. These are dated as Tremadoc to Early Arenig and are tentatively referred to the arandaspids. The best preserved vertebrate fauna has, however, been found in the Stairway Sandstone Formation, with *Arandaspis* and *Porophoraspis* (Ritchie and Gilbert-Tomlinson 1977), which are of Early Llanvirn age. Arandaspid remains occur there up to the Late Ordovician (Ashgill).

The third major Ordovician vertebrate localities are those in the Anzaldo Formation of central Bolivia, around the town of Cochabamba. Here the arandaspid *Sacabambaspis* occurs quite abundantly, in association with marine invertebrates. This genus, as well as the still poorly known genus *Andinaspis* (Gagnier 1992) are almost similar to the Australian arandaspids, and may even occur also at some levels of the Amadeus Basin (Stokes Siltstone, Llanvirn–Caradoc). The age of the Bolivian forms is still debated, although a Caradoc age is favoured.

A few isolated scales or denticles resembling thelodont scales have been described as conodonts (*Evencodus, Stereoconus*) by Moskalenko (1973) from the Ordovician of the Siberian Platform (Fig. 4.3E). They may actually be conodonts, but their histological structure deserves to be studied. Astraspid tesserae, as well as scales of acanthodians and thelodonts, have been mentioned in the Late Ordovician of Siberia, but detailed descriptions of the specimens is awaited.

In sum, the Ordovician vertebrates fall into at least two major taxa, the Arandaspida (*Arandaspis, Porophoraspis, Sacabambaspis*) and the Astraspida (*Astraspis*, and possibly *Eriptychius* and *Pycnaspis*). In addition, there are possibly Ordovician thelodonts (mentioned but undescribed) and some still undetermined jawless or jawed vertebrate taxa with cellular dermal bone.

Beside these undoubted Ordovician vertebrates, represented by the arandaspids, astraspids, and probably other taxa known only from isolated scales and tesserae, a number of enigmatic fossils made up of apatite have been recorded from the Cambrian and Lower Ordovician and referred to, with reservations, as vertebrates or craniates. These remains have been reviewed by Blieck (1992), who showed that most of them are now referred to various non-vertebrate groups. The latest debate in this connection concerns the conodonts, recently claimed by Sansom *et al.* (1992) and Briggs (1992) to be true craniates, if not vertebrates. As will be discussed in Chapter 7 (p. 283), this conclusion, based on anatomical and histological arguments, needs further corroboration.

Anatolepis (Bockelie and Fortey 1976) deserves special comment. It is represented by small phosphatic carapace and spine fragments ornamented with minute tubercles, the structure of which recalls vaguely the dentine of the vertebrate exoskeleton. These fragments occur from the Late Cambrian to the Ordovician and have been regarded as possible vertebrate remains during the past decade. They have, however, been claimed by invertebrate palaeontologists (Peel and Higgins 1977) as belonging to a group of extinct arthropods, the aglaspidids. Others have tried to demonstrate that the carapace of the aglaspidids had a totally different structure, and that *Anatolepis* could be regarded as more closely related to the vertebrates (Briggs and Fortey 1982). The structure of *Anatolepis* is admittedly different from that of the dentine and aspidine of true vertebrates (Ørvig 1989), but it is equally different from that of any arthropod. The specimens figured by Repteski (1978) clearly show a basal laminated layer, a cancellar layer, and a superficial layer, which look more similar to the structure of a heterostracan carapace. Although the vertebrate affinities of *Anatolepis* sound less and less probable, it is not yet entirely ruled out.

Further reading

Aldridge *et al.* (1993); Blieck (1992); Briggs and Fortey (1982); Darby (1982); Denison (1967); Elliott (1987): Elliott *et al.* (1991); Gagnier (1992, 1993*a,b*); Gagnier *et al.* (1986); Peel and Higgins (1977); Ørvig (1958, 1989); Repetski (1978) Smith (1991); Ritchie and Gilbert-Tomlinson (1977); Shergold (1991); Stensiö (1964).

4.2 FOSSIL HAGFISHES (HYPEROTRETI)

The only undoubted fossil hagfish known to date is *Myxinikela* (Bardack 1991) from the late Carboniferous of Illinois, USA (Fig. 4.1B). It is preserved in carbonaceous shales deposited in an anoxic environment that permitted the preservation of soft tissues in the form of an imprint. *Myxinekela* clearly displays the tentacles (Fig. 4.1, 3) and long prenasal sinus (Fig. 4.1, 4) that are characteristic of the group, and resembles the Recent hagfishes, despite its very small size (about 50 mm in length). It differs from the latter by the more anterior position of the branchial apparatus (Fig. 4.1B2, 1), which is situated just behind the braincase, by the less elongated body shape, and by the more developed caudal fin(Fig. 4.1B1). Another form from the same locality, *Gilpichthys* (Bardack and Richardson 1977) may also be a hagfish, but its tentacles are not clearly visible (Fig. 4.1A). It displays, however, hagfish-like non-mineralized teeth (Fig. 4.1A2, 2). These two Carboniferous hagfishes lived in a marine near-shore environment.

Again, the affinities of the conodonts with hagfishes (see Chapter 7, p. 283) advocated by some authors may enlarge considerably the stratigraphical range of the group. It still rests, however, on very tenuous and dubious arguments and will not be considered here.

Further reading

Bardack (1991); Bardack and Richardson (1977); Janvier (1993).

4.3 ARANDASPIDA

The Arandaspida are characterized by the extreme anterior position of the eyes, at the tip of the head (Fig. 4.2, 9). The eyes and possibly the nostrils (Fig. 4.2, 11), separated

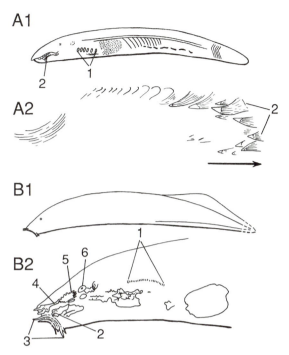

Fig. 4.1. Fossil Hypetrotreti (hagfishes). A, *Gilpichthys*, Late Carboniferous of North America, reconstruction of the entire animal (A1, about ×1) and detail of the non-mineralized teeth (A2, the arrow points forward, ×15); B, *Myxinikela*, Late Carboniferous of North America, reconstruction of the entire animal (B1, ×1) and imprints of internal structures in the head (B2). 1, presumed trace of gill pouches; 2 horny teeth; 3, tentacles; 4, prenasal sinus; 5, olfactory organ; 6, eyes. (A, redrawn from Bardack and Richardson (1977); B, redrawn from Bardack (1991).)

by a T-shaped bone (Fig. 4.2, 8), are situated inside a large terminal embayment of the anterior margin of the dorsal shield. Most other features of the Arandaspida occur in other vertebrate groups. The group at present includes four genera: *Arandaspis*, *Porophoraspis*, *Sacabambaspis* and *Andinaspis*, which seem to differ only in minor features of the dermal ornamentation (Fig. 4.2.H).

The head armour of arandaspids, so far as known from *Arandaspis* (Fig. 4.2A) and particularly *Sacabambaspis* (Fig. 4.2B–G) consists of two large median plates, or shields, covering the dorsal and ventral side respectively. Each shield is roughly oval in shape, the dorsal one being rather flattened (Fig. 4.2, 1), and the ventral one more convex (Fig. 4.2, 2). Both are ornamented with minute and closely set, drop-shaped (Fig. 4.2H1) or oak-leaf-shaped tubercles (Fig. 4.2H3) that are somewhat reminiscent of those of some early heterostracans (Fig. 4.10, 5).

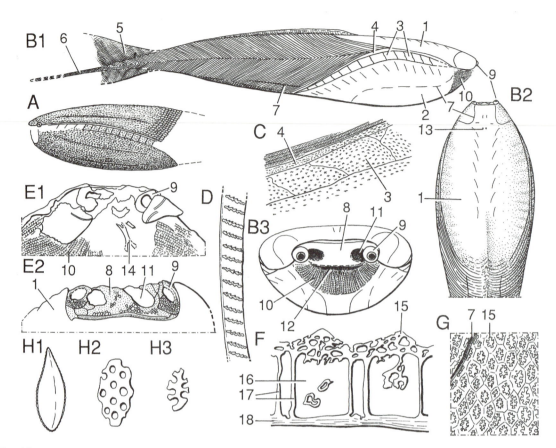

Fig. 4.2. Arandaspida. A, *Arandaspis*, Early Ordovician of Australia, reconstruction of the head in lateral view (×0.5); B–G, *Sacabambaspis*, Ordovician of Bolivia; B, reconstruction in lateral (B1, ×0.5), dorsal (B2), and anterior (B3) view; C, detail of the branchial platelets of the right side; D, detail of a portion of flange scale; E, sketch of the anterior part of the head in two of the best-preserved specimens known so far, in ventral (E1) and dorsal (E2) view respectively; F, vertical section through the exoskeleton of the dorsal shield (×30); G, aspect of the external surface of the dorsal shield, showing the grooves that mark the limits between the exoskeletal units; H, tubercles of the dermal ornamentation of *Arandaspis* (H1), *Porophoraspis* (H2), and *Sacabambaspis* (H3), brought to the same scale (×20). 1, dorsal shield; 2, ventral shield; 3, branchial platelets; 4, epibranchial plate; 5, caudal fin; 6, chordal lobe; 7, lateral-line grooves; 8, T–shaped plate; 9, presumed sclerotic ring; 10, oral plates or scales; 11, presumed nostrils; 12, mouth; 13, pineal and parapineal openings; 14, dumbell-shaped bone or calcified cartilage; 15, tubercle; 16, cancellae; 17, double walls of the cancellae; 18, laminar basal layer. (A, H1, H2, from Ritchie and Gilbert-Tomlinson (1977); B–H3, based on Gagnier (1993a) and original material.)

The two shields are separated laterally by a series of diamond-shaped platelets, the branchial plates (Fig. 4.2, 3), lined dorsally with an elongated marginal, or epibranchial plate (Fig. 4.2, 4). Anteriorly, and behind the embayment containing the eyes and nostrils, the dorsal shield is pierced by a pair of closely set foramina for the pineal and parapineal 'eyes' respectively (Fig. 4.2, 13). Arandaspids are thus the only vertebrates with paired pineal and parapineal openings. The eyes seem to be encapsulated in an endoskeletal ossification, possibly covered with an exoskeletal sclerotic ring (Fig. 4.2, 9).

The position of the nostrils remains uncertain, yet Gagnier (1993a) suggested that they may have opened medially to each eye, and were separated by a median T-shaped dermal bone (Fig. 4.2, 11, 8). The mouth (Fig. 4.2, 12) is situated ventrally to this dermal bone and its lower 'lip' was covered with a fan-shaped bundle of minute scale rows that could probably expand as the mouth opened (Fig. 4.2, 10).

The body scales (Fig. 4.2D) of arandaspids are elongated, rod-shaped, and arranged in chevrons on each flange, much as in anaspids (Fig. 4.11). There are no ventral or dorsal median ridge scales. Arandaspids possess no paired or unpaired fin, except the caudal, which consists of a moderately large web (Fig. 4.2, 5), covered with minute scales, and expands ventrally and dorsally to a slender median axial lobe (Fig. 4.2, 6).

The lateral-line system is well developed over the head and body, and was housed in narrow grooves lined with rows of minute tubercles (Fig. 4.2, 7).

Although very little is known of the internal anatomy of arandaspids, some faint impressions on the shields (Fig. 4.2B2) suggest the presence of at least ten branchial units or pouches. There may have been more (up to twenty) if one admits that the number of branchial plates reflects that of the branchial openings. The latter opened between the branchial plates, through a minute foramen.

There seems to have been some endoskeletal mineralization in arandaspids, at least in the anterior part of the head and around the eyes, but it is still uncertain whether it is simply calcified cartilage or bone. A peculiar, dumbell-shaped and supposedly endoskeletal element has been observed inside the mouth cavity of two specimens (Fig. 4.2, 14), and may belong to some form of 'lingual' apparatus comparable to that of modern jawless craniates (Gagnier 1993*a*).

The structure of the arandaspid exoskeleton is still poorly known, despite some data on *Sacabambaspis*. It is three-layered, like that of heterostracans, with a laminar basal layer (Fig. 4.2, 18), a middle cancellar layer (Fig. 4.2, 16), and a superficial spongy layer that forms the tubercles (Fig. 4.2, 15). The cancellar layer, when seen in horizontal section, displays a honeycomb aspect, as in heterostracans, but the walls of each cancella are double, suggesting some separation between the adjacent dermal units, or tesserae. In the rearmost part of the shields, there is a network of grooves between the tubercles, which mark the limits between these tesserae (Fig. 4.2G). Each of them is surmounted by a layer of spongy bone bearing a single tubercle, which seems to contain no dentine. Whether the exoskeleton is made up of cellular bone or aspidine is undecided, but some peculiar vacuities recall cell spaces (Gagnier 1993*a*).

Further reading

Gagnier (1989, 1992, 1993*a,b*); Gagnier and Blieck (1992); Janvier and Blieck (1993); Ritchie and Gilbert-Tomlinson (1977).

4.4 ASTRASPIDA

The Astraspida are characterized by a thick, glassy enameloid cap on the tubercles of the ornamentation (Fig. 4.3C1, 9). The dermal head armour is made up of loosely attached polygonal units of spongy aspidine or tesserae. The latter bear numerous small tubercles surrounding a larger median one (Fig. 4.3A1, 1).

Astraspis is known from only two partly articulated specimens, on which are based the reconstructions in Fig. 4.3A. The head armour consists, as in arandaspids, of a relatively flat dorsal shield and a presumably bulging ventral shield, united laterally by a series of branchial platelets (Fig. 4.3, 4) which separate at least eight, horizontally aligned external gill openings (Fig. 4.3, 5). In contrast to the arandaspids, the gill openings are relatively large and clearly distinct, with no cover. The orbits are small and laterally placed (Fig. 4.3, 6). almost nothing is known of the snout and oral regions. In *Astraspis*, the surface of the dorsal shield is marked with five longitudinal ridges (Fig. 4.3, 3). The sensory-line grooves can be traced with difficulty and pass through the centre of some of the tesserae (Fig. 4.3, 2). The scales of *Astraspis* are large, diamond-shaped, and ornamented with tubercles. The shape of the caudal fin is still unknown, but there is no evidence of any other median fin, nor is there any evidence of paired fins.

Astraspis and *Pycnaspis* (if the latter really is a different genus) alone can be included in the Astraspida with certainty. The general morphology of the other genus from the Harding Sandstone Formation, *Eriptychius*, is still poorly known, and its histological characters (namely the presence of very large dentine tubules; Fig. 4.3C3) are regarded by Ørvig (1989) as closer to those of heterostracans than to those of *Astraspis*. However, an isolated branchial platelet referred to as this genus shows a branchial notch which suggests that it possessed separate branchial openings, like *Astraspis* (Fig. 4.3B2, 5).

Thanks to the excellent preservation of the hard tissues in the Harding Sandstone Formation, the histological characters of the Astraspida have been studied in detail (see review by Ørvig 1989). The tubercles of *Astraspis* display a thick amorphous enameloid layer (Fig. 4.3C1, 9), which rests directly on the spongy aspidine (Fig. 4.3C1, 11). In *Pycnaspis*, there is an intervening layer of mesodentine-like tissue (Fig. 4.3C2, 10) between the enameloid cap and the aspidine. Finally, in *Eriptychius*, this dentinous tissue, with long branching tubules, forms most of the tubercle (Fig. 4.3C3, 10). Since the aspidine of astraspids is somewhat different from that of heterostracans (it is thicker, with dense bundles of fibre, and somewhat resembles dentine), Halstead (1987*b*) coined the name 'astraspidine' for it.

The tubercles of the dermal skeleton of astraspids show traces of regeneration and replacement. The 'old' tubercles are buried under the younger ones, and become progressively necrosed.

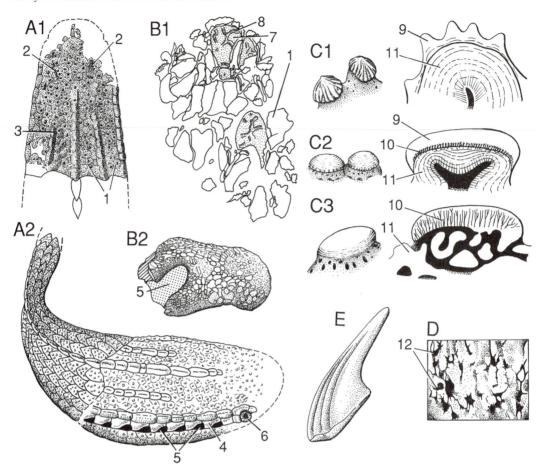

Fig. 4.3. Astraspida and other Ordovician vertebrates. A, *Astraspis*, Late Ordovician of North America, dorsal shield in dorsal view (A1, ×0.7) and reconstruction in dorsolateral view (A2); B, *Eriptychius*, Late Ordovician of North America, partially articulated snout in dorsal view (B1, ×1), showing masses of globular calcified cartilage with canals (stippled), and isolated branchial plate (B2, ×2); C, comparison of the aspect and structure (vertical sections) of the tubercles of the dermal ornamentation in *Astraspis* (C1), *Pycnaspis* (C2), and *Eriptychius* (C3) from the Ordovician of North America; D, cell spaces (lacunae) in the basal bone of an undetermined tubercle from the Ordovician of North America (about ×150); E, *Stereoconus*, Ordovician of Siberia, a possible vertebrate dermal denticle (×50). 1, tesserae; 2, supraorbital sensory line; 3, longitudinal ridges; 4, branchial plates; 5, branchial openings; 6, eye; 7, calcified endoskeleton (globular calcified cartilage); 8, grooves for subaponevrotic vessels; 9, enameloid; 10, dentinous tissue (meso- or metadentine); 11, aspidine; 12, cell space. (A1, from Halstead (1973*b*); A2, from Elliott (1987); B1, from Denison (1967); B2, C, based on Ørvig (1958, 1989); D, from M. M. Smith (1991); E, from Moskalenko (1973).)

Nothing is known of the internal anatomy of the Astraspida, except that they may have possessed at least eight gill-pouches or gill units. The masses of globular calcified cartilage described by Denison (1967) in a presumed snout of *Eriptychius* contain a number of vascular canals (Fig. 4.3, 8). Usually, cartilage is not—or is only poorly—vascularized, except at its surface. This condition is thus rather different from what is met with in other extant or fossil vertebrates whose endoskeleton is known. Another possibility is that these canals are mainly superficial, and represent the subaponevrotic canal network that lies between the exo- and endoskeletons in many early vertebrates.

The Astraspida are known only in the Ordovician of North America and Siberia. There are some records from the Early Silurian of Siberia, but these remain to be confirmed on histological grounds.

Further reading

Denison (1967); Elliott (1987); Elliott *et al.* (1991); Halstead (1973*b*, 1987*b*); Halstead Tarlo (1967*b*); Ørvig (1951, 1958, 1965, 1989); Smith (1991); Stensiö (1964)

4.5 HETEROSTRACI

Heterostracans are characterized by a pair of common external branchial openings on either side of the head armour (Fig. 4.4, 16). They have long been known only in the form of isolated plates of the head armour. This fragmentary state has engendered controversy about the affinities of the group. Since the dermal plates of hetero-

stracans are made up of dentine and aspidine and generally display a peculiar honeycomb-like structure (Fig. 4.4C2), their vertebrate nature was not clear at first, and they have been variously referred to crustaceans or squids. T. H. Huxley (1858*a*) was the first to suggest the piscine nature of heterostracan shields, and this was later confirmed by E. R. Lankester (1864, 1868; see Chapter 8), who also pointed out a character that made them different from all other known Silurian and Devonian jawless vertebrates: the single common branchial opening. Even now, there is no better unique character of this group, although one may argue that, among the gnathostomes, holocephalans and osteichthyans do possess single, slit-shaped, common gill opening and that,

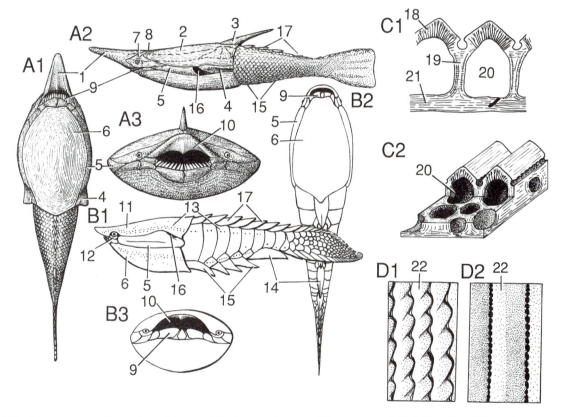

Fig. 4.4. Heterostraci. External morphology and histology. A, the pteraspidiform *Errivaspis*, Early Devonian of Europe, in ventral (A1, ×0.5), lateral (A2), and anterior (A3) view; B, the cyathaspidiform *Anglaspis*, Early Devonian of Europe, in lateral (B1, ×1), ventral (B2), and anterior (B3) view; C, generalized structure of the heterostracan exoskeleton in vertical section (C1, ×25) and in block diagram, showing the cancellar middle layer (C2); D, external aspect of the dermal bone ornamentation in pteraspidiform heterostracans (D1, ×50) and cyathaspidiform heterostracans (D2, ×50). 1, rostral plate; 2, dorsal disc; 3, dorsal spinal plate; 4, cornual plate; 5, branchial plate; 6, ventral disc; 7, orbital plate; 8, pineal plate; 9, oral plates; 10, prenasal groove; 11, dorsal shield (fused rostral, pineal, orbital epitega + dorsal disc); 12, suborbital plate; 13, sensory-line pores; 14, postanal plate; 15, preanal ridge; 16, common external branchial opening; 17, median dorsal ridge; 18, dentine; 19, aspidine; 20, cancella; 21, basal layer; 22, dentine ridge. (A, from E. I. White (1935) and Blieck (1984); B, from Kiaer (1932) and Blieck and Heintz (1983); C1 from Novitskaya (1971); D, from Blieck (1982*a*).)

among extant hagfishes, *Myxine* also possesses a common branchial opening.

The earliest known heterostracans are Early Silurian (Early Wenlock) in age, and the youngest ones are presumably Late Frasnian (367 Ma). They share with the arandaspids and astraspids the oblong, fusiform overall body shape and the acellular structure of the exoskeleton. The latter character was first used to include astraspids in the heterostracans, but it is irrelevant if regarded as just a generalized type of vertebrate dermal skeleton (see Chapter 6, p. 276). However, arandaspids and astraspids also share with heterostracans the large median ventral and dorsal shields covering the head (Fig. 4.4, 2, 6), the fan-shaped arrangement of the oral plates (Fig. 4.4, 9, Fig. 4.5), and (at least for arandaspids and some early heterostracans), the oak-leaf-shaped tubercles of the dermal ornament (Fig. 4.10, 5). These characters are unique to this ensemble of three groups.

The question of the affinities of heterostracans with other vertebrate taxa (beside arandaspids and astraspids) has been a much-debated subject over the past hundred years, and the controversies will be dealt with in Chapter 5 (p. 240). Suffice it to say here that these taxa were first referred to as chondrichthyans, then as 'agnathans', and, in an evolutionary context, as ancestral to either hagfishes or the gnathostomes.

There are now some instances of excellently preserved heterostracans, in particular in the Early Silurian of the Canadian Arctic or in the Early Devonian of Germany (Hunsrück Shale) and Wales, but in none of them is there any trace of the endoskeleton, apart from some poorly informative impressions on the internal surface of the dermal plates of the head armour (Fig. 4.5A). Most of the discussions on their phylogeny, classification, or affinities therefore, rest on the structure of their exoskeleton.

4.5.1 Exoskeleton and internal anatomy

In all heterostracans, the head is covered with a dermal armour that is generally composed of several plates containing the canals of the sensory-line system (Fig. 4.4, 13). The body is covered with large scales, and the tail, which is either pad-shaped or diphycercal, is covered with smaller scales, some of which are slightly larger than others and arranged in radial rows (Fig. 4.4B2). The rows probably correspond to the position of large underlying cartilaginous radials. There is always a series of large median dorsal and ventral ridge scales (Fig. 4.4, 15, 17) but there is no median fin, except the caudal, and there are no paired fins. The tail and body squamation is known only in a few species and does not seem to display a great diversity of shape, apart from the size and shape of the flange scales. In contrast, the head-shield may be

fusiform, flattened, or ventrally bulged, and may bear dorsal or lateral expansions.

Despite differences in the proportions, arrangement, the number of the plates, the head armour of heterostracans almost always comprises large median ventral and dorsal shields and a pair of lateral, elongated branchial plates (Fig. 4.4, 5). The only exception is in the Amphiaspidida and the genus *Boothiaspis*, where all the plates of the armour are fused into a rigid 'muff' (Fig. 4.6C, D), and in *Cardipeltis*, in which the ventral shield is replaced by numerous platelets. The other plates of the armour may vary in number or be fused with others in various ways.

The phylogeny and the classification of heterostracans are still much debated, and are largely based on the pattern of the plates and lateral-line canals. The most currently accepted classification recognizes two major clades, the Cyathaspidiformes (Fig. 4.8, 4) and the Pteraspidiformes (Fig. 4.9, 1), and a number of minor groups of uncertain affinities, such as the Lepidaspidida (Fig. 4.8A), Tesseraspidida, Traquairaspidiformes (Fig. 4.8B, C), Corvaspidida, Cardipeltida, and Tolypelepidida (Fig. 4.8D). The data that are used here to illustrate heterostracan morphology are taken from the two major clades.

4.5.1.1 Exoskeleton

In most textbooks the external morphology of heterostracans is exemplified by a cyathaspidiform (usually *Poraspis*, but here *Anglaspis*, Fig. 4.4B) and a pteraspidiform (usually *Pteraspis* or *Errivaspis*, Fig. 4.4A). This is due partly to the fact that these are among the best-known taxa, and partly to the difficulty of determining which, of the cythaspidiform or pteraspidiform structures, is the most generalized condition for the group. Despite recent discoveries of supposedly primitive Early Silurian heterostracans (e.g. *Athenaegis*, Soehn and Wilson 1990; Fig. 4.10A), I still feel compelled to stick to this tradition.

The cyathaspidiform *Anglaspis* has a fusiform head armour, in which the dorsal shield is a single plate covering the entire head (Fig. 4.4, 11). It is embayed laterally by notches for the orbits and the common external branchial openings (Fig. 4.4B1, 16). The superficial ornamentation of the exoskeleton of *Anglaspis* and typical cyathaspids consists of parallel, more or less longitudinal dentine ridges with finely crenulated margins (Fig. 4.4D2, 22). From the orientation of the ridges, one may distinguish different areas of the dorsal shield, or 'epitega', which are supposed to arise from separate growth centres. There are rostral, pineal, and branchial epitega, which do not correspond exactly with the individual plates of the pteraspidiforms. The orbit is lined ventrally by a small, crescentiform suborbital plate (Fig. 4.4, 12). The position of the pineal organ (Fig. 4.5A) is indicated by a slight

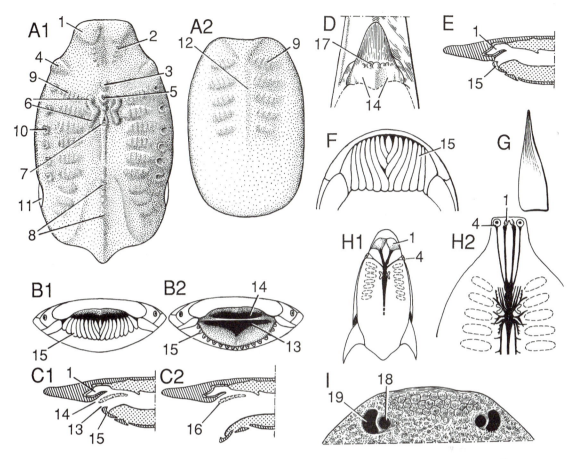

Fig. 4.5. Heterostraci. Internal anatomy and interpretations. A, internal cast of the head armour of a cyathaspidiform in dorsal (A1) and ventral (A2) view, showing impressions of internal organs; B, heterostracan head reconstructed in anterior view to illustrate the position of the oral plates when the mouth is closed (B1) and open (B2), according to the theory that a prenasal sinus and a nasopharyngeal duct were present; C, sagittal section through a heterostracan head to show the position of the oral plates when the mouth is closed (C1) and open (C2), according to the same theory (exoskeleton hatched, endoskeleton and soft parts stippled); D, *Rhinopteraspis*, Early Devonian of Europe, ventral side of the rostrum showing the outgrowth interpreted as upper tooth plates; E, sagittal section through a heterostracan head, according to the theory that the oral plates could bite against upper tooth plates of the rostrum (same symbols as in C). F, *Protopteraspis*, Early Devonian of Europe, head in ventral aspect, showing oral plates in natural position; G, *Errivaspis*, Early Devonian of Europe, isolated oral plate in internal aspect (×6); H, attempted reconstruction of the brain and nerves (black) in a generalized pteraspidiform (H1) and a long-snouted amphiaspid (H2); I, *Gabreyaspis*, Early Devonian of Siberia, anterior part of head in dorsal aspect (×0.7). 1, olfactory organ; 2, olfactory tract; 3, pineal recess; 4, eye; 5, cerebellum; 6, vertical semicircular canals; 7, medulla oblongata; 8, arcualia; 9, gill lamellae; 10, possible gill-arch elements; 11, common external branchial opening; 12, ventral groove; 13, mouth; 14, prenasal sinus and nasopharyngeal duct; 15, oral plates; 16, nasobuccal shelf; 17, outgrowth or tooth plate; 18, orbit; 19, adorbital opening. (A, B, C, E, from Janvier (1993); D, from Halstead (1973*b*); F, from Heintz (1962); G, from E. I. White (1935); H, from Novitskaya (1983); I, from Novitskaya (1971).)

elevation with no pineal foramen proper, although the exoskeleton is thin and translucent at this particular place. The snout, or rostral region, is embayed ventrally by a broad notch, sometimes divided into two furrows by a slight median longitudinal ridge (Fig. 4.4B3, 10). The mouth is lined ventrally by a few small oral plates (Fig. 4.4B3, 9), which do not abut against the rostrum but probably protected the lower lip or acted as a scoop for scraping food. The dorsal and ventral shields are separated laterally by the elongated branchial plate (Fig. 4.4B1, 5), which lines ventrally the branchial opening (Fib. 4.4B1, 16). The ventral shield consists of a single epitegum and is more convex than the dorsal one.

The dermal head armour also contains canals of the

lateral sensory-line system, which open to the exterior by large pores (Fig. 4.4B1, 13). These canals and pores occur also in the trunk scales, as far back as the tail. The pattern of the lateral-line canals in *Anglaspis* consists of two pairs of dorsal and one pair of ventral longitudinal lines with a number of transverse commissural lines (Fig. 4.7A). The supraorbital line forms a V in the anterior part of the dorsal shield, but in cyathaspidiforms the two branches never meet behind the pineal region (Fig. 4.7A, 1).

The body squamation of *Anglaspis* consists of very large and deep flange scales, and dorsal and ventral ridge scales, which diminish in size toward the tail. This is a characteristic of the Cyathaspidida, a subgroup of the Cyathaspidiformes. The anus is supposed to have been situated just behind a large median ventral scute (Fig. 4.4B, 15). The leading edges of the tail are covered with elongated median ridge scales, and the web is pad-

shaped, sometimes regarded as hypocercal; that is, the lobe containing the notochord is pointing posteroventrally. The only undistorted and complete cyathaspidiform tail hitherto recorded is in *Torpedaspis* (Fig. 4.6A1), a large form with elongate head armour. Its tail is pad-shaped and slightly pointed posteriorly. The tail of another cyathaspidid *Nahanniaspis* (Fig. 4.10B), resembles that of *Anglaspis*, but with a longer dorsal part. In contrast, the tails of *Ctenaspis* (Fig. 4.6B) and *Athenaegis* (Fig. 4.10A), are diphycercal, with a few series of larger scales that may possibly correspond to the position of the underlying radials, as in pteraspidiforms. The pad-shaped tail of *Anglaspis* and some other cyathaspidiforms can thus be regarded as derived from to the fan-shaped (diphycercal) tail of other heterostracans.

The morphology of pteraspidiform *Errivaspis*, a close relative to the classical genus *Pteraspis*, displays a

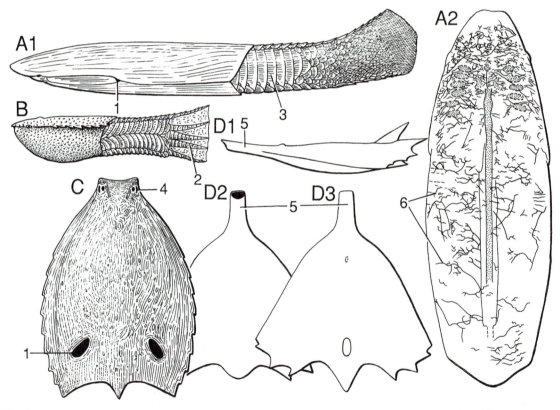

Fig. 4.6. Heterostraci. Morphology of the Cyathaspidiformes. A, the cyathaspidid *Torpedaspis*, Early Devonian of Canada, reconstruction in lateral view (A1, ×0.5) and internal cast of the dorsal shield (A2) showing the impressions of the subaponevrotic vascular network (impressions of brain, labyrinth, and gills stippled); B, *Ctenaspis*, Early Devonian of Canada reconstruction in lateral view (×0.7); C, the amphiaspidid *Kureykaspis*, Early Devonian of Siberia, dorsal view of head armour (×1); D, the amphiaspidid *Eglonaspis*, Early Devonian of Siberia, head armour in lateral (D1), ventral (D2), and dorsal (D3) view (×0.3). 1, external branchial opening; 2, dermal 'fin rays' or zonation; 3, flanges scales; 4, adorbital opening; 5, oral tube; 6, impressions of the subaponevrotic vessels. (A, from Broad and Dineley (1973); B, from Dineley (1976); C, D, from Novitskaya (1971).)

slightly more complex pattern of head armour plates. The dorsal shield consists of several separate plates: the median dorsal 'disc' (Fig. 4.4A, 2), the orbital plates (Fig. 4.4A, 7), the pineal plate (Fig. 4.4A, 8) and the rostral plate (Fig. 4.4A, 1). Each of these plates has its own growth centre. The superficial ornamentation of these plates consists of very thin, concentric dentine ridges with a distinctly serrated edge (Fig. 4.4D1, 22). The dorsal disc is embayed posteriorly by a notch for a long median spinal plate (Fig. 4.4A, 3), which is a modified dorsal median ridge scale. Laterally, as in cyathaspidiforms, there is an elongated branchial plate (Fig. 4.4A, 5), but the common external branchial opening (Fig. 4.4A, 16) is bounded posteriorly by a small cornual plate (Fig. 4.4A, 4). In other pteraspidiform genera, the cornual process of the cornual plate may be rather long (Fig. 4.9B2, C1). The orbits are very small and pierce the orbital plate, and the ventral side of the rostral plate is embayed by two large furrows for the pre-nasal sinus (Fig. 4.4A3, 10; see discussion below). The ventral margin of the mouth is covered with oral and postoral plates (Fig. 4.4A, 9). The oral plates are small and pointed; they could probably expand in a fan-like manner as the skin of the lip evaginated (Fig. 4.5C). The ventral side of the head armour consists of a single ventral disc (Fig. 4.4A, 6). The lateral-line canals open to the exterior by elongated slits.

The body is covered with large diamond-shaped scales, but the median dorsal and ventral ridge scales are smaller than in most cyathaspidiforms (Fig. 4.4A, 15, 17). The tail is diphycercal and covered with minute scales, some of which are larger than others, and arranged in radial digitations.

The pattern of the sensory-line canals of *Errivaspis* and most other pteraspidiforms (except the anchipteraspids, where it is more similar to that in cyathaspidiforms, thus generalized) consists of two pairs of dorsal longitudinal lines (Fig. 4.7C, 2, 3), the medial one being closely set, and three oblique commissural lines that radiate from the growth centre of the dorsal disc (Fig. 4.7C, 4). There is no evidence, so far, of sensory lines on the body. Characteristic of higher pteraspidiforms are the two supraorbital lines meeting behind the pineal region (Fig. 4.7C, 1, 5).

4.5.1.2 *Internal anatomy*

The internal anatomy of heterostracans is inferred from impressions on the internal surface of the head-armour plates observed in a few taxa (Fig. 4.5A). These impressions are always well-marked in cyathaspidiforms and in some pteraspidiforms, in particular the Protopteraspididae.

Among the most general of these impressions are those of the two vertical semicircular canals (Fig. 4.5, 6), the

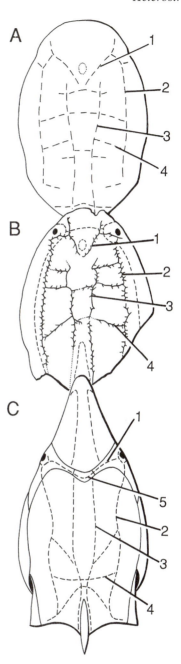

Fig. 4.7. Heterostraci. Sensory-line patterns. Sensory-line canals in the dorsal shield of the Cyathaspidida (A), Anchipteraspididae (B), and Pteraspididae (C). 1, supraorbital line; 2, lateral longitudinal line; 3, medial longitudinal line; 4, transverse lines; 5, pineal line. (A, C, from Blieck (1984); B, from Elliott (1984).)

brain (Fig. 4.5, 5, 7), the eyes (Fig. 4.5, 4), the pineal organ (Fig. 4.5, 3),and the olfactory organs (Fig. 4.5, 1), on the internal surface of the dorsal shield. The position of the branchial apparatus is often indicated by a paired series of eight to ten transversally elongated impressions, which are sometimes marked with parallel longitudinal grooves left by the gill lamellae (Fig. 4.5, 9). In addition, and laterally to each of these 'gill-pouch' impressions, there are U- or Y-shaped impressions (Fig. 4.5, 10), which may be attributed to some external component of the gill arches or the extrabranchial atria. On the internal surface of the ventral disc there is sometimes a median impression (Fig. 4.5, 12), referred to either as the trace of an endostyle or a 'lingual' apparatus of hagfish or lamprey type. It is flanked on either side by the impressions of the ventral part of the 'gill-pouches'.

In a few specimens, the impression of the brain and spinal chord is continued posteriorly by a median series of rounded depressions that may have housed the arcualia of the vertebral column (Fig. 4.5, 8). The outline of the brain impression itself fits that of most other early jawless vertebrates, such as osteostracans or galeaspids (see Figs 4.16, 4.21, and 6.11). It shows at least three distinct areas: the pineal recess (Fig. 4.5, 3), the paired impression attributed to the metencephalon (cerebellum, Fig. 4.5, 5), and the impression of the medulla oblongata, with a median ridge corresponding to the fossa rhomboidea (Fig. 4.5, 7).

The impressions of the eyeballs are semi-conical in shape, and the orbits are generally very small (Fig. 4.5, 4). The impression of the olfactory organ (Fig. 4.5, 1), is generally paired and connected to the brain impression by two convergent grooves, which may correspond to the olfactory tract (Fig. 4.5, 2). In some specialized amphiaspidids, the mouth, eyes, and olfactory organs are situated at the anterior end of a long tube-like expansion of the head armour, and very far in front of the pineal region. This implies that in these forms the olfactory and optic tracts were very long (Fig. 4.5H2, 1, 4).

The olfactory organ certainly opened either ventrally into the mouth cavity (Fig. 4.5E, 1), or anteroventrally into a common prenasal sinus (Fig. 4.5C, 1, 14, Fig. 6.6). A notch observed on the snout margin of a few cyathaspid specimens has been claimed to be either a posterior nostril, comparable to that of the gnathostomes (Novitskaya 1983), or a passage for a tentacle (Stensiö 1964), but these interpretations remain debated.

The internal surface of the armour plates is generally quite smooth and shiny, with a few nutrient canals, even at places marked with internal organ impressions. This may indicate that no endoskeleton, whether cartilaginous or calcified, was directly attached to the exoskeleton. The fact that the brain and labyrinth could leave such an accurate impression on the exoskeleton suggests that they

were enclosed only in fibrous sheath, and that there was no intervening cartilage. Some blood-vessel impressions may occur, but a subaponevrotic vascular network is known only in *Torpedaspis* (Fig. 4.6A2, 6). Heterostracans may thus have possessed subcutaneous lymphatic sacs, like hagfishes, which are therefore 'floating in their own skin'. In all other vertebrates possessing an ossified or cartilaginous endoskeleton attached to the exoskeleton (Galeaspida, Osteostraci, Placodermi, and Osteichthyes), the internal surface of the exoskeleton is roughened or spongy in aspect, or the basal layer shows the impression of a vascular subaponevrotic plexus.

There has been much discussion about the interpretation of these internal impressions in heterostracans, largely triggered by Stensiö's (1927, 1964, 1968) suggestion that heterostracans were ancestral—or related—to hagfishes, whereas many palaeontologists regarded them as somehow related to the gnathostomes. In fact, most researchers agreed on the interpretation of the brain, gill, labyrinth impressions, but there have been divergent interpretations of the snout. Some have regarded the olfactory organ as paired and opening either inside the mouth cavity (Fig. 4.5E) or directly to the exterior by means of external nares, whereas others have considered that it opened into a separate duct comparable to the prenasal sinus and nasopharyngeal duct of hagfishes (Fig. 4.5B, C, 14). The second interpretation is supported by the fact that the ventral side of the rostrum in most heterostracans is channelled by two grooves leading posteriorly towards the olfactory impressions (Fig. 4.5D, 14). These grooves resemble those in the roof of the prenasal sinus of the embryonic hagfishes. In addition, a comparison with the Galeaspida (see p. 118), where the actual endoskeleton is known (Fig. 4.21, 2), suggests that the condition of the nasal region in the latter is quite similar to that in heterostracans, given that the opening of the prenasal sinus has been shifted dorsally. The presence of a prenasal sinus, and possibly of a nasopharyngeal duct (Fig. 4.5B, C, 14) in heterostracans, both separated from the oral cavity by a nasobuccal shelf (Fig. 4.5B, C, 16), would, however, not be a unique character shared with the galeaspids and hagfishes, but rather a generalized craniate condition (see Chapter 6, p. 263). In the Amphiaspidida, a group of cyathaspidiforms, the orbit is lined laterally by a larger, bean-shaped opening, the adorbital opening (Fig. 4.5I, 19), of unknown function. It has been interpreted as a secondarily paired and dorsally displaced opening of the prenasal sinus, a spiracle, or a prespiracle, but none of these interpretations is fully satisfactory. An inhalent function remains probable, however, since amphiaspids were benthic forms.

This interpretation of the nasal region has bearings on that of the mouth and oral plates. Some have claimed that the latter could bite against the ventral side of the rostrum, and have described 'upper tooth plates' (Fig. 4.5D, 17), which are probably mere pathological outgrowths of the dentinous covering. Accurate reconstructions of the oral plates in various heterostracans have shown that these could hardly abut against the rostrum and were more probably covering the ventral lip of the mouth (Fig. 4.5F, 15), as in arandaspids. It is interesting that the mouth of a hagfish, when retracted, displays ventral skin folds that match exactly the pattern of the oral plates in arandaspids and heterostracans (Fig. 3.4A1). This could mean that the fan-shaped pattern of the oral plates is linked with a more or less similar feeding mechanism. It is also probable that, in pteraspidiforms, the anterior tip of the oral plates was free and could be used to scrape or even grasp the food (Figs 4.5B2, 4.5G, 15).

The interpretation of heterostracans as 'biting agnathans', that is with lower and upper tooth plates reflects the deep-rooted idea that they are related to the gnathostomes. Halstead (1973*a,b*), for example tried to interpret the heterostracan internal anatomy in terms of gnathostome anatomy. To him, the 'gill pouch' impressions were in fact muscular units, and the U- or Y-shaped lateral impressions (Fig. 4.5, 10) were those of forked gills of gnathostome type. He also thought that heterostracans may have possessed a horizontal semicircular canal, like the gnathostomes, but this is unlikely since one of the gill-pouch impressions penetrates between the two vertical semicircular canal impressions (Fig. 4.5, 6).

Another question is how the gills could have been organized in heterostracans. A comparison between the 'gill-pouch' impressions of heterostracans and those on the endoskeleton-covered oralobranchial chamber of osteostracans (Fig. 4.15E) shows that the gill lamellae impressions observed in some heterostracans are most probably those of the posterior hemibranch of each gill unit. But how these gills were suspended, and whether they were partly enclosed in pouches or free, remains unknown. The best model for heterostracan (and osteostracan) gills is probably that of extant larval lampreys, with a somewhat larger posterior hemibranch (Fig. 3.6C1).

Virtually nothing is known of the axial skeleton of heterostracans, apart from the slight 'vertebral' impressions (Fig. 4.5A, 8) of the dorsal shield, and the possible presence of a few large radials in the tail, which are suggested by the digitations of larger scales in various genera (*Athenaegis*, Fig. 4.10A; *Drepanaspis*, Fig. 4.9E1; *Ctenaspis*, Fig. 4.6B; *Errivaspis*, Fig. 4.4A). The number of the presumed caudal rays is from four to ten, and this scarcity suggests that they were not movable and did not possess radial muscles. In *Athenaegis* (Fig. 4.10A) or *Cardipeltis*, the digitations even seem to be free, giving the tail the aspect of a hand.

4.5.2 Histology

The structure of the heterostracan exoskeleton generally consists of three layers: a basal laminar layer (Fig. 4.4, 21), a middle cancellar layer with honeycomb-shaped cavities (Fig. 4.4, 19, 20), and a superficial layer of dentine ridges or tubercles (Fig. 4.4, 18). However, in a group of pteraspidiforms, the Psammosteidae, the cancellar pattern is presumed to be lost, the exoskeleton consists of thick, spongy aspidine. Histologically, the middle and basal layers of the exoskeleton of heterostracans is made up of aspidine, a type of acellular hard tissue showing incremental growth zones, as well as some peculiar fibre-like lineaments. There has been debates about the nature of aspidine (see Chapter 6, p. 276), in particular whether it is closely similar to dentine or bone, and whether it is primitive or advanced. Clearly, aspidine is not formed like dentine, since no structures comparable to dentine tubules are observed in it. Its mode of growth is still obscure, but it is probable that it was produced by a single retreating cell front, because no trace of pycnotic (dying) cells has ever been observed near the surface of the aspidine layers.

4.5.3 Diversity and phylogeny

The two major heterostracan groups, the Cyathaspidiformes and Pteraspidiformes, are regarded as monophyletic. The Cyathaspidiformes can, however, be characterized only if such problematic genera as *Athenaegis*, *Tolypelepis*, *Corvaspis*, *Davelaspis*, *Listraspis*, and *Traquairaspis* are excluded. This is the solution adopted here (Fig. 4.8).

4.5.3.1 Cyathaspidiformes

The Cyathaspidiformes are characterized by an ornamentation of longitudinal, more-or-less parallel dentine ridges. The latter have finely crenulated margins and are separated by a groove devoid of dentine (Fig. 4.4D2). The dorsal shield is a single plate formed by the fusion of several epitega. The two supraorbital canals do not meet behind the pineal region (Fig. 4.7A, 1). The body scales are quite large, with flange scales either arranged in chevrons or vertical (Fig. 4.8E, G, H, I).

The Cyathaspidiformes comprise two major clades: the Cyathaspidida (Fig. 4.10, 8) and Amphiaspidida (Fig. 4.10, 6). The Cyathaspidida have a fusiform—and even cigar-shaped—head armour (*Irregulareaspis*, *Anglaspis*, *Poraspis*, and *Torpedaspis*) and are characterized by deep, vertical flange scales (Figs 4.4B1, 4.6A1). *Nahanniaspis* retains chevron-shaped flange scales (Fig. 4.10B), but shares with the Cyathaspidida a reduced and slender caudal fin. In the Siberian Amphiaspidida, all the plates of the head armour are fused in a single 'muff', and the entire head is much

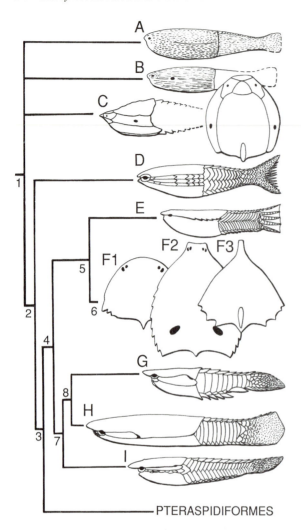

Fig. 4.8. Heterostraci. Interrelationships. *Terminal taxa*: A, *Lepidaspis*; B, *'Traquairaspis' mackenziensis*: C, *Toombsaspis* (lateral and dorsal views); D, *Athenaegis*; E, *Ctenaspis*; F, Amphiaspidida (head shields in dorsal view; F1, *Prosarctaspis*; F2, *Kureykaspis*; F3, *Eglonaspis*); G, *Anglaspis*; H, *Torpedaspis*; I, *Nahanniaspis. Nested taxa and synapomorphies*: 1, Heterostraci (single external branchial opening); 2, single dorsal shield, large scales, loss of oak leaf-shaped tubercles; 3, 'higher heterostracans' (single branchial or branchio-cornual plate not pierced by the branchial opening); 4, Cyathaspidiformes (ornamentation of longitudinal dentine ridges with finely crenulate margins, flange scales arranged in chevrons); 5, head armour plates fused into a single unit; 6, Amphiaspidida (flattened armour, adorbital opening); 7, caudal fin pad-shaped, without distinct zonation; 8, Cyathaspidida (large, deep flange scales). *Age*: Early Silurian (D); Early Devonian (A–C, E–I). (C, from Tarrant (1991); D, from Soehn and Wilson (1990); E, from Dineley (1976); F, from Novitskaya (1971); G, from Blieck and Heintz (1983); H, from Broad and Dineley (1973); I, from Dineley and Loeffler

PTERASPIDIFORMES

flattened, suggesting benthic or even burrowing habits (Fig. 4.6C, D). In some genera, the anterior part of the armour is elongated into a tube-shaped structure, and the eyes disappear (*Empedaspis, Pelurgaspis, Eglonaspis*, Fig. 4.6D, 5). In others (*Prosarctaspis*, Fig. 4.8F1; *Kureykaspis*, Fig. 4.8F2; *Olbiaspis, Gabreyaspis*), there is a peculiar opening, the adorbital opening (Fig. 4.5, 19; Fig. 4.6, 4), situated anterolaterally to the orbit, which has been interpreted as a spiracle, a 'pre-spiracle', or a dorsally displaced nostril. It is probably an inhalent opening and indicates that amphiaspids lived half-buried in the sediment, more or less like extant rays (seen Chapter 7, p. 301). The closest relative of the amphiaspids outside Siberia is probably *Ctenaspis*, which, however, differs from other cyathaspidiforms by its ornamentation of stel-

late tubercles (Figs 4.6B, 4.8E). The fusion of the plates into a single unit occurs also, by homoplasy, outside the amphiaspids, in the cyathaspidid genus *Boothiaspis*, which is probably a close relative of *Torpedaspis* (Fig. 4.6A).

4.5.3.2 *Pteraspidiformes*

The Pteraspidiformes are characterized by a dorsal shield made up of several independent plates, ornamented with concentric, laterally serrated dentine ridges (Fig. 4.4D1), except in the Psammosteidae, which have secondarily tuberculate ornamentation. The dermal plate pattern is fairly constant, the only exceptions being the two orbital plates in the Psammosteidae (Fig. 4.9E1, E2), and the single branchio-cornual plate fused to the dorsal shield in the Anchipteraspididae (Fig. 4.9A). The supraorbital canals meet behind the pineal region (Fig. 4.7B, C, 1), and the body scales are relatively small and diamond-shaped (Fig. 4.4A1; this is probably a plesiomorphous condition, since it occurs also in such non-pteraspidiform heterostracans as *Athenaegis*).

Pteraspidiforms comprise five major taxa: the Anchipteraspididae, Protopteraspididae, Pteraspididae, Protaspididae, and Psammosteidae. The Anchipteraspididae (Fig. 4.9A) retain some features that are widespread among cyathaspidiforms, such as a single branchio-cornual plate, the fused plates (or epitega) of the dorsal shield, including the branchio-cornual plate, the transverse orientation of the dorsal commissural sensory lines (Fig. 4.7B, 4), and the pineal recess included in the dorsal epitegum. The ornamentation, however, is concentric, as in most other pteraspidiforms. The anchipteraspidids (*Anchipteraspis*, *Ulutitaspis*, *Rhachiaspis*) are small forms from the Late Silurian and Early Devonian of the Canadian Arctic. All other pteraspidiforms, i.e. the Pteraspidida, share separate branchial and cornual plates and separate plates in the dorsal shield (Fig. 4.9, 2). The Protopteraspididae and Pteraspididae are quite similar in overall morphology (Fig. 4.9B1, C2), but some protopteraspidids, such as *Doryaspis* (Fig. 4.9B2), display a peculiar morphology, with a rostrum-like median oral plate which extents ventrally to the mouth. The Protopteraspididae are the sister-group of the Pteraspidina, which are characterized by the supraorbital canal passing through the pineal plate (Fig. 4.7C, 5; Fig. 4.9, 3). The Pteraspidina comprise the Pteraspididae (Fig. 4.9C), Protaspididae (Fig. 4.9D), and Psammosteidae (Fig. 4.9E). Among the Pteraspididae, *Rhinopteraspis* (Fig. 4.9C3) is remarkable in having a very long rostrum and spinal plate. The Psammosteidae (Fig. 4.9E) have long been regarded as quite primitive—if not the most primitive—heterostracans because of their partly 'tessellate' head armour, which recalls that of the Ordovician arandaspids and astraspids. However, they share unique characters with pteraspidiforms, in particular the shape of the dorsal disc, the separate plates, and the orbit piercing the anterior orbital plate, even though the pattern of their sensory lines remains unknown (the sensory lines were presumably too superficial to leave any trace on the exoskeleton). In adult psammosteids, the major plates of the armour are more-or-less separated by fields of polygonal platelets (wrongly referred to as 'tesserae'), but juvenile stages apparently did not possess such platelets and had a head armour composed of large jointed plates which recalls that of, say, the protopteraspidids (Fig. 2.7A2). Moreover, the platelets of psammosteids differ from the arandaspids tesserae in being able to grow like other plates and having no central tubercle. The psammosteids clearly form a clade, characterized

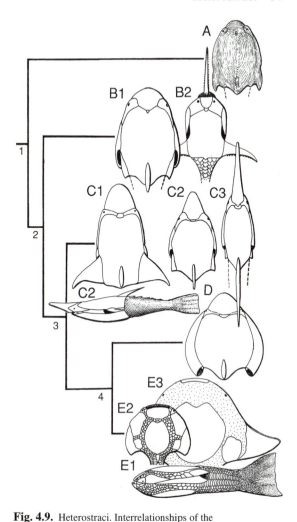

Fig. 4.9. Heterostraci. Interrelationships of the Pteraspidiformes. *Terminal taxa*: A, Anchipteraspididae (*Rhachiaspis*); B, Protopteraspididae (B1, *Protopteraspis*; B2, *Doryaspis*); C, Pteraspididae (C1, *Unarkaspis*; C2, *Errivaspis* in dorsal and lateral views; C3, *Rhinopteraspis*); D, Protaspididae (*Cyrtaspidichthys*); E, Psammosteidae (E1, *Drepanaspis* in lateral view; E2, *Psammolepis*; E3, *Pycnosteus*). *Nested taxa and synapomorphies*: 1, Pteraspidiformes (ornamentation of concentric dentine ridge with serrated margins, supraorbital sensory-line canals meeting behind pineal plate or pineal area, dorsal spinal plate); 2, Pteraspidida (separate cornual plates, dorsal shield made up of several separate units [homoplasy with Fig. 4.8C]); 3, Pteraspidina (pineal sensory-line canal passing through pineal plate); 4, Protaspidoidea (enlarged and flattened branchial plates, reduced cornual plates). *Age*: Late Silurian (A); Early Devonian (B–D, E1); Middle to Late Devonian (E2, E3). (A, from Elliott (1984); B, C, modified from Blieck (1984), Elliott (1983), and Heintz (1968); D, from Denison (1970); E1, from Gross (1963*a*); E2, 3, from Obruchev and Mark-Kurik (1965) and Tarlo (1964).)

by two pairs of orbital plates, reduced cornual plates, a spongy structure of their aspidine (the honeycomb layer is supposed lost), and an ornamentation of stellate, shiny tubercles. They are the youngest known heterostracans, since they survive until the end of the Frasnian. The best-known psammosteid is the Early Devonian *Drepanaspis* (Figs 2.7A, 4.9E1), which is of moderate size, but the Late Devonian forms, such as *Pycnosteus* (Fig. 4.9E3), *Psammolepis* (Fig. 4.9E2), *Schizosteus*, or *Tartuosteus*, could reach a breath of 1.5 metres, with laterally expanded and ventrally curved branchial plates. The Protaspididae (Fig. 4.9D) are probably the sister-group of the Psammosteidae, since they share with them a much enlarged branchial plate and a much reduced, even minute, cornual plate.

4.5.3.3 Taxa of uncertain affinities

In addition to the groups mentioned above, a number of taxa of uncertain affinities are classified as heterostracans because of their overall shape, the (sometimes only presumed) presence of a common external branchial opening, and the histological structure of their exoskeleton. None of these forms can, however, be 'shoe-horned' into the two major taxa defined above. This may be due to the fact that these generally early forms are stem-group heterostracans, but it also may mean that one or both of the major 'higher heterostracan' taxa described above is not monophyletic.

There is an ensemble of 'tessellate heterostracans', such as the Silurian genera *Tesseraspis*, *Oniscolepis*, and *Kallostracon*, that are known only from fragments, but none of them is clearly a heterostracan. *Tesseraspis*, with its longitudinal ridges, is strikingly similar to *Astraspis* but its histological structure is quite different, and much more heterostracan-like. Another type of tessellate form is represented by *Lepidaspis* (Figs 4.10F, 4.8A) or *Aserotaspis*, in which the tesserae are elongated in shape and bear an oak-leaf-shaped tubercle (Fig. 4.10, 5). Strangely enough, the degree of fusion of the tesserae in *Lepidaspis* varies from one individual to another, some specimens having a completely solid shield. Although the oak-leaf-shaped type of tubercle is common among early heterostracans (and also among arandaspids), there is no clear evidence of a common branchial opening in these forms.

Other taxa resemble the cyathaspidiforms in having the dorsal and ventral shields made up of a single unit respectively, but their ornamentation looks as if it consisted of many overlapping, separate scales. This is the case of the corvaspids (*Corvaspis*) and tolypelepids (*Athenaegis*, *Tolypelepis*), for example. *Athenaegis* possesses three series of small branchial plates on each side (Fig. 4.10A, 2), and shows no evidence of a common branchial opening. Nor is there any clear evidence of this opening in

corvaspids. If this type of ornamentation is regarded as primitive, because it suggests a growth from discrete scale-like units, then all these enigmatic forms, as well as 'lepidaspids', may represent either stem-group heterostracans (Fig. 4.8) or close relatives of heterostracans, not to be included in this taxon (if devoid of a common gill opening). *Athenaegis*, however, possesses rather large flange scales, which points towards particular relationships with 'higher heterostracans', i.e. cyathaspidiforms and pteraspidiforms.

Nahanniaspis, from the Early Devonian of the Canadian Arctic, is a typical cyathaspidiform, regarded here as the sister-group of the Cyathaspidida (Fig. 4.8, 7), but it shows a number of additional branchial plates, ventral to the normal, elongated branchiocornual plate (Fig. 4.10B, 2). Whether this is due to mere individual variation or represents a primitive condition for the group is still unanswered.

The Traquairaspidiformes are clearly heterostracans, by virtue of their common branchial opening, and they are usually recognized by their typical ornamentation of lacrymiform or oak-leaf-shaped tubercles arranged in longitudinal rows (Fig. 4.10C–E). However, they display a wide range of diversity in the structure of the armour, from completely fused plates (Fig. 4.10E), to a pteraspidiform-like pattern (Fig. 4.10C, D), with orbits piercing the orbital plates. This variable condition throws serious doubt on the monophyly of the group, yet there seems to be a common pattern of sensory lines that supports monophyly. They are, however, tentatively placed here outside the 'higher heterostracans', because of the position of their branchial opening, which pierces the branchial plate (Fig. 4.10C, D, 1; Fig. 4.8B, C).

The Cardipeltida (*Cardipeltis* and perhaps *Obruchevia*) are relatively large heterostracans with a dorsoventrally depressed head armour that is composed of large plate units and smaller platelets, and devoid of a ventral shield. The common branchial openings are dorsally placed, in a lateral notch in the dorsal disc. The sensory-line pattern looks vaguely cyathaspidiform-like, with transverse commissural lines. The tail is diphycercal, with at least four large digitations devoid of web. The partly 'tessellate' structure of the armour of *Cardipeltis* is suggestive of psammosteids, yet no clear character of this group can be found in this genus.

The interrelationships of heterostracans depicted in Fig. 4.8 and 4.9 show two major monophyletic sister-groups, the Cyathaspidiformes and Pteraspidiformes, which represent the 'higher heterostracans' (Fig. 4.8, 3) and are characterized by a single branchial or branchiocornual plate, which is not pierced by the branchial opening. *Athenaegis* (or all tolypelepids) may be regarded as the sister-group of the 'higher heterostracans', since it

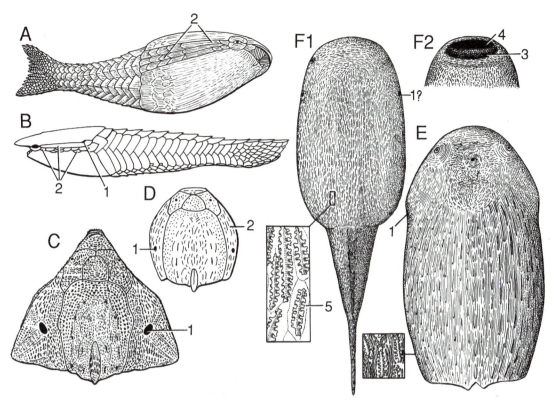

Fig. 4.10. Heterostraci. Problematical taxa. A, *Athenaegis*, Early Silurian of Canada, reconstruction in ventrolateral view (×1.3); B, *Nahanniaspis*, Early Devonian of Canada, reconstruction in lateral view (×1); C–E, 'Traquairaspidiformes', head armours in dorsal view of *Phialaspis* (C, ×0.5), Early Devonian of Britain, *Toombsaspis* (D, ×1), Early Devonian of Britain, and *'Traquairaspis' mackenziensis* (E, ×0.7), Early Devonian of Canada, with detail view of the ornamentation; F, *Lepidaspis*, Early Devonian of Canada, attempted reconstruction of the entire animal in dorsal view (F1, ×0.5), and the snout in ventral view (F2), with detail view of the oak leaf-shaped tubercles of the ornamentation. 1, external branchial opening; 2, branchial plates; 3, mouth; 4, ?prenasal sinus and nasopharyngeal duct; 5, dermal unit. (A, from Soehn and Wilson (1990); B, from Dineley and Loeffler (1976); C, D, from Tarrant (1991); E, F, based on photographs in Dineley and Loeffler (1976).)

shares with them a single dorsal shield and rather large, diamond-shaped flange scales. The other taxa mentioned above are left in an undecided position, as a basal polytomy, but two major types of Traquairaspidiformes (with either a single dorsal shield or separate plates; Fig. 4.8B, C) have been distinguished.

The classification in Table 4.1 is derived from the cladograms in Figs. 4.8 and 4.9.

4.5.4 Stratigraphical and geographical distribution, habitat

The earliest known 'heterostracan-like' forms, such as *Athenaegis*, are Early Silurian in age, but 'higher heterostracans' are not known before the Late Silurian. The Silurian heterostracans are essentially cyathaspidiforms or traquairaspidiforms, or belong to a number of 'tessellate' taxa of uncertain affinity. However, anchipteraspidid pteraspidiforms had already appeared at the end of the Silurian. By early Devonian times, the cyathaspidiforms are less numerous, and are progressively replaced (possibly outcompeted) by the pteraspidiforms, except in western Siberia, where the amphiaspidids survive until the Pragian. By the end of Middle Devonian times, the only surviving heterostracans are the psammosteids, which seem to have been confined to sandy deltas or even flood-plains. The latest known occurrence of a psammosteid is in the Late Frasnian.

Heterostracans proper are known chiefly from North America, northern Europe, Russia (including Siberia), and the Ukraine. No heterostracan is known from the southern hemisphere or from China or Indochina.

Table 4.1. Classification of the Heterostraci (based on cladograms in Figs 4.9 and 4.10)

Heterostraci
  Lepidaspis sedis mutabilis
  '*Traquairaspis' mackenziensis sedis mutabilis*
  Toombsaspis sedis mutabilis
    Athenaegis
    'Higher heterostracans' (unnamed taxon)
      Cyathaspidiformes
        (unnamed taxon)
          Ctenaspis
          Amphiaspidida
        (unnamed taxon)
          Nahanniaspis
          Cyathaspidida
      Pteraspidiformes
        Anchipteraspididae
        Pteraspidida
          Protopteraspididae
          Pteraspidina
            Pteraspididae
            Protaspidoidea
              Protaspididae
              Psammosteidae

Most heterostracans lived in a marine, shallow-water environment, but some of them may have inhabited river systems or deltas (see Chapter 7, p. 293).

Further readings

Anatomy, histology, and interpretation: Bendix-Almgreen (1986); Blieck and Janvier (1991); Denison (1961, 1971*b*); Janvier (1974, 1975*b*, 1981*a*); Janvier and Blieck (1979, 1993); Gross (1961*a*, 1963*a*); Halstead (1973*a*,*b*); Halstead Tarlo and Whiting (1965); Heintz (1962); Kiaer (1928, 1932); Kiaer and Heintz (1935); Novitskaya (1971, 1983, 1992); Soehn and Wilson (1990); Stensiö (1927, 1964, 1968); Tarlo (1964); White (1935, 1973); Tarrant (1991). *Interrelationships and systematics*: Blieck (1982*a*,*b*, 1984, 1985); Blieck and Heintz (1983); Blieck *et al.* (1991); Denison (1953, 1964, 1966, 1970, 1973); Dineley (1976); Dineley and Loeffler (1976); Elliott (1983, 1984); Elliott and Loeffler (1988); Obruchev and Mark-Kurik (1965, 1968); Tarlo (1964, 1965).

4.6 ANASPIDA

Anaspids are among the few early jawless vertebrates that do not possess a dermal head armour, hence their name ('shield-less'). They are slender in shape (Fig. 4.11A), laterally compressed, and characterized by one or several triradiate spines behind the series of external branchial openings (postbranchial spines; Fig. 4.11, 2). Another feature of the Anaspida, which may not be unique to this group, is their strongly hypocercal tail, in which the lobe

containing the notochord (Fig. 4.11, 6) bends postero-ventrally and bears a large dorsal fin web (Fig. 4.11, 5). Anaspids are exclusively Silurian in age, but there are a few atypical forms in the Early Silurian and Late Devonian that are traditionally referred to this taxon, although they do not display any unique anaspid characters.

4.6.1 Exoskeleton and external morphology

The head of anaspids is fusiform in shape, and is generally covered with minute scales (Fig. 4.11A2, 3) or a few larger dermal plates (Fig. 4.11, 9, 13). The eyes are large, laterally placed, and surrounded by a ring of dermal plates, but these are not a sclerotic ring. Between the orbits is a pineal foramen (Fig. 4.11, 11), and in front of it a T-shaped or keyhole-shaped opening, which is currently interpreted as a dorsal nasohypophysial opening (Fig. 4.11, 10), similar to that of lampreys and osteostracans (see p. 107). The mouth is known in only a few genera, and seems to have been more-or-less rounded in shape (Fig. 4.11, 8). The upper and lower lips are armed with dermal oral plates (Fig. 4.11, 9), which could probably bite vertically against each other, much in the same way as gnathostome jaws. Behind the eyes, there is a series of six to fifteen external branchial openings arranged in slanting line (Fig. 4.11, 1). These openings either pierce a single branchial plate or are lined with crescentiform scales.

The body is generally covered with elongated, rod-shaped scales arranged in chevrons (Fig. 4.11D1), as in arandaspids, but there is always a median dorsal series of large ridge scales (Fig. 4.11, 7). In one form (*Lasanius*) the squamation is reduced to these large median dorsal scales, and the postbranchial and branchial scales (Fig. 4.11C). In *Birkenia* and *Lasanius*, the median dorsal ridge scales can be considerably enlarged and bicuspidate (Fig. 4.12D3).

The paired fins are found only in *Pharyngolepis* and *Rhyncholepis*, but may have been present in most—if not all—anaspids (Fig. 4.11, 3). They extend ventrolaterally from the postbranchial spines to the anal region, and their length depends on the position of the latter. In *Pharyngolepis* (Fig. 4.11A1) the body is very elongated in shape and the anal region situated far back. As a consequence, the paired fins are extremely elongated and ribbon-like. In contrast, *Rhyncholepis* (Fig. 4.11B) has a very short preanal region, and rather short-based paired fins (Fig. 4.11B, 3). The paired fins of anaspids are covered with minute scales arranged in thin rows which suggest the presence of numerous underlying radials. The presence of radial muscles is also indicated by the presence of a narrow basal lobe covered with larger scales.

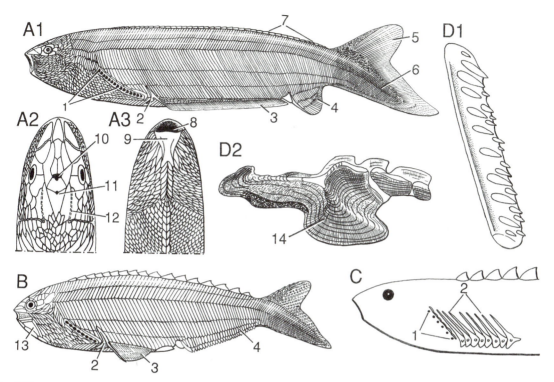

Fig. 4.11. Anaspida. External morphology and histology. A, *Pharyngolepis*, Early Silurian of Norway, reconstruction in lateral view (A1, ×1), and head in dorsal (A2) and ventral (A3) view; B, *Rhyncholepis*, Early Silurian of Europe, reconstruction in lateral view (×1); C, *Lasanius*, Early Silurian of Scotland, head in lateral view (×2); D, flange scale in external view (D1) and transverse section (D2, ×70). 1, external branchial openings; 2, tri-radiate postbranchial spine; 3, paired fins; 4, anal fin; 5, epichordal lobe of caudal fin; 6, chordal lobe of caudal fin; 7, median dorsal scutes; 8, mouth; 9, oral plate; 10, presumed nasohypophysial foramen; 11, pineal foramen; 12, sensory-line groove; 13, ventral 'gular' plate; 14, laminar aspidine. (A, combined from I. C. Smith (1957), Ritchie (1964), and Stensiö (1964); B, from Ritchie (1980); C, from Parrington (1958); D, from Gross (1958).)

There is no dorsal fin in anaspids. The only unpaired fins are the dorsal lobe, or epichordal lobe (Fig. 4.11, 5), of the tail and the anal fin, which, however, is reduced or lacking in several genera (e.g. *Birkenia*, *Lasanius*, *Rhyncholepis*; Figs 4.11B, 4, 4.12D2–4). Whether the epichordal lobe of anaspids belongs to the caudal fin or is the posterior dorsal fin is undecided.

There may be a lateral-line system, at least on the head, as suggested by very thin grooves (Fig. 4.11, 12). This, however, is still debated.

4.6.2 Internal anatomy

Very little is known of the internal anatomy of anaspids, apart from vague imprints of soft structures preserved under particular conditions in four genera (*Jamoytius*, *Endeiolepis*, *Euphanerops*, and *Legendrelepis*) which are often referred to as anaspids on the basis of overall resemblance, in particular their strongly hypocercal tail (Fig. 4.12A–C, D5–7). Being virtually naked, these forms do not display either postbranchial spines or median dorsal ridge scales, and therefore cannot be referred with certainty to the Anaspida as defined here. One may, however, concede that their internal anatomy may have been quite similar to that of the Anaspida.

Jamoytius (Fig. 4.12A, D7), from the Early Silurian (Llandovery) of Scotland, is an eel-shaped form with no mineralized skeleton, which may have possessed elongated paired fins. Some specimens show tarry imprints of the eyes (Fig. 4.12A, 3), possibly the olfactory organ or an annular cartilage (Fig. 4.12A, 2), and a branchial 'basket' with about twenty branchial units or openings (Fig. 4.12A, 1). *Jamoytius* was first regarded as a very primitive craniate, but it is currently interpreted either as a primitive lamprey or as a naked anaspid.

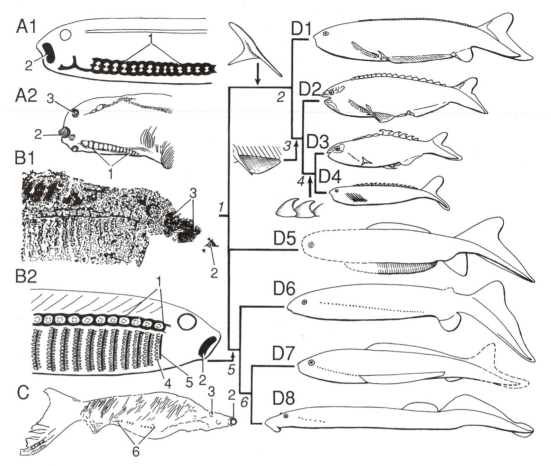

Fig. 4.12. Anaspid-like fossil craniates and interrelationships of the Anaspida and Hyperoartia. A, *Jamoytius*, Early Silurian of Scotland, reconstruction of the head skeleton in lateral view (A1, ×1) and sketch of a squashed head (A2); B, *Legendrelepis*, Late Devonian of Quebec, head in lateral aspect (B1, ×2.5) and attempted reconstruction (B2); C, *Euphanerops*, Late Devonian of Quebec (×1). D, interrelationships of the Anaspida and Hyperoartia; *Terminal taxa*: D1, *Pharyngolepis*; D2, *Rhyncholepis*; D3, *Birkenia*; D4, *Lasanius*; D5, *Endeiolepis*; D6, *Legendrelepis* or *Euphanerops*; D7, *Jamoytius*; D8, Petromyzontiformes (Carboniferous to Recent lampreys). *Nested taxa and synapomorphies*: *1*, strongly hypocercal tail, posteroventrally tilted series of external branchial openings, ?dorsal nasohypophysial opening, epi- and hypobranchial myomeres; *2*, Anaspida (tri-radiate postbranchial spine); *3*, Rhyncholepidida (shortened paired fins, reduced anal fin); *4*, Birkeniidae (enlarged, hook-shaped median dorsal scutes); *5*, Hyperoartia (annular cartilage, loss of mineralized exoskeleton); *6*, elongate, eel-shaped body. *Age*: Silurian (D1–D4, D7); Late Devonian (D5, D6); Carboniferous to Recent (D8). 1, branchial 'basket'; 2, annular cartilage; 3, eye; 4, gill lamellae; 5, gill arch; 6, external branchial openings or trematic rings. (A, D7, combined from Ritchie (1968, 1984*a*) and Forey and Gardiner (1981); B1, C, D5, D6, from Arsenault and Janvier (1991); D1, D2, from Ritchie (1964, 1980); D3, from Stetson (1928); D4, from Parrington (1958).)

Euphanerops and *Legendrelepis* (which may be synonyms) are from the late Devonian of Miguasha, Quebec and have an admitedly anaspid-like aspect, although the characteristic tri-radiate postbranchial spines cannot be observed (Fig. 4.12B, C). They are also poorly mineralized, and thanks to their reduced exoskeleton, part of their internal anatomy is preserved in imprint. They seem to have possessed an annular cartilage surrounding the mouth (Fig. 4.12B, C, 2), as in lampreys, and a long ventral series of about thirty gill units that extends back to the anal region, thereby recalling the branchial apparatus of the cephalochordates (Fig. 4,12B, C, 4–6). Whether this condition is primitive or derived is difficult to decide but, according to the phylogenetic position given to these forms here (Fig. 4.12D), we would rather view it as unique to the ensemble Anaspida + Hyperoartia (lampreys), with reversions in either group. There are also large cartilaginous radials in the caudal fin.

Endeiolepis is from the same age and locality as *Euphanerops* and *Legendrelepis*, but it differs in having a peculiar ventrolateral series of scales that form a kind of undulated crest (Fig. 4.12D5). Whether this crest is a modified paired fin or a neoformation is undecided. Some specimens of *Endeiolepis* display traces of the myomeres and the imprint of either the intestine or the stomach.

On the basis of these admittedly sparse data on the internal anatomy of anaspid-like forms, one may believe that such 'true anaspids' as *Pharyngolepis* or *Pterygolepis* did in fact possess a cartilaginous endoskeleton more or less comparable to that of lampreys, with a small braincase and a branchial basket extending far posteriorly. The shape of the oral plates would, however, indicate that it is unlikely that the true anaspids possessed an annular cartilage and a sucker.

4.6.3 Histology

All the plates and scales of anaspids are made up by a kind of acellular, laminar bone that strongly resembles aspidine (Fig. 4.11, 14). There is no evidence of dentine, even in the tubercles of the body scales. This condition is reminiscent of the condition in astraspids (Fig. 4.3C1), although the tubercles in anaspids do not seem to have been capped with enameloid (Gross 1958; Märss 1986*b*).

4.6.4 Diversity and phylogeny

The Anaspida have often been regarded as the sister-group of lampreys since they share with them a dorsal nasohypophysial opening (or so we suppose), a slender body shape, and gill openings arranged in slanting line (Kiaer 1924, Stensiö 1927). Moreover, if the arrangement of the body scales in anaspids reflects that of the underlying myomeres, one may infer from it that, as in lampreys, their body musculature extended far forward above—and perhaps beneath—the branchial apparatus. The fact that the epibranchial musculature reached the eye even suggests that anaspids possessed the same visual accommodation device as lampreys, i.e. a corneal muscle spanning the cornea (Fig. 3.6I, 23). Finally, the tail is known to be slightly hypocercal in lampreys, and displays some resemblance to that of anaspids, if the second dorsal fin is considered as homologous to the epichordal lobe of the anaspid tail. *Jamoytius* has been regarded as a possible link between the anaspid and lamprey morphologies because of its naked skin, slender body shape, and possible annular cartilage. Nevertheless, since lampreys have

no dermal skeleton and since true anaspids do not show much of their endoskeleton, the use of lampreys as an out-group to reconstruct the phylogeny of the Anaspida (as defined here) is limited, and a comparison with any other craniate group is more appropriate. In addition, true anaspids are so rare and their occurrences so close in time (they are all Silurian, in particular Wenlock), that the geological age of the taxa is of no help either. Anaspids have also been placed as the closest relatives to the gnathostomes (Maisey 1986), because their body musculature suggests the presence of a horizontal septum and their paired fins extend to the anal region (a prerequisite for the formation of pelvic fins). The argument for this theory remains tenuous, however (see Chapter 5, p. 237)

When considering the Anaspida proper, that is, those forms in which a triradiate postbranchial spine has actually been observed (Fig. 4.12D, *2*), *Pharyngolepis* appears as the most generalized form because of the small size of its dorsal ridge scales, the large scale-covered areas on its head, and its well-developed anal fin (Fig. 4.12D1). Conversely, its elongated, ribbon-like paired fins and numerous gill openings could be assessed as unique in relation to other vertebrates, but general for the group, because such elongated paired fins and numerous gill openings also occur in *Jamoytius*, here regarded as more closely related to lampreys (Fig. 4.12D7).

All other anaspids show a tendency toward the shortening of the paired fins (Fig. 4.12D, *3*), enlargement of the median dorsal ridge scales and head plates, reduction of the number of gill openings, and reduction or loss of the anal fin. *Rhyncholepis* (Fig. 4.12D2) has large median dorsal scutes, fewer gill openings, and a reduced anal fin. It is unique in possessing large dermal plates ('gulars') on the ventral side of the head (Fig. 4.11, 13). In *Birkenia* (Fig. 4.11D3) and *Lasanius* (Fig. 4.11D4), there are only eight to six gill openings, no anal fin, and the dorsal scutes are considerably enlarged and hook-shaped (Fig. 4.12D, *4*). *Lasanius* has lost most of its scales, except the dorsal scutes, the postbranchial spines, and the branchial scales (Fig. 4.11C).

As already mentioned, the status of *Jamoytius*, *Endeiolepis*, *Euphanerops*, and *Legendrelepis* is still obscure. In the phylogeny proposed here, *Jamoytius* (Fig. 4.12D7) is regarded as the sister-group of lampreys (Fig. 4.12D8), and *Euphanerops* (Fig. 4.12D6, interchangeable with *Legendrelepis*) as the sister-group of *Jamoytius* + lampreys. All these forms seem to have possessed an annular cartilage (Fig. 4.12D, *5*). It is undecided whether *Endeiolepis* (Fig. 4.12D5) is more closely related to these lamprey-like forms or to the Anaspida. Nevertheless, its long ventrolateral crest, if derived from elongated paired fins, reflects a general condition for the ensemble lampreys + Anaspida.

The classification of the Anaspida shown in Table 4.2 is derived from the cladogram in Fig. 4.12D1–4. For the classification of *Endeiolepis*, *Euphanerops*, *Legendrelepis*, and *Jamoytius*, see section 4.7.

Table 4.2. Classification of the Anaspida (See also Fig. 4.12 D1–4)

Anaspida
    Pharyngolepis
    Rhyncholepidida
        Rhyncholepis
    Birkeniidae
        Birkenia
        Lasanius

4.6.5 Stratigraphical and geographical distribution, habitat

The Anaspida occur from the Early Silurian (Llandovery) to the latest Silurian in North America and Europe. There is a single record of a 'birkeniid' squamation from the Silurian of China. *Jamoytius* is Early Silurian in age, but *Jamoytius*-like forms occur in the Early Devonian of the USA. *Endeiolepis*, *Euphanerops*, and *Legendrelepis* all come from the Late Devonian locality of Miguasha (Quebec). Most of what is known of anaspid morphology comes from the remarkably well-preserved Early Silurian material of Ringerike (Norway), with *Pharyngolepis*, *Pterygolepis*, and *Rhyncholepis*. All known anaspids, including the atypical Late Devonian forms, were marine, and most probably epipelagic animals.

Further reading

Arsenault and Janvier (1991); Forey and Gardiner (1981); Gross (1958); Heintz (1958); Janvier (1987); Jarvik (1959); Kiaer (1924); Märss (1986*b*); Parrington (1958); Ritchie (1964, 1968, 1980, 1984*a*); Stensiö (1964); Stetson (1928); Traquair (1899); Wickstead (1969).

4.7 FOSSIL LAMPREYS (HYPEROARTIA, PETROMYZONTIFORMES)

Leaving aside *Jamoytius*, which may be closely related to lampreys because of the presence of an annular cartilage, but not a true lamprey (Petromyzontiformes), as defined by an oral sucker and piston cartilage, the only fossil lampreys recorded so far are Carboniferous in age. *Mayomyzon* (Fig. 4.13B) from the Late Carboniferous of Mazon Creek, Illinois, is by far the best known of all, since it shows imprints of the cartilaginous endoskeleton, which appears remarkably similar to that of extant

lampreys. Although the presence of a sucker is uncertain in *Mayomyzon*, there was certainly a piston cartilage (Fig. 4.13, 8), which provides evidence for the same type of 'lingual' device as in modern lampreys. The main difference between *Mayomyzon* and extant lampreys rests in the position of the gill pouches, which are small, closely set, and situated closer to the eyes (Fig. 4.13D2). There may have been only five or six gill pouches, instead of seven in extant lampreys. The unpaired fins of *Mayomyzon* also differ from those of extant lampreys in forming a continuous web, with no distinct dorsal and caudal fins (Fig. 4.13B1, 1, 2).

In the same locality and layers as *Mayomyzon* occurs a second genus of a presumed lamprey, *Pipiscius*, which displays an oral sucker armed with presumably horny plates (Fig. 4.13C1, C2, 10,11). The pattern of these plates is somewhat reminiscent of that in the Recent genus *Ichthyomyzon*, which is regarded as the most primitive extant lamprey (Fig. 4.13C3, 11).

Hardistiella (Fig. 4.13A1), from the Bear Gulch locality (see Chapter 1), is another Carboniferous lamprey, somewhat older and less well preserved than *Mayomyzon*. The tail of *Hardistiella* seems to be slightly hypocercal (if undistorted), and it may retain a small anal fin with radials (Fig. 4.13A1, 1, 3).

The small size of all these Carboniferous lampreys (about 5 cm) has been invoked to suggest that they are juvenile individuals and did not undergo a larval development and metamorphosis. Since a piston cartilage is present in *Mayomyzon*, and a sucker exists in *Pipiscius*, one may rather regard them as adults. A specimen of *Hardistiella* displays imprints of cartilaginous bars that are strikingly similar to the trabecles and nasal capsule of extant larval lampreys, and may be an indication of a larval development in Carboniferous lampreys (Fig. 4.13A2, A3, 5, 6).

All these Carboniferous lampreys were marine, and the presence of a sucker and piston cartilage in some of them suggests that they already had developed parasitic habits.

The classification (by phyletic sequencing) of the Hyperoartia and Anaspida in Table 4.3 is derived from the cladogram in Fig. 4.12:

Further reading

Arsenault and Janvier (1991); Bardack and Zangerl (1971); Bardack and Richardson (1977); Forey and Gardiner (1981); Janvier (1983); Janvier and Lund (1983); Lund & Janvier (1985); Ritchie (1984*a*).

4.8 OSTEOSTRACI

Osteostracans are armoured jawless vertebrates that possess a perichondrally ossified endoskeleton and an

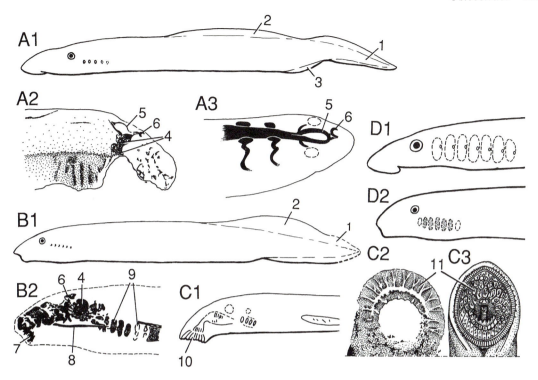

Fig. 4.13. Fossil Hyperoartia (lampreys). A, *Hardistiella*, Early Carboniferous of Montana (USA), reconstruction in lateral view (A1, ×1) and sketch of the head of a supposedly larval specimen showing the trabecles (A2, ×7) in the same position as in extant larval lampreys (A3); B, *Mayomyzon*, Late Carboniferous of Illinois (USA), reconstruction in lateral view (B1, ×2), and imprints of the cartilages in the head (B2); C, *Pipiscius*, Late Carboniferous of Illinois (USA), sketch of the head in lateral view (C1, ×1), and oral sucker (C2), compared to that of a primitive extant lamprey (C3, *Ichthyomyzon*). D, outline of the head and gill pouches in an extant lamprey (D1, *Lampetra*) and a Carboniferous lamprey (D2, *Mayomyzon*), showing the relative position and size of the gill pouches. 1, caudal fin; 2, dorsal fin; 3, anal fin; 4, eye; 5, trabecle; 6, nasal capsule; 7, annular cartilage; 8, piston cartilage; 9, gill pouches; 10, sucker; 11, horny plates on sucker. (A, D, from Janvier and Lund (1983) and Lund and Janvier (1986); B, from Bardack and Zangerl (1971); C1, 2, from Bardack and Richardson (1977).)

Table 4.3.   Classification of the Hyperoartia and Anaspida (See also Fig. 4.12)

Anaspida *sedis mutabilis*
Endeiolepis sedis mutabilis
Hyperoartia *sedis mutabilis*
  Euphanerops/Legendrelepis
  Jamoytius
  Petromyzontiformes (lampreys)
    Hardistiella
    Mayomyzon
    Pipiscius
    All extant lampreys

exoskeleton of cellular bone. They are characterized by peculiar median and lateral shallow depressions in the dorsal surface of the head shield, which are referred to as 'cephalic fields' (Fig. 4.14, 6, 7; Fig. 4.15, 29). These fields are covered with free polygonal dermal bone platelets and are connected to the labyrinth cavity by means of large, dichotomously ramified canals (Fig. 4.16, 11). They have been variously interpreted as having housed either electric organs, dynamosensory expansions of the labyrinth, or other kinds of sensory organs linked with the lateral-line system. None of these interpretations is as yet corroborated.

Another unique character of the Osteostraci is a peculiar horizontal lobe lining ventrally the caudal fin, which may possibly be a modified anal fin (Fig. 4.14, 4). This character is, however, assumed to be lost in some osteostracan taxa (Fig. 4.14C1).

Most osteostracans have a horseshoe-shaped head, on which attach a pair of pectoral fins; this represents the generalized condition for the group (Fig. 4.14A, B, D, 5).

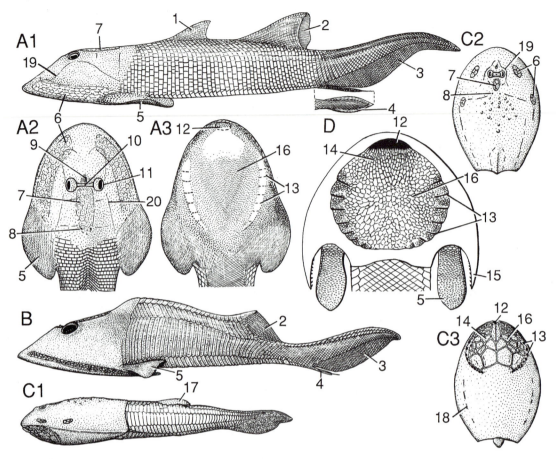

Fig. 4.14. Osteostraci. External morphology. A, *Ateleaspis*, Early Silurian of Scotland, reconstruction in lateral view (A1, ×0.8), and head in dorsal (A2) and ventral (A3) view. B, zenaspidid osteostracan, Early Devonian of Scotland, reconstruction in lateral view (×0.5); C, *Tremataspis*, Late Silurian of Estonia, reconstruction in lateral view (C1, ×1), and head in dorsal (C2) and ventral (C3) view; D, generalized cornuate osteostracan, head and paired fins in ventral view. 1, anterior dorsal fin; 2, posterior dorsal fin; 3, caudal fin; 4, horizontal caudal lobe; 5, pectoral fin; 6, lateral fields; 7, median field; 8, endolymphatic duct; 9, nasohypophysial opening; 10, pineal foramen; 11, sclerotic ring; 12, mouth; 13, external branchial openings; 14, oral plates; 15, cornual process; 16, scales or dermal plates covering of oralobranchial chamber; 17, median dorsal scute; 18, ventral trunk sensory line; 19, infraorbital sensory line; 20, main lateral sensory line. (A, from Ritchie (1967); C, from Janvier (1985*b*); D, partly based on specimens of '*Cephalaspis*' *cradleyensis* in White and Toombs (1983).)

Various modifications of this pattern occur, by development of processes (Fig. 4.18F2–4), reduction of the cephalic fields (Fig. 4.14C2, 6, 7), or loss of the paired fins (Fig. 4.14C). The body shape is more conservative, yet there is a tendency towards the reduction of the dorsal fins (Fig. 4.14A–C, 1, 2) and the loss of the horizontal lobe of the caudal fin.

4.8.1 Exoskeleton and external morphology

The external aspect of osteostracans will be exemplified here by three different types, ranging from the most

generalized to the most derived. The first type is represented by the Early Silurian genus *Ateleaspis* (Fig. 4.14A), from Scotland; the second by *Zenaspis* (Fig. 4.14B), from the Early Devonian of Europe and Spitsbergen; and the third by the Late Silurian genus *Tremataspis* (Fig. 4.14C), from Estonia.

Ateleaspis has the typical dorsoventrally depressed head-shield, with broad-based paired fins whose squamation passes progressively to the tesserae of the shield and to the squamation of the flanges (Fig. 4.14A, 5). The lateral and median 'fields' are large (Fig. 4.14A, 6, 7), and the endolymphatic duct opens to the exterior inside the median field (Fig. 4.14A, 8). The orbits are closely set, and the eyeballs bear a sclerotic ring. They are separated by a rod-shaped pineal plate pierced by the pineal foramen (Fig. 4.14A, 10). Anteriorly, the keyhole-shaped nasohypophysial opening is situated in a shallow heart-shaped depression (Fig. 4.14A, 9). The mouth (Fig. 4.14A, 12) and gill openings (Fig. 4.14A, 13) are situated on the ventral side of the head, which is partly covered with minute scales (Fig. 4.14A, 16).

The sensory-line system, which is fairly constant in osteostracans, consists of canals and grooves in the exoskeleton of the head-shield, but it is uncertain whether there were sensory lines on the trunk. The pattern of the sensory lines consists, as in most vertebrates, of longitudinal and transverse lines, but because of the extreme dorsal position of the orbits, there is no supraorbital canal. Instead, the main lateral line continues anteriorly into a long infraorbital canal that forms a loop on the lateral fields (Fig. 4.14A, 19).

As in most other osteostracans (Fig. 4.17B), the exoskeleton of the head-shield is made up of polygonal tesserae ornamented with tubercles of mesodentine, which are firmly attached to the underlying endoskeleton, except at the level of the cephalic 'fields'.

The body is covered with rows of small, elongated scales. There are no distinct median dorsal ridge scales, but there are ventrolateral ones. Two dorsal fins are present, but the anterior has no fin web and is a mere pointed elevation covered with scales (Fig. 4.14A, 1, 2). The tail is epicercal, with a large ventral web (Fig. 4.14A, 3) and the characteristic horizontal lobe (Fig. 4.14A, 4).

Zenapis has much the same shape as *Ateleapis*, but its paired fins are narrow-based (Fig. 4.14B, D, 5) and are situated just behind a spine-shaped process of the head-shield, the cornual process (Fig. 4.14D, 15). The cornual process is a widespread character among osteostracans, and defines a major clade, the Cornuata (Fig. 4.18, 4). The lateral fields are narrower, but the rest of the shield remains quite similar to that of *Ateleapis*. Ventrally, the mouth (Fig. 4.14D, 12) is distinctly lined with elongated oral plates (Fig. 4.14D, 14), and each gill opening is covered with small scaly flap (Fig. 4.14D, 13). The rest of the ventral side is covered with larger dermal plates (Fig. 4.14D, 16), which are separated from the scales of the trunk by a solid bridge of dermal bone belonging to the shield.

The body is covered with larger scales than those of *Ateleaspis*, and there is a distinct row of median dorsal ridge scales. The anterior dorsal fin is lost and only the posterior one remains (Fig. 4.14B, 2).

Tremataspis is an example of an osteostracans with many unique characters, although it has long been regarded as 'primitive' because of its Silurian age and relatively simple morphology. The paired fins and cornual processes are absent and the head-shield is prolonged posteriorly by a long abdominal division (Fig. 4.14C). The 'fields' are small and the lateral ones are divided into

anterior and posterior parts (Fig. 4.14C2, 6). Because of the reduction of the median field, the endolymphatic ducts open outside it (Fig. 4.14C2, 8). There are indications that the pineal plate could move slightly back and forth and that the pineal lumen could be obturated in this way. Ventrally, the mouth is very small (Fig. 4.14C3, 12) and is armed with two large oral plates (Fig. 4.14C3, 14). Its margin bears minute rounded denticles. The ventral side of the head is covered with relatively large platelets (Fig. 4.14C3, 16). The body is covered with large scales and there is no dorsal fin at all. A large median dorsal scute is assumed to be a remnant of one of the dorsal fins (Fig. 4.14C1, 17). The tail, too, is much reduced and its web is covered with polygonal scales, instead of the elongated dermal fin rays of other osteotracans, a condition that suggest poor mobility. There is no horizontal lobe in the caudal fin.

The aspect of the exoskeleton of *Tremataspis* is also unique, since no tesserae are visible, except on the 'fields'. Instead of tubercles, there is a shiny layer of mesodentine, covered with enameloid, and containing a network of canals, the 'pore-canal system' (Fig. 4.17C, 5), and flask-shaped cavities (Fig. 4.17C, 8) which open to the exterior by numerous minute pores (Fig. 4.17C, 4). This type of exoskeleton, called 'cosmine', is quite similar, by convergence, to that of some fossil sarcopterygian osteichthyans (see p. 205). The network it contains seems to have been linked with the lateral sensory-line system and may have contained electrosensory organs. The canals of the pore-canal system and the ordinary lateral lines in *Tremataspis* are both divided horizontally by a very thin perforated bone lamina of unknown function, which clearly suggests a common origin for these two structures (Fig. 4.17C, 9). It is however, probable, that all osteostracans possessed a similar network of sensory canals (Fig. 4.17B, 3, 5), but this was situated more superficially, between the tubercles of the ornamentation, and is thus untraceable. In addition to the pore-canal system, *Tremataspis* has a normal lateral-line system, with two ventral lines continuing far back on the abdominal division of the shield (Fig. 4.14C3, 18). Since this part of the shield is derived from fused flange scales, one may infer that all osteostracans also possessed lateral-line canals on the trunk.

Tremataspis belongs to a clade, the Thyestiida, whose most generalized members have the classical cornuate osteostracan morphology (Fig. 4.18H1, 2). This shows that the peculiar and apparently simple morphology of this genus is in fact derived, and may represent an adaptation to burrowing habits.

4.8.2 Internal anatomy

The endoskeletal head-shield of osteostracans is a single mass of cartilage lined with a thin layer of perichondral bone, and pervaded by canals and cavities for the brain, sensory capsules, cranial nerves, and major blood vessels, It is hollowed ventrally by a large, hemispherical oralobranchial chamber (Fig. 4.15, 1), which contained the branchial apparatus and probably also a water-pumping device similar to the velum of larval lampreys (Fig.

3.6C1; Fig. 4.15E, 32). This chamber is closed posteriorly by the postbranchial wall, which also encloses the heart (Fig. 4.15, 10; Fig. 4.16, 31, 32) and is pierced by openings for the oesophagus and dorsal aorta (Fig. 4.15B, 20), the ventral aorta, and the hepatic vein (Fig. 4.15A, 11). Beyond the postbranchial wall the abdominal cavity

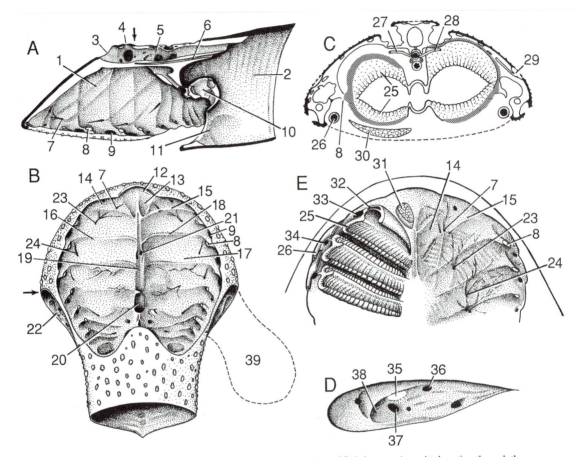

Fig. 4.15. Osteostraci. Internal anatomy. A–D, *Norselaspis*, Early Devonian of Spitsbergen; A, sagittal section through the head-shield (section of exoskeleton black, endoskeleton white; ×6); B, head-shield in ventral view; C, transverse section through the head-shield at the levels indicated by arrows in A and B (exoskeleton black, endoskeleton white), showing alternative reconstructions of the gills and gill arches (gill arches partly incorporated in endoskeletal shield on the right-hand side and independent in the left-hand side; gill lamellae stippled, gill arches grey); D, area of attachment for the paired fins in lateral view (×14). E, *Scolenaspis*, Early Devonian of Spitsbergen, roof of the oralobranchial cavity, with velum and gills reconstructed on left-hand side (×1). 1, oralobranchial chamber; 2, abdominal cavity; 3, nasohypophysial opening; 4, pineal foramen; 5, cranial cavity; 6, canal for the notochord; 7, canal for the trigeminus nerve; 8, medial ventral process for attachment of branchial arches; 9, canal for extrabranchial veins; 10, pericardic (intramural) cavity; 11, foramen for the hepatic vein; 12, supraoral field; 13, depressions for oral musculature; 14, prebranchial fossa, housing the velum; 15, prebranchial ridge; 16, interbranchial ridge; 17, branchial fossa; 18, groove for extrabranchial artery; 19, groove for dorsal aorta and oesophagus; 20, foramen for dorsal aorta and oesophagus; 21, internal carotids; 22; area for attachment of pectoral fin; 23, foramen for the facial nerve; 24, foramen for the glossopharyngeus nerve; 25, gill lamellae; 26, external branchial duct; 27, extrabranchial artery; 28, efferent branchial artery; 29, lateral field; 30, branchial musculature; 31, oral musculature; 32, velum and velar skeleton; 33, hyoid arch; 34, first branchial arch; 35, articular area for paired fin skeleton; 36, brachial vein; 37, brachial artery; 38, insertion area for fin muscles; 39, outline of pectoral fin. (A–D, from Janvier (1981*b*); E, based on Wängsjö (1952).)

extends posteriorly (Fig. 4.15A, 2), which is covered dorsally by the occipital endoskeleton.

Thanks to its perichondral bone lining, and to the exceptional conditions of preservation of the material from the Late Silurian of Saaremaa (Estonia) and Early Devonian of Spitsbergen, the endoskeletal head-shield of osteostracans has preserved natural casts of many internal organs. These were first revealed by Stensiö in 1927,

using Sollas's section-grinding method (see Chapter 8, p. 315).

The brain cavity provides a fairly accurate cast of the brain itself (Fig. 4.15A, 5; Fig 4.16A, 6); the divisions for the telencephalon, mesencephalon, metencephalon, and medulla oblongata can easily be distinguished (Fig. 4.16C, 26–30). Behind the pineal canal (Fig. 4.15A, 16A, 4), there is an asymmetrical swelling which suggests

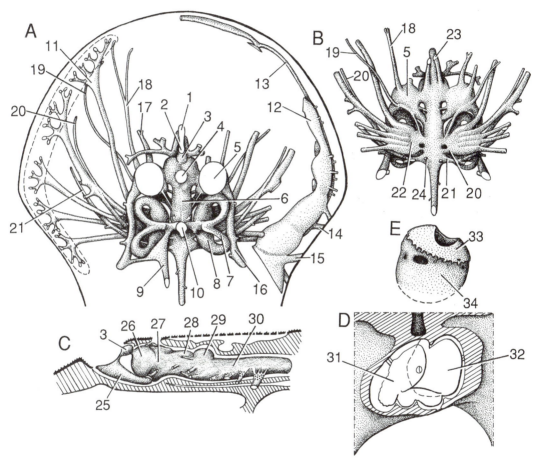

Fig. 4.16. Osteostraci. A–D, *Norselaspis*, Early Devonian of Spitsbergen; A, B, reconstruction of the internal cast of the cavities and canals in the head-shield, in dorsal (A) and ventral (B) view (×10); C, attempted reconstruction of the brain and nasohypophysial complex in lateral view (section of endoskeleton hatched, exoskeleton in black); D, attempted reconstruction of the heart in the pericardic cavity, in dorsal view (section of endoskeleton hatched). E, *Tremataspis*, Late Silurian of Estonia, eye capsule in anterior view (×20). 1, hypophysial division of nasohypophysial opening; 2, nasal division of nasohypophysial opening; 3, nasal cavity; 4, pineal foramen; 5, orbit; 6, brain cavity; 7, vertical semicircular canal; 8, ampulla; 9, endolymphatic duct; 10, 'sel' canal to the median field; 11, 'sel' canal to lateral fields; 12, marginal vein; 13, marginal artery; 14, brachial artery; 15, brachial vein; 16, dorsal jugular vein; 17, anterior division of jugular vein and profundus branch of trigeminus nerve; 18, maxillaris and mandibularis branches of trigeminus nerve; 19, facial nerve; 20, glossopharyngeus nerve; 21, vagus nerve; 22, otic capsule; 23, internal carotid; 24, occipital artery; 25, hypophysial tube; 26, telencephalon; 27, diencephalon; 28, mesencephalon; 29, metencephalon; 30, medulla oblongata; 31, atrium; 32, ventricle; 33, sclerotic ring; 34, ossified sclera. (A, B, E, modified from Janvier (1981*b*, 1985*a,b*); D, from Janvier *et al.* (1991).)

that the right habenular ganglion was larger than the left one, as in lampreys. The brain cavity is continued anteriorly by the ethmoid cavity, which contained a small, pear-shaped olfactory organ (Fig. 4.16, 3), and probably also a large hypophysial tube extending ventrally to the hypophysial region of the diencephalon, much as in lampreys (Fig. 4.16, 25). Beneath the brain cavity is a thin median canal for the notochord (Fig. 4.15, 6), which ends anteriorly just behind the hypophysial region. In the posterior part of the occipital region, this canal becomes much enlarged, showing that, in the body, the notochord must have been as large as in hagfishes and lampreys.

A number of canals branched off laterally from the brain cavity. Some housed cranial nerves (Fig. 4.16, 17–21), and other cerebral veins and arteries (Fig. 4.16, 16, 23, 24). There are large fenestrae leading toward the orbits and the labyrinth cavity, for the optic tract and acoustic nerve respectively (Fig. 4.16B).

The orbital cavities display a number of distinct recesses, or mydomes, to which the extrinsic eye muscles were attached, and which are almost in the same position as in lampreys (and placoderms among the gnathostomes); that is, the superior oblique muscle attached in the posterior part of the orbit, a condition that was once regarded as linked with the dorsal migration of the naso-hypophysial complex, but which may be general for the vertebrates. Thin canals for the oculomotor and trochlear nerves confirm the presence of extrinsic eye muscles.

The eyes are ossified in some osteostracans. In *Tremataspis*, for example, there is a distinct scleral (perichondral), cup-shaped ossification (Fig. 4.16, 34), and a dermal sclerotic ossification (Fig. 4.16, 33).

The labyrinth has two vertical semicircular canals, with their respective ampullae, which join into a crus commune (Fig. 4.16, 7, 8). From the latter arises an endolymphatic duct (Fig. 4.15, 9), which opened to the exterior inside or behind the median field (Fig. 3.14, 8). The most intriguing feature of the osteostracan labyrinth is the ensemble of five lateral canals and one dorsal canal of large diameter, which arise from the medial part of the labyrinth cavity and whose numerous distral ramifications open into the lateral and dorsal fields respectively (Fig. 4.16, 10, 11). These canals, here termed 'sel' canals, as Stensiö (1927) first referred to them in abbreviations, have engendered a number of controversies. Their aspect and large size is suggestive of electromotor nerves, as in torpedoes for example, but they seem to have a proximal swelling that suggests the presence of a ganglion and, thus, a sensory function (Fig. 4.16B, 22). Another interpretation is that these 'sel' canals did not transmit nerves, but instead housed prolongations of the labyrinth, comparable or even homologous to the ciliated chambers of the lamprey labyrinth (Fig. 3.6, 22). Such diverticles filled with endolymph would have transmitted to the labyrinth the pressure or vibrations of the surrounding water, by means of the flexible membrane covering the fields. All these interpretations are linked with those of the fields, since these canals are connected to them, but none of them is fully satisfactory.

The cranial nerve canals and their relationships to the roof of the oralobranchial cavity have been the source of considerable disputes over the past 65 years in connection with the premandibular arch theory (see Chapter 6, p. 254). It is now clear that the third nerve canal leading to the oralobranchial cavity emerges from the brain cavity just behind the labyrinth, and then bends forward, passing dorsally to the labyrinth, (Fig. 4.16, 20). It is thus the canal for the glossopharyngeus nerve. As a consequence, the two nerve canals anterior to it (Fig. 4.15, 7, 23; Fig. 4.16, 18, 19) housed three branchial or visceromotor and viscerosensory nerves: the facial nerve, and the mandibular and maxillar branches of the trigeminus nerve. It still cannot be decided whether the mandibular branch passed through the foremost canal, together with the maxillar branch, or passed through the second canal, together with the facial nerve. The latter solution would be more compatible with what is met with in most extant vertebrates, including lampreys, where the mandibular branch of the trigeminus nerve often runs close to the facial nerve. The other cranial nerves (trochlear, oculomotor, acoustic, and vagus) and the spinal or spino-occipital nerves pose no major problem in their interpretation.

The blood circulation can be reconstructed on the basis of the vascular canals preserved in the endoskeletal shield. The heart was housed in a cavity inside a posterior swelling of the postbranchial wall, the pericardic cavity (Fig. 4.15, 10; Fig. 4.16D). The shape of this cavity suggests that there was a ventricle and an atrium, lying side by side and quite similar to those of lampreys (Fig. 4.16D, 31, 32). From the ventricle, the blood was sent through a large foramen to the ventral aorta and the afferent branchial arteries. After oxygenation, it passed, via the efferent branchial arteries, to the dorsal aorta, which was unpaired and housed with the oesophagus in a large median groove in the roof of the oralobranchial cavity (Fig. 4.15, 19, 20). Then, from the dorsal aorta, the blood was sent: (1) to the brain and eyes via the carotids (Fig. 4.16, 23) and also an occipital artery (probably a modified segmental artery; Fig. 4.16, 24); (2) to the gill musculature, velum, and oral musculature via the extrabranchial arteries (Fig. 4.15, 27); (3) to the paired fins via the subclavian and brachial arteries (Fig. 4.15D, 37; Fig. 4.16A, 14); and (4) to the body via the posterior continuation of the dorsal aorta.

Finally, the deoxygenated blood returned to the atrium of the heart through at least two large venous trunks: the dorsal jugular vein (Fig. 4.16, 16), which drained the blood from the dorsal side of the head, and the marginal vein (Fig. 4.16, 12), lying in the marginal part of the shield, which drained the blood from the extrabranchial veins and the brachial veins of the paired fins (Fig. 4.15, 36; Fig. 4.16, 15). These very large marginal veins, which are unique to osteostracans, played the same role as the anterior cardinal veins, and one may wonder if they are not actually these veins, yet in a different position. The blood from the body probably reached the atrium of the heart (Fig. 4.16, 31) through the posterior cardinal veins and ductus cuvieri; and a large hepatic vein is thought, by analogy with lampreys, to have reached the heart through the ventral foramen of the postbranchial wall (Fig. 4.15, 11).

The roof of the oralobranchial cavity displays a number of paired depressions, the branchial fossae (Fig. 4.15, 17), separated by transverse or slightly oblique interbranchial ridges (Fig. 4.15, 18), which housed the gills, as shown by impressions of the gill lamellae on their surface (Fig. 4.15E, right-hand side). Anteriorly, there is a triangular supraoral field (Fig. 4.15, 12), which represents the roof of the oral cavity, and is sometimes hollowed by pits for the oral musculature (Fig. 4.15, 13, 31), or bears minute dermal denticles.

Stensiö (1927, 1964) first believed that the interbranchial ridges, which separate the branchial fossae, were the dorsal part of the branchial arches incorporated into the endoskeletal shield (Fig. 4.15C, right-hand side). The ventral part of the arches was supposed to be cartilaginous and attached to small processes that prolong ventrally the ridges (Fig. 4.15, 8). Another possibility is that the pattern we see in the roof of the oralobranchial cavity is a mere 'cast' of the entire branchial apparatus, and that the entire cartilaginous gill arches were situated inside this cavity, being attached to the shield by means of the processes mentioned above (Fig. 4.15C, left-hand side). This latter interpretation makes it easier to explain a number of features, in particular the blood vascular system.

The interpretation of the cranial nerves of osteostracans has important bearings on the interpretation of the interbranchial ridges, since the rank of the nerves indicates that of the branchial arches (for example, the facial nerve innervates the hyoid arch). Stensiö, for example, considered that the foremost canal leading to the roof of the oralobranchial cavity (Fig. 4.15, 7) transmitted a visceromotor branch of the profundus nerve (later regarded as the maxillar branch of the trigeminus nerve). The ridge that is closest to the opening of this canal was therefore regarded by Stensiö as the premandibular ridge (and thus as the premandibular gill arch). The current interpretation of the cranial nerves of osteostracans makes the oralo-branchial cavity quite similar to the condition of larval lampreys (Fig. 3.6C1). The visceromotor branches of the trigeminus nerve (maxillar and mandibular) innervated the oral region and the velum (Fig. 4.15E, 7, 32), and the facial nerve went to the hyoid arch (Fig. 4.15E, 23), which was closely associated with the velum and lacked an anterior hemibranch, as in larval lampreys (Fig. 4.15E, 25, 33). This interpretation implies that, as in larval lampreys, the mandibular arch, which bears the velum, and the hyoid arch are closely associated or perhaps fused into a single arch (Fig. 4.15E, 33). Consequently, the foremost gill pouch in osteostracans was that situated between the hyoid and the first branchial arch (Fig. 4.15E, 33, 34), and there is no longer any reason to assume that they possessed a prespiracular or even a spiracular gill-pouch. In fact, the unusually anterior position of the branchial apparatus—relatively to the brain—in osteostracans is due to a forward shift that brought the foremost gill arch (innervated by the glossopharyngeus nerve, Fig. 4.15E, 34) into a preorbital position.

The paired fins contained a complex musculature, as is shown by multiple depressions on the area of the shield where they attached (Fig. 4.15, 22, 38). There are some indications of an endoskeleton in the paired fins, in the form of a possible articulation area (Fig. 4.15, 35).

Little is known of the axial skeleton, but a median series of small patches of endoskeleton attached to the dorsal wall of the abdominal cavity behind the occipital region of *Tremataspis* suggests that osteostracans possessed at least ossified dorsal arcualia.

Although osteostracans are by far the best-known fossil jawless vertebrates, many details of their internal anatomy remain frustratingly enigmatic, such as the 'sel' canals and the cephalic 'fields'. The ethmoid cavity, for example, is unanimously accepted since Stensiö's work as having housed a nasohypophysial complex, which comprised a blind, dorsally opening hypophysial tube (Fig. 4.16C, 25). However, the only evidence for this interpretation is the resemblance of the ethmoid cavity with the nasohypophysial complex in lampreys, and the lack of a buccohypophysial canal. This resemblance is further reinforced by that of the keyhole-shaped nasohypophysial foramen of osteostracans (Fig. 4.16A, 1, 2) and the internal collaret in the nasohypophysial duct of lampreys. This derived state of the structure of the ethmoid region, shared only by osteostracans and lampreys (the condition in anaspids remains poorly known), is incongruent with a number of other characters, such as the epicercal tail, sclerotic ring, or true bone, which are unique to osteostracans and gnathostomes (see Chapter 5, p. 239). The heart, too, which is enclosed in a pericardic cavity (the intramural cavity in osteostracans), seems to have been remarkably similar in osteostracans and lampreys, in the close association of the atrium and ventricle, which lie side by side (Fig. 4.16D). But this resemblance is probably a general character of the Vertebrata.

4.8.3 Histology

The histology of osteostracan hard tissues is much more diverse and complex than that of all other jawless vertebrates. The exoskeleton, that is, the scales and the tesserae of the shield, are made up of true bone, with osteocyte lacunae (Fig. 4.17A, 2), though a secondarily acellular exoskeleton occurs in the Late Devonian osteostracan *Escuminaspis*, and it is covered with a more or less continuous layer of a dentinous tissue, the mesodentine (Fig. 4.17A, 1). The mesodentine is sometimes (e.g. in thyestidians) lined with a thin layer of enameloid. In the mesodentine of osteostracans, the cells that produced that dentinous matter, the odontoblasts, did not occupy a single large pulp cavity, but were instead grouped into smaller cavities inside the dentine, which were connected by vascular canals. Mesodentine was once regarded as 'intermediate' between bone and true dentine, or orthodentine, but it is now considered as derived from orthodentine (see Chapter 6, p. 276). The middle layer of the osteostracan exoskeleton (Fig. 4.17, 11) is bony and contains a horizontal network of canals often referred to as the 'mucous canal' network. It consists of larger inter-areal canals forming a polygonal pattern along the limits of the tesserae (Fig. 4.17B, 3), and smaller intra-areal canals within the tesserae (Fig. 4.17B, 5). In osteostracans whose superficial, dentinous layer is reduced or lacking these canals are turned into 'mucous' grooves. It is probable that these 'mucous' grooves and canals are identical to the pore-canal system of *Tremataspis* (Fig. 4.17C, 5) and other thyestidians, and may actually belong to the lateral sensory-line system.

Below this network of 'mucous' canals, there are one or more layers of thin vascular canals that radiate from the centre of each tessera (Fig. 4.17B, 6). Finally, and still deeper, there is a basal, laminar layer of bone that rests on the endoskeleton (Fig. 4.17, 12). Between the exo- and endoskeleton there is an intervening network of sub-aponevrotic vascular canals, from which arise the vessels that supply the exoskeleton (Fig. 4.17, 7).

The endoskeleton of the head was a mass of cartilage lined with perichondral bone, which also contained osteocyte lacunae. In some instances the perichondral bone layer sends off projections or trabeculae inside the cartilage; the most extreme condition is displayed by the Boreaspididae (Fig. 4.18F2), where the endoskeleton has a spongy structure that mimics the endochondral bone of osteichthyans. Globular calcified cartilage also occurs in

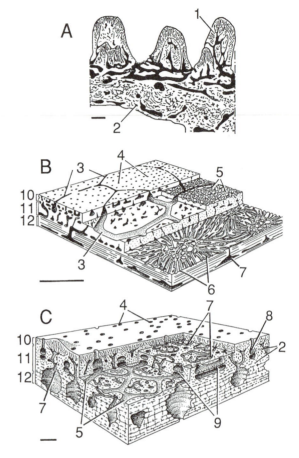

Fig. 4.17. Osteostraci. Histology. A, *Procephalaspis*, Silurian of Estonia, vertical section through the exoskeleton (scale bar: 0.1 mm); B, block diagram of the exoskeleton of a generalized cornuate osteostracan with continuous superficial layer (scale bar: 1 mm); C, block diagram of the exoskeleton of *Tremataspis*, Late Silurian of Estonia, showing the structure of the cosmine and the pore-canal system (scale bar: 0.1 mm). 1, mesodentine in tubercle; 2, osteocyte (bone cell) space; 3, circum-areal canals; 4, pores of the intra-areal canals (or cosmine pores); 5, intra-areal canals; 6, radiating vascular canals; 7, ascending vascular canal; 8, flask-shaped cavity of pore-canal system; 9, perforated horizontal lamina of pore-canal system; 10, superficial (dentinous) layer; 11, middle (spongy bone) layer; 12, basal (laminar bone) layer. (A, from Ørvig (1951); B, from Wängsjö (1952); C, modified from Denison (1947).)

osteostracans, generally lining internally the perichondral bone, or filling narrow cartilaginous processes, such as the cornual processes. Apart from the major arterial and venous trunks (jugular vein, marginal vein and artery), the endoskeleton was not vascularized and all the vascular network was perichondral in position.

The nature of the osteostracan head endoskeleton is still unclear. Stensiö (1927) regarded it as possibly homologous to the muco-cartilage of larval lampreys (Fig. 3.6M), and Johnels (1950) noticed that the dermis of the head in adult lampreys is thickened and strongly vascularized, thereby recalling vaguely the shape of the osteostracan head-shield. None of these interpretations can, however, be clearly tested. Although the osteostracan endoskeletal shield may not have incorporated gill arches, as Stensiö (1927) suggested, it was certainly continuous with the endoskeletal support of the paired fins (scapular area) and even surrounded the heart. This peculiar distribution of the endoskeleton is, however, not incompatible with the interpretation of the endoskeletal shield as being essentially the neurocranium. After all, such a broadly expanded neurocranium also occurs in some gnathostomes, such as the placoderms (Fig. 4.42A), where it is prolonged laterally over the branchial apparatus in the form of processes or blade-like expansions.

Although it is unknown whether they went through a larval stage, it is quite clear that osteostracans with a tessellate exoskeleton could grow, but the solidity of their shield, which ensured fossilization, was largely dependent on the degree of ossification of their endoskeleton. The latter depends on the taxon in consideration, and probably also the age of the individual. White and Toombs (1983), for example, have noticed differences in size ranging up to 200 per cent in a single population of osteostracans (*Zenaspis*). There are nevertheless instances where determinate growth in osteostracans may be envisaged. Large populations of Boreaspididae or Tremataspididae, for example, are now well known, and almost no variation in size has ever been noticed in each species. In these taxa, in which the exoskeleton has lost the tessellate pattern, one can accept that the entire shield became ossified once the animal reached a definitive size. Stages of development of the exoskeleton in *Tremataspis*, for example, show that the dentinous superficial layer formed first, and that the rest of the exoskeleton grew centripetally (Fig. 7.11B).

4.8.4 Diversity and phylogeny

The Osteostraci comprise a large clade, the Cornuata (Fig. 4.18, 4), characterized by the cornual processes in front of the paired fins (Fig. 4.14, 15) and the posterior closure of the dermal rim of the oralobranchial cavity. In addition to this group, there are five or six genera of 'non-cornuate' osteostracans, some of which are more closely related to the cornuates than others.

As we have seen, *Ateleapis* (Figs 4.14A, 4.18A) is regarded as the most generalized osteostracan, lacking cornual processes, ventral platelets, and median dorsal ridge scales. *Hirella* (Fig. 4.18B) has narrow-based paired fins, and ventral platelets covering the ventral side of the head. *Hemicyclaspis* (Fig. 4.18C) has in addition narrower lateral fields, and its anterior dorsal fin is reduced to a large median scute.

Five major clades can be recognized within the Cornuata: (1) the Cephalaspidida (Fig. 4.18D), with a long prehypophysial region and broad cornual processes; (2) the Zenaspidida (Fig. 4.18E), with narrow and thick cornual processes, posteriorly enlarged lateral fields, and enlarged hypophysial division of the nasohypophysial opening; (3) the Benneviaspidida (Fig. 4.18F), with no radiating canals in the exoskeleton and a depressed shield; (4) the Kiaeraspidida (Fig. 4.18G), with reduced cornual processes and enlarged supraoral field; and (5) the Thyestiida (Fig. 4.18H), with a perforated lamina dividing the pore canals and lateral-line canals, and with an infra-orbital sensory line passing medially to the lateral fields.

Within each of these groups, some remarkable morphologies occur. Among cephalaspidids, *Parameteoraspis* (Fig. 4.18D2) is represented by the largest known osteostracan species, reaching 40 cm in shield breadth, some of them having a crescent-shaped shield.

Zenaspidids are also generally large forms, and are perhaps the most conservative cornuates. However, some of them, like *Machairaspis* (Fig. 1.5, 12), develop a very high and anteriorly tilted spinal process over the occipital region. Others, like *Diademaspis* (Fig. 4.18E2), are very large and have a very broad median 'field'.

Benneviaspidids, such as *Benneviaspis* (Fig. 4.18F1), have a very flat head-shield. In some of them, the Boreaspididae, the head-shield is produced anteriorly into a long rostral process (*Belonaspis*, Fig. 4.18F2), and in *Hoelaspis* (Fig. 4.18F3), the cornual processes are curved forward. *Tauraspis* (Fig. 4.18F4) is unique in having two long anterolateral 'horns'.

Kiaeraspidids are very small forms, barely larger than a finger-nail. *Kiaeraspis* or *Norselaspis* (Fig. 4.18G1) have a rather elongated abdominal division; all others have no cornual processes and show a tendency toward the reduction or subdivision of the fields. *Gustavaspis* (Fig. 4.18G2) is the most derived genus of this group, with much reduced 'fields', and a dorsally placed mouth (Fig. 1.5, 8).

Thyestiids include the most derived osteostracans, the Tremataspididae, which have lost the paired fins and cornual processes and have gained an elongated, almost olive-shaped head-shield. Generalized thyestiids, like *Procephalaspis* (Fig. 4.18H1), display a typical cornuate shield pattern. In more advanced forms, such as *Thyestes* (Fig. 4.18H2), the cornual processes are reduced, the paired fins minute, and a number of rows of body scales

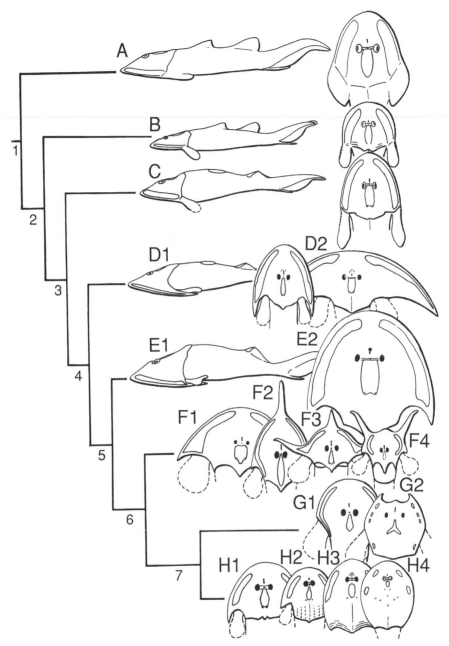

Fig. 4.18. Interrelationships of the Osteostraci. *Terminal taxa*: A, *Ateleaspis* in lateral and dorsal views; B, *Hirella* in lateral and dorsal views; C, *Hemicyclaspis* in lateral and dorsal views; D, Cephalaspidida (D1, *Cephalaspis* in lateral and dorsal views; D2, *Parameteoraspis*); E, Zenaspidida (E1, *Zenaspis* in lateral view; E2, *Diademaspis*); F, Benneviaspidida (F1, *Benneviaspis*; F2, *Belonaspis*; F3, *Hoelaspis*; F4, *Tauraspis*); G, Kiaeraspidida (G1, *Norselaspis*; G2, *Gustavaspis*); H, Thyestiida (H1, *Procephalaspis*; H2, *Thyestes*; H3, *Dartmuthia*; H4, *Tremataspis*). *Nested taxa and selected synapomorphies*: 1, Osteostraci (cephalic 'fields', horizontal caudal lobe); 2, narrow-based paired fins, large dermal plates closing oralobranchial fenestra; 3, anterior dorsal fin reduced to a large scute, paired fins separated from trunk musculature; 4, Cornuata (cornual processes, oralobranchial fenestra posteriorly closed by a bar of dermal bone); 5, loss of anterior dorsal scute; 6, first canal to lateral field bifurcates at a point close to the orbit; 7, loss or reduction of posterior ventral myodome. *Age*: Silurian (A, B, H); Early Devonian (C–G; E, possibly Early to Late Devonian). (A, from Ritchie (1967); B, from Heintz (1939); C, from Stensiö (1932); D–F3, G, from Wängsjö (1952) and Janvier (1985a); F4, from Mark-Kurik and Janvier (1995); H, from Janvier (1985b).)

fuse to the rear of the head-shield. The Tremataspididae, such as *Dartmuthia* (Fig. 4.18H3) or *Tremataspis* (Fig. 4.18H4) show congruent tendencies towards the elongation of the shield, reduction of the 'fields', and development of a continuous layer of cosmine. *Sclerodus* is the most derived tremataspid, since its elongated head-shield is, in part, secondarily reduced, and it is unique among osteostracans in having large fenestrations in the shield margin.

In addition to these five major monophyletic groups, there are some genera of uncertain affinities, such as *Waengsjoeaspis*, *Tannuaspis*, *Ilemoraspis*, *Didymaspis*, and *Escuminaspis* the last-named being unique among cornuate osteostracans in having no lateral 'fields' (probably owing to secondary loss) and an acellular exoskeleton.

The phylogeny of the osteostracans is relatively well corroborated so far as the non-cornuate forms or the major cornuate groups are concerned. The relationships of the latter however, are, still only tenuously supported. The thyestiids and kiaeraspidids, for example, share only the lack of a medially directed posterior ventral myodome (Fig. 4.18, 7). Noteworthy is the fact that all these major cornuate taxa, except the cephalaspidids, which retain a large median dorsal scute (Fig. 4.18D1), do not show any trace of the anterior dorsal fin (Fig. 4.18, 5).

The classification (by phyletic sequencing) of the Osteostraci derived from the cladogram in Fig. 4.18 is shown in Table 4.4.

Table 4.4. Classification of the Osteostraci (See also Fig. 4.18)

Osteostraci
Ateleaspis
Hirella
Hemicyclaspis
Cornuata
Cephalaspidida
Zenaspidida
Benneviaspidida
Kiaeraspidida
Thyestiida

4.8.5 Stratigraphical and geographical distribution, habitat

The earliest known osteostracans are Early Silurian (Wenlockian) in age. They comprise generalized, primitive forms, like *Ateleaspis*, as well as derived cornuates, like tremataspids. This suggests that most of osteostracan history and diversification occurred earlier, in Llandoverian or perhaps Ordovician times. Osteostracans remain rare until the Early Devonian, when a wide range of morphologies suddenly appear in the record, in association with the extension of the Old Red Sandstone facies. By Middle Devonian times, most of the major cornuate monophyletic groups have disappeared, and only cephalaspidid or zenaspidid-related forms survive until the Frasnian. The latest known osteostracan is *Escuminaspis*, from the Frasnian of Miguasha, Quebec.

Osteostracans are known exclusively in North America, Europe, the Ukraine, and Russia, the easternmost forms being the tannuaspids from Tuva in Central Asia. They display well-defined patterns of endemism. The thyestiids, for example, are known only in Britain, the Baltic, Timan, and the Urals. The Boreaspididae occur only in Spitsbergen and Severnaya Zemlya.

Most osteostracans lived in quiet marine environments, preferably in lagunas, tidal flats, or deltas, but it is not ruled out that some of them could live in rivers. The shield shape and ornamentation in some taxa (olive-shaped in tremataspids, roof-shaped in zenaspidids) may reflect adaptations to resist frequent flows of sand or mud. Some osteostracans probably lived half-buried in the sediment, but the absence of any dorsal incurrent opening suggests that they could not spend much time in this position without damaging their gills. The long and fragile rostral and cornual processes of the Boreaspididae are more suggestive of nectonic habits, the effect of these processes being perhaps to make the head appear larger to a predator. Osteostracans were microphagous, possibly suspension-feeders. They may have been outcompeted by the antiarchs (see Chapter 7, p. 228), which occupied very much the same niche by Middle Devonian times.

Further readings

Anatomy, histology, and interpretation: Adrain and Wilson (1994); Allis (1931); Damas (1954); Denison (1947, 1951); Gross (1956, 1961a, 1968c); Heintz (1939, 1967); Janvier (1975a, 1978a, 1981a,b, 1984, 1985a–c, 1993); Janvier *et al.* (1991); Kiaer (1928); Ørvig (1951, 1968a); Ritchie (1967); Stensiö (1927, 1932, 1964, 1968); Watson (1954); Wängsjö (1952); White and Toombs (1983); Whiting (1977). *Interrelationships and systematics*: Afanassieva (1991); Afanassieva and Janvier (1985): Forey (1987a); Heintz (1939); Janvier (1985a,b); Mark-Kurik and Janvier (1995); Novitskaya and Karatayute-Talimaa (1988); Wängsjö (1952); Stensiö (1932).

4.9 GALEASPIDA

The Galeaspida are armoured jawless vertebrates with a massive endo-and exoskeletal head-shield, like osteostracans. They are characterized by a large median dorsal opening in the anterior part of the head-shield, which communicates ventrally with the underlying oralo-branchial cavity (Fig. 4.19, 1). Another galeaspid charac-teristic is the festooned pattern of the main and infraorbital lateral-line canals on the dorsal side of the head-shield (Fig. 4.19, 8, 11).

Most galeaspids have an elongated, oval head-shield (Fig. 4.19A), but some have a shorter, horseshoe-shaped shield which recalls that of osteostracans (Fig. 4.19E); others have long rostral and cornual processes (Fig. 4.19C, D, 13, 14).

4.9.1 Exoskeleton and external morphology

Complete articulated galeaspids are so far virtually unknown, or are so poorly described that it is not yet possible to decipher the precise aspect of the body and tail. The body was covered with minute, button-shaped scales with a bulging base and a single external tubercle (Fig. 4.19, 15). Galeaspids had no paired fins, but had instead a ventrolateral ridge of scales, as in osteostracans. The tail seems to be preserved in a specimen of *Sanqiaspis*, and displays scaly digitations like that of heterostracans, but it is not known whether it is diphycer-cal, hypo- or epicercal (Fig. 4.19, 21).

The head-shield morphology will be exemplified here by three genera, *Hanyangaspis*, *Polybranchiaspis*, and *Eugaleaspis*, which illustrate the diversity of head shape in the group.

Hanyangaspis (Fig. 4.19B), from the Silurian of the Wuhan province, in China, is regarded as the most

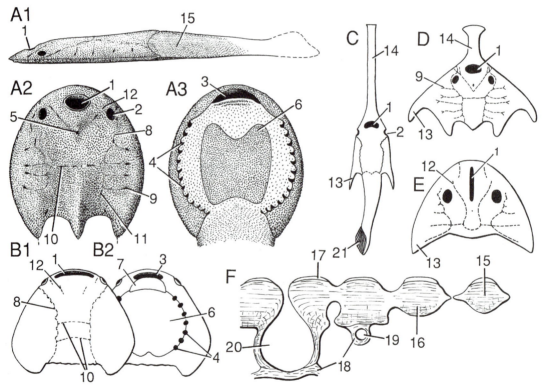

Fig. 4.19. Galeaspida. External morphology. A, *Polybranchiaspis*, Early Devonian of China, reconstruction in lateral view (A1), head in dorsal (A2) and ventral (A3) view (×1); B, *Hanyangaspis*, Silurian of China, head in dorsal (B1) and ventral (B2) view (×0.3); C–E, various morphologies of the head-shield in galeaspids in dorsal view: *Sanqiaspis* (C, ×0.5), *Sanchaspis* (D, ×0.5) and *Eugaleaspis* (E, ×0.5), Early Devonian of China; F, *Bannhuanaspis*, Early Devonian of Vietnam, vertical section through the exoskeleton in the posterior part of the head-shield (×50). 1, median dorsal opening; 2, orbit; 3, mouth; 4, external branchial openings; 5, pineal foramen; 6, median ventral plate; 7, oral plate; 8, infraorbital sensory line; 9, lateral transverse sensory lines; 10, commissural transverse sensory line; 11, main lateral sensory line; 12, supraorbital sensory line; 13, cornual process; 14, rostral process; 15, scale; 16, laminar aspidine; 17, enameloid layer; 18, perichondral bone; 19, subaponeurotic blood vessel; 20, lumen of a lateral-line canal; 21, dermal 'fin rays' or zonation. (A, based on Y. H. Liu (1975) and Tong-Dzuy *et al.* (1995); B, from Pan (1984) and S. T. Wang (1986); C, E, from Y. H. Liu (1975); D, from P'an and S. T. Wang (1981); F, from Janvier (1990) and new observations.)

generalized galeaspid, because of its almost terminal and transversally elongated median dorsal opening (Fig. 4.19B1, 1). The shield is roughly trapezium-shaped, with laterally placed orbits but no pineal opening. The dorsal surface of the shield shows no other detail than a uniform ornamentation of stellate tubercles. The canals of the sensory-line system lie against the basal layer of the exoskeleton, and some large pores or slit-like openings on the external surface indicate their position. The lateral-line canals consist of short supraorbital canals that converge towards the pineal region (Fig. 4.19B, 12), and a long, festooned main lateral-line canal which gives off many transverse lateral canals (Fig. 4.19B, 8). The main lateral lines of either sides are united by two commissural transverse lines (Fig. 4.19B, 10), and are continued anteriorly by an infraorbital portion. No externally opened endolymphatic duct is observed in galeaspids, except in *Xiushuiaspis* (Fig. 4.21, 23; Fig. 4.22B). It is probable that this character is more widespread but difficult to observe.

The ventral side of the shield shows a large oralo-branchial fenestra covered with two median plates: a large posterior one (Fig. 4.19B, 6), and a smaller anterior one that lines the mouth (Fig. 4.19, 7) and plays the role of an oral plate. The seven pairs of branchial openings are situated at the junction between the large posterior plate and the rim of the shield (Fig. 4.19B, 4).

Polybranchiaspis (Fig. 4.19A), which is widely distributed in the Early Devonian of South China and northern Vietnam, has a roughly oval head-shield, with a bean-shaped median dorsal opening, which is more dorsally placed than that of *Hanyangaspis* (Fig. 4.19B). The orbits, too, are dorsal in position (Fig. 4.19A2, 2). There is a small, rounded pineal opening (Fig. 4.19A2, 5). The ornamentation consists of minute, closely set tubercles arranged in radiating rows, giving the impression of larger, stellate tubercles. The pattern of the lateral-line canal is much as in *Hanyangaspis*, but the supraorbital lines are longer and meet posteriorly (Fig. 4.19A2, 12), and there is only one transverse commissural line (Fig. 4.19A, 10). Lateral transverse lines arise from the main lateral line (Fig. 4.19A2, 9). As in all other galeaspids, the sensory-line canals are very large (Fig. 4.19F, 20) and open to the exterior by means of a few large slits or pores, generally situated in the distal part of the canals. Ventrally, the oralobranchial chamber is closed by a single large median plate (Fig. 4.19A3, 5), surrounded by minute scales. There are about 13 pairs of branchial notches, which lodged the external branchial openings (Fig. 4.19A3, 4).

Eugaleaspis (Fig. 4.19E), which commonly occurs together with *Polybranchiaspis*, has a horseshoe-shaped shield, with a slit-shaped median dorsal opening that

makes it resemble osteostracans (Fig. 4.19E, 1). For this reason, it was put in this group by Liu (1965) when first discovered. However, the median dorsal opening of *Eugaleapis* extends from the anterior shield margin, up to near the pineal region, and is thus much larger than the osteostracan nasohypophysial opening. Also, as in other galeaspids, it communicates ventrally with the branchial cavity. The pattern of the lateral-line canals differs from that of *Polybranchiaspis* and *Hanyangaspis* by the fewer lateral transverse canals and the junction between the supra-orbital and transverse commissural lines (Fig. 4.19E, 12). Ventrally, the oralobranchial fenestra is rounded in shape and marked by seven or eight branchial notches.

Various galeaspid genera (*Pentathyraspis, Micro-hoplonaspis*) have been claimed to possess large dorsal fenestrations in the dorsal shield (Fig. 4.20, 2). These have been interpreted either as dorsally displaced branchial openings (Fig. 4.20C, 6) or as lateral fields similar to those of osteostracans. There is no evidence in favour of the latter interpretation, and in only one genus, *Microhoplonaspis* (Fig. 4.20B, 2), does this fenestration seem to be lined with perichondral bone, and thus to be a real duct. The most likely interpretation is that these fenestrations are in most instances mere artefacts that are due to the fragility of the exoskeleton in this area of the shield (over the branchial fossae) where the underlying endoskeleton is extremely thin (Fig. 4.20C, 7).

4.9.2 Internal anatomy

The internal anatomy of galeaspids is known in a small number of forms, essentially the Devonian genera *Polybranchiaspis, Duyunolepis* (Fig. 4.21A) and *Paradyunaspis*, and the Silurian *Xiushuiaspis* (Fig. 4.21E) and *Sinogaleaspis*. Many details are still poorly known. The gross morphology of the endoskeletal shield is basically similar to that of osteostracans, since it consists of a single mass of cartilage lined with perichondral bone (hatched in Fig. 4.21D), which surrounds the brain (Fig. 4.21, 11), sensory capsules (Fig. 4.21, 2, 12, 17), nerves, and major vascular trunks (Fig. 4.21, 15, 7). As in osteostracans, too, there is a large oralobranchial cavity (Fig. 4.21D1, 14), which, however, does not seem to be completely closed posteriorly by a postbranchial wall, except perhaps in some Silurian forms. The major difference between the internal anatomy of galeaspids and that of osteostracans is the large median dome-shaped depression of the oralobranchial cavity, which opens dorsally to the exterior by the median dorsal opening and into which open the two separate nasal cavities (Fig. 4.21, 1, 2). The latter also differ from the single median ethmoid cavity of osteostracans in being well separated and connected to the brain cavity by means of two large canals for the

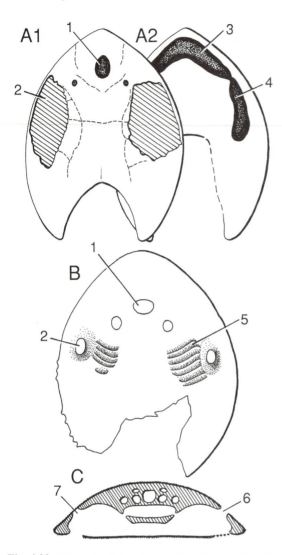

Fig. 4.20. Galeaspida. External morphology. Interpretation of the dorsal fenestrations. A, *Pentathyraspis*, Early Devonian of China, head-shield in dorsal (A1) and ventral (A2) aspect (dorsal fenestrations hatched, ×1); B, *Microhoplonaspis*, Early Devonian of China, head-shield in dorsal view, with cast of oralobranchial chamber partly exposed (×2); C, transverse section through a 'fenestrate' galeaspid to show alternative interpretations: true dorsal branchial opening on the right-hand side; and artefact on the left-hand side (exoskeleton black, endoskeleton hatched). 1, median dorsal opening; 2, dorsal fenestration; 3, mouth; 4, fenestra for the series of external branchial openings; 5, branchial fossae; 6, dorsal external branchial opening; 7, area devoid of endoskeleton. (A, B, from Pan (1992).)

olfactory tracts (Fig. 4.21, 18). The orbital cavities are small, with no clear myodomes. The labyrinth cavity (Fig. 4.21, 17) has a small vestibular division, two vertical semicircular canals and ampullae (Fig. 4.21C, 24, 25), and a dorsal endolymphatic duct (Fig. 4.21, 23), which generally does not open to the exterior. There is no evidence of the large 'sel' canals that, in osteostracans, lead to the cephalic fields. The brain cavity is roughly similar to that of osteostracans, with a paired dorsal swelling attributed to the metencephalon (Fig. 4.21E, 22). The cranial nerves are still poorly known and their identification (Fig. 4.21, 13, 19) is to be taken with reservations. Finally, galeaspids have no marginal vein or artery, and no subclavian artery, since they have no paired fin.

The two genera, *Duyunolepis* and *Xiushuiaspis*, illustrated in Fig. 4.21 show two different states of the oralobranchial cavity. *Xiushuiaspis* is one of the most generalized galeaspids and has only eight pairs of branchial fossae (Fig. 4.21E, 3), whereas *Duyunolepis* has about 20 pairs (Fig. 4.21A, 3), and the closely related genus *Paraduyunaspis* has about 24 pairs. As in osteostracans, the branchial fossae are separated by interbranchial ridges (Fig. 4.21, 4); their lateral ends are produced into a small process for the attachment of the cartilaginous branchial arches, as in osteostracans (Fig. 4.21, 5). The branchial nerves entered the oralobranchial cavity in the same way as in osteostracans, but more medially. The extrabranchial veins emptied into the dorsal jugular vein (Fig. 4.21, 8)—a condition which suggests that there was no anterior cardinal veins. The efferent branchial arteries (Fig. 4.21, 9) emptied into a median dorsal aorta, which was housed in a canal overlying the oralobranchial cavity (Fig. 4.21, 15). The dorsal jugular veins drained two large venous sinuses (Fig. 4.21, 6), situated behind the eyes.

The major question about the internal anatomy of galeaspids is that of the nature of the median dorsal opening and the duct by which it communicated with the oralobranchial cavity. The dorsal position of this median dorsal opening suggests immediately a nasohypophysial opening, like that of osteostracans, lampreys, or anaspids, and this is how it was first interpreted, before details of its internal structure were known. It is, however, more closely comparable to the nasopharyngeal duct of hagfishes because it communicates with the gill chamber. The surface of this dorsal duct in galeaspids is covered with minute sharp tubercles that point toward the exterior (Fig. 4.21D1, 1; Fig. 4.25, 9). Since the orientation of such tubercles or denticles in water-conducting ducts of fishes is always against the main water flow (supposedly to repel ectoparasites), we can assume that the median dorsal duct of galaspids was inhalent in function, as is the

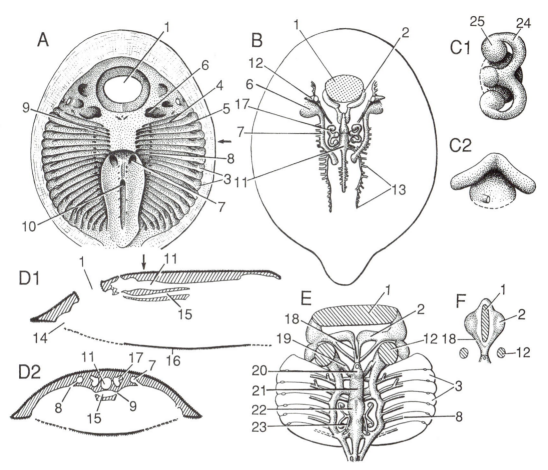

Fig. 4.21. Galeaspida. Internal anatomy. A–E, *Duyunolepis*, Middle Devonian of China; A, head-shield in ventral view (×0.7); B, internal cast of the internal canals and cavities of the shield, in dorsal view; C, labyrinth cavity in dorsal (C1) and lateral (C2) view (×4); D, sagittal (D1) and transverse (D2) section through the head at levels indicated by arrows (exoskeleton black, endoskeleton hatched). E, *Xiushuiaspis* (*Changxingaspis*), Early Silurian of China, internal cast of the cavities and canals in the head-shield and of the underlying cast of the oralobranchial cavity, in dorsal view (×2.5); F, *Eugaleaspis*, Early Devonian of China, position of the nasal cavities relatively to the median dorsal opening. 1, median dorsal opening; 2, nasal cavities; 3, branchial fossae; 4, interbranchial ridge; 5, processes for the attachment of the branchial arches; 6, sinus of the dorsal jugular vein; 7, dorsal jugular vein; 8, foramen for extrabranchial veins; 9, foramen for efferent branchial artery; 10, canal for the spinal cord; 11, brain cavity; 12, orbital cavity; 13, canal for the vagus nerve; 14, mouth; 15, canal for the dorsal aorta; 16, median ventral plate; 17, labyrinth cavity; 18, olfactory tract; 19, trigeminal cavity; 20, habenula; 21, mesencephalon; 22, metencephalon; 23, endolymphatic duct; 24, vertical semicircular canal; 25, ampulla. (A, B, C, modified from P'an and S. T. Wang (1978) and Janvier (1984); E, F, from N. Z. Wang (1991).)

nasopharyngeal duct of hagfishes. Whatever may have been the embryonic origin of this dorsal duct (endo- or ectodermal; see Chapter 3, p. 47), it can be considered as a prenasal sinus, since the olfactory organs open into its posterior wall (Fig. 4.21, 2).

The shape of the cavities for the olfactory organs and the canals for the olfactory tracts is accurately known in the two Silurian genera *Xiushiuaspis* and *Sinogaleaspis*

(Fig. 4.21E, F, 2). In the former, they display a remarkable resemblance to the olfactory impressions of the heterostracan dorsal shield (Fig. 4.5A, 1). This leads quite naturally to the hypothesis that the median dorsal opening and duct of *Xiushuiaspis* or *Hanyangaspis*, which is almost terminal in position, would be readily adaptable to the condition in heterostracans, which may thus support the reconstruction of an inhalent prenasal sinus in heterostracans (Fig. 4.5B, C, 14). As will be discussed in the next chapter (p. 264), this condition can be regarded as general for the craniates, although the presence of two separate nasal cavities is possibly a derived condition, unique to the vertebrates.

4.9.3 Histology

The entire exoskeleton of galeaspids consist of more-or-less fused, minute units, that are totally different from the larger tesserae of osteostracans. In the head-shield, these units are fused side by side into a continuous layer and display, in vertical section, a laminar structure (Fig. 4.19, 16). There is no indication of dentine or cellular bone, but in some genera the external elevation, or tubercle, in the centre of each unit is covered with a thin layer of a more transparent tissue that looks like enameloid (Fig. 4.19, 17). This structure, with an enameloid cap that rests directly on aspidine-like acellular bone, closely resembles that of the Ordovican *Astraspis*. In some taxa, such as *Polybranchiaspis*, the exoskeleton is hollowed by large, polygonal cavities that mimic the honeycomb-like structure of the heterostracan exoskeleton, yet are probably not homologous to it. At the boundary between the exo- and endoskeleton, there is, as in osteostracans, a dense subaponevrotic vascular network (Fig. 4.19, 19), and the large canals of the lateral sensory-line system, which are partly lined with perichondral bone (Fig. 4.19, 18, 20). The structure of the endoskeleton seems to have been identical to that of osteostracans, that is, composed of cartilage lined with a thin layer of perichondral bone (which, however, has not yet been shown to be cellular).

The scales have the same structure as the units of the exoskeleton of the shield, and in many galeaspids, such as *Hanyangaspis* or *Bannhuanaspis*, the posterior part of the shield passes progressively into the body squamation, with no clear break (Fig. 4.19F, 15).

4.9.4 Diversity and phylogeny

If one considers *Hanyangaspis* (Fig. 4.22A) as the most generalized galeaspid, because of its broad and almost terminal median dorsal opening, and laterally placed eyes, then galeaspids may fall into two major clades, the Eugaleaspidiformes (Fig. 4.22C) and the Huananaspidiformes (Fig. 4.22G), and a probably paraphyletic ensemble, often referred to as the Polybranchiaspidiformes (Fig. 4.22D, E, F).

The Huananaspidiformes are characterized by a long rostral process and slender cornual processes, which parallels the condition in the boreaspid osteostracans. *Sanquiaspis* (Fig. 4.22G1), *Gantarostrataspis* (Fig.

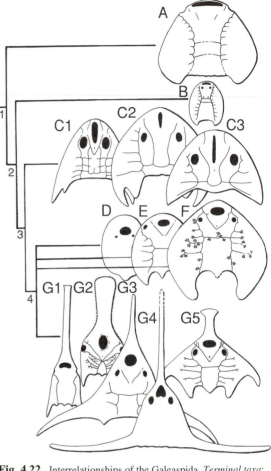

Fig. 4.22. Interrelationships of the Galeaspida. *Terminal taxa*: A, *Hanyangaspis*; B, *Xiushuiaspis* (*Changxingaspis*); C, Eugaleaspidiformes (C1, *Sinogaleapis*; C2, *Yunnanogaleaspis*; C3, *Eugaleaspis*); D, *Duyunolepis*; E, *Polybranchiaspis*; F, *Dongfangaspis*; G, Huananaspidiformes (G1, *Sanqiaspis*; G2, *Gantarostrataspis*; G3, *Asiaspis*; G4, *Lungmenshanaspis*; G5, *Sanchaspis*). *Nested taxa and synapomorphies*: 1, Galeaspida (large median dorsal inhalent opening, festooned pattern of main lateral-line canal); 2, narrower median dorsal opening, dorsally placed orbits (reversion in some Huananaspidiformes); 3, 'higher galeaspids' (single transverse commissural sensory-line canal); 4, Polybranchiaspidida (more than 10 branchial fossae). *Age*: Silurian (A); Early Silurian (B, C1); all other taxa Early Devonian. (A, from Pan (1984) and N. Z. Wang (1986); B, from N. Z. Wang (1991) and Pan and S. T. Wang (1983); C1, C2, from P'an and S. T. Wang (1980); C3, E–G1, from Y. H. Liu (1975); D, from P'an and S. T. Wang (1978); G2, from J. Q. Wang and N. Z. Wang (1992); G3, G4, from P'an *et al.* (1975); G5, from Pan and S. T. Wang (1981).)

4.22G2), and *Sanchaspis* (Fig. 4.22G5) have a slightly spatulate rostral process, and *Lungmenshanaspis* (Fig. 4.22G4) has forwardly recurved cornual processes. They also display a very large number of branchial fossae and gill opening, a character shared with 'polybranchiaspidiforms'. Their median dorsal opening is bean-shaped, rounded, or heart-shaped. In some forms, such as *Wumengshanaspis*, the ventral dermal covering is completely fused with the shield, leaving holes only for the mouth and branchial openings.

The Eugaleaspidiformes have a pear- or slit-shaped median dorsal opening, and a shortened head-shield with small cornual processes, which vaguely recalls that of osteostracans. They have only seven or eight external branchial openings. In *Sinogaleaspis* (Fig. 4.22C1) and *Yunnanogaleaspis* (Fig. 4.22C2), there are distinct cornual processes, which are reduced in *Eugaleaspis* (Fig. 4.22C3).

The 'Polybranchiaspidiformes', such as *Polybranchiaspis* (Fig. 4.22E), *Duyunolepis* (Fig. 4.22D), or *Dongfangaspis* (Fig. 4.22F) have an oval to rounded head-shield, with an oval or bean-shaped median dorsal opening, and numerous branchial fossae and gill openings as in the Huananaspidiformes. This shield shape may, however, be a general character, for it occurs in some of the most generalized galeaspids, such as *Xiushuiaspis* (Fig. 4.22B). One may nevertheless gather most—but not all—polybranchiaspiforms in a group characterized by the star-shaped pattern of the lateral extremity of the transverse sensory-line canals, such as *Dongfangaspis* (Fig. 4.22F). The largest-known galeaspids, reaching about 20 cm in shield length, are found among the Devonian 'polybranchiaspidiforms'.

These three major groups all share the presence of a single commissural sensory-line canal on the shield (Fig. 4.19, 10; Fig. 4.22, 3), whereas two such lines occur in *Hanyangaspis*, *Xiushuiaspis*, and *Dayongaspis*, all Silurian in age. The presence of numerous commissural transverse lines is regarded here as general for the vertebrates, since it occurs in the most generalized heterostracans (Fig. 4.7A, 4) and in thelodonts. Their interrelationships are, however, still obscure. Eugaleaspidiforms and huananaspidiforms share cornual processes, but 'polybranchiaspidiforms' and huananaspidiforms share a large number of branchial fossae and gill openings (Fig. 4.22, 4). Stratigraphy would rather suggest that the eugaleaspidiforms diverged quite early, before the two other groups appeared. In fact, no Silurian galeaspid displays the 'polybranchy' (more than eight to ten gill openings) that is the most common condition in Devonian galeaspids (except eugaleaspidiforms).

The cladogram of the Galeaspida in Fig. 4.22 is based on the assumption that a homoplasy on the cornual process in the eugaleaspidiforms and huananaspidiforms is more likely than on the 'polybranchial' condition, which implies important modification in the branchial blood circulation and innervation. Therefore, 'polybranchiaspidiforms' and huananaspidiforms are regarded as a taxon, the Polybranchiaspidida (Fig. 4.22, 4). The classification that can be derived (by phyletic sequencing) from this cladogram is shown in Table 4.5.

Table 4.5.   Classification of the Galeaspida
(See also Fig. 4.22)

Galeaspida
   Hanyangaspis
   Xiushuiaspis
   'higher galeaspids' (unnamed taxon)
      Eugaleaspidiformes
      Polybranchiaspidida
         'Polybranchiaspidiformes'
         Huananaspidiformes

4.9.5 Stratigraphical and geographical distribution, habitat

The earliest known galeaspid is *Dayongaspis*, from the Early Silurian (Llandovery) of Hunan. *Hanyangaspis*, *Xiushuiaspis*, and *Sinogaleaspis* are from the Early Silurian (Wenlock) of Wuhan and Jiangxi. Most other galeaspids are from the Early Devonian of Yunnan, Guizhou, Guangxi, and Sichuan in China, and the Bac Bo (Tonkin) in Vietnam. The youngest known galeaspid is a large, unnamed form from the red sandstone of Ningxia, northern China, which is found in association with the antiarch *Remigolepis* and is thus dated as Early Famennian or Late Frasnian.

Galeaspids are known exclusively from China and northern Vietnam. Most occurrences are in the South China Block, but some are in the North China block and even in western Tibet (Xinjiang). Their distribution seems, on the present evidence, to have been endemic to these areas.

Most galeaspids were benthic forms, living on sandy or muddy substrates in marginal, marine environments (deltas, lagoons). Some forms (*Duyunolepis*, *Paraduyunaspis*) occur in frankly marine deposits, in association with articulate brachiopods (S. T. Wang 1991). The shape of the shield in some huananaspidiforms, such as *Sanqiaspis* or *Lungmenshanaspis*, may be interpreted in the same way as that in boreaspid osteostracans (see Chapter 7, p. 300), i.e. suggestive of more nectonic habits.

Further reading

Anatomy, histology, and interpretation: Halstead (1979); Halstead *et al.* (1979); Janvier (1975*a*, 1984, 1990, 1993); Janvier *et al.* (1993); Y. H. Liu (1975, 1986); Pan (or P'an (1984, 1992); Pan and Chen (1993); Pan and S. T. Wang (1978); Tong-Dzuy *et al.* (1994); N. Z. Wang (1991). *Interrelationships and systematics*: Janvier (1984); Liu (1965, 1973, 1975, 1986); Pan (or P'an) (1984, 1988, 1992); Pan and Dineley (1988); Pan and S. T. Wang (1980, 1981, 1983); Pan *et al.* (1975); N. Z. Wang (1986, 1991); J. Q. Wang and N. Z. Wang (1992); Zhu (1992).

4.10 PITURIASPIDA

The Pituriaspida are poorly known armoured jawless vertebrates from the Lower or Middle Devonian of Australia. The group is so far known by only a few speci-

mens preserved in impression in sandstone, and recorded from a single locality of the Georgina Basin in western Queensland. They have been referred to two different genera, *Pituriaspis* (Fig. 4.23A) and *Neeyambaspis* (Fig. 4.23B), the former being by far the best known. The two genera look quite different, despite the presence of a long rostral process in both (Fig. 4.23, 7). *Neeyambaspis* seems to have possessed a large median opening (Fig. 4.22, 14) which, if dorsal in position, strongly recalls that of galeaspids.

Pituriaspis has an elongated shield that includes a long abdominal division, with a ventral oralobranchial fenestra (Fig. 4.23, 4), short cornual processes (Fig. 4.23, 8), and a distinct area for the attachment of paired fins (Fig. 4.23, 6). In ventral aspect, it is thus quite similar to an osteostracan. In contrast, the dorsal side of the head-shield is totally different from that of osteostracans. The orbits

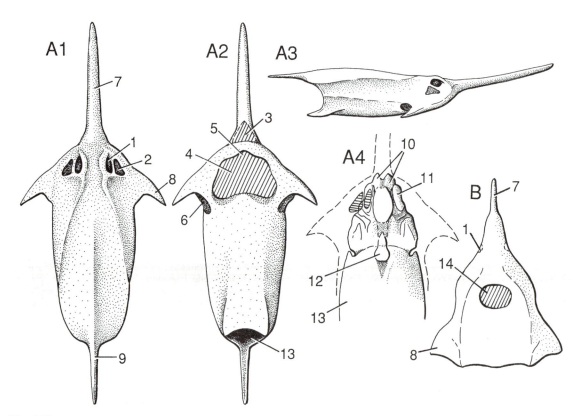

Fig. 4.23. Pituriaspida. Morphology and internal anatomy.
A, *Pituriaspis*, Early Devonian of Australia, head-shield in dorsal (A1, ×0.6), ventral (A2) and lateral (A3) view, and reconstruction of the cast of the internal cavities in dorsal view (A4). B, *Neeyambaspis*, Early Devonian of Australia, as interpreted by Young (1991*a*) (×0.5). 1, orbit; 2, adorbital depression, possibly for some sense organ; 3, presumed position of olfactory organ; 4, oralobranchial cavity; 5, oral notch; 6, area of attachment for the paired fins; 7, rostral process; 8, cornual process; 9, spinal process; 10, ?telecephalon or olfactory tract; 11, orbital cavity; 12, ?brain cavity; 13, abdominal cavity; 14, presumed mouth opening. (Based on Young (1991*a*).)

are situated well apart and face almost laterally (Fig. 4.23, 1), with a peculiar adorbital pit of unknown function (Fig. 4.23, 2). There is neither a pineal foramen nor a naso-hypophysial opening, and the nasal apertures may have been situated ventrally, near the base of the rostral process (Fig. 4.23, 3). The abdominal division (Fig. 4.23, 13) is prolonged posteriorly by a long horizontal spinal process (Fig. 4.23, 9) and ventrally by an expanded subanal lamina.

The fact that some internal structures are preserved as natural casts suggests that there was a perichondrally ossified or calcified endoskeleton, as in osteostracans and galeaspids, but these vague indications of the internal anatomy of the head are difficult to interpret. There is only an indication that there were two separate olfactory bulbs (Fig. 4.23, 10), and the metencephalic division of the brain cavity, if correctly interpreted, may have been as in osteostracans, galeaspids, and heterostracans (Fig. 4.23, 12). The exoskeleton is ornamented with minute rounded tubercles, but nothing is known of its histology.

Pituriaspis shares with the osteostracans and gnathostomes the well-defined pectoral fins, and with the latter two and the galeaspids the perichondrally ossified endoskeleton. It is thus provisionally placed close to these three groups (see Chapter 5, p. 240).

The two pituriaspid genera occur together in supposedly deltaic or freshwater deposits, in association with placoderms (e.g. *Wuttagoonaspis*) and thelodonts.

Further reading

Young (1991*a*)

4.11 THELODONTI

The thelodonts are presumably jawless vertebrates that possessed an almost entirely micromeric and micro-squamose exoskeleton. They can be characterized by their growing scale base (Fig. 4.26, 1), which often develops anchorage devices in the form of outgrowths or processes. Although thelodont scales are fairly abundant in Silurian and Early Devonian marine sediments, only a few genera are known as complete individuals. Even in such cases, the absence of ossified endoskeleton and of large dermal plates makes them difficult to reconstruct in three dimensions. They are generally dorsoventrally flattened when preserved, with the tail turned on one side (Fig. 4.24D), and this suggests that they were in fact rather depressed in shape. Recent discoveries of excellently preserved thelodonts from the Early Silurian and Early Devonian of Canada have, however, shown that some of them could be deep-bodied forms (Fig. 4.24E, F). Despite some

common—yet not general—features of scale morphology, it is not certain that thelodonts, as currently defined, represent a clade. Some of them may be more closely related to some of the major vertebrate clades than others.

4.11.1 Exoskeleton and external morphology

The morphology of the thelodonts will be exemplified here by three of the best-preserved forms known at present: *Lanarkia* (Fig. 4.24C) *Loganellia* (Fig. 4.24A), *Phlebolepis* (Fig. 4.24B), and some unnamed forms from Canada (Fig. 4.24E, F).

Lanarkia, from the Silurian of Europe, is a small thelodont with broadly separated eyes, small triangular lateral flaps, and a forked tail. The scale covering consists of spine-shaped scales, some of which are very large and aligned in longitudinal series (Fig. 4.24C, 10). The gill openings are supposed to have been situated on the ventral side, medially to the paired flaps.

Loganellia, also from the Silurian of Europe, has very much the same aspect as *Lanarkia*, but the scales are smaller, with a bulging base. It clearly displays seven or eight external gill openings (Fig. 4.24A, 7), which are situated ventrally to the paired flaps or fins (Fig. 4.24A, 6). The eyes are surrounded by two larger, crescentiform circumorbital scales (Fig. 4.24A, 1). There are one dorsal and one anal fin (Fig. 4.24A, 2, 3). The tail is also forked, with a somewhat longer ventral lobe, which contained the notochord (Fig. 4.24A, 5). The caudal fin web (epichordal lobe) shows radiating zones of enlarged scales that are comparable with the 'digitations' in the heterostracan tail (Fig. 4.24A, 4).

Phlebolepis, from the Silurian of Estonia, is a small form with boat-shaped scales hollowed by a very large pulp cavity (Fig. 4.25F, 3). The eyes are also surrounded by two large circumorbital scales. The unpaired fins and tail are as in *Loganellia*. *Phlebolepis* is the only form in which the pattern of the sensory-line system has been described in detail, although this sensory system was certainly present in most, if not all, thelodonts. It consisted of closed canals that opened to the exterior by minute pores piercing the scales (Fig. 4.24B, 8). As in most other vertebrates, there were divergent supraorbital lines and roughly parallel longitudinal lines united by transverse commissures. The sensory lines were present on both the dorsal and ventral side, and extended to the tail.

The newly described—as yet unnamed—forms from the Silurian and Early Devonian of the Canadian Arctic seem to deep-bodied. They display a wide diversity in the shape and distribution of the unpaired fins. Some have no dorsal or no anal fin, and the epichordal lobe of the tail is considerably enlarged, with traces of thick and scarce fin

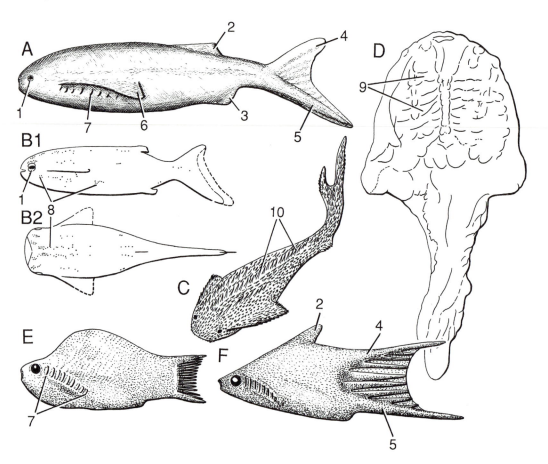

Fig. 4.24. Thelodonti. External morphology and internal anatomy. A, *Loganellia*, Late Silurian of Scotland, reconstruction in lateral view (about ×0.5); B, *Phlebolepis*, Late Silurian of Estonia, reconstruction in lateral (B1) and dorsal (B2) view (about (×0.5); C, *Lanarkia*, Early Silurian of Scotland, reconstruction in dorsolateral view (about ×1); D, *Turinia*, Early Devonian of Scotland, complete specimen in dorsal view, showing impressions of some internal structures (×0.5); E, F, unnamed thelodonts from the Late Silurian and Early Devonian of Canada, reconstruction in lateral view (about ×1). 1, circumorbital scales; 2, dorsal fin; 3, anal fin; 4, epichordal lobe; 5, chordal lobe; 6, paired flaps or fins; 7, external branchial openings; 8, pores of the sensory-line canals; 9, ?gill pouches; 10, spine-shaped scales. (A, modified from Turner (1970); B, from Märss (1979); C, from Turner (1992); D, from Turner (1991); E, F, from Wilson and Caldwell (1993).)

rays (Fig. 4.24F, 4). In the earliest of these forms, the tail is diphycercal and quite similar in shape to that of many heterostracans (Fig. 4.24E). The form of the tail in these two Canadian forms is likely to be a particular condition, relatively to that of other thelodonts in which the chordal lobe (Fig. 4.24A, 5) is longer than the epichordal lobe.

4.11.2 Internal anatomy

Virtually nothing is known of the internal anatomy of the thelodonts, apart from some imprints observed in a few specimens. Some specimens of *Loganellia*, for instance,

are known to display imprints of internal structures such as the eyes (Fig. 4.25, 2), olfactory organ (Fig. 4.25, 1), braincase, and 'branchial basket', the last-named being quite similar to that observed in *Jamoytius* (Fig. 4.12A, 1). Laterally placed gill openings, as well as large eyes and a peculiar cylindrical 'stomach' occur also in the Canadian forms referred to above (Fig. 4.24E; Fig. 7.9A).

An exceptionally well-preserved large specimen of *Turinia* from the Silurian of Scotland shows faint impressions of the branchial apparatus (gill pouches?), in the form of paired undulations of its surface (Fig. 4.24D, 9). Nevertheless, none of these data on thelodont internal anatomy provides important information about their

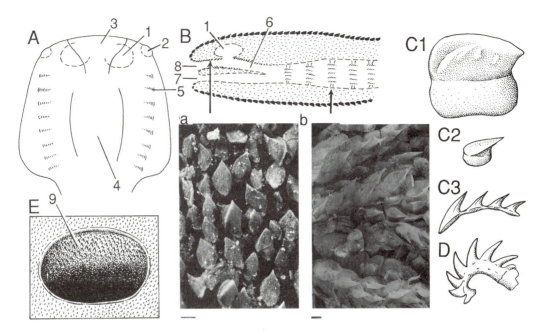

Fig. 4.25. Thelodonti. Internal anatomy and interpretation. A–C, *Loganellia*, Late Silurian of Scotland; A, outline of the head, showing the position of the three types of internal denticles (3–5); B, hypothetical parasagittal section through the head, showing the position of the forward-pointing denticles (a) and the denticle whorls and platelets (b) (exoskeleton black, endoskeleton and soft parts stippled; scale bar: 0.1 mm); C, comparative size of an external scale (C1), a forward-pointing denticle (C2), and a denticle whorl (C3), brought to the same scale (about ×10). D, branchial tooth whorl of the Carboniferous chondrichthyan *Hamiltonichthys* (×12). E, detail of the median dorsal opening of the galeaspid *Polybranchiaspis*, showing the forward-pointing tubercles on the walls of the canal (×3). 1, olfactory organs; 2, eyes; 3, field of forward pointing denticles; 4, field of backward-pointing denticles; 5, denticle whorls associated to gill units; 6, nasopharyngeal duct (hypothetical); 7, mouth; 8, prenasal sinus; 9, forward-pointing tubercles. (A, from Van der Brugghen and Janvier (1993), a, b, courtesy of W. Van der Brugghen, Almere Stad); D, from Maisey (1989*b*).)

affinities. All that can be said is that the endoskeleton was not ossified and did not expand over the branchial apparatus in the form of a shield. The braincase, if correctly interpreted from imprints, may have been more developed than in lampreys, and its outline is more suggestive of that of a shark or a placoderm. The olfactory organ seems to have been paired. Acid preparation of well-preserved specimens of *Loganellia* has recently, yielded a variety of internal denticles, the size of which is about a quarter of that of the external scales (Fig. 4.25B, C2). The organization of these denticle-bearing areas may provide some information on the internal structure of the thelodonts. In the snout (Fig. 4.25, 3), these internal denticles point forward (Fig. 4.25Ba), an orientation that is incompatible with teeth or pharyngeal denticles, which usually point backward to convey the food particles towards the digestive tract. These forward-pointing denticles recall the outward-pointing tubercles of the median dorsal duct of galeaspids (Fig. 4.25E, 9) and suggest that at least some thelodonts possessed a similar median terminal inhalent duct (Fig. 4.25B, 8) above the mouth. If this interpretation

is correct, then the large oval opening in the anterior part of the head of many articulated thelodonts would not be the mouth, as is usually thought, but the opening of the prenasal sinus. Further posteriorly, in the pharynx (Fig. 4.25, 4), the minute denticles point backwards, and laterally there are assemblages of slender denticles (Fig. 4.25A, 5), sometimes fused into curved series or platelets (Fig. 4.25Bb, C3). These recall the pharyngeal tooth whorls of various early gnathostomes (Fig. 4.25D), and may thus be associated with either the gill septa or the gill covers. *Loganellia* is the only known jawless craniate in which such an extensive internal dermal skeletal covering occurs. This condition is suggestive of the gnathostomes and may be considered to favour the theory that some thelodonts may be more closely related to the gnathostomes than to any other vertebrate taxon.

The question whether the nasal cavities opened into the median prenasal sinus or directly outside by nostrils (as in gnathostomes) is not answered. Both conditions may have been present if thelodonts are non-monophyletic and if a median nasopharyngeal duct is regarded as a general

concern essentially the suspension of the mandibular arch and its relationship to the hyoid arch, the structure of the gill arches, the neurocranial fissures, the structure of the vertebrae and fins, and the histological structure of the skeleton.

4.12.1 The neurocranial fissures

There are two major types of braincase, or neurocranium, in the gnathostomes. That of osteichthyans and the fossil

acanthodians (Fig. 4.27C–E) is rather narrow and deep, with ventrally open orbital cavities, and two transverse fissures: the otico-occipital fissure posteriorly, and the ventral fissure anteriorly (Fig. 4.27, 3, 4). The otico-occipital fissure may be continued further ventrally as far as the ventral otic notch. In sarcopterygians there is generally a complete separation of the braincase into two halves, at a level that is believed by some to be that of the ventral fissure (Fig. 4.27E, 3). In contrast, the braincase of chondrichthyans is a single, broad mass of cartilage, and

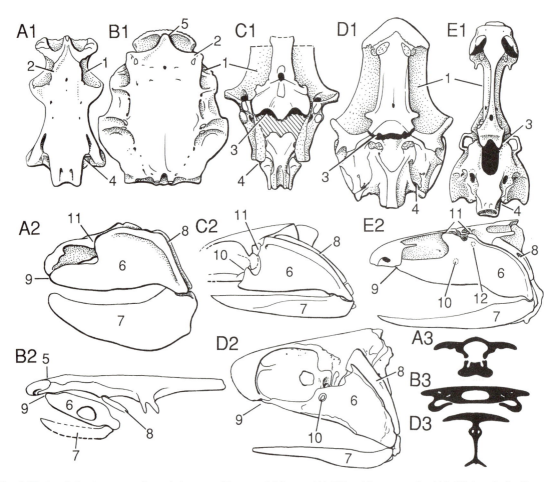

Fig. 4.27. Fossil Gnathostomata. General characters. Neurocranial fissures (A1–E1) and jaw suspension (A2–E2) in early fossil gnathostomes. A1–E1, outline of the braincase of early gnathostomes in ventral view, showing the position of the major neurocranial fissures; A2–E2, braincase, mandibular, and hyoid arches in lateral view; A3, B3, D3, transverse sections through the orbital region of the braincase, showing the difference between the broad-based, chondrichthyan, and placoderm type (A3, B3) and the narrow-based osteichthyan type (D3); A, Chondrichthyes (*Xenacanthus*, Permian); D, Placodermi (*Dicksonosteus*, Devonian); C, Acanthodii (*Acanthodes*, Permian); D, Actinopterygii (*Mimia*, Devonian); E, Sarcopterygii (*Eusthenopteron*, Devonian). 1, orbital cavity; 2, ectethmoid process; 3, ventral fissure; 4, otico-occipital fissure; 5, cranio-ethmoidian (ethmo-otic) fissure; 6, palatoquadrate; 7, Meckelian cartilage or bone; 8, hyomandibula; 9, ethmoid articulations; 10, basal (palatobasal, basipterygoid) articulation; 11, dorsal (postorbital, otic, or suprapterygoid) articulations; 12, paratemporal articulation. (A, from Schaeffer (1981); B, from Goujet (1975, 1984a); C, from Miles (1973b); D, from Gardiner (1984a), E, from Jarvik (1980).)

shows no such fissure, except in the fossil xenacanths, where an otico-occipital fissure is present (Fig. 4.27A, 4). In the fossil placoderms (Fig. 4.27B), the posterior part of the braincase is also unfissured and very broad, but some forms had a separate rhinocapsular ossification containing the olfactory capsules (Figs 4.42, 4.46). Here, the rhinocapsular ossification is separated from the rest of the braincase by the cranio-ethmoidian fissure (Fig. 4.27B, 5). This condition is now regarded as general to all placoderms, but lost in some taxa. In both the chondrichthyans and the placoderms the orbital cavity is lined ventrally by an orbital shelf (Fig. 4.27A, B, 2).

Which of these two types of braincase is general for the gnathostomes may be decided either by ontogenetic inferences or by out-group comparison with extant and fossil jawless vertebrates. The neurocranium of all embryonic gnathostomes shows a zone of 'fragility' at the level of the junction between the trabecles and the otico-occipital region, which is marked by the polar cartilage and lateral commissure (Fig. 4.28, 2, 3). This zone corresponds approximately to the position of the ventral fissure. The otico-occipital fissure is always present as a separation between the embryonic pila occipitalis and the otic capsule. Extant lampreys, as an out-group, have no ventral fissure and no occipital region; i.e. their braincase ends posteriorly before the level of the otico-occipital fissure of the gnathostomes. The consideration of fossil jawless taxa is perhaps more fruitful. Osteostracans and galeaspids, which share the largest number of unique features with the gnathostomes and thus may well be their

closest relatives (see Chapter 5), have an unfissured braincase and a well-developed occipital region, somewhat comparable to that of the chondrichthyans and placoderms. It may thus be concluded that the morphotypic gnathostome braincase is an unfissured mass of cartilage lined with perichondral bone, and that the otico-occipital fissure of the xenacanth sharks on the one hand, and of osteichthyans and acanthodians on the other, is homoplasic. The ventral fissure would be unique to osteichthyans and acanthodians, and the cranio-ethmoidian fissure may be unique to placoderms as well. Young (1986b) has suggested, however, that the rhinocapsular ossification of placoderms is homologous to the nasal capsule of lampreys, which is separated from the rest of the braincase (Fig. 3.3B2), and thus represents a general condition for the gnathostomes. This is inconsistent with the fact that, among the jawless vertebrates which possess an ossified endoskeleton (galeaspids, osteostracans), the nasal capsule is never independent of the rest of the braincase.

As will be summed up in Chapter 6 (p. 259), Bjerring (1977, 1978) proposed a more complex theory for the homologies of the various types of ventral fissure in the gnathostomes. This was based on the assumption that what is commonly termed the 'ventral fissure' may occur between different primordial segments of the craniate skull. According to this theory, the ventral fissure would be homoplasic between actinopterygians and sarcopterygians, and even between various taxa of actinopterygians and sarcopterygians, since it corresponds to different intermetameric joints. However well supported this theory may look in figures, it is largely untestable, because of the unavailability of embryonic data on the fossil taxa or on the extant Actinistian *Latimeria*, the only living vertebrate with an intracranial joint. The conjunction test rather points toward a global homology of the ventral fissure throughout all the osteichthyan taxa, since no known gnathostome clearly possesses several ventral fissures at the same time.

4.12.2 Jaw suspension and teeth

The endoskeleton of the gnathostome jaws comprises the palatoquadrate and Meckel's cartilage (Fig. 4.27, 6, 7). The latter is generally rod-shaped, with a depression for the adductor muscles; the palatoquadrate is generally b-shaped, unless modified by fusion to the braincase or reduced, with a large lateral depression for the adductor muscles. One major exception to this morphology is in the fossil placoderms, where the palatoquadrate is more or less omega-shaped, with a *medial* depression for the adductor muscles (Fig. 4.27B2, 6; Fig. 4.42F). This quite different aspect has been used as an argument for con-

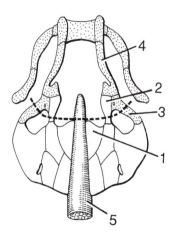

Fig. 4.28. Embryonic skull of a 8 mm embryo of the extant actinopterygian *Amia*, in ventral view (neural crest derivatives shaded). The dashed line indicates the possible position of the ventral fissure in early osteichthyans. 1, anterior parachordal; 2, polar cartilage; 3, lateral commissure; 4, trabecle; 5, notochord. (Modified from Bjerring (1978).)

sidering this group as the sister-group of all other gnathostomes (Schaeffer 1975; Young 1986*b*). It is, however, clear that this condition is merely due to the fusion of the palatoquadrate to a large dermal bone, the suborbital plate, which leaves no space for a laterally placed jaw musculature. This exceptional condition in placoderms is thus likely to be unique to this group.

The suspension of the palatoquadrate to the braincase is effected by means of various processes linking the palato-quadrate to the braincase (Fig. 4.27, 9–12), and also by means of the hyoid arch (Fig. 4.27, 8). Fusion of the palatoquadrate to the braincase in, e.g., holocephalans, lungfishes, and tetrapods clearly represents secondary conditions, often due to durophagous diet or to the loss of the suspensory function of the hyomandibula.

The palatoquadrate can be connected to the braincase in three areas: anteriorly (rostral, ethmoid articulations; Fig. 4.27, 9), dorsally (orbital, postorbital, paratemporal, suprapterygoid articulations; Fig. 4.27, 11, 12), and ven-trally (basipterygoid or palatobasal articulation; Fig. 4.27, 10). In no chondrichthyan do all three articulations coexist. Placoderms have only one or two articulations, which may be the ethmoid (usually rather complex and tripartite) and possibly a more posteriorly placed one, which could be compared to the basipterygoid articula-tion (Fig. 4.27B2). Acanthodians have no ethmoidal articulation (Fig. 4.27C2), and actinopterygians have no postorbital and paratemporal articulation (Fig. 4.27D2). In the latter, the palatoquadrate is attached to the epihyal, which plays the major role in the dorsal articulation of the mandibular arch to the braincase. In acanthodians, only the postorbital and basipterygoid articulations usually occur. There is a double postorbital articulation in *Acanthodes* (Fig. 4.27C2, 11), which has been interpreted in several ways. However, considering the fact that *Acanthodes* is the only acanthodian providing reliable data on internal anatomy, and that some sarcopterygians may well have possessed a true postorbital articulation, one can assume that the ethmoid and postorbital articulation are general (morphotypic) gnathostome char-acters, and that the basipterygoid articulation is unique to osteichthyans and acanthodians. Conversely, there may be a far-fetched argument to support the idea that the basipterygoid articulation is general for the gnathostomes, if we assume that the medial velar skeleton of larval lampreys is homologous to the mandibular arch of the gnathostomes (see Chapter 6, p. 258). This part of the velar skeleton and the basipterygoid process are both connected in the same way to the trabecles (Fig. 6.3, 8).

The suspensory function of the hyomandibula (Fig. 4.27, 8), or epihyal, which is connected to the quadrate part of the palatoquadrate by ligaments or inter-vening skeletal elements, is a general gnathostome feature

and occurs in all primitive members of every major group, except perhaps the placoderms, where both the epihyal and the palatoquadrate are firmly attached to the exoskeleton (Fig. 4.42, 10, 14). The connection between the hyomandibula and the palatoquadrate by means of an interhyal is unique to the osteichthyans and acanthodians.

The articulation between the hyomandibula and the braincase has been considered by some to have developed independently several times, by incorporation of hypo-thetical pharyngohyal elements to the braincase, forming the lateral commissure (Fig. 4.28, 3) and the otical shelf (see Chapter 6, p. 257). Such a complicated process is, however, unnecessary to explain the absence of pharyn-gohyals, if one considers that the hyoid arch was initially devoid of such a component and was not constructed exactly like a branchial arch. It now seems that, from the beginning, the hyoid arch had no respiratory function and essentially served the jaw mechanism.

Despite the fact that the oral plates of some jawless craniates (anaspids, osteostracans) may have had a biting function, true teeth, that is dermal elements preformed in a dental lamina, seem to be unique to the gnathostomes. The elasmobranch dentition, i.e. independent teeth that are constantly produced in the dental lamina, and shed when they reach the lingual side of the jaw (Figs 3.13I, 6.16), are generally regarded as the primitive condition for the gnathostomes, because it is 'simple' and can easily be derived from the placoid scale pattern (see chapter 6, p. 278). Small, independent teeth arranged in 'families' (with old and young teeth) are in fact found in most pri-mitive chondrichthyans and acanthodians, and may well be the general gnathostome condition, but the way in which the old teeth are shed in modern sharks (at a very high rate) is a character that appeared only later (Patterson 1992). All early chondrichthyans and acanthodians (climatiids) show a very slow replacement rate, and each tooth family consists of many functional teeth. This is in fact exactly what could be expected from the theory that teeth arose from a microsquamose shagreen such as in the skin of the lip of the thelodonts. Reif (1982*a*) suggested that placoderms did not possess any dental lamina, since they have no true teeth, but denticle-bearing dermal plates that grow by apposition, as denticles are worn. If this is true, then the dental lamina would be a character of all gnathostomes except placoderms (see Chapter 5).

4.12.3 Gill arch structure

Gnathostome gill arches are generally composed of four paired elements: pharyngo-, epi-, cerato-, and hypo-branchials, the last-named articulating with a single element, the basibranchial in osteichthyans, and with sep-arate basibranchials in holocephalans and some elasmobranchs. The pharyngobranchials may be modified, lost, or fused to the epibranchials, in particular in osteichthyans.

The orientation of the hypobranchials deserves some comment: in osteichthyans and acanthodians, the hypobranchials prolongate anteriorly the ceratobranchials and converge toward the basibranchial plate or basibranchial series (Fig. 4.59, 10; Fig. 4.66, 18). In contrast, the hypobranchials 4 to 5 of holocephalans are oriented towards a posteriorly pointed basibranchial copula (Fig. 3.15, 15); in modern elasmobranchs, all but the first hypobranchial are posteriorly directed (Fig. 3.13, 9) towards the basibranchial copula. The latter condition is probably unique to elasmobranchs, and the condition in acanthodians and osteichthyans, with separate basibranchials and forwardly directed hypobranchials, may well be the most generalized gnathostome condition.

The gill rays, which support the interbranchial septa in elasmobranchs and some fossil stem-group chondrichthyans (*Cladoselache*) are extremely long (Fig. 4.30, 16), but less so in holocephalans. In osteichthyans, the interbranchial septa are shorter than the gill lamellae and are divided in sarcopterygians (Fig. 3.10D–F). Out-group comparison with lampreys (Fig. 3.10A), in which the interbranchial septa separate the gill chambers, and with most fossil jawless craniates would suggest that the condition in elasmobranchs is the general gnathostome condition and, consequently, that separate gill slits is general for the gnathostomes. Five such gill slits and corresponding gill arches may also be regarded as the general condition for the gnathostomes, since it occurs in the earliest fossil chondrichthyans (see p. 136). Larger or smaller numbers of gill arches and gill slits in some gnathostomes should be regarded as a particular condition.

4.12.4 Skeletal histology and structure

The exoskeleton of the gnathostomes consists either of minute dermal denticles or scales with a single pulp cavity (chondrichthyans; Fig. 4.29A2) or of large bony plates (osteichthyans, placoderms; Fig. 4.29A3), with intermediate conditions in acanthodians and some fossil chondrichthyans, where larger scutes or spines occur together with minute growing scales (Fig. 4.29A4). Out-group comparison with fossil armoured agnathans, such as osteostracans, would tend to suggest that the large, compound (macromeric) scales and dermal plates are general for the gnathostomes, but the ontogeny of the osteichthyan dermal skeleton shows that the large dermal plates form by accretion of small dentinous dermal units, the odontodes, to a base of dermal bone. Moreover, there is no close resemblance, and presumably no homology, between the pattern of the tesserae of the dermal shield of an osteostracan and that of the dermal bones of osteichthyans, except perhaps in the sclerotic ring. One may thus consider that the macromeric condition in the gnathostomes arose several times in the placoderms, some acanthodians (climatiids), and osteichthyans respectively, and that the micromeric condition of chondrichthyans (if one excepts the fin spines) is general for the gnathostomes. As was pointed out by Reif (1982*a*), the widely used terms 'micro-', 'meso-', and 'macromeric' are vague, and most palaeontologists use the term 'micromeric' as a synonym of 'small scales'. Instead, he proposed the terms 'microsquamose' (single odontodes), 'mesosquamose' (dermal elements growing by addition of new odontodes, but which may be shed during life, such as growing scales), and 'macrosquamose' (dermal elements that are never shed, such as dermal head bones). The terms 'micro-', 'meso-', and 'macromeric' should be used only to refer to the relative size of the macrosquamose dermal bones.

Very few Recent and fossil gnathostomes are entirely microsquamose, which is believed to be the generalized condition. Only some chondrichthyans possess an almost entirely microsquamose exoskeleton (placoid scales), which recalls that of the thelodonts. In the frame of his 'lepidomorial theory' (see Chapter 6, p. 278), Stensiö (1962) suggested that the placoid scale is not 'primitive' but is derived from mesosquamose (growing) scales and is the result of the fusion of several odontodes at the papillary level. He therefore, considered that all primitive chondrichthyans had growing, compound scales. In contrast, Reif regards the placoid scale as a primitively microsquamose condition for the gnathostomes. This could be supported by the fact that almost typical chondrichthyan placoid scales are now known to occur as early as the Silurian (e.g. *Elegestolepis*, see p. 150, Fig. 4.31B). As will be discussed below, this poses a real problem, because the current phylogeny of the chondrichthyans would imply a multiple appearance of the mesosquamose condition in this group. Out-group comparison with the thelodonts, as well as the mode of growth of compound scales by addition of odontodes, would thus imply that the microsquamose condition is general for the gnathostomes (Fig. 4.29A1), but, in this case, the mesosquamose condition would have appeared independently several times (e.g. in some Palaeozoic chondrichthyans and acanthodians) in the early history of the gnathostomes.

In most gnathostomes, the superficial layer of the scales, the crown of the teeth or the tubercles or the dermal bones, are made up of metadentine or orthodentine. Exceptions are the scale crown and dermal ornament of some acanthodians (Fig. 6.63B), which consists of mesodentine, and the tubercles and jaw-bones of the placoderms, which consist of semidentine (Fig. 4.43B).

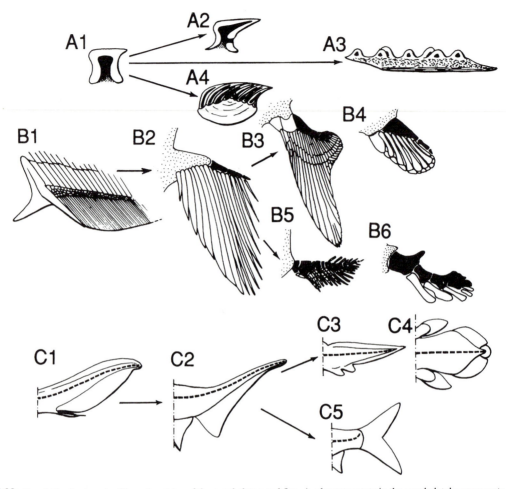

Fig. 4.29. Fossil Gnathostomata. Character states of the exoskeleton and fins. A, character state in the exoskeletal components (vertical section, pulp cavity in black): A1, microsquamose and micromeric state (single odontode, thelodonts as an out-group); A2, microsquamose and micromeric state (placoid scale with neck canals of modern elasmobranchs); A3, macrosquamose and macromeric state (odontodes not shed, and attached to large dermal bones; placoderm, osteichthyans, some acanthodians); A4, mesosquamose and micromeric state (scales of some Palaeozoic chondrichthyans, acanthodians, and osteichthyans). B, character states in the pectoral fin endoskeleton (metapterygial axis in black): B1, eurybasal paired fin (Anaspida as an out-group, radials unknown); B2, eurybasal pectoral fin of the Devonian chondrichthyan *Cladoselache*); B3, B4, stenobasal pectoral fins of a modern shark (B3) and a modern actinopterygian (B4); B5, B6, monobasal pectoral fin of a Palaeozoic xenacanthiform shark *Expleuracanthus* (B5) and the Devonian sarcopterygian *Eusthenopteron* (B6). C, character states in the tail morphology (notochordal axis in dashed line): C1, epicercal tail (osteostracans as an outgroup); C2, epicercal tail in a gnathostome (acanthodian); C3, C4, true diphycercal (isocercal) tail in a chondrichthyan (*Expleuracanthus*, C3) and an actinistian (*Latimeria*, C4); C5, apparent diphycercal tail in a teleost actinopterygian. (B1, from Ritchie (1964); B2, B3, from Bendix-Almgreen (1975); B4, B6, from Jarvik (1980); B5, from Zangerl (1981).)

The teeth of some chondrichthyans have a core of trabecular dentine (osteodentine); in holocephalans, lungfishes, and rays the teeth are strengthened by vertical
bundles of tubate orthodentine arranged into denteons
around vascular canals and embedded in pleromic (filling)
dentine. Osteostracans, the presumed sister-group of the

gnathostomes (see p. 238), have only mesodentine. Out-
group comparison would thus imply that mesodentine is
general for the gnathostomes, semidentine is unique to the
placoderms, and gnathostome orthodentine is homoplasic
with that of heterostracans. If, however, out-group comparison is made with the thelodontid thelodonts, then

metadentine or orthodentine would be general for the gnathostomes. All gnathostomes possess some kind of enameloid, yet true enamel (monotypic enamel) is met with only in osteichthyans.

The dermal bone in most early gnathostomes is cellular, as in osteostracans, and this may represent the generalized condition. The acellular bone met with in some acanthodians, in the base of the placoid scales of modern chondrichthyans, or in some osteichthyans (teleosts), is thus secondary.

The histological structure of the fossil and extant gnathostome endoskeleton falls into three major types: that of chondrichthyans, which is made up of cartilage lined with small tesserae of prismatic calcified cartilage; that of placoderms and acanthodians, which consists of cartilage lined with perichondral bone; and that of osteichthyans, which is made up of cartilage, perichondral bone, and endochondral bone (Fig. 6.12A). These types may not be clear-cut, since a thin coating of perichondral bone sometimes occurs outside the layer of prismatic claified cartilage in some chondrichthyans, and endochondral-like spongy bone has been recorded in the rhinocapsular bone of some placoderms (Fig. 4.42E, 34). Ontogeny shows that in the endoskeleton of osteichthyans, cartilage forms first, then the perichondral bone appears, and finally the endochondral bone invades centripetally the cartilage. Out-group comparison shows that only cartilage and perichondral bone occur in fossil jawless vertebrates (galeaspids and osteostracans). The occurrence of a spongy endoskeleton in boreaspid osteostracans is unique to that family, owing to the appearance of trabecular perichondral bone. It can thus be assumed that a cartilaginous endoskeleton lined with perichondral bone is general for the gnathostomes. Chondrichthyans have gained prismatic calcified cartilage and have lost most of its perichondral bone, whereas osteichthyans have gained endochondral bone. The condition in placoderms and probably acanthodians is identical to or very close to the generalized gnathostome condition. Both the perichondral and endochondral bone of the gnathostomes is cellular.

4.12.5 Paired fins and girdles

In the gnathostomes, the paired fins arise during ontogeny from a ventrolateral fold of the ectoderm, called the Wolff's crest. This feature has long been the principal support of the 'fin-fold' theory which assumes that ancestral vertebrates had a long ventrolateral fin-fold. The fact that some of the earliest-known gnathostomes, the acanthodians, possess a ventrolateral series of large intermediate scutes or spines (Fig. 4.64A) between the pectoral and pelvic fins has often been regarded as giving palaeontological support to this theory. The long, ribbon-shaped paired fins of some anaspids (Fig. 4.11A, 3; Fig. 4.29B1) can also be explained in the light of this fin-fold theory. However, all gnathostomes possess only two distinct paired fins, pectoral and pelvic, and there is no evidence that their morphotype includes a continuous ventrolateral fin (Coates 1993).

The endoskeletal pectoral girdle displays a wide range of diversity, and the comparison with the jawless vertebrates that have paired fins (osteostracans, anaspids, pituriaspids) is of poor value in assessing the morphotypic gnathostome condition. The endoskeletal support of the paired fins in osteostracans is not independant of the braincase, contrary to what is found in all gnathostomes, but the area for the attachment of the musculature and fin skeleton in advanced osteostracans shows some striking resemblance to that in placoderms (Figs 4.15D, 4.41C). In the gnathostomes, one can distinguish stenobasal and eurybasal paired fins. The former are narrow-based and attach on a short girdle by means of a few basal elements (Fig. 4.29B3–B6). The latter are broad-based and attach on a long girdle; each radial articulates independently on the girdle, with the exception of the metapterygial radials (Fig. 4.29B2). These different types apparently occur in various major gnathostome taxa. In osteostracans, the narrow-based paired fin is a derived condition, and the broad-based fin of *Ateleaspis* is the primitive condition, yet nothing is known of the girdle and radials (if any) in this genus. In Anaspids (Fig. 4.29B1), the paired fin rays may have attached independently to the flange, if the dermal scale rows reflect their orientation. It is thus tempting to consider the broad-based fins with independent radials (Fig. 4.29B2) as the generalized condition for the gnathostomes. One must, however, be aware that this condition may easily reappear from a narrow-based condition, as exemplified by the advanced brachythoracid arthrodires (see p. 153).

It is currently assumed that the most generalized type of gnathostome girdle and fin is found in the fossil (Devonian) chondrichthyan *Cladoselache* (Fig. 4.29B2). The shoulder girdle has a dorsal shaft and a ventral plate-shaped part on which unjointed radials articulate separately. Posteriorly, the girdle is prolonged by a long fin element, the metapterygial axis (Fig. 4.29B2, black), which is composed of a variable number of mesomeres on which metapterygial radials are attached. The pelvic girdle is smaller and plate-shaped, but the fin skeleton consists of the same elements as in the pectoral fin, i.e. unjointed radials, metapterygial axis, and metapterygial radials. The reason for the choice of *Cladoselache* as the morphotypic structure of the paired fins and girdles is merely that it displays the most simple condition, and resembles most closely that met with in the embryo of most fishes. In some osteichthyans (teleosts, amniotes), the shoulder girdle consists of at least two ossifications, scapular and coracoid, but in the majority of the early gnathostomes, whether chondrichthyans, placoderms, or osteichthyans, there is a single scapulocoracoid element. The same occurs in the pelvic girdle, which tends to become divided into three ossification (pelvis, ilium, and ischium) in amniotes.

The subsequent modifications of the girdles and paired fin endoskeleton show the same 'trends' in various gnathostome groups. When the scapulocoracoid is firmly attached to large dermal plates (placoderms, osteichthyans), it shows a tendency toward the reduction of its dorsal process and ventral blade (coracoid). When the fins become narrow-based, the radials become reduced in number, and their proximal segments fused into basipterygial plates; the propterygium and mesopterygium (Fig. 4.29B3, B4; most chondrichthyans, some actinopterygians, some placoderms). Finally, in some groups, such as the xenacanth chondrichthyans, some holocephalans, and the sarcopterygian osteichthyans, all the radials situated in front of the metapterygial axis disappear, and the fin skeleton consists of a long metapterygial axis with preaxial and even postaxial metapterygial radials (monobasal fin; Fig. 4.29B5, B6).

The girdles are covered with large dermal bones in the placoderms, climatiid acanthodians, and osteichthyans. If it is assumed that the primitive condition of the gnathostome exoskeleton is micromery or even microsquamosity, then one has to admit that the dermal shoulder girdle has appeared independently in these three groups. Homologies in the dermal bone pattern of placoderms and osteichthyans have been proposed, but they are far from unanimously accepted. A dermal pelvic girdle is exceptional in the gnathostomes. Only in some placoderms does an enlarged scale cover the pelvic area.

The dermal fin rays are of two kinds: the ceratotrichiae, which prolong the radials inside the fin web, and the more or less modified scales that cover the entire fin. In fossil and Recent chondrichthyans, the latter do not differ markedly from the other body scales, and the ceratotrichiae are well developed even when the radials are long. In acanthodians, also, the scales covering the fins are slightly more elongated in shape than the other scales, and there are long ceratotrichiae. This state probably represents the generalized gnathostome condition. In contrast, the paired fins of osteichthyans are covered with rows of tile-shaped, elongated scales, the lepidotrichiae, which tend to sink into the skin and replace the ceratotrichiae.

4.12.6 Unpaired fins

There is a consensus that the general condition for the gnathostome paired fins is the presence of two dorsal fins

and an epicercal caudal fin (Fig. 4.29C2). This is decided on the basis of out-group comparison with osteostracans, which are assumed to have had in their primitive form two dorsal fins (although the anterior one is a mere hump, devoid of fin web) and an epicercal tail (Figs 4.18A, B, 4.29C1)). A single dorsal fin occurs in some chondrichthyans, some acanthodians, and actinopterygians. The condition in placoderms is poorly known, but they seem to have possessed a single, large dorsal fin. The caudal fin is clearly epicercal in osteostracans, chondrichthyans, placoderms, acanthodians, and most piscine osteichthyans. Both ontogeny and phylogeny show that the apparently diphycercal tail of cladistians and teleosteans is a derived condition. The true diphycercal tail of actinistians is also regarded as derived on account of the presence of an epicercal tail in the early actinistian *Miguashaia* (Fig. 4.75A). The question of the anal fin is more problematical. It is usually assumed that the anal fin is present in the morphotypic gnathostome, because it exists in some jawless vertebrates, such as anaspids and thelodonts. Most Palaeozoic chondrichthyans lack the anal fin, however, and there is no clear evidence that the horizontal lobe of the osteostracans is a modified anal fin. Here, I accept that an anal fin is a general character for the gnathostomes.

4.12.7 Vertebrae and ribs

Early fossil gnathostomes largely confirm that the vertebrae are basically compounded of two dorsal and two ventral elements (basi- and interdorsal; basi- and interventral; Fig. 3.3C), and that the formation of a centrum has occurred more than once, by fusion or close association of these elements with the calcified or ossified notochordal tissues. It is thus quite certain that vertebral centra would be lacking in the morphotype of the gnathostomes. In fact, among the early (Devonian) gnathostome groups, virtually none displays centra, with the exception of some Devonian lungfishes, osteolepiforms, panderichthyids and tetrapods. Ossified ribs are also extremely rare among the early gnathostomes; they occur in Late Devonian and later lungfishes and osteolepiforms. Calcified ribs occur in hybodontiform sharks.

4.12.8 Sensory-line pattern

Much has been written about the primitive condition of the gnathostome sensory-line pattern. The most substantiated out-group comparison has been made with heterostracans, because it is assumed that their sensory-line pattern has not been greatly modified by the development of the prehypophysial region of the head (unlike osteostracans, for example). If this comparison is justified, then the supraorbital, infraorbital, and main lateral lines are general to all gnathostomes. The mandibular line seems to be unique to the gnathostomes, yet there is a small postoral line in most heterostracans and osteostracans, which may have been innervated by lateralis fibres accompanying the mandibular branch of the trigeminus nerve.

The pattern of the sensory lines on the head and body is largely bound to the course of the lateralis-nerve fibres, which in turn are fascicularized with the cranial and spinal nerves. The frequent V-shaped pattern of the supraorbital lines (Fig. 4.41, 26), for example, is linked with the course of the profundus nerve, with which are associated the corresponding lateralis fibres. Similarly, the infraorbital and mandibular lines are linked with the course of the maxillaris and mandibularis branches of the trigeminus nerve. On the body, the sensory lines tend to be organized in longitudinal and transverse lines, although there is usually only one main lateral line, and this probably represents the general gnathostome condition.

The sensory-line canals are closed in most fossil jawless vertebrates (heterostracans, osteostracans, galeaspids, thelodonts), and this is assumed to represent the general condition for the gnathostomes also.

Further readings

Neurocranium: Bjerring (1977, 1978, 1985); Gardiner (1984a); Goujet (1984a); Jarvik (1980, 1981a); Rosen *et al.* (1981); Schultze (1987, 1993); Young (1986b). *Jaw suspension and teeth*: Arratia and Schultze (1991); Jarvik (1980); Lauder (1980a); Reif (1982a); Schaeffer (1975); *Gill arch morphology*: Nelson (1969); Maisey (1986, 1989a); Rosen *et al.* (1981); *Histology and structure of skeleton*: Nelson (1970); Ørvig (1967a); Reif (1982a); Schaeffer (1977); Stensiö (1959); *Paired fins and girdles*: Jarvik (1965a, 1981a); Rosen *et al.* (1981); Stensiö (1959); Zangerl (1981); *Unpaired fins*: Jarvik (1980); Maisey (1986, 1988); Zangerl (1981). *Vertebrae and ribs:* Gardiner (1983, 1984a); Rosen *et al.* (1981); *Sensory-line pattern*: Bjerring (1979, 1984); Jarvik (1980); Ørvig (1975); Säve-Söderbergh (1941).

4.13 CHONDRICHTHYES

When fossils are considered, chondrichthyans remain characterized by the prismatic calcified cartilage lining the endoskeletal elements. However, the second characteristic of Recent Chondrichthyans, i.e. the pelvic metapterygium modified into a mating organ, the mixipterygium or pelvic clasper, may no longer be valid when some Devonian chondrichthyans are considered, since some of them do not display this feature of sexual

dimorphism. This is the case of the Late Devonian *Cladoselache* (Fig. 4.30D1, 10), where all known specimens show an unmodified pelvic metapterygium. If this absence of pelvic claspers in *Cladoselache* is regarded as general for chondrichthyans (Fig. 4.39, 1), it would imply that the claspers of other chondrichthyans and some placoderms are homoplasic (see Chapter 5, p. 243). However, it is not ruled out that the pelvic claspers have been lost in *Cladoselache* as a consequence of a reduction of the pelvic fins, which culminates with the complete loss of these fins in the Eugenodontida (see below).

We saw in Chapter 3 that the interrelationships of Recent chondrichthyans are far from satisfactorily elucidated. The situation is even worse when fossil, and in particular Palaeozoic, chondrichthyans are considered. Until the early 1980s, the chondrichthyans, fossil and extant, were divided into elasmobranchs (all shark-like and ray-like forms) and holocephalans (typical chimaeroids and the fossil bradyodonts, regarded as holocephalans on the basis of tooth histology). The position of the batomorphs (sawfishes, torpedoes, skates, and rays) has been little debated by palaeontologists, since the earliest members of the group are virtually similar to the Recent forms and do not add much to the elucidation of this question (see Chapter 3). All other shark-like chondrichthyans, regarded as elasmobranchs, were arranged into a series of grades, based essentially on tooth, scale, or fin spine morphology: the 'cladodonts', 'ctenacanthids', 'xenacanthids', and 'hybodonts'. In general, all modern elasmobranchs were regarded as a clade, the neoselachians (Fig. 4.39K), possibly derived from some hybodontid ancestor (Fig. 4.39J). It has, however, long been claimed, on the basis of tooth morphology and various anatomical features, that the extant frilled shark, *Chlamydoselachus*, is more primitive than most other fossil and recent elasmobranchs and would more probably be of 'cladodont' grade; it has also been claimed that the Port Jackson shark, *Heterodontus*, is a survivor of the 'hybodont' grade. These views are no longer considered in the current literature.

The first attempts at analysing chondrichthyan phylogeny on the basis of extensive cladistic character analysis date back to the early 1980s, in the work of Schaeffer (1981), Young (1982a), Maisey (1984a, 1986), Mader (1986), and more recently Gaudin (1991). There are important disagreements between these works, but only Maisey (1984a, 1986) considers *Cladoselache* to be the sister-group of all other chondrichthyans, Recent and fossil, on the basis of the lack of pelvic mixipterygium. In addition, it is now possible to recognize a number of higher fossil clades, even though their phylogenetic position remains obscure. There is no gnathostome group whose phylogeny is as poorly elucidated as the chon-

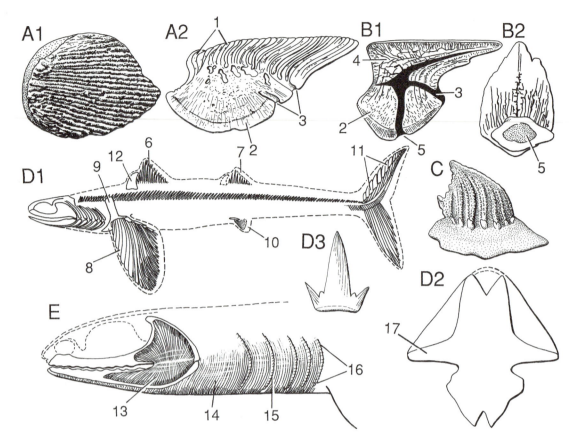

Fig. 4.30. Fossil Chondrichthyes. Silurian chondrichthyans and Devonian Cladoselachidae. A, *Mongolepis*, Early Silurian of Central Asia, scale in external view (A1, ×15) and vertical section (A2). B, *Elegestolepis*, Early Silurian of Central Asia, scale in vertical section (B2) and basal view (B3, ×40). C, *Pilolepis*, Early Silurian of Canada, dermal scute in lateral view (×30); D, *Cladoselache*, Late Devonian of the USA, reconstruction of the skeleton in lateral view (D1, ×0.2), outline of the braincase in ventral view (D2) and tooth in labial view (D3). E, Cladoselachidae, Late Devonian of Tennessee (USA), reconstruction of the musculature, based on a specimen with preserved muscle impressions. 1, odontodes; 2, base; 3, neck canals; 4, microsquamose crown; 5, pulp cavity and basal canal; 6, anterior dorsal fin; 7, posterior dorsal fin; 8, pectoral fin radials; 9, scapulocoracoid; 10, pelvic fin; 11, large calcified plates (radials); 12, dorsal fin spine; 13, adductor jaw muscles; 14, visceral constrictor gill musculature; 15, gill slit; 16, gill rays; 17, postorbital process. (A, from Karatayute-Talimaa *et al.* (1990); B, from Karatayute-Talimaa (1973); C, from Thorsteinsson (1973); D1, D2, after Bendix-Almgreen (1975) and Zangerl (1981); D3, from Schaeffer and Williams (1977); E, from Maisey (1989a).)

drichthyans, and in particular the elasmobranchs. The situation is slightly better for the holocephalans, thanks to the discovery of remarkable Carboniferous localities such as Bear Gulch and Mecca Quarry in the USA, which have yielded complete, articulated specimens. Its now seems quite clear that bradyodonts are stem-group holocephalans (Fig. 4.37). More puzzling is the case of some shark-like chondrichthyans that share with the holocephalans a number of unique characters and may thus be quite generalized holocephalans. This is for example the case of the so-called 'stethacanthid sharks',

which are here put—albeit questionably—with holocephalans (Figs 4.33, 4.37).

In addition to the major taxa discussed below, there are many Palaeozoic chondrichthyan taxa that are represented by 'ichthyoliths', i.e. isolated teeth, scales, and fin spines. These will not be considered here, except for the earliest known chondrichthyan (or supposedly so) remains, from the Silurian of Canada, Russia, and Central Asia (Fig. 4.30A–C; Fig. 4.40), as well as some enigmatic Devonian and Carboniferous teeth and scales that resemble those of neoselachians (Fig. 4.40D–G).

4.13.1 Cladoselachidae

The Cladoselachidae, and *Cladoselache* (Fig. 4.30D) in particular, are among the best-studied Devonian chondrichthyans. They are known by complete specimens in which some traces of soft tissues (kidneys, musculature; Fig. 4.30E) are preserved. They are shark-like chondrichthyans, but differ from all other members of the group in lacking pelvic claspers. As mentioned above, the paired fins contain unjointed radials that articulate separately on the girdle or the metapterygial axis (Fig. 4.30, 8), and also ceratotrichiae. The braincase is short, with very long postorbital processes (Fig. 30D2, 17). The tail is large, lunate in shape, and the chordal lobe is strengthened by large cartilaginous plates, which are enlarged neural arches and dorsal radials (Fig. 4.30, 11). The anal fin is supposed to have been lost. Anteriorly to the first dorsal fin there is a short, bony spine (Fig. 4.30, 12).

In the Cladoselachidae, the mouth is subterminal in position, and there are five gill slits (Fig. 4.30, 15). The interbranchial septa are strengthened by very long gill rays (Fig. 4.30, 16). The teeth are small and tricuspidate (the so-called 'cladodont type'), and are considered by some as the most generalized tooth type for chondrichthyans (Fig. 4.30D3). Although cladoselachids are known exclusively from the Devonian, this type of tooth may occur much later, until the Late Permian, and in other taxa, such as symmoriids. Tooth replacement is assumed to have been very slow, rows of teeth being functional at the same time. Although the body is almost naked, scales of the head and fins of *Cladoselache* are compound, or mesosquamose.

4.13.2 Eugeneodontida

Eugeneodontids (also referred to as edestids *sensu lato*) are known from the Carboniferous to the Early Triassic. They are characterized by a median series of large teeth on the symphysis of the lower jaw (Fig. 4.31, 1). This tooth series is simply arched and low in the most generalized taxa, such as *Bobbodus* (Fig. 4.31F, 1), *Fadenia* (Fig. 4.31B1), or *Sarcoprion* (Fig. 4.31C, 1), but is scrolled into a tooth 'spiral' of high, sharp teeth in the more advanced edestids (*Helicoprion*; Fig. 4.31E, 1). When the mouth was closed, this symphysial tooth series was housed in a cavity of the upper jaw (Fig. 4.31, 3), between two similar series of upper teeth (Fig. 4.31, 2). The other teeth of the jaws are small, multicuspidate, and elongated in shape (Fig. 4.31, 4). The scales are generally compound, with several flame-shaped odontodes attached on a single base, but there are also some isolated odontodes, with a single pulp cavity, like ordinary placoid scales.

In the generalized eugeneodontids, such as *Caseodus* (Fig. 4.31A), the palatoquadrate is free, whereas in edestids it is much reduced or fused to the braincase (Fig. 4.31B1). They have therefore sometimes been compared to holocephalans, and hence reconstructed with a single gill slit. At present the number of gill slits remains unknown. The pectoral fins, when known, show nearly the same structure as those of cladoselachids, with long, unsegmented radials articulating directly on the girdle (Fig. 4.31B2, 6). In others, such as *Caseodus* (Fig. 4.31A) or *Romerodus*, each radial displays at least a short proximal and a long distal segment. The anal and pelvic fins are lacking in all forms of which the body is preserved (hence the impossibility of knowing if they possessed a pelvic clasper). The neural and haemal arches of the chordal lobe of the tail tend to fuse into large calcified plates, in a way that recalls the structure of the cladoselachid tail (Fig. 4.31, 7).

The actual body shape of eugeneodontids is known in only a few taxa. Some were huge and shark-like, with a single prominent dorsal fin, whereas others were eel-shaped. *Ornithoprion* is a peculiar form, in which the symphysial part of the lower jaw was produced anteriorly into a long beak covered with denticles (Fig. 4.31D, 5). They were no doubt a highly diversified group, probably to be placed outside the ensemble formed by the elasmobranchs and holocephalans (crown-group chondrichthyans), although Lund (1990) puts them in the Paraselachimorpha, i.e. stem-group holocephalans.

4.13.3 Petalodontida

Like eugeneodontids, petalodontids are a small group of Carboniferous and Permian chondrichthyans that can hardly be classified in any of the two major extant taxa. They are characterized by petal-shaped teeth (Fig. 4.32B) and possessed, like edestids, a symphysial tooth series (Fig. 4.32, 2). They were for long known only from isolated teeth and from poorly preserved specimens of *Janassa*, which gave the false impression of their being ray-like fishes. Complete and well-preserved petalodontids were later discovered in the Carboniferous of Bear Gulch, which provided more precise information on their body shape. *Belantsea* (Fig. 4.32C), for example, has very large and rounded pectoral (Fig. 4.32, 4) and dorsal fins. Whether it has pelvic fins is unclear, since the fin referred to as a pelvic lies very close to the caudal, and therefore may be the anal fin. Pelvic fins have, however, been described in *Janassa*. The caudal fin is small, with a short chordal lobe, and resembles the tail of teleosts. The body was deep, and was covered with scattered star-shaped denticles. Whether the palatoquadrate was free or was fused to the braincase is still not known, but there

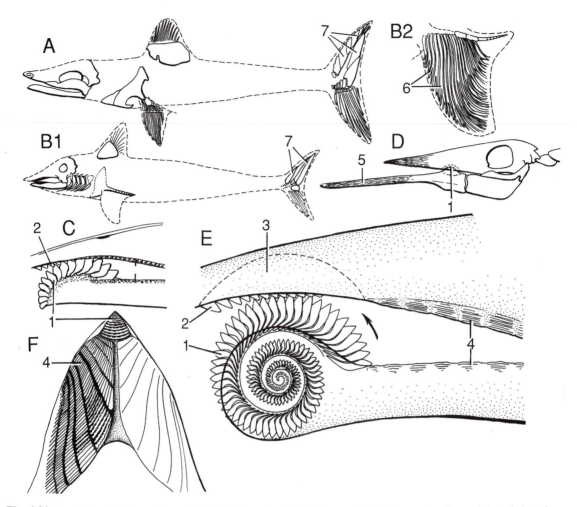

Fig. 4.31. Fossil Chondrichthyes. Eugeneodontida. A, *Caseodus*, Carboniferous of the USA, reconstruction of skeleton in lateral view (about ×0.1); B, *Fadenia*, Permian of Greenland, reconstruction of the skeleton in lateral view (B1, about ×0.06), and skeleton of pectoral fin (B2); C, *Sarcoprion*, Permian of Greenland, symphysial region of lower jaw and snout in lateral view (×0.1); D, *Ornithoprion*, Late Carboniferous of the USA, skull in lateral view (×0.5); E, *Helicoprion*, Permian, attempted restoration of the symphysial region of the lower jaw and snout, in lateral aspect (×0.2); F, *Bobbodus*, Carboniferous of the USA, lower jaw dentition (×0.3). 1, symphysial tooth spiral or tooth series; 2, upper, parasymphysial tooth series; 3, space between the upper parasymphysial teeth to accomodate the symphysial tooth spiral; 4, marginal teeth; 5, symphysial process covered with denticles; 6, unsegmented radials; 7, large calcified plates (neural and haemal spines). (A, F, from Zangerl (1981); B1, from Zangerl (1981); B2, from Bendix-Almgreen (1975); C, from Nielsen (1952); D, from Zangerl (1966),

was probably a single gill slit (Fig. 4.32, 3). The radials of the fins have never been clearly observed; in *Netsepoye*, they seem to have been segmented and articulated on basal plates.

Petalodontids had the overall aspect of the extant scorpaenid or cyclopterid teleosts, and their parrot-bill-like teeth may have served in eating corals or shells. They occur in marine environments, and their teeth are often

found in Carboniferous to Permian limestones, in association with corals, brachiopods, crinoids, and fusulines. Some isolated teeth referred to *Megactenopetalus* (Fig. 4.32D) reach 10 cm in crown breadth and belonged to very large individuals. The histology of the petalodontid teeth shows a massive core of trabecular dentine, coated with a layer of orthodentine, but the surface of the boundary between these two types of tissues is like a

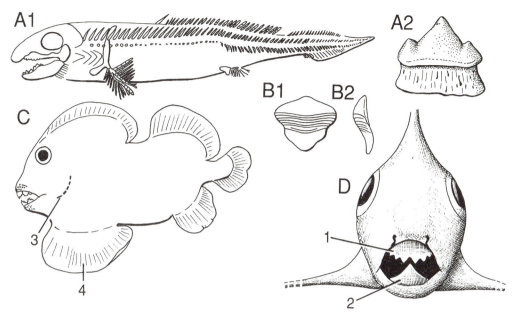

Fig. 4.32. Fossil Chondrichthyes. Petalodontida and *Heteropetalus*. A, *Heteropetalus*, Early Carboniferous of Montana, reconstruction of skeleton in lateral view (A1, ×1.5) and isolated tooth (A2); B, *Petalodus*, Carboniferous, isolated tooth in labial (B1) and lateral (B2) view; C, *Belantsea*, Early Carboniferous of Montana, reconstruction in lateral view (×0.3); D, *Megactenopetalus*, Permian, attempted reconstruction of the head in anterior view, showing the presumed position of the large symphysial teeth (×0.2). 1, upper median tooth; 2, lower median tooth; 3, gill slit; 4, enlarged pectoral fin. (A1, from Lund (1977*a*); A2, B, from Hansen in Zangerl (1981); C, from Lund (1989); D, based on reconstruction by Hansen (1978).)

waffle-iron. This is interpreted as incipient tubate dentine (orthotrabeculine; Zangerl *et al.* 1993); it recalls the tooth structure of orodontids and eugeneodontids (Taylor and Adamec 1977, Lund 1983).

The Carboniferous genus *Heteropetalus* (Fig. 4.32A, 1) was first regarded by Lund (1977*a*) as a primitive petalodontid and later (Lund 1990) as a hybodontiform. Zangerl (1981) considers that its teeth accord better with *Desmiodus*, which is known only as isolated teeth, but its tooth histology is quite different from that of the petalodontids (Fig. 4.32A2).

4.13.4 Symmoriida: Symmoriidae and Stethacanthidae

Symmoriids are characterized by a very long metapterygial rod (Fig. 4.33, 3) in the pectoral fins and a peculiar arrow-shaped endoskeletal element posteriorly to the dorsal fin supports (Fig. 4.33, 1). The radials of the paired fins are segmented into proximal and distal parts (Fig. 4.33, 4, 5); those of the median fins articulate with separate supports. The pelvic metapterygium in males is modified into a pelvic clasper or mixipterygium. The

teeth are of cladodont type, with a large, median cusp (Fig. 4.33C). They are known from the Late Devonian to the Carboniferous. Scales are lacking, apart from some compound denticles on the head and fins.

Typical genera are *Cobelodus* (Fig. 4.33A), *Denaea* (Fig. 4.33B), and *Symmorium*. Some peculiar Carboniferous forms are also referred to this group: the Stethacanthidae (*Stethacanthus*, *Orestiacanthus*, *Falcatus*; Fig. 4.33 D–F). In these forms the males have a peculiar brush-shaped or club-shaped device on the back. They also share with symmoriids a long metapterygial rod in the pectoral fins and the characteristic arrow-shaped element in the dorsal fin support. Their dorsal 'brush' or club is probably a modified anterior dorsal fin; it is preceded by a massive dermal spine (Fig. 4.33, 6). It consists of bundles of presumably calcified radials or fibrous tubules (Fig. 4.33, 8), surmounted by a shagreen of spiny denticles (Fig. 4.33, 7). Williams (1985) and Lund (1985) suggested that this 'brush' could abut against a patch of denticles covering the head, and could thus have served during mating or/for sexual display behaviour (see Chapter 7, p. 305). Zangerl (1984) also suggested that the ensemble of the 'brush' and head denticles may have been used in a

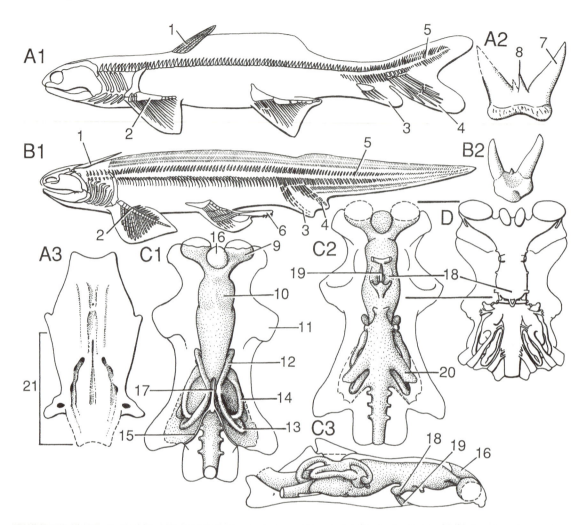

Fig. 4.34. Fossil Chondrichthyes. Elasmobranchii, Xenacanthiformes. A, *Antarctilamna*, Middle Devonian of Antarctica, attempted reconstruction based on the Carboniferous genus *Diploselache* (A1, about ×0.2), isolated tooth (A2) and braincase in dorsal view (A3); B, *Expleuracanthus*, Carboniferous to Permian, reconstruction in lateral view (B1, about ×0.2) and isolated tooth (B2). C, *Xenacanthus*, Permian of Texas, outline of the braincase, with internal cast of the internal cavities, in dorsal (C1), ventral (C2), and lateral (C3) view (×0.5). D, *Squalus*, Recent, outline of the braincase and internal cavities in ventral view, to be compared with the proportions of corresponding structures in *Xenacanthus*. 1, dorsal spine; 2, metapterygium; 3, anal fin; 4, hypochordal lobe of tail; 5, chordal lobe of tail; 6, mixipterygial (pelvic) clasper; 7, lateral cusps; 8, median cusps; 9, nasal cavities;10, brain cavity; 11, postorbital process; 12, anterior semicircular canal; 13, posterior semicircular canal; 14, horizontal semicircular canal; 15, otico-occipital fissure; 16, precerebral fontanelle; 17, endolymphatic fossa; 18, hypophysial pit; 19, buccohypophysial canal; 20, vagus nerve canal; 21, otico-occipital region of braincase. (A, from Young, 1982*a*, 1991*b*; B, from Schaeffer and Williams (1977); C, D, from Schaeffer (1981).)

their elongated and often low teeth, which are ornamented with sinuous ridges and pierced by numerous nutrient canals (the 'hybodont' tooth type; Fig. 4.35C3, D). Their fin spines are rounded in section, with two posterior rows of denticles, and are ornamented with smooth ridges (Fig. 4.35B, 1). They share with neoselachians the non-growing, placoid scale type (Fig. 4.35C2), although some early hybodontiforms possess some growing scales. Their monophyly is now favoured by a majority of chondrichthyan experts, and is supported by at least four characters: the calcified pleural ribs (Fig. 4.35B, 8); two pairs of large recurved denticles on the head (Fig. 4.35, 9,);

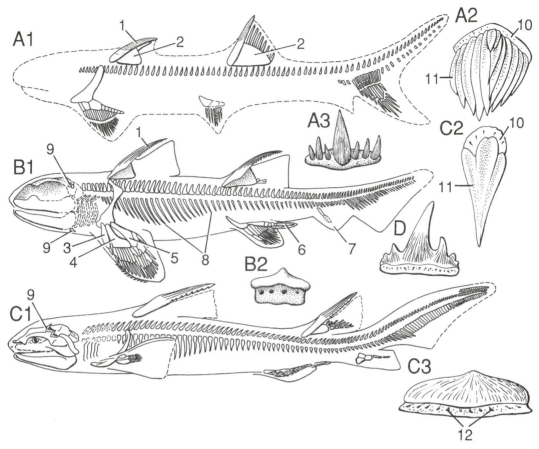

Fig. 4.35. Fossil Chondrichthyes. Elasmobranchii. Ctenacanthiformes and Hybodontiformes. A, the ctenacanthiform *Ctenacanthus*, Devonian to Carboniferous, reconstruction of skeleton in lateral view (A1, ×0.2), tooth in labial view (A2), and isolated growing scale in external view (A2). B, D, Hybodontiformes; B, *Onychoselache*, Early Carboniferous of Scotland, reconstruction of the skeleton in lateral view (B1, ×0.5) and tooth in labial view (B2); C, *Hamiltonichthys*, Late Carboniferous of Kansas (USA), reconstruction of skeleton in lateral view (C1, ×0.5), non-growing scale in external view (C2) and tooth in labial view (C3); D, *Hybodus*, Cretaceous, tooth in labial view. 1, fin spine; 2, basal plate; 3, propterygium; 4, mesopterygium; 5, metapterygium; 6, mixipterygial (pelvic) clasper; 7, anal fin; 8, large calcified ribs; 9, cephalic spines; 10, scale base; 11, odontode; 12, nutrient foramina. (A1, from Moy-Thomas (1936*b*) B, from Dick and Maisey (1980); C, from Maisey, (1989*b*), D, from Maisey (1983).)

smooth ribs of dentine on fin spines; and low-crowned teeth (Fig. 4.35C3); but some late hybodontiforms, such as *Hybodus*, have relatively high cusps (Fig. 4.35D). Among the best-known Palaeozoic hybodontiforms are the Carboniferous genera *Tristychius, Onychoselache* (Fig. 4.35B), and *Hamiltonichthys* (Fig. 4.35C). Early Devonian isolated scales referred to as *Arauzia* resemble hybodontiform scales and may be the earliest known evidence of the group (Fig. 4.40F).

The Neoselachii are characterized by at least one layer of haphazardly fibred enameloid on teeth and scales, paired occipital condyles (Fig. 4.40B, 3), closed buccohy-

pophysial canal, large endolymphatic fossa (Fig. 4.40, 2), and a constricted and septate notochordal canal (linked with the formation of calcified centra). They comprise all extant elasmobranchs and the fossil representatives of their included taxa. The earliest articulated neoselachian, *Palaeospinax* (Fig. 4.40A), is Early Jurassic in age; a tooth fragment from the Early Triassic of Turkey has been referred to this genus (Thies 1982). However, the skull of the Cretaceous *Synechodus* (Fig. 4.40B) is the earliest well-preserved Mesozoic neoselachian skull known so far; it indicates that this genus is the sister-group of either all other neoselachians, or galeomorphs only (Maisey 1985). Batomorphs, such as *Spathobatis* (Fig. 4.40C) are known first in the Early Jurassic, but their rhinobatid-like aspect is probably a generalized batomorph condition. Some Late Palaeozoic and Early Mesosoic taxa, such as *Hopleacanthus* and *Acronemus*, possess non-growing placoid scales and incipient calcified vertebral centra which suggest closer affinities to the neoselachians. In addition, isolated placoid scales and teeth of neoselachian type (but without the typical fibrous enameloid) are recorded from Permian and Carboniferous sediments (*Anachronistes*, Fig. 4.40G); and the Silurian chondrichthyan scales referred to as *Elegestolepis* (Fig. 4.30B) are strikingly neoselachian-like. For various theories on neoselachian interrelationships, including fossils, see Maisey (1984*b*, 1985), Thies and Reif (1985), and Cappetta (1987).

4.13.6 Holocephali

A number of chondrichthyans with large, crushing teeth or tooth plates made up of columnar (or 'tubate') dentine and interstitial pleromin have been recorded, mostly in the form of isolated teeth, from Late Devonian to Late Permian rocks. These forms, called 'bradyodonts', because of their supposedly slow tooth replacement, had been regarded as related to extant chimaeroids on the basis of tooth histology (tubate dentine, or orthotrabeculine) and some features of cranial morphology (e.g. holostyly). The discovery of numerous articulated bradyodonts in the Carboniferous locality of Bear Gulch has considerably strengthened this interpretation. It is now widely accepted that bradyodonts are stem-group holocephalans, some of which are more closely related to chimaeroids than others (Fig. 4.37). In addition, thanks to this excellently preserved material, it has been possible to understand how the peculiar, scrolled teeth of some bradyodonts (cochliodonts) have formed by fusion of tooth families. One may thus make a transformation series leading from a condition with massive, but separate, teeth made up of tubate dentine, such as in *Helodus* (Fig. 4.36E, G1, 10), to a few series of large, plate-like teeth, such as in *Cochliodus* (Fig. 4.36F1, 9) or *Deltodus* (Fig. 4.36G2, 9), and finally large plates of tubate dentine, which also form lingually in holocephalans (Fig. 4.36F2, G3, 9). Moreover, Patterson (1992) pointed out that the mechanism of tooth growth in the extant chimaeroid *Callorhinchus* is virtually the same as in cochliodontid 'bradyodonts'.

'Bradyodonts' display an amazing diversity of morphology, in particular by the development of peculiar appendages on the head, which may have had the same function as the frontal claspers (tenaculum) of male extant chimaeroids (Fig. 4.36A, 1). These claw-shaped, horn-shaped, spine-shaped, or whip-shaped frontal appendages of bradyodonts were, however, generally paired, whereas the frontal clasper of chimaeroids are unpaired, a difference that suggests homoplasy, though probably linked with rather similar mating habits. Paired series of frontal spines comparable to those of most 'bradyodonts' (Fig. 4.36, 2) occur in juvenile individuals of extant chimaeroids. In cochliodontids (Fig. 1.19, 4), the frontal spines were long and slender in males, and formed a brush. In chondrenchelyids (e.g. *Harpagofututor*; Fig. 4.36A), the paired frontal claspers were articulated and ended with two branches covered with denticles. One of these branches was prolonged with a whip-shaped structure covered with denticles. In chochliodontids, there were also horn-shaped nuchal and mandibular spines (Fig. 4.36, 3; 4.37D). In echinochimaerids, the frontal spines were antler-shaped (Fig. 4.36D, 2).

In chondrenchelids, the paired pectoral fins are monobasal, i.e. reduced to the metapterygium only (Fig. 4.36A), with preaxial and postaxial radials, and thereby parallel those of pleuracanthids among xenacanthiforms.

One possible theory of holocephalan interrelationships is illustrated by the cladogram in fig. 4.37 (combined from Lund 1986*b,c*). It is based essentially on transformation from independent teeth into tooth plates, as well as on some features of the postcranial skeleton. The closest relative of modern chimaeroids may be the Carboniferous genus *Echinochimaera* (Fig. 4.37F), which is strikingly *Chimaera*-like in overall aspect. It probably possessed a prepelvic clasper (Fig. 4.36, 6) and a toothplate structure essentially similar to that of extant chimaeroids. Patterson (1992) has, however, suggested that chondrenchelyids (Fig. 4.37C) are more similar to extant chimaeroids in tooth structure and growth process. Some cochliodontids (Fig. 4.36B, C; Fig. 4.37E) share with *Echinochimaera* and chimaeroids a large synarcual (Fig. 4.36, 5), on which articulate the dorsal fin spine (Fig. 4.36, 4) as well as a long dorsal process of the pelvic girdle and dorsal scutes in the tail (Fig. 4.36, 8). These forms, as well as the menaspids (Fig. 4.37D), have extremely long frontal spines and large mandibular spines, which are lacking in modern chimaeroids and *Echinochimaera*, and are here assumed to have been lost or reduced in these taxa.

Both bradyodonts and typical chimaeroids have a holostylic jaw suspension, i.e. the palatoquadrate is reduced and fused to the braincase (Fig. 4.36E); a characteristic that is probably linked with durophagous diet. However, as we have seen above, the 'shark-like' symmoriids, which retain a free palatoquadrate and a suspensory hyomandibula, and whose branchial apparatus lies largely behind the braincase, display at least one char-

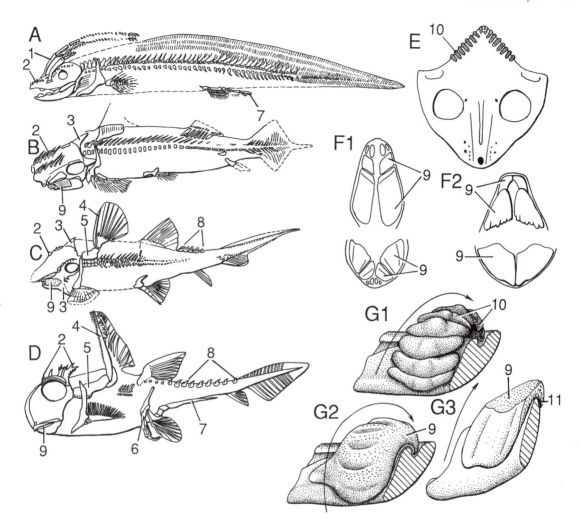

Fig. 4.36. Fossil Chondrichthyes. Holocephali. A, *Harpagofututor*, Early Carboniferous of Montana (USA), reconstruction of skeleton of a male in lateral view (×0.8); B, C, two unnamed cochliodonts, Early Carboniferous of Montana, reconstruction of skeleton in dorso-lateral view (×0.6); D, *Echinochimaera*, Early Carboniferous of Montana, reconstruction of a male in lateral view (×0.5); E, *Helodus*, Carboniferous of Europe, braincase and teeth in ventral view (×0.8); F, upper and lower dentition of a Carboniferous cochliodont (F1, *Cochliodus*) and a Recent holocephalan (F2, *Chimaera*); G, lower jaw and teeth of a generalized holocephalan (G1, based on *Helodus*), a cochliodont (G2, based on *Deltodus*) and a Recent chimaeroid (G3, *Callorhinchus*); the arrow indicates the direction of the tooth development or tooth plate growth. 1, paired frontal claspers; 2, frontal scutes or spines; 3, nuchal and mandibular spines; 4, mobile dorsal fin spine; 5, synarcual; 6, prepelvic clasper; 7, mixipterygial (pelvic) clasper; 8, dorsal scutes; 9, tooth plates; 10, tooth family; 11, descending lamina. (A, from Lund (1982); B–D, from Lund (1986*b*); E, from Patterson (1965) and Moy-Thomas (1936*a*); F, from Zangerl (1981); G2, from Bendix-Almgreen (1983); G3, based on Patterson (1992) and Didier *et al.* (1994).)

acter that is unique to holocephalans: the calcified rings strengthening the sensory-line canals. They also share with holocephalans the development of large denticles on the head and perhaps a tenaculum-like mating device. They are therefore here tentatively placed as the sister-group of holocephalans (Fig. 4.37A)

4.13.7 Iniopterygia

Iniopterygians are a group of Carboniferous chondrichthyans that display a relatively large number of unique characteristics, such as the almost dorsal position of the pectoral fins and a possibly diphycercal tail

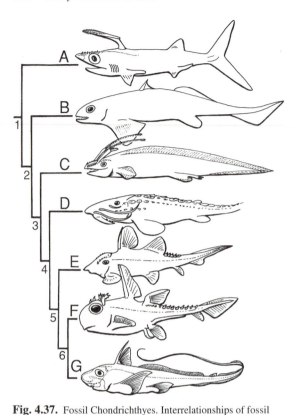

Fig. 4.37. Fossil Chondrichthyes. Interrelationships of fossil and extant Holocephali. *Terminal taxa*: A, Symmoriida (outgroup); B, Helodontidae (*Helodus*); C, Chondrenchelyidae (*Harpagofututor*); D, Menaspidae (*Menaspis*); E, Cochliodontidae (unnamed 'cochliodont'); F, Echinochimaeridae (*Echinochimaera*); G, Chimaeroidei (Mesozoic and extant holocephalans). *Nested taxa and synapomorphies*: 1, calcified rings surrounding the sensory-line canals, frontal denticles, reduced body squamation; 2, Holocephali (palatoquadrate fused to braincase, durophagous diet, teeth containing 'tubate dentine', or denteons, and columns of pleromin); 3, large, recurved tooth plates growing from lingual to labial side of jaw, nuchal and frontal spines, synarcual; 4, mandibular spines (subsequently lost); 5, high dorsal pelvic process, mobile dorsal spine; 6, prepelvic clasper in males. *Age*: Carboniferous (A–C, E, F); Permian (D); Triassic to Recent (G). (A, C, E, F, from Lund (1985*b*); B, from Moy-Thomas (1936*a*) and Patterson (1965); D, based on Patterson (1968) and Bendix-Almgreen (1971).)

(Fig. 4.38). They resemble holocephalans in the position of their branchial apparatus, which lies below the braincase, and have therefore been gathered with holocephalans in the taxon Subterbranchialia (Zangerl 1981). Some iniopterygians have a free palatoquadrate (Fig. 4.38, 8), whereas others seem to be holostylic (Stahl 1980). The teeth are either arranged in families of separate denticles or form various types of peculiar whorls. There was prob-

ably had a single gill opening, covered by an opercular flap that was strengthened by long rays (Fig. 4.38, 5). In *Sibyrhynchus* and *Inioptera*, the snout is adorned with large, pointed scutes (Fig. 4.38B, 7). The pectoral fin skeleton has a large basal plate (Fig. 4.38, 1), which articulates with the dorsal part of a large, crescentiform scapulocoracoid (Fig. 4.38, 2). The ventral end of the latter articulates in turn with a basibranchial element. The foremost radial is very large in males and is covered with

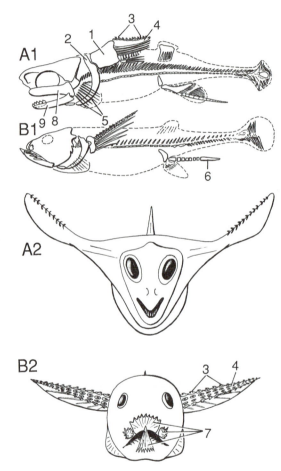

Fig. 4.38. Fossil Chondrichthyes. Iniopterygia, Carboniferous of North America. A, *Iniopteryx*, reconstruction of skeleton in lateral view (A1, ×0.3) and of the head and pectoral fins in anterior view (A2); B, *Sibyrhynchus*, reconstruction of skeleton in lateral view (B1, ×1.2), and of the head and pectoral fins in anterior view (B2). 1, pectoral basipterygium; 2, scapulocoracoid; 3, pectoral fin denticles (in males only); 4, enlarged radial; 5, opercular rays; 6, mixipterygial (pelvic) clasper; 7, rostral and symphysial scutes; 8, palatoquadrate; 9, Meckelian cartilage. (A1, B1, from Zangerl and Case (1973), A2, modified from Zangerl and Case (1973).)

a shagreen of spiny denticles, which give it the aspect of a large fin spine (Fig. 4.38, 3, 4). The pelvic metapterygium in males is modified into a rod-shaped pelvic clasper (Fig. 4.38, 6). Peculiar pouch-like organs are associated with the pelvic fins and have been interpreted as a poisonous gland. The trace of a spiral valve has been observed in the intestine of some specimens.

Although the overall aspect of iniopterygians recalls that of holocephalans, i.e. a large head, single gill opening, and slender tail, there is no clear evidence that the two groups are closely related, apart from the similar subcranial position of the branchial apparatus (Fig. 4.39, 12). If, however, one accepts the taxon Subterbranchialia, then the calcified rings of the lateral line must be assumed to have been lost—or not yet found—in iniopterygians.

4.13.8 Phylogeny and classification

We have seen that in addition to the prismatic calcified cartilage and pelvic claspers, all extant chondrichthyans (neoselachians and chimaeroids) share the presence of a basibranchial copula (Fig. 3.13, 13; Fig. 3.15, 15) and basal plates supporting the radials of paired and unpaired fins. Leaving aside the basibranchial copula, which is rarely observed in fossils, a glance at Devonian and Carboniferous chondrichthyans (referred to this group on the basis of the prismatic calcified cartilage) shows that some taxa do not display either pelvic claspers or basal plates (cladoselachids), whereas others have either no basal plates (symmoriids) or have them in the unpaired fins only (eugeneodontids). This suggests that some Palaeozoic chondrichthyan groups fall outside the ensemble formed by the fossil and extant elasmobranchs and holocephalans (crown-group chondrichthyans). It is also possible that some of the characters shared by these two groups, in particular the basal plates in fins, have appeared twice, as suggested by the fact that stem-group holocephalans, such as some bradyodonts or possibly symmoriids, have no basal plate in either unpaired or paired fins.

The cladogram in Fig. 4.39 illustrates two possible theories of relationships of extant and Palaeozoic chondrichthyans, combined to various extents from Schaeffer (1981), Zangerl (1981), and Young (1982a), Maisey (1984a), and Gaudin (1991). However, both rest on very tenuous synapomorphies. Only the relationships within the Holocephali (Fig. 4.37) and Elasmobranchii (Fig. 4.39, 6) respectively are reasonably reliable. There are different phylogenies, also based on cladistic analyses, but most of them differ in the position of the xenacanthiforms, cladoselachids, and symmoriids. Eugeneodontids and petalodontids are rarely considered, but Zangerl (1981) gathered them with the cladoselachids and symmoriids

into a clade. In fact, cladoselachids share at least with eugeneodontids the enlarged radials and neural spines in the tail (Fig. 4.39, 10). But petalodontids and eugeneodontids share the symphysial tooth series, as well as some histological characters (Fig. 4.39, 2). Schaeffer and Williams (1977) and Mader (1986) consider xenacanthiforms as the sister-group of all other elasmobranch, in which they include symmoriids and cladoselachids, whereas Young (1982a) assigns this position to the symmoriids. Mader (1986) considers cladoselachids as most closely allied to symmoriids (and stethacanthids). Gaudin (1991) made the most extensive, computer-assisted analysis of chondrichthyan phylogeny. He recognizes three clades, holocephalans, iniopterygians, and elasmobranchs, in an unresolved trichotomy. He puts cladoselachids, symmoriids, and elasmobranchs (as defined here) in a clade Elasmobranchii, characterized, for example, by the precerebral fontanelle, gill arches extending beyond the neurocranium, posteriorly directed hypobranchials, and large hypochordal lobe. The first two characters also occur in lampreys; the third is not clearly known in cladoselachids and symmoriids. Lund (1985a), however, described anteriorly directed hypobranchials in stethacanthids.

Moreover, there are many chondrichthyan taxa that are left aside from phylogenetic consideration, with the excuse that they are too poorly known, but which, in fact, are hard to place anywhere. This is, for example the case of the Devonian and Carboniferous 'Orodontida', with hybodont-like, low-crowned teeth made up of tubate dentine, eel-shaped body, and reduced paired fins. The histological resemblance between orodontid and bradyodonts teeth is difficult to assess. Lund (1990) puts them among holocephalans in a large clade Paraselachimorpha, which also includes iniopterygians, petalodontids, and eugeneodontids. The Carboniferous genus *Polysentor*, which is a holocephalan-like but seems to have been hyostylic is another 'non-aligned' taxon. Yet another Carboniferous form, *Squatinactis*, displays some ray-like features (forwardly expanded pectoral fins, caudal spine), but its teeth are of 'cladodont' type. There are also a large number of 'ichthyoliths', that is, isolated spines, scales, and teeth of chondrichthyans, which do not clearly belong to any of the clades defined here. For example, chondrichthyan ichthyoliths from the Middle and perhaps Early Devonian of Bolivia include peculiar triangle-shaped or recurved and spiny plates, which have turned out to be fused radials embedded in a mass of calcified cartilage (*Zamponiopteron*). They are found together with endoskeletal elements (*Pucapampella*) that recall the synarcual of holocephalans (Janvier and Dingerkus 1991). These ichthyoliths are therefore referred to the holocephalans (in this case, the earliest known evidence of the

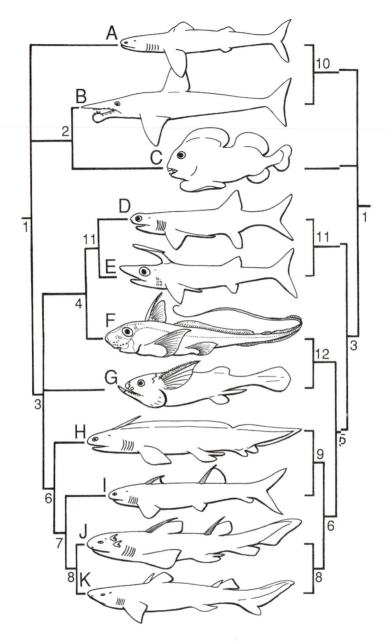

group). The peculiar, yet poorly preserved, Early Devonian Stensioellida and Pseudopetalichthyida, which are classically referred to as placoderms (see p. 171), somewhat resemble the 'bradyodont' *Menaspis* and may also be early holocephalans.

A classification of the Chondrichthyes, based on the cladogram in Figs 4.37 and 4.39 (left-hand side) would be as shown in Table 4.6.

4.13.9 Stratigraphical and geographical distribution, habitat

The earliest known fossils that are thought to be chondrichthyan remains are isolated scales referred to as *Mongolepis* (Fig. 4.30A), *Udalepis*, *Sodolepis*, *Elegestolepis* (Fig. 4.30B), and *Niualepis*, from the Early Silurian of Mongolia and Siberia (Karatajute-Talimaa

Fig. 4.39. Fossil Chondrichthyes. Two theories of interrelationships of the fossil and extant Chondrichthyes (some Palaeozoic taxa omitted). *Terminal taxa*: A, Cladoselachidae; B, Eugeneodontida; C, Petalodontida; D, Symmoriidae; E, Stethacanthidae; F, Holocephali; G, Iniopterygia; H, Xenacanthiformes; I, *Ctenacanthus*; J, Hybodontiformes; K, Neoselachii. *Nested taxa and selected synapomorphies*: 1, Chondrichthyes (prismatic calcified cartilage, ?loss of anal fin); 2, large median symphysial teeth in lower and upper jaw, incipient tubate dentine covered with orthodentine; 3, mixipterygial (pelvic) claspers; 4, calcified rings in sensory-line canals, frontal spines or denticles; 5, crown-group Chondrichthyes (all premetapterygial radials of the pectoral fin articulate with compound basal cartilages, pelvic metapterygium articulates with all but the first few radials); 6, Elasmobranchii (posteriorly directed hypobranchials, basibranchials separated from basihyal, tribasal paired fins, ?anal fin reappeared, two fin spines [except in xenacanthiforms, by reversion]); 7, fin spine base deeply embedded between the myomeres; 8, Euselachii (shoulder joint with strong propterygial support, enlarged pelvic propterygium); 9, Ctenacanthiformes (large fin spines with pectinate ornamentation, lost in advanced xenacanthids, broad and expanded occipital region of braincase); 10, neural arches and dorsal radials of chordal lobe enlarged in the form of calcified plates, reduced pelvic fin, unsegmented pectoral radials; 11, Symmoriida (arrow-shaped posterior dorsal fin plate, enlarged pelvic plate, very long metapterygial axis of pectoral fin); 12, Subterbranchialia (branchial arches situated ventrally to the braincase, large opercular cartilage and rays). *Age*: ?Early Devonian to Cretaceous (J); Middle Devonian to Triassic (H); Middle Devonian to Recent (F); Late Devonian (A); Late Devonian to Carboniferous (E, I); Carboniferous (D, G); Carboniferous to Permian (C); Carboniferous to Triassic (B); Triassic to Recent (K).

Table 4.6. **Classification of the Chondrichthyes** (See also Figs 4.37 and 4.39)

Chondrichthyes
  Cladoselachidae *sedis mutabilis*
  (unnamed taxon) *sedis mutabilis*
    Eugeneodontida
    Petalodontida
  Crown-group Chondrichthyes *sedis mutabilis*
    (unnamed taxon) *sedis mutabilis*
      Symmoriida
      Holocephali *sedis mutabilis*
        Helodontidae
        Chondrenchelyidae
        Menaspidae
        Cochliodontidae
        Echinochimaeridae
        Chimaeroidei
      Iniopterygia *sedis mutabilis*
      Elasmobranchii *sedis mutabilis*
        Xenacanthiformes
        Ctenacanthiformes
        Euselachii
          Hybodontiformes
          Neosalachii

1973, 1992; Karatajute Talimaa *et al.* 1990). The apparently compound scales of *Mongolepis* bear only a vague resemblance to some Devonian and Carboniferous compound chondrichthyan scales (Fig. 4.36A2). However, sections of the scales clearly show that the crown consists of closely set odontodes (Fig. 4.30A, 1) attached to a growing base (Fig. 4.30A, 2), with several basal and neck canals (Fig. 4.30A, 3), which are characteristic of chondrichthyans. These odontodes are made up of 'lamelline', a unique kind of dentine that does not contain dentine tubules and has a lamellar structure. More convincingly chondrichthyan-like are the scales referred to as *Elegestolepis* (Karatayute-Talimaa 1973; Fig. 4.30B),

from the Early Silurian of Central Asia (Tuva). These are very small scales, with a basal canal (Fig. 4.30B, 5), a neck canal (Fig. 4.30B, 3), and a pulp cavity. The base is bulged or flattened, and they resemble the placoid scales of euselachians (hybodontiforms and neoselachians). Karatayute-Talimaa (1992) has recognized nine different types of supposedly chondrichthyan scale structures in the Silurian and Devonian, all derived from individual odontodes of lamelline or orthodentine. *Pilolepis* (Fig. 4.30C), from the Early Silurian of Arctic Canada, is represented by isolated scutes that vaguely resemble those of menaspid holocephalans (Thorsteinsson 1973) but are regarded by some authors as climatiid acanthodian scutes. Although these scales may well belong to chondrichthyans, the only clear evidence of the group is the presence of prismatic calcified cartilage (the unique character of the clade), which first occurs at the end of Early Devonian times. This may be due to mere hazards of preservation, but may also mean that this type of hard tissue did not appear in chondrichthyans before this time. Nevertheless, undoubted chondrichthyan diplodont and cladodont teeth occur as early as the Late Pragian (late Early Devonian). It is worthy noting that chondrichthyan teeth are very rarely found before the Middle Devonian, and are so far completely unknown, in the Silurian. This suggests either that tooth replacement was very slow in early chondrichthyans (Maisey 1984a, Patterson 1992) or that because of very small size they have gone unnoticed (Turner and Young 1987).

The idea put forward by Mader (1986) that the so-called 'diplodont' teeth, which are met with in xenacanthiforms, represent the primitive tooth type for all chondrichthyans is based only on stratigraphical arguments and is not consistent with the phylogenies proposed here. It could however, be supported by the fact that the branchial denticles in most chondrichthyans and osteichthyans, which can be regarded as a generalized type of pharyngeal denticle, often resemble 'diplodont' teeth.

The earliest known chondrichthyan from which substantial and more-or-less articulated endoskeletal remains are known is *Antarctilamna*, from the late Early and Middle Devonian (Fig. 4.34A). It has 'diplodont' teeth and an elongated braincase and probably belongs to the Xenacanthiformes. By Late Devonian times, complete, articulated chondrichthyans become more and more abundant and, in the Early Carboniferous, most major chondrichthyan monophyletic groups are represented. Apart from euselachians and chimaeroids, virtually no other group survived the Permian–Triassic boundary. Only a few eugeneodontids (*Helicampodus*, *Parahelicampodus*) and possibly some xenacanthiforms survived into the Early Triassic. Supposedly Mesozoic ctenacanthiform teeth and spines (e.g. *Pheobodus*) bear only general elasmobranch characters.

Special mention should be made of the so-called 'anachronic' neoselachian remains. Teeth of 'neoselachian type', i.e. with one or two large nutrient foramina piercing the base anteriorly, are now known to occur in the Carboniferous (*Anachronistes*, Duffin and Ward 1983; Fig. 4.40G), and even the early Middle Devonian (*Mcmurdodus*, Turner and Young 1987; Fig. 4.40D1), although the earliest articulated neoselachians are Early Jurassic in age. *Mcmurdodus* is even more surprising by its remarkable resemblance to the teeth of extant hexanchiform neoselachians (Fig. 4.40D3). Does this mean that neoselachians existed as early as the Middle Devonian? The histology of these 'anachronic' teeth has not yet been properly investigated to check whether they have the characteristic haphazardly fibred enameloid, which is regarded as a unique neoselachian character. Only *Anachronistes* has been shown to lack this enameloid (Thies and Reif 1985).

Some Palaeozoic placoid-like scales and fin spines have also been regarded as of neoselachian character. This is the case, for instance, for the Early Devonian scales referred to as *Ellesmereia* (Fig. 4.40E) and, to some extent, the Silurian *Elegestolepis* (Fig. 4.30B). Such a very early age for neoselachians may perhaps be consonant with certain elasmobranch phylogenies based only on Recent forms, in particular those in which *Chlamydoselachus* is regarded as the sister-group of all other Recent and fossil elasmobranchs, and those in which *Heterodontus* is regarded as closely related to the hybodontiforms. Given the present state of knowledge, such a change in elasmobranch relationships would imply too many homoplasies or reversions (e.g., in the structure of the enameloid, paired fins, and braincase), and it is therefore regarded here as unlikely (see for review Maisey 1984a,b). Only the Early Devonian scales of *Arauzia* (Fig. 4.40F) are strikingly hybodontiform-like,

and suggest that euselachians, at least, can occur rather early and, consequently, that the diversification of the crown-group chondrichthyans occurred either in the Early Devonian or the Silurian.

Among Recent chondrichthyans, all holocephalans and the majority of elasmobranchs are bound to marine environments. Only a few rays can live permanently in rivers, and a few marine sharks may survive and even mate in freshwater. The question of the habitat of Palaeozoic chondrichthyans follows, as usual a circular line of argument. Since this group is regarded as marine, the localities in which it occurs are said to be marine or brackish-water. However, when chondrichthyans occur in large numbers with typically continental tetrapods, for example, then they are said to be freshwater (e.g. many Mesozoic hybodontiforms). In the Palaeozoic, the only likely instance of a freshwater chondrichthyan is *Expleuracanthus*, which occurs widely in Carboniferous and Permian intramontane basins.

Some Palaeozoic chondrichthyan taxa seem to have a restricted geographical distribution: iniopterygians in North America, *Zamponiopteron* in South America, or *Antarctilamna* in Gondwanan regions. This may, however, be due to mere chance of preservation. It is likely that most Palaeozoic chondrichthyans, being marine epipelagic animals, had a rather wide distribution.

Further reading

Silurian chondrichthyans: Karatayute-Talimaa (1973, 1992); Karatayute-Talimaa *et al.* (1990). *Cladoselachidae*: Bendix-Almgreen (1975); Dean (1909); Maisey (1984a, 1989a). *Eugeneodontida*: Nielsen (1952); Zangerl (1966, 1981). *Petalodontida*: Lund (1977a, 1983, 1989); Didier *et al.* (1994); Zangerl (1981); Zangerl *et al.* (1993). *Symmoriida*: Lund (1984, 1985a,b, 1986a, 1990), Zangerl (1984), Zangerl and Case (1976); Zangerl and Williams (1975); Williams (1985). *Elasmobranchii*: Cappetta (1987); Maisey (1975, 1982, 1983, 1984a, 1989b); Poplin and Heyler (1989); Schaeffer (1981); Zangerl (1981); Young (1982a). *Holocephali (including bradyodonts)*: Bendix-Almgreen (1968, 1971); Lund (1977b,c, 1982, 1986b,c, 1990); Moy-Thomas (1936a); Patterson (1965, 1968, 1992). *Iniopterygia*: Zangerl and Case (1973); Stahl (1980). *Early Neoselachii and enigmatic taxa*: Duffin and Ward (1983); Maisey (1976, 1980, 1984b, 1985); Reif (1973); Thies and Reif (1985); Turner and Young (1987). *Interrelationships and systematics*: Capetta (1987); Gaudin (1991); Lund (1985a); Mader (1986); Maisey (1984a,b); Schaeffer (1981); Schaeffer and Williams (1977); Thies and Reif (1985); Young (1982a); Zangerl (1973, 1981).

4.14 PLACODERMI

The placoderms are an extinct clade of gnathostomes characterized by a dermal armour which is made up of two

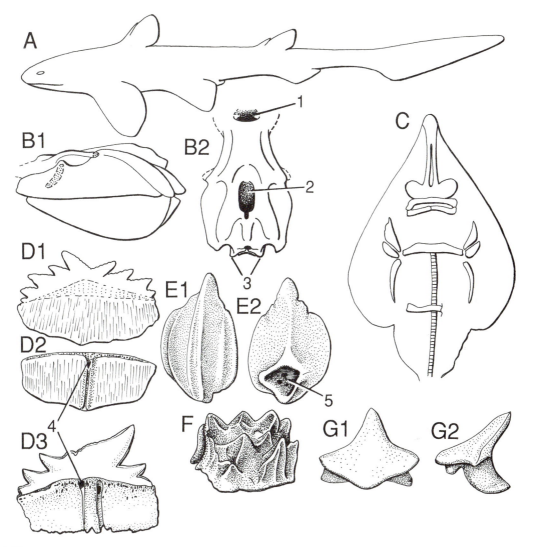

Fig. 4.40. Fossil Chondrichthyes. Early Euselachii and Neoselachii, or Palaeozoic neoselachian-like taxa. A, *Palaeospinax*, Early Jurassic, reconstruction in lateral view (about ×0.3); B, *Synechodus*, Late Cretaceous, skull in lateral (B1) and dorsal (B2) view (×0.7); C, *Spathobatis*, Late Jurassic, outline of the head and anterior part of the body in ventral view (about ×0.3); D, *Mcmurdodus*, Early to Middle Devonian of Antarctica and Australia, tooth in anterior (D1) and basal (D2) view (×7), compared to a tooth of the modern squalomorph *Echinorhinus* (D3); E, *Ellesmereia*, Early Devonian of Canada, scale in external (E1) and basal (E2) views (×60); F, *Arauzia*, Early Devonian of Spain, scale in lateral view (×30); G, *Anachronistes*, Early Carboniferous of Great Britain, tooth in labial (G1) and lateral (G2) view (×15). 1, anterior (precerebral) fontanelle; 2, endolymphatic fossa; 3, occipital condyles; 4, nutrient canal; 5, pulp cavity. (A, based on Dean (1909); B, from Maisey (1985); C, D3, from Cappetta (1987); D1, 2, from Turner and Young (1987); E, from Vieth (1980); F, redrawn from a photograph in Mader (1986); G, from Duffin and Ward (1983).)

parts, and covers the head and the anterior part of the trunk respectively. There is a single exception in *Synauchenia* (see p. 159 Fig. 4.47C), in which the two divisions are fused into a rigid box. Placoderms are known from the late Early Silurian (Wenlock) to the Late Devonian (Famennian), when they seem to undergo a sudden extinction.

4.14.1 General characters

The head armour is primitively not articulated with the thoracic armour, but an articulation device has developed independently in four groups (the arthrodires, antiarchs, petalichthyids, and ptyctodonts). This articulation is always borne by the anterior dorsolateral plate of the thoracic armour and the paranuchal plate of the skull-roof (Fig. 4.41, 11, 16). Even in placoderms with a reduced and very short thoracic armour, the dermal shoulder girdle can be distinguished from that of osteichthyans by the presence of at least one median dorsal plate that links the dermal bone series of either sides (Fig. 4.41, 15). Another placoderm characteristic, assumed to have been lost independently, in several taxa, is a special type

of dermal hard tissue called semidentine (Fig. 4.43, 1). This is a type of dentine in which the odontocytes were enclosed and unipolar in shape. The cell spaces in semidentine are thus drop-shaped (Fig. 4.43, 3), with a single, outwardly directed prolongation (Fig. 4.43, 4). This tissue occurs in the superficial part of the tubercles that cover the dermal bones of the armour (unless there is reduction of the superficial ornamentation), as well as in the jawbones or gnathal plates.

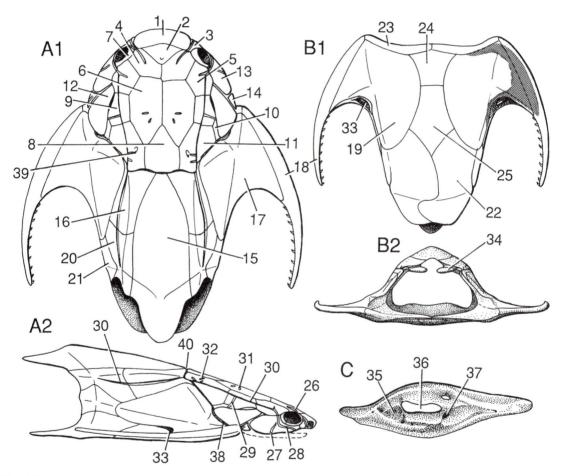

Fig. 4.41. Placodermi. Arthrodira. External morphology. A–C, *Dicksonosteus*, Early Devonian of Spitsbergen. A, head and thoracic armour in dorsal (A1) and lateral (A2) view (×1.5); B, thoracic armour in ventral (B1) and anterior view (B2) (outline of scapulocoracoid stippled); C, pectoral fenestra (×3). 1–25 *dermal plates of armour*. 1, rostral; 2, pineal; 3, postnasal; 4, preorbital; 5, postorbital; 6, central; 7, sclerotic ring; 8, nuchal; 9, marginal; 10, postmarginal; 11, paranuchal; 12, submarginal; 13, suborbital; 14, postsuborbital; 15, median dorsal; 16, anterior dorsolateral; 17, anterolateral; 18, spinal; 19, anterior ventrolateral; 20, posterior dorsolateral; 21, posterolateral; 22, posterior ventrolateral; 23, interolateral; 24, anterior median ventral; 25, posterior median ventral; 26–32: *sensory-line grooves*: 26, supraorbital; 27, supraoral; 28, postorbital and infraorbital; 29, postmarginal; 30, main lateral line (thoracic and cephalic parts); 31, middle pit line; 32, posterior pit line. *Others*: 33, pectoral fenestra; 34, condyle for cranio-thoracic articulation; 35, muscle insertion areas; 36, articulation area for fin skeleton; 37, canal for brachial blood vessel; 38, gill opening; 39, endolymphatic duct; 40, limit between the head and trunk armours. (A–C, from Goujet (1984*a*).)

The head and thoracic armours consist of several bony plates, the pattern of which is characteristic of each major placoderm group. In some placoderms (petalichthyids, rhenanids, some acanthothoracids), there are, in addition to the larger plates, a number of minute platelets often referred to as 'tesserae'. These are superimposed on the large dermal plates, and probably represent a derived condition (Figs 4.50A, 4.55A, 4.57H).

The head armour comprises a skull-roof and a variable number of cheek plates (up to three), which form a mobile operculum covering the gill chamber (Fig. 4.41, 12, 13, 14). There was a single branchial opening, situated between the cheek and the thoracic armour (Fig. 4.41, 38). The orbit is generally large and the eye is protected by a sclerotic ring made up of four plates (Fig. 4.41, 7). As usually reconstructed, the nostrils are anteroventral in position, as in sharks, but in some forms (e.g. rhenanids and antiarchs; Figs 4.52B, 4.55) they were situated dorsally. Where gnathal plates are known, there is one pair in the lower jaw (inferognathals; Fig. 4.42, 38) and one or two pairs in the upper jaw (superognathals; Fig. 4.42, 16, 17). In some placoderms the palate bears a small parasphenoid surrounding the openings for the carotid arteries and the buccohypophysial canal, which is sometimes paired (Fig. 4.42, 3, 4). Whether this dermal bone is homologous to the parasphenoid of osteichthyans is still a matter of debate.

The skull-roof, which is firmly attached to the braincase, consists of several plates, which overlap or are united by complex sutures. A pineal foramen is sometimes present, and pierces a separate pineal plate (Fig. 4.41, 2). In the posterior part of the skull-roof, the endolymphatic ducts open to the exterior by a minute foramen that pierces either the nuchal or paranuchal plates (Fig. 4.41, 39).

The thoracic armour is also composed of overlaping or sutured plates, the pattern of which is somewhat more constant than those of the head armour. There are always one or two median dorsal plates, and also paired anterior dorsolateral, anterolateral, and anterior ventrolateral plates (Fig. 4.41, 15, 16, 17, 19).

When found, the tail is epicercal and there is one dorsal fin (Figs 4.47A, 4.52A). The pectoral fins are generally well developed, and are covered with minute scales (Fig. 4.50A, 17), except in a specialized group, the antiarchs, where they are covered with dermal plates and modified into peculiar appendages (Fig. 4.52, 23–8). The pelvic fins are small. In only one group, the ptyctodonts (Fig. 4.51, 23, 24), the pelvic fins of males bear hook-shaped dermal scutes on the metapterygium. These are reminiscent of the claspers of chondrichthyans, but this homology is still debated and cannot be generalized to all placoderms. There is no clear evidence of an anal fin in placoderms.

The unarmoured parts of the placoderm body is sometimes covered with scales, which are extremely variable in shape, size, and thickness. When present, the scales are often diamond-shaped, and tuberculated. The arthrodire *Sigaspis* (Fig. 4.45A), for example, and the antiarch *Pterichthyodes* (Fig. 4.53E) have very large scales, whereas the petalichthyid *Lunaspis* (Fig. 4.50A) has smaller ones, and the rhenanid *Gemuendina* (Fig. 4.55A) has polygonal scales over most of the body. The majority of the placoderms seem, however, to have had an almost naked trunk and tail.

The endoskeleton of placoderms consists of cartilage lined with perichondral bone, as in osteostracans or galeaspids among jawless vertebrates. This condition is regarded as general for the gnathostomes. The endocranium consists of a flattened braincase with long lateral processes, sometimes with a separate rhinocapsular ossification housing the nasal capsules (Fig. 4.42, 18), a palatoquadrate (attached laterally to the cheek plates, in particular the suborbital plate; Fig. 4.42, 10), a Meckelian bone (Fig. 4.42, 39), a hyoid arch, and probably five branchial arches, which are very rarely ossified. In the hyoid arch, the epihyal (Fig. 4.42, 14) is expanded and attached to the internal surface of the submarginal plate of the cheek (Fig. 4.42, 12). Strangely, the ceratal and basal elements of the visceral arches are always poorly ossified; even the Meckelian bone is ossified only in its anterior (mentomandibular) and posterior (articular) parts (Fig. 4.47, 18; Fig. 4.48, 4). The sclera of the eyeballs is sometimes perichondrally ossified (Fig. 4.42D, 36). In probably all placoderms, the eyes were attached to the braincase by means of a cartilaginous eye-stalk (Fig. 4.42, 31, 35), a character that is also shared by elasmobranchs among chondrichthyans (Fig. 3.13G, 20).

The axial skeleton consists only of neural and haemal arches (basidorsals and basiventrals respectively; Fig. 4.47A1). Anteriorly, the vertebral elements are fused into a synarcual (Fig. 4.55, 18) on which rest the median dorsal plates. The endoskeletal shoulder girdle is a single element, the scapulocoracoid, which lies against the internal surface of the thoracic armour (Fig. 4.41B1, grey; Fig. 4.47, 5; Fig. 4.48, 9; Fig. 4.55, 15). The pectoral fin endoskeleton consists of unsegmented radials that articulate either on basal plates, or separately on the scapulocoracoid. The latter condition, met with in some brachythoracid arthrodires, is regarded as derived. In advanced antiarchs (euantiarchs), the articulation device of the pectoral fin was essentially dermal. The pelvic girdle, when present, is a very small element in the body wall (Fig. 4.47, 1).

It is virtually impossible to imagine a hypothetical generalized placoderm, or to select one particular taxon that would approach this morphotype, because there is still

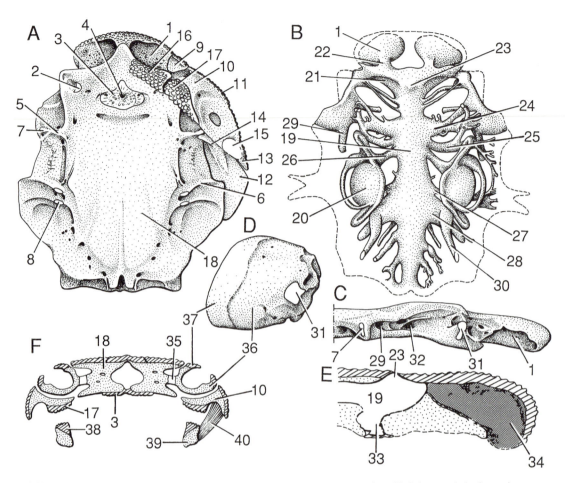

Fig. 4.42. Placodermi. Arthrodira. Internal anatomy. A–F, *Dicksonosteus*, Early Devonian of Spitsbergen. A, braincase in connection with the palatoquadrate, and the suborbital and postnasal plates on the right-hand side (anterior superognathal restored) (×2); B, outline of the braincase, with internal cast of the cavities and canals in dorsal view; C anterior part of the braincase in lateral view, showing the orbital cavity (×6); D, eye capsule in ventral view (×6); E, sagittal section of the snout (exoskeleton obliquely hatched, perichondral bone black, endoskeleton of braincase stippled, endoskeleton of rhinocapsular bone grey); F, reconstructed transverse section through the head at the level of the eyes, with jaw musculature added on the right-hand side. 1, nasal cavity; 2, area of attachment for the palatoquadrate; 3, parasphenoid; 4, buccohypophysial foramen; 5, anterior postorbital process; 6, posterior postorbital process; 7, area of attachment for hyomandibula; 8, area of attachment for gill arches; 9, postnasal plate; 10, palatoquadrate; 11, suborbital plate; 12, submarginal plate; 13, postsuborbital plate; 14; hyomandibula; 15, articulation for the Meckelian cartilage; 16, anterior superognathal; 17, posterior superognathal; 18, braincase; 19, brain cavity; 20, labyrinth cavity; 21, optic tract; 22, olfactory tract; 23, pineal recess; 24, trigeminus nerve; 25, facial nerve; 26, acoustic nerve; 27, glossopharyngeus nerve; 28, vagus nerve; 29, jugular vein; 30, endolymphatic duct; 31, area for attachment of eye-stalk; 32, posterior myodome; 33, hypophysial fossa; 34, rhinocapsular ossification; 35, eye-stalk; 36, scleral ossification; 37, sclerotic ring; 38, inferognathal; 39, Meckelian cartilage; 40, adductor jaw muscle. (A–E, from Goujet (1984*a*).)

much controversies about the polarity and distributions of various character states in the group. As we shall see below, one of the current theories of phylogenetic interrelationships of the placoderms implies that the Early Devonian 'acanthothoracids' (also known as 'palae-acanthaspids', Fig. 4.56) are an assemblage of generalized

members of various clades (Fig. 4.57D, F, H). If this is true, 'acanthothoracids' such as *Radotina* or *Romundina* may display a rather generalized placoderm armour. However, recent discoveries of Early Silurian placoderms—the earliest representatives of the group—which display a mixture of characters formerly believed to be

unique to other taxa, will lead to radical changes in placoderm phylogeny and character-state polarities. For this
reason, we prefer to describe here the major placoderm
clades on the basis of what are presumed to be
generalized representatives of each of them.

4.14.2 Arthrodira

Arthrodires are characterized by two gnathal plates in the
upper jaw: the smaller anterior superognathal, and the
larger posterior superognathal (Fig. 4.42, 16, 17), which
both bite against the inferognathal of the lower jaw
(Fig. 4.42, 38). The arthrodires are by far the most diverse
and abundant of all placoderms, their size ranging from a
few centimetres up to five or six metres in the largest Late
Devonian forms such as *Dunkleosteus* or *Titanichthys*
(Fig. 1.15, 1, 2).

The arthrodires are divided into three major groups: the
Actinolepidoidei (Fig. 4.49A, B), probably including the
Phyllolepida (Fig. 4.49B), which were once regarded as
not being arthrodires; the Phlyctaenii (Fig. 4.49C); and
the Brachythoraci (Fig. 4.49, 4). Only the second and third
of these groups possess an articulation device between the
skull-roof and the thoracic armour. It consists of a condyle
on the anterior dorsolateral plate (Fig. 4.41, 34) of the
thoracic armour, and a corresponding fossa in the
paranuchal plate of the skull-roof. In order to work, this
dermal articulation has to be in the same horizontal plane
as that between the braincase and the synarcual. In
actinolepids, there is a simple sliding contact between the
skull-roof and the thoracic armour.

4.14.2.1 Exoskeleton and general morphology

Both the actinolepids and phlyctaeniids have a rather
elongated thoracic armour (Figs 4.41, 4.45A). The pectoral fenestra (Fig. 4.41, 33), through which the pectoral
fin attaches on the scapulocoracoid, is closed posteriorly.
In contrast, many brachythoracids have a comparatively
short thoracic armour and, in the most derived states, a
posteriorly open pectoral fenestra (Fig. 4.45F, 3; Fig.
4.48, 7). If phlyctaeniids are regarded as more closely
related to brachythoracids than to actinolepids, as is
indicated by the shared dermal neck joint, then the
elongated thoracic armour is a general condition for the
arthrodires (Fig. 4.49).

In most textbooks the arthrodires are exemplified by
the genus *Coccosteus* (Fig. 4.47A), a Middle Devonian
brachythoracid. I have chosen here the Early Devonian
phlyctaeniid *Dicksonosteus* (Figs 4.41, 4.42), in which the
endoskeleton is known to a large extent.

Dicksonosteus, from the Early Devonian of
Spitsbergen, was a decimetre-long fish with a dermal

armour ornamented with rounded tubercle of semidentine
(Fig. 1.5, 1). The head is rather depressed dorsoventrally,
and the thoracic armour is armed with long, denticulated
spinal plates (Fig. 4.41, 18). The skull-roof consists of
several dermal plates (see Fig. 4.41 for the names of these
plates), which do not overlap. The cheek consist of three
separate plates; the submarginal plate, which served as an
operculum, is attached to the epihyal (Fig. 4.42, 12, 14).
The plates of the thoracic armour (again see Fig. 4.41 for
the names of these plates) have narrow overlap areas. The
cranio-thoracic articulations (Fig. 4.41, 34) are situated
close to the midline, which is regarded as a general condition for phlyctaenioids (phlyctaeniids + brachythoracids;
Fig. 4.49, 3). The jaws are armed with tuberculated
gnathal plates. The anterior superognathal (Fig. 4.42, 16)
is attached on the ectethmoid process of the braincase.
The posterior superognathal (Fig. 4.42, 17) rests on the
palatoquadrate, which, in turn, expands over the medial
surface of the suborbital plate and articulates with the
braincase by means of the ethmoidian articulation
(Fig. 4.42, 2, 10). The inferognathal is a long rod that
rests on the Meckelian bone and bears numerous denticles
(Fig. 4.42, 39). The parasphenoid is small, triangle-
shaped, ornamented with tubercles, and pierced by a large
buccohypophysial canal (Fig. 4.42, 3, 4).

The sensory-line system was housed in open grooves
or pit lines on the surface of the dermal bones (see
Fig. 4.41A2 for the names of the sensory lines).

4.14.2.2 Endoskeleton

The braincase is much flattened and sends off laterally
two long postorbital processes (Fig. 4.42, 5, 6). The
anterior portion of the braincase (Fig. 4.42, 34), which
contains the nasal capsules, is separated from the rest of
the braincase by a perichondrally ossified wall that marks
the position of a discontinuity known as the cranio-
ethmoidian fissure. In several other arthrodires, this
fissure is still present and results in a total separation
between the braincase and the rhinocapsular ossification
(e.g. *Buchanosteus*, Fig. 4.46B; for discussion, see
Chapter 5, p. 243). The cavities for the olfactory organs
are large and open ventrally (Fig. 4.42, 1). The orbital
cavity is hollowed by mydomes, in particular a deep
posterior myodome for the recti eye muscles (Fig. 4.42,
32). There was a cartilaginous eye-stalk linking the
eyeball to the orbital wall (Fig. 4.42, 31, 35). The ventral
surface of the braincase bears articular facets for the
hyoid and gill arches (Fig. 4.42, 7, 8). The ossified sclera
of the eye is cup-shaped and is pierced by canals for the
optic nerve and ophthalmic blood vessels (Fig. 4.42, 36).

As in osteostracans, the perichondral bone lamella
closely lines the canals and cavities inside the braincase,
and thus provides a rather accurate 'cast' of the internal

organs, such as the brain, labyrinth nerves, and blood vessels (Fig. 4.42B). Since Stensiö's (1969a) first descriptions of the neurocranial anatomy of arthrodires, which were based on the actinolepid *Kujdanowiaspis*, it has been traditional to compare the neurocranial anatomy with that of elasmobranchs. In fact, most of the features of the labyrinth cavity or cranial nerve canals are simply general gnathostome features, or even vertebrate features. In phlyctaeniid arthrodires, the position of the large labyrinth cavities (Fig. 4.42, 20), well apart from the brain cavity (Fig. 4.42, 19), is probably linked with the flattened shape of the braincase.

In the thoracic armour, the scapulocorcoid is an elongated mass of cartilage pierced by vascular and nerve canals (grey in Fig. 4.41B1). The surface for the attachment of the pectoral fin bears a keyhole-shaped articular faced for the fin skeleton (Fig. 4.41, 36), and depressions for the muscle insertions (Fig. 4.41, 35). The pectoral fin was thus clearly stenobasal in structure, although very little is known of the pectoral fin endoskeleton.

Nothing is known of the axial skeleton of *Dicksonosteus*, but the internal surface of the median dorsal plate bears a median groove, which probably housed the synarcual. In other plyctaeniids (*Groenlandaspis*) and in actinolepidoids (phyllolepids), the vertebral column consists of ventral and dorsal arcualia.

4.14.2.3 Histology

The exoskeleton of the arthrodires consists of thick plates of cellular bone, covered with costate tubercles. In the tubercles, the semidentine occupies the external layer (pallial semidentine; Fig. 4.43A). In many brachythoracids, the external ornamentation has, however, disappeared, and the dermal plates were probably covered by skin.

4.14.2.4 Diversity and phylogeny

The diversity of arthrodires is perhaps the most important among Devonian vertebrates. The actinolepidoids (actinolepids, phyllolepids, and wuttagoonaspids; Fig. 4.49, 2) are regarded as the sister-group of the phlyctaenioids (phlyctaeniids and brachythoracids), which share the cranio-thoracic dermal articulation (Fig. 4.49, 3).

The Actinolepida are a small group of Early Devonian arthrodires, with a rather elongated thoracic armour, moderately long spinal plates, and often ornamented with large tubercles arranged in concentric rows. Among the best-known actinolepids are *Bryantolepis*, from the Early Devonian of the USA, and *Sigaspis* (Fig. 4.45A), from the Early Devonian of Spitsbergen, the latter being known to have possessed large body scales. A peculiar placoderm, *Wuttagoonaspis* (Fig. 4.44C), from the Middle Devonian of Australia, is regarded as closely related to actinolepids and phyllolepids. Its suborbital plate is fused

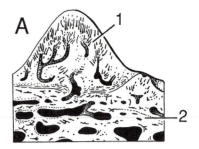

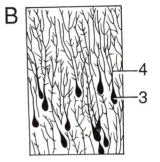

Fig. 4.43. Placodermi. Arthrodira. Histology of exoskeleton. A, vertical section through a tubercle of the dermal bone ornamentation (×35); B, detail of the organization of the odontocyte spaces in the semidentine (×350). 1, semidentine; 2, dermal bone; 3, odontocyte space; 4, odontocyte process (semidentine canalicule). (A, from Ørvig (1957); B, from Ørvig (1967a).)

to the skull-roof and only its submarginal plate is free (Fig. 4.44C, 15).

The Phyllolepida (Fig. 4.44A, B) have long been regarded as a separate placoderm taxon, of the same rank as the arthrodires. It is now suggested that they form, together with the actinolepids and wuttagoonaspids, a clade, the Actinolepidoidei, in which they are the sister-group of the wuttagoonaspids. This clade is, however, poorly supported. Phyllolepids have a dorsoventrally depressed armour, and are characterized by a very large nuchal plate which occupies two-thirds of the surface of the skull roof (Fig. 4.44A, 1). There is no cranio-thoracic articulation, and the thoracic armour lacks posterior lateral and dorsolateral plates. The latter character was the major argument in excluding them from the Arthrodira. Like arthrodires, they possess a large parasphenoid (Fig. 4.44, 12), but they seem to lack the anterior superognathal. The ornamentation of the dermal bones in most phyllolepids consists of sharp, concentric ridges of bone, which can be derived from the concentric rows of large tubercles of actinolepids; the jaws bear one upper and one lower pair of tuberculated gnathal plates armed with

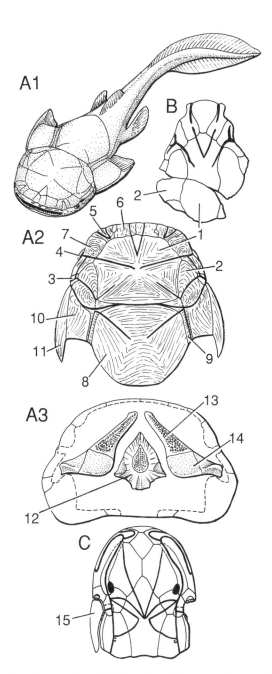

Fig. 4.44. Placodermi. Arthrodira. Phyllolepida and Wuttagoonaspida. A, *Austrophyllolepis*, Middle Devonian of Australia, reconstruction (A1, ×0.3), armour in dorsal view (A2) and restored palate (A3); B, *Antarctaspis*, Middle Devonian of Antarctica, skull-roof (×0.3); C, *Wuttagoonaspis*, Middle Devonian of Australia, skull in dorsal view (×0.3). 1, nuchal plate; 2, paranuchal plate; 3, marginal plate; 4, suborbital plate; 5, postnasal plate; 6, preorbital plate; 7, postorbital plate; 8, median dorsal plate; 9, anterior dorsolateral plate; 10, anterolateral plate; 11, spinal plate; 12, parasphenoid; 13, superognathal; 14, palatoquadrate; 15, submarginal plate. (A, from Long (1984); B, from E. I. White (1968); C, from Ritchie (1973).)

pointed cusps (Fig. 4.44, 13). The tail is rather long, and the pelvic fins are well developed, possibly with a metapterygial clasper (Fig. 4.44A1; see also Chapter 5 p. 244). *Antarctaspis* (Fig. 4.44B), from the Middle Devonian of Antarctica, is regarded as a rather generalized phyllolepid, on the basis of its very large nuchal plate and the pattern of its sensory-line grooves.

The Phlyctaenii (Fig. 4.49C) generally have an elongated trunk armour, with a long and narrow median dorsal plate (Fig. 4.41, 15). The spinal plate may be very long, slender, and denticulated (*Dicksonosteus*, *Arctolepis*). In one phlyctaeniid family, the Groenlandaspididae (*Tiaraspis*, *Groenlandaspis*), the thoracic armour is shortened, and the median dorsal plate is formed into a very high median crest (Fig. 4.45B, 1). Most phlyctaeniids are Early or Middle Devonian in age, and only the Groenlandaspididae remain abundant until the very end of the Late Devonian.

The Brachythoraci (Fig. 4.49, 4), which are particularly abundant in the Middle and Late Devonian, are easily recognizable by their rather thick dermal bones with overlapping or sutured margins. The inferognathal comprises a biting part and a large posterior, non-biting blade (Fig. 4.47, 17; Fig. 4.48, 3). The median dorsal plate has a well-developed ventral keel (Fig. 4.48, 13), and the internal surface of the nuchal plate has two deep pits in which are inserted a paired process of the neurocranium. They comprise a major group, the eubrachythoracids, which includes some very large forms (Fig. 4.49, 7), and several minor monophyletic groups, the gemuendenaspids (Fig. 4.49D), holonematids (Fig. 4.49E), homostiids (Fig. 4.49F), and buchanosteids (Fig. 4.46). The ensemble of the four minor groups has been generally seen as para-phyletic (Gardiner and Miles 1990; Young 1979) but Lelièvre (1984, 1991) and Carr (1991) have suggested that homostiids and buchanosteids formed a monophyletic group, the Migmatocephalia, which is defined by some unique features, such as the separate supraoral and infra-orbital sensory lines on the cheek (Fig. 4.45, 2). As the study of Early Devonian brachythoracids progresses, the Migmatocephalia emerge as a quite highly diversified taxon, and which may have been outcompeted by the eubrachythoracids during the Middle Devonian.

Gemuendenaspids are represented by a single genus, *Gemuendenaspis* (Fig. 4.45C), which is phlyctaenaspid-like in its very long and crested median dorsal plate, but whose skull-roof structure is of brachythoracid type.

Holonematids (*Holonema*; Fig. 4.45D) have an elongated trunk armour, and their dermal bones are ornamented with sinuous, parallel bony ridges. Their gnathal plates have a peculiar, tubular structure. They appear at the beginning of the Middle Devonian and survive,

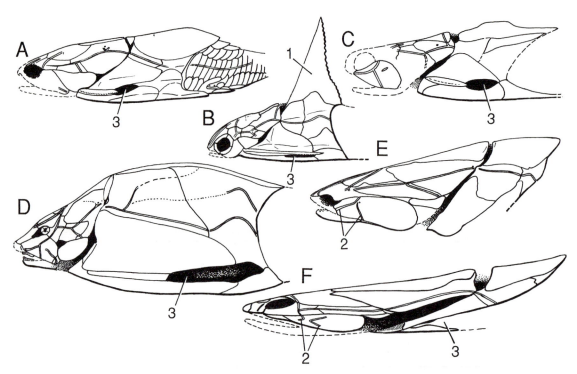

Fig. 4.45. Placodermi. Arthrodira. Actinolepida (A), Phylctaenii (B), primitive Brachythoraci (C, D) and Migmatocephalia (E, F). Reconstruction of armour in lateral view. A, *Sigaspis*, Early Devonian of Spitsbergen (×0.6); B, *Tiaraspis*, Early Devonian of Germany (×1); C, *Gemuendenaspis*, Early Devonian of Germany (×0.5); D, *Holonema*, Middle to Late Devonian (×0.3); E, *Antineosteus*, Early Devonian of Morocco (×0.5); F, *Homostius*, Middle Devonian of Estonia (×0.06). 1, median dorsal plate; 2, disconnected supraoral and infraorbital sensory-lines; 3, pectoral fenestra or notch. (A, from Goujet (1973); B, from Schultze (1984b); C, from Miles (1962); D, from Miles (1971), modified according to Lelièvre *et al.* (1990); E, from Lelièvre (1984), F, from Heintz (1934).)

almost unchanged, until the end of the Devonian. Because of their long dermal armour and very shallow keel on the median dorsal plate, holonematids were once regarded as allied to the phlyctaeniids. They may be more closely related to the migmatocephalians and eubrachythoracids.

Homostiids (Fig. 4.45E, F) are rather large forms, with a short thoracic armour and elongated skull. In the most advanced forms, the eyes are almost frontal in position and the pectoral fenestra is widely open posteriorly (Fig. 4.45F, 3). Homostiids occur from the late Early Devonian (*Antineosteus*, *Goodradigbeeon*, *Taemasosteus*) to the Middle Devonian (*Homostius*). They are dorso-ventrally depressed forms, which suggests a benthic habit.

Buchanosteids (Fig. 4.46) have the overall aspect of eubrachythoracids, such as the Coccosteidae (Fig. 4.47A), in particular in the shape of the skull-roof. They are, however, distinguished by their entirely separate rhinocapsular component of the braincase, which can be detached from the rest of the head (Fig. 4.46, 13). This peculiar structure is regarded as a general placoderm (and perhaps gnathostome) feature. *Buchanosteus* is the only brachythoracid in which the structure of the braincase is known extensively; it shows a surprisingly long occipital region (Fig. 4.46A).

The Eubrachythoraci (Fig. 4.49, 7) are characterized e.g. by a palatoquadrate with two separate ossifications (autopalatine and pterygoquadrate), a posterior thickening of the skull-roof, a rod-shaped submarginal plate (Fig. 4.47,6), and paired lateral thickenings or crests on the ventral surface of the skull-roof. They comprise two major groups, the coccosteomorphs (Fig. 4.47; Fig. 4.49, 8) and pachyosteomorphs (Fig. 4.48; Fig. 4.49, 9), which are particularly diversified in Middle and Late Devonian times. The coccosteomorphs are characterized by an anterior embayment of the central plates of the skull-roof, to accomodate a process of the preorbital plate (Fig. 4.47A2, 10), and a pointed posterior process of the median dorsal plate (Fig. 4.47A2, 19). One of the most generalized coccosteomorphs is the Middle Devonian genus *Coccosteus* (Fig. 4.47A; Fig. 4.49G), a moderately large arthrodire with a closed pectoral fenestra (Fig. 4.47A, 5). There are, however, some very derived coccosteomorphs by Late Devonian times, in particular the long-snouted camuro-piscids *Camuropiscis*, *Tubonasus*, and *Rolfosteus* (Fig. 4.47D) and the laterally compressed brachydeiroids (*Brachydeirus*, *Leptosteus*; Fig. 4.47B). The brachy-deiroids also include long-snouted forms (*Oxyosteus*) that parallel the camuropiscids, and the only placoderm with fused cranio-thoracic armour, *Synauchenia* (Fig. 4.47C). In these advanced coccosteomorphs the cheek (suborbital and postsuborbital plates) is fixed to the skull-roof (Fig. 4.47B, C, D, 8, 9). Pachyosteomorphs are characterized by various unique features of the skull-roof pattern,

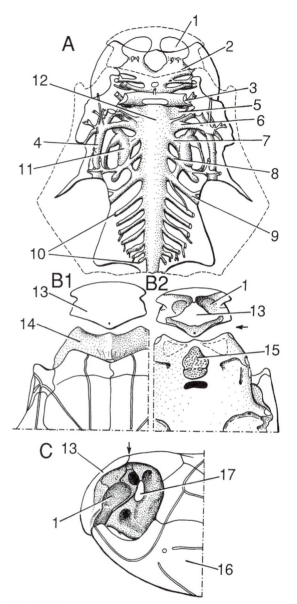

Fig. 4.46. Placodermi. Arthrodira, Brachythoraci. Anatomy of *Buchanosteus*, Early Devonian of Australia. A, outline of the braincase (continuous line) and skull-roof (dashed line), superimposed to the internal cast of the cavities in the braincase, in ventral view (×0.5). B, anterior part of the skull in dorsal (B1) and ventral (B2) view, showing the free rhinocapsular ossification; C, snout in left lateral view, showing the orbital cavity and cranio-ethmoidian fissure (arrow). 1, nasal cavity; 2, optic tract; 3, pituitary vein; 4, jugular vein; 5, trigeminus nerve; 6, facial nerve; 7, acoustic nerve; 8, glossopharyngeus nerve; 9, vagus nerve; 10, spino-occipital nerves; 11, labyrinth cavity; 12, brain cavity; 13, rhinocapsular ossification; 14, braincase; 15, parasphenoid; 16, suborbital plate; 17, eye-stalk. (from Young (1979, 1986*b*).)

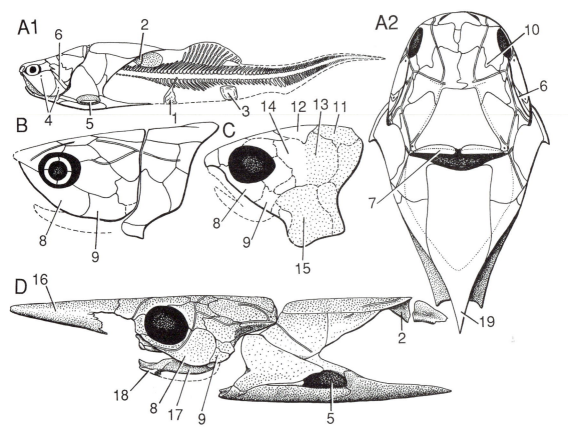

Fig. 4.47. Placodermi. Arthrodira. Eubrachythoraci, Coccosteomorphi. A, *Coccosteus*, Middle to Late Devonian, reconstruction of skeleton in lateral view (A1, about ×0.3) and armour in dorsal view (A2); B, *Brachydeirus*, Late Devonian of Germany, armour in lateral view (×1); C, *Synauchenia*, Late Devonian of Germany, armour in lateral view (part corresponding to the thoracic armour stippled) (×0.3); D, *Rolfosteus*, Late Devonian of Australia, armour in lateral view (×0.6). 1, pelvic girdle; 2, ventral 'keel' of median dorsal plate; 3, anal plate; 4, pits for sense organs; 5, pectoral fenestra; 6, rod-shaped submarginal plate; 7, extrascapular plates; 8, suborbital plate; 9, postsuborbital plate; 10, embayment in central plate; 11, median dorsal plate; 12, nuchal plate; 13, anterior dorsolateral plate (fused to paranuchal plate); 14, paranuchal (fused to anterior dorsolateral plate); 15, anterolateral plate; 16, rostral plate; 17, inferognathal; 18, anterior portion of Meckelian bone (mentomandibular); 19, pointed posterior process of median dorsal plate. (A, from Miles and Westoll (1968); B, from Gross (1932); C, from Stensiö (1963); D, from Dennis and Miles (1979).)

such as a reduced contact between the paranuchal and central plates. They have also lost the posteroventral branch of the main lateral-line groove on the anterior dorsolateral plate of the thoracic armour (Fig. 4.48). They comprise a number of generally small forms that have lost the spinal plate (aspinothoracids); many of them have very large eyes (leiosteids, selenosteids, pachyosteids, *Heintzichthys*, *Gorgonichthys*; Figs 4.48C, 4.49J). The other pachyosteomorphs, in particular the Dinichthyidae, include very large forms, such as *Dunkleosteus* (Fig. 4.48B) or *Eastmanosteus* (Fig. 4.48A). Their gnathal plates develop long pointed cusps (Fig. 4.48, 5), unless they are secondarily adapted to a durophagous diet. In

dinichthyids, the nuchal gap between the skull-roof and the thoracic armour (Fig. 4.48, 8) is very large, which suggests greater mobility of the intracranial joint and, consequently, a larger gape. Some authors, however, consider that the mechanics of this joint is incompatible with such a large movement. The palate of dinichthyids shows a peculiar notch in the anterior superognathals (Fig. 4.48, 11), which has been interpreted as intrabuccal posterior nostrils, thereby parallelling holocephalans and lungfishes.

There are several brachythoracid taxa of uncertain phylogenetic position. The Heterosteidae, for example, (*Heterosteus*, *Herasmius*) are large Middle Devonian brachythoracids with a very short thoracic armour and

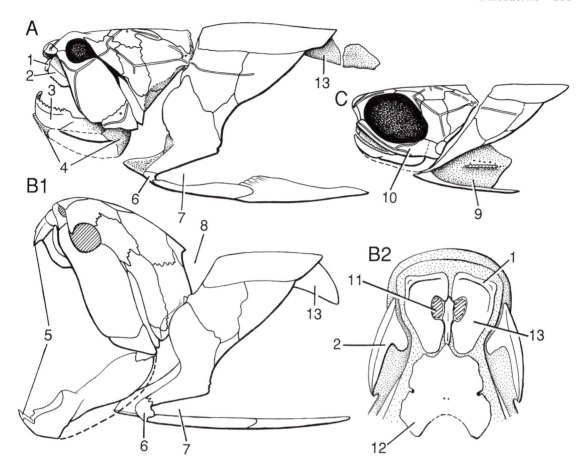

Fig. 4.48. Placodermi. Arthrodira. Eubrachythoraci, Pachyosteomorphi. A, *Eastmanosteus*, Late Devonian, armour in lateral view (×0.2); B, *Dunkleosteus*, Late Devonian, armour in lateral view, showing the possible gape (B1, about ×0.08), and reconstructed palate in ventral view, showing the possible position of the posterior nostrils (B2, reconstructed parts stippled); C, *Rhinosteus*, Late Devonian of Germany, armour in lateral view (×0.8). 1, anterior superognathal; 2, posterior superognathal; 3, inferognathal; 4, Meckelian bone; 5, cusp; 6, spinal plate; 7, pectoral notch; 8, nuchal gap; 9, scapulocoracoid; 10, suborbital plate; 11, possible position of posterior nostrils; 12, parasphenoid; 13, ventral 'keel' of median dorsal plate. (A, from Dennis-Bryan (1987); B, partly from Dunkle and Bungart (1946), Heintz (1931), and unpublished data provided by H. Lelièvre, Paris; C, from Stensiö (1963).)

very broad skull-roof. They recall homostiids in having frontally placed eyes, but are clearly eubrachythoracids, as is shown by the presence of a pronounced posterior keel process of the median dorsal plate. *Titanichthys* (Fig. 1.15, 2) is another huge brachythoracid from the Late Devonian, and is perhaps the largest known placoderm. It has a depressed armour, with a thoracic part resembling that of heterosteids.

The current theory of arthrodire interrelationships is depicted in Fig. 4.49. The phlyctaeniids and brachythoracids are grouped into the taxon Phlyctaenioidei (Fig. 4.49, 3), which is characterized by the craniothoracic ball-and-socket articulation. The Brachythoraci

(Fig. 4.49, 4) include the gemuendenaspids, holonematids, migmatocephalians, and eubrachythoracids. This phylogeny shows a number of parallel trends within the latter, two groups, such as the tendency towards the posterior opening of the pectoral fenestra in homostiids and pachyosteomorphs (Fig. 4.49F, I, J)

4.14.3 Petalichthyida

Petalichthyids are characterized by dorsally placed orbits, which are usually in contact with the central plates, and a large, elongated nuchal plate on which the supraorbital

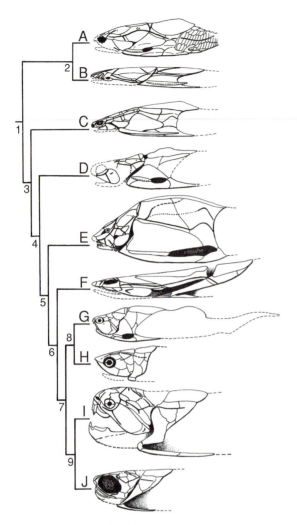

Fig. 4.49. Placodermi. Interrelationships of the Arthrodira. *Terminal taxa*: A, Actinolepida; B, Phyllolepida and Wuttagoonaspida; C, Phlyctaenii; D, *Gemuendenaspis*; E, Holonematidae; F, Migmatocephalia (Homostiidae, Buchanosteidae); G, Coccosteidae and Camuropiscidae; H, Brachydeiroidea, I, Dinichthyidae; J, Aspinothoraci. *Nested taxa and selected synapomorphies*: 1, Arthrodira (endolymphatic pore opening in single paranuchal plate, anterior superognathal, skull-roof made up of three unpaired plates, and seven paired plates); 2, Actinolepidoidei (radiation centre of paranuchal plate near its topographic centre, ornamentation consisting of concentric ridges); 3, Phlyctaenioidei (dermal cranio-thoracic articulation consisting of a condyle on thoracic plate, and a fossa on skull-roof, dorsal occipital process on braincase); 4, Brachythoraci (nuchal plate trapezoid in shape, overlapping dermal plates in skull-roof, keel, or short carinal process on ventral surface of median dorsal plate); 5, loss of denticles on occlusal surface of anterior superognathal, lateral lobe of central plates reduced; 6, cranio-thoracic articulation laterally displaced, lateral thickening of the skull-roof; 7, Eubrachythoraci (trilobate central plate, posterior thickening of skull-roof, inferognathal with a blade-like posterior portion, dorsal process of anterior superognathal, submarginal plate rod-shaped, palatoquadrate consisting of two separate ossifications); 8, Coccosteomorphi (embayment in central plate for the preorbital plate, posteriorly pointed median dorsal plate); 9, Pachyosteomorphi (loss of the posteroventral lateral-line groove on the anterior dorsolateral plate). *Age*: Early Devonian (A, D); Early to Middle Devonian (C, F); Middle to Late Devonian (B, E, G, I); Late Devonian (H, J). (A–J, same sources as for Figs 4.44–48)

sensory line and the posterior pit line meet to form a X-shaped pattern (Fig. 4.50, 9). In most petalichthyids, the orbits are dorsally placed and bounded ventrolaterally by the postorbital plate (Fig. 4.50A, 3), but in the Chinese quasipetalichthyids (e.g. *Eurycaraspis*) they are still open ventrally and are separated from the central plates (Fig. 4.50B). There are two paranuchal plates on each side, and the endolymphatic duct opens in the anterior one (Fig. 4.50, 7, 8, 16). The sensory lines are enclosed in canals, which open to the exterior by means of large pores.

The cranio-thoracic articulation consists, as in phlyctaenioid arthrodires, of a condyle on the anterior ventrolateral plate of the thoracic armour and a depression in the posterior paranuchal plate, but these are shaped dif-

ferently than those of arthrodires and rather resemble those of ptyctodontids (see below).

The thoracic armour is moderately long to short, with relatively long spinal plates (Fig. 4.50A, 11). The body and tail (known only in *Lunaspis*; Fig. 4.50A) is covered with small diamond-shaped scales, and a few very large median dorsal scutes. The pectoral fins are covered with relatively large scales (Fig. 4.50A, 17).

The braincase is well ossified and provides detail information on the internal anatomy. The Late Devonian *Macropetalichthys*, in particular, provided the basis for some of Stensiö's (1925) early work on placoderm anatomy.

Nothing is known of the jaws in petalichthyids, and there is no evidence of gnathal plates, even in complete

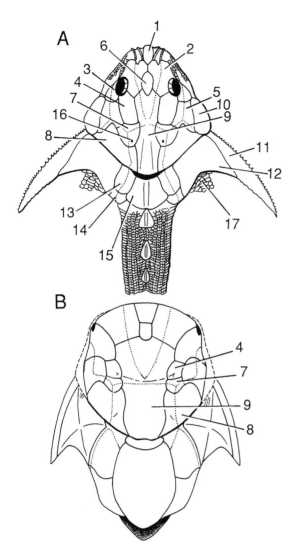

Fig. 4.50. Placodermi. Petalichthyida. A, *Lunaspis*, Early Devonian of Germany, reconstruction in dorsal view (×0.5); B, *Eurycaraspis*, Middle Devonian of China, armour in dorsal view (×1). 1–15, *dermal plates*: 1, rostral; 2, preorbital; 3, postorbital; 4, central; 5, marginal; 6, pineal; 7, anterior paranuchal; 8, posterior paranuchal; 9, nuchal; 10, submarginal; 11, spinal; 12, anterolateral; 13, anterior dorsolateral; 14, posterior dorsolateral; 15, median dorsal; 16–17, *others*: 16, endolymphatic duct; 17, pectoral fin. (A, from Cross (1961*b*); B, from Y. H. Liu (1991).)

specimens of *Lunaspis*. The jaws may thus have been armed with minute separate denticles.

Petalichthyids show a very low diversity; most of them display an overall morphology that is quite comparable to

that of extant guernards, which suggests benthic habits. The overall shape of quasipetalichthyids (Fig. 4.50B) is more indicative of that of generalized arthrodires.

4.14.4 Ptyctodontida

Ptyctodontids have a reduced dermal armour, and are charaterized by the trabecular structure of their gnathal plates, which recalls the columnar, or tubate, dentine of holocephalans and contains columns of pleromic dentine (Fig. 4.51D, 21). Ptyctodontid tooth plates occur in great abundance in Middle to Late Devonian sediments.

The skull roof (Fig. 4.51A2) is poorly ossified in front of the pineal region, and the orbits are very large. The nuchal plate is elongated and at its centre the supraorbital sensory-line canal and the posterior pit line meet to form an X-shaped pattern, as in petalichthyids (Fig. 4.51, 15, 16). Also, as in the latter, the sensory lines are enclosed in canals that open through large pores. In all ptyctodontids, the endolymphatic duct is closed. The cheek is covered only by the submarginal plate, which is rod-shaped (Fig. 4.51, 8).

The thoracic armour is very short and forms a mere 'ring' of dermal plates, with a large anterolateral plate (Fig. 4.51, 11) and a median dorsal plate, which may bear one to three separate spines (modified median dorsal ridge scales; Fig. 4.51C, 9, 12). There is a spinal plate, which is often armed with recurved denticles. The craniothoracic articulation consists of a condyle on the anterior dorsolateral plates and a fossa on the paranuchal plate of the skull-roof.

The dermal bones of most ptyctodontids have a spongy surface and were covered with skin. However, in Early Devonian forms from Siberia (*Tollodus*), there is still a tuberculate ornamentation.

The braincase consists of several ossifications (occipital, orbital). The palatoquadrate is made up of three ossifications (quadrate, autopalatine, and metapterygoid; Fig. 4.51, 20) and articulates on the braincase. There is only one gnathal plate on each jaw (superognathal and inferognathal; Fig. 4.51, 13).

The scapulocoracoid is short and deep, with a very narrow surface for the articulation of the pectoral fin skeleton, which comprises at least three large basal elements (Fig. 4.51, 19). The pelvic fins bear specialized, hook-shaped or spiny dermal elements that vaguely resemble the pelvic and prepelvic claspers of male holocephalans (Fig. 4.51, 14, 23, 24). This sexual dimorphism has been best observed in *Rhamphodopsis* (Fig. 4.51B) and *Ctenurella* (Fig. 4.51A3–5); it has been regarded by Ørvig (1962) as a synapomorphy of the ptyctodontids and holocephalans (Fig. 3.15G). The prepelvic scute is dermal and is thus quite different from the prepelvic clasper of

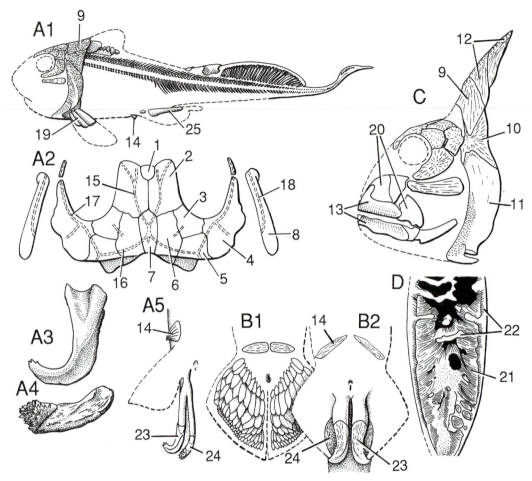

Fig. 4.51. Placodermi. Ptyctodontida. A, *Ctenurella*, Late Devonian, reconstruction in lateral view (A, ×1), skull-roof (A2), dermal elements of pelvic claspers in ventral view (A3, A4, ×4), and attempted reconstruction of the claspers (A5); B, *Rhamphodopsis*, Middle Devonian of Europe, reconstruction of the pelvic fins in a female (B1) and a male (B2) (×3); C, *Campbellodus*, Late Devonian of Australia, armour in lateral view (×0.5); D, vertical section through a ptyctodontid tooth plate (×20). 1–14, *dermal plates*: 1, pineal; 2, preorbital; 3, postorbital; 4, marginal; 5, paranuchal; 6, central; 7, nuchal; 8, submarginal; 9, median dorsal; 10, anterior dorsolateral; 11, anterolateral; 12, scutes forming the dorsal spine; 13, gnathal tooth plates; 14, prepelvic scute; 15–18, *sensory lines*: 15, supraorbital; 16, posterior pit line; 17, infraorbital; 18, submarginal; 19–25, *others*: 19, basipterygium; 20, palatoquadrate; 21, columnar pleromin; 22, bony hard tissue; 23, hook-shaped clasper scute; 24, spiny clasper scute; 25, pelvic metapterygium. (A1, from Ørvig (1960); A2, from Ørvig (1962), modified; A3, A4, from Miles and Young (1977); B, from Miles (1967); C, from Long (1988*a*); D, from Ørvig (1980).)

chimaeroids (Fig. 3.15, 5), but the two dermal elements of the pelvic clasper (Fig. 4.51, 23, 24) may well fit on the extremities of a bifid mixipterygium, similar to that of some modern chimaeroids (Fig. 4.51A5). The axial skeleton and body shape are well known from the Late Devonian genus *Ctenurella* (Fig. 4.51A1). The overall body shape of *Ctenurella*, with a large, deep head, and a long, whip-shaped tail, is also quite reminiscent of that in extent chimaeroids, and may be linked to similar habits.

Ptyctodontids could certainly reach a fairly large size, as is illustrated by isolated gnathal plates of *Palaeomylus* up to 15 cm long.

4.14.5 Antiarcha

Antiarchs are characterized by an elongated trunk armour with two median dorsal plates (Fig. 4.52, 11, 14) and pectoral fins modified into jointed appendages covered with

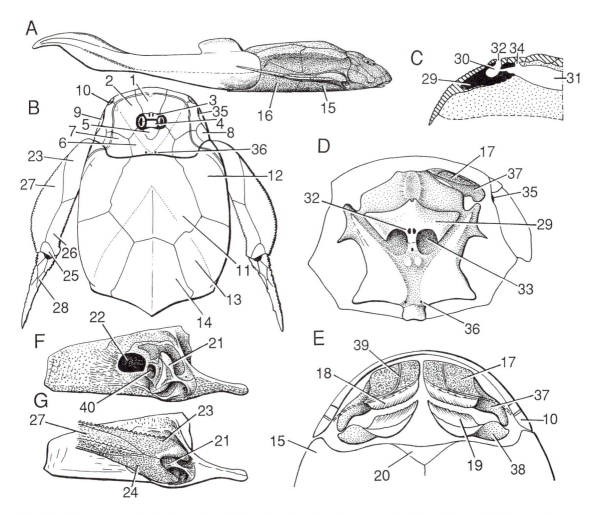

Fig. 4.52. Placodermi, Antiarcha. *Bothriolepis*, Middle to Late Devonian; A, reconstruction in lateral view (×0.5); B, armour in dorsal view; C, sagittal section through the skull (dermal bone hatched, hypothetical braincase stippled, hypothetical rhinocapsular cartilage black); D, attempted reconstruction of braincase and right palatoquadrate in dorsal view; E, reconstruction of the jaws in ventral view; F, right anterior ventrolateral plate and brachial process, in lateral view; G, the same when articulated with the proximal pectoral fin plates. 1–28, *dermal plates*: 1, premedian; 2, lateral; 3, rostral; 4, pineal; 5, postpineal; 6, nuchal; 7, paranuchal; 8, postmarginal; 9, submarginal; 10, prelateral; 11, anterior median dorsal; 12 anterior dorsolateral; 13, mixilateral (posterior dorsolateral + posterolateral); 14, posterior median dorsal; 15, anterior ventrolateral; 16, posterior ventrolateral; 17, suborbital; 18, cutting edge of suborbital (serving as a superognathal); 19, inferognathal; 20, semilunar; 21, brachial process; 22, axillary foramen; 23, proximal dorsal central plate of pectoral fin; 24, proximal ventral central plate of pectoral fin; 25, distal dorsal central plate of pectoral fin; 26, marginal medial plate of pectoral fin; 27, marginal lateral plate of pectoral fin; 28, distal segment of pectoral fin 29–40, *others*: 29, rhinocapsular cartilage; 30, nasal cavity; 31, brain cavity; 32, nostrils; 33, orbits; 34, pineal recess or foramen; 35, spiracle; 36, endolymphatic ducts; 37, palatoquadrate; 38, Meckelian bone; 39, infraorbital sensory-line groove; 40, funnel pit. (A, B, from Stensiö (1948); C–E, based on Young (1984); F, G, based on Young (1988) and Stensiö (1948).)

dermal plates (Fig. 4.52, 23–28). The head is short, and the eyes, pineal opening, and nostrils are housed in an oval orbitonasal fenestra in the centre of the skull-roof (Fig. 4.52B). There is no superognathal, but the suborbital plate has a cutting edge which bites against the infero-

gnathal (Fig. 4.52, 17, 19). A spiracular opening is present between the cheek and the skull-roof (Fig. 4.52, 35). The cranio-thoracic articulation involves a condyle on the paranuchal plate and a fossa on the anterior dorso-lateral plate, that is the reverse of the condition in arthrodires, petalichthyids, and ptyctodontids. The nuchal plate is almost pentagonal and pierced by the openings of the endolymphatic ducts (Fig. 4.52, 6, 36); it bears an X-shaped commissure between the posterior pit line and the central sensory line. Anterior to the orbitonasal fenestra there is a large premedian plate (Fig. 4.52, 1), which forms the anterior edge of the skull-roof; in contrast, the rostral plate is a minute bone lining the nostrils (Fig. 4.52, 3). The body and tail is either covered with very large scales (*Asterolepis*, *Pterichthyodes*; Fig. 4.53E, F) or almost naked (*Bothriolepis*; Fig. 4.52A). There is only one dorsal fin; the anal and pelvic fins are lacking.

The dermal plates and scales of antiarchs are ornamented either with tubercles or with vermiculations. Virtually nothing is known of their internal anatomy, apart from the palatoquadrate, which is of typical placoderm type (Fig. 4.52, 37), i.e. with a medial depression for the jaw muscles. The braincase has been reconstructed on the basis of its impression on the internal surface of the skull-roof. Young (1984) considered that a median recess of the latter housed a separate rhinocapsular ossification (Fig. 4.52, 29).

Natural casts of some soft internal structures in *Bothriolepis* have been described by Denison (1941). These have been used for the reconstruction the overall shape of the gill chamber and intestine with a spiral valve. Peculiar lateral evaginations of the pharynx have been interpreted as 'lungs', though with great reservations.

The only known antiarchs have long been the Middle and Late Devonian euantiarchs, such as *Bothriolepis* (Fig. 4.52A), *Asterolepis* (Fig. 4.53F), or *Pterichthyodes* (Fig. 4.53E), with a relatively homogeneous morphology and a characterictic ball-and-socket type of dermal articulation of the pectoral appendage (Fig. 4.52, 21). In the late 1960s and in the 1970s, however, a number of peculiar antiarchs were uncovered from the Early Devonian of southern China (Yunnan), showing what may be regarded as the primitive morphology of the group. In these 'Yunnanolepiformes' (*Yunnanolepis*, Fig. 4.53A), the premedian plate is quite short, and the orbitonasal fenestra more anterior in position, and lined anteriorly by a peculiar triangular preorbital depression (Fig. 4.53A, 1, 2). The latter may have served in protecting the nostrils when the animal was half-buried in sand or mud. The most striking feature of these early antiarchs is the absence of any complex dermal articular device for the pectoral appendage (Fig. 4.53, 3). As field research progressed in Yunnan, and also in northern Vietnam, a variety of other Early Devonian antiarchs turned up, showing different types of pectoral joint, some approaching the euantiarch structure. The Procondylolepiformes (*Chuchinolepis*, Fig. 4.53B), for example, have a rudimentary articular process on the anterior ventrolateral plate, but this does not resemble that of euantiarchs (Fig. 4.53B, 5). In contrast, the

Sinolepidae (*Sinolepis*, *Grenfellaspis*, Fig. 4.53C, D), a clade characterized by a large ventral fenestration in the thoracic armour (Fig. 4.53, 13), display a variety of incipient brachial processes that foreshadow the helmet-shaped brachial process of euantiarchs (Fig. 4.53, 5). In both, the brachial process is hollowed by a funnel-shaped pit housing some endoskeletal remnant of the scapulocoracoid (Fig. 4.52, 40; Fig. 4.53, 7). In addition, sinolepids share with the euantiarchs an axillary foramen, i.e. an opening in the anterior ventrolateral plate of the thoracic armour, through which passed the brachial vessels and nerves (Fig. 4.52, 22; Fig. 4.53, 6). The complex brachial articulation of the euantiarchs thus appeared stepwise, as is suggested by a transformation series with the yunnanolepiforms (Fig. 4.53A2), procondylolepiforms (Fig. 4.53B), sinolepids (Fig. 4.53D2), and euantiarchs (Fig. 4.52F, G).

The phylogeny of the antiarchs shown in Fig. 4.54 depicts the current consensus. The Euantiarcha (Fig. 4.54, 4) fall into two major clades, the bothriolepidoids (Fig. 4.54, 5) and asterolepidoids (Fig. 4.54, 6), which can be distinguished on the basis of skull-roof features. A jointed pectoral appendage is a general feature for euantiarchs, since it occurs also in sinolepids (Fig. 4.53C, 4.54C). However, in one asterolepidoid genus, *Remigolepis* (Fig. 4.53G, 4.54H), it becomes secondarily unjointed. The preorbital depression (stippled in Fig. 4.54) is present in the yunnanolepiforms (Fig. 4.54A), procondylolepiforms (Fig. 4.54B), and sinolepids (Fig. 4.54C), as well as in some supposedly primitive bothriolepidoids (*Microbrachius*, Fig. 4.54D), thereby indicating that its absence from advanced bothriolepidoids and from asterolepidoids is homoplasic. The diversity of the antiarchs is not very great. Some species may develop spines or crests on the median dorsal plates, and the ornamentation turns into radiating riges in some taxa.

The peculiar morphology of antiarchs, which is probably linked with benthic or burrowing habits, gave rise to much discussion among the early students of Devonian vertebrates. In the early nineteenth century because of their jointed pectoral appendages and their carapace, the antiarchs were first regarded as arthropods until Hugh Miller and John Malcolmson (see Chapter 8, p. 310) provided evidence that they were fishes. Since the late nineteenth century the antiarchs have been regarded as relatives of the other placoderms, although Cope (1889) and, more recently, Novitskaya and Karatayute-Talimaa (1989) have placed them among the jawless vertebrates. Their position within the placoderms is, however, still debated; Young (1984, 1986*b*) considers them as the closest relatives of arthrodires, whereas Goujet (1984*b*) relates them to rhenanids and some 'acanthothoracids' (Fig. 4.57, 5) on the basis of their dorsally placed nostrils. The antiarchs seem always to have been confined to

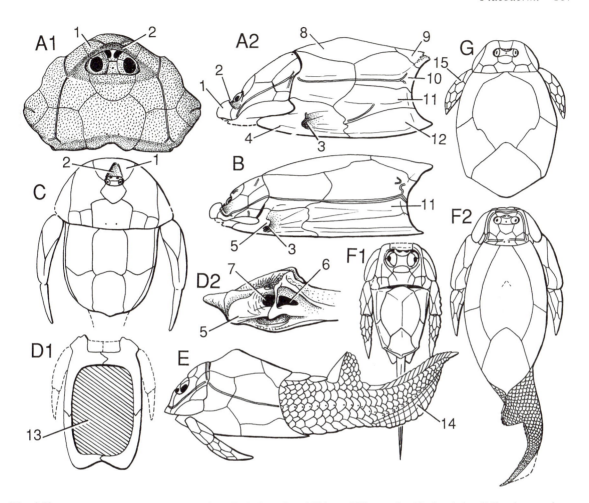

Fig. 4.53. Placodermi. Antiarcha. A, *Yunnanolepis*, Early Devonian of China and Vietnam, head in dorsal view (A1) and armour in lateral view (A2, ×2); B, *Chuchinolepis*, Early Devonian of China and Vietnam, armour in lateral view (×1); C, *Sinolepis*, Late Devonian of China, armour in dorsal view (×0.8); D, *Grenfellaspis*, Late Devonian of Australia, thoracic armour in ventral view (D1, ×0.3) and detail of brachial process in lateral view (D2); E, *Pterichthyodes*, Middle Devonian of Scotland, reconstruction in lateral view (×0.5); F, *Asterolepis*, Late Devonian, reconstruction in dorsal view of a juvenile (F1, ×2.5) and an adult (F2, ×0.2); G, *Remigolepis*, reconstruction in dorsal view (×0.5). 1, premedian plate; 2, preorbital depression; 3, pectoral fenestra; 4, prepectoral corner of anterior ventrolateral plate; 5, brachial process; 6, axillary foramen; 7, funnel pit; 8, anterior median dorsal plate; 9, posterior median dorsal plate; 10 posterior dorsolateral plate; 11, posterolateral plate; 12, posterior ventrolateral plate; 13, ventral fenestration of thoracic armour; 14, caudal fin web; 15, unjointed pectoral appendage. (A1, from Zhang (1980); A2, B, from Tong-Dzuy and Janvier (1990); C, from Liu and P'an (1958); D, from Ritchie *et al.* (1992); E, from Hemmings (1978); F1, from Upeniece and Upenieks (1992); F2, from Lyarskaya (1981); G, from Panteleyev (1992).)

marginal marine or deltaic environments, and since they often occur in arenaceous facies (e.g. 'Old Red Sandstone'), they are often said to be freshwater fishes. They generally occur in large quantities and are probably the most common fishes in the Late Devonian.

The function of the pectoral appendages is still obscure, but they certainly could not have enabled the fish to walk on land. The most likely interpretation is that they were used by the animal to throw sand or mud over its back, and thereby bury itself quickly, as do some tropical crabs (see Chapter 7, p. 302).

Juvenile individuals of various antiarchs (*Bothriolepis*, *Asterolepis*) have been described; they show an important allometry (figs 4.53F1, 7.11C). In particular, the young have large eyes (like most juvenile gnathostomes), a large head, and longer pectoral appendages.

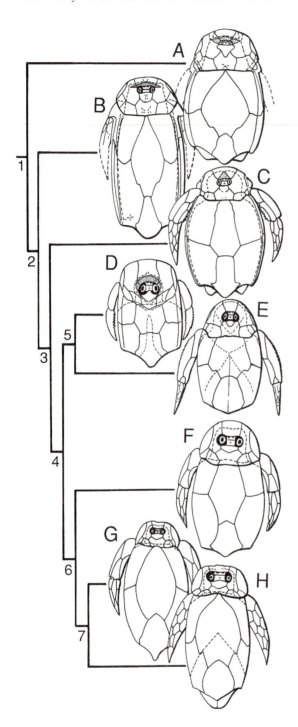

Fig. 4.54. Placodermi. Interrelationships of the Antiarcha. *Terminal taxa* (preorbital depression stippled): A, Yunnanolepiformes (*Yunnanolepis*); B, Procondylolepiformes (*Chuchinolepis*); C, Sinolepidae (*Dayaoshania*); D, *Microbrachius*; E, Bothriolepidae (*Bothriolepis*); F, *Pterichthyodes*; G, *Asterolepis*; H, *Remigolepis. Nested taxa and selected synapomorphies*: 1, Antiarcha (two median dorsal plates, plate covered pectoral fins, endolymphatic ducts in nuchal plate); 2, incipient brachial process; 3, axillary foramen; 4, Euantiarcha (helmet-shaped brachial process); 5, Bothriolepidoidei (unpaired semilunar plate, posterior oblique cephalic pit line extending from nuchal plate onto central plates); 6, Asterolepidoidei (posterolaterally expanded postmarginal plate, narrow lateral plate, short nuchal plate); 7, Asterolepidae (anterior notch on premedian plate). *Age*: Early Devonian (A, B); Early to Late Devonian (C); Middle to Late Devonian (D–G); Late Devonian (H). (A, B, based on Zhang (1980) and Tong-Dzuy and Janvier (1990); C, from Ritchie *et al.* (1992); D, F, from Hemmings (1978); E, from Stensiö (1948); G, from Lyarskaya (1981); H, from Stensiö (1931).)

4.14.6 Rhenanida

Rhenanids are characterized by a reduction of the thoracic armour and considerably enlarged pectoral fins (Fig. 4.55A, 6), which simulate those of batomorphs (skates and rays). Many other features of their internal anatomy (shape of the braincase, labyrinth, synarcual, mandibular arch) are also suggestive of rays; Stensiö (1969) even suggested a close relationship between the two groups. However, rhenanids are clearly placoderms, as is shown by the pattern of their dermal plates; in particular by the presence of a median dorsal plate (Fig. 4.55, 5), the structure of the palatoquadrate (Fig. 4.55, 9), and the histological structure. Moreover, they differ from the

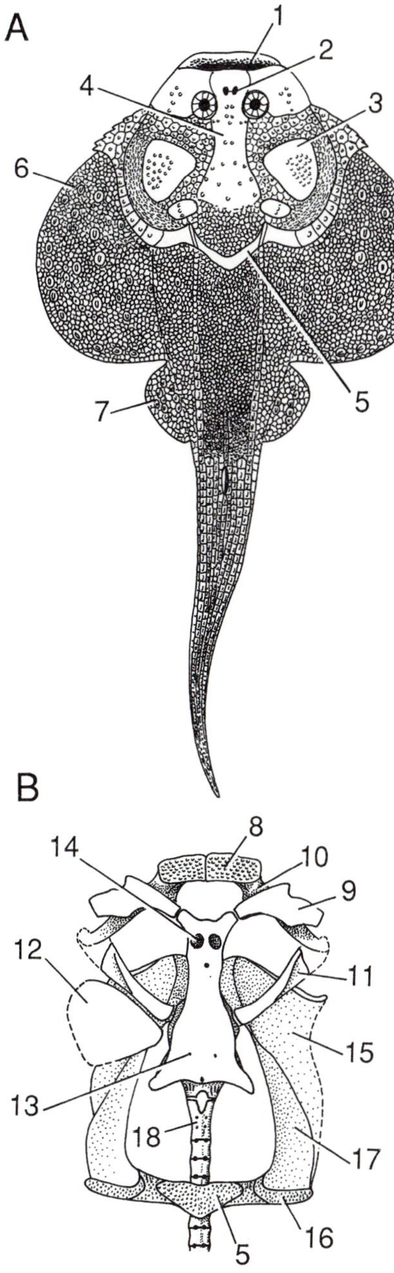

Fig. 4.55. Placodermi. Rhenanida. A, *Gemuendina*, Early Devonian of Germany, reconstruction in dorsal view (×0.8); B, *Jagorina*, Late Devonian of Germany, skull and shoulder girdle in dorsal view (dermal plates of skull-roof removed, ×0.3). 1, mouth; 2, nostrils; 3, submarginal plate; 4, skull-roof; 5, median dorsal plate; 6, pectoral fin; 7, pelvic fin; 8, inferognathal (denticle-bearing plate); 9, palatoquadrate; 10, Meckelian bone; 11, epihyal (or anterior postorbital process); 12, opercular cartilage (or epihyal); 13, endolymphatic duct; 14, nasal cavity; 15, scapulocoracoid; 16, anterior dorsolateral plate; 17 anterolateral plate; 18, synarcual. (A, from Gross (1963*b*); B, combined from Denison (1978) and Young (1986*b*).)

batomorphs by their dorsally placed nostrils and a single gill slit (Fig. 4.55, 2). The fins, part of the head, and the body of rhenanids are covered with minute platelets, which fill the space between the larger dermal units and even cover some of them. This condition is regarded as derived. Of the head armour, only the skull-roof (Fig. 4.55, 4), suborbital plates, and submarginal plates (Fig. 4.55, 3) can still be recognized. In the thoracic armour, the median dorsal, anterior dorsolateral, anterolateral, and anterior ventrolateral plates are still present (Fig. 4.55, 5, 16, 17), though reduced and sometimes covered with a secondary layer of polygonal platelets. The gnathal plates of rhenanids are flattened, and are covered with a shagreen of minute denticles (Fig. 4.55, 8).

Rhenanids are quite rare, and are known by only four monospecific genera: *Asterosteus*, *Gemuendina* (Fig. 4.55A), *Jagorina* (Fig. 4.55B), and *Bolivosteus*, distributed throughout the Middle and Late Devonian in marine environments. This apparent scarcity may, however, be due to the fact that their exoskeleton is fragile, and many rhenanid platelets are now more and more frequently found as 'ichthyoliths'.

4.14.7 Acanthothoraci

The acanthothoracids, or 'palaeacanthaspids', are represented by an ensemble of small Early Devonian forms. Their thoracic armour is short and the skull-roof is relatively elongated, but they cannot be defined by any unique characteristic. The internal anatomy of the braincase of some acanthothoracids is well known; for example in *Brindabellaspis* (Fig. 4.56B, left-hand side). Acanthothoracids show a striking diversity of skull-roof morphology, with either anteriorly or dorsally placed nostrils (Fig. 4.56, 18, Fig. 4.57D, F). Some of them have a partly tessellate skull-roof (Fig. 4.57H). Young (1980, 1986*b*) regarded acanthothoracids as primitive rhenanids on the basis of the dorsal position of their nostrils, whereas Goujet (1984*b*) considered them as paraphyletic. According to Goujet, the species '*Radotina*' *prima* (Fig. 4.57D) would be more closely related to the arthrodires, phyllolepids, petalichthyids, and ptyctodontids, because of the histologic structure of its tubercles, which are rounded and have a layer of pallial semidentine (i.e. the semidentine layer is restricted to the superfical part of the tubercles; Fig. 4.57, 2). Another ensemble comprising the acanthothoracid genera *Brindabellaspis* (Fig. 4.56B), *Romundina* (Figs 4.56A, 4.57F), and *Palaeacanthaspis* would be more closely related to antiarchs and would share with them a large premedian plate with a preorbital depression (Fig. 4.56, 1, 30), as well as a cranio-thoracic joint of antiarchan type and a single anterior ventrolateral plate unit (formed by the anterolateral, spinal, and anterior ventrolateral plates of other

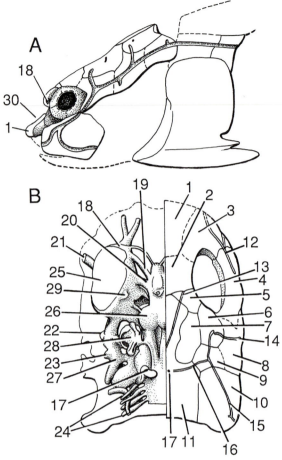

Fig. 4.56. Placodermi, Acanthothoraci. A, *Romundina*, Early Devonian of Canada, reconstruction of armour in lateral view (×3); B, *Brindabellaspis*, Early Devonian of Australia, internal cast of the cavities in the braincase (left-hand side) and skull-roof (right-hand side), in dorsal view (×0.7). 1–11, *dermal plates*: 1, premedian; 2, rostropineal; 3, postnasal; 4, postorbital; 5, preorbital; 6, marginal; 7, central; 8, postmarginal; 9, anterior paranuchal; 10, posterior paranuchal; 11, nuchal; 12–16, *sensory lines*: 12, infraorbital; 13, supraorbital; 14, postmarginal; 15, main lateral line; 16, posterior pit line; 17–29, *others*: 17, endolymphatic duct; 18, nasal cavity; 19, olfactory tract; 20, optic tract; 21, facial nerve; 22, glossopharyngeus nerve; 23, vagus nerve; 24, spinal nerves; 25, orbital cavity; 26, cerebellum (metencephalon); 27, dorsal jugular vein; 28, labyrinth cavity; 29, myodome for superior oblique eye muscle; 30, preorbital depression. (A, from Goujet (1984*b*), modified after Ørvig (1975); B, based on Young (1980).)

placoderms). finally, the other acanthothoracids *Radotina kossorensis* (Fig. 4.57H), *Kossoraspis*, and *Kimaspis* would be more closely allied to the rhenanids on the basis of their

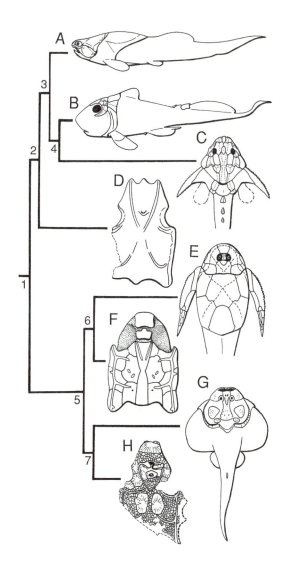

Fig. 4.57. Placodermi. Interrelationships of the Placodermi (the Acanthothoraci being supposed paraphyletic, following Goujet (1984*b*). *Terminal taxa*: A, Arthrodira (including Phyllolepida and Wuttagoonaspida); B, Ptyctodontida; C, Petalichthyida; D, *'Radotina' prima*; E, Antiarcha; F, *Romundina*; G, Rhenanida; H, *Radotina kosorensis*. *Nested taxa and selected synapomorphies*: 1, Placodermi (body armour forming a complete ring, with a median dorsal plate, double cervical joint with endoskeletal and exoskeletal components, semidentine in dermal ornamentation, omega-shaped palatoquadrate associated with cheek plates, epihyal fused with submarginal plate, ?rhinocapsular ossification separated from the rest of the braincase by a crano-ethmoidian fissure); 2, pallial (superficial) layer of semidentine in tubercles; 3, anterior and posterior median ventral plates, endolymphatic duct (lost in ptyctodonts) in the paranuchal plate which bears the commissure between the posterior pit line and the main lateral line; 4, sensory lines enclosed in canals situated below the internal surface of the dermal plates, and opening by large pores; 5, premedian plate, backward-shifted nasal capsules, dorsally placed nostrils; 6, large premedian plate with a preorbital depression, dermal cranio-thoracic joint with a condyle on paranuchal and a fossa on anterior dorsolateral plate; 7, numerous tesserae covering larger dermal bones on head and shoulder girdle. *Age*: Silurian–Devonian (E); Early Devonian (D, F, H); others Devonian. (A, based on Miles and Westoll (1968); B, based on Ørvig (1962); C, based on Gross (1961*b*); D, from von Koenen (1895); E, from Stensiö (1948); F, from Ørvig (1975); G, based on Gross (1963*b*); H, from Gross (1959).)

partly tessellated skull-roof and stellate tubercles (Fig. 4.57, 7). Seen in this way, the acanthothoracids would be an ensemble of generalized placoderms. It is interesting that the internal anatomy of the acanthothoracid braincase (Fig. 4.56B) is strikingly similar to that some fossil jawless vertebrates, in particular galeaspids and osteostracans (Figs 4.16A, 4.21E). This suggests that it features the most generalized gnathostome condition (see Chapter 6 p. 273).

4.14.8 Stensioellida and Pseudopetalichthyida

Two groups, the Stensioellida and Pseudopetalichthyida, are represented by articulated, but poorly preserved, specimens from the Early Devonian Hunsrück Shales of Germany. They have been referred to as placoderms, for lack of clearer affinities. Stensioellids (*Stensioella*) and pseudopetalichthyids (*Pseudopetalichthys*, *Paraplesiobatis*) were regarded by Gross (1962*a*) as placoderms because they seem to have possessed a skull-roof and a short shoulder girdle armed with dermal plates, but they show no evidence of a median dorsal plate, in spite of the presence of a synarcual. There is in fact no evidence that they are placoderms. The striking overall resemblance between *Stensioella* and the Carboniferous and Permian 'bradyodonts' *Menaspis* or *Deltoptychius* suggests that it may be an early holocephalan.

4.14.9 Placoderm phylogeny

The interrelationships of the placoderms are still much debated and there are very few points of consensus. There is now a tendency to consider phyllolepids either as the closest relatives of arthrodires or as highly derived members of this group. They may be most closely related to actinolepids and wuttagoonaspids (Fig. 4.49, 2). In any case, they share with arthrodires an interolateral plate in the thoracic armour. Ptyctodontids share a number of unique features with the petalichthyids, e.g. the enclosed sensory-line canals with large pores and the X-shaped commissure between the supraorbital canal and the posterior pit line (Fig. 4.57, 4). The antiarchs and arthrodires (and the phyllolepids) are closely related, according to several authors (Miles and Young 1977; Young 1986*b*; Denison 1978; Gardiner 1984*b*), because they share a single pair of paranuchal plates. In contrast, Goujet (1984*b*) puts antiarchs with rhenanids and some acanthothoracids in a clade characterized by a premedian plate and dorsally placed nostrils (Fig. 4.57, 5). Acanthothoracids (Fig. 4.57D, F, H), as seen above, are regarded either as primitive rhenanids (Young 1986*b*) or as stem-group placoderms (Goujet 1984*b*). The latter theory is illustrated in Fig. 4.57. The classification of the Placodermi (by phyletic sequencing) derived from the cladograms in Fig. 4.49, 4.54, and 4.57 is set out in Table 4.7.

4.14.10 Stratigraphy, distribution, habitat

Placoderms have long been regarded as exclusively Devonian in age, although their sudden occurrence in the earliest Devonian clearly suggests that they had existed long before. This has been confirmed by the discovery of relatively abundant placoderm material in the Silurian of China, bringing the distribution of the group down to the Early Silurian (Wenlock). These Silurian placoderms are still poorly known, although some of them, from the Silurian of Yunnan, are undoubtedly antiarchs. Others, such as the Early Silurian *Wangolepis* (Pan 1988), cannot be placed in any of the major Devonian groups defined above (Goujet, personal communication). Placoderms are thus probably more diverse than currently believed and their absence from the Silurian localities outside China is due to unsuitable environmental conditions.

The majority of the placoderms known at present are Devonian in age. Their diversity depends on the taxon, the environment, and the region of the world which is considered. For example, the Early Devonian of China and Vietnam is remarkable for its diversity of antiarchs. In contrast, the Early Devonian of North America and Europe has yielded a large variety of arthrodires, but no antiarch. The most impressive diversity of placoderms is

Table 4.7.   Classification of the Placodermi (See also Figs 4.49, 4.54, and 4.57)

Placodermi
  (unnamed taxon)
    'Radotina' prima ('Acanthothoraci')
  (unnamed taxon)
    Arthrodira
      Actinolepidoidei
          Actinolepida *sedis mutabilis*
          Wuttagoonaspida *sedis mutabilis*
          Phyllolepida *sedis mutabilis*
      Phlyctaenioidei
        Phlyctaenii
        Brachythoraci
          Gemuendenaspis
          Holonema
          Migmatocephalia
          Eubrachythoraci
             Coccosteomorphi
             Pachyosteomorphi
  (unnamed taxon)
    Petalichthyida
    Ptyctodontida
  (unnamed taxon)
    (unnamed taxon)
      Romundina ('Acanthothoraci')
      Antiarcha
        Yunnanolepiformes
        Procondylolepiformes
        Sinolepidae
        Euantiarcha
          Bothriolepidoidei
          Asterolepidoidei
  (unnamed taxon)
    Radotina kosorensis ('Acanthothoraci')
    Rhenanida

probably shown by the famous Late Devonian inter-reef fauna of Gogo, Australia, where more than twenty different species (most of them belonging to a monospecific genus) have been recorded (see p. 10 and Fig. 1.12). The large Late Devonian arthrodires, such as *Dunkleosteus* or *Titanichthys*, which may have reached up to six metres in length, were probably the largest animals of their time (Fig. 1.15, 1, 2). The latest known placoderms are Late Famennian in age and represented essentially by antiarchs, ptyctodonts, phyllolepids, and one plyctaeniid arthrodire family, the Groenlandaspididae. Unlike many invertebrate groups, placoderms show no clear decrease in diversity at the Frasnian–Famennian boundary, which is reputedly a 'mass extinction' period (see Chapter 7, p. 290), and very large arthrodires are still abundant in Famennian seas. However, no placoderm survives with certainty into the Carboniferous. On the information at present available, the extinction of the placoderms thus seems to be quite sudden.

Like many groups of jawless vertebrates, the Early Devonian placoderms seem to display more patterns of endemism than Middle and Late Devonian ones. There are taxa of antiarchs (yunnanolepiforms, procondylolepiforms), for example, that are unique to China and Vietnam, and wuttagoonaspid arthrodires are unique to Australia. Phyllolepids occur suddenly, and in great abundance, in Europe and North America by Famennian times, but are found as early as the Middle Devonian in Gondwanan regions; this implies a process of rapid expansion resulting from changes in geography (Young 1990).

Most placoderms were marine, but in Early Devonian times, a majority of placoderms were bound to deltaic or lagoonal facies (actinolepids, plyctaeniid, antiarchs). By Middle and Late Devonian times, many arthrodires, ptyctodonts, and rhenanids lived in frankly marine environments, in particular around coral reefs, whereas antiarchs and some arthrodires (groenlandaspids, phyllolepids) remained bound to sandy shores. Most arthrodires had a crushing-shearing dentition and may have preyed on small crustaceans, brachiopods, or molluscs; advanced brachythoracid arthrodires had cutting gnathal plates and therefore probably preyed on other fishes (some, such as *Mylostoma*, secondarily gained a durophagous diet). Ptyctodontids had a crushing dentition which is somewhat convergent in structure to that of modern chimeras. The ray-like aspect of the rhenanids clearly suggests that they had a benthic mode of life.

Further reading

Arthrodira: Carr (1991); Denison (1978); Dennis and Miles (1979, 1981); Dennis-Bryan (1987); Gardiner and Miles (1990); Goujet (1975, 1984*a*); Gross (1932); Heintz (1931, 1934); Lelièvre (1984, 1991); Lelièvre *et al.* (1990); Miles (1962, 1971); Miles and Dennis (1979); Miles and Westoll (1968); Ørvig (1957, 1967*b*) Stensiö (1959, 1963, 1969); Young (1979. *Phyllolepida and Wuttagoonaspida*: Long (1984); Ritchie (1973, 1984*b*). *Petalichthyida*: Gross (1961*b*); Liu (1991); Stensiö (1964); Young (1978). *Ptyctodontida*: Forey and Gardiner (1986); Mark-Kurik (1977); Mark-Kurik *et al.* (1991); Miles (1967); Miles and Young (1977); Ørvig (1960, 1962, 1980, 1985). *Antiarcha*: Hemmings (1978); Janvier and Pan (1982); Lyarskaya (1981); Ritchie *et al.* (1992); Stensiö (1931, 1948); Tong-Dzuy and Janvier (1990); Wang (1989); Young (1984, 1988); Young and Zhang (1992); Zhang (1978, 1980). *Rhenanida*: Goujet *et al.* (1985); Gross (1963*b*); Stensiö (1969); Young (1986*b*). *Acanthothoraci*: Goujet (1984*b*); Gross (1959); Young (1980). *Interrelationships and systematics*: Denison (1978, 1983); Gardiner (1984*b*); Gardiner and Miles (1990); Miles and Young (1977); Forey and Gardiner (1986); Goujet (1984*a,b*); Stensiö (1963, 1969); Young (1980, 1986*b*).

4.15 ACANTHODII

Acanthodians are a fossil group of gnathostomes characterized by the presence of large spines in front of all fins (Fig. 4.58, 8, 9) except the caudal fin. They also have small scales of a particular type with a bulging base, whose crown grows by adjunction of new odontodes that surround and completely overly the preceding ones. This results in a unique onion-like structure of the scale crown, with successive layers of orthodentine or mesodentine (Fig. 4.63B, C). The head and shoulder girdle of acanthodians is essentially covered with scales or minute bony platelets, although some larger dermal plates may occur on the gill covers and shoulder girdle (Fig. 4.61), as well as around the orbits and on the cheek (Figs 4.58, 33, 4.64C). None of these larger dermal elements is clearly homologous to those of osteichthyans. Acanthodians are often described as 'spiny sharks', and are still regarded by some as being closely related to or even belonging to elasmobranchs. Their vague shark-like aspect is more likely to be a generalized gnathostome condition, with two dorsal fins, separate gill slits (at least in some of them; Fig. 4.60A, 3) and an epicercal tail (Fig. 4.58, 6). In contrast, the shape of their head, with very large and anteriorly placed eyes rather recalls the aspect of some early actinopterygians (see Fig. 4.65C1).

Many acanthodians, in particular the climatiiforms, have a series of ventrolateral 'intermediate spines', which extend between the pectoral and pelvic fins (Fig. 4.64A). These spines have been traditionally regarded as giving palaeontological support for the fin-fold theory (see p. 133). Although this series of spines or scutes occurs mainly in the earliest acanthodians, it may be a unique character either of the climatiiforms or of acanthodians in general.

The morphology of acanthodians looks relatively homogeneous at first glance (Fig. 4.64), but their micromeric exoskeleton and poorly ossified endoskeleton make them rare as fossils even in good conditions. Most of what we know of acanthodian internal anatomy comes from the Carboniferous and Permian genus *Acanthodes*, and from some other acanthodiforms (Figs 4.58, 4.59), which are all probably derived forms. However, for want of better material, we are compelled to refer again to this source of data, and in particular the studies made on the excellently preserved material from the Early Permian of Lebach, Germany. The fact that there are great differences in the organization of the gill openings and teeth (Figs 4.60, 4.62) between the major acanthodian taxa suggests that acanthodians may reveal a wider range of diversity in their internal anatomy.

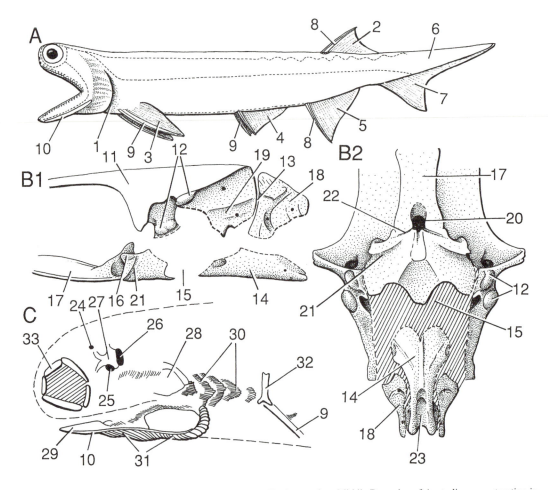

Fig. 4.58. Acanthodii. External morphology and braincase. A, *Howittacanthus*, Middle Devonian of Australia, reconstruction in lateral view (×0.5); B, *Acanthodes*, Carboniferous in Permian, reconstruction of the braincase in lateral (B1) and ventral (B2) view (×1.2; interpretation according to Miles 1973b); C, *Utahacanthus*, Carboniferous of Utah (USA), sketch of head in lateral view (×5). 1, gill slit; 2, dorsal fin; 3, pectoral fin; 4, pelvic fin; 5, anal fin; 6, chordal lobe of caudal fin; 7, caudal fin web; 8, median fin spine; 9, paired fin spine; 10, mandibular dermal bone; 11, dorsal part of braincase; 12 otic condyles for articulation of palatoquadrate; 13, otico-occipital fissure; 14, basioccipital ossification; 15, ventral fissure; 16, basipterygoid process; 17, basisphenoid ossification; 18, occipital ossification; 19, groove for jugular vein; 20, buccohypophysial foramen; 21, spiracular groove; 22, groove for internal carotids; 23, groove for epaxial musculature; 24, utricular otolith; 25, lagenar otolith; 26, saccular otolith; 27, labyrinth; 28, palatoquadrate; 29, Meckelian bone; 30, gill rays; 31, branchiostegals; 32, scapulocoracoid; 33, circumorbital plates. (A, from Long (1986a); B, from Miles (1973b); C, from Schultze (1990).)

4.15.1 Exoskeleton and external morphology

Acanthodes (Fig. 4.64H), like other acanthodiforms such as *Howittacanthus* (Fig. 4.58A), is a slender-bodied acanthodian with no intermediate spines (they are presumed lost) and only one dorsal fin (Fig. 4.58, 2); the single dorsal fin is regarded as an acanthodiform character. *Acanthodes* is devoid of teeth, and its gill arches are armed with long gill-rakers, which suggest that it was a

macrophagous suspension-feeder (Fig. 4.59A2, B, 8). This is, however, a derived condition in the group, since teeth and tooth-bearing jaw-bones are known in other acanthodians (Fig. 4.62). It has a single gill cover and probably a single gill opening (Fig. 4.58, 1).

The dermal skeleton of *Acanthodes* consists essentially of minute scales with an unornamented crown of dentine and a base of acellular bone (Fig. 4.63C); again presumably a derived condition. The scales may be very scarce

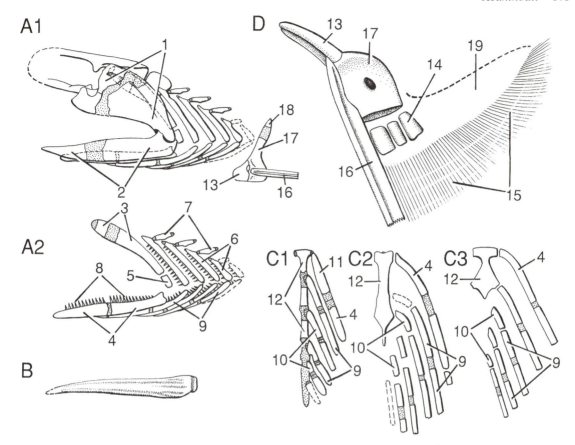

Fig. 4.59. Acanthodii. Visceral skeleton, shoulder girdle, and pectoral fin of *Acanthodes*, Carboniferous to Permian. A, reconstruction of the skull and shoulder girdle (cartilage stippled) in lateral view (A1) and separate hyobranchial apparatus with gill rakers (A2); B, isolated hyoid gill-raker (×10); C, three reconstructions of the ventral part of gill arches, according to Nelson (1968; C1), Miles (1973*b*; C2), and Gardiner (1984*a*; C3); D, shoulder girdle and pectoral fin skeleton in ventral view. 1, palatoquadrate (with separate metapterygoid and quadrate ossifications); 2, Meckelian bone; 3, epihyal; 4, ceratohyal; 5, interhyal; 6, epibranchials; 7, pharyngobranchials; 8, gill-rakers; 9, ceratobranchials; 10, hypobranchials; 11, hypohyal; 12, basibranchial series or basibranchial plate; 13, procoracoid; 14, radials; 15, ceratotrichiae; 16, fin spine; 17, scapulocoracoid; 18, suprascapular ossification; 19, presumed metapterygial division of the fin. (A, combined from Miles (1973*b*), Jarvik (1977) and Schultze (1993), B, from Zidek (1992); C1, from Nelson (1968), C2, from Miles (1973*b*); C3, from Gardiner (1984*a*); D, combined from Miles (1973*a*) and Forey and Young (1983).)

or absent on the head. Growth series show, however, that the scale covering extends forwards as the individuals grow, with a more rapid progression along the main lateral line (Fig. 7.11D). Larger plates occur around the eyes (circumorbital plates, Fig. 4.58, 33) and below the lower jaw (mandibular bone; Fig. 4.58, 10), and the gill cover is strengthened by dermal branchiostegal rays (Fig. 4.58, 31, Fig. 4.60C, 1). There is no dermal shoulder girdle, apart from the fin spine, but a number of large dermal plates strengthen the shoulder girdle in other acanthodians, in particular climatiiforms (Fig. 4.61A).

The sensory lines pass between two scale rows, which may be modified into minute bony tubes (see Fig. 4.60, 5–10 for the names of the lines).

The fin spines are slender, ornamented with only two rounded ridges of dentine (Fig. 4.63D). The pectoral fin spines are inserted in the scapulocoracoid and were probably not movable (Fig. 4.59, 13, 16). The fins are covered with minute elongated scales, which may fuse into longer units, thereby resembling the lepidotrichs of osteichthyans. The distal part of the fin web contains numerous, very thin ceratotrichiae (Fig. 4.59, 15).

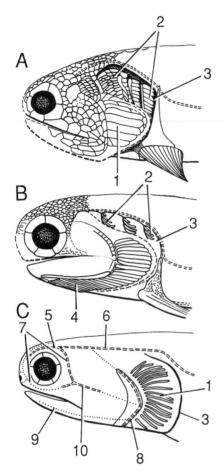

Fig. 4.60. Acanthodii. Head reconstructions showing the extent of the hyoidean gill cover in the Climatiiformes (A) and Acanthodiformes (B, C). A, *Brachyacanthus*, Early Devonian (×2); B, *Mesacanthus*, Early to Middle Devonian (×4); C, *Homalacanthus*, Late Devonian (×3). 1, hyoidean gill cover; 2, subsidiary gill covers; 3, gill slit; 4, branchiostegals; 5, supraorbital line; 6, main lateral line; 7, infraorbital line; 8, preopercular line; 9, mandibular line; 10, supramaxillar line. (A, B, from Watson (1937); C, from Miles (1966).)

4.15.2 Endoskeleton

The braincase of *Acanthodes* consists of several ossifications, which have been observed in a few specimens preserved as a natural mould. They have thus been assembled and interpreted in different ways, according to the presumed affinities of acanthodians. Jarvik (1980), for example, interprets the braincase of *Acanthodes* in the light of chondrichthyan anatomy, whereas Miles (1973b) does it in the light of osteichthyans (Fig. 4.58B). As currently reconstructed, the braincase of *Acanthodes* com-

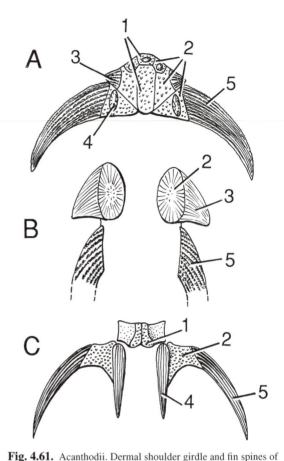

Fig. 4.61. Acanthodii. Dermal shoulder girdle and fin spines of the Climatiiformes in ventral view. A, *Climatius*, Early Devonian (×1); B, *Gyracanthides*, Devonian to Carboniferous (×0.2); C, *Diplacanthus*, Middle to Late Devonian (×1). 1, lorical plate; 2, pinnal plates; 3, prepectoral spine; 4, intermediate spine; 5, fin spine. (A–C, combined from Miles (1973a) and Long (1986b).)

prises two median ventral ossifications, a posterior, basioccipital one (Fig. 4.58, 14), and an anterior, basisphenoid one (Fig. 4.58, 17). The latter is pierced by the median buccohypophysial foramen (Fig. 4.58, 20) and produced laterally into a large basipterygoid process (Fig. 4.58, 16), which is lined posteriorly by a spiracular groove (Fig. 4.58, 21). Dorsally, there is a supraoccipital ossification (Fig. 4.58, 18), whose position is unclear, and a large component that includes the otic region and part of the orbital region (Fig. 4.58, 11). Nothing is known of the ethmoid region. The otic cavity contains three large otoliths, or statoliths (lagenar, saccular, and utricular; Fig. 4.58C, 24–26) made up of calcite with traces of dolomite (Miles 1973b; Schultze 1988, 1990). This condition, which is confirmed by observations on other acanthodians, such as *Utahacanthus* (Fig. 4.58C), is similar to that in extant actinopterygians and is therefore regarded by Schultze (1988) as the general osteichthyan condition. Coates (1993) rejects this interpretation and considers that the three statoliths arose more than once (see p. 187).

There seems to have been a wide gap between the two ventral ossifications, and this suggests the presence of a ventral fissure (Fig. 4.58, 15), as in generalized osteichthyans (Fig. 4.27).

The palatoquadrate consists of two or three ossifications, with a large quadrate part (Fig. 4.59, 1), and the Meckelian bone of two ossifications (Fig. 4.59, 2), lined ventrally by the dermal mandibular bone (Fig. 4.58, 10). The palatoquadrate articulates with the braincase by means of two articulations, a double postorbital one (otic, Fig. 4.58, 12) and the palatobasal one (Fig. 4.58, 16). Both the palatoquadrate and Meckelian bone are hollowed laterally by a depression for the jaw musculature. The hyoid arch consists of large epal and ceratal elements, also made up of two separate ossifications (Fig. 4.59, 3, 4), and an interhyal (Fig. 4.59, 5). It does not seem to have played any role in the jaw suspension but is closely associated to the posterior margin of the palatoquadrate. There are five gill arches with pharyngobranchials (presumably oriented posteriorly; Fig. 4.59, 7). Various reconstructions have been proposed for the ventral part of the hyobranchial apparatus of *Acanthodes* (Fig. 4.59C), but it seems now that the hyoid and branchial arches are connected basally to a single median basibranchial ossification (Miles 1973*b*; Gardiner 1984*a*; Fig. 4.59C2, C3, 12), although Nelson (1968) reconstructs separate basibranchial elments (Fig. 4.59C1, 12). There seems to be no hypohyal, and no hypobranchial in the first gill arch. The hyoid and branchial arches bear long dermal gill rakers housing an endoskeletal core (Fig. 4.59A2, B, 8). Jarvik (1980) suggests that the branchiostegal rays of *Acanthodes*, which look as if they were attached to the hyoid arch, are in fact endoskeletal gill rays, comparable to those of the extant frilled shark (*Chlamydoselachus*). Although this has not yet been clearly refuted, the organization of these rays is so similar to the series of dermal branchiostegal plates in other acanthodians that they are probably of the same nature (Fig. 4.60, 1, 4). However, true gill-rays, which armed the interbranchial septum and supported the gill lamellae, are also present in acanthodids (Fig. 4.58, 30) and seem to have been rather short, as in osteichthyans.

The axial skeleton consists of only neural and haemal arches; some radials have been observed in the tail. Two basal plates seem to support the dorsal fin.

The endoskeletal shoulder girdle consists of three ossifications: scapulocoracoid, procoracoid, and suprascapular (Fig. 4.59, 13, 17, 18). This condition seems to be derived, as the scapulocoracoid of climatiiforms is undivided. Three large but short radials or basal elements articulate with the girdle by a small articular area (Fig. 4.59, 14).

4.15.3 Histology

The scales of acanthodians show a bulging base of acellular bone and a crown of dentine arranged in nested layers (Fig. 4.63A, B1, C1). This condition is not, however, general for all acanthodians, and the group displays other types of histological structure. The scales of climatiiforms generally have a cellular base and a crown of mesodentine (Fig. 4.63B). Other acanthodians have an acellular base and a crown of orthodentine. Some acanthodians (*Poracanthodes*) have a pore-canal system in the scale crown, although it does not have the same detailed structure as that in osteostracans (see p. 112) and sarcopterygians (see p. 205). This network of canals may be linked with the lateral-line system. In only one genus, *Machaeracanthus*, the successive dentine layers of the scale crown do not completely cover the preceding ones; in this respect they resemble the structure of some Palaeozoic chondrichthyans with growing scales (Fig. 4.63E1).

The fin spines of acanthodians are essentially made up of orthodentine or mesodentine, which forms the external ornamentation and covers a core of trabecular dentine and bone (cellular or acellular), often surrounding a central canal (Fig. 4.63B3, C3). Some climatiid fin spines can resemble superficially the median fin spines of some Palaeozoic elasmobranchs (e.g. ctenacanthiforms). However, the spines of the latter have a much larger central canal (which is always widely open basally), a long basal insertion, a median ridge on their posterior surface, and a thin enameloid layer on the surface of the dentine of the external ornamentation.

4.15.4 Diversity and phylogeny

Three major acanthodian clades are currently recognized: the Climatiiformes (Fig. 4.64, 3), Ischnacanthiformes (Fig. 4.64, 9), and Acanthodiformes (Fig. 4.64, 5). The Diplacanthida (Fig. 4.64, 4) and Gyracanthidae (Fig. 4.64B) are generally referred to the Climatiiformes, though with some doubt (Fig. 4.64, left-hand side).

Climatiiforms can be recognized by their rather broad-based fin spines, ornamented with numerous, often nodose, ridges, as well as by their numerous intermediate spines and well-developed dermal shoulder girdle (e.g. *Climatius*, *Euthacanthus*, *Brachyacanthus*). The shoulder girdle consists of ventral pinnal and lorical plates, with paired prepectoral spines attached to the pinnal plates (Fig. 4.61A, 1, 2). The head has a large hyoid gill cover (Fig. 4.60A, 1), but four or five smaller gill covers are visible above it (Fig. 4.60A, 2), providing evidence for separate gill slits (Fig. 4.60A, 3). The teeth are minute, often petal-shaped, and entire tooth families may be fused into a spiral (Fig. 4.62A, 1), but they do not seem to have been shed. In *Ptomacanthus* or *Latviacanthus* (Fig. 4.62A, B, 1) for example, the dentition somewhat resembles that of a shark, with numerous families of blunt or pointed teeth lining the palatoquadrate and Meckelian bone. Very little is known of the internal anatomy of the climatiiforms, except that the palatoquadrate is elongated and has no otic (postorbital) articulation (Fig. 4.62B, 3).

Many of these climatiiform characters can be interpreted as general for acanthodians or even for gnathostomes. If the numerous intermediate spines are regarded as a primitive acanthodian feature, and the five gill slits or separate teeth as general gnathostome features, then climatiiforms are defined only by their dermal shoulder girdle and the mesodentine of their scales and spine ornamentation (Fig. 4.63B); this, however, has been evidenced in only a few genera. As a consequence, we cannot be sure that the group is a clade. Diplacanthids are often regarded as an advanced climatiiform group (Fig. 4.64, right-hand side), because of their well-developed dermal shoulder girdle and relatively large dermal bones on the head (Fig. 4.61C). However, they retain only one pair of intermediate spines. They are characterized by a large dermal plate on the cheek (e.g. *Culmacanthus* Fig. 4.64C) and lack teeth, like acanthodiforms. Gyracanthids (Fig. 4.64B) are also considered as climatiiforms because of their large dermal shoulder girdle and prepectoral spines (Fig. 4.61B). They are characterized by large recurved spines ornamented ridges of dentine which form chevrons on the leading edge. Gyracanthids would thus be the only climatiiforms that straddle the Frasnian–Famennian boundary and survived until the Late Carboniferous.

Ischnacanthiforms (Fig. 4.64J) are characterized by large tooth-bearing jaw-bones on the palatoquadrate and Meckelian bone (Fig. 4.62C, 5). Large triangle-shaped teeth are firmly attached to these jaw bones which grow anteriorly by apposition of larger and larger teeth (Fig. 4.62C, 6). In the symphysial region of the jaws were large separate tooth spirals. The dermal covering of the head and shoulder girdle is reduced and there are only a few larger platelets on the gill cover. The gill cover is large and covers all the subsequent gill slits. The fin spines are slender and are covered with orthodentine, as is the crown of the scales. The mandibular arch is the only endocranial element ever described in this group (Fig. 4.62C, 3, 4). The palatoquadrate is elongated, as in climatiiforms, and does not seem to have an otic (post-

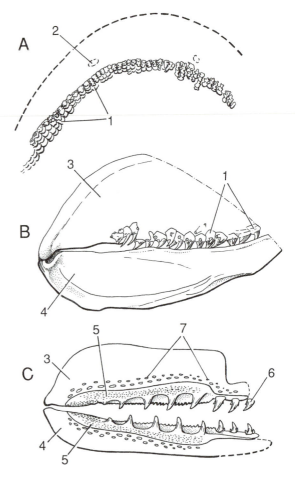

Fig. 4.62. Acanthodii. Dentition. A, *Ptomacanthus*, Early Devonian of Great Britain, upper dental arch in ventral view (×1.5); B, *Latviacanthus*, Early Devonian of Latvia, mandibular arch in lateral view (×4); C, *Poracanthodes*, Early Devonian of Severnaya Zemlya, mandibular arch in internal view (×3). 1, tooth 'families'; 2, external nostril; 3, palatoquadrate; 4, Meckelian bone; 5, dentigerous jaw-bone; 6, free tooth (youngest tooth); 7, toothed platelets. (A, redrawn from a photograph in Miles (1973*a*); B, from Schultze and Zidek (1982); C, reconstruction based on Valiukevicus (1992) and the interpretation of jaw-bone growth by Ørvig (1973).)

orbital) articulation. *Ischnacanthus*, *Uraniacanthus*, *Persacanthus*, and *Poracanthodes* are moderately large forms, but some isolated jaw-bones of *Xylacanthus* are very large and suggest a total body length of about 2 metres.

Acanthodiforms have only one dorsal fin and spine, a very large hyoidian gill cover and they are devoid of teeth. Most of them have no intermediate spine; the pelvic fin

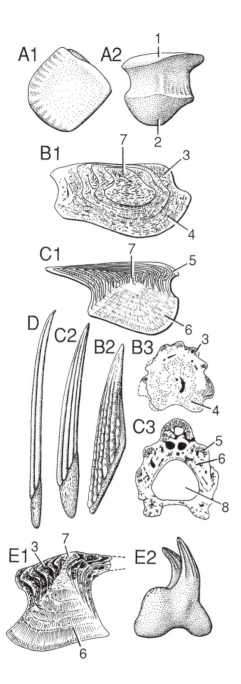

Fig. 4.63. Acanthodii. Exoskeleton and histology.
A, *Gomphonchus*, Silurian, scale in lateral (A1), and external
(A2) view (×40). B, The Silurian climatiiform *Nostolepis*,
vertical section through a scale (B1, ×50), fin spine in lateral
view (B2, ×3), and transverse section through a fin spine
(B3, ×20). C, the Silurian ischnacanthiforme *Gomphonchus*,
vertical section through a scale (C1, ×50), fin spine in lateral
view (C2, ×3), and transverse section through a fin spine
(C3, ×20). D, *Acanthodes*, Carboniferous to Permian, fin spine
in lateral view. E, *Machairacanthus*, Early Devonian, vertical
section through a scale (F1, ×35) and isolated tooth in lateral
view (F2, ×20). 1, crown; 2, base; 3, mesodentine; 4, cellular
bone; 5, orthodentine; 6, acellular bone; 7, primordial odontode;
8, central canal. (A, from Gross (1947); B1, from Ørvig
(1967*b*); C, B3, from Gross (1971*a*), E1, from Goujet (1976);
E2, redrawn from a photograph in Mader (1986).)

is elongated in shape in the most advanced forms
(Fig. 4.64H). Some Carboniferous acanthodiforms have an
elongated, almost eel-shaped body (*Pseudacanthodes*,
Traquairichthys; Fig. 4.64I) and probably lose the pelvic
spines and even the pelvic fins. The acanthodiforms comprise
three families, the Mesacanthidae (*Mesacanthus*, Figs 4.60B,

4.64E), Cheiracanthidae (*Cheiracanthus*, Fig. 4.64F), and
Acanthodidae (*Acanthodes*, *Homalacanthus*, *Traquairichthys*;
Figs. 4.58A, 4.64G–I). Only the Acanthodidae survived until
the Early Permian.

Acanthodian phylogeny is still poorly corroborated.
There is a majority consensus on a cladogram in which

the climatiiforms are the sister-group of the ischnacanthiforms and acanthodiforms (Fig. 4.64, left-hand side), but with diplacanthids in an undeterminate position. The problem with this theory is that the synapomorphies shared by the acanthodiforms and ischnacantiforms are essentially losses or reductions: reduction of the dermal shoulder girdle, or of the head scutes; reduction or loss of intermediate spines; and loss of cellular bone (Fig. 4.64, 10).

◀ **Fig. 4.64.** Acanthodii. Interrelationships of the Acanthodii. Left-hand side, according to the classification of Miles (1973*a*); right-hand side, according to Long (1986*b*). *Terminal taxa*: A, Climatiidae (*Climatius*); B, Gyracanthidae (*Gyracanthides*); C, Culmacanthidae (*Culmacanthus*); D, Diplacanthidae (*Diplacanthus*); E, Mesacanthidae (*Mesacanthus*); F, Cheiracanthidae (*Cheiracanthus*); G, *Homalacanthus*; H, *Acanthodes*; I, *Traquairichthys*; J, *Ischnacanthus*. *Nested taxa and selected synapomorphies*: 1, Acanthodii (paired fin spines, onion-like structure of scale crown, at least one intermediate spine); 2, double mandibular joint, numerous branchiostegals; 3, Climatiiformes (numerous ventral dermal bones in shoulder girdle or broad-based spines); 4, Diplacanthida (large cheek plate, deep body shape, dorsal fin spines very long and slender); 5, Acanthodiformes (single dorsal fin); 6, Acanthodida (loss of intermediate spine); 7, Acanthodidae (reduction of branchiostegals, pelvic fins close to pectoral fins); 8, Acanthodinae (anal and dorsal fins at the same level and very close to caudal fin); 9, Ischnacanthida (dentigerous jaw-bone); 10, loss of dermal shoulder girdle; 11, Climatiida (broad-based fin spines, large and numerous intermediate spines). *Age*: Silurian to Early Devonian (A); Early Devonian to Carboniferous (B); Devonian (D–H, J); Late Devonian (C); Carboniferous (I); Carboniferous to Early Permian (H). (Reconstructions based on Denison (1979), except C, from Long (1983).)

Only the enlargement of the hyoid gill cover may be a positive character, yet it seems to have occurred independently, since the most generalized acanthodiforms (mesacanthids) display a condition which resembles that in climatiiforms (Fig. 4.60A, B). Maisey (1986) therefore suggested a reverse character polarity, in which the acanthodiform condition is regarded as general for acanthodians, and a gain of intermediate spines, teeth, and dermal shoulder girdle occurs in the climatiiforms and ischnacanthiforms. This solution is, however, unsatisfactory, since out-group comparison with most other gnathostome groups shows that two dorsal fins and teeth are general for the gnathostomes. Moreover, the theory favoured by the majority is more consistent with the earlier occurrence (in the Silurian) of the climatiiforms and ischnacanthiforms. A third possibility has been suggested by Long (1986*b*), who puts ischnacanthiforms as the sister-group of all other acanthodians (Fig. 4.64, right-hand side), on the basis of a supposedly derived type of double jaw articulation in acanthodiforms and climatiiforms. Again, this phylogeny implies that the dermal shoulder girdle and intermediate spines in climatiiforms are derived conditions.

The major problem in acanthodian phylogeny concerns the status of climatiiforms, which may well be paraphyletic, as they are currently defined. One may even question the monophyly of the acanthodians as a whole. The structure of the acanthodian scales appears to be unique, but somewhat similar structures or morphologies can be found elsewhere. The nested structure of the dentine layers of the crown is comparable to the ganoin of early actinopterygians (Fig. 4.65E3), and the bulging base is also met with in some Palaeozoic chondrichthyans (Fig. 4.30A2). The paired fin spines are indeed unique among the gnathostomes, although the spinal plates of placoderms can be regarded as a fin spine incorporated in the dermal shoulder girdle. The case of *Machaeracanthus* is one that throws doubt on acanthodian monophyly. This Devonian genus is known only from isolated scales, spines, and teeth. The scales look like climatiid scales, although the mesodentine layers do not entirely cover the central part of the crown (Fig. 4.63E1); this condition

recalls the structure of some Palaeozoic chondrichthyan scales (see p. 143). The teeth have a peculiar 'goat-head' shape (Goujet 1993) and look more like diplodont chondrichthyan teeth (Fig. 4.63E2); and the knife-shaped spines, though known in hundreds, are all asymmetrical, and thus paired. If it is an acanthodian, *Machairacanthus* is thus atypical in having no unpaired fin spines. If it is a chondrichthyan, it is equally atypical in having paired fin spines, or it indicates that chondrichthyans are derived from some acanthodian-like gnathostomes. It is thus probable that acanthodian phylogeny is doomed to undergo profound reworkings in the future, and will need to be reconsidered in the general framework of gnathostome phylogeny.

The classification of the Acanthodii shown in Table 4.8 is derived from the cladogram in Fig. 4.64 (left-hand side):

4.15.6 Statigraphical and geographical distribution, habitat

The earliest known acanthodian remains (scales and spines) are Early Silurian in age, and there are records of

Table 4.8. **Classification of the Acanthodii** (See also Fig. 4.64)

Acanthodii
  Climatiiformes
    Climatiida
      Climatiidae
      Gyracanthidae
    Diplacanthida
      Culmacanthidae
      Diplacanthidae
  (unnamed taxon)
    Ischnacanthiformes
    Acanthodiformes
      Mesacanthidae
      Acanthodida
        Cheiracanthidae
        Acanthodidae

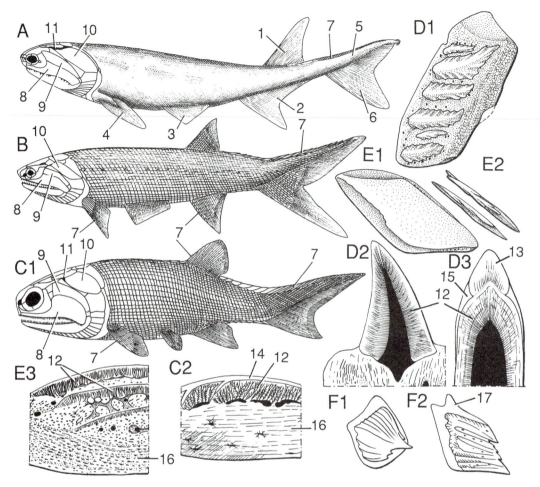

Fig. 4.65. Fossil Actinopterygii. Morphology of the Silurian and Devonian Actinopterygii. A, *Cheirolepis*, Middle to Late Devonian of Europe and North America, reconstruction in lateral view (×0.2); B, *Howqualepis*, Middle Devonian of Australia, reconstruction in lateral view (×0.1); C, *Moythomasia*, Late Devonian, reconstruction in lateral view (C1, ×1), and vertical section through a scale (C2, ×100); D, *Lophosteus*, Late Silurian Estonia, scale in lateral view (D1, ×25) and vertical section through a tooth (D2, ×90), compared with that of a typical actinopterygian tooth (the Mesozoic genus *Semionotus*, D3); E, *Andreolepis*, Late Silurian of Sweden, flange scale (E1, ×25), and possible fulcral scales (E2, ×20) in lateral view, and vertical section through a scale (E3, ×100); F, comparative outline of the scales of *Cheirolepis* (F1) and *Moythomasia* (F2), brought to the same size. 1, dorsal fin; 2, anal fin; 3, pelvic fin; 4, pectoral fin; 5, chordal lobe of caudal fin; 6, caudal fin web; 7, fringing fulcra; 8, maxilla; 9, preopercular; 10, opercular; 11, spiracular opening; 12, dentine; 13, acrodine; 14, ganoine; 15, collar enamel; 16, bony scale base; 17, dorsal peg of scale. (A, F1, from Pearson and Westoll (1979); B, from Long (1988*b*); C, F2, from Jessen (1968); C2, from Gardiner (1984*a*); D1, from Gross (1969); D2, from Gross (1971*b*); D3, from Olsen and McCune (1991); E1, from Gross (1968*b*); E2, from Janvier (1978*b*).)

Late Ordovician occurrences, as yet undescribed. The youngest ones are Early Permian, but acanthodians are most abundant from the Late Silurian to the Late Devonian. Climatiiforms and ischnacanthiforms are widespread in the Silurian and Early Devonian; climatiiforms become very rare after the Middle Devonian, although gyracanthids (which may belong to the climatiiforms)

occur until the Later Carboniferous. Ischnacanthids occur until the Late Carboniferous, but most Permian and Carboniferous acanthodian remains belong to the Acanthodidae.

Since they are nectonic fishes, most acanthodians have a world-wide distribution. There are, however, some apparently endemic taxa, such as the diplacanthid

Culmacanthus, which is known exclusively in the Devonian of Australia and Antarctica. The gyracanthid *Gyracanthides* also seems to occur in much greater abundance in Gondwanan regions (i.e. Australia, Antarctica, the Middle East, and South America) than in Euramerica or China. How far these patterns are due to mere sampling chance is unknown. Long (1986a) has suggested, on the basis of stratigraphical occurrences, that acanthodiforms have dispersed world-wide from a centre situated somewhere in Euramerica.

The acanthodians are essentially marine, although species associated with Old Red Sandstone facies have often been regarded as freshwater. The Carboniferous and Permian acanthodids commonly occur in intramontane basins and indeed probably lived in freshwater. The diet of the acanthodians can be guessed from the morphology of their dentition. It is clear that the ischnacanthiforms and most climatiiforms were predators, although the latter may have preyed on small invertebrates. The acanthodiforms, being devoid of teeth and possessing long gill-rakers, were clearly microphagous suspension-feeders. The stomach contents of some Late Devonian acanthodids show large amounts of small crustaceans.

Further reading

Anatomy and histology: Denison (1979); Gross (1947); Heyler (1969); Jarvik (1980); Long (1983, 1986a,b); Miles (1966, 1968, 1973a,b); Nelson (1968, 1969); Ørvig (1967a,b, 1973); Reis (1896); Schultze (1990); Schultze and Zidek (1982); Valiukevicius (1992); Watson (1937); Zidek (1985). *Interrelationships and systematics*: Denison (1979); Long (1983, 1986a,b); Maisey (1986); Miles (1973a).

4.16 OSTEICHTHYES: FOSSILS AND CHARACTERS

The addition of early fossil osteichthyans to a phylogeny based on Recent taxa does not greatly modify the definition of osteichthyans and the structure of their hypothetical ancestral morphotype. Many osteichthyan characteristics concern soft tissues and cannot be observed or inferred from fossils. Even such skeletal structures as the gill arch components or the fin radials are rarely preserved in the earliest representatives of the major osteichthyan taxa. Of the long list of osteichthyan characteristics cited by Rosen *et al.* (1981) or Lauder and Liem (1983), only a few, such as the sclerotic rings of four plates, appear to be more generalized when other fossil gnathostomes are considered (e.g. placoderms and acanthodians). The question of the degree of generality of endochondral bone is still debated; it has been claimed to be present in some arthrodires (Forey 1980, Gardiner

1989c, but Goujet (1984a,b) considers that the few patches of spongy bone observed in the snout of some forms (Fig. 4.42, 34) are not really comparable to the widespread endochondral bone of true osteichthyans. As to the dermal bones of the head, the absence of a maxilla in actinistians and lungfishes remains valid when fossils are considered, but phylogenies based on many other characteristics imply that this is a convergent loss. Apart from a few Silurian forms generally referred to as actinopterygians (see below), there is no fossil osteichthyan that falls outside the crown-group osteichthyans (i.e. the group including all the extant osteichthyans and their respective fossil relatives). This perhaps means that the diversification of osteichthyans took place long before the earliest evidence of the group.

There are virtually no significant osteichthyan remains before the Devonian. A few isolated scales, teeth, and dermal bone fragments have been recorded from the Late Silurian (Ludlow) of Sweden, Estonia, Russia, and China, and have been referred to either stem-group actinopterygians or stem-group osteichthyans. Some possible lungfish remains have been recorded from the Late Silurian of China. In contrast, the earliest stages of the Devonian have yielded typical actinopterygian and sarcopterygian remains.

4.17 ACTINOPTERYGII

The actinopterygians are one of the largest and most diverse extant vertebrate groups, but they did not gain this great diversity before Triassic or Jurassic times, and they first became relatively abundant by Early Carboniferous times. Among Palaeozoic fossil actinopterygians known from articulated specimens, only the Middle to Late Devonian genus *Cheirolepis* (Fig. 4.65A) seems to lack the acrodine cap on its teeth and has atypical scales (Fig. 4.65F1). It remains, however, referred to this group because of its single dorsal fin, large maxilla, and some other features apparently unique to actinopterygians, such as ganoin on scales or fulcral scales on the leading edge of the tail (Fig. 4.65A, 7), nasals with notches for the nostrils (Fig. 4.69B), and a spiracular opening situated between the dermosphenotic and the intertemporal and supratemporal bones (Fig. 4.65A, 11). The definition of an actinopterygian given in Chapter 3 (p. 68) for extant forms thus remains valid for the majority of the Palaeozoic actinopterygians.

The earliest remains referred to the actinopterygians are isolated scales, teeth, and dermal bone fragments of *Andreolepis* (Fig. 4.65E) and *Lophosteus* (Fig. 4.65D1), from the Later Silurian of the Baltic area. All the teeth associated with these scales are made up of orthodentine but show no collar enamel and no acrodine cap (Fig. 4.65D2).

The scales are diamond-shaped (*Andreolepis*) or somewhat elongated (*Lophosteus*) and are ornamented with confluent ridges of orthodentine. A thin layer of ganoine seems to be present in *Andreolepis*, above each dentine layer (Fig. 4.65E3, 12). None of these scales has a characteristic peg-and-socket articulation, but some scales of *Andreolepis* are extremely elongated and rod-shaped (Fig. 4.65E2); they resemble the fulcral scales on the leading edge of the caudal fin of *Cheirolepis* (Fig. 4.65A, 7). The few dermal bones referred to *Andreolepis* are ornamented with rounded tubercles, whereas those referred to *Lophosteus* bear star-shaped tubercles. In some of them, open sensory-line grooves are observed (Märss 1986*b*). Peculiar spine-like elements with large tubercles are also referred to *Lophosteus* (Otto 1993). Admittedly, the few complete dermal plates known from these Silurian forms are not clearly actinopterygian-like, and the ornamentation of the dermal bones of *Lophosteus* recalls that of 'acanthothoracid' placoderms. Nevertheless, their histology does not show any evidence of semidentine. Schultze (1977) suggested that *Andreolepis* is a stem-group actinopterygian and *Lophosteus* may possibly be a stem-group osteichthyan or a stem-group sarcopterygian. More recently (Schultze 1992), he has referred to both genera as actinopterygians. The affinities of these two forms are thus far from clear, although they are here regarded as stem-group actinopterygians (Fig. 4.67A, B).

In contrast, the isolated scales and dermal bones of the Later Silurian *Naxilepis* (Fig. 4.67C) and the Early Devonian *Orvikuina* (Fig. 4.67D), *Ligulalepis* (Fig. 4.67E), and *Dialipina* (Fig. 4.67F) clearly belong to the actinopterygians, since they are covered with typical ganoine and some of them possess a distinct, narrow-based dorsal peg. Their ornamentation consists of separate dentine and ganoine ridges, a pattern that occurs widely in Devonian actinopterygian scales, although there is a tendency for the ridges to fuse into a continuous ganoine layer in Later Devonian and later forms. Schultze (1992) has suggested a morphocline in scale structure, leading from *Lophosteus* to *Cheirolepis* or *Moythomasia* (Fig. 4.67). It involves the progressive thickening of the ganoine, superimposition of ganoine layers, and development of an anterodorsal process into a narrow-based peg.

Apart from the late Middle Devonian *Cheirolepis* and *Howqualepis* (Fig. 4.65B), the earliest known articulated actinopterygians are Late Devonian in age, and are represented essentially by *Moythomasia* (Fig. 4.65C) and *Mimia* (Fig. 2.6) from the Frasnian of Europe and Australia. The diversity of the Carboniferous actinopterygians is foreshadowed by the still poorly known Famennian forms, with *Tegeolepis*, *Osorioichthys*,

Stegotrachelus, and some remains of platysomids. It thus seems that the so-called actinopterygian 'radiation' took place by the end of the Frasnian. Before this time, actinopterygians are rare and rather homogeneous in aspect.

4.17.1 General morphology and exoskeleton

Mimia (Fig. 2.6, 4.66) is a small fish (about 10 cm in length) that displays what is regarded as a rather generalized actinopterygian morphology, by comparison with *Cheirolepis* whose elongated body and minute scales are now considered as unique to this genus. *Mimia* is more derived that *Cheirolepis* and the extant cladistians by at least six characteristics, the most conspicuous of which is the presence of fringing fulcrae (arrow-shaped scales, Fig. 3.17, 5) on the leading edge of all fins, paired and unpaired (Fig. 4.70, 3). In the head, the snout extents slightly further than the lower jaw, and the eye is very large. The teeth show a distinct acrodine cap, the maxilla is posteriorly expanded, and the mandibular sensory-line canal runs through the dentary, as in most other early actinopterygians. In the palate, all the basic dermal bones of osteichthyans are present: small, paired vomers (Fig. 4.66, 1), a very broad parasphenoid (Fig. 4.66, 2) pierced by a buccohypophysial canal that does not extend posteriorly beyond the ventral fissure (Fig. 4.66, 4), large ento- and ectopterygoids, accessory vomerine plates, and numerous dermopalatines. The lower jaw shows a large prearticular and several coronoids, and the dentary encloses the mandibular sensory-line canal. Numerous minute tooth plates are associated with the hyoidean and branchial arches. In the snout, the rostral bone is shield-shaped and reaches the mouth margin, where it bears a few teeth, and the nasals bear notches for anterior and posterior nostrils. The orbit is lined ventrally and posteriorly by two large infraorbital dermal bones (lachrymal and jugal). A minute dermal bone, the dermohyal (Fig. 4.66, 15), is attached to the lateral surface of the epihyal (Fig. 4.66, 13). As in most other early actinopterygians, the preopercular (Fig. 4.65, 9) is strongly inclined and attached to the palatoquadrate, the spiracular opening is situated between the dermosphenotic, intertemporal, and supratemporal (Fig. 4.65, 11), and, posteriorly to the skull-roof, there are one or two pairs of extrascapulars.

The dermal shoulder girdle of *Mimia* and other early actinopterygians comprises large cleithrum and clavicula, and is linked to the skull-roof by a supracleithrum and presupracleithrum. The body is covered with typical actinopterygian rhombic scales, with a peg-and-socket articulation and ganoine ridges. There are large median dorsal and ventral (post-anal) ridge scales that pass posteriorly to fulcra on the leading edge of the tail (Fig. 4.65, 7). There is a single dorsal fin and a large anal fin (Fig. 4.65, 1,

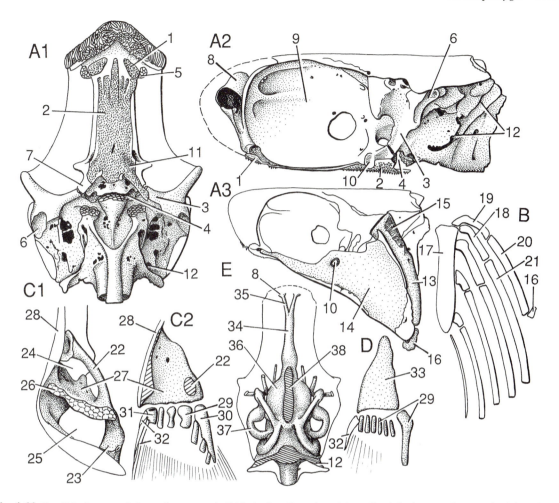

Fig. 4.66. Fossil Actinopterygii. Internal anatomy. A–C *Mimia*, Late Devonian of Australia; A, braincase and some palatal dermal bones in ventral (A1, ×5) and lateral (A2) view, and with palatoquadrate and epihyal added (A3); B, reconstruction of the ventral part of the right hyoid and branchial arches; C, scapulocoracoid in posterior (C1) and dorsal view (C2, dorsal buttress removed, radials added) (×5); D, *Moythomasia,* pelvic girdle and fin in dorsal view (×10). E, *Kansasiella,* Carboniferous of Kansas, internal cast of the cavities in braincase, in dorsal view. 1, vomer; 2, parasphenoid; 3, lateral commissure; 4, ventral fissure; 5, facet for the autopalatine part of the palatoquadrate; 6, facet for epihyal; 7, descending process of sphenethmoid; 8, nasal capsule; 9, orbital cavity; 10, basipterygoid process and basal articulation; 11, spiracular groove; 12, otico-occipital fissure; 13, epihyal; 14, palatoquadrate; 15, dermohyal; 16, interhyal; 17, basibranchial; 18, hypobranchial; 19, hypohyal; 20, ceratohyal; 21, ceratobranchial; 22, dorsal buttress; 23, coracoid plate; 24, supraglenoid canal; 25, supracoracoid canal; 26, articulation area for radials; 27, horizontal plate; 28 cleithrum; 29, radials; 30, metapterygium; 31, propterygium; 32, fringing fulcra; 33, pelvic plate (metapterygial axis); 34, telencephalon; 35 olfactory tract; 36, mesencephalon; 37, semicircular canals; 38, dorsal fontanelle. (A–D, from Gardiner (1984*a*); E, from Poplin (1974).)

2). The caudal fin is epicercal, and the squamation shows a distinct change in pattern at the root of the tail ('hinge line'). The paired fins are rather small, and the pelvic fin is broad-based, yet less so than in *Cheirolepis* or *Howqualepis* (Fig. 4.65A, B, 3). All the fins are covered with thin, elongated dermal fin rays made up of lepidotrichiae.

4.17.2 Endoskeleton

Most of the endoskeleton of *Mimia* is perichondrally ossified, but some trabecles of endochondral bone occur. The braincase is a single mass of bone hollowed laterally by very large orbital cavities separated by a very thin and

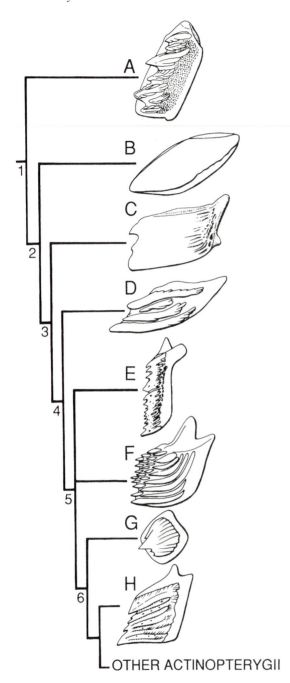

Fig. 4.67. Fossil Actinopterygii. Interrelationships of stem-group Actinopterygii (cladistians omitted) based on scale structure and morphology, according to Schultze (1992). *Terminal taxa*: A, *Lophosteus*; B, *Andreolepis*; C, *Naxilepis*; D, *Orvikuina*, E, *Ligulalepis*; F, *Dialipina*; G, *Cheirolepis*; H, *Moythomasia*. *Nested taxa and synapomorphies*: 1, Actinopterygii (scales rhombic in shape, with overlying sequence of dentine denticles and cancellous bone; Fig. 4.65, E3); 2, occasional ganoine layers over dentine of denticles; 3, anterodorsal process, ganoine on dentine denticles; 4, overlying sequences of dentine denticles without bony base (Fig. 4.65, C2); 5, 'peg-and-socket' articulation of scales, peg with narrow base; 6, ganoine layer directly overlying other ganoine layer (Fig. 4.65, C2, 14). *Age*: Late Silurian (A–C); Early Devonian (E, F); Middle Devonian (D); Middle–Late Devonian (G); Late Devonian (H). (A, from Gross (1969); B, from Gross (1968*b*); C, from Wang and Dong (1989); D–F, from Schultze (1968); G, from Pearson and Westoll (1979); H, from Jessen (1968).)

lacunar interorbital septum (Fig. 4.66, 9), and bounded posteriorly by strong postorbital ridges. In ventral aspect, the braincase shows two major fissures: the ventral fissure (Fig. 4.66, 4), and the otico-occipital fissure, which is continued ventrolaterally by the vestibular fontanelle (Fig. 4.66, 12). The ventral fissure was filled by cartilage and may correspond to part of the intracranial joint of sarcopterygians (see Chapter 7, p. 259), whereas the otico-occipital fissure was free and gave passage to the branches of the vagus nerve. An important feature of the braincase of *Mimia* is the large canal for the jugular vein straddled by the lateral commissure (Fig. 4.66, 3), which bears the groove for the spiracle (Fig. 4.66, 11) and meets ventrally the descending process of the sphenethmoid region (Fig. 4.66, 7) in a way that is strikingly similar to that in some dipnomorph sarcopterygians (Fig. 4.84, 6).

No reconstruction of the brain and labyrinth cavities has been made for *Mimia*, but that of the slightly more derived Carboniferous genus *Kansasiella* (Fig. 4.66E) can be regarded as being quite similar in outline. As in most other early actinopterygians, it shows a large midbrain and elongated olfactory tracts (Fig. 4.66, 34, 36).

The palatoquadrate is attached to the braincase by means of two articulations, ethmoid and basipterygoid (Fig. 4.66, 5, 10). The hyomandibula lies against the posterior edge of the palatoquadrate and articulates with the braincase (Fig. 4.66, 6, 13). It is thus likely that most of the movement of the palate was due to the cranio–hyoidean articulation. A small interhyal lies against the quadrate part of the palatoquadrate (Fig. 4.66, 16). There are five branchial arches (Fig. 4.66B), the fifth one being much reduced, and they are connected basally to a large median basibranchial bone (Fig. 4.66, 17), a condition that may be primitive for osteichthyans if similar to that in acanthodians (Fig. 4.59C2, C3., 12).

The axial endoskeleton consists of arcual and haemal arches that are presumed to represent basidorsals and basiventrals, but there is no ossified centrum (Fig. 2.6, 14). Unpaired supraneurals are present in the anterior part of the body. The dorsal and anal fins are supported by several large rods, or by basal plates (Fig. 2.6 15).

The endoskeletal pectoral shoulder girdle, or scapulocoracoid, is a single element attached to the internal surface of the cleithrum (Fig. 4.66C). It consists of a horizontal plate-shaped part (Fig. 4.66, 27), on which the radials articulate posteriorly (Fig. 4.66, 26, 29) and large dorsal and ventral buttresses (Fig. 4.66, 22, 23) which bounded the spaces for the dorsal and ventral fin musculature respectively (Fig. 4.66, 24, 25). These large passages may be regarded as homologous to the supraglenoid and supracoracoid foramens in the scapulocoracoid of sarcopterygians (Fig. 4.82, 4, 5). The radials of the pectoral fin are short, with a perforated propterygium (Fig. 4.66, 31; a character shared by *Mimia* and all other actinopterans), and a metapterygial axis formed by a single element (Fig. 4.66C2, 30). The pelvic girdle is a single triangle-shaped plate on which short radials attach (Fig. 4.66, 33, 29). As in all actinopterygians, the pelvic girdle is likely to be formed by part of the metapterygium, and thus is not homologous with that of chondrichthyans and sarcopterygians.

4.17.3 Histology

There are two types of hard tissues that are unique to actinopterygians, with a few exceptions among the earliest known forms: ganoine and acrodine. Ganoine is a shiny enamel layer covering the dentine of the scales and dermal bones in actinopterygians, except *Lophosteus*. Only traces of ganoine seem to occur, however, in *Andreolepis*. Ganoine was long believed to be different from enamel and to have been produced, like dentine, by ectomesenchymal cells, but it is now clearly shown to be of epidermal origin, like enamel: it displays virtually the same crystallite organization as monotypic enamel (that is, enamel originating from a single source of cells), which is classically regarded as a sarcopterygian synapomorphy. Considering ganoine as enamel would imply that it appeared independently in actinopterygians and sarcopterygians, unless *Lophosteus* is not an actinopterygian. Another possibility is that enamel is a general gnathostome character that has been lost many times (Smith 1992). What, in fact, is unique to actinopterygians is the way in which the dentine and ganoine layers overlap each other (Fig. 4.65E3, C2). Acrodine is also unique to actinopterygians, except for *Lophosteus*, *Andreolepis* and *Cheirolepis*. It is a translucent cap of hypermineralized dentine on the tip of the tooth, which is classified as bitypic enamel, that is, it originates from both mesenchymatous and ectodermal cells (Smith 1992). In contrast to monotypic enamel, which is clearly separated from the underlying dentine, the dentine tubules may penetrate acrodine for a short distance (Fig. 4.65, 13).

Actinopterygians possess, like acanthodians, three large otoliths, or statoliths (saccular, lagenar, and utricular). This is regarded by Schultze (1988) as the general osteichthyan condition, because the same occurs in acanthodians. In Recent actinopterygians, the otoliths consist of aragonite and vaterite. Palaeozoic actinopterygian otoliths are extremely rare. Talimaa (in Nolf 1985) described isolated Early Devonian otoliths that were referred to as actinopterygians, despite the fact that they are made up of calcium phosphate (apatite), which is known only in extant lamprey statoconiae. Coates (1993) pointed out that the few Palaeozoic actinopterygians with otoliths *in situ* have only a pair of large calcitic statoliths with a marginally placed groove (sulcus). These statoliths resemble the single utricular statoliths of extant lungfishes and *Latimeria*, which suggests that this represents the general osteichthyan condition. The three otoliths in actinopterygians are thus regarded by Coates as having appeared independently in cladistians, actinopterans, and acanthodians.

4.17.4 Diversity

Silurian and Devonian actinopterygians are few, and most Devonian forms are quite similar in general aspect, apart from the peculiar genus *Cheirolepis*, which differs from all other Palaeozoic actinopterygians in its minute ganoin-covered scales (Fig. 4.65F1). Carboniferous and Permian actinopterygians display a wider range of diversity, with eel-shaped or deep-bodied forms (Fig. 4.70 P–R); this evokes, though to a lesser extent, the diversity of modern teleosts.

If *Andreolepis* and *Lophosteus* are accepted as actinopterygians, then their difference in scale and bone ornamentation may reflect some differences in general morphology and habitat. The same applies to *Cheirolepis* with regard to most other Devonian actinopterygians. Its large size, slender body-shape and wide gape suggest predacious habits, whereas other forms, such as *Moythomasia*, *Mimia*, or *Stegotrachelus* were comparatively small and may have formed large populations around coral reefs. By Middle Devonian times, in particular in the Givetian, the diversity of actinopterygians increases, with other large forms, such as *Howqualepis*, which could reach 70 cm in length and possessed peculiar symphysial toothed plates. In the Famennian appear advanced forms, probably actinopterans, such as gonatodontids and the deep-bodied platysomids (Fig. 4.70Q) that, in addition, possessed a crushing dentition. Some highly specialized forms, such as the eel-shaped tarrasiids occur as early as the Carboniferous (Fig. 1.19, 8)

4.17.5 Phylogeny

Early actinopterygians, i.e. mainly Devonian to Permian forms, have long been included in a taxon Palaeonisciformes, ranked in the chondrostean grade, together with extant sturgeons and paddlefishes. Historically, this tradition stems from Agassiz's (1833–44) concept of 'ganoids' (including sturgeons, cladistians, gars, and Palaeozoic actinopterygians, as well as fossil lungfishes and even osteostracans and some acanthodians), and from Müller's (1846) early classification of actinopterygians into chondrosteans, holosteans, and teleosteans. Cope (1898) finally accepted Müller's classification and coined the name 'Actinopteri' (Fig. 4.70, 3) for all actinopterygians except the Cladistia, which remained among the 'crossopterygians'. Woodward (1891) coined the name 'Actinopterygii' for higher actinopterans, a name that Goodrich (1909) later extended to actinopterans plus cladistians (Fig. 4.70, 2). The name 'palaeoniscids' was first used by Vogt (1845) to include certain 'ganoids', among which was the Permian genus *Palaeoniscum* (Fig. 4.71D). Huxley (1861) and Traquair (1877) referred the palaeoniscids to the 'acipenseroids' (sturgeons and paddlefishes), within the 'ganoids'. This classification was retained by Woodward (1891) and Goodrich (1909) who, however, decided to leave aside the ambiguous term 'ganoids'.

Although Stensiö (1921) and Watson (1928) showed that the endoskeleton, in particular the braincase, of early actinopterygians could provide a large amount of information, exoskeletal characteristics have always been preferred for reconstructing the relationships and systematics of this group. As new taxa were discovered, in particular in the Late Palaeozoic and Triassic, it appeared that some 'palaeonisciforms' shared a number of supposedly advanced characteristics with 'holosteans' (extant bowfins and gars, and many Mesozoic taxa). With the later works of Gardiner (1967), Schaeffer (1973) and Patterson (1982*a*), it became clear that chondrosteans, when including the 'palaeonisciforms', were paraphyletic, as were

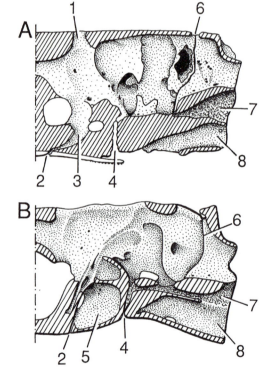

Fig. 4.68. Fossil Actinopterygii. Sagittal section of the posterior part of the braincase, showing the development of the posterior myodome. A, *Mimia*, Devonian of Australia; B, *Kansasiella*, Carboniferous of Kansas. 1, pineal foramen; 2, buccohypophysial foramen; 3, hypophysial pit; 4, ventral fissure; 5, posterior myodome; 6, otico-occipital fissure; 7, canal for the notochord; 8, canal for the dorsal aorta. (A, from Gardiner (1984*a*); B, from Poplin (1974).)

holosteans (Patterson 1973). Early attempts at resolving the relationships of the 'palaeonisciforms' on the basis of the structure of the neurocranium were unsatisfactory, because of the scarcity of taxa in which internal anatomy can actually be observed (e.g. *Moythomasia*, *Mimia*, *Kansasiella*, *Kentuckia*, *Cosmoptychius*, *Pteronisculus*, *Boreosomus*, and *Birgeria*). Moreover, the braincase of these early actinopterygians turned out to be very conservative, apart from variations in the number of ossification centres and a trend towards the enlargement of the posterior myodome (for the recti eye muscles), which is absent in *Cheirolepis*, *Polypterus*, and *Mimia* (Fig. 4.68A); well-developed but still paired in *Kansasiella* (Fig. 4.68B, 5) or *Kentuckia*; and finally unpaired in *Pteronisculus*, *Birgeria*, and *Saurichthys*. The posterior myodome is lacking in chondrosteans (by loss), but extends very far backwards in most neopterygians, and pushed back the ventral fissure (Fig. 4.68, 4; see, however, Schaeffer and Dalquest 1978). It became thus clear that the phylogenetic relationships of early actinopterygians had to be considered essentially on the basis of the pattern of the dermal bones of the head, despite the risk of homoplasies in such highly fragmented patterns. The earliest known actinopterygians in which at least part of the dermal skull-roof can be reconstructed show a trend toward an elongation of the parietals and shortening of the postparietal (Fig. 4.69, 2, 3); only *Dialipina* (Fig. 4.69A) and *Cheirolepis* (Fig. 4.69B) retain a pineal plate (Fig. 4.69, 1).

The differences that can be observed in the dermal bone patterns of the skull-roof among 'palaeonisciforms' are quite difficult to interpret and use in reconstructing a phylogeny. These patterns are often not clear-cut, with incongruent distributions, and they hardly define major taxa. For this reason, evolutionary classifications of early actinopterygians have long been based on a few 'trends', such as the increasingly vertical 'suspensorium' of the jaw (i.e. the hyomandibula, palatoquadrate, and adjoining preopercular, Fig. 4.65, 9) and the reduction of the notochordal lobe of the caudal fin (Fig. 4.65, 5). With the advent of the cladistic method, and in particular with the

help of cladistic computer programs, it became possible to analyse a large number of characteristics, taken from the dermal bones, fin structure, or endoskeleton, in a large number of taxa, and search for parsimonious distribution, regardless of the geological age of the taxa (Lauder and Liem 1983; Gardiner 1984a). Gardiner and Schaeffer (1989) defined 27 monophyletic non-neopterygian

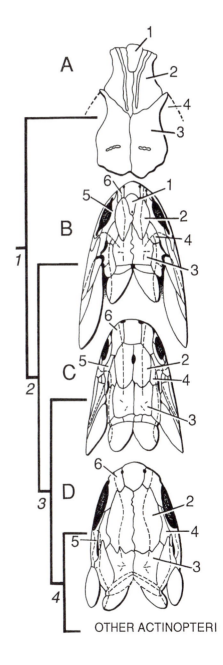

Fig. 4.69. Fossil Actinopterygii. Trends in skull-roof morphology among early actinopterygians (cladistians omitted). A, *Dialipina*; B, *Cheirolepis*; C, *Moythomasia*; D, *Kentuckia*. *Synapomorphies*: *1*, postparietals slightly larger than parietals; *2*, postparietals shorter, intertemporals in contact with parietals, dermosphenotics in contact with nasals; *3*, loss of pineal plate, larger parietals, larger rostral; *4*, very large parietals, reduced postparietals, intertemporal in contact with nasals, loss of pineal foramen. 1, pineal plate; 2, parietal, 3, postparietal; 4, intertemporal; 5, dermosphenotic; 6, nasal. (A, from Schultze (1992); B, from Pearson and Westoll (1979); C, from Jessen(1968); D, from Rayner (1951).)

actinopterygian taxa, and proposed a pattern of relationships that differs in many respects from previous phylogenies. These authors admit that their theory is perhaps far from the truth, but it may be regarded as better than others by the number of characters (in particular the pattern of the circumorbital dermal bones) simultaneously taken into consideration, and the arguments used in support of each node. I therefore chose this phylogeny to be illustrated here (Figs 4.70, 4.71). Parts of it have been recently criticized by Coates (1993).

The terminal taxa defined by Gardiner and Schaeffer, to which is added *Howqualepis* as a separate group (Long 1988*b*), are briefly listed below, with some of the included genera and their respective ages:

1 '*Cheirolepis* Group'. *Cheirolepis* (Fig. 4.70A, Devonian).
2 Cladistia. *Polypterus* (Fig. 4.70B), *Erpetoichthys* (?Cretaceous to Recent).

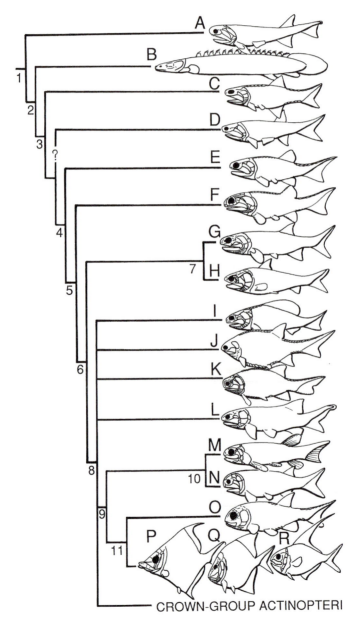

CROWN-GROUP ACTINOPTERI

3 '*Mimia* Group'. *Mimia* (Fig. 4.70C), *Tegeolepis* (all Devonian).

4 '*Howqualepis* Group'. *Howqualepis* (Fig. 4.70D, Devonian).

5 '*Moythomasia* Group'. *Moythomasia* (Fig. 4.70E), *Stegotrachelus* (all Devonian).

6 '*Kentuckia* Group'. *Kentuckia* (Fig. 4.70F), *Elonichthys* (all Carboniferous).

7 '*Pteronisculus* Group'. *Nematoptychius*, *Paramblypterus* (all Carboniferous), *Acrolepis* (Permian), *Pteronisculus* (Fig. 4.70G), *Tarseodus*, *Dicellopyge* (all Triassic).

8 '*Boreosomus* Group'. *Boreosomus* (Fig. 4.70H), *Ptycholepis* (all Triassic).

9 '*Australichthys* Group'. *Australichthys* (Fig. 4.70I), *Namaichthys*, *Mentzichthys*, *Cycloptychius*, '*Rhadinichthys*' *canobiensis*, *Phanerorhynchus* (all Carboniferous).

10 '*Watsonichthys* Group'. *Watsonichthys*, *Mesonichthys*, *Cosmoptychius*, *Willomorichthys*, *Strepheoschema* (Fig. 4.70J) (all Carboniferous), *Rhabdolepis* (Permian).

11 '*Belichthys* Group'. *Aetheretmon* (Fig. 4.70K), '*Rhadinichthys*' *carinatus*, *Cornuboniscus*, *Gonatodus*, *Commentrya*, *Phanerosteon* (all Carboniferous), *Belichthys* (Triassic).

12 '*Amblypterus* Group'. *Rhadinichthys ornatissimus*, *Pseudogonatodus*, *Gyrolepidotus*, *Oxypteriscus* (all Carboniferous), *Amblypterus* (Fig. 4.70L, Permian).

13 'Haplolepid Group'. *Haplolepis* (Fig. 4.70M), *Pyritocephalus* (all Carboniferous).

14 'Redfieldiid Group'. numerous genera, including *Redfieldius* (Fig. 4.70N), *Brookvalia* and *Helichthys* (all Triassic).

15 '*Mesopoma* Group'. *Mesopoma*, *Canobius*, *Styracopterus* (all Carboniferous).

16 '*Aeduella* Group'. *Aeduella* (Fig. 4.70O), *Decazella*, *Igornella* (Carboniferous to Permian).

17 '*Platysomus* Groups'. *Platysomus* (Fig. 4.70Q, Carboniferous–Permian), *Plectrolepis*, *Paramesolepis*, *Adroichthys*, *Proteurynotus*, *Amphicentrum*, *Cheirodopsis*, *Chirodus* (all Carboniferous).

18 '*Bobasatrania* Group'. *Ebenaqua* (Fig. 4.70P. Permian), *Bobasatrania*, *Ecrinesomus* (all Triassic).

19 '*Dorypterus* Group'. *Dorypterus* (Fig. 4.70R, Permian).

20 '*Saurichthys* Group'. *Saurichthys* (Fig. 4.71C, Triassic).

21 'Chondrostean Group' (Fig. 4.71, 3). *Gyrosteus*, *Chondrosteus* (Fig. 4.71B, Jurassic), all other Mesozoic to Recent Chondrosteans (Fig. 4.71A).

22 '*Palaeoniscum* Group'. *Palaeoniscum* (Fig. 4.71D, Permian), *Cosmolepis*, *Scanilepis* (all Jurassic).

23 '*Birgeria* Group'. *Birgeria* (Fig. 4.71E, Triassic).

24 '*Australosomus* Group'. *Australosomus* (Fig. 4.71F), *Pholidopleuron*, *Macroaethes* (all Triassic).

25 '*Cleithrolepis* Group'. *Chleithrolepis* (Fig. 4.71G), *Dipteronotus*, *Hydropessum* (all Triassic).

26 '*Perleidus* Group'. *Perleidus* (Fig. 4.71H), *Pseudobeaconia*, *Mendocinichthys*, *Dollopterus*, *Meridensia*, *Colobodus*, *Manlietta*, *Meidiichthys*, *Procheirichthys* (all Triassic).

27 '*Peltopleurus* Group'. *Peltopleurus* (Fig. 4.71I), *Habroichthys* (Triassic), *Platysiagum* (Jurassic).

28 '*Luganoia* Group'. *Luganoia* (Fig. 4.71J), *Besania* (all Triassic).

29 Crown-group Neopterygii (extant neopterygians and their in-group fossil relatives).

The consensus tree produced by Gardiner and Schaeffer shows that only a few of these terminal groups can be arranged into larger clades. The *Pteronisculus* and

---

Fig. 4.70. Fossil and Recent Actinopterygii. Interrelationships of the Actinopterygii, according to the phylogeny proposed by Gardiner and Schaeffer (1989); some terminal taxa omitted). *Terminal taxa*: A, *Cheirolepis*; B, *Polypterus*; C, *Mimia*; D, *Howqualepis*; E, *Moythomasia*; F, *Kentuckia*; G, *Pteronisculus*; H, *Boreosomus*; I, *Australichthys*; J, *Strepheoschema*; K, *Aetheretmon*; L, *Amblypterus*; M, *Haplolepis*; N, *Redfieldius*; O, *Aeduella*; P, *Ebenaqua*; Q, *Platysomus*; R, *Dorypterus*. *Nested taxa and selected synapomorphies*: 1, Actinopterygii (single dorsal fin, peg-and socket articulation of scales, narrow-based peg on scales, otoliths in part composed of vaterite, dermohyal on hyomandibula, shield-shaped rostral with ethmoid commissure, T-shaped dermosphenotic contacting nasals, two infraorbitals, postorbital and squamosal absent, one or two pairs of extrascapulars, ?posteriorly expanded maxilla, dentary with enclosed mandibular sensory-line canal, tail with hinge-line in squamation, basal fulcra on dorsal edge of caudal fin); 2, crown-group Actinopterygii (acrodine cap on teeth, postcleithrum differentiated from body scales, pineal plate lost); 3, Actinopteri (lateral cranial canal in braincase, fringing fulcra on leading edge of all fins); 4, ascending process of parasphenoid in spiracular groove, intertemporal bone meets nasal, supra-angular; 5, spiracular canal, paired posterior myodome, suborbital; 6, single median posterior myodome, distinct fossa Bridgei, prismatic ganoine (enamel) on teeth and scales; 7, dermal basipterygoid process; 8, dermopterotic present, overlaps or abuts dermosphenotic; 9, less than 12 or 13 branchiostegals; 10, one branchiostegal ray, single narial opening for both external nostrils, enlarged postcleithrum; 11, almost vertical suspensorium (homoplasy with node 6 in Fig. 4.71), small, sickle-shaped preopercular, no suborbital, premaxillo-antorbital toothless and almost excluded from jaw margin. *Age*: Devonian (A, C, D, E); Carboniferous (F, I–M, O, Q); Permian (P, R); Triassic (G, H); Cretaceous–Recent (B). (Reconstructions based on Campbell and Le Duy Phuoc (1983), Gardiner (1969, 1984b), Heyler (1969), Jessen (1968); Lehman (1966), Long (1988b), Moy-Thomas and Miles (1971), and Pearson and Westoll (1979).)

Boreosomus groups are sister-groups and are united on the basis of three characters, including a dermal basipterygoid process and a robust opercular process on the hyomandibula (Fig. 4.70, 7). The *Saurichthys* group and chondrosteans are gathered on the basis of six synapomorphies, including the posterior prolongation of the parasphenoid beyond the occipital condyle, the lengthened ethmoid region, and the enlarged craniospinal process (Fig. 4.71, 2). In addition, there is a large clade, including the haplolepid, *Mesopoma*, redfieldiids, *Aeduella*, *Platysomus*, *Bobasatrania*, and *Dorypterus*

groups, which is characterized by only two synapomorphies: less than twelve branchiostegal rays, and the dermopterotic never overlapping more than one-third of the dermopterotic (Fig. 4.70, 9). Coates (1993) provides a different phylogeny for this clade. There is also a major polytomy which includes the latter ensemble, the *Watsonichthys*, *Australichthys*, *Belichthys* and *Amblypterus* groups, as well as all other actinopterygians above this node, which is defined by only one synapomorphy, the presence of a dermopterotic that overlaps or abuts the dermosphenotic (Fig. 4.70, 8). This cladogram

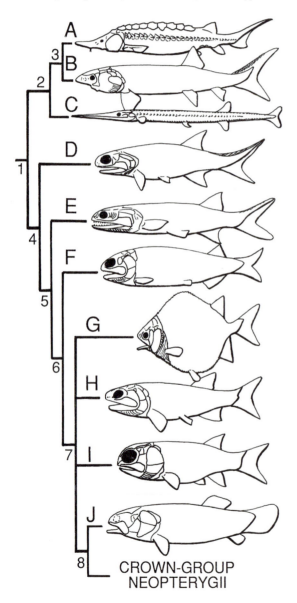

Fig. 4.71. Fossil and Recent Actinopterygii. Interrelationships for the crown-group Actinopteri, according to the phylogeny proposed by Gardiner and Schaeffer 1989; some terminal taxa omitted). *Terminal taxa*: A, *Acipenser*; B, *Chondrosteus*; C, *Saurichthys*; D, *Palaeoniscum*; E, *Birgeria*; F, *Australosomus*; G, *Cleithrolepis*; H, *Perleidus*; I, *Peltopleurus*; J, *Luganoia. Nested taxa and selected synapomorphies*: 1, crown-group Actinopteri (keystone-shaped dermosphenotic, rudimentary post-temporal fossa); 2, enlarged craniospinal process, parasphenoid extends posteriorly beyond occipital condyle; 3, Chondrostei (posterior myodome lost, palatoquadrates meet anteriorly, reduced body scaling, mandibular sensory-line excluded from dentary); 4, numerous anamestic supraorbitals; 5, separate premaxillae and antorbitals; 6, large supratemporal fossa, hyomandibula nearly vertical (homoplasy with node 11 in Fig. 4.70), uncinate process on first two epibranchials; 7, Neopterygii (dermal fin rays equal number of radials in dorsal and anal fin, ascending process on premaxilla, dermal fin rays of dorsal caudal lobe extend far beyond the tip of the chordal lobe); 8, free maxilla, coronoid process of surangular. *Age*: Permian (D); Triassic (C, E–J); Jurassic (B); Cretaceous to Recent (A). (Reconstructions based on Griffith (1977) and Lehman (1966).)

CROWN-GROUP
NEOPTERYGII

is clearly inconsistent with the majority of the classical 'trends' recognized in actinopterygian evolution, in particular the increasingly vertical 'suspensorium' and the reduction of the notochordal lobe of the tail. The vertical suspensorium and, to some extent, the reduction of the notochordal lobe occurs at least twice (in clade 9 of Fig. 4.70 and the *Dorypterus* group, and in clade 5 of Fig. 4.71, towards the crown-group neopterygians).

Groups 3 to 19 in the list above are Palaeozoic or Triassic in age and represent stem-group actinopterans. Groups 20 to 29 are crown-group actinopterans and are essentially Mesozoic to Recent.

Neopterygian phylogeny is not the subject of this book; it concerns Mesozoic to Recent forms, which cannot be regarded as 'early vertebrates'. However, Fig. 4.72 and 4.73 show one of the phylogenies of this now considerable group (Patterson and Rosen 1977). Fossils generally do not overturn the phylogeny of neopterygians based on Recent taxa, although Olsen and McCune (1991) have used fossils to suggest that gars were more closely related to teleosts than halecomorphs. Within the halecomorphs, *Caturus* (Fig. 4.72E) is the sister group of fossil and extant bowfins. There are a number of Mesozoic neopterygians, formerly referred to as 'holosteans', that are still difficult to place in this cladogram. This is, for example, the case for the pycnodonts (Fig. 4.72B) and semionotids (Fig. 4.72C). Others, such as parasemionotids (Fig. 4.72D), may be more closely related to the halecomorphs than to any other taxon. The phylogeny of the Teleostei in Fig. 4.73 is one of those consistent with Patterson and Rosen's (1977) cladogram. It is intended to show that a number of small, 'sardine-shaped' Mesozoic teleosts often referred to as 'leptolepids' (*Pholidophorus*, *Leptolepis*, *Leptolepides*, *Tharsis*, or *Anaethalion*; Fig. 4.73A, B, C, E, H, I, J) are in fact an assemblage of generalized teleosts, some of which are more closely related to one particular extant teleost group—osteoglossomorphs, (Fig. 4.73F), elopomorphs (Fig. 4.73G), and euteleosteans (Fig. 4.73K)—than others, despite an overall similarity in head morphology. The structure of the caudal skeleton has proved to be the most character-bearing and informative part of the skeleton of these fishes, and has largely served in reconstructing this phylogeny. (See also Arratia (1991) for a different theory of the relationships.)

The classification of actinopterygians (by phyletic sequencing), with special reference to Palaeozoic taxa, in Table 4.9 is derived from the cladograms in Figs 4.70–3:

4.17.6 Stratigraphical distribution and habitat

Isolated actinopterygian scales and teeth are quite abundant in marine sediments after the Givetian, and they may

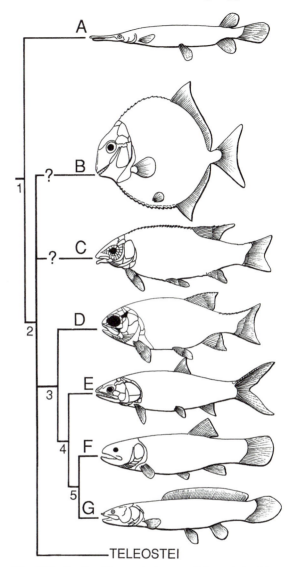

Fig. 4.72. Fossil and Recent Actinopterygii. Interrelationships of the crown-group Neopterygii, partly according to the phylogeny proposed by Patterson (1973, 1982*a*). *Terminal taxa*: A, Ginlymodi (*Lepisosteus*); B, Pycnodontidae (debated position); C, Semionotidae (debated position); D, Parasemionotidae; E, Caturidae (*Caturus*); F, *Amiopsis*; G, *Amia*. *Nested taxa and selected synapomorphies*: 1, crown-group Neopterygii (consolidated upper pharyngeal dentition, clavicle much reduced or lost; 2, Halecostomi (mobile maxilla, intraopercular, median neural spines, reduction of the quadratojugal); 3, Halecomorphi (both symplectic and quadrate participate to the jaw articulation); 4, loss of quadratojugal; 5, Amiidae (perichordal vertebral centra). *Age*: Triassic (D); Triassic to Cretaceous (B, C, E); Jurassic (F); Cretaceous to Recent (A, G). (Reconstructions based on Lehman (1966).)

Table 4.9. Classification of the Actinopterygii (See also Figs 4.70–3)

Actinopterygii
  Cheirolepis
  Crown-group Actinopterygii
    Cladistia
    Actinopteri
      Moythomasia Group
      Howqualepis Group
      Mimia Group
      Kentuckia Group
      (unnamed taxon)
        Pteronisculus Group
        Boreosomus Group
      (unnamed taxon)
        Australichthys Group *sedis mutabilis*
        Watsonichthys Group *sedis mutabilis*
        Belichthys Group *sedis mutabilis*
        Amblypterus Group *sedis mutabilis*
        (unnamed taxon, *sedis mutabilis*)
          Haplolepid Group
          Redfieldiid Group
          (unnamed taxon)
            Bobasatrania Group
            Platysomus Group
            Dorypterus Group
    Crown-group Actinopteri *sedis mutabilis*
    (unnamed taxon)
      Saurichthys Group
      Chondrostei
    (unnamed taxon)
      Palaeoniscum Group
      Birgeria Group
      Australosomus Group
      Neopterygii
        Cleithrolepis Group *sedis mutabilis*
        Perleidus Group *sedis mutabilis*
        Peltopleurus Group *sedis mutabilis*
        (unnamed taxon) *sedis mutabilis*
          Luganoia Group
          Crown-group Neopterygii
          Ginglymodi
          Halecostomi
            Pycnodontidae *sedis mutabilis*
            Semionotidae *sedis mutabilis*
            Halecomorphi *sedis mutabilis*
              Parasemionotidae
              Caturidae
              Amiidae
            Teleostei *sedis mutabilis*

provide some information. There is, for instance, an overall trend toward an extensive ganoine covering on scales, from the Frasnian to the Carboniferous. The acrodin cap on teeth is unknown before the Givetian, and teeth of durophagous type, that is, onion-shaped teeth (of platysomids, for example), do not occur before the latest Famennian. However, although the characteristic, minute scales of *Cheirolepis* are easily recognized in Givetian–Frasnian marine sediments, the determination of isolated actinopterygian remains is extremely difficult. The actual stratigraphical range of the taxa as defined from complete specimens is therefore difficult to assess. The preservation of articulated actinopterygians requires very quite conditions of deposition, such as Carboniferous coal basins or lagoons, which are exceptional. Consequently, it is probable that most of the genera cited in the above list, many of which are known only from a single locality, have a much wider stratigraphical distribution. Some classical Carboniferous forms, such as platysomids or *Canobius*, are now recorded from the Late Famennian, in the form of isolated bones; and, if we accept the cladogram in Fig. 4.70, these are rather derived forms. This implies that a large part of actinopteran history took place in the Devonian, but is poorly recorded.

In sum, apart from stratigraphically restricted forms, with a well-defined scale and tooth morphology or ornamentation (e.g. *Cheirolepis*, platysomids), actinopterygian remains are at present of little stratigraphical interest.

All Devonian actinopterygians are marine, and probably pelagic. Only *Cheirolepis* occurs in more confined environments, such as the Escuminac Formation of the Orcadian Basin. *Moythomasia* or *Mimia* remain associated with typically marine facies, such as Late Devonian brachiopod or coral-bearing limestones. By Late Carboniferous times some presumably freshwater actinopterygian faunas appear. *Aeduella*, for example, seem to have been restricted to Carboniferous and Permian intramontane basins of Europe. However, until the Late Mesozoic, actinopterygians remain essentially marine.

Further reading

Anatomy and histology: Bjerring (1978, 1985, 1986a); Coates (1993); Gardiner (1967, 1984a); Gardiner and Schaeffer (1989); Gross (1968b, 1969, 1971b); Janvier (1978b); Jessen (1968, 1972); Long (1988b); Nielsen (1942); Otto (1993); Patterson (1975); Pearson (1982); Pearson and Westoll (1979); Poplin (1974); Sire *et al.* (1987); Schultze (1968, 1977, 1992); Watson (1928). *Interrelationships and systematics*: *Actinopterygii in general*: Gardiner (1984a); Gardiner and Schaeffer (1989); Lauder and Liem (1983); Lehman (1966); Long (1988b); Patterson (1982a); Schaeffer (1973); Schultze (1992). *Neopterygii*: Patterson (1973); Olsen and McCune (1991). *Teleostei*: Arratia (1991); Lambers (1992); Patterson (1975, 1977b, 1982a); Patterson and Rosen (1977).

4.18 SARCOPTERYGII

Sarcopterygians, fossil and extant, are characterized chiefly by the monobasal articulation of the paired fins

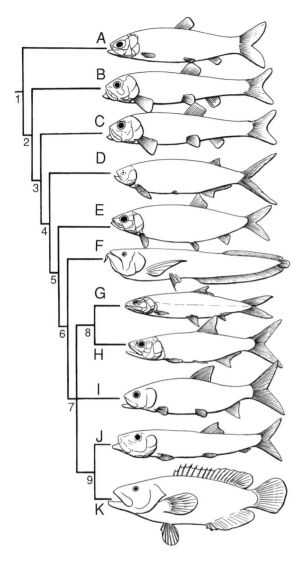

Fig. 4.73. Fossil and Recent Actinopterygii. Interrelationships of the Teleostei, according to the phylogeny proposed by Patterson and Rosen (1977). *Terminal taxa*: A, *Pholidophorus bechei*; B, *Proleptolepis*; C, *Leptolepis coryphaenoides*; D, Pachycormidae (debated position); E, *Tharsis dubius*; F, Osteoglossomorpha (*Osteoglossum*); G, *Elops*; H, *Ichthyemidion* ('*Anaethalion*') *vidali*; I, other *Anaethalion* species; J, *Leptolepides sprattiformis*; K, crown-group Clupeocephala (clupeomorphs and euteleosteans). *Nested taxa and selected synapomorphies*: 1, Teleostei (mobile premaxilla, unpaired basibranchial tooth plates, loss of enamel on most skull bones, cycloid scales); 2, loss of separate prearticular in lower jaw, partial closure of otico-occipital fissure; 3, complete closure of otico-occipital fissure, spiracular canal reduced, posterior myodome open posteriorly, three epurals and horizontal, elongated uroneurals in caudal skeleton; 4, loss of spiracular canal, sutures between cartilage bones of skull retained throughout life; 5, loss of suborbitals, strongly constricted vertebral centra, epipleural intermuscular bones in middle part of trunk; 6, crown-group Teleostei (seven or less hypurals, caudal axis of vertebral column turned sharply upward at level of the first pre-ural centrum); 7, Elopocephala (two uroneurals extending forward beyond the second ural centrum); 8, Elopomorpha (fused angular and retroarticular in lower jaw); 9, Clupeocephala (retroarticular excluded from joint surface for the quadrate). *Age*: Jurassic (A–E, J); Jurassic to Cretaceous (I); Cretaceous (H); Cretaceous to Recent (F, G, K). (Reconstructions based on Forey (1973), Gaudant (1968), Lambers (1992), Lehman (1966), and Taverne (1981).)

with fin musculature extending in a more or less developed basal lobe (see Chapter 3, p. 77). Other characters are the endoskeletal urohyal (Fig. 4.78), the rearmost gill-arch articulating with the preceding one (Fig. 4.78, 33), more than four sclerotic plates (Fig. 4.74, 10), and an ascending process of the palatoquadrate (Fig. 4.78, 25). True enamel has long been claimed to be unique to sarcopterygians, but actinopterygian ganoine and collar enamel is now regarded as similar to true enamel, since it is also monotypic. The presence of enamel on the entire tooth remains unique to sarcopterygians, however. There are several other characters (subdermal anocleithrum, intracranial dermal or endoskeletal joint, double-headed

hyomandibula) that have a more restricted distribution among sarcopterygians, but are regarded by some as general for the group and secondarily lost in some sarcopterygian taxa. These characters will be discussed below (Chapter 5, p. 246) in connection with the question of sarcopterygian interrelationships.

The earliest known sarcopterygian remains are Early Devonian in age and are represented by lungfishes and their close relatives, the porolepiforms, as well as by the genera *Youngolepis* and *Powichthys*, of debated affinities. Some isolated remains referred to either *Youngolepis* or lungfishes have been recorded from the Late Silurian of China, though undescribed. Strangely, the supposedly

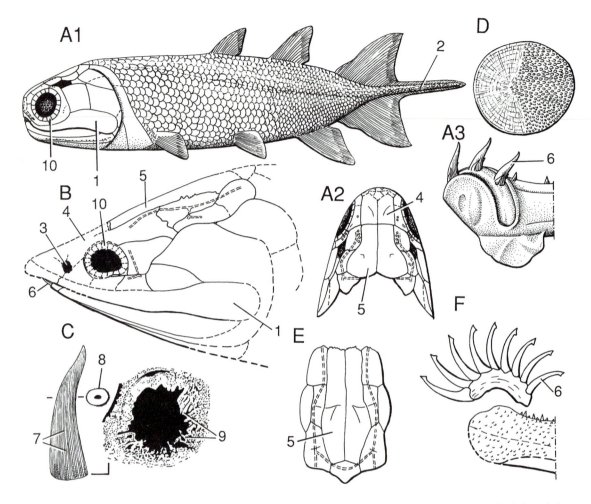

Fig. 4.74. Fossil Sarcopterygii. Onychodontiformes. External orphology. A, *Strunius*, Late Devonian, reconstruction in lateral view (A1, ×2), head in dorsal view (A2) and symphysial region of lower jaw in internal view (A3); B, *Grossius*, Middle Devonian of Spain, head in lateral view (×0.3); C–F, *Onychodus*, Middle–Late Devonian, parasymphysial tooth in lateral view, with apical and basal horizontal sections (C), scale in external view (D), posterior part of skull-roof in dorsal view (E) and parasymphysial toothed plate of a form with hamate (hooked) teeth (F, ×0.8). 1, maxilla; 2, chordal lobe of tail; 3, nasal opening (anterior, posterior, or common); 4, parietal; 5, postparietal; 6, parasymphysial teeth; 7, striae on enamel; 8, enamel; 9, basal dentine folds; 10, sclerotic ring. (A, F, from Jessen (1966); B, C, from Schultze (1969, 1973); D, from Aquesbi (1988); E, from Andrews (1973).)

more primitive sarcopterygian taxa, such as the ony-chodontiforms and actinistians, appear somewhat later, in the late Early Devonian and the Late Middle Devonian respectively. Osteolepiforms first appear at the beginning of Middle Devonian times; the rhizodontiforms, panderichthyids, and tetrapods do not occur before the Late Devonian. It is thus clear that the stratigraphical distribution of sarcopterygians does not match closely the currently accepted pattern of their interrelationships (Fig. 5.3). This indicates that most of the history of the group occurred much earlier than the earliest fossil record, i.e. in Silurian times.

Sarcopterygians display a wider range of diversity than Palaeozoic actinopterygians, even at the level of their internal anatomy. The anatomically best-known Devonian sarcopterygians were for long the porolepiforms *Glyptolepis* (Fig. 4.78), the osteolepiform *Eusthenopteron* (Fig. 4.88C), and the actinistian *Diplocercides* (Fig. 4.75B), which had been investigated by the grinding section method, as well as, to some extent, the tetrapod *Ichthyostega* (Fig. 4.95).

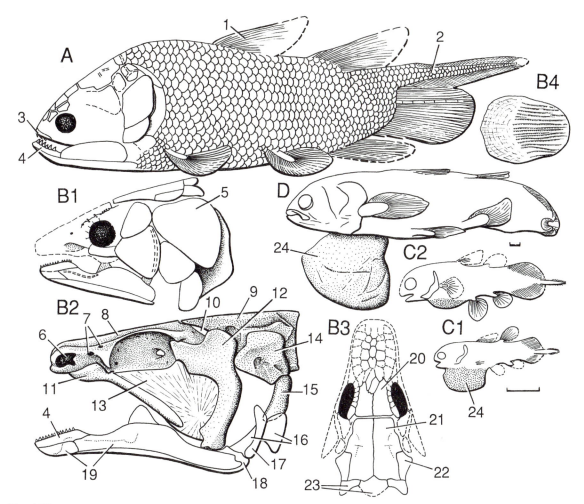

Fig. 4.75. Fossil Sarcopterygii. Actinistia. External morphology and internal anatomy. A, *Miguashaia*, Late Devonian of Quebec, reconstruction in lateral view (×2); B, *Diplocercides*, Late Devonian, head in lateral view (B1, ×1.2), braincase with mandibular and hyoid arches in lateral view (B2), skull in dorsal view (B3) and scale in external view (B4, ×5); C, *Rhabdoderma*, Carboniferous, two juvenile individuals at successive stages of reduction of the yolk sac (C2, C1; scale bar: 10 mm); D, *Latimeria*, Recent, foetus with large yolk sac (to be compared with C1; scale bar: 10 mm). 1, anterior dorsal fin; 2, chordal lobe of tail; 3, premaxilla; 4, dentary; 5, opercular; 6, nasal cavity; 7, openings of the rostral organ; 8, ethmosphenoid; 9, otoccipital; 10, intracranial joint; 11, autopalatine part of palatoquadrate; 12, pterygoquadrate part of palatoquadrate; 13, entopterygoid; 14, hyomandibula; 15, stylohal (or interhyal); 16, interhyal and ceratohyal; 17, symplectic; 18, retroarticular process of articular; 19, infradentaries (splenials); 20, parietal; 21, postparietal; 22, tabular; 23; extrascapulars; 24, yolk sac. (A, from Schultze (1973); B1, B4, from Jessen (1973); B3, from Stensiö (1937); B2, from Jarvik (1980); C, from Schultze (1980); D, from Smith *et al.* (1975).

However, in the 1970s, a large amount of new material from the Early and Late Devonian of Australia and from the Early Devonian of China has increased considerably our knowledge of early sarcopterygians. In particular, acid preparations of the Australian lungfishes (*Speonesydrion*, *Dipnorhynchus*, *Gryphognathus*, *Chirodipterus*, and *Holodipterus*) and osteolepiform (*Gogonasus*) have provided much new information. In addition, further studies on the dipnomorphs *Powichthys*, *Youngolepis*, and *Diabolepis*, have helped in answering the question of lungfish affinities. More extensive descriptions of the panderichthyids (*Panderichthys*, *Elpistostege*) have provided support for their closer affinities with the tetrapods. Only the onychodontiforms and rhizodontiforms are still poorly known.

In contrast to actinopterygians, it is difficult to decide which sarcopterygian shows the most generalized condition and could exemplify the entire group. As for placoderms, each major sarcopterygian taxon will be described here on the basis of one of the best-known and most generalized forms, when available. The apparent heterogeneity of sarcopterygians is due to two major groups, the lungfishes and tetrapods. The former have, even in early forms, a strongly modified skull structure that has made zoologists doubt their belonging to the sarcopterygians or even the osteichthyans. Their body, at least in Palaeozoic forms, is, however, quite clearly of sarcopterygian type, with a monobasal fin skeleton. Early tetrapods, in contrast, retain a skull pattern with many generalized osteichthyan or sarcopterygian characters, but their paired fins are modified into limbs with digits and they have lost most of their scales and unpaired fins. Other sarcopterygians, formerly referred to as 'crossopterygians', generally share a peculiar feature, the intracranial joint, as in the extant actinistian *Latimeria* (Fig. 3.21, 12). In these forms, the braincase is made up of two halves, ethmosphenoid and otoccipital, which articulate behind the level of the hypophysial region. This endoskeletal joint is sometimes coupled with a joint in the skull-roof, between the parietals and postparietals. In some taxa (e.g. actinistians, holoptychiid porolepiforms), however, the position of this dermal joint does not match that of the endoskeletal joint, and the intracranial articulation was non-functional (Fig. 6.4).

This joint in the sarcopterygian braincase has variously been interpreted as a general gnathostome or even a craniate feature, or a sarcopterygian specialization, linked with predation and feeding, yet its mobility is very slight, if not lacking. There are two competing theories about the intracranial joint: some regard it as homologous to the ventral fissure of actinopterygians (e.g. Gardiner 1984*a*; Ahlberg 1991*a*), whereas others consider that it is a remnant of a series of joints between the different cranial metameres of an hypothetical craniate ancestor (Bjerring 1978; Jarvik 1980; see Chapter 6, p. 258). It sounds, however, more realistic to assume that all osteichthyans have a propensity to develop a gap or a joint between the tip of the notochord and the hypophysis, i.e. at the level where the two halves of the embryonic braincase are united by the sole polar cartilage and lateral commissure (Figs 4.27, 4.28). The development of the intracranial joint may vary considerably. In the dipnomorph *Youngolepis*, for example, there is a wide individual variation in the development of the joint, which may be reduced to a mere ventral fissure (Fig. 6.4D). This shows that the joint has been progressively ankylosed in early lungfishes. In the other 'jointless' sarcopterygian group, the tetrapods, there remains a ventral fissure behind the parasphenoid (Fig. 4.96, 12) in the earliest known forms.

The interrelationships of sarcopterygians are still much debated and will therefore be treated in the text chapter (p. 246). Some of the fossil sarcopterygians, which have folded dentine in teeth (plicidentine) and were thought to possess choanae, have long been referred to a particular taxon, the Rhipidistia. This taxon is now defined on the basis of plicidentine only (Ahlberg 1991*a*); it includes lungfishes, which are assumed to have lost this character.

4.18.1 Onychodontiformes

Onychodontiforms, also known as 'struniiforms', are represented by only a few genera, among which the Late Devonian *Strunius* (Fig. 4.74A) and the Middle Devonian *Grossius* (Fig. 4.74B) are the only well-documented forms. Some data on the well-preserved *Onychodus*-specimens from the Late Devonian of Gogo, Australia, have also been distilled by Andrews (1973). The Onychodontiformes are characterized by a single row of large sigmoid teeth on the parasymphysial plates, which rest on either side of the lower jaw symphysis (Fig. 4.74, 6). When the mouth is closed these teeth are housed inside deep internasal pits of the palate. The premaxilla is slender, bearing no sensory-line canal, and the maxilla is large (Fig. 4.74, 1), extending posteriorly beyond the oral commissure and reaching the preopercular, as in actinopterygians. There are probably two separate external nostrils, but only the anterior one has been located with certainty (Fig. 4.74, 3). The teeth are simple, with no infolded dentine, except at the very base where some slight folds occur (Fig. 4.74C, 9). They are covered with a characteristic finely striated enamel (Fig. 4.74C, 7, 8). In some *Onychodus* species, the parasymphysial teeth are slightly hook-shaped (Fig. 4.74F, 6)

Apart from these common features, *Grossius* and *Strunius* display quite different head morphologies: *Grossius* is a very large form, with small eyes and a pointed snout (Fig. 4.74B), whereas *Strunius* is small, with a blunt snout and very large eyes (Fig. 4.74A). The body is known in *Strunius* only and is covered with large rounded scales. The basal lobe of the fins is short and lacking in the dorsal fins (as in early actinistians; see next paragraph). The tail of *Strunius* is diphycercal, with a long axial notochordal lobe (Fig. 4.74, 2), as in most actinistians (except *Miguashaia*, Fig. 4.75A), but that of *Onychodus* is said to be epicercal–the general gnathostome condition.

Grossius and *Onychodus* have remarkably long postparietals (Fig. 4.74B, E, 5) and short parietals (Fig. 4.74B, 4), which probably represent a derived condition. As in other sarcopterygians, the mandibular sensory-line canal passes through the infradentary, which is single, but there is sometimes a secondary line in the dentary, as in actinopterygians. The dermal bones and scales of the onychodontiformes are ornamented with ridges or horseshoe-shaped tubercles of dentine (Fig. 4.74D). The presence of patches of cosmine (Jessen 1966) remains uncertain.

Almost nothing is known of the internal anatomy of onychodontiforms, except that they possessed a large notochordal canal in the otico-occipital region of the braincase, like most other sarcopterygians. Strangely, this little-known sarcopterygian group is fairly abundant in marine Devonian sediments, in the form of teeth, which are quite characteristic, or scale fragments. They occur from the Early Devonian (Pragian) to the Late Famennian. (Carboniferous occurrences are now discarded.) Onychodontiforms were probably active predators, and could reach a fairly large size (about a metre in length).

Further reading

Andrews (1973); Aquesbi (1988); Jessen (1966); Long (1995); Schultze (1969, 1973); Zhu and Janvier (1994).

4.18.2 Actinistia

Actinistians have been defined and described on the basis of the extant coelacanth, *Latimeria* (see Chapter 3, p. 71). Fossil actinistians, in particular Devonian forms such as *Euporosteus*, *Miguashaia* (Fig. 4.75A), or *Diplocercides* (Fig. 4.75B), do not differ markedly from the extant species, except in their rather small size. Some of the actinistian characters met with in *Latimeria* do not, however, occur in these early forms. In *Miguashaia*, which is the best known amongst the earliest actinistians, the dermal fin rays of the anterior dorsal fin are thin and numerous, with unfused lepidotrichs (Fig. 4.75, 1), and the caudal fin is epicercal (Fig. 4.75, 2), though with a large, rounded ventral web. Most characters of the extant actinistian are, however, present in the Devonian forms: the dorsal fins each supported by a single basal element; the endoskeleton of second dorsal and anal fin resembling that of paired fins; the large rostral organ opening by pores (Fig. 4.75, 7); the large triangular pterygoids in the palate (Fig. 4.75, 13), the short and stout hyomandibula (Fig. 4.75, 14); the lack of a maxilla; the small premaxilla (Fig. 4.75, 3); the very short dentary (Fig. 4.75, 4); the two enlarged infradentaries (Fig. 4.75, 19), the articulation between the retroarticular process and the symplectic (Fig. 4.75, 17, 18); the very large opercular (Fig. 4.75, 5); the postcleithrum; and the lack of branchiostegal rays. The spiny ornamentation of the scales on Mesozoic and Recent actinistians seems to have occured at a rather late stage. The Devonian coelacanth scales are ornamented either with horseshoe-shaped tubercles (as in onychodontiforms) or with thin dentinous ridges (Fig. 4.75B4).

In the earliest known actinistians (Late Devonian and Carboniferous forms), the two halves of the braincase are well ossified (Fig. 4.75, 8, 9), whereas in later forms each of them tends to break up into several ossifications. This trend towards forming several ossification centres in the braincase is met with in other osteichthyans, such as lungfishes and actinopterygians (see Chapter 6, p. 275; Fig. 6.12B, C).

Some Mesozoic actinistians (*Undina*, *Macropoma*, *Mawsonia*; Fig. 4.76G, H) show in their abdominal cavity a peculiar elongated and calcified sac, which is currently interpreted as a swim bladder or a lung. This organ, which has no equivalent in *Latimeria* (except perhaps the 'fatty lung'), has a multi-layered structure and is extremely hard.

We now know that *Latimeria* is ovoviviparous, and that fully formed young can be found in the oviduct of the female. This had been suggested by Watson (1927) long before *Latimeria* was discovered, on the basis of the presence of young inside a specimen of the Jurassic actinistian *Holophagus*. However, this interpretation was later rejected, and cannibalism was invoked, because other specimens of *Holophagus* showed some non-actinistian fishes (thus ingested) in the same position (Schultze 1972). In 1972, Schultze described very young individuals of the Carboniferous actinistian *Rhabdoderma* as possessing a large yolk sac (Fig. 4.75C, 24) which is reduced in larger individuals (Fig. 4.75C1). Three years later, the ovoviviparity of *Latimeria* was demonstrated, and the young in the oviduct of a female were also found to possess a large yolk sac (Fig. 4.75D, 24; Smith *et al.* 1975). The young of *Latimeria*, like those of *Rhabdoderma*, have a somewhat longer chordal lobe of the caudal fin than the adults.

4.18.2.1 Diversity and phylogeny

The most generalized actinistian known at present is the Late Devonian *Miguashaia* (Fig. 4.76A), which displays traces of cosmine on dermal bones, retains an epicercal tail, and has no preorbital bone, and in which the anterior opening of the rostral organ is situated outside the premaxilla (Forey 1991*b*). *Diplocercides* is the second well-known Devonian actinistian, and the only one in which the endoskeleton has been well described, although the body is still virtually unknown. Its braincase and mandibular arch provide a remarkable example of morphological stability from the Devonian to Recent (Fig. 4.75B2). There are, however, also examples of strongly divergent morphologies among actinistians, in particular in the Carboniferous genus *Allenypterus* (Fig. 4.76C), which is deep-bodied and mimics some extant teleosts. In contrast, the Carboniferous *Caridosuctor* (Fig. 4.76D) and the Permian *Coelacanthus* (Fig. 4.76F) have a slender body shape.

The phylogeny in Fig. 4.76 (simplified from Forey 1991*b*) shows that the extant *Latimeria* (Fig. 4.76I) is most closely related to the Cretaceous genus *Macropoma* (Fig. 4.76H). Both share a considerable forward extension of the basicranial muscle on to the parasphenoid (Fig. 4.76, 8). These and other Mesozoic coelacanths, such as *Mawsonia* (Fig. 4.76G) share a shortened otico-occipital region of the braincase, relatively to the ethmosphenoid region (Fig. 4.76, 7). This contrasts with the condition in Palaeozoic forms, where the two regions are almost equal in size (Fig. 4.75B3). The typical actinistian structure of the anterior dorsal fin, with large, rigid dermal fin rays, occurs first in the clade that includes all Mesozoic forms, as well as the Carboniferous *Rhabdoderma* and *Caridosuctor* (Fig. 4.76, 4). Unbranched fin rays occur in all actinistians except *Miguashaia* (Fig. 4.76, 2).

4.18.2.2 Stratigraphical distribution and habitat

The earliest known actinistian remains are Givetian to Frasnian in age, with *Euporosteus*, *Miguashaia*, and *Diplocercides*, all from Europe and the Middle East. *Chagrinia* is a poorly known Late Famennian form from North America. The number of actinistian species in the fossil record shows an increase from the Devonian to the Carboniferous, a clear decrease in the Permian, and a peak in the Triassic. The number of recorded species then decreased until the end of the Cretaceous. There is no actinistian record from the Tertiary. Forey (1991*b*) analysed the changes of the ratio between the marine and presumed freshwater actinistian species, showing that the largest number of freshwater species lived in Carboniferous and Cretaceous times. The habitat of fossil actinistians (as well as of many other fishes) is, however, often inferred from sedimentological data that are now known to be unreliable or, in turn, inferred from palaeontological data (e.g. the abundance of land vertebrates or the lack of marine invertebrates). It is probable that the number of fossil freshwater actinistians has been over-estimated.

Further reading

Anatomy of Palaeozoic forms: Bjerring (1973, 1985, 1986*b*); Cloutier (1991*a*); Forey (1981); Jarvik (1980); Jessen (1966, 1973); Lund and Lund (1984); Schultze (1973); Stensiö (1937). *Interrelationships and systematics*: Cloutier (1991*a*,*b*), Forey (1991*b*).

4.18.3 Dipnomorpha

The dipnomorphs comprize two major taxa, the Porolepiformes (also referred to as holoptychiids) and the Dipnoi (lungfishes), as well as some genera of debated position (*Powichthys*, *Youngolepis*, and *Diabolepis*; Fig. 4.84). Porolepiforms, which are exclusively Devonian, have been variously regarded either as relatives of the osteolepiforms (with which they share folded dentine in teeth) or as ancestral only to the Caudata (urodeles) (see Chapter 5, p. 246); they are here considered to be the closest relatives of lungfishes.

Dipnomorphs are generally stout-bodied fishes with a blunt snout, small eyes, epicercal tail, and posteriorly placed dorsal fins (Figs 4.77, 4.80). They are characterized by elongated, leaf-shaped pectoral fins (Fig. 4.77, 33; Fig. 4.80, 1), in which the endoskeleton consists of numerous mesomeres with pre-and postaxial radials (Fig. 4.79A2). In addition, they have in the snout a network of branching canals, the rostral tubules, which are interpreted as transmitting either nerve fibres to sensory organs on the surface of the snout, or capillary blood vessels. The rostral tubules are few in porolepiforms, but quite numerous in *Youngolepis* and lungfishes, where they also occur in the lower jaw (Figs 4.81D, 4.84B2).

There are other dipnomorph characteristics, such as the supraneurals articulating with the neural arches, the branched posterior radials in the second dorsal fin (Fig. 4.79, 11), the pineal region anterior to the parietals (Fig. 4.78C2, 20, 21), the lack of contact between the parietals and supraorbitals, the preoperculo-submandibular bone, and the *crista rostrocaudalis* in the nasal cavity (Ahlberg 1991*a*).

4.18.3.1 Porolepiformes

Porolepiforms are characterized by strongly folded teeth filled with attachment bone ('dendrodont' type; Fig. 4.77D) and by the otic part of the main lateral-line canal passing through the centre of the postparietals (Fig. 4.77A3, 18). Other characters may be the lack of radials in the anterior dorsal fin (Fig. 4.79A5) and large processes on the mesomeres of paired fins (Fig. 4.79A2), but these have so far been observed only in a few porolepiforms. They are known essentially from the genera *Porolepis* (Fig. 4.77A1, C1), *Glyptolepis* (Fig. 4.77C2), *Laccognathus* (Fig. 1.11, 2), and, to some extent, *Holoptychius* (Fig. 4.77B, C3), and *Quebecius*. Three of the best-known of these genera are the Early Devonian *Porolepis*, and the Middle to Late Devonian *Holoptychius* and *Glyptolepis*, which may serve to illustrate the group.

General morphology and exoskeleton   *Holoptychius* and *Glyptolepis* are rather large and massive fishes with rounded scales (Fig. 4.77B2). Their snout is blunt, with small, anteriorly placed eyes. In *Holoptychius*, the anterior and posterior nostrils are closely set and separated by a small dermal bone, the nariodal (Fig. 4.77C3, 23). The posterior nostril is only slightly more ventral in position than the anterior one, but in other porolepiforms, such as

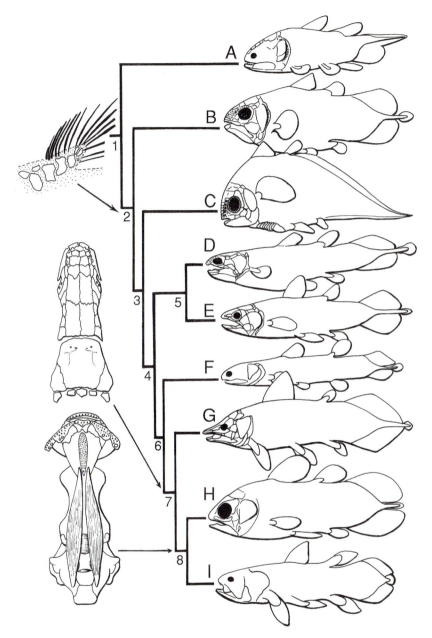

Fig. 4.76. Fossil Sarcopterygii. Interrelationships of fossil and Recent Actinistia, according to the phylogenies proposed by Forey (1991*b*) and Cloutier (1991) (some taxa omitted). *Terminal taxa*: A, *Miguashaia*; B, *Hadronector*; C, *Allenypterus*; D, *Caridosuctor*; E, *Rhabdoderma exiguum*; F, *Coelacanthus*; G, *Mawsonia*; H, *Macropoma*; I, *Latimeria. Nested taxa and selected synapomorphies*: 1, Actinistia (rostral organ, reduced dentary, maxilla lost); 2, diphycercal tail, basal lobe in second dorsal and anal fins, unbranched fin rays, scales with longidudinal ridges only, large sensory-line pores, loss of basipterygoid process, pelvic fins in abdominal position; 3, five extrascapulars; 4, sensory-line pore present on dentary, basal plate of second dorsal fin with an anterior process, lepidotrichs of anterior dorsal fin fused into rods; 5, frontals of both sides similar in shape; 6, ornamentation of lower jaw restricted to the angular, several lateral-line pores per scale; 7, postparietal shield length half or less than half that of the parieto-ethmoidal shield; 8, basicranial muscle extending over more than 50 per cent of parasphenoid length. *Age*: Late Devonian (A); Carboniferous (B–E); Permian (F); Cretaceous (G, H); Recent (I). (A, from Schultze (1973); B–D, from Lund and Lund (1984); E–H, based on Forey (1981, 1991*b*).)

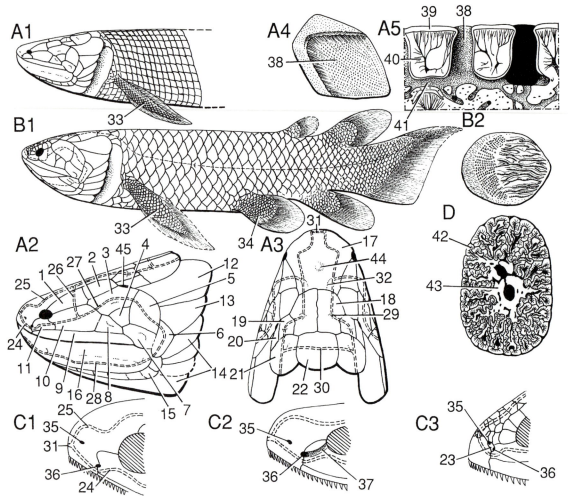

Fig. 4.77. Fossil Sarcopterygii. Dipnomorpha, Porolepiformes. External morphology and histology. A, *Porolepis*, Early Devonian of Europe, reconstruction of the anterior part of the body in lateral view (A1, ×0.5); head in lateral (A2) and dorsal (A3) view; scale in lateral view (A4); vertical section through the cosmine of exoskeleton (A5, about ×70). B, *Holoptychius*, Middle to Late Devonian, reconstruction in lateral view (B1, ×0.5) and scale in lateral view (B2). C, detail of the snout, in lateral view, of *Porolepis* (C1), *Glyptolepis* (C2), and *Holoptychius* (C3); D, *Hamodus*, Middle Devonian of Estonia, horizontal section through a tooth. 1–23, *dermal bones*: 1, postorbital; 2, prespiracular; 3, upper squamosal; 4, accessory squamosal; 5, 6, preoperculars; 7, quadratojugal; 8, lower squamosal; 9, maxilla, 10, jugal, 11, lachrymal; 12, opercular; 13, subopercular; 14, branchiostegals; 15, submandibulars; 16, dentary and infradentaries; 17, parieto-ethmoidal shield (fused parietals and dermal bones of the snout); 18, postparietals; 19, tabular; 20, extratemporal; 21, lateral extrascapular; 22, median extrascapular; 23, nariodal; 24–32, *sensory lines*: 24, infraorbital line; 25, supraorbital line; 26, postorbital line; 27, jugal and preopercular line; 28, mandibular line; 29, postparietal pit lines; 30, supratemporal commissure; 31, ethmoid commissure; 32, parietal pit line; 33–45, *others*: 33, pectoral fin; 34, pelvic fin; 35, anterior nostril; 36, posterior nostril; 37, groove of posterior nostril; 38, pore and flask-shaped cavity of cosmine; 39, enamel; 40, dentine; 41, horizontal canal network of cosmine; 42, dentine (plicidentine); 43, attachment bone or osteodentine filling pulp cavity; 44 pineal region; 45, spiracle. (A1–4, combined from Jarvik (1980) and original observations; B1, C, from Jarvik (1972); B2, from Ørvig (1957); A5, from Gross (1956); D, from Schultze (1970), after Bystrov.)

the Early Devonian *Porolepis*, it is quite ventral in position and is situated near the mouth margin (Fig. 4.77C1, 36). In *Glyptolepis*, the posterior nostril is prolonged by a groove that meets the orbital margin (Fig. 4.77C2, 37). This groove has been regarded as an incipient nasolacrymal duct.

The skull-roof shows very short parietals and longer postparietals. In *Porolepis*, the thick cosmine covering masks the sutures in the anterior half of the head, but the parietals were probably quite short, and the snout is covered with a mosaic of small bones. The median extrascapular overlaps the lateral extrascapulars (Fig. 4.77, 21, 22).

The cheek displays a few extra bones, not met with in other osteichthyans: the preoperculo-submandibular (Fig. 4.77, 6) and subsidiary squamosals (Fig. 4.77, 4, 8).

The operculo-gular series of dermal bones comprises a very small opercular (Fig. 4.77, 12), a feature common to all porolepiforms, but probably general for dipnomorphs.

The subopercular series consists of several elongated branchiostegals (Fig. 4.77, 14).

The lower jaw is massive, with large coronoid tusks (Fig. 4.78, 9) and parasymphysial toothed plates that bear several series of large folded teeth (Fig. 4.78, 7), in contrast to those of onychodontiforms. These parasymphysial teeth are housed in paired internasal pits of the palate (Fig. 4.78, 1). The mandibular sensory-line canal passes through the infradentaries (Fig. 4.77, 28), when these are distinct from the dentary.

The dermal bones of the palate display the general osteichthyan pattern, but the parashenoid (Fig. 4.78, 5) is

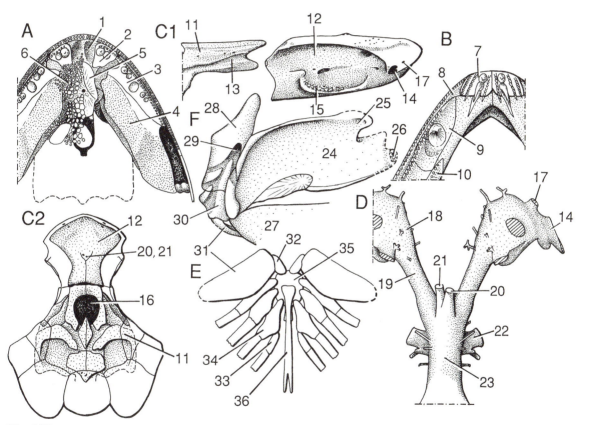

Fig. 4.78. Fossil Sarcopterygii. Dipnomorpha, Porolepiformes. Internal anatomy. A–E *Glyptolepis*, Middle Devonian; A, reconstruction of the palate in ventral view (about ×1); B, anterior part of the lower jaw in dorsal view (about ×1.5); C, braincase in lateral view (C1) and dorsal view with skull-roof superimposed (C2); D, internal cast of the anterior part of brain cavity and nasal cavities in dorsal view; E, ventral division of hyobranchial skeleton in ventral view; F, *Porolepis*, Early Devonian, part of the hyoid arch and palatoquadrate in lateral view (about ×2). 1, internasal fossae; 2, vomer; 3, dermopalatine; 4, entopterygoid; 5, parasphenoid; 6, palatal platelets; 7, parasymphysial toothed plates; 8, dentary; 9, coronoid; 10, prearticular; 11, otoccipital; 12, ethmosphenoid; 13, groove for jugular vein; 14, ventrolateral fenestra (opening in endoskeleton for the posterior nostril); 15, basipterygoid process; 16, dorsal fontanelle; 17, anterior nostril; 18, nasal cavity; 19, olfactory tract; 20, pineal canal; 21, parapineal canal; 22, optic tract; 23, brain cavity; 24, palatoquadrate; 25, processus ascendens; 26, ethmoid articulation of palatoquadrate; 27, Meckelian bone; 28, epihyal; 29, canal for facial nerve; 30, symplectic; 31, ceratohyal; 32, hypohyal; 33, fifth hypobranchial; 34, fourth hypobranchial; 35, basibranchial; 36, urohyal. (A–E, from Jarvik (1972) and Bjerring (1991); F, from Véran (1988) and original observations.)

condition, and also that they display some resemblances to the tetrapods, in characters that are now known to have appeared independently within the group. Devonian lungfishes do not differ basically from extant forms, although their skeletons are much more heavily ossified. Devonian lungfishes are, however, much more diverse that post-Devonian ones. In order to illustrate this diversity, the group is exemplified here by two Late Devonian genera, *Chirodipterus* (Figs 4.80C, 4.81A) and *Griphognathus* (Figs 4.80B, 4.81B), which are now well

known thanks to the excellently preserved material from Gogo, Australia. The former is a short-snouted form, with typical radiating ridges of dentine on the tooth plates (Fig. 4.81A, 1, 4); the second is a long-snouted form, in which the tooth plates are secondarily modified into large denticle-bearing plates (Fig. 4.81B, 1, 4).

General morphology and exoskeleton Devonian lungfishes are generally stout fishes with elongated pectoral and pelvic fins (Fig. 4.80, 1, 2) and two dorsal fins in

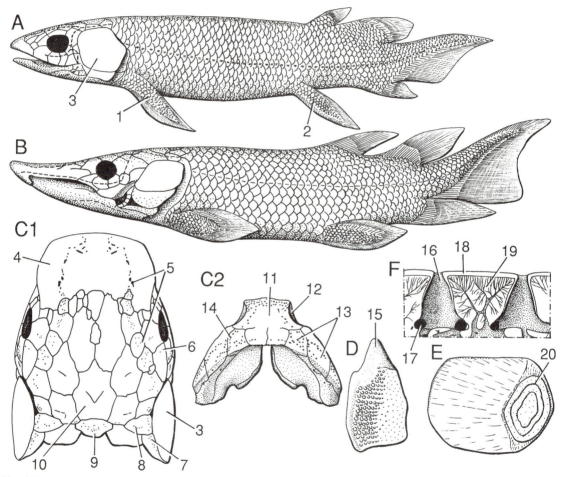

Fig. 4.80. Fossil Sarcopterygii. Dipnomorpha, Dipnoi. External morphology and histology. A, *Dipterus*, Middle to Late Devonian, reconstruction in lateral view (about ×1); B, *Griphognathus*, Late Devonian, reconstruction in lateral view (about ×0.2); C, *Chirodipterus*, Late Devonian, skull in dorsal view (C1, ×1) and lower jaw in ventral view (C2); D, *Dipnorhynchus*, Early Devonian, scale in external view (cosmine stippled); E, *Rhinodipterus*, Late Devonian, scale in external view (cosmine stippled); F, *Dipterus*, Late Devonian, vertical section through the cosmine of a dermal bone (×140). 1, pectoral fin; 2, pelvic fin; 3, opercular; 4, ethmoidal shield (fused dermal bones of the snout); 5, supraorbital sensory-line pores; 6, infraorbital sensory-line pores; 7, main lateral-line pores; 8, pores of the supratemporal commissure; 9, median extrascapular; 10, 'B'-bone; 11, dentary; 12, labial pit; 13, infradentaries; 14, mandibular sensory line; 15, dorsal peg; 16, flask-shaped cavity; 17, canal of the pore-canal system; 18, enamel; 19, dentine; 20, Westoll lines. (A, B, from Campbell and Barwick (1988*b*); C, from Miles (1977); D, from Thomson and Campbell (1971); E, F, from Gross (1956), modified.)

a posterior position. The body of *Chirodipterus* has not yet been reconstructed, but it is probably quite similar to that of *Dipterus* (Fig. 4.80A), another Devonian genus. The body scales of the earliest Devonian lungfishes are more or less diamond-shaped and cosmine-covered (Fig. 4.80D), like those of *Porolepis*, but in later forms they tend to be more rounded in shape (Fig. 4.80E).

The dermal skull of Devonian lungfishes displays a mozaic-like pattern of the dermal bones (Fig. 4.80C1), which renders difficult any bone-to-bone comparison with other osteichthyans. The presence of a large median bone at the back of the skull-roof (Fig. 4.80, 10), the five extrascapulars, and the series of small bones enclosing the supraorbital sensory-line canal (Fig. 4.80, 5) are, for example, among the most striking unique lungfish features. Also, lungfishes lack a maxilla (the presence of the premaxilla is disputed; see comments on *Diabolepis* below), and their lower jaw is so strongly modified that the dentary and infradentaries are difficult to distinguish (Fig. 4.80, 11, 13). The vomers may still be present (Fig. 4.81, 6), but the palatine, ectopterygoids, and coronoids are either lost or fused to the tooth plates (Smith and Chang 1990). The parasphenoid does not completely separate the entopterygoids, but extends posteriorly under the otico-occipital region (Fig. 4.81, 2). As will be discussed below, I have accepted here the theory according to which lungfishes are the sister-group of the porolepiforms. I also accept that some taxa, such as *Youngolepis* (Fig. 4.84B, 4.85C) and *Diabolepis* (Figs 4.84C, 4.85D) are stem-group lungfishes that provide a 'link' between the general osteichthyan and the lungfish dermal bone patterns. Thanks to these forms, some of the lungfish dermal bones, such as the parietals, postparietals, extrascapulars, dentary, and infradentaries, can be reliably homologized with those of other osteichthyans and can be given the same names (Fig. 4.83, 4, 5). The supernumerary bones, or those to which no precise homology can be assigned are named with letters (Forster-Cooper 1937). This if for example the case of anamestic 'B' bone (Fig. 4.80, 10; Fig. 4.83, 6), which becomes increasingly large and repels the parietals and postparietals laterally. In all Devonian lungfishes, the opercular (Fig. 4.80, 3) and subopercular are remarkably large, but there is no branchiostegal series. Instead of the gular plates, there are several pairs of rounded, scale-like plates.

Owing to the migration of the nostrils inside the mouth, the ethmoid commissure of the infraorbital sensory-lines is interrupted. The dermal skeleton of the snout and jaw symphysis is pitted by numerous openings, which are larger than the pores of the cosmine, and may correspond to the openings of special sense-organs (possibly the Fahrenholz organs). Characteristic of all lungfishes is also a deep lateral depression, or labial pit, in the lower jaw

(Fig. 4.80, 12), which is probably linked with the reduction of the dentary and the lack of maxilla.

The dermal shoulder girdle comprises the clavicle, cleithrum, anocleithrum, and supracleithrum, the latter meeting a bone of the skull-roof (possibly a posttemporal; Fig. 4.80, 7). The anocleithrum is covered by the skin (subdermal), as in other sarcopterygians except the onychodontiforms and osteolepiforms.

Endoskeleton The braincase of Devonian lungfishes is a massive bony structure in which the palatoquadrate is presumably incorporated. It shows no trace of an intracranial joint or ventral fissure (except in *Diabolepis*, Fig. 4.84C1, 5). This condition is probably linked with durophagous habits; it parallels that in holocephalans. The dorsal surface of the braincase is firmly attached to the skull-roof anteriorly, but posteriorly the two skull components are connected only by narrow median and lateral buttresses or crests, between which passed the epaxial musculature and some of the large jaw muscles. The notochord penetrates a short distance into the braincase, but does not extend as far forward as in generalized sarcopterygians (Fig. 4.81, 27). In palatal view, the braincase shows the large cavities for the olfactory organs (Fig. 4.81, 3). In *Griphognathus*, the two nostrils are surrounded by anamestic dermal bones (Fig. 4.81, 8, 9). In *Chirodipterus*, only the notch for the anterior nostrils can be distinguished.

The internal anatomy of the braincase has been reconstructed in *Chirodipterus* and some other dipnoans, such as the Early Devonian *Dipnorhynchus* (Fig. 4.81C). Here again, despite some unique features, it shares some features with other sarcopterygians, such as the supraotic cavity, which contained derivatives of the endolymphatic ducts. Also, like porolepiforms, lungfishes have separate pineal and parapineal canals (Fig. 4.81, 21, 22). Ventrally, there is a long, posteriorly directed buccohypophysial canal (Fig. 4.81, 25), which opens at the junction between the parasphenoid and entopterygoid tooth plates. The endoskeleton of the snout and lower jaw is pervaded by bundles of large tubules lined with perichondral bone, the rostral and symphysial tubules (Fig. 4.81, 28), which have been interpreted as transmitting nerve fibres toward the electroreceptors and mechanoreceptors of the snout. Bemis and Northcutt (1992), however, regard them as homologous to the dense vascular network in the snout of *Neoceratodus*.

The hyoid arch comprises a small, stout epihyal (Fig. 4.81, 11), which can be regarded as double-headed in some Devonian forms. The ventral shaft of the epihyal is oriented anteroventrally and is in contact with the hyosuspensory process of the quadrate (Fig. 4.81, 12). The ceratohyal (Fig. 4.81, 13) was probably also attached

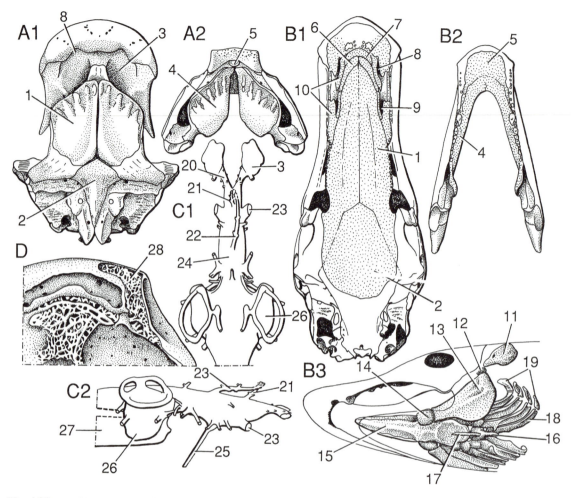

Fig. 4.81. Fossil Sarcopterygii. Dipnomorpha, Dipnoi. Palate and internal anatomy. A, *Chirodipterus*, Late Devonian, palate in ventral view (A1, ×1) and lower jaw in dorsal view (A2); B, *Griphognathus*, Late Devonian, palate in ventral view (B1, denticle-covered area stippled), lower jaw in dorsal view (B2, denticle-covered area stippled), and hyobranchial apparatus in ventrolateral view (B3) (B1, B2, ×0.5); C, *Dipnorhynchus*, Early Devonian of Australia, internal cast of cranial cavities in dorsal (C1) and lateral (C2) view (×1). D, *Speonesydrion*, Early Devonian of Australia, symphysial tubule system in dorsal view (×1). 1, entopterygoid tooth plate or denticle-covered plate; 2, parasphenoid; 3, nasal cavity; 4, prearticular tooth plate or denticle-covered plate; 5, median tooth plate; 6, vomerine tooth plate; 7, premaxilla; 8, anterior nostril; 9, posterior nostril; 10, anamestic bones or dermopalatines; 11, hyomandibula; 12, hyoid process of quadrate; 13, ceratohyal; 14, hypohyal; 15, sublingual rod, basihyal and anterior basibranchial; 16, posterior basibranchial; 17, urohyal; 18, ceratobranchials; 19, epibranchials; 20, olfactory tract; 21, parapineal canal; 22, pineal canal; 23, optic tract; 24, brain cavity; 25, buccohypophysial canal; 26, labyrinth cavity; 27, canal for the notochord; 28, symphysial tubules. (A, B1, B2, from Miles (1977); B3, from Campbell and Barwick (1987); C, from Campbell and Barwick (1982); D, from Campbell and Barwick (1984*b*).)

to the quadrate by means of a hyosuspensory ligament, as in extant lungfishes. In both *Griphognathus* and *Chirodipterus*, the basibranchial series consists of two elements (Fig. 4.81, 15, 16), the foremost of which is the basihyal, fused to the first basibranchial. There are four, well-developed gill arches (Fig. 4.81, 18, 19), which suggest an extensive branchial respiration (Campbell and

Barwick 1988*b*, 1990). A ventral element is assumed to be the urohyal (Fig. 4.81, 17).

The axial skeleton of most Devonian lungfishes consists of vertebrae with an ossified centrum that derives from the chordal sheath. The neural arches are situated between adjacent centra, whereas the haemal arches are attached to the respective centra. In all post-Devonian lungfishes, the chordal centra have disappeared (the centra in the extant *Protopterus* are neomorphs of perichordal origin), but some forms display rather elongated ribs, as in the extant *Neoceratodus*.

The scapulocoracoid of *Chirodipterus* (Fig. 4.82, 2) is massive, and is attached to the cleithrum (Fig. 4.82, 1) by buttresses that straddle the supracoracoid and supraglenoid canals (Fig. 4.82, 4, 5). It is thus built basically as in osteolepiforms (Fig. 4.91, 8). The pelvic girdle consists of two large pelvic plates (Fig. 4.82, 6), probably united by a symphysis.

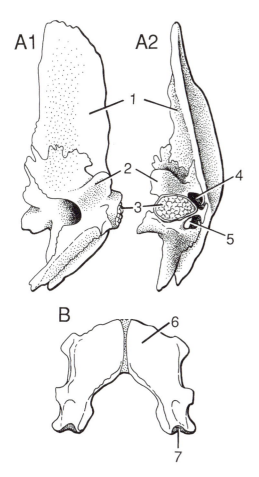

Fig. 4.82. Fossil Sarcopterygii. Dipnomorpha, Dipnoi. Girdles of *Chirodipterus*, Late Devonian. A, cleithrum and scapulocoracoid in medial (A1) and posterior (A2) view; B, pelvic girdle in ventral view. 1, cleithrum; 2, scapulocoracoid; 3, articular area for pectoral fin skeleton; 4, supraglenoid foramen, 5, supracoracoid foramen; 6, pelvic girdle; 7, articular area for pelvic fin skeleton. (A, from Janvier (1980*a*); B, from Young *et al.* (1989).)

The skeleton of both paired and unpaired fins is poorly known, for it is rarely ossified, even in otherwise well-ossified Devonian forms. The radials of the unpaired fins seem, however, to articulate on a large basal plate, and the rearmost radials of the second dorsal fin form a branching structure, as in porolepiforms. From the overall shape of the pectoral fins, it may be inferred that the endoskeleton was similar to that of porolepiforms and extant lungfishes, i.e. with a long series of mesomeres, and post- and preaxial radials.

Histology The endoskeleton of lungfishes, Devonian ones in particular, is strongly ossified, both peri- and endochondrally, as in all other osteichthyans. The exoskeleton of most Devonian lungfishes is covered with a type of cosmine resembling that of porolepiforms, i.e., the enamel generally lines the bottle-shaped spaces (Fig. 4.80, 16, 18). In addition, the cosmine of lungfishes shows more or less concentric lines of discontinuity, the Westoll lines (Fig. 4.80, 20; Fig. 4.83, 7), which are interpreted as being due to cyclic resorbtion and redeposition during growth (Ørvig 1969) and are a unique feature of this group. From the Late Devonian onwards, lungfishes show a tendency towards bone reduction and loss of the cosmine. This leads to the condition in Recent forms (Fig. 3.22D, E, F) in which only part of the dermal skull-roof remains, in the form of a few bones covered by the skin, and where the endoskeleton is largely cartilaginous. As early as the Carboniferous, lungfishes were almost as poorly ossified as Recent forms.

Diversity and phylogeny Devonian dipnoans display a relatively high diversity involving differences in the skull proportions and tooth-plate histology. The long-snouted condition exemplified by *Griphognathus* is derived and probably arose several times in different ways. The differences in the aspect and growth of the dermal bones of the mouth (tooth plates and parasphenoid) have led to animated discussions relating to phylogeny. Campbell and Barwick (1990) have considered that the shagreen-covered entopterygoids, parasphenoid, and prearticulars of the Early Devonian *Uranolophus* (Fig. 4.83B2, 9, 10) and of *Griphognathus* appeared independently of the typical tooth plates found in most other lungfishes (e.g. *Dipnorhynchus*, *Chirodipterus*, and *Dipterus*). They also consider and that among the other lungfishes there is a fundamental difference between tooth plates that grow by adding denticles to the margins of the tooth plates, and those that grow by deposition of successive dentine layers. The entopterygoid and prearticular of many non-dipnoan sarcopterygians, for example, are covered with a shagreen of minute denticles, sometimes arranged in radiating rows. It is thus likely that the most generalized con-

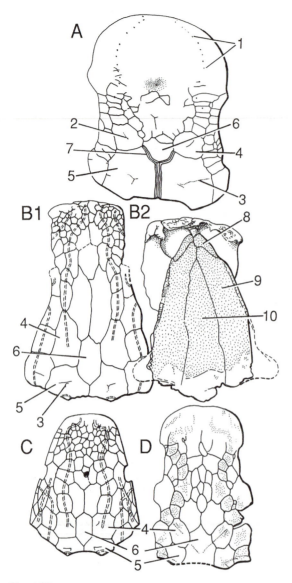

Fig. 4.83. Fossil Sarcopterygii. Dipnomorpha, Dipnoi. A, *Diabolepis*, Early Devonian of China, skull-roof in dorsal view (×2); B, *Uranolophus*, Early Devonian of Wyoming, skull-roof in dorsal view (B1) and palate in ventral view (B2) (×0.5); C, *Dipnorhynchus*, Early Devonian, skull-roof in dorsal view (×0.5); D, *Dipterus*, Middle to Late Devonian, skull-roof in dorsal view (×1.5). 1, pores of the supraorbital sensory line; 2, anterior (parietal) pit line; 3, middle (postparietal) pit line; 4, parietal; 5, postparietal; 6, 'B'-bone; 7, Westoll lines; 8, vomer; 9, entopterygoid; 10, parasphenoid. (A, from Chang and Yu (1984); B, from Denison (1968) and Schultze and Campbell (1987); C, from Campbell and Barwick (1987); D, from Jarvik (1980).)

dition in lungfishes is close to that in the still-debated genus *Diabolepis* (Fig. 4.84C, 17, 20; see below), i.e. with numerous small denticles arranged in radiating rows. The condition in *Uranolophus* or *Griphognathus* would thus be derived, independently, by the enlargement of the denticulated plates and the loss of the radiating organization of the denticles; and that in other lungfishes would be derived by a reduction of the denticle rows and their trend towards a fusion into crests or a more or less continuous dentinous covering. The phylogeny favoured here (Fig. 4.85, left-hand side; for more details see Schultze and Marshall (1993)) assumes that *Diabolepis* (Fig. 4.85D) is the sister-group of all dipnoans in which the posterior nostril is situated inside the palate (Fig. 4.85, 8). Then comes *Uranolophus* (Fig. 4.85E), in which the parasphenoid extends far forward, but the two entopterygoids are in contact anteriorly (Fig. 4.83, 9, 10). In all other lungfishes, the parasphenoid is short and the entopterygoids are largely in contact along the midline of the palate (Fig. 4.81, 1, 2; Fig. 4.85, 9). Other primitive lungfishes are probably *Dipnorhynchus* (Figs 4.83C, 4.85F) and *Speonesydrion*. The skull-roof of *Uranolophus* and *Dipnorhynchus* also displays a transformation series towards the typical lungfish pattern. In these two genera, the anamestic 'B' bone (incipient in *Diabolepis*) is still small (Fig. 4.83, 6) and completely separates only the parietals (Fig. 4.83, 4), while a median contact is retained between the two postparietals (Fig. 4.83A–C, 5). The postparietals are completely separated by the 'B' bone in all other lungfishes (Fig. 4.83, 5).

A number of forms generally referred to as 'dipterids' (including *Chirodipterus* and *Dipterus*, Fig. 4.85G) probably represent a paraphyletic ensemble of cosmine-covered forms. The long-snouted rhynchodipterids (including *Griphognathus*, Fig. 4.85H) are probably more derived than *Dipterus* and form a clade. The Late Devonian *Scaumenacia* (Fig. 4.85I) shows an increased elongation of the paired fins, a considerable enlargement of the posterior dorsal fin, and a reduction of the anal fin, which forshadows the morphology of post-Devonian forms. During the Late Palaeozoic and up to the Cretaceous occur a number of taxa, such as *Phaneropleuron* (Fig. 4.85J), *Uronemus*, *Sagenodus*, *Megapleuron*, *Gnathorhiza*, *Ctenodus*, and *Ceratodus*, that display a progressive reduction of the endo- and exoskeletal ossification and an increasing proportion of the characters found in the extant taxa, *Neoceratodus* and the Lepidosirenidae (Fig. 4.85, 14).

Stratigraphical distribution and habitat   At present the earliest known undisputed lungfishes are *Uranolophus*, *Dipnorhynchus*, and *Speoneshydrion*, all from the late Early Devonian (Emsian). All three genera are marine, as

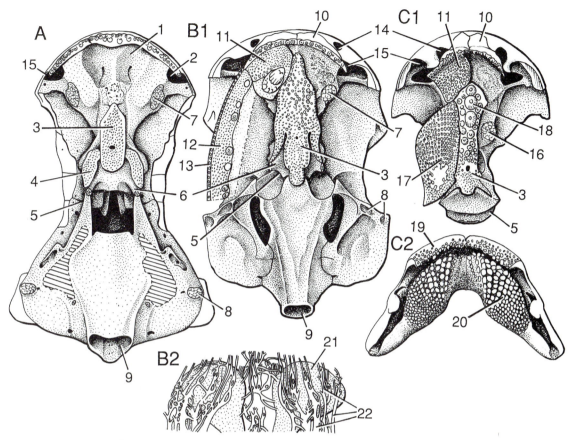

Fig. 4.84. Fossil Sarcopterygii. Dipnomorpha. A, *Powichthys*, Early Devonian of Canada, braincase in ventral view (×1.5); B, *Youngolepis*, Early Devonian of China, braincase and part of the palate (reconstructed) in ventral view (B1, ×1.5) and rostral tubules in dorsal view (B2); C, *Diabolepis*, Early Devonian of China, ethmosphenoid part of the braincase and part of the palate (reconstructed) in ventral view (C1, ×4) and lower jaw in dorsal view (C2). 1, apical or internasal fossa (to accomodate parasymphysial toothed plates of lower jaw); 2, ventrolateral fenestra of nasal cavity; 3, parasphenoid; 4, parasphenotic plates; 5, intracranial joint; 6, processus descendens of ethmosphenoid; 7, ethmoid articulation for palatoquadrate; 8, facet for hyomandibula; 9, canal for the notochord; 10, premaxilla; 11, vomer; 12 dermopalatine and ectopterygoid; 13, maxilla; 14, anterior nostril; 15, posterior nostril; 16, attachment area for palatoquadrate; 17, entopterygoid; 18, parasphenoid tooth with folded base; 19, dentary; 20, prearticular; 21, nasal capsule; 22, rostral tubules. (A, from Jessen (1980); B1, based on Chang (1982, 1991a); B2, from Cheng (1992); C, based on Chang and Yu (1984) and Smith and Chang (1990).)

are most—if not all—other Devonian dipnoans. Biomechanical analysis of the hyoid and branchial arches in Devonian lungfishes shows that they were entirely dependent on branchial respiration, and were probably unable to gulp air, as extant lepidosirenids do. They lived in near-shore marine environments, preferably among coral reefs, and did not invade true freshwater ecosystems before the Carboniferous. Late Devonian lungfishes may, however, have been found in estuarine or fluvial environments. Aestivation burrows, comparable to those of modern lepidosirenids, containing remains of the lungfish

Gnathorhiza (Fig. 7.14B) first occur in Early Permian times (McAllister 1992). The young of the Carboniferous lungfish *Megapleuron* show the same aspect as those of the modern *Neoceratodus*, with very large, blade-shaped pectoral fins. It thus seems that the shaping of the modern dipnoan morphology and mode of life took place in Carboniferous to Permian times.

18.3.3 Dipnomorph phylogeny

The question of the relationships of lungfishes with other sarcopterygians was partly answered during the 1980s by the discovery of forms that, at various times, have been regarded as stem-group porolepiforms, stem-group dipnoans, or stem-group dipnomorphs. These are the genera *Powichthys* (Jessen 1980; Fig. 4.84A), *Youngolepis* (Chang 1982; Fig. 4.84B), and *Diabolepis* (Chang and Yu 1984; Fig. 4.84C). These forms were discovered after a long period of stability in the systematics of fossil sarcopterygians, during which most phylogenetic debates

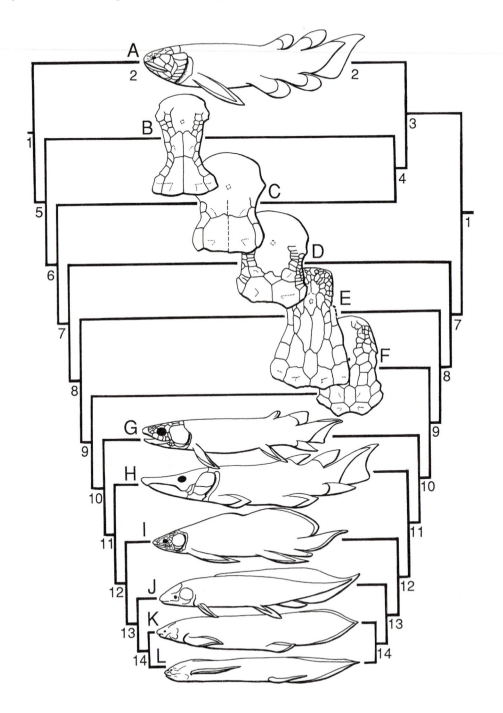

were based on the major taxa defined by Jarvik (1942). The new forms did not fit the clear-cut distinctions established between these taxa, either because they did not display *all* their defining characters, or because they showed assemblages of characters of several taxa.

Diabolepis, from the Early Devonian of Yunnan, China, is perhaps the most fascinating of these forms: it combines some lungfish characters (tooth plates, Westoll lines; Fig. 4.83, 7; Fig. 4.84C, 17, 20) and other characters found either in porolepiforms, in *Youngolepis*, or more generally in other osteichthyans (intracranial joint or ventral fissure, external nostrils, premaxilla, separate entopterygoids; Fig. 4.84C, 5, 10, 14, 15, 17). Its skull-roof, for example, has an incipient 'B' bone between the postparietals, and a pronounced tendency towards the fragmentation of the rest of the skull-roof into small bones (Fig. 4.83A, 6, 7). The palate and the braincase show a mix of lungfish-like and generalized osteichthyan or sarcopterygian features. There seems to be a ventral fissure (Fig. 4.84C, 5), and the parasphenoid is in contact with the vomers (Fig. 4.84C, 3, 11). The vomers are covered with minute teeth and lie along the premaxillae (Fig. 4.84C, 10). There are two external nostrils, resembling those of *Youngolepis*, but the posterior nostril lies exactly on the oral margin, showing again an incipient lungfish-like condition (Fig. 4.84C, 14, 15). Typically lungfish-like are the isolated entopterygoid tooth plates and lower jaws that Chang and Yu (1984) have referred to this taxon. Although those who refuse to include *Diabolepis* among lungfishes (e.g. Campbell and Barwick 1984*a*, 1987; Schultze 1987) argue that the skull-roof and tooth plates belong to two different animals, the tooth plates match perfectly the lateral margins of the parasphenoid, as well as the large contact area for the palatoquadrate (Fig. 4.84C, 16), thereby allowing a reconstruction of entire palate. I here treat *Diabolepis* as the sister-group of all other lungfishes (Fig. 4.85, 7), thus bridging a morphological gap between them and other dipnomorphs (in particular porolepiforms and *Youngolepis*). It is also worth noting that, despite their modified state, some of the teeth of *Diabolepis* (in particular those on the parasphenoid; Fig. 4.84C, 18) retain a slightly folded base, like the folded teeth of other rhipidistians.

When this position is accepted for *Diabolepis*, *Youngolepis*, which shares many resemblances with *Diabolepis* (ventrally placed external nostrils, proportions of the skull, tendency to produce large, rounded denticles in the mouth) and comes from the very same Chinese locality and layer, can be placed as the sister-group of lungfishes (including *Diabolepis*; Fig. 4.85, 6). *Youngolepis* has, however, some general dipnomorph or particular porolepiform characteristics, such as the tubules in the rostral region only (Fig. 4.84, 22), the structure of the cosmine, the median extrascapular overlapping lateral ones, and the small scapulocoracoid with reduced supracoracoid and supraglenoid foramina (Fig. 4.79B, 2). It also shares some features with osteolepiforms (and tetrapods), such as the processus dermintermedius in the anterior nostril (Fig. 4.88, 10), a large ventrolateral fenestra of the nasal cavity (Fig. 4.84B1, 2), the paired postparietals flanked by the supratemporals and tabulars (Panchen and Smithson 1987), and the polyplocodont structure of the folded teeth. Some of these characters are thus either homoplasies, or more probably general rhipidistian characters, but uncertainties remain as to which character should be allocated to which category.

Powichthys, from the Early Devonian of Arctic Canada, is more typically porolepiform-like in the

Fig. 4.85. Interrelationships of the Dipnomorpha. Left-hand side, according to Ahlberg (1991*a*), right-hand side, according to Chang and Smith (1992). Relationships above node 9 according to Schultze and Marshall (1993). *Terminal taxa* (B–F, represented by the skull-roof only): A, Porolepiformes; B, *Powichthys*; C, *Youngolepis*; D, *Diabolepis*; E, *Uranolophus*; F, *Dipnorhynchus*; G, *Dipterus*; H, *Griphognathus*; I, *Scaumenacia*; J, *Phaneropleuron*; K, *Neoceratodus*; L, Lepidosirenidae. *Nested taxa and selected synapomorphies*: 1, Dipnomorpha (more than four mesomeres in pectoral fin skeleton, pineal reached but not surrounded by parietals, no parietal–supraorbital contact, infraorbital sensory-line canal follows premaxilla suture, some rostral tubules in snout, enamel entering the pores of the cosmine, branching posterior radials in posterior dorsal fin); 2, Porolepiformes (dendrodont tooth structure, otic part of main lateral line passing through the centre of postparietals, no radials in anterior dorsal fin); 3, three large sensory pits on lower jaw; 4, Youngolepiformes (parieto-ethmoidal shield very long, reaching back to the level of the exit for trigeminus nerve); 5, no sensory-line canal in premaxilla, marginal tooth pavement, more rostral tubules in snout; 6, no complete intracranial joint, parietals do not reach pineal foramen or pineal region, loss of postorbital junction between supraorbital and infraorbital sensory lines, postotic division of lateral-line canal along skull-roof margin, no basicranial fenestra, enlarged prearticular, posterior nostril very close to mouth margin; 7, Dipnoiformes (entopterygoid and prearticular tooth plates made up of highly mineralized dentine or petrodentine, median 'B'-bone in skull-roof, tubules in snout and symphysial part of lower jaw, loss of maxilla); 8, Dipnoi (posterior nostril inside palate, loss of basal folding in teeth or denticles); 9, parasphenoid short, entopterygoid plates meeting along mid-line; 10, postparietals completely separated by 'B'-bone, pelvic fins elongated in the same manner as pectorals, rounded scales, large ossified vertebral centra (subsequently lost at node 12); 11, loss of anterior furrow (between dentary and tooth plates) in lower jaw; 12, enlarged posterior dorsal fin, loss of cosmine, lozenge-shaped parasphenoid, large ossified ribs; 13, caudal fin coalescent with dorsal fin, loss of dentary, cheek bones reduced; 14, no separate anal fin. *Age*: Devonian (A–I), Carboniferous–Permian (J), Cretaceous–Recent (K, L). (B–H, same sources as for Figs 4.80–4; I–L, based on Jarvik (1980).)

position of the nostrils, elongate postorbital bones, and large internasal cavities to accomodate the parasymphysial teeth (Fig. 4.84, 1). However, it shares with *Youngolepis* the broad and toothed parasphenoid (Fig. 4.84A, 3), which is probably a generalized osteichthyan feature, and peculiar sensory pits in the lower jaw. The supraorbital and main lateral line of *Powichthys* pass, as in *Youngolepis* and lungfishes, through a series of small dermal bones, laterally to the parietals and postparietals (Fig. 4.85B, C). Both *Powichthys* and *Youngolepis* have a descending process of the sphenethmoid (Fig. 4.84, 6), as in the Devonian actinopterygian *Mimia* (Fig. 4.66, 7). This process, which meets the lateral commissure of the otoccipital may be a general osteichthyan feature, but is unknown in other sarcopterygians. If its presence in these two taxa is not due to a reversion, then it may suggest that these are stem-group sarcopterygians rather than dipnomorphs. Chang and Smith (1992) have put *Powichthys* and *Youngolepis* on the side of the porolepiforms (Fig. 4.85, 3), and only *Diabolepis* on that of the lungfishes. Certain characters, such as the large sensory pits on the lower jaw, are admittedly unique to porolepiforms, *Powichthys*, and *Youngolepis*, but one may wonder if their absence in lungfishes cannot be due to the strongly modified jaw structure in the latter. I am rather inclined toward accepting an earlier phylogeny (Janvier 1986; Ahlberg 1991*a*), in which *Powichthys*, *Youngolepis*, and lungfishes form a morphocline showing an increased fusion of the two endocranial halves, loss of the sensory-line canal in the premaxilla, increased marginal tooth 'pavement' (on dentary, premaxilla, and maxilla), and increasing number of rostral tubules (Fig. 4.85, 5–7).

The phylogenetic classification (by phyletic sequencing) of the Dipnomorpha drawn from the cladogram in Fig. 4.85 (left-hand side) is shown in Table 4.10.

Further reading

Porolepiformes, Powichthys, and Youngolepis: Ahlberg (1989, 1991*a*); Bjerring (1991); Chang (1982, 1991*a*); Jarvik (1942, 1966, 1972, 1980); Jessen (1980); Kulczycki (1960); Véran (1988). *Dipnoi*: Campbell and Barwick (1982, 1984*a*,*b*, 1987, 1988*a*,*b*); Chang and Yu (1984); Denison (1968); Forster-Cooper (1937); Jarvik (1980); Miles (1977); Schultze and Campbell (1986); Smith and Chang (1990); Thomson and Campbell (1971). *Histology*: Bemis and Northcutt (1992); Borgen (1992); Gross (1956); Meinke (1984); Meinke and Thomson (1983); Ørvig (1969); Peyer (1968); Schultze (1969, 1970); Smith and Campbell (1987). Thomson (1975, 1977). *Interrelationships and systematics of the Dipnomorpha*: Ahlberg (1991*a*); Chang and Smith (1992); Schultze (1987, 1994). *Relationships, interrelationships, and systematics of the Dipnoi*: Campbell and Barwick (1990); Forey (1987*b*); Holmes (1985); Marshall (1987); Marshall and Schultze (1992); Miles (1975); Panchen and Smithson (1987); Rosen

Table 4.10.   Classification of the Dipnomorpha (See also Fig. 4.85)

Dipnomorpha
Porolepiformes
Porolepidae
Holoptychiidae
Powichthys
Youngolepis
Dipnoiformes
Diabolepis
Dipnoi
Uranolophus
Dipnorhynchidae
'Dipteridae'
Rhynchodipteridae
Scaumenacia
Phaneropleuron
Crown-group Dipnoi
Neoceratodus
Lepidosirenidae

et al. (1981); Schultze (1987, 1994); Schultze and Marshall (1993).

18.4 Rhizodontiformes

Although recorded from the Coal Measures as early as the 1830s, rhizodontiforms are still poorly known. This is largely due to the fact that their skull dermal bones are thin and loosely attached to each other, and were dispersed after decay. In addition, most rhizodontiforms are very large (some lower jaws can be over a metre in length), a condition that favours scattering during fossilization. Rhizodontiforms are characterized by the fact that the dermal fin-rays (lepidotrichiae) have a long unsegmented portion (Fig. 4.86, 25), which is covered by a layer of rounded scales, like those of the fin lobe (Fig. 4.86, 24, 31). The cleithrum has a widely expanded ventral blade that has both lateral and medial overlap areas for the clavicle (Fig. 4.87B, 5). The dermal bones are ornamented with bony vermiculations, and there is no cosmine. The folded teeth are of polyplocodont type (Fig. 4.86D), like osteolepiforms and tetrapods, though with only slightly sigmoid primary folds. In the lower jaw, the parasymphysial tooth-bearing plates are much reduced, as in osteolepiforms (Fig. 4.86, 21). The scales are large, rounded, ornamented with thin ridges, and bear a rounded knob on their internal surface (Fig. 4.86, 34). Other rhizodontiform characters are the very large extrascapulars (Fig. 4.86, 10, 11), which overlap in the same way as in osteolepiforms, and the elongated postparietals (Fig. 4.86, 7).

The braincase is partly known in *Notorhizodon*, where it clearly shows a complete intracranial joint and a large

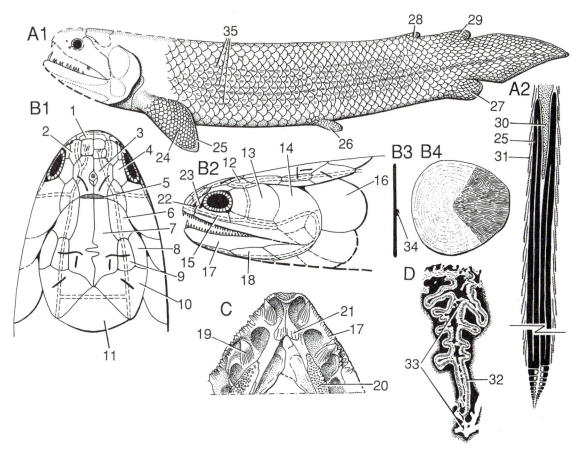

Fig. 4.86. Fossil Sarcopterygii. Rhizodontiformes. External morphology. A, *Strepsodus*, Early Carboniferous, reconstruction in lateral view (A1, ×0.5) and vertical section through a fin (A2, radials stippled, lepidotrichiae black); B, *Barameda*, Early Carboniferous of Australia, head in dorsal (B1) and lateral (B2) view (×0.7), and scale in transverse section (B3) and external view (B4); C, *Notorhizodon*, Late Devonian of Antarctica, lower jaw symphysis in dorsal view (×0.3); D, *Rhizodus*, Early Carboniferous, horizontal section through a fold of a tooth (×50). 1–21, *dermal bones of head*: 1, premaxilla; 2, nasal; 3, parietal; 4, supraorbital; 5, dermal intracranial gap; 6, supratemporal; 7, postparietal; 8, extratemporal; 9, tabular; 10, lateral extrascapular; 11, median extrascapular; 12, intertemporal; 13, postorbital; 14, squamosal; 15, lachrymal; 16, opercular; 17, dentary; 18, infradentaries; 19, coronoids; 20, prearticular; 21, parasymphysial plates; 22–34, *others*: 22, anterior nostril; 23, posterior nostril; 24, scale-covered part of pectoral fin lobe; 25, lepidotrichiae; 26, pelvic fin; 27, anal fin; 28, anterior dorsal fin; 29, posterior dorsal fin; 30, radial; 31, external scales extending over unsegmented division of lepidotrichs; 32, dentine (plicidentine); 33, attachment bone; 34, internal boss on scales; 35, accessory lateral lines. (A, from Andrews (1985); B, from Long (1989); C, from Young *et al.* (1992); D, from Schultze 1969).)

parasphenoid (Fig. 4.87A, 1, 2). The pectoral fin endoskeleton is known only in *Sauripterus* (Fig. 4.87C), *Rhizodus*, and *Barameda* (Fig. 4.87D). It differs radically from that of the dipnomorphs and shows a dichotomous branching pattern which recalls that of osteolepiforms (Fig. 4.91A), although here the radials also are branched, so that the pectoral basal lobe is further expanded into a solid paddle. The humerus, however, is similar to that of osteolepiforms, with a

rounded caput humeri and a large entepicondylar process (Fig. 4.87, 7, 8).

The anatomy of the nasal openings of the rhizodontiforms is still uncertain. There is a strong probability that there were two external nasal openings on each side (Fig. 4.86, 22, 23), but we do not know whether there was a choana. Current theories predict that there was none (if the posterior nostril is regarded as homologous to the choana).

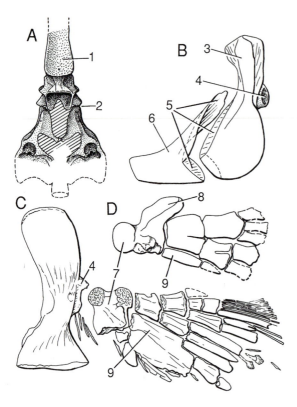

Fig. 4.87. Fossil Sarcopterygii. Rhizondontiformes. Internal anatomy. A, *Notorhizodon*, Late Devonian of Antarctica, ventral part of braincase and parasphenoid in ventral view (×0.3); B, *Rhizodus*, Early Carboniferous, shoulder girdle in lateral view (×0.2); C, *Sauripterus*, Late Devonian, shoulder girdle and pectoral fin in lateral view (×0.5); D, *Barameda*, Early Carboniferous of Australia, endoskeleton of pectoral fin in lateral view (×0.5). 1, parasphenoid; 2, intracranial joint; 3, cleithrum; 4, glenoid fossa of scapulocoracoid; 5, cleithroclavicular overlap; 6, clavicle; 7, proximal mesomere (humerus); 8, entepicondyle; 9, foremost preaxial radial (radius). (A, B, from Young *et al.* (1992); C, from Andrews and Westoll 1970*b*); D, from Long (1989).)

The interrelationships of the Rhizodontiformes have been recently worked out by Young *et al.* (1992), although the few genera hitherto recorded look quite similar, and are generally known only from very fragmentary material. *Notorhizodon* would be the sister-group of all other rhizodontiforms.

As pointed out by Andrews (1985), the rhizodontiforms were certainly large predators, and their stout body suggests that they waited for their prey while ambushed among sunken roots and twigs. Rhizodontiforms are Late Devonian to Carboniferous in age.

Further reading

Andrews (1985); Andrews and Westoll (1970*b*); Long (1989), Young *et al.* (1992).

4.18.5 Osteolepiformes

The Osteolepiformes, as defined here, are a probably monophyletic group of fossil sarcopterygians with folded teeth, and are characterized by an enlarged axillary scale, or scute, on the posterior margin of the basal lobe of paired and unpaired fins (Fig. 4.88, 1). Osteolepiforms have, however, long been regarded as ancestral to tetrapods, because of the structure of the endoskeleton of their paired fins, and also because they are supposed to have possessed choanae (Fig. 4.89A1).

The best-known osteolepiform is the Late Devonian *Eustenopteron* (Fig. 4.88C), thanks to the detailed anatomical investigations by Jarvik (1942, 1980). There are numerous other osteolepiform genera in which some internal anatomical structures are preserved (e.g. *Osteolepis*, *Ectosteorhachis*, *Megalichthys*, *Sterropterygion*, and *Gogonasus*), but none of them has yielded an extensive, articulated endoskeleton. From what we know of this group, it seems to display a wider diversity than porolepiforms and rhizodontiforms. However, like porolepiforms, it comprises cosmine-covered and non-cosmine-covered forms, the latter condition being assumed to be derived from the former. The classical genera *Osteolepis*, *Ectosteorhachis*, and *Eusthenopteron* still remain the best models to illustrate the varied morphology of osteolepiforms, but the discussion on relationships will include a number of recently discovered and quite different taxa.

4.18.5.1 Exoskeleton and external morphology

In overall aspect, the differences between *Osteolepis* and *Eusthenopteron* are mainly in the proportions of the body and fins (Fig. 4.88A1, C1). *Osteolepis* is stout, with rounded fins and a broad and blunt snout, whereas *Eusthenopteron* is slender, with fins that are more triangular, posteriorly placed dorsal fins, and a more pointed snout. Juvenile specimens of *Eusthenopteron* (Fig. 4.88C2) have a shorter snout, which indicates that the adult morphology is derived in relation to that of *Osteolepis*. The tail of *Osteolepis* is epicercal, whereas that of *Eusthenopteron* tends toward an apparent diphycercy by enlargement of the web on the dorsal edge of the notochordal lobe (Fig. 4.88, 4). The tail of *Eusthenopteron* clearly represents a derived condition, which also occurs in other forms, such as *Gyroptychius* (Fig. 4.88B). The diamond-shaped scales and dermal bones of *Osteolepis* are coated with a shining layer of cosmine (Fig. 4.88A2, A3) and have a small dorsal peg (Fig. 4.88A2, 13), whereas the

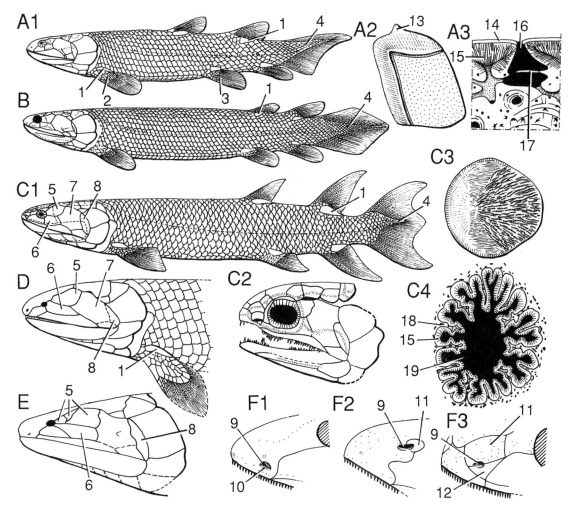

Fig. 4.88. Fossil Sarcopterygii. Osteolepiformes. External morphology. A, *Osteolepis*, Middle Devonian, reconstruction in lateral view (A1, ×0.5), scale in external view (A2), and vertical section of the cosmine (A3, ×100); B, *Gyroptychius*, Middle Devonian, reconstruction in lateral view (×0.5); C, *Eusthenopteron*, Late Devonian, reconstruction in lateral view (C1, ×0.3), head of a juvenile individual in lateral view (C2, ×3), scale in external view (C3), and horizontal section through a tooth (C4, ×30); D, *Koharalepis*, Late Devonian of Antarctica, reconstruction of head and pectoral fin in lateral view (×0.2); E, *Canowindra*, Late Devonian of Australia, reconstruction of head in lateral view (×0.5); F, lateral view of the snout of *Gyroptychius* (F1), *Megalichthys* (F2) and *Eusthenopteron* (F3). 1, basal, or axillary, fin scute; 2, pectoral fin; 3, pelvic fin; 4, chordal lobe of caudal fin; 5, postorbitals; 6, jugal; 7, squamosal; 8, preopercular; 9, anterior nostril; 10, dermintermedial process; 11, anterior tectal; 12, lateral rostral; 13, anterodorsal process of scale; 14, enamel; 15, dentine or plicidentine; 16, flask-shaped cavity; 17, horizontal septum; 18, attachment bone; 19, pulp cavity. (A1, A2, B, C1, C3, from Jarvik (1948a, 1980); A3, from Gross (1956); C2, from Schultze (1984a); C4, from Schultze (1969); D, from Young *et al.* (1992); E, from Long (1985a); F, from Jarvik (1972).)

rounded scales and dermal bones of *Eusthenopteron* have a tuberculate or vermiculate ornamentation (Fig. 4.88C3). There are also some minor differences in the organization of the dermal bones of the head, such as the presence of an 'extratemporal' (Fig. 4.92, 1) and a lateral rostral in contact with the supraorbito-tectal bone in *Osteolepis*. In both

fishes, the snout is covered with numerous small dermal bones, but their external sutures are not visible in *Osteolepis* because of the extensive cosmine covering. The cheek of these osteolepiforms is also characteristic in showing relatively short postorbitals and jugals, with a large squamosal, a quadratojugal, and a bar-shaped preop-

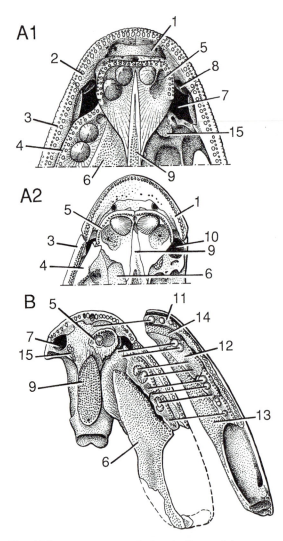

Fig. 4.89. Fossil Sarcopterygii. Osteolepiformes. Palate.
A, *Eusthenopteron*, Late Devonian, two different re
constructions of the anterior part, assuming that the choanae do
(A1) or do not (A2) exist. B, *Gogonasus*, Late Devonian of
Australia, palate and lower jaw showing the position of the pits
that housed the respective tusks. 1, apical fossa; 2, premaxilla;
3, maxilla; 4, dermopalatine; 5, vomer; 6, entopterygoid;
7, choana; 8, intermedial process of nasal cavity;
9, parasphenoid; 10, ventral fenestra of nasal cavity;
11, dentary; 12, coronoid; 13, prearticular; 14, parasymphysial
toothed plate; 15, ethmoid articulation for palatoquadrate.
(A1, from Bjerring (1987); A2, from Rosen *et al.* (1981);
C, from Long (1988*a*).)

ercular (Fig. 4.88A1, C1; to locate these bones, use the
pattern in porolepiforms, Fig. 4.77A2). However, the
canowindriids, which are regarded here as generalized

osteolepiforms, retain elongated postorbitals and jugals
(Fig. 4.88D, E, 5, 6), like *Porolepis*. The opercular and
subopercular are relatively large in all osteolepiforms.

In the skull-roof, the pineal foramen lies well between
the parietals (Fig. 4.92). The contact between the ethmosphenoid and otico-occipital divisions is straight but
shows no indication of pronounced mobility (Fig. 4.92,
9). In some taxa (e.g. *Gyroptychius*), it displays an asymmetry which suggests that there was no dermal joint at all.
The median extrascapular is overlapped by the lateral
ones (Fig. 4.92, 7, 8).

Osteolepiforms have a single pair of external nostrils,
currently interpreted as the anterior nostril of other fishes
(Fig. 4.88, 9). Rosen *et al.* (1981) have suggested, however,
that the two (anterior and posterior) external nostrils may
have been lodged inside the same opening, which is often
slit-shaped (Fig. 5.4A). Each nostril is lined internally by a
dermintermedial process (Fig. 4.88, 10). The dermal bones
covering the internal side of the mouth display the general
osteichthyan pattern (Fig. 4.89). In *Osteolepis* and the
closely related *Gogonasus*, the parasphenoid displays the
generalized condition met with in *Youngolepis* or *Mimia*,
i.e. it is broad and denticle-covered (Fig. 4.89B, 9). In contrast, that of *Eusthenopteron* is thin, slender, and penetrates
between the two posterior processes of the vomers (Fig.
4.89A, 9). The space traditionally referred to as the choana
(Fig. 4.89, 7) is situated at the junction of four dermal
bones, the vomer, dermopalatine, premaxilla and maxilla,
and lies ventrally to a large opening in the nasal cavity (Fig.
4.89A2, 10; Fig. 4.90, 19). Rosen *et al.* (1981) considered
that this space is not a choana, but a mere hinge between the
palate and the snout, and was probably closed by soft
tissues (as exemplified by some actinopterygians) or acco-
modated one of the coronoid fangs of the lower jaw.
Although this interpretation has been rejected as an offence
to one of the greatest conquests of evolutionary palaeontology (e.g. Jarvik 1981*b*; Holmes 1985; Schultze 1991*a,b*),
an uncertainty remains by the virtue of the nature of fossils
(the hypothesis that a space in a fossil was closed by soft
tissue cannot be refuted). The position of the anterior coronoid fangs in an excellently preserved specimen of
Gogonasus shows, however, that they fit medially to the
presumed choana, and not in it (Fig. 4.89B).

In the lower jaw, the parasymphysial tooth-bearing
plates (Fig. 4.89B, 14) are much reduced in osteolepiforms, and, consequently, there is no deep internasal
fossa. Only a short apical fossa (Fig. 4.89A, 1) accommo-
dates the foremost dentary teeth.

The dermal shoulder girdle is slender (Fig. 4.91A, 1, 2)
and has a rather broad cleithrum with an ornamented
anocleithrum (which is thus not subdermal). Its structure
is that of generalized osteichthyans, i.e. with a clavicle
and interclavicle.

4.18.5.2 *Endoskeleton*

In *Osteolepis*, *Ectosteorhachis*, and *Eusthenopteron* the neurocranium looks much the same. The differences between the three genera in the breadth of the snout are reflected by the position of the nasal cavities, which are more closely set in *Eusthenopteron* than in *Osteolepis* or *Ectosteorhachis* (Fig. 4.90E, 28). The brain cavity shows a single pineal canal (Fig. 4.90, 30) and a deep pituitary cavity. The basicranial muscle (Fig. 4.90B, 22) was prob-

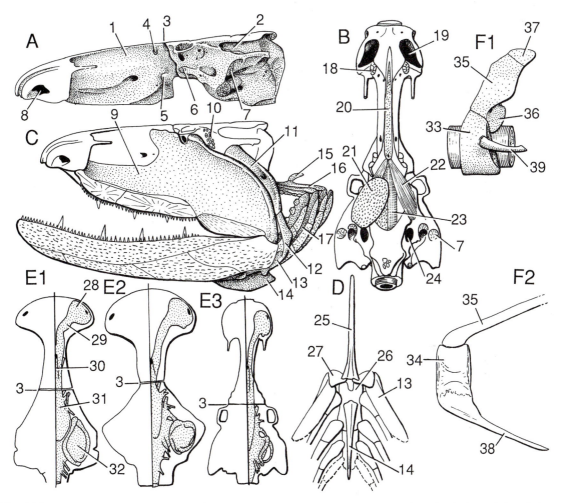

Fig. 4.90. Fossil Sarcopterygii. Osteolepiformes. Internal anatomy. A–D, *Eusthenopteron*, Late Devonian, braincase in lateral view (A) and ventral view (B, parotic plate, notochord, and basicranial muscle reconstructed), braincase, palatoquadrate, lower jaw, and hyobranchial apparatus in lateral view (C), and ventral part of hyobranchial apparatus in ventral view (D); E, dorsal outline of the braincase and its internal cavities (stippled) in *Osteolepis* (E1), *Ectosteorhachis* (E2), and *Eusthenopteron* (E3); F, anterior vertebra of *Eusthenopteron* (F1, notochord reconstructed) and posterior vertebra of *Megalichthys* (F2) in lateral view. 1, ethmosphenoid; 2, otoccipital; 3, intracranial joint; 4, suprapterygoid process; 5, basipterygoid process; 6, paratemporal articulation; 7, articular facet for hyomandibula; 8, anterior nostril; 9, palatoquadrate; 10, spiracular platelets; 11, hyomandibula; 12, interhyal; 13 ceratohyal; 14, urohyal; 15, pharyngobranchial; 16, epibranchial; 17, ceratobranchial; 18, ethmoid articulation for palatoquadrate; 19, presumed endochoanal opening; 20, parasphenoid; 21, parotic plate; 22, basicranial muscle; 23, notochord; 24, vestibular fontanelle; 25, sublingual rod; 26, basibranchials; 27, hypohyal; 28, nasal cavity; 29, olfactory tract; 30, pineal canal; 31, brain cavity; 32, labyrinth; 33, ventral vertebral arch (intercentrum); 34, enlarged ventral vertebral arch (intercentrum and ? pleurocentrum); 35, neural arch (basidorsal); 36, interdorsal (pleurocentrum); 37, supraneural; 38, haemal arch (basiventral); 39, rib. (A, C, D, F1, from Jarvik (1980); B, from Bjerring (1967); E, from Thomson (1965); F2, from Andrews and Westoll (1970*a*).)

ably much shorter than in the extant coelacanth and was covered by large parotic dermal plates (Fig. 4.90B, 21). The palatoquadrate is more expanded posteriorly in *Osteolepis* than in *Eusthenopteron*, but the organization of the visceral skeleton is basically the same in both genera (Fig. 4.90C, D). The palatoquadrate is attached to the braincase by means of at least four major articulations, the usual ethmoid and basipterygoid (basal) articulations (Fig. 4.89, 15; Fig. 4.90, 5, 18) and also by the suprapterygoid and paratemporal articulations (Fig. 4.90A, 4, 6), the last-named being unique to osteolepiforms. The double-headed hyomandibula lines the posterior edge of the palatoquadrate and is articulated on the lateral commissure, as in most generalized osteichthyans (Fig. 4.90, 7, 11). The spiracular canal passed between the hyomandibula and the palatoquadrate, and was lined with minute dermal bones (Fig. 4.90, 10). There are five branchial arches that attach on two basibranchial elements (Fig. 4.90, 15–17, 26). The basibranchial series is prolonged anteriorly by a thin sublingual rod (Fig. 4.90, 25), and is underlain by a blade-shaped urohyal (Fig. 4.90, 14). For an extensive description of the endoskeleton of the head of osteolepiforms, see Jarvik (1980) for *Eusthenopteron*, Romer (1937) for *Ectosteorhachis*, and Thomson (1965) for *Osteolepis*. Additional information on other osteolepiform taxa is given in Vorobjeva (1977), Long (1985*a–c*), and Young *et al.* (1992).

Each vertebra consists of the two basic dorsal arcualia, a small interdorsal (or pleurocentrum), and a large basidorsal (Fig. 4.90, 35, 36), which is connected to a hemicylindrical perichordal centrum, or ventral vertebral arch (Fig. 4.90F1, 33). In *Osteolepis*, *Megalichthys*, and *Ectosteorhachis*, the basidorsal and basiventral are firmly fused to a ring-shaped centrum (Fig. 4.90F2, 34). The ribs are short and articulate with the centrum (Fig. 4.90, 39). Supraneurals are present only in the anterior part of the vertebral column (Fig. 4.90F1, 37).

The endoskeleton of the unpaired fins consists of radials that articulate separately on basal plates or rods which, in turn, are in contact with the vertebral column.

The scapulocoracoid (Fig. 4.91A, 8) is tri-radiate in shape, with large supraglenoid and supracoracoid foramina (Fig. 4.91, 9, 10). In early osteolepids (Fig. 4.91B, 8), the body of the scapulocoracoid is dorsoventrally flattened, with a transversely elongated glenoid fossa, thereby recalling the condition in porolepiforms and *Youngolepis* (Fig. 4.79A, B, 2). In contrast, the area for the articulation of the pectoral fin skeleton (glenoid fossa) is pear-shaped in *Eusthenopteron* (Fig. 4.91A, 11). The endoskeleton of the pectoral fin has been described at length and seems to be quite homogeneous among all osteolepiforms in which it is preserved (Jarvik 1980). Its axis consists of four mesomeres, to which are connected

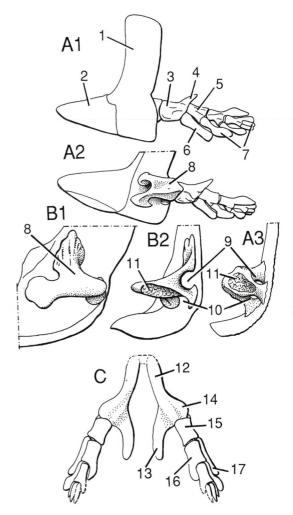

Fig. 4.91. Fossil Sarcopterygii. Osteolepiformes. Shoulder and pelvic girdles, paired fin skeleton. A, *Eusthenopteron*, Late Devonian, shoulder girdle and pectoral fin skeleton in lateral (A1), medial (A2) and posterior (A3) view (×1). B, scapulocoracoid of an early and supposedly generalized osteolepiform, from the Early or Middle Devonian of Iran, in medial (B1) and posterior (B2) view (×3). C, pelvic girdle and pelvic fin skeleton of *Eusthenopteron* in dorsal view (C2, ×1). 1, cleithrum; 2, clavicle; 3, proximal pectoral mesomere (humerus); 4, entepicondylar process; 5, second pectoral mesomere (ulna); 6, first pectoral metapterygial radial (radius); 7, distal radials; 8, scapulocoracoid; 9, supraglenoid foramen; 10, supracoracoid foramen; 11, glenoid fossa; 12, pubic process; 13, ischiadic process; 14, iliac portion; 15, proximal pelvic mesomere (femur); 16, second pelvic mesomere (fibula); 17, first pelvic metapterygial radial (tibia). (A, C, from Jarvik (1980); B, from Janvier (1980*a*).)

at least four or five preaxial radials. All the mesomeres, except the second one, are produced into large postaxial processes, which may be interpreted as postaxial radials, fused to the metapterygial axis (Fig. 4.91, 4). The proximal mesomere, or humerus (Fig. 4.91, 3), is quite similar in the rhizondontiforms, osteolepiforms, panderichthyids, and tetrapods. Its various crests and foramina have accordingly been interpreted in the light of those in the tetrapod humerus. It has a bulged caput humeri, and its large dorsal process is referred to as an entepicondylar process (Fig. 4.91, 4); other ridges are interpreted as supinator and ectepicondylar processes. The second mesomere and first preaxial radial are currently regarded as homologous to the ulna and radius of the tetrapods (Fig. 4.91, 5, 6).

The pelvic girdle (Fig. 4.91, 12–14) is a more or less tri-radiate structure with a striking resemblance to a proximal mesomere; hence Panchen and Smithson's (1990) suggestion that it is serially homologous to the humerus, not to the scapulocoracoid (see discussion, Chapter 6, p. 268). Its has pubic and ischiadic processes (Fig. 4.91, 12, 13), and a shallow iliac portion (Fig. 4.91, 14), which is not connected to the vertebral column. The pelvic fin skeleton is basically similar to that of the pectoral fin, but here the proximal mesomere has no postaxial process (Fig. 4.91, 15). The second mesomere and first preaxial radial are also interpreted as homologues of the tibia and fibula of tetrapods (Fig. 4.91, 16, 17).

4.18.5.3 Histology

In cosmine-covered osteolepiforms ('osteolepids', megalichthyids, and some canowindriids), the cosmine layer, which covers the dermal bones and scales, looks smoother than in dipnomorphs, since its pores are not bounded by a crescent-shaped ridge (Fig. 4.88A2). The pores of the cosmine are smaller and more scattered than in the dipnomorphs, and the flask-shaped cavities are generally not lined internally with enamel (Fig. 4.88A3, 14, 16). In addition, each cavity is divided by a horizontal septum (Fig. 4.88A3, 17). In megalichthyids, the resorbtion of the cosmine can be considerable, leaving only islets of cosmine. In all non-cosmine-covered osteolepiforms, the dermal bones and scales are ornamented with bony tubercles, ridges, or vermiculations.

The folded teeth are of either polyplocodont or eusthenodont type. In both types the bone of attachment penetrates between the primary folds of the dentine (Fig. 4.88, 21), but in the eusthenodont type (*Eusthenodon*), the pulp cavity is also largely filled with bone or osteodentine.

4.48.5.4 Diversity and phylogeny

Although they comprise a relatively large number of species, the osteolepiforms are rather homogenenous in overall morphology (Fig. 4.93). Two major clades are recognized: the Megalichthyidae (Fig. 4.93, 4) and the Tristichopteridae (or Eusthenopteridae; Fig. 4.93, 5). The Megalichthyidae (*Megalichthys*, *Ectosteorhachis*, *Mahalalepis*, and *Megistolepis*) are large, cosmine-covered forms with a more dorsally placed external nostril (Fig. 4.88F2, 9), large tusks on the premaxillae, transversally elongated vomers, and closed pineal foramen. The Tristichopteridae (*Marsdenichthys*, *Eusthenodon*, *Eusthenopteron*, and *Jarvikina*) are characterized by the lack of contact between the lateral rostral (Fig. 4.88F3, 12) and the supraorbito-tectal, or posterior tectal, and they are devoid of cosmine. Some tristichopterids, such as *Eusthenodon* (Fig. 4.93I) could reach a very large size (over two metres in length). All other cosmine-covered osteolepiforms generally referred to as 'osteolepids', such as *Osteolepis*, *Thursius*, *Gyroptychius*, and *Glyptopomus* (Fig. 4.93B–E), cannot be united into a clade and are probably paraphyletic, although their precise relationships with the megalichthyids and tristichopterids are not yet elucidated. Long (1985b) has tried to define 'trends' in osteolepiform evolution, in particular as to the processes involved in the elongation of the head. In tristichopterids, the ethmosphenoid undergoes a considerable elongation from the condition in *Marsdenichthys* (Fig. 4.92C1) to that in *Eusthenodon* (Fig. 4.92C2), whereas in 'osteolepids', it is the otoccipital that becomes elongated, from the condition in *Gyroptychius* (Fig. 4.92B1) to that in *Glyptopomus* (Fig. 4.92B2). How much these trends reflect the phylogenetic pattern is not clear.

The interrelationships of osteolepiforms have been recently enlightened by the discovery of some peculiar froms, referred to as the 'canowindriids' (*Canowindra*, *Beelarongia*, and *Koharalepis*), in the Middle and Late Devonian of Australia and Antarctica (Long 1985c, 1987; Young *et al.* 1992). These forms have very small orbits, large postorbitals and jugals (Fig. 4.88D, E, 5, 6), and broad postparietal shield, with an extratemporal (Fig. 4.92A, 1). These features, which recall porolepiforms (*Canowindra* had been first referred to the porolepiforms by Thomson (1973)), are most probably general for rhipidistians (i.e., all sarcopterygians with folded teeth). However, Young *et al.* (1992) regard the canowindriids as a clade, and put them as the sister-group of all other osteolepiforms plus panderichthyids and the tetrapods. Since the pectoral fin of the canowindriid *Koharalepis* is reconstructed with a large axillary scute (Fig. 4.88D, 1), I here regard this group as the sister-group of only all other osteolepiforms (Fig. 4.93, 1). Vorobyeva (1977) described a number osteolepiforms, such as *Thysanolepis*, *Viliuichthys*, and *Lamprotolepis*, which are still poorly known and therefore cannot be considered in this phylogeny. Finally, Chang and Zhu (1993) described a

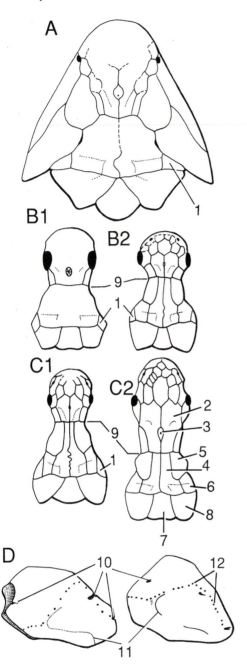

Fig. 4.92. Fossil Sarcopterygii. Osteolepiformes. Diversity in skull-roof morphology. A, *Koharalepis*, Middle–Late Devonian of Antarctica, head in dorsal view; B, Osteolepididae: *Gyroptychius* (B1), Middle Devonian, and *Glyptopomus* (B2), Middle Devonian. C, Tristichopteridae: *Marsdenichthys* (C1), Late Devonian of Australia, and *Eusthenodon* (C2), Late Devonian of Europe and North America; note the supposedly reverse 'trend' in the two groups, with the elongation of the sphenethmoid in tristichopterids, and of the otoccipital in osteolepids. D, 'cheek plate' (fused squamosal, quadratojugal, and preopercular) in the early osteolepiform *Kenichthys* (left), Middle Devonian of China, and the dipnomorph *Youngolepis* (right), Early Devonian of China. 1, extratemporal; 2, parietal; 3, pineal foramen; 4, postparietal; 5, supratemporal; 6, tabular; 7, median extrascapular; 8, lateral extrascapular; 9, dermal intracranial 'joint' or parietal-postparietal suture; 10, large sensory pits; 11, pit line; 12, pores of the jugal sensory line. (A, from Young *et al.* (1992); B, C, from Long (1985*b*); D, from Chang (1991*a*) and Chang and Zhu (1993).)

Middle Devonian form from China, *Kenichthys*, which, strangely, allies typical osteolepiform and *Youngolepis*-like features. It possesses a single, slit-shaped external nostril, an osteolepiform type of skull-roof (with the postorbital straddling the dermal intracranial joint), and small

parasymphysial (adsymphysial) toothed plates. In contrast, it shares with *Youngolepis* a free premaxilla, dorsally lined by the intrasutural infraorbital canal, and the cheek bones (squamosal, jugal, quadratojugal) fused into a single plate (Fig. 4.92D). It seems to have possessed an axillary scute, which would make it an osteolepiform, as defined here. If its *Youngolepis*-like characters are regarded as general rhipidistian characters, then it may be a very generalized ostolepiform, quite close morphologically to the common ancestor of the dipnomorphs and tetrapodomorphs (see Chapter 5).

The classification shown in Table 4.11 can be derived from the cladogram in Fig. 4.93:

Table 4.11. Classification of the Osteolepiformes
(See also Fig. 4.93)

Osteolepiformes
  Canowindridae
  (unnamed taxon)
    'Osteolepididae' *sedis mutabilis*
    Megalichthyidae *sedis mutabilis*
    Tristichopteridae *sedis mutabilis*
      Marsdenichthys
      Tristichopterinae
        Eusthenopteron
        Eusthenodon

4.18.5.5 Stratigraphical and geographical distribution, habitat

The earliest known osteolepiforms are of latest Early Devonian or early Middle Devonian in age (Emsian or Eifelian), and are represented by cosmine-covered 'osteolepids'. The greatest diversity of osteolepiforms had been reached by the end of the Middle Devonian and the beginning of the Late Devonian with a relatively large number of 'osteolepid' and tristichopterid species. In the Carboniferous and Early Permian, only the large megalichthyids (*Megalichthys*, *Ectosteorhachis*; Fig. 1.22, 1) and the poorly known rhizodopsids survive (Schultze and Heidke 1986). It is noteworthy that the canowindriids, which are probably the sister-group of all other osteolepiforms, are so far known only in the late Middle and Late Devonian of East Gondwana (Australia, Antarctica). Long (1990) has therefore suggested that this area was a centre of origin for all osteolepiforms, and even for panderichthyids and tetrapods. Most osteolepiforms were marine fishes, although those found in arenaceous sediments or coal deposits have long been regarded as freshwater.

Further reading

Anatomy: Andrews and Westoll (1970*a*), Bjerring (1987); Chang and Zhu (1993); Janvier (1980*a*); Jarvik (1942, 1944, 1948*a*, 1952, 1954, 1964, 1966, 1980, 1981*a,b*); Jessen (1973); Long (1985*a–c*, 1987); Rackoff (1980); Romer (1937); Thomson (1964, 1965, 1973); Vorobyeva

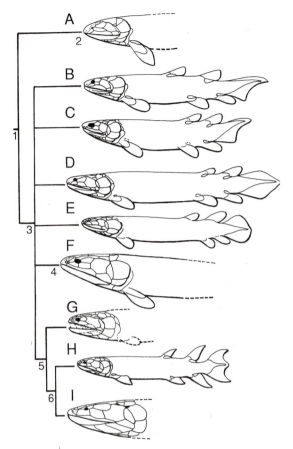

Fig. 4.93. Interrelationships of the Osteolepiformes (in part according to Young *et al.* (1992).) *Terminal taxa*: A, Canowindridae (*Koharalepis*); B, *Osteolepis*; C, *Thursius*; D, *Gyroptychius*; E, *Glyptopomus*; F, Megalichthyidae (*Ectosteorhachis*); G, *Marsdenichthys*, H, *Eusthenopteron*; I, *Eusthenodon*. *Nested taxa and selected synapomorphies*: 1, Osteolepiformes (large basal, or axillary fin scutes, large dermal anocleithrum between post-temporal and cleithrum); 2, Canowindridae (jugal at least twice as long as high, postorbital bones excluded from orbital margin); 3, 'higher' osteolepiforms (similar size of lachrymal, jugal and postorbital bones, smaller lateral extrascapulars); 4, Megalichthyinae (slit-like external nares, posterior nasals notched in frontals, loss of pineal foramen); 5, Tristichopteridae (lateral rostal not in contact with supraorbito-tectal); 6, Tristichopterinae (vomers with very long posterior processes extending posteriorly at the same level as the parasphenoid). *Age*: Middle Devonian (A–E); Late Devonian (G–I); Late Devonian to Early Permian (F). (A, after Young *et al.* (1992); B–E, after Jarvik (1948*a*); F, based on Thomson (1975); H, I, after Jarvik (1952, 1980).)

(1977); Young *et al.* (1992). *Histology*: Gross (1956); Schultze (1969, 1970); Thomson (1975). *Interrelationships and systematics*: Jarvik (1980); Long (1985*b*); Vorobyeva (1977); Young *et al.* (1992).

4.18.6 Panderichthyida

Panderichthyids are a small group of Late Devonian sarcopterygians, characterized by peculiar thickenings of the postfrontals ('eyebrows') medially to the orbits (Fig. 4.94, 41), a large median rostral that separates the premaxillae (Fig. 4.94, 15), paired posterior postrostrals, a very large median gular plate, and a lateral recess in the nasal capsule. They are represented by only two or three genera: *Panderichthys*, *Elpistostege*, and perhaps *Obruchevichthys* (regarded by some as a tetrapod). Their overall aspect, known only in *Panderichthys* from the Late Devonian of Latvia (Fig. 4.94A), is quite different from that of most other sarcopterygians. This is due to the fact that they share with the tetrapods a number of uniquely derived features, such as the flattened skull, posterodorsally placed orbits, paired frontals (Fig. 4.94, 2), jugal meeting the quadratojugal (Fig. 4.94, 10, 11), 'labyrinthodont' plicidentine (Fig. 4.94E), no dorsal and anal fin, and dermal bones ornamented with vermicula-

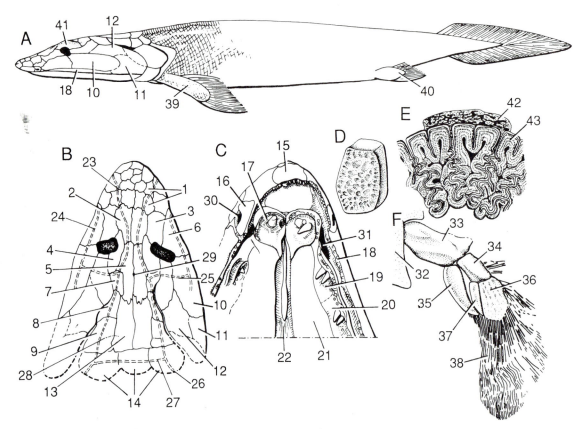

Fig. 4.94. Fossil Sarcopterygii. Panderichthyida. *Panderichthys*, Late Devonian of Latvia. A, reconstruction in lateral view (×0.2); B, skull in dorsal view; C, anterior part of the palate in ventral view; D, isolated scale in external view; E, portion of a horizontal section through a main tusk (×40); F, skeleton of the pectoral fin in ventral view. 1–22, *dermal bones*: 1, nasals; 2, frontal; 3, prefrontal; 4, postfrontal; 5, parietal; 6, lachrymal; 7, intertemporal; 8, supratemporal; 9, tabular; 10, jugal; 11, quadratojugal; 12, squamosal; 13, postparietal; 14, extrascapulars; 15, median rostral; 16, lateral rostral; 17, vomer; 18, maxilla; 19, dermopalatine; 20, ectopterygoid; 21, entopterygoid; 22, parasphenoid; 23–8, *sensory-line canals*; 23, supraorbital; 24, infraorbital; 25, postorbital; 26, main lateral line; 27, supratemporal commissure; 28, middle (postparietal) pit line; 29–43, *others*: 29, pineal foramen; 30, external (anterior) nostril; 31, choana; 32, scapulocoracoid; 33, proximal mesomere (humerus); 34, second mesomere (ulna); 35, foremost radial (radius); 36, distal plate (fused distal mesomeres and radials); 37, second radial; 38, lepidotrichs; 39, pectoral fin; 40, pelvic fin; 41, 'eyebrows' (ridges) in postfrontal; 42, attachment bone; 43, dentine (plicidentine). (A–C, based on Vorobyeva and Schultze (1991); D, from Vorobyeva and Obruchev (1964); E, from Pander (1860); F, from Vorobyeva (1992).)

tions. These resemblances, pointed out earlier by Gross (1964), have since been further emphasized (Schultze and Arsenault 1985; Schultze 1994; Vorobyeva and Schultze 1992) in support of the hypothesis that panderichthyids are the closest fossil relatives of tetrapods. As will be seen in the next chapter (p. 246), this interpretation now seems well corroborated, in spite of some anatomical data that are not consistent with the classical view on the fish–tetrapod transition.

4.18.6.1 Exoskeleton and external morphology

The head of *Panderichthys* is strikingly tetrapod-like in the position of the orbits (which have a peculiar elliptical shape), the depressed snout, the presence of four median pairs of bones in the skull-roof (including frontals, parietals and postparietals; Fig. 4.94, 2, 5, 13), the lack of a dermal intracranial joint, the vermiculate ornamentation, and the jugal-quadratojugal contact (Fig. 4.94, 10, 11). However, it remains fish-like in having a well-developed operculo-gular bone series, extrascapulars (Fig. 4.94A, B, 14), a spiracular slit, and sensory-line canals enclosed in the bones. The palate is quite similar to that of *Eusthenopteron*, with posteriorly pointed vomers, a narrow parasphenoid, and elongated openings regarded as choanae (Fig. 4.94, 17, 22, 31). The external nostrils lie close to the oral margin, as in most early tetrapods (Fig. 4.94, 30; Fig. 4.95, 38). (See Fig. 4.94B, C for the names of other dermal bones and sensory lines).

The dermal shoulder girdle comprises a clavicle and a large, dorsally expanded cleithrum connected to the skull-roof by means of the anocleithrum, supracleithrum, and post-temporal, as in other piscine sarcopterygians. The scales are comparatively small, diamond-shaped (Fig. 4.94D), and bear the same vermiculate ornamentation as the head. The fins are covered with lepidotrichiae arranged in branching rows (Fig. 4.94, 38).

18.6.2 Endoskeleton

Unfortunately, little is known of the endoskeleton of panderichthyids. In the head, the palatoquadrate, hyomandibula, and urohyal seem to have been quite similar to those in the osteolepiform *Eusthenopteron*. The braincase is still virtually unknown, but the nasal cavities are small and elongated in shape, with a dermintermedial process, as in osteolepiforms. The scapulocoracoid is large and massive (Fig. 4.94, 32), with three buttresses, as in osteolepiforms and lungfishes, but these converge laterally to form a large bony plate against the internal surface of the cleithrum. The supraglenoid and supracoracoid foramina are thus completely surrounded by the endoskeleton, as in tetrapods. The endoskeleton of the paired fins is known only in the pectoral fin. It resembles basically that of osteolepiforms, with preaxial radials only,

but the humerus is comparatively more elongated in shape (Fig. 4.94, 33), and the distal mesomeres and radials are fused into a large bony plate (Fig. 4.94, 36). The vertebrae comprise paired neural arches (basidorsals), which are sutured to a large perichordal ventral arch. Large ribs are also fused to both the ventral and neural arches.

4.18.6.3 Histology

The histological structure of the teeth of *Panderichthys* resembles that of the early tetrapod teeth (in particular *Ichthyostega*) in being of 'labyrinthodont' type, that is, with a polyplocodont type of folding and almost no attachment bone between the folds (Fig. 4.94, 42).

18.6.4 Diversity, distribution, and habitat

The two best-known panderichthyids, *Panderichthys* and *Elpistostege*, are similar in overall aspect (general dermal bone pattern, position of orbits, flattened snout, 'eyebrows'), but the orbits are small and rounded in *Elpistostege*. The two genera also differ slightly in the proportions and pattern of the dermal bones of the skull, but it is reasonable to consider the group as a clade. Panderichthyids are known exclusively from the Late and perhaps Middle Devonian of the Baltic and Quebec.

Further reading

Anatomy: Schultze and Arsenault (1985); Vorobyeva and Schultze (1991); Vorobyeva (1992); Vorobyeva and Kuznetzov (1992). *Histology*: Pander (1860); Schultze (1969, 1970).

4.18.7 Tetrapoda

The definition of the tetrapods is not much modified when fossils are added to a phylogeny based on extant taxa. The most spectacular of these modifications concern the number of digits and the dermal bones of the cheek. When Jarvik (1965a) suggested that the foot of the Late Devonian tetrapod *Ichthyostega* had six, or perhaps seven, digits (Fig. 4.95F, 56; including the prehallux and postminimus), and that its dermal skull retained a preopercular, subopercular, and lateral rostral (Fig. 4.95, 15, 19, 20), he received sceptical comments from many other early tetrapod specialists. Later, the hands of two other Late Devonian tetrapods, *Tulerpeton* (Fig. 4.97B) and *Acanthostega* (Fig. 4.96D), turned out to have six and eight digits respectively. In addition, the Early Carboniferous tetrapod *Crassigyrinus* also retains a preopercular in the cheek (Fig. 4.99, 4). The tail of *Ichthyostega* is also different from all other known tetrapod tails in possessing, like the fish tail, endoskeletal radials, scales, and dermal fin rays (lepidotrichiae) (Fig. 4.95, 45, 46). These features not only show that the tetrapods are osteichthyans, but

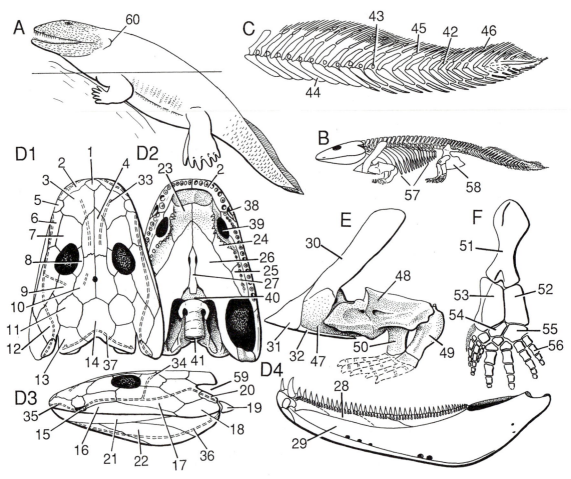

Fig. 4.95. Fossil Sarcopterygii, Tetrapoda. *Ichthyostega*, Late Devonian of Greenland. A, reconstruction, assuming that a gill slit was present (×0.1); B, reconstruction of the skeleton in lateral view; C, tail skeleton in lateral view; D, skull in dorsal (D1), ventral (D2), and lateral (D3) view, and lower jaw in internal view (D4); E, shoulder girdle and fore limb in lateral view; F, hind limb in anterior view. 1–32, *dermal bones*: 1, median rostral; 2, premaxilla; 3, nasal; 4, frontal; 5, tectal; 6, lachrymal; 7, prefrontal; 8, postfrontal; 9, postorbital; 10, parietal; 11, intertemporal + supratemporal; 12, squamosal; 13, tabular; 14, fused postparietals; 15, lateral rostral; 16, maxilla; 17, jugal; 18, quadratojugal; 19, subopercular; 20, preopercular; 21, dentary; 22, infradentaries; 23, vomer; 24, dermopalatine; 25, ectopterygoid; 26, entopterygoid; 27, parasphenoid; 28, coronoid; 29, prearticular; 30, cleithrum; 31, clavicle; 32, interclavicle; 33–7, *sensory line canals*: 33, supraorbital line; 34, postorbital line; 35, infraorbital line; 36, mandibular line; 37, supratemporal commissure; 38–60, *others*: 38, external nostril; 39, choana; 40, ventral fissure or reduced intracranial joint; 41, canal for notochord; 42, neural arch (basidorsal); 43, interdorsal (pleurocentrum); 44, haemal arch (basiventral); 45, radial; 46, dermal fin ray; 47, scapulocoracoid; 48, humerus; 49, ulna; 50, radius; 51, femur; 52, fibula; 53, tibia; 54, indermedium; 55, fibular; 56, phalanges; 57, ribs; 58, pelvic girdle; 59, otic notch; 60, gill slit. (A, modified from Bjerring (1985); B–E, D4, from Jarvik (1980); F, from Coates and Clack (1990).)

also that pentadactyly (five digits) is not a character of the tetrapods. Further investigations on *Ichthyostega* (Jarvik 1980) and more recently on *Acanthostega* (Clack 1988) have confirmed that some of these early tetrapods retain a number of generalized osteichthyan or sarcopterygian features that have been modified or lost in all other tetrapods,

such as a ventral fissure (Fig. 4.95, 40; Fig. 4.96, 12), a relatively mobile cheek and palatoquadrate (Fig. 4.96, 9), and sensory lines partly enclosed in canals and opening by means of pores (Fig. 4.96, 1).

Ichthyostega is certainly the best known of all these early tetrapods, yet its hand remains underscribed.

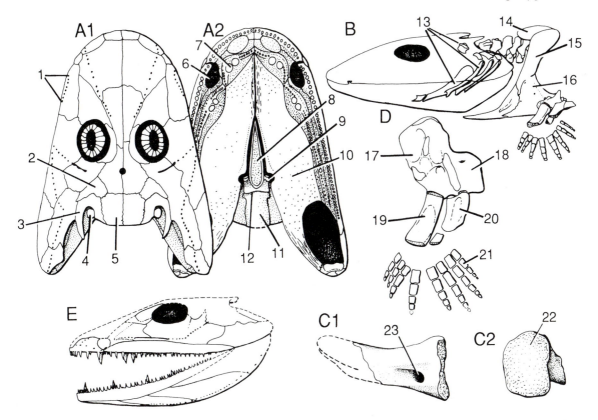

Fig. 4.96. Fossil Sarcopterygii. Tetrapoda. A–D, *Acanthostega*, Late Devonian of Greenland; A, skull in dorsal (A1) and palatal (A2) views (×0.5); B, reconstruction of the head and fore limb in lateral view, showing the presumed position of the gill arches; C, stapes in lateral (C1) and medial (C2) view; D, reconstruction of the fore limb in dorsal view. E, *Ventastega*, Late Devonian of Latvia, skull in lateral view (×0.5). 1, pores of infraorbital sensory line; 2, supratemporal; 3, tabular; 4, stapes; 5, postparietal; 6, choana; 7, vomer; 8, parasphenoid; 9, basipterygoid process; 10, entopterygoid; 11, part of braincase enclosing the notochord; 12, ventral fissure or reduced intracranial joint; 13, hyobranchial skeleton; 14, anocleithrum; 15, cleithrum; 16, scapulocoracoid; 17, humerus; 18, entepicondylar process; 19, radius; 20, ulna; 21, digits; 22, footplate; 23, stapedial foramen. (A, B, from Clack and Coates (1993); C, from Clack (1989, 1993); D, from Coates and Clack (1990); E, from Ahlberg *et al.* (1994).)

Equally detailed data on *Acanthostega* are to be published by J. Clack and M. Coates (Cambridge). Because of its early age, *Ichthyostega* has often been regarded as the most generalized, or primitive, tetrapod. However, it displays a number of characteristics that are likely to be early specializations, such as its large and overlapping ribs (Fig. 4.95, 57), the complex structure of its humerus (Fig. 4.95, 48), the intertemporal fused with the supratemporal (Fig. 4.95, 11), the premaxilla excluded from the margin of the choana (Fig. 4.95D2, 2, 39), and the fusion of its two postparietals into a median bone (Fig. 4.95, 14). Notwithstanding these unique ichthyostegid characters, *Ichthyostega* will be used here to exemplify the general tetrapod structure.

4.18.7.1 *Exoskeleton and external morphology*

Ichthyostega was about 1.5 m in length and probably had a rather stout overall body shape (Fig. 4.95A). Its limbs are relatively short, the hind limbs being larger than the fore limbs, and with a limited mobility. Its digits were probably partly embedded in a large fleshy pad, more or less like those of a dolfin or a seal. The skin of its tail and belly was covered with thin scales. The fin web of the tail is narrow and extended mostly on the dorsal edge of the notochordal lobe (Fig. 4.95C). The head is a massive structure, with no indication of a cranial mobility, and it is possible that a small gill slit was present. The nostrils are ventrally placed, lying along the oral margin (Fig. 4.95, 38). At the back of the skull, a well-marked otic notch may have been occupied by a tympanic membrane (Fig. 4.95, 59); see Chapter 6, p. 272, for a different interpretation).

The bones of the head of *Ichthyostega* are ornamented with vermiculations that recall those of panderichthyids (Fig. 4.94B), and the skull-roof agrees in many respects with that of the latter, besides the fusion of the two postparietals. The nostril is unlike that of most other tetrapods; it is lined ventrally by a small lateral rostral bone enclosing the infraorbital sensory-line canal (Fig. 4.95, 15), as in osteolepiforms and panderichthyids (Figs 4.88, 12; 4.94, 16). As in panderichthyids also, the jugal meets the quadratojugal to separate the squamosal from the maxilla (Fig. 4.95, 17, 18). The preopercular is a small bone in the back of the cheek, through which passes the preopercular sensory line (Fig. 4.95, 20); the subopercular is a small triangular bone that is separated from the cheek and probably lay free in the skin (Fig. 4.95, 19). There is no extrascapular, and the sensory line that pass through these bones in other osterichthyans (the supratemporal commissure, Fig. 4.95, 37) passes here through the rearmost bones of the skulltable, i.e. the postparietals and tabulars (Fig. 4.95, 13, 14). There have been lengthy debates about this major difference between fishes and tetrapods, which bears on dermal bone homologies, but the most likely explanation is that the extrascapulars have become reduced (to the same extent as in advanced actinistians, for example; Fig. 4.76, 7), and then fused to the skull-table. There is no opercular, submandibulo-branchiostegal bone series, or gular plates.

The palate is solid and displays the generalized osteichthyan pattern (vomer, dermopalatine, ectopterygoid, entopterygoid, and parasphenoid; Fig. 4.95, 23–7), with a large anterior palatal fenestra to accommodate the large anterior tusks of the lower jaw, and a pair of large, welldefined choanae (Fig. 4.95, 39). However, in contrast to the condition in all piscine osteichtyans, the dermal bones of the palate are not covered with a shagreen of small denticles, except for rows of smaller teeth on the ectopterygoid, dermopalatine, and vomers, which line medially the choanae. This is regarded as a derived ichthyostegid condition, but paralleling that in many other tetrapod groups (temnospondyls, amniotes). The parasphenoid is short and the two entopterygoids meet anteriorly (Fig. 4.95, 26, 27). In both the upper and lower jaws, the largest teeth are borne by the external dental arcade, i.e. the premaxilla, maxilla, and dentary (Fig. 4.95, 2, 16, 21), whereas the vomer, dermopalatine, ectopterygoids, and coronoids bear only small teeth (Fig. 4.95, 23, 24, 25, 28). This character is an important difference between *Ichthyostega* and piscine osteichthyans, where the largest teeth occur on the inner dental arcade. However, many other early tetrapods (Figs 4.96, 4.101 and 4.102) retain a few large coronoid tusks, even though the teeth of the outer arcade are large. The mandibular sensory-line canal

passes through the series of infradentaries (Fig. 4.95, 22, 36), as in piscine sarcopterygians.

The dermal shoulder girdle is reduced, with a narrow cleithrum, but there is a relatively large clavicle and interclavicle, the latter being produced posteriorly into a long median process (Fig. 4.95, 30–2). The cleithrum is no longer connected to the skull-roof by a series of dermal bones.

Very thin and rounded scales covered the chordal lobe of the tail. The lepidotrichiae were elongated in shape and probably unjointed.

4.18.7.2 *Endoskeleton*

The endoskeleton of the head of *Ichthyostega* is still poorly known. The braincase probably had a ventral fissure behind the posterior limit of the parasphenoid (Fig. 4.95, 40), and the otico-occipital region was rather short, as in all other tetrapods. The notochord seems to have penetrated into the braincase in the same way as in most sarcopterygian fishes (Fig. 4.95, 41), and there was no occipital condyle. It is probable that the occipital region included a number of vertebrae, since cranial ribs are present. The stapes, or hyomandibula modified into a sound-conducting element, is an important feature of the tetrapods, but it is still unknown in *Ichthyostega*. It has, however, been described in *Acanthostega* (Clack 1988), where it appears to be a stout, massive element, probably connecting the fenestra ovalis of the braincase to the quadrate (Fig. 4.96C). The fact that the stapes in early tetrapods does not seem to be oriented toward the otic notch had led Clack (1989) to suggest that the latter was not closed by the tympanic membrane but instead represented a functional spiracular opening that served for the intake of air when the head was partly emerged (Fig. 6.10C). *Acanthostega* retains complete gill arches (Fig. 4.96, 13).

The axial skeleton closely resembles that of osteolepiforms, with large centra (ventral vertebral arch), haemal and neural arches, and small interdorsals (Fig. 4.95, 42–4). The neural arches develop pre- and postzygapophyses for the intervertebral articulations. The ribs are very large, double-headed, and articulate with both the centra and neural arches. They are produced posteriorly into a large blade that overlaps one or two of the subsequent ribs, thereby forming a solid thoracic cage (Fig. 4.95, 57). This structure may be linked with the partly terrestrial habits of *Ichthyostega*, and would have prevented the animal from being squashed by its own weight.

The scapulocoracoid is a large and massive structure attached to the ventral end of the cleithrum (Fig. 4.95, 47). It displays a large coracoid plate and a laterally facing glenoid fossa. The supracoracoid and supraglenoid canals are relatively small. The humerus bears a number of prominent ridges, including a large entepicondylar process (Fig. 4.95, 48). The articulations for the radius and ulna are well separated, and that for the radius is ventral in position. The radius and ulna are short, stout bones (Fig. 4.95, 49, 50), and the latter has a well-developed olecranon, which does not exist in any piscine sarcopterygian. The reconstruction of the fore limb, as implied by the structure of these bones (the hand is still unknown), shows that the shoulder joint may have been quite mobile, whereas the elbow was almost incapable of any antero-posterior flection. Rotary movements of the elbow were, however, possible to some extent. The same probably applies also to *Acanthostega*, although its humerus is shorter and more osteolepiform-like (Fig. 4.96D, 17).

The pelvic girdle is a single large bony plate, fused to its contralateral by a pubic synchondrosis, and connected to the vertebral colum by a sacral rib (Fig. 4.95B, 58). The pubo-ischiadic portion is a large plate for the attachment of a powerful musculature, but the iliac portion is small, with a long posterior iliac process. The acetabular fossa, for the articulation of the femur, is oblique and elongated in shape. The femur is relatively elongated, and the tibia and fibula are large, plate-shaped bones (Fig. 4.95, 51–3). The knee certainly permitted ample anteroposterior movements. The ankle bones (intermedium and fibulare; Fig. 4.95, 54, 55) are also plate-like, and there are clearly seven, and possibly eight, toes as in *Acanthostega* (Fig. 4.95, 56; Fig. 4.96, 21). All the bones distal to the femur seem to have formed a stiff pad, and there is no clear indication of an ankle joint allowing anterior projection of the foot.

4.18.7.3 Histology

The teeth of *Ichthyostega* have a folded structure of advanced polyplocodont type ('labyrinthodont' type), i.e. without attachment bone between the folds of dentine, and the axes of the primary folds are undulating (Fig. 4.98A). As in panderichthyids, however, the dentine folds retain side branches, unlike all other tetrapods with advanced labyrinthodont teeth (except perhaps loxommatids; Fig. 4.98B). At the level of the primary folds, the dentine of *Ichthyostega* displays dense bundles of very long dentine tubules that recall the 'dark dentine' of anthrocosauroids (Fig. 4.98A, C, 1). This peculiar structure is probably due to a more rapid retreat of the odontoblasts at the level of the folds.

4.18.7.4 Diversity and phylogeny

Until recently, very few tetrapods were known before the end of Early Carboniferous times, and ichthyostegids were regarded as the sister-group of all other tetrapods,

or Neotetrapods (Gaffney 1979; Gardiner 1983). Now, better knowledge of the Late Devonian *Acanthostega* (Fig. 4.96A–C) and the Early Carboniferous *Crassigyrinus* (Fig. 4.99), and the discovery of the Late Devonian *Tulerpeton* (Fig. 4.97) and *Ventastega* (Ahlberg *et al.* 1994, Fig. 4.96E) have complicated the problem of early tetrapod interrelationships, because some of these early taxa display assemblages of generalized tetrapod characters and derived characters of one or another major tetrapod subgroup. Hence the classical theories about 'mosaic' evolution that are invoked when parsimony is neglected. For example, *Tulerpeton* shares some unique features with the Carboniferous anthracosaurs in its skull structure (though this is poorly known) and the morphology of its humerus, but retains six digits (all anthracosaurs have only five) and a distinct anocleithrum, lost in *Ichthyostega* (Lebedev 1984,

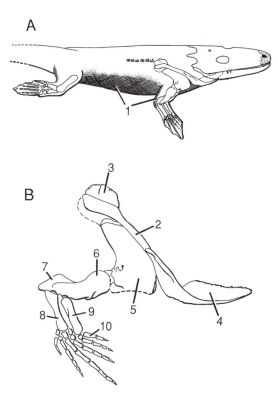

Fig. 4.97. Fossil Sarcopterygii. Tetrapoda. *Tulerpeton*, Late Devonian of Russia. A, attempted reconstruction of the skeleton in dorsolateral view (the outline of the skull is based on anthracosaurs, ×0.2); B, reconstruction of the shoulder girdle and fore limb, anterolateral view). 1, scale-covered areas; 2, cleithrum; 3, anocleithrum; 4, clavicle; 5, scapulocoracoid; 6, humerus; 7, entepicondylar process; 8, ulna; 9, radius; 10, digits. (A, B, from Lebedev (1985, 1990).)

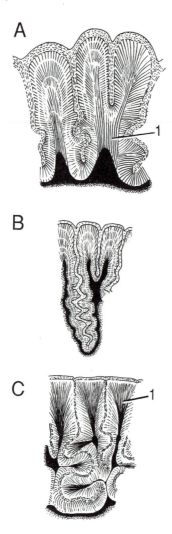

Fig. 4.98. Fossil Sarcopterygii. Tetrapoda. Tooth histology. Portion of horizontal sections through teeth of the ichthyostegid *Ichthyostega* (A, ×30), the loxommatid *Loxomma* (B, ×30), and the anthracosaur *Eogyrinus* (C, ×20). 1, 'dark dentine'. (A–C, from Schultze (1969).)

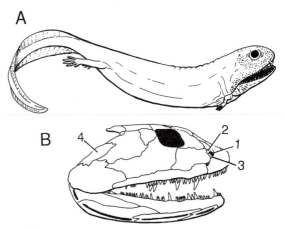

Fig. 4.99. Fossil Sarcopterygii. Tetrapoda. *Crassigyrinus*, Early Carboniferous of Scotland. A, reconstruction in lateral view (×0.05); B, skull in lateral view. 1, external nostril; 2, anterior tectal; 3, lateral rostral; 4, preopercular. (A, from Milner *et al.* (1986), reconstruction by M. I. Coates; B, from Panchen (1985).)

less, loss of preopecular and subopercular) have been twice acquired independently.

In the following discussion, we consider the following tetrapod terminal clades:

(1) *Acanthostega* (Devonian; Figs 4.96A–C, 4.105A)
(2) *Ichthyostega* (Devonian; Figs 4.95, 4.105B)
(3) *Tulerpeton* (Devonian; Figs 4.97, 4.105C)
(4) Loxommatidae (*Loxomma*, *Megalocephalus*, Carboniferous; Figs 4.100, 4.105C)
(5) *Crassigyrinus* (Early Carboniferous; Figs 4.99, 4.105E)
(6) Anthracosauroidea (*Anthracosaurus*, *Eoherpeton*, Carboniferous to Triassic; Figs 4.101, 4.105F)
(7) Amniota (a large clade comprising the mammals, birds, 'reptiles', and a number of extinct Palaeozoic taxa; Figs 4.102, 4.105G)
(8) Aistopoda (Carboniferous to Permian limbless tetrapods; Fig. 4.105H)
(9) Nectridea (Carboniferous to Permian; Fig. 4.105I)
(10) Microsauria (Carboniferous to Permian)
(11) Colosteidae (*Pholidogaster*, *Greererpeton*, and *Colosteus*, all Carboniferous; Fig. 4.105J)
(12) Temnospondyli (a large clade comprising Carboniferous to Recent taxa (Fig. 4.103) e.g. *Dendrerpeton*, *Eryops*, the Trematosauria, Capitosauroidea (Fig. 4.105K), 'Dissorophoidea', and Lissamphibia (Fig. 4.105L)

The interrelationships of these taxa, and a few others not considered here, have been the subject of considerable debate in the past fifty years. A wide variety of opinions have been expressed, in particular about the relationships

Lebedev and Clack 1993). In a similar way, *Crassigyrinus* has the 'dark dentine' (Fig. 4.98B) and the 'mobile cheek' of anthracosaurs but retains a preopercular (Fig. 4.99, 4); lost in all other tetrapods, except *Ichthyostega* and *Acanthostega*) and has a pelvic girdle that is regarded by Panchen (1985) as more primitive than that of any other tetrapod. Panchen and Smithson (1988) have proposed a hand-made phylogeny of a number of early tetrapod taxa, which assumes that Gaffney's 'neotetrapod' characteristics (five digits or of modern amphibians (Lissamphibia). This question is far from settled, and the recent discoveries of several new Devonian tetrapods may reveal further contradictory characters. In the cladogram in Fig. 4.105, the tetrapods comprise a large crown-group, the Neotetrapoda (Fig. 4.105, 4), which includes the Recent Amniota and Lissamphibia and all their respective fossil relatives. The Neotetrapoda are characterized by a maximum number of five digits (with the exception of secondary cases of poly-dactyly in some amniotes), as well as by the closure of the gill slit. This makes *Acanthostega*, *Ichthyostega*, *Tulerpeton* and *Ventastega* fall outside the Neotetrapoda. *Ventastega*, from the Late Devonian of Latvia is perhaps more generalized than *Acanthostega* (Ahlberg *et al.* 1994). Some authors claim, however, that *Tulerpeton* should be more closely related to the Anthracosauroidea, on account of various characters such as the ornamenta-tion of dermal bones and shape of humerus. In the same way, *Crassigyrinus* (Fig. 4.105E) is put with the anthra-cosauroidea and amniotes on the basis of the closed post-temporal fossa and 'dark dentine' (the latter may be a plesiomorphous feature, since it occurs in *Ichthyostega*), but it retains a preopercular, which is lost in all other neotetrapods. The Aistopoda, which are regarded as limb-less tetrapods, have an indeterminate position, but are ten-tatively put here among the Neotetrapoda. They may be close relatives of the Microsauria, a small tetrapod group possibly belonging to the Amphibia (Fig. 4.105, 8). The Loxommatidae is yet another taxon of erratic position: it has variously been regarded as the sister-group of the Neotetrapoda (Gardiner 1983), the Anthracosauroidea, and the Amniota (Panchen and Smithson 1988), or the Amphibia (Trueb and Cloutier 1991). Loxommatids are characterized by bean-shaped orbits (Fig. 4.100, 1). All other tetrapod taxa fall within the two major taxa of the Neotetrapoda, the Reptiliomorpha (Fig. 4.105, 5–7) and the Amphibia (or Batrachomorpha; Fig. 4.105, 8).

One of the classical differences between the Reptiliomorpha and the Amphibia is that the tabular is in contact with the parietal in the former (Fig. 4.101, 4; Fig. 4.102, 3) but not in the latter (Fig. 4.103, 3). This charac-ter is, however, to be taken with caution, since the latter state (tabular separated from the parietal) occurs also in loxommatids (Fig. 4.100, 4) *Ichthyostega* (Fig. 4.95, 13) and panderichthyids (Fig. 4.94, 9) and is thus plesiomor-phous for the tetrapods. It also occurs in some seymouri-amorphs (Panchen 1980b). The same reasoning applies also to the difference in vertebral structure, where the rha-chitomous condition (Fig. 4.103B2, 14, 15; large inter-centrum, and small interdorsal, or pleurocentrum) of temnospondyls is likely to be the general tetrapod con-dition, whereas the embolomerous (or diplospondylous) condition (Fig. 4.101E, 15, 16; Fig. 4.102B, 10, 11; large

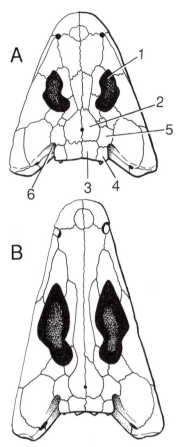

Fig. 4.100. Fossil Sarcopterygii. Tetrapoda, Loxommatidae. Skull in dorsal view of *Loxomma* (A, ×0.2) and *Megalocephalus* (B, ×0.1). 1, anterior orbital notch; 2, parietal; 3, postparietal; 4, tabular; 5, supratemporal, 6, otic notch. (A, B, from Beaumont (1977).)

or reduced intercentrum and large centrum) of anthra-cosaurs and amniotes is derived, as are also the tube-shaped vertebrae of aistopods, nectrideans, and lissamphibians. Nevertheless there remains a widespread tendency among early tetrapod specialists to consider that there is something valid in this major dichotomy, as did for example Gauthier (1989).

The Reptiliomorpha are now best characterized by the embolomerous vertebral structure, with an increasing en-largement of the centrum (and reduction of the inter-centrum), and by the closure of the post-temporal fossa, which normally lies, in fishes and other tetrapods, between the opisthotic, postparietal, and tabular (Fig. 4.101D, 3, 4, 11). Another possible reptiliomorph character is the pres-ence of a pit in the braincase, for the attachment of the re-tractor muscles of the eye (Fig. 4.101, 9). This character

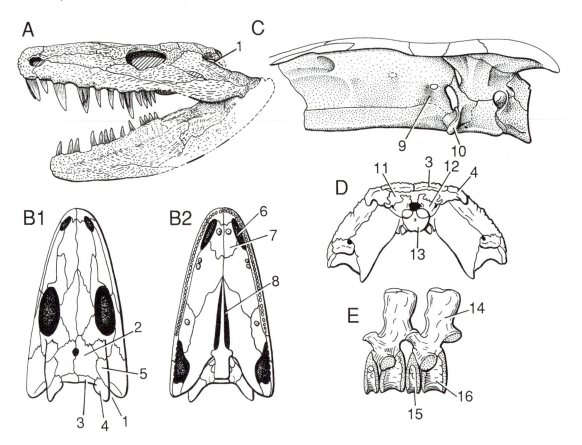

Fig. 4.101. Fossil Sarcopterygii. Tetrapoda, Anthracosauroidea. A, *Anthracosaurus*, Carboniferous of Scotland, skull in lateral view (×0.2). B, *Gephyrostegus*, Early Carboniferous, skull in dorsal (B1) and ventral (B2) view (×1); C, *Archeria*, Permian of the USA, braincase in lateral view (×1); D, *Palaeoherpeton*, Carboniferous of Scotland, skull in posterior view; E, *Eogyrinus*, Carboniferous of Scotland, vertebrae in lateral view. 1, otic notch; 2, parietal; 3, postparietal; 4, tabular; 5, supratemporal; 6, choana; 7, vomer; 8, interpterygoid vacuity; 9, retractor pit; 10 basal articulation (basipterygoid process); 11, otic capsule (opisthotic); 12, exoccipital; 13, occipital condyle of basioccipital; 14, neural arch (basidorsal); 15, intercentrum (ventral vertebral arch); 16, centrum. (A, from Panchen (1981); B, from Heaton (1980), after Carroll; C, from Clack and Holmes (1988); D, E, from Panchen (1970, 1980*b*).)

is present in loxommatids, but the condition in *Ichthyostega* and *Acanthostega* in not known. A retractor pit of this kind may already be present in osteolepiforms (Bjerring 1987). The presence of mobile basal articulation for the palatoquadrate (Fig. 4.101, 10) is often regarded as an important reptiliomorph character, but it seems to occur also in such non-neoterapods as *Acanthostega* (Fig. 4.96, 9). The solid palate of *Ichthyostega* (Fig. 4.95D2) would thus be a convergence with the condition in temmospondyls. The Reptiliomorpha comprise essentially the Anthracosauroidea (Fig. 4.101, 4.105F) and Amniota (Fig. 4.105G). The former are characterized by a special type of dermal ornamentation and the 'dark dentine' (Fig. 4.98C, 1; this, however, may possibly be already present in *Ichthyostega*). The amniotes are char-

acterized by separate scapular and coracoid ossifications, pterygoid flanges of the entopterygoid (Fig. 4.102, 7), a reduced intercentrum (Fig. 4.102B, 11) and a number of characters in the soft anatomy, such as the amnion. The Seymouriamorpha (Fig. 4.102) can either be included in the Amniota (as is done here) or be regarded as the sister-group of the Amniota.

The Amphibia are characterized by e.g. a four-digit hand (Fig. 4.103B1) and large interpterygoid vacuities (Fig. 4.103, 6). They are essentially represented by the Temnospondyli (Fig. 4.105, 10), which comprise the Lissamphibia (apodans, salamanders, and frogs; Fig. 4.105L) and a large number of Palaeozoic and Mesozoic forms, such as the Capitosauroidea (Fig. 4.105K). The Temnospondyli are characterized by a

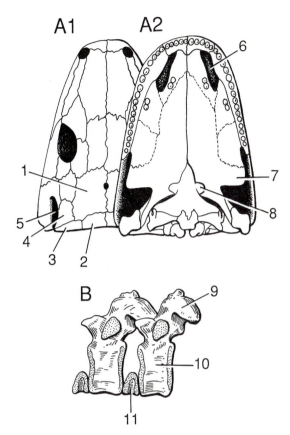

Fig. 4.102. Fossil Sarcopterygii. Tetrapoda, Amniota. *Seymouria*, Permian of the USA; A, skull in dorsal (A1) and ventral (A2) view (×0.5); B, vertebrae in lateral view. 1, parietal; 2, postparietal; 3, tabular; 4, supratemporal; 5, otic notch; 6, choana; 7, pterygoid flange; 8, basal articulation of palatoquadrate; 9, neural arch (basidorsal); 10, centrum; 11, intercentrum. (A, B, from T. E. White 1939).)

'waisted' humerus (Fig. 4.103B1) and a strap-shaped occipital condyle (Fig. 4.103D1). One of the most generalized temnospondyls is the Carboniferous *Dendrerpeton* (Fig. 4.103A), but the group comprises large and highly specialized forms such as the Permian and Mesozoic capitosauroids and trematosaurs. The latter were marine, piscivorous forms that retain a well-developed sensory-line system (Fig. 4.103C, 16) and scaly skin (Fig. 4.103C, 17). A series of small Permian temnospondyls often referred to as 'dissorophoids' links the modern Lissamphibia to these heavily ossified forms. The Early Permian dissorophid *Doleserpeton* (Fig. 4.103D1) is one of these forms in which the pterygoid is strikingly lissamphibian-like (Fig. 4.103D1, 7) and the teeth are pedicellate, as in

Lissamphibians (Fig. 4.103D2, 10). Other Early Permian forms, the 'branchiosaurs', are yet more closely related to lissamphibians; they looked like newts, with large external gills (Fig. 4.103F). The earliest known Gymnophiona, the Early Jurassic *Eocoecilia* (Fig. 104), retains small limbs and is therefore regarded as the sister-group of all other gymnophiones, or Apoda (Jenkins and Walsh 1993). Gymnophiones have been regarded as derived from the Microsauria (Carroll and Currie 1975, Carroll 1990), a theory which implies that the gymnophiones have diverged before all other temnospondyls. This theory is, however, not unanimously accepted (see review in Milner 1993b). The earliest known Urodela is the Jurassic *Karaurus* (Fig. 4.103E), which is the sister-group of all other urodeles, or Caudata (Ivachnenko 1978). The earliest known Salientia is the Early Triassic *Triadobratrachus*, which is the sister-group of all other salientia, or Anura. *Triadobatrachus* has many frog specializations, such as the elongated tarsals (Fig. 4.103, 19), but retains a number of caudal vertebrae (Fig. 4.103, 18), as well as separate radius, ulna, tibia, and fibula (Rage and Rocek 1989).

In addition to the Temnospondyli, other minor taxa are also included in the Amphibia, such as the Carboniferous Colosteidae (Fig. 4.105J) and the newt-like Nectridea (Fig. 4.105I), which share with temnospondyls the large interpterygoid vacuities and four-digit manus.

From this account, it appears that the interrelationships of the Tetrapoda are largely unresolved. Contrary to the opinion of Panchen and Smithson (1988; see also Panchen 1992), I consider that only a thorough computerized cladistic analysis of a large number of characters could help in providing a parsimonious cladogram. This has been done in detail by Trueb and Cloutier (1991) for the amphibian clade. It is probable that tetrapod phylogeny will be deeply reworked in the near future.

The classification of the Tetrapod in Table 4.12 is derived from the cladogram in Fig. 4.105 and from the classification of the Temnospondyli by Trueb and Cloutier (1991).

4.18.7.5 *Stratigraphical and geographical distribution, habitat*

The first findings of Devonian tetrapods were geographically restricted to Greenland. Following the classical practice of evolutionary palaeontology, the 'cradle' of the tetrapods has been regarded as being the 'Old Red Sandstone continent'. Later, some doubtful tetrapod remains (*Metaxygnathus*) but undoubted tetrapod trackways were recorded from the Late Devonian of Australia (Fig. 7.14A), providing evidence for a world-wide distribution (Campbell and Bell 1977, Warren and Wakefield 1972). Late Devonian (Famennian) tetrapods have been

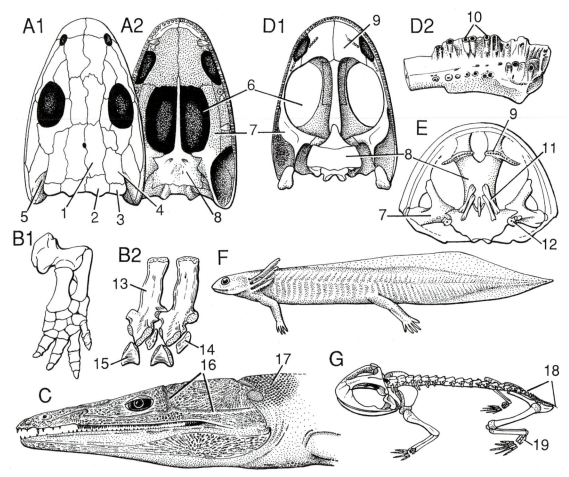

Fig. 4.103. Fossil Sarcopterygii. Tetrapoda, Temnospondyli. A, *Dendrerpeton*, a Carboniferous generalized temnospondyl, skull in dorsal (A1) and ventral (A2) view (×0.6); B, *Eryops*, Permian, anterior limb (B1, ×0.3) and vertebrae (B2) in lateral view ; C, *Trematosaurus*, a Triassic marine temnospondyl, reconstruction of the head and anterior part of the trunk in lateral view (×0.5); D, *Doleserpeton*, a Permian generalized dissorophoid, skull in ventral view (D1, ×3) and anterior part of the lower jaw in medial view (D2, ×1.5); E, *Karaurus*, a Jurassic Urodela, skull in ventral view (×1); F, *Branchiosaurus*, a Permian branchiosaurid, reconstruction in lateral view (×1.2); G, *Triadobatrachus*, a Triassic Salientia, reconstruction of the skeleton (×0.8). 1, parietal; 2, postparietal; 3, tabular; 4, supratemporal; 5, otic notch; 6, interpterygoid vacuity; 7, entopterygoid; 8, parasphenoid; 9, vomer; 10, pedicles of teeth; 11, hyobranchial skeleton; 12, stapes; 13, neural arch (basidorsal); 14, interdorsal (pleurocentrum); 15, intercentrum (ventral vertebral arch); 16, sensory-line grooves; 17, scales; 18, caudal skeleton; 19, elongated tarsals. (A, from Carroll (1967); B, from W. K. Gregory (1911), modified; C, from Janvier (1992); D, from Bolt (1991); E, from Ivakhnenko (1978); F, from Milner (1982); G, from Rage and Rocek (1989).)

since recorded from North America, Scotland, Latvia, southern Russia, and Australia (Ahlberg and Milner 1994). The earliest-known records are the Australian trackway and the Scottish form (Ahlberg 1991*b*), which may be Frasnian in age. A possibly Early Devonian trackway has been described in Australia (Warren *et al.* 1986), but both its age and its nature need confirmation.

The current theory is that tetrapods arose in the Frasnian and diversified rapidly in the Famennian and Tournaisian. The Tournaisian and Early Visean (Earliest Carboniferous) represent a gap of about 20 Ma, during which we still have no tetrapod remain, and which corresponds to the period of diversification of all the major neotetrapod taxa mentioned in the preceding section. As early as the Late Visean, 335 Ma ago, all these taxa are already present, including the amniotes (Smithson 1989).

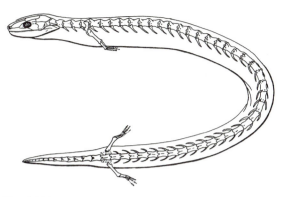

Fig. 4.104. Fossil Sarcopterygii. Tetrapoda, Temnospondyli, Gymnophiona. *Eocoecilia*, Early Jurassic of USA, reconstruction of the skeleton (×1.5). (From Jenkins and Walsh (1993).)

The earliest tetrapods have long been regarded as freshwater animals because extant amphibians are essentially freshwater, and also because the Old Red Sandstone and Coal Measures have been long regarded as strictly freshwater deposits. However, most fishes associated with the earliest tetrapods (antiarchs, groenlandaspids, lungfishes, and osteolepiforms) have a wide distribution and commonly also occur in marine deposits. It is therefore probable that Devonian tetrapods lived in vast tidal flats, lagoons, and deltas. The locality of Andeyevka, near Tula (Russia), where *Tulerperton* occurs, is clearly marine. But the environment of *Ichthyostega* and *Acanthostega* in Greenland is interpreted as large, meandering rivers, though not very far from deltas. It is now admitted that most tetrapod features, including the dactylic limbs, appeared in essentially aquatic animals, and served swimming rather than walking on land.

Further reading

Devonian tetrapods: Ahlberg (1991*b*); Ahlberg and Milner (1994); Ahlberg *et al.* (1994); Campbell and Bell (1977); Clack (1988, 1989, 1993); Clack and Coates (1993); Coates and Clack (1990); Jarvik (1952, 1980); Lebedev (1984, 1985, 1990); Lebedev and Clack (1993); Leonardi (1983); Warren and Wakefield (1972); Warren *et al.* (1986). *Other Palaeozoic tetrapods, tetrapod interrelationships and systematics (includ-*

ing fossils): Ahlberg and Milner (1994); Benton (1988); Bolt (1991); Carroll (1987); Gaffney (1979); Gardiner (1982, 1983); Gauthier (1989); Gauthier *et al.* (1988); Milner (1988, 1993*a*); Milner *et al.* (1986); Panchen (1970, 1980*b*, 1985); Panchen and Smithson (1987, 1988); Schultze and Trueb (1991); Trueb and Cloutier (1991).

Table 4.12. Classification of the Tetrapoda (See also Fig. 4.105)

Tetrapoda
  Acanthostega
  Ichthyostegalia
    Ichthyostega
  (unnamed taxon)
    Tulerpeton
  Neotetrapoda
    Reptiliomorpha *sedis mutabilis*
      Loxommatidae
      (unnamed taxon)
        Crassigyrinus
      (unnamed taxon)
        Anthracosauroidea
        Amniota (incl. Seymouriamorpha, Diadectomorpha)
    Aistopoda *sedis mutabilis*
    Amphibia *sedis mutabilis*
      Nectridea
      (unnamed taxon)
        Colosteidae
        Temnospondyli
          Edopoidea
          Eryopoidea
          Capitosauroidea
          Dissorophoidea
            'Dissorophidae'
            'Branchiosauridae'
          Lissamphibia
            Gymnophiona
              Eocoecilia
              Apoda
            Batrachia
              Urodela
                Karaurus
                Caudata
              Salientia
                Triadobatrachus
                Anura

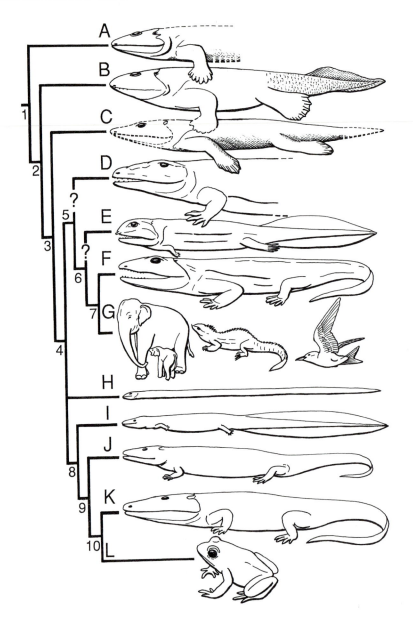

Fig. 4.105. Interrelationships of the Tetrapoda (partly based on the phylogeny proposed by Panchen and Smithson (1988). *Terminal taxa*: A, *Acanthostega; B, Ichthyostega;* C, *Tulerpeton*; D, Loxommatidae; E, *Crassigyrinus*; F, Anthracosauroidea; G, Amniota (including Seymouriamorpha); H, Aistopoda; I, Nectridea; J, Colosteidae; K, Capitosauroidea; L, Lissamphibia. *Nested taxa and selected synapomorphies*: 1, Tetrapoda (limbs with digits, sacrum, fenestra ovalis, hyomandibula modified into a stapes); 2, reduced gill apparatus; 3, cleithrum free from scapulocoracoid, five digits in hind limb); 4, Neotetrapoda (five digits or less in both fore and hind limbs); 5, Reptiliomorpha (retractor pit for eye muscles); 6, loss of post-temporal fossa; 7, incipient pterygoid flange, phalangeal formula of pes is 2, 3, 4, 5, 4–5, diplospondylous vertebrae, stapes sound-conducting; 8, Amphibia (four-digit manus); 9, large interpterygoidian vacuities; 10, Temnospondyli ('waisted' humerus, strap-like occipital condyle, stapes sound-conducting). *Age*: Late Devonian (A–C); Carboniferous (D–E, H, J); Carboniferous to Permian (I); Carboniferous to Triassic (F); Carboniferous to ?Cretaceous (K); Carboniferous to Recent (G); Triassic to Recent (L). (A, based on Clack and Coates (1993); B, based on Jarvik (1980); C, based on Lebedev (1985); D–F, H, J, based on Milner *et al.* (1986).)

Interrelationships of the major craniate taxa: current phylogenetic theories and controversies

The major taxa described in the preceding chapter are all supposed to be clades, with perhaps one exception, the thelodonts, which are regarded by some palaeontologists as an ensemble of generalized craniates or vertebrates. However, the relationships between these taxa remain the subject of sometimes heated discussion, even if such higher clades as the gnathostomes or osteichthyans are now relatively well corroborated.

Until the 1970s, discussions on interrelationships of major taxa were relatively simple to sum up, since they were based on the consideration of a small number of characters. Most of them were 'pet characters' of one or another authority, and the 'minor' characters that contradicted them were regarded as homoplasies or simply not considered. With the advent of phylogenetic systematics, or cladistics, an increasingly large number of characters were taken into consideration for phylogeny reconstruction, but it was still possible to compare and discuss contradictory cladograms, since these were few. Since efficient cladistic computer programs (see Chapter 2, p. 40) came on the scene, the phylogenies, and the characters on which they are based have become so numerous that the task of comparing in detail the merits of respective phylogenies is now virtually impossible at the level of a general book such as this. This method of analysis of the characters, which no doubt will be in general use for a long time, is based on parsimony, and the only point that can be discussed by the average reader is the criterion of choice among a number of equally parsimonious trees (see, for example, Schultze 1994). The consensus tree of these equally parsimonious trees produced by the computer is often the only result published by phylogeneticists, and it is generally not very informative. There are even instances of authors who have published only one of the 400 or 500 equally parsimonious trees they obtained, but without telling the reader! Moreover, the options offered by current cladistic computer programs are numerous and the same data matrix may produce widely different sets of trees, depending on character weights, ordering of character states, etc. Less than ten years after the publication of the first computer-generated phylo-

genies of the major craniate groups, the discussion of the various competing phylogenies, character by character, has become quite difficult. To some, this may sound like the end of phylogenetics, but such is not the case. First, this way of considering character-state distributions prevents—to some extent—writers from falling in the pitfall of 'pet characters'. Second, the rapidity of such character analyses enables a large number of research workers to propose phylogenies based on various sets of characters. Over a period of less than ten years, it appears that the trees produced for the same taxa (extant ones in particular) are more and more stable, that is, better and better corroborated. One may thus expect that, with time, there may be a wide consensus on one particular tree for the craniates, the gnathostomes, or the osteichthyans, for example, although a consensus is not the truth. The interrelationships of major extant craniate taxa are probably not likely to be deeply modified in the future, yet some changes may occur in the position of hagfishes or the coelacanth–lungfish–tetrapod problem. In contrast, trees involving fossils are doomed to be largely unresolved or to be modified as knowledge of fossil forms (both in their structure and diversity) progresses.

5.1 INTERRELATIONSHIPS OF THE CRANIATA

Let us first consider the interrelationships of the Craniata, with the Gnathostomata (Fig. 5.1H) as a terminal taxon. As we have seen in Chapter 3, the question of the relationships of the two extant jawless craniate taxa, hagfishes and lampreys, and the gnathostomes is still far from settled, in spite of an impressive number of characters shared only by lampreys and gnathostomes. Molecular data still provide ambiguous answers that depend largely on the nature of the out-group (tunicates or cephalochordates) used for sequence analyses and, of course, on whether distance or parsimony methods are used (Stock and Whitt 1992). Only a hagfish–gnathostome sister-group relationship can be reasonably ruled out. However, considering the number and the nature of the characteristics that support the monophyly of extant Vertebrata (in the sense used in this book, that is lampreys and gnathostomes), we are inclined toward preferring this theory, rather than the monophyly of the Cyclostomi (hagfishes and lampreys). Fossil hagfishes and lampreys are of no help in resolving this dilemma, since they are virtually similar to the extent forms. Only

Jamoytius, if regarded as a lamprey (Forey and Gardiner 1981), may provide evidence for the secondly loss of the paired fins in Carboniferous to Recent lampreys. Also, some anaspid-like forms, such as *Euphanerops*, may show that the annular cartilage is a more general character than was previously believed (Fig. 4.12B, 2). The fossil lamprey *Hardistiella* perhaps retains a hypocercal tail and an anal fin, which recall the condition in anaspids

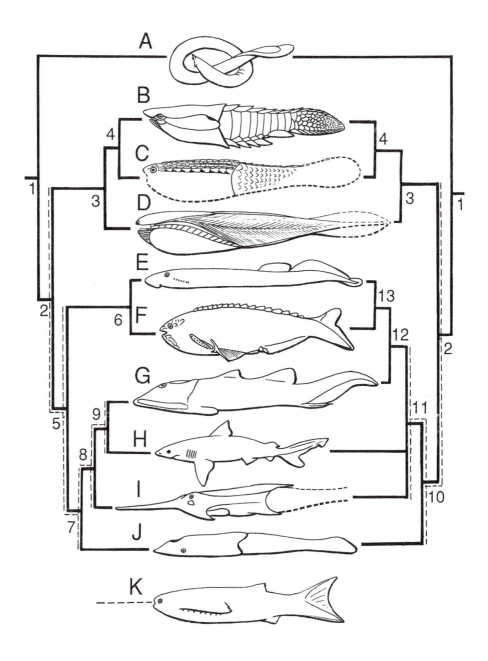

(Fig. 4.13, 1, 3). The fossil hagfishes show that the posterior position of the gill pouches and the slender body shape of extant forms are derived characters, which have appeared since the Carboniferous (Fig. 4.1B, 1).

How the exclusively fossil clades, such as arandaspids, astraspids, heterostracans, osteostracans, anaspids, pituriaspids, and galeaspids, relate to these extant groups is a different question. Notwithstanding the fact that the internal anatomy of some of them is fairly well known, they display features that are either generalized craniate or vertebrate characters, or unique to each of them, which tell us nothing about their relationships. Only a few shared features display a more restricted pattern, and may serve in reconstructing a phylogeny.

5.1.1 Monophyly versus paraphyly of the Cephalaspidomorphi

To the first students of jawless craniate phylogeny, such as Kiaer (1924) and Stensiö (1927), the most conspicuous of these shared characters was the presence of a dorsally placed nasohypophysial opening in lampreys, osteostracans, and anaspids, although in the latter no endoskeleton could help in demonstrating that it comprised a blind hypophysial tube. These three taxa have thus been regarded as forming a clade, the Cephalaspidomorphi (Fig. 5.1, 12; Goodrich 1909), regardless of the otherwise very different characters in the rest of the external morphology and internal anatomy, which were regarded as 'adaptive', and thus neglected. When galeaspids were first discovered in the 1960s, it was thus logical to relate them to the cephalaspidomorphs, because of their median dorsal opening, which 'had' to be a dorsal nasohypophysial opening (Janvier 1975b). Later, this was shown to be wrong in two respects: the median dorsal opening communicates basally with the oralobranchial chamber, and galeaspid phylogeny implies that the general condition for the entire group is an almost terminal median dorsal opening, such as in *Hanyangaspis* (Fig. 4.22A). Leaving aside the poorly known anaspids, most studies on osteostracans tended to interpret their internal anatomy in the light of that of adult or larval lampreys. This worked to some extent, probably because most of the comparable structures of both groups displayed a general vertebrate condition. In the 1970s, the use of cladistic analysis gave a cold look at all possible characteristics, even those said to be 'adaptive', such as the shape of the caudal fin. This has led to the strange conclusion that osteostracans, despite their dorsal nasohypophysial opening, shared more unique characters with the gnathostomes than with lampreys and anaspids. These characters are the epicercal tail, the large dorsal jugular vein (vena capitis lateralis), the sclerotic rings, the endoskeletal scleral ossification of eye capsule, the perichondral bone, the cellular dermal bone, the externally open endolymphatic ducts, and the pectoral fins with muscles attached to the endoskeleton (Janvier 1981a, 1984a; Forey 1984; Forey and Janvier 1993; Fig. 5.1, 9). Even if anaspids could be left with lampreys (Fig. 5.1, 6), for the lack of more characters, the Cephalaspidomorphi had to be either para- or diphyletic, depending on whether the dorsal nasohypophysial opening was regarded as lost in the gnathostomes or acquired independently in lampreys (+ anaspids) and osteostracans. Janvier (1981a) suggested that the gnathostomes could have derived from an ancestor with a nasohypophysial complex similar to that of lampreys, but opening in a more terminal position. In contrast, Schaeffer and Thomson (1980) considered the possibility that the condition in osteostracans is merely a convergence with lampreys, and is due to the forward shift of the branchial apparatus in the former. Improved knowledge of galeaspid internal anatomy does not greatly change this pattern of character distribution, and they can be regarded as the sister-group of osteostracans and

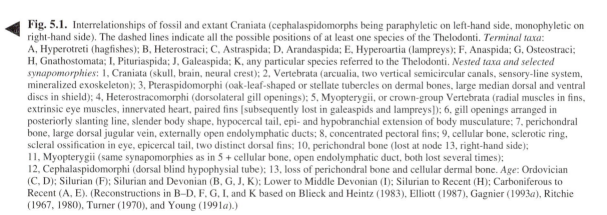

Fig. 5.1. Interrelationships of fossil and extant Craniata (cephalaspidomorphs being paraphyletic on left-hand side, monophyletic on right-hand side). The dashed lines indicate all the possible positions of at least one species of the Thelodonti. *Terminal taxa*: A, Hyperotreti (hagfishes); B, Heterostraci; C, Astraspida; D, Arandaspida; E, Hyperoartia (lampreys); F, Anaspida; G, Osteostraci; H, Gnathostomata; I, Pituriaspida; J, Galeaspida; K, any particular species referred to the Thelodonti. *Nested taxa and selected synapomorphies*: 1, Craniata (skull, brain, neural crest); 2, Vertebrata (arcualia, two vertical semicircular canals, sensory-line system, mineralized exoskeleton); 3, Pteraspidomorphi (oak-leaf-shaped or stellate tubercles on dermal bones, large median dorsal and ventral discs in shield); 4, Heterostracomorphi (dorsolateral gill openings); 5, Myopterygii, or crown-group Vertebrata (radial muscles in fins, extrinsic eye muscles, innervated heart, paired fins [subsequently lost in galeaspids and lampreys]); 6, gill openings arranged in posteriorly slanting line, slender body shape, hypocercal tail, epi- and hypobranchial extension of body musculature; 7, perichondral bone, large dorsal jugular vein, externally open endolymphatic ducts; 8, concentrated pectoral fins; 9, cellular bone, sclerotic ring, scleral ossification in eye, epicercal tail, two distinct dorsal fins; 10, perichondral bone (lost at node 13, right-hand side); 11, Myopterygii (same synapomorphies as in 5 + cellular bone, open endolymphatic duct, both lost several times); 12, Cephalaspidomorphi (dorsal blind hypophysial tube); 13, loss of perichondral bone and cellular dermal bone. *Age*: Ordovician (C, D); Silurian (F); Silurian and Devonian (B, G, J, K); Lower to Middle Devonian (I); Silurian to Recent (H); Carboniferous to Recent (A, E). (Reconstructions in B–D, F, G, I, and K based on Blieck and Heintz (1983), Elliott (1987), Gagnier (1993a), Ritchie (1967, 1980), Turner (1970), and Young (1991a).)

gnathostomes, with which they share perichondral bone, a large dorsal jugular vein, and perhaps open endolymphatic ducts (Fig. 5.1, 7; N. Z. Wang 1991). Galeaspids indicate only that the common hypothetical ancestor to osteostracans and the gnathostomes had widely separate nasal cavities, and that the reduction of the olfactory organ in osteostracans is yet one more convergence with lampreys. Pituriaspids are still too poorly known to be considered seriously, but they had perichondral bone and narrow-based paired fins, two features that would place them closer to the osteostracans and gnathostomes than to the galeaspids (Fig. 5.1, 8; Young 1991a). Galeaspids differ from osteostracans (and perhaps pituriaspids) in two respects: their exoskeleton is acellular and contains no dentine, and they have no paired fins. The former feature is regarded as general for the vertebrates; the second may be unique to the group, and to secondary loss, since anaspids do have paired fins. Instead of the cephalaspidomorphs, we have now a monophyletic group of vertebrates that is characterized by the ability to produce perichondral bone, and which includes three jawless taxa, the galeaspids, osteostracans and pituriaspids, and the gnathostomes (Fig. 5.1, 7). The former three taxa may thus be called 'stem-group gnathostomes'.

5.1.2 The Pteraspidomorphi

Having put together lampreys, anaspids, and osteostracans, Stensiö (1927) was left with heterostracans and thelodonts. (*Palaeospondylus*, considered by Stensiö in his analysis, is here ignored, despite Forey and Gardiner's (1981) and Jarvik (1980) suggestions that it may be a larval lungfish or osteolepiform.) Since he regarded jawless craniates as a clade, e.g. because of their gill structure Stensiö rejected Kiaer's (1924) suggestion that heterostracans were related to gnathostomes because they have a paired olfactory organ ('Diplorhina'). He was thus inclined to consider heterostracans as closely related to hagfishes, and he included both groups in the taxon Pteraspidomorphi, which had been erected by Goodrich (1909) solely for heterostracans and thelodonts. In Stensiö's phylogeny, the Pteraspidomorphi (including hagfishes) are the sister-group of the Cephalaspidomorphi (including lampreys). Subsequently, Stensiö (1932, 1964, 1968) further emphasized this theory by interpreting the internal impressions of heterostracans in the light of hagfish anatomy. In contrast, Halstead (see review in Halstead 1973a,b) and Novitskaya (1983) argued in favour of a closer relationship between heterostracans and gnathostomes, on the basis of the paired nasal sacs, possibly paired external nostrils, and presumed outwardly directed gills. Thanks to our knowledge of the anatomy of galeaspids, which matches fairly well the internal organ

impressions of heterostracans (compare Figs 4.5A1 and 4.21E), we can assume that the nasal cavities in heterostracans were actually paired, but opened ventrally either into the mouth or into a median nasopharyngeal duct, homologuous to that of hagfishes and to the median dorsal duct of galeaspids. This 'diplorhiny' is now regarded as a general vertebrate condition, and not as a unique character shared only with gnathostomes (Forey and Janvier 1993).

Heterostracans are clearly more advanced than hagfishes, because they possess two vertical semicircular canals, a well-developed sensory-line system, a cerebellum, a mineralized skeleton, and possible arcualia or other vertebral elements (Fig. 5.1, 2). However, they look less advanced than the anaspids, pituriaspids, osteostracans, and gnathostomes, in lacking paired fins (Fig. 5.1, 5). In any case, they are clearly less derived than the osteostracans and gnathostomes, in lacking perichondral bone, cellular bone in the exoskeleton, large eyes with sclerotic rings, and open endolymphatic ducts (Fig. 5.1, 7–9). The caudal fin of heterostracans, which is the only unpaired fin, shows fan-shaped digitation of larger scales (Denison 1971b; Soehn and Wilson 1990), which suggests the presence of a few, large underlying radials. Such a condition resembles that in hagfishes, where the caudal radials are devoid of musculature. This contrasts with the tail in lampreys and gnathostomes, and presumably also anaspids and osteostracans, where the radials are thin and numerous, and are provided with radial muscles that allow undulations of the fin web. For this reason, these taxa were grouped by Janvier (1978a) into the Myopterygii ('muscular fins'). The taxon Myopterygii (Fig. 5.1, 5 or 11) is redundant with Vertebrata when extant taxa are considered, but heterostracans may be non-myopterygian vertebrates, since they had vertebral elements.

The status of the thelodonts is still obscure, but light has been shed here by the discovery of some internal structures (internal denticles, branchial 'basket', stomach). Thelodonts are regarded by Turner (1991) as a clade by the virtue of the anchoring device of their growing scales. In contrast, Ørvig (1968), Karatayute-Talimaa (1978), and Janvier (1981a) have suggested that they may be generalized vertebrates, since their entirely microsquamose exoskeleton is just what we would expect to find in the most generalized members of each major vertebrate group. As far as their internal organization is concerned, we have seen that they probably had paired olfactory organs and a series of gill units with separate external openings. The recent discovery of an internal covering of forward-pointing denticles in the snout, and of whorl-like toothed platelets in the branchial region, in the Silurian thelodont *Loganellia* (Fig. 4.25), suggests that they could have possessed a median, terminal inhalent duct separated from the mouth, like hagfishes and galeaspids (and probably heterostracans), and a branchial organization comparable to that of the gnathostomes, with slender denticles, or even tooth whorls, on the pharyngeal surface of the interbranchial septa. This assemblage of features would thus imply that *Loganellia* is more closely related to the gnathostomes than to any other jawless group.

In conclusion, the pteraspidomorphs, as defined by Stensiö are probably paraphyletic and do not cohere better than the cephalaspidomorphs when several characters are taken into consideration in the cladistic analysis. Gagnier (1993*b*) has, however, recently revived the name 'Pteraspidomorphi' to include heterostracans, astraspids, and arandaspids (Fig. 5.1, 3), which all share large median dorsal and ventral shields, as well as typical oak-leaf-shaped tubercles on dermal bones. It is also probable that stem-group pteraspidomorphs were thelodont-like, i.e., with a microsquamose exoskeleton. This means that this new definition of the Pteraspidomorphi is, after all, very near to that of Goodrich (1909).

5.1.3 A cladogram of the Craniata

The hand-made—and admittedly subjective—cladistic analysis of the major craniate taxa, including the soft anatomy and physiological characters of extant forms has led to two cladograms (Janvier 1984; Forey 1984*a*) which are congruent in many respects, except for the thelodonts, which are regarded as monophyletic by Forey (1984*a*). A computer analysis using HENNIG 86 (Farris 1988) produces a far less resolved consensus tree, although the galeaspids and osteostracans remain more closely related to the gnathostomes than they are to any other jawless group. Anaspids, lampreys, and the *incertae sedis* genera *Jamoytius* and *Euphanerops* remain together in the majority of the most parsimonious trees. Figure 5.1 shows two cladograms of the Craniata. On the left-hand side, the Cephalaspidomorphi are paraphyletic (or diphyletic), with osteostracans more closely related to the gnathostomes than to any other taxon. On the right-hand side, the Cephalaspidomorphi are a clade, thereby implying that such characters as cellular dermal bone, open endolymphatic duct, and epicercal tail have either appeared twice, or have been lost in various other taxa (anaspids, galeaspids, lampreys). A sclerotic ring is assumed to be present in arandaspids (Gagnier 1993*a*; Fig. 4.2, 9), whereas it is lacking in all other craniates except osteostracans and gnathostomes. The nature of this dermal bone ring in *Sacabambaspis* needs, however, to be carefully checked. The eye of osteostracans and gnathostomes (placoderms and osteichthyans) is unique in having both a sclerotic

ring and a perichondrally ossified sclera (Figs 4.16E, 4.42D). The thelodonts are considered here as paraphyletic, i.e. any craniate branching off from the dashed line in Fig. 5.1 may be called a 'thelodont', by virtue of its microsquamose scale morphology.

Many other theories of craniate relationships will certainly be proposed in the future, including some in which all 'bony' jawless craniates ('ostracoderms') are regarded as more closely related to the gnathostomes than to hagfishes or lampreys (Gagnier1993*b*, Forey and Janvier 1994). This latter solution would avoid many reversions in lampreys (loss of paired fins, sensory-line canals, etc.). Nevertheless, considering the available data, I cannot see any character unambiguously supporting a clade Agnatha (jawless craniates). Moreover, despite the fact that the gnathostome skull looks more 'primitive' (i.e. more metameric in organization; see Chapter 6, p. 257) than that of any jawless craniate, I doubt whether the gnathostomes will ever be shown to be paraphyletic, i.e. agnathans being a sub-group of the gnathostomes.

5.2 INTERRELATIONSHIPS OF THE GNATHOSTOMATA

Fossils do not refute the monophyly of the two major extant gnathostome taxa, the Chondrichthyes and Osteichthyes. We have, however, pointed out in the preceding chapter (p. 149) that there is no evidence of prismatic calcified cartilage (the major chondrichthyan synapomorphy) before the end of the Early Devonian; and chondrichthyan-like scales—whether compound (growing) or placoid (non-growing)—occur at least as early as the Early Silurian. This may be interpreted either as merely due to sampling chance, or as an indication that there are stem-group chondrichthyans that lack this type of hard tissue—unless these are in fact stem-group gnathostomes. This question may be answered when articulated specimens of these early forms are found.

Where fossils are involved, the major questions of gnathostome interrelationships concern, as one could expect, the two exclusively fossil higher taxa, the Placodermi (Fig. 5.2A) and the Acanthodii (Fig. 5.2C).

5.2.1 Relationships of the Placodermi

Stensiö (1963, 1969) regarded placoderms as a paraphyletic group from which arose, independently, various chondrichthyan groups. He considered authrodires to be more closely related to elasmobranchs than to other placoderms because of some similarity in the internal anatomy of the braincase (e.g. the shape of the otic capsule) and the scapulocoracoid. Despite differences in

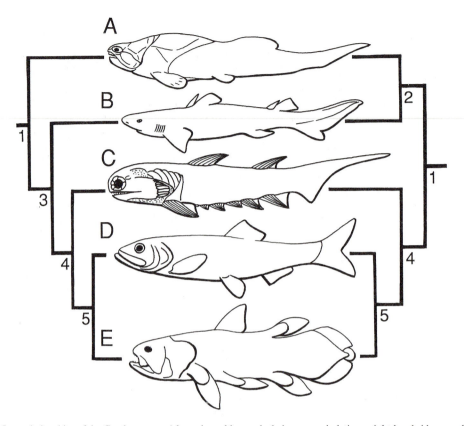

Fig. 5.2. Interrelationships of the Gnathostomata (elasmobranchiomorphs being monophyletic on right-hand side, paraphyletic on left-hand side). *Terminal taxa*: A, Placodermi; B, Chondrichthyes; C, Acanthodii; D, Actinopterygii; E, Sarcopterygii. *Nested taxa and selected synapomorphies*: 1, Gnathostomata (jaws, horizontal semicircular canal, medial gill arches, pelvic fins); 2, Elasmobranchiomorphi (pelvic claspers or internal fertilization, eye-stalk); 3, crown-group Gnathostomata (ossified ventral division of gill arches, cranio-ethmoidian fissure lost, dental lamina, anteriorly placed myodome for the superior oblique eye muscle); 4, Teleostomi (mouth in terminal position, deep and narrow braincase, ventral fissure, otico-occipital fissure, large hyoid gill cover armed with dermal plates or branchiostegals, large otoliths, or statoliths, separate anterior and posterior nostrils); 5, Osteichthyes (endochondral bone, large dermal units on head and shoulder girdle, main teeth borne by premaxilla, maxilla, and dentary, lepidotrichiae). *Age*: Silurian and Devonian (A); Silurian to Early Permian (C); Silurian to Recent (B, D); Early Devonian to Recent (E). (Reconstruction in A and B based on Miles and Westoll (1968) and Watson (1937).)

the position of the nasal openings, he suggested a relationship between the rhenanids and batomorphs. Later, Ørvig (1962) pointed out striking resemblances between the ptyctodontids and holocephalans (similar columnar pleromic dentine in tooth plates, prepelvic clasper, overall morphology), which may account for close relationships. Chondrichthyans and placoderms were thus gathered by Stensiö into the Elasmobranchiomorphi, a clade to which he also added the acanthodians (see also Jarvik 1980). This theory of placoderm paraphyly is no longer held: it would imply too many homoplasies in chondrichthyans, and too many independent losses of characters in the various placoderm taxa. There is now a consensus

over the monophyly of the placoderms and of the chondrichthyans.

There are, however, three competing theories of the relationships of the placoderms. They may be regarded as the sister-group of either all other gnathostomes, of chondrichthyans, or of osteichthyans.

5.2.1.1 Placoderms as the sister-group of all other gnathostomes (Fig. 5.2, left-hand side)

This theory implies that there are unique characters shared by osteichthyans and chondrichthyans, which never occur in placoderms. Although this idea was foreshadowed by Dean (1909), it was propounded by

Schaeffer (1975), who considered that there is a major difference between placoderms and all other gnathostomes in the shape of the palatoquadrate, which is roughly b-shaped in chondrichthyans and osteichthyans, but omega-shaped in placoderms (Fig. 4.27). In addition, the jaw musculature is attached on the external surface of the palatoquadrate in chondrichthyans and osteichthyans, but on its medial surface in placoderms (Fig. 4.42, 40). Later, Schaeffer (1981) considered that the presence of a postorbital (otic) articulation between the palatoquadrate and the braincase is a more reliable character of crown-group gnathostomes (chondrichthyans and osteichthyans), despite its secondary loss in some teleostomian taxa (e.g. some acanthodians and actinopterygians; Fig. 4.27D2). There is now, however, some evidence for such an articulation in arthrodires, although Young (1986b) doubts its homology with that of other gnathostomes. Rosen *et al.* (1981) added one more chondrichthyan–osteichthyan character: the presence of a metapterygium in paired fins. There again, and despite the poor preservation of the fin skeleton in placoderms, the articular surface for the pectoral fin skeleton shows an enlarged posterior area that may correspond to the articulation of the metapterygium (Young 1986b). The difference in palatoquadrate shape between placoderms on the one hand, and other gnathostomes on the other may simply be due to a unique condition in the former. The passage of the adductor muscles of the jaw in a medial position can easily be explained by the close fusion of the palatoquadrate and the dermal bones of the cheek, at least in arthrodires, antiarchs, rhenanids, and some 'acanthothoraci' (Fig. 4.42, 10). The thin bone layer (vinculum) between the autopalatine and the pterygoquadrate tends to disappear and the jaw musculature had no area for attachment other than the dorsomedial part of the palatoquadrate. Subsequently, a new endoskeletal lamina of the palatoquadrate expanded against the suborbital plate, laterally to the musculature (Fig. 4.42F).

Young (1986b) added four more unique chondrichthyan–osteichthyan characters: (1) the internal rectus eye muscle inserting in a posterior position in the orbit; (2) the superior oblique eye muscle inserting in the anterior part of the orbit; (3) the fusion of the nasal capsule (rhinocapsular bone) to the rest of the braincase; and (4) the presence of a dental lamina which is a prerequisite for the formation of true teeth (see also Chapter 6, p. 278). For the first three characters, the condition in placoderms is admittedly more similar to that in lampreys (internal rectus muscle in somewhat anterior position, superior oblique muscle in posterior position (Fig. 4.56, 29), nasal capsule separated from the braincase (Fig. 4.42, 34), as long as the attribution of the myodomes in placoderms is correct. If, however, osteostracans are regarded as the

out-group of the gnathostomes, then a separate nasal capsule does not hold as a general gnathostome character, since no such a separate capsule occurs in osteostracans. The fourth character (absence of a dental lamina) is more difficult to assess, since gnathal plates are not known in all placoderms, in particular not in the supposedly primitive 'Acanthothoraci'.

In sum, there remain some arguments to justify the theory that placoderms are the sister-group of all other gnathostomes (Fig. 5.2, left-hand side). I would consider that the position of the extrinsic eye muscles is the most persuasive. This theory nevertheless remains poorly supported.

5.2.1.2 Placoderms and chondrichthyans as the sister-groups (Fig. 5.2, right-hand side)

The affinity of the placoderms with chondrichthyans was first suggested by Quenstedt (1838), and was first developed in detail by Stensiö (1925, 1963, 1969). Goujet (1982) proposed three unique characters of placoderms and chondrichthyans: (1) the presence of an eye-stalk, uniting the eye to the braincase (Fig. 4.42, 35), (2) the presence of processes for muscle attachment to the eyeball (Fig. 4.42D), and (3) the presence of pelvic claspers derived from the pelvic metapterygium in males, implying similar reproductive biology with internal fertilization and horny egg-cases (23, 24, Fig. 4.51; see Chapter 7, p. 305). Young (1986b) added to these synapomorphies: (4) the loss of otico-occipital and ventral fissures, and (5) the extensive subocular shelf of the braincase. Other resemblances in internal anatomy between placoderms and chondrichthyans (Stensiö 1963) can be interpreted as general gnathostome characters.

The eye-stalk is present in most placoderm groups and in most elasmobranchs (Fig. 3.13, 20), but not in holocephalans. Its presence in fossil stem-group chondrichthyans (that is, chondrichthyans which are probably neither elasmobranchs, nor holocephalans) suggests, however, that is has been lost in holocephalans. Gardiner (1984c) regarded the eye-stalk as a gnathostome character because he considered it to be present, though in the form of a ligament, in some osteichthyans, but there is no clear support for this homology. Consequently, the eye-stalk still remains as a possible chondrichthyan–placoderm synapomorphy.

The question of the claspers in ptyctodonts has long been debated, but there is as yet no satisfactory clue (Ørvig 1962; Miles 1967; Miles and Young 1977; Goujet 1982; Young 1986b). We know only that in ptyctodonts two dermal elements (one hook-shaped and the other spiny) are associated with the pelvic fin, and were probably borne by endoskeletal elements of the basipterygium (Fig. 4.51, 25). There elements would thus be comparable to the fields of denticles on the claspers in male chondrichthyans. Miles and Young (1977) have shown that these elements match fairly well a bifid clasper as found in some Recent Holocephalans (Fig. 3.15G). If ptyctodonts had pelvic claspers, then they had the same type of internal fertilization as chondrichthyans, and perhaps the capacity to secrete horny egg capsules. The problem with pelvic claspers is that they are apparently lacking in all other placoderms, although very few pelvic fins have actually been observed. Long (1984) has described a peculiar club-shaped endoskeletal (probably metapterygial) element in the pelvic fin of the phyllolepid *Austrophyllolepis*; this may suggest the presence of a pterygopodium that supported the clasper. One of the current chondrichthyan phylogenies implies that pelvic claspers are lacking in the morphotype of this group, because they are lacking in *Cladoselache*, which can be regarded as the sister-group of all other chondrichthyans on the basis of such characters as the unsegmented radials (Fig. 4.39, left-hand side). It is, however, possible that this is linked with a reduction of the pelvic fins in the *Cladoselache*, if the latter are regarded as the sister-group of the Eugeneodontida, where these fins are totally lacking (Fig. 4.39, 10).

The other placoderm–chondrichthyan characters, i.e. the unfissured braincase and extensive suborbital shelf, have been discussed in the preceding chapter (p. 128) in connection with the gnathostome characteristics. By comparison with jawless vertebrates, such as osteostracans or galeaspids, these characteristics would appear to be general for the gnathostomes, since they both occur in osteostracans (given that the ventral endoskeletal lining of the orbits is the homologue of the suborbital shelf).

5.2.1.3 Placoderms and osteichthyans as the sister-groups

The idea that placoderms may be related to osteichthyans because they possess large dermal bones was widespread among early students of the group. The placoderms have been classified as siluriforms (Agassiz 1835; Huxley 1861), ganoids (McCoy 1848), sturgeons (Jaekel 1911), or even relatives of lungfishes (Woodward 1891). Although he was convinced that placoderms were more closely related to chondrichthyans, Stensiö (1959, 1963) proposed homologies between their dermal bones and those of osteichthyans, thereby implying that the macromeric condition of the exoskeleton was primitive for the gnathostomes. He therefore refers to the gnathal plates of arthrodires to as 'vomers', 'palatopterygoid' and 'mixicoronoid', for example. For the skull-roof, the cheek, or the thoracic armour, he used osteichthyan terminology only when it was clear from the underlying structures, such as the 'parietals' covering the otic capsule, the

'opercular' attached to the hyomandibula, or the 'cleithroclavicular' lining the scapulocoracoid. All other bones that did not match osteichthyans were given non-committal names, such as 'nuchal' or 'paranuchal' plates. Gross (1962b) rejected this partial homology, and most subsequent authors have used noncommittal names for placoderm plates. Forey (1980) and Gardiner (1984c) have, however, revived this homology, but in advocating the theory that placoderms and osteichthyans are sister-groups. They support this relationship by ten unique characters: (1) neurocranium and branchial arches protected by a series of large dermal plates that interlock and overlap one another; (2) dermal plates covering the lateral face of the scapulocoracoid; (3) a parasphenoid; (4) endochondral bone; (5) descending bony laminae of the skull-roof, which embrace the neurocranium; (6) postbranchial lamina ornamented with multicuspid denticles; (7) supracoracoid foramen; (8) dermohyal (submarginal plate) fused to the head of hyomandibula; (9) basihyal and urohyal; and (10) several perichondral ossifications in the shoulder girdle, palatoquadrate, Meckelian bone, and basibranchial series. Most of these synapomorphies have been discussed and refuted by Goujet (1982) and Young (1986b). The resemblance in dermal bone pattern between placoderms and osteichthyans (characters 1, 2, 3, 5, 6, 8) would be plausible if only arthrodires—and in particular advanced arthrodires—are concerned, but it becomes largely irrelevant when the supposedly generalized placoderm pattern (exhibited by the 'acanthothoracids') is considered. Only the dermal shoulder girdle of primitive placoderms may show some similarity with that of osteichthyans. The vague resemblance between arthrodires and osteichthyans in the pattern of the dermal bones of the palate (gnathal plates, parasphenoid, character 3) may be explained as a general tendency in macromeric gnathostomes, where bones occur primarily in certain areas where dermal denticles are more concentrated. Nelson (1970) has shown that in sharks these areas are situated around the buccohypophysial canal (parasphenoid) and at places that match the position of the vomers, palatine, and entopterygoids of osteichthyans. The presence of endochondral bone in placoderms (character 4) is uncertain, and is based on a few spongy trabecles in the rhinocapsular bone (Fig. 4.42, 34). However, most of the endoskeleton, including the braincase, is only perichondrally ossified. The multicuspid denticles (character 6) of the postbranchial lamina are similar to the independant multicuspid scales that occur around the gill openings of many vertebrates, including perhaps thelodonts, and may well be a general device for preventing ectoparasites from entering the gill chamber (Patterson 1977a) in all craniates with a mineralized exoskeleton. A supracoracoid-like foramen (character 7) occurs only in ptyctodonts (Mark-

Kurik *et al.* 1991) and some 'acanthothoracids' and also in acanthodians. It remains as a valid synapomorphy, only if the deep scapulocoracoid (of ptyctodont type) is regarded as general for all placoderms. The dermohyal (in fact, the submarginal plate; (Fig. 4.42, 12) fused with the head of the hyomandibula (character 8) is also cited by Gardiner and Schaeffer (1989) as a synapomorphy of actinopterygians. There is also uncertainty about the nature of what is called the hyomandibula in placoderms. Goujet (1984*a*) sees it as a true hyomandibula, whereas Young (1986*b*) regards it as the opercular cartilage, i.e. a derivative of the hyoidean gill rays. The presence of a basihyal and urohyal (character 9), inferred from a disarticulated ptyctodont, remains uncertain. Finally, the fact that there are several perichondral ossifications (character 10) in various endoskeletal elements is difficult to assess, since chondrichthyans have lost perichondral bone. Osteostracans and galeaspids have a single perichondral ossification for the entire head, but they have no palatoquadrate and Meckelian bone.

In sum the placoderm-osteichthyan sister-group relationships remains poorly supported, or supported by characters that are difficult to assess, because they are too vaguely defined. We are thus inclined towards choosing here between the two earlier theories (Fig. 5.2, 2, 3). The sister-group relationship between placoderms and chondrichthyans remains supported by more discrete and well-defined characteristics, such as the eye-stalk and pelvic claspers (or internal fertilization) (Fig. 5.25, 2). The name 'Elasmobranchiomorphi' can be retained for the group including chondrichthyans and placoderms, although an ancestor–descendant relationship between these two groups is ruled out.

5.2.2 Relationships of the Acanthodii

The question of the relationships of acanthodians has been less debated than that of the placoderms, perhaps because they provide fewer anatomical data than the latter. Many early authors placed them in a group 'of their own', with no clear relationships to any other particular group of gnathostomes. In contrast, Huxley (1861) and Traquair (1890) related them to the 'ganoids', i.e., actinopterygians or osteichthyans in general. This opinion prevails today, in spite of a long tradition of placing the acanthodians with the chondrichthyans, which stems from Quenstedt (1851). Woodward (1935) and Watson (1937) regarded acanthodians as most closely related to the placoderms, both taxa being supposed to share a functional gill slit between the mandibular and hyoid arches (Watson's 'aphetohyoidy', now rejected, except by Zangerl and Williams 1975). They were thus considered to be the most primitive gnathostomes. Most of the con-

clusions on the affinities of acanthodians are based on anatomical data that are taken from a few specimens of one of the latest known species, *Acanthodes bronni*. As we have seen above (Chapter 4, p. 176), the interpretation of the internal anatomy depends largely on the way in which the braincase is reconstructed, as shown by the analyses of Miles (1973*b*) and Jarvik (1977, 1980).

Postcranial characters have little bearing on the phylogenetic position of acanthodians. We shall consider here the three possible theories of the relationship of acanthodians advocated by various authors during the past fifteen years: as the sister-group of all other gnathostomes (including placoderms), of the chondrichthyans, or of the osteichthyans.

5.2.2.1 Acanthodians as the sister-group of all other gnathostomes

Although foreshadowed by Watson (1937), this idea was resurrected by Rosen *et al.* (1981), on the basis of the presumed absence of a metapterygium in acanthodians, in contrast to the presence of this endoskeletal fin element in both chondrichthyans and osteichthyans (and probably also in placoderms). This status would be consistent with the very early (Early Silurian) age of the earliest known acanthodians, but it rests on a single, poorly evidenced character. In fact, the endoskeletal support of the acanthodian paired fins is very poorly known and can hardly be used with confidence in a character analysis.

The arrangement of the ceratotrichs in the pectoral fin of *Acanthodes* (Forey and Young 1983), which extend far posteriorly, strongly suggests the presence of a large metaptergygium (Fig. 4.59D).

5.2.2.2 Acanthodians and chondrichthyans sister-groups

Acanthodians have long been regarded as closely related to the chondrichthyans, and particularly the elasmobranchs, mostly on the basis of their overall morphology (which is largely generalized), small scales, the interscale position of the sensory lines, fin spines, and structure of the branchial skeleton (posterior orientation of the pharyngobranchials; Fig. 4.59, 7), which is supposed to have extended far behind the braincase. Jarvik (1980) has emphasized the resemblances between *Acanthodes* and elasmobranchs, in particular the 'orbitostylic sharks' (Fig. 3.14, 5) because of the presence of a prominent orbital process on its palatoquadrate. In any case, since they possess neither prismatic calcified cartilage nor pelvic claspers, acanthodians cannot be regarded as a subgroup of the chondrichthyans. At best, they may be their sister-group. In fact, the presence of median fin spines might be regarded as unique to the two groups.

5.2.2.3 Acanthodians and osteichthyans as sister-groups

The largest number of unambiguous characters links acanthodians with osteichthyans. The presence of three otoliths, for example, is regarded as unique to osteichthyans, and well-preserved acanthodians clearly display the lagenar, saccular, and utricular otoliths (Fig. 4.58, 24–6). Recent studies on the otoliths of early actinopterygians (Coates 1993*a*) suggest, however, that three otoliths are not general for osteichthyans. Instead, a single, large saccular otolith is a more likely general condition. Other characters shared only with osteichthyans are the long branchiostegals in the main (hyoidean) gill cover (Fig. 4.60, 1, 4), the spiracular groove in the neurocranium (Fig. 4.58, 21), an otico-occipital fissure (Fig. 4.58, 13), a possible ventral fissure (Fig. 4.58, 15) separate anterior and posterior nostrils, and a narrow-based braincase (Fig. 4.27C). The small scales that cover the fins of acanthodians are elongated in shape, and are somewhat suggestive of the osteichthyan lepidotrichs (Heyler 1969), yet a similar condition is also met with in osteostracans. The presence of endochondral bone is still uncertain in acanthodians, although some bony trabecles have been observed internally to the perichondral bone sheath. In sum, the relationships of acanthodians with osteichthyans seems to be supported by the best-defined and the most reliably assessed characters (Fig. 5.2, 4). If the placoderms are considered to be the sister-group of osteichthyans, acanthodians would fall closer to the latter, because of the presence of large otoliths and a fissured braincase, which are lacking in placoderms, but this would imply that acanthodians have lost the large dermal plates. The name 'Teleostomi' is now often used for the group that includes acanthodians and osteichthyans (Schultze 1993).

5.3 INTERRELATIONSHIPS OF THE SARCOPTERYGII

The question of sarcopterygian interrelationships includes that of the affinities, or 'origins' of the tetrapods. As we come closer to Man, relationships become more difficult to unravel because characters are often assessed in a less objective manner: each palaeontologist tries to demonstrate that his pet fossil is the closest to the ancestry of a major taxon, by the virtue of its age or its morphology. This interesting behaviour is particularly illustrated by the attempts to show that *Ichthyostega* was Carboniferous, instead of Devonian, in age, and to demonstrate that *Elpistostege*, from the Late Devonian of Miguasha, was the earliest (see Jarvik 1948*b*). There are also several other early 'tetrapod' remains, which have later turned out to be fish, but also fish remains which are now referred to tetrapod. The strong reactions against the suggestion by Rosen *et al.* (1981) that lungfishes were more closely related to the tetrapods than to any other extant or fossil taxon can also be explained by this appeal to ancestors. If Rosen *et al.* were right, then osteolepiforms, which are such good fossils for constructing transformation series, would mean nothing, and the common ancestor to lungfishes and tetrapods would be unknown. In a totally different, and refreshing way, Long (1985*b*, see also Young *et al.* 1992) has treated osteolepiforms as any other fish clade, with no reference to the tetrapods (possibly to avoid entering a still-hot debate), and by using such 'trivial' characteristics as the axillary scutes, instead of endoskeletal features, classically assessed by reference to the tetrapods. Long's work had the merit of defining the osteolepiforms; with the consequence of excluding the panderichthyids, which now appear as more closely related to the tetrapods than the osteolepiforms.

In the preceding chapter, a number of sarcopterygian clades were defined. The monophyly of onychodontiforms and porolepiforms is still tenuously supported. In contrast, that of actinistians, lungfishes, and rhizodontiforms seems to hold up well. Osteolepiforms and panderichthyids are regarded here as monophyletic taxa, but this could be overturned with only a slight loss of parsimony.

Schultze (1991) reviewed the current theories concerning the 'origin of tetrapods'; his discussion extends, in fact, to the interrelationships of all sarcopterygians. He recognized four theories: (1) the diphyletic origin of tetrapods (e.g. Holmgren 1949; Jarvik 1942, 1981*a*); (2) the lungfish–tetrapod sister-group relationship (e.g. Rosen *et al.* 1981; Forey *et al.* 1991); (3) the osteolepiform (or panderichthyid)–tetrapod sister-group relationship (Cope 1892 and most authors since then; e.g. Holmes 1985; Schultze 1991, 1994); and finally (4) the sarcopterygian–tetrapod sister-group relationship (Chang 1991*b*). I shall not discuss here these different theories, which have been debated at length through piles of pages. The most constroversial one has long been the diphyletic origin of tetrapods, which Jarvik (1981*a*) now tends to transform into a tri- or quadriphyletic origin (with the amniotes appearing independently from the anurans, and perhaps also the apodans, from osteolepiform ancestors). The first idea, that the urodeles (or caudates) arose independently from all other tetrapods from fish ancestors stems from Holmgren's (1933) observation of limb development, and his subsequent suggestion that the urodele limb was more readily derived from the pectoral fin skeleton of lungfishes. Since lungfishes turned out to be unsuitable as ancestors (because of their numerous specializations and 'false' choanae), they were replaced in this role by the fossil porolepiforms, which were generalized enough, and might have possessed small choanae (Jarvik 1942). A long period then followed during which both porolepiforms and osteolepiforms—or rather, two species *Glyptolepis groenlandica* and *Eusthenopteron foordi*—were meticulously studied to match urodeles and anurans respectively. When integrated into a fully segmentalist and transcendental view of craniate evolution (see Chapter 6, p. 259), this theory became even more immune from criticism, because every missing character could be inferred from its presumed presence in any ideal metamere. The fact that the choana in urodeles forms from a gut-process (endoderm), and not from the stomodeum (ectoderm) as in amniotes and apodans, is explained, for example, as a derivation from a once-existing prespiracular gill slit that gave rise to the nasal sacs (Bjerring 1977).

The polyphyletic origin of tetrapods is now considered as unparsimonious with regard to the number of tetrapod, amphibian, and lissamphibian synapomorphies (see Chapter 3, p. 76 and Chapter 4, p. 225), and the peculiarities in the development of the urodele limbs are now interpreted as a unique derived condition (see Chapter 6, p. 270).

All the other theories imply tetrapod monophyly, but differ in the way in which the characters shared by some fishes and the tetrapods are assessed. The goal is, of course, to find the largest number of these characters. As we have seen in Chapter 3 (p. 81), there are still uncertainties about the interrelationships of extant sarcopterygians (actinistians, lungfishes, and tetrapods). Many of the tetrapod-like features of lungfishes are discarded as results of parallelism, yet some of them remain less ambiguous, such as the sigmoid aspect of the arterial cone of the heart (Fig. 3.22). However, most of the physiological and molecular evidence points towards a closer relationship between lungfishes and tetrapods. Thus, contrary to the problem of the cyclostomes versus vertebrates discussed above, the Recent taxa do not help much in providing an initial hypothesis. Let us now see how fossils can help.

5.3.1 Sarcopterygian monophyly

Most palaeontologists now agree on the monophyly of the Sarcopterygii, as including the tetrapods (Fig. 5.3, 1). They are characterized essentially by the monobasal articulation of paired fins, numerous sclerotic plates, and a processus ascendens in the palatoquadrate; but only the first of these characters is observed in all taxa (except some limbless tetrapods). Most palaeontologists agree also on the monophyly of the Rhipidistia (sarcopterygians with folded teeth), but only a few include lungfishes in this taxon. Chang's (1991*b*) theory that tetrapods might

be the sister-group of all other sarcopterygians was something of a surprise. She considers that all sarcopterygyans, except the tetrapods, share five synapomorphies: cosmine (lost in several taxa independently); intracranial joint (lost in lungfishes); submandibular series of dermal bones; anocleithrum in shoulder girdle; median extrascapular and extensive hyomandibular facet. Chang's theory would imply that folded teeth, and perhaps also choanae, have appeared twice (unless choanae are unique to the tetrapods).

Sarcopterygians comprise a major clade, the Rhipidistia (Fig. 5.3, 3), which includes the extant lungfishes and tetrapods as well as a number of fossil taxa (see below), and two taxa of still-debated affinities, the Onychodontiformes and Actinistia. I accept that actinistians are more closely related to rhipidistians than onychodontiforms. In fact, the latter retain poorly developed basal fin lobes and a few other characters assessed as plesiomorphous, since they are present in actinopterygians (e.g. the large, posteriorly expanded maxilla; (Fig. 4.74, 1).

5.3.2 Rhipidistian monophyly

Rhipidistians, including the tetrapods, can be defined by the presence of plicidentine, i.e. folded teeth, notwithstanding the fact that this tooth structure also occurs independently in other, younger vertebrate groups (ginglymods, ichthyosaurs), and may be incipient in the basal part of onychodontiform tusks (Fig. 4.74, 9). I have assumed here that lungfishses have lost this character, although some of the larger teeth on the parasphenoid of *Diabolepis* have a folded base (Fig. 4.84, 18). The general dipnomorph condition of this character is exhibited by porolepiforms (Fig. 4.77, 42), although the dendrodont (highly folded) structure of their teeth is probably derived. Another rhipidistian character may be the presence of a postorbital junction between the supraorbital and infraorbital sensory-line canals (Ahlberg 1991*a*; Fig. 4.77A2, 25, 26). This junction does not exist in early actinopterygians, onychodontiforms, and actinistians, but is present in porolepiforms, osteolepiforms, and panderichthyids (the condition in the rhizodontiforms is poorly known). It is, however, also lacking in lungfishes and *Youngolepis*, probably as the result of a reversion, as suggested by the persistence of a connection between the infraorbital line and the otic part of the main lateral line (contrary to the general osteichthyan condition represented by actinopterygians).

The Rhipidistia include the Dipnomorpha (porolepiforms and lungfishes), the Rhizodontiformes and the Choanata (Osteolepiformes, Panderichtyida, and Tetrapoda). The Rhizodontiformes are regarded as more closely related to the choanates on the basis of their similar uniserial paired fins (doubtfully unique to this group, if regarded as the primary metapterygial structure) and, in particular, of their quite similar humerus (Fig. 4.87, 7; Fig. 4.91, 3; Fig. 4.96, 17). Rhizodontiforms and choanates form the Tetrapodomorpha (Fig. 2.3. 5).

5.3.3 Choanate monophyly

The taxon Choanata, i.e. vertebrates with choanae, has been erected by Säve-Söderbergh (1934) to include the osteolepiforms, lungfishes, and tetrapods. Yet, he regarded tetrapods as diphyletic, with lungfishes and urodeles as sister-groups, and osteolopiforms and all other

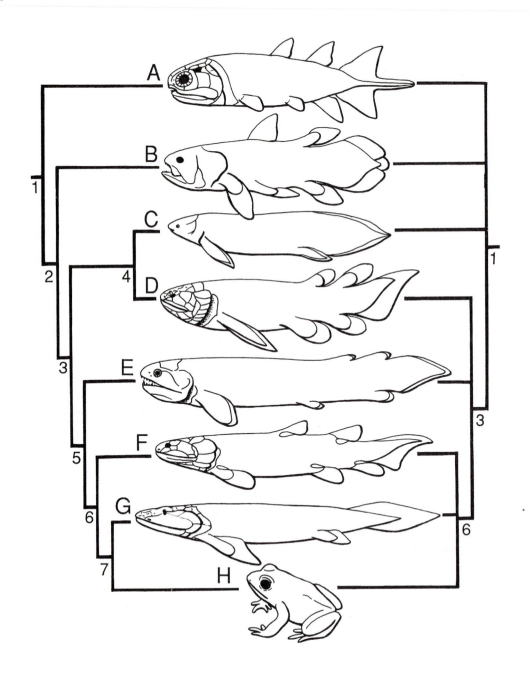

tetrapods as sister-groups. The lengthy debate about the lungfish and tetrapod choanae will be briefly outlined below (Chapter 6, p. 266). It culminated with the publication of a paper by Rosen *et al.* (1981) and various subsequent responses to it (e.g. Jarvik 1981*b*, Holmes 1985, Panchen and Smithson 1987). The highly probable presence of two external nostrils, associated with typical lungfish characters (tooth plates with radiating ridges, Westoll lines, 'B' bone) in *Diabolepis* (Fig. 4.84C1, 14, 15) supports the long-held idea that the internal nostrils of other lungfishes have been acquired *within* the group, independently from those of tetrapods and their presumed piscine relatives, osteolepiforms and panderichthyids. Since choanae were first defined in tetrapods (Man), I see no reason to reject the taxon Choanata because lungfishes are now excluded. There are many major taxa that have changed in content, but not in name. In this case, it should include only the tetrapods, panderichthyids, and osteolepiforms (Fig. 5.3, 6), since no other sarcopterygian clearly possesses choana-like palatal openings surrounded by the maxilla, premaxilla, vomer, and dermopalatine (Fig. 4.89, 7; Fig. 4.94, 31; Fig. 4.95, 39). This choanate condition is congruent with the presence of only one external nostril, ventrally lined by a lateral rostral. The condition in rhizodontiforms is still unknown, but they seem to retain two external nostrils (Fig. 4.86, 22, 23). The major problem now is perhaps with the osteolepiforms and panderichthyids. They admittedly form a neat morphocline leading to the tetrapods and, like the latter, possess a single external nostril (Fig. 4.88, 9; Fig. 4.94, 30). However, as pointed out by Rosen *et al.* (1981) and Forey (1987, see also Forey *et al.* 1991), most of the resemblances between the tetrapods and these fishes is due to shared general osteichthyan, sarcopterygian, or rhipidistian characteristics. In addition, the resemblance pointed out by Jarvik (1942, 1981*b*) between

Eusthenopteron and a Triassic capitosauroid is clearly the result of a convergence, and the use of a more generalized temnospondyl, an anthracosaur, or *Ichthyostega* would provide a quite different pattern (Fig. 4.101B, 4.103A). Ahlberg (1991*a*) considers that only two characteristic are unique to osteolepiforms, panderichthyids, and tetrapods (single external nostrils, dermintermedial process in the nasal cavities), and that only two are unique to tetrapods and panderichthyids (frontals and loss of the intracranial joint; Fig. 5.3, 7). To him, the choana is a character of rhipidistians, and may coexist with two external nostrils in dipnomorphs and rhizodontiforms. Strangely, the long, rod-shaped hyomandibula of osteolepiforms and panderichthyids does not foreshadow the stout stapes of early tetrapods (Fig. 4.96C; Clack 1988). In contrast, the stapes resembles more closely the stout hyomandibula of the dipnomorphs (Fig. 4.78, 28; Fig. 4.81, 11) or even actinistians (Fig. 4.75, 14).

Many of the unique characters that support the osteolepiform–panderichtyid–tetrapod morphocline are easily overturned or shown to occur elsewhere (Rosen *et al.* 1981; Forey *et al.* 1991). The uniserial paired fins can be interpreted as a general sarcopterygian characteristic, when considered as the primitive structure of the gnathostome metapterygium (Fig. 4.29B6). The dermintermedial process also occurs in *Youngolepis* and may be a general rhipidistian character. Paired dermal bones anterior to the parietals (frontals) occur in some actinistians (Fig. 4.76, 7) and also in lungfishes (Fig. 4.80C), and the loss of intracranial joint occurs also in lungfishes. The single external nostril occurs in lungfishes (admittedly by convergence), but Rosen *et al.* (1981) have shown that the exoskeletal nasal fenestra of osteolepiforms may well have contained two closely-set external nostrils, anterior and posterior (Fig. 5.4A). This latter interpretation has been largely discarded by subsequent

◀ **Fig. 5.3.** Interrelationships of the Sarcopterygii. One of the fully resolved cladograms on left-hand side, essentially from Ahlberg (1991*a*), strict consensus of most of the current phylogenies on right-hand side (except the phylogenies proposed by Jarvik (1981*a*), Rosen *et al.* (1981), Chang (1991*a*), and Forey *et al.* (1991).) *Terminal taxa*: A, Onychodontiformes; B, Actinistia; C, Dipnoi; D, Porolepiformes; E, Rhizodontiformes; F, Osteolepiformes; G, Panderichthyida; H, Tetrapoda. *Nested taxa and selected synapomorphies*: 1, Sarcopterygii (monobasal paired fins, ventral fissure developed dorsally into an intracranial joint, more than four sclerotic plates); 2, maxilla reduced or lost, large basal lobe in fins; 3, Rhipidistia (folded dentine and enamel in teeth, no maxilla–preopercular contact, large quadratojugal, postorbital junction of supraorbital and infraorbital sensory-lines [lost in lungfishes], only three coronoids, nasal capsule with lateral recess, ?dermintermedial process [lost in porolepiforms]); 4, Dipnomorpha (elongated pectoral fin, with numerous mesomeres, rostral tubules, branched posterior radials in posterior dorsal fins, enamel entering the flask-shaped cavities of cosmine, pineal reached but not surrounded by parietals, no parietal–supraorbital contact, infraorbital sensory-line canal follows premaxillar suture); 5, Tetrapodomorpha (concave glenoid fossa, perforated humeral process, large entepicondylar process, deltoid and supinator processes on humerus); 6, Choanata (single external nostril, choanae, dermintermedial and tectal processes in external nostril [unless at node 3]); 7, large frontals meeting along the midline, elongated humerus, jugal meeting quadratojugal, dorsally placed orbits, ventrally placed external nostril near oral margin, loss of dorsal and anal fins). *Age*: Devonian (A, D); Devonian to Carboniferous (E); Late Silurian or Early Devonian to Recent (C); ?Middle to Late Devonian (G); Middle Devonian to Early Permian (F); Middle Devonian to Recent (B); Late Devonian to Recent (H). (Reconstructions of A–G based on Andrews (1985), Jarvik (1980), Jessen (1966), and Vorobyeva and Schultze (1991).)

authors. The external nostril in a well-preserved snout of the osteolopiform *Shirolepis* (Middle Devonian of Siberia; holotype, Palaeontological Institute of the Russian Academy of Science) shows, however, two ventral grooves, almost devoid of cosmine, and separated by the cosmine-covered area of the dermintermedial process (Fig. 5.4B, 5, 6). This suggests the presence of two narinal ducts. In addition, a generalized Middle Devonian osteolepiform, *Kenichthys*, displays a long, slit-like external nostril that may be compatible with the presence of two narial tubes (Chang and Zhu 1993). By assessing the tetrapod-like characters of osteolepiforms

and panderichthyids as general for osteichthyans or sarcopterygians, and by regarding their choanae as a mere palatal hinge or a pit for the lower tusks (Fig. 4.89A2), Rosen *et al.* (1981) and Forey *et al.* (1991) were forced to regard these taxa as sister-groups to all other sarcopterygians. However, and whatever doubts there may be about the characteristics mentioned above, the description of the panderichthyids by Vorobyeva and Schultze (1991) has provided additional unique characters shared by the panderichthyids and tetrapods, e.g., the jugal meeting the quadratojugal (Fig. 4.94, 10, 11); 'labyrinthodont' tooth structure (with no attachment bone between the folds; Fig. 4.94E); posteriorly extended nasal capsule; and loss of median fins, except the caudal (Fig. 4.94A). The sister-group relationships between panderichthyids and tetrapods thus seem better and better corroborated (Fig. 5.3, 7).

The presence of diamond-shaped scales and a ventrally placed external nostril in panderichthyids suggests that the common ancestor they may have shared with osteolepiforms and tetrapods (Fig. 5.3, 6) looked more like an 'osteolepid' (such as *Thursius* or *Gyroptychius*; Fig. 4.93B–F) or a canowindrid rather than like *Eusthenopteron*. Also, the polyplocodont type of tooth folding may well be a general rhipidistian characteristic, since it occurs also in *Youngolepis*. (The dendrodont type in porolepiforms would thus be unique to this group.)

In sum, almost all conceivable relationships have been assigned to the tetrapods within the Sarcopterygii or, at least, within the Rhipidistia. The theory illustrated by the cladogram on the left-hand side of Fig. 5.3 is derived from Ahlberg (1991*a*) and is the most easily acceptable in terms of transformation series and character distribution. It is also relatively consistent with the stratigraphical record (except for actinistians and rhizodontiforms). The poorly resolved cladogram on the right-hand side depicts the consensus on which most palaeontologists would agree (except Jarvik 1981*a*, Chang 1991*b*, and Forey *et al.* 1991).

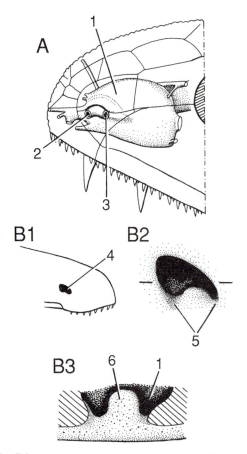

Fig. 5.4. A, Heterodox interpretation by Rosen *et al.* (1981) of the external nostril in the osteolepiform *Eusthenopteron*; B, *Shirolepis*, Late Devonian of Siberia (specimen figured by Vorobyeva (1977, pl. 2:3)), outline of the snout in anterolateral view (B1, ×2), right external nostril in lateral view (B2) and dorsal view of its floor (B3, section at the level indicated in B2). 1, nasal cavity; 2, anterior nostril; 3, posterior nostril; 4, external nostril; 5, ventral grooves devoid of cosmine; 6, dermintermedial process. (A, combined from Jarvik (1980).)

5.4 CRANIATE CLASSIFICATION

The classification of the fossil and extant Craniata in Table 5.1 is derived from the cladograms in Fig. 5.1 (left-hand side), 5.2 (right-hand side), and 5.3 (left-hand side), which show the here preferred theories of the relationships.

Further reading

Craniate interrelationships: Bjerring (1984); Forey (1984*a*); Forey and Janvier (1993, 1994); Gagnier (1993*b*); Halstead (1973*a*, 1982); Janvier (1978*a*, 1981*a*, 1984, 1986, 1993);

Table 5.1. Classification of fossil and extant Craniata
(See also Figs 5.1–3)

Craniata
  Hyperotreti
  Vertebrata
    Pteraspidomorphi
      Arandaspida
      Heterostracomorphi
        Astraspida
        Heterostraci
    Myopterygii (crown group Vertebrata)
      (unnamed taxon)
        Anaspida
        Hyperoartia
      (unnamed taxon)
        Galeaspida
        (unnamed taxon)
          Pituriaspida
          (unnamed taxon)
            Osteostraci
            Gnathostomata
              Elasmobranchiomorphi
                Placodermi
                Chondrichthyes
              Teleostomi
                Acanthodii
                Osteichthyes
                  Actinopterygii
                  Sarcopterygii
                    Onychodontiformes
                    (Crown-group Sarcopterygii)
                      Actinistia
                      Rhipidistia
                        Dipnomorpha
                          Porolepiformes
                          Dipnoi
                      Tetrapodomorpha
                        Rhizodontiformes
                        Choanata
                          Osteolepiformes
                          (unnamed taxon)
                            Panderichthyida
                            Tetrapoda

Janvier and Blieck (1979, 1993); Jarvik (1964, 1981*a*); Jollie (1968); Karatayute-Talimaa (1978); Kiaer (1924); Maisey (1986, 1988); Moy-Thomas and Miles (1971); Novitskaya (1983, 1992); Novitskaya and Karatayute-Talimaa (1989); Schaeffer and Thomson (1980); Stensiö (1927, 1964, 1968); Turner (1991); N. Z. Wang (1991); Wängsjö (1952); Young (1991*a*). *Placoderm relationships*: Denison (1978); Forey (1980); Forey and Gardiner (1986); Gardiner (1984*b*); Goujet (1982); Gross (1962*b*); Jarvik (1980); Maisey (1986); Miles and Young (1977); Moy-Thomas and Miles (1971); Ørvig (1962, 1980, 1985); Schaeffer (1975); Stensiö (1963, 1969); Young (1986*b*). *Acanthodian relationships*: Denison (1979); Forey (1980); Jarvik (1977, 1980); Maisey (1986); Miles (1937*b*), Moy-Thomas and Miles (1971); Nelson (1968, 1969); Rosen *et al.* (1981); Schaeffer (1975); Schultze (1988); Watson (1937). *Sarcopterygian interrelationships*: Ahlberg (1991*a*); Andrews (1973); Bonde (1977); Campbell and Barwick (1984*a*); Chang (1991*b*); Forey (1984*b*); Forey *et al.* (1991); Gross (1964); Holmes (1985); Janvier (1980*a,b*, 1986); Jarvik (1942, 1964, 1980, 1981*a,b*); Maisey (1986); Miles (1975, 1977); Moy-Thomas and Miles (1971) Panchen (1967); Panchen and Smithson (1987); Patterson (1980); Rage and Janvier (1982); Rosen *et al.* (1981); Säve-Söderbergh (1934); Schultze (1987, 1991, 1993, 1994); Szarski (1977); Thomson (1964); Vorobyeva (1977); Vorobyeva and Schultze (1991); Westoll (1943*a,b*).

6

Anatomical philosophy: homologies, transformations, and character phylogenies

In 1809 Lamarck published his *Zoological philosophy*, and in 1818 and 1822 Etienne Geoffroy Saint-Hilaire published his *Anatomical philosophy*. Lamarck expressed ideas on how organisms have transformed with time; Geoffroy Saint-Hilaire speculated on how organs transformed or arose from a pattern that was once common to all animals. To Geoffroy, anatomical philosophy was a way of finding order and sense in the distribution of similarities among organisms. It included 'analogies' (homologies in the modern sense) and theories about transformations. This sounds like idealistic morphology, but it is just what people have done—and still do—when trying to find out how fins have transformed into limbs, for example. I therefore take the risk of resurrecting this old term for the title of this chapter.

I expect that readers will find in the present chapter what most students are accustomed to hearing, for they live in the evolutionary world of transformation series and anamorphose-like description of Nature: gills transform into jaws, fins into limbs, arms into wings, etc. All that follows is essentially what scientists have believed and still believe about craniate evolution. Some of these inferences, generally based on palaeontology, are certainly quite close to the truth (they are consistent with character distributions), but all of them are beyond reality in the sense that they incorporate a number of assumptions on processes, which represent the bulk of the palaeontological literature. These processes have led to the common practice of giving or retaining different names to things when they change, whether they be taxa (fishes and tetrapods) or characters (fins and limbs) and, ultimately, to a gradulastic view of life. Even present-day cladists, who reject ancestor–descendent relationships between taxa, continue to admit this kind of relationship between characters that bear different names (Nelson 1994), thereby supporting—often unconsciously—the idea that taxa are defined by their common ascent rather than by their characters.

Cuvier discarded Geoffroy's 'philosophy', as he did Larmarck's, because he considered that it was foolish to try to go beyond the facts to such a wide extent. The same is true of cladists, and in particular 'pattern cladists' (a mythical category erected to frighten evolutionary biologists and make them reject all cladistics), with evolutionary stories about transformations. However, many of the theories, in particular theories of homology, which arose from anatomical philosophy, are linked with ontogenetic inferences, and therefore have some value in assessing the degree of generality of characters. Knowing about these classics of anatomical philosophy is important in understanding why some of the theories about the relationships or evolution of early vertebrates are so deep-rooted, although some of them have been clearly refuted. It also shows how new theories may arise and are selected by scientists according to their fitness for these processes. This is the obligatory part of the mythology in modern science. To many, the question whether the tetrapods are sarcopterygians or otherwise sounds spurious in comparison to the extraordinary 'mystery' of the transformation of a fin into a limb, and even more so if selective pressure, or any other process, comes into play.

6.1 THE SEGMENTED VERTEBRATE

6.1.1 The three historical periods of the segmental theory

The idea that the vertebrates, and particularly their head, were once entirely composed of serially homologous segments is said to stem from Goethe (1820; see Goethe 1890–1906) and Oken (1807) (Fig. 6.1A). It became a real debate, however, when E. Geoffroy St Hilaire compared this presumed vertebrate segmentation with that of arthropods. This triggered the now-famous controversy between Geoffroy and Cuvier, in 1832, about the common *Bauplan* of the arthropods, molluscs, and vertebrates. At that time, Goethe had never published anything detailed on that particular subject, but when he heard of Geoffroy's interest in the recognition of vertebral components in the vertebrate skull he published an old manuscript which he had written in the late eighteenth century.

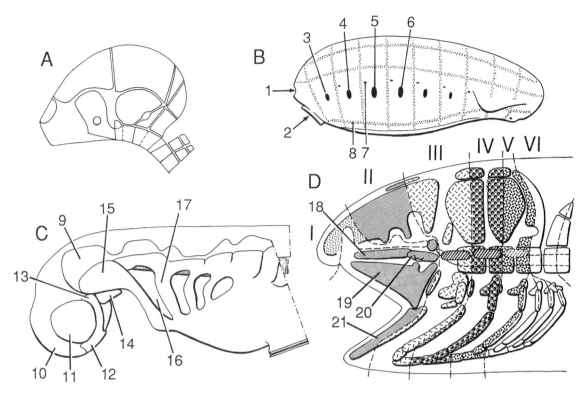

Fig. 6.1. The segmented vertebrate head. A, the Goethe–Oken model of the 'vertebral skull', as reconstructed by Jollie (1984); B, the ideal ancestor of the Craniata, as reconstructed by Bjerring (1984); C, the mesodermal segments and brain in the head of a 5.5-mm *Squalus* embryo, according to Bjerring's (1977) interpretation; D, the cranial segments of a gnathostome, according to Jollie's (1984) interpretation (dashed lines indicate the limit between the segments, myotomal components hatched). 1, terminal, inhalent gill opening (part of the nasopharyngeal duct or Rathke's pouch + Sessel's pouch); 2, sucker, corresponding to the stomodeum; 3, first paired gill opening (part of nasal sac, external nostrils, and choanae); 4, second paired, or prespiracular, gill opening (giving rise to part of the teleostome mouth, eye lens, and nasolacrymal duct); 5, spiracular gill opening (spiracle, auditory duct in tetrapods); 6, fourth paired gill slit (foremost functional gill slit in extant vertebrates); 7, nephric opening; 8, lateral-line neuromasts; 9, sclerotomic mesenchyme derived from somites (giving rise to most of the braincase and vertebrae); 10, brain; 11, optic vesicle (eye and optic tract); 12, Chiarugi's vesicle (parasomitic part of the first, or terminal segment according to Bjerring); 13, Platt's vesicle (somitic part of the first, or terminal segment, according to Bjerring); 14, premandibular mesoderm (second segment of Bjerring); 15, mandibular somite (third segment of Bjerring); 16, spiracular cleft; 17, hyoid somite (fourth segment of Bjerring); 18, trabecle; 19, palatoquadrate; 20, polar cartilage; 21, Meckelian cartilage; I, premandibular segment (devoid of visceral component); II, mandibular segment; III, hyoid, or antotic segment; IV–VI, metotic segments. (A, D, from Jollie (1984); B, from Bjerring (1984); C, from Bjerring (1977).)

Neal and Rand (1936) considered that the history of ideas on head segmentation passed through three periods: transcendental, anatomical, and embryological. The early models of Oken, Goethe, and Geoffroy St Hilaire belong to the transcendental period, when the idea—the search for a common plan for all metazoans—was regarded as more important than observation of the actual facts. The anatomical period began with Cuvier, but it was mainly Huxley (1858b) and Gegenbaur (1872) who carried it forward. Huxley regarded both Goethe's and Oken's arguments as ridiculous, but he did not comment upon

Geoffroy's. Huxley addressed two major questions: (1) are all vertebrate skulls constructed on a common plan? and (2) is this plan identical in development and pattern with that of vertebrae? (Jollie 1984). He answered the first question in the positive, and the second in the negative. Gegenbaur (1872), who was strongly impressed by Haeckel's idea that a metamere could be regarded as a morphological individual of a colony (Clark 1964), was the first to convert the problem of the segmental origin of the skull into the problem of the segmentation of the entire head (i.e. including the nervous system, musculature, etc.) The embryological period began with the works of Vogt (1842) on the development and amphibians, and later gained momentum with the masterpieces of Goette (1901), von Kupffer (1892), Balfour (1878), and Van Wijhe (1882), until De Beer (1937) published a synthesis on the structure of the vertebrate skull. The most spectacular evidence of head segmentation was the arrangement of the head mesoderm into series of 'somites', which are one source of the muscular and skeletogenic tissues. The modern period begins with the works of Holmgren and Stensiö (1936), which draws on a large body of palaeontological data. Present-day ideas on morphology-based segmental theory are best illustrated by the works of Jollie (1971, 1981, 1984), Bjerring (1977, 1984), and Jarvik (1980), but it is probable that they represent only one step in the embryological period, which will continue under the considerable challenge of developmental genetics and molecular biology (Thorogood 1993). The segmental theory has not been universally accepted, and some embryologists, morphologists, and even palaeontologists (see review in Jefferies 1986) have regarded the apparent metamery of the vertebrates as secondary. Their statements are based on various grounds, such as comparison with the tunicates, the development of cephalochordates and hagfishes, or some supposedly vertebrate fossil ancestors (see Chapter 7, p. 281). As we shall see below, there may be the possibility of a compromise between the segmentalist and non-segmentalist views, at least for the foremost part of the head.

6.1.2 How many head segments?

If we admit that there is evidence for a metameric pattern in the craniate head, whether in the embryo or the adult, the first question that arises is: how many segments, or metameres, are there? For the braincase alone, Oken, Owen, Huxley, and Goette saw four segments, Goethe six, Balfour eight, and Van Wijhe nine. Moreover, when the braincase and visceral skeleton are combined, Gegenbaur considered nine or ten metameres, Kingsbury eight to fourteen, Bjerring nine, and Jollie five and a half (Fig. 6.1D). (The half is due to the fact that the sclerotomic segments producing the skeleton alternate with the myotomic segments producing the musculature and part of the skeleton (Jollie 1984).) Clearly, a consensus is not yet reached. This lack of agreement is largely due to divergent interpretations in two regions: the snout (in front of the mandibular segment) and the branchial apparatus. For example, and considering only the current theories, Jollie (1984) regards the third and all subsequent gill arches as belonging to post-cephalic metameres (Fig. 6.1D, VI), whereas Bjerring (1984) includes them in the head.

6.1.3 A segmented snout?

I shall not rehearse in detail the embryological arguments that have led to the present-day segmentalist models. For a detailed review of that question, see, for example, Jollie (1962, 1968, 1971, 1981, 1984), Bjerring (1977, 1984), Jarvik (1981*a*) or Hanken and Hall (1993). As an example of the combination of embryological inferences and palaeontological data, I shall consider here the case of the formation of the snout (the 'rostralization' of Bjerring 1977) in craniates.

In the frame of a segmentalist model, the anteriormost embryonic mesodermal segment that is capable of producing both braincase and visceral skeleton components is the mandibular segment (Fig. 6.1, 15). In the lamprey embryo, for example, it comprises a somitic vesicle and a parasomitic blade (fused with the rest of the 'lateral plate' in the gnathostomes), like the subsequent mesodermal segments. In various craniates, however, there are smaller additional mesodermal vesicles *in front* of the mandibular somite, which were noticed by early embryologists, and have long been believed to possess no parasomitic component. These vesicles have been termed the premandibular somite (Fig. 6.1, 14) and the Platt's vesicle (Fig. 6.1, 13) respectively. They are known to be the source of the tissues that later contribute to the formation of the some of the extrinsic eye muscles, the eye-stalk, and the skeleton of the anterior part of the braincase. It was thus tempting to early segmentalists to see these vesicles as remnants of a once-existing complete series of mesodermal segments that were similar to the mandibular and subsequent segments, and could thus have produced visceral arches. This idea seemed to be supported by a number of anatomical arguments, in particular the fact that the trigeminus nerve complex in hagfishes and lampreys comprises two ganglia and may thus be composite. In some instances (sharks), the premandibular somite is prolonged ventrally by a mesodermal bar that meets the parasomitic part of the mandibular mesoderm (Fig. 6.1, 14); this and a ventral prolongation of the Platt's vesicle (Chiarugi's vesicle; Fig. 6.1, 12) have been claimed by Bjerring (1977) to be the 'visceral tube' (parasomitic part) of these vesicles, the two components being respectively premandibular and 'terminal', but they have never been proved to include neural crest cells. Nevertheless, the idea that craniates once possessed a complete premandibular arch has progressively developed (Allis 1914) and was soon exploited by palaeontologists who were working on early vertebrates (e.g. Stensiö 1927). Enthusiasm for the premandibular arch theory was not shared by all morphologists and, given that only a premandibular somite without visceral bar is present in the lamprey embryo, it was generally believed that the premandibular somites in craniate embryos were merely forward extensions of the mandibular somite, tending to provide ventral support to the brain (Schaeffer 1975; Gans and Northcutt 1983, Gans 1993). The only possible indication of a contribution from the premandibular somite to the visceral skeleton is the part it is supposed to take in the formation of the anterior part of the palatoquadrate of the gnathostomes—hence the view that the palatoquadrate is a composite visceral element, in which the autopalatine is premandibular and the pterygoquadrate is mandibular (see e.g. Bjerring 1977; Jarvik 1954).

In sum, the question whether the snout of craniates was once segmented into one (premandibular) or two (premandibular and terminal) metameres with somitic and visceral components, or is merely an unsegmented forward prolongation of the mandibular somite derivatives is largely a matter of conception; the choice depends on whether one believes in laws of geometry in biology (serial homology) or prefers constructs based on the actual fate of the embryonic cells.

The contribution of palaeontology to this question is ruled by the same methodological constraint. As we shall see below, the best palaeontological support for full segmentalist theories, in particular those for the premandibular arches, has been seen in the sarcopterygian *Eusthenopteron* (Fig. 6.2A; Jarvik 1981*a*). In contrast, the earliest known craniates, such as the fossil jawless craniates, provide no clear support for a segmented snout. Their pattern of phylogenetic relationships nevertheless strongly supports the view that a median, terminal, inhalent nasopharyngial duct is primitive for the craniates (see below, and Fig. 6.6), thereby meeting one of the predictions of Bjerring's terminal arch theory (Bjerring 1989*b*).

6.1.4 The anatomy of a segmentalist theory

Stensiö (1927, 1968) has given considerable weight to the premandibular arch theory and, subsequently, to the idea that the best way of understanding the homologies and transformations in the vertebrate skull is to assign each character to a particular segment, the rank of which is defined by that of the corresponding cranial nerve. He was followed here by Jarvik (see, for a full presentation of the theory, Jarvik 1981*a*), and this philosophy culminated with Bjerring's works on head segmentation (in particular Bjerring 1977). Bjerring even recognizes a pre-premandibular, or terminal, segment; i.e. two segments in front of the mandibular segment. The gill slits that once existed between these segments are supposed by Bjerring (1977) to have given rise to the eye lens (Fig. 6.1, 4), the nasal and choanal openings (Fig. 6.1, 3), and a

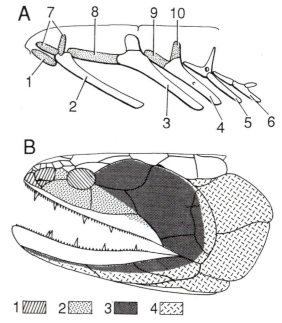

Fig. 6.2. Head segmentation as applied to the osteolepiform *Eusthenopteron*. A, Jarvik's (1954, 1981*a*) interpretation of the dorsal visceral arch components (stippled) incorporated into the braincase. B, Bjerring's (1979) interpretation of the external dermal bones of the head as gill covers. 1, terminal segment; 2, premandibular segment; 3, mandibular segment; 4, hyoid segment; 5, 6, first and second posthyoid segments; 7, 'supra- and infrapharyngopremandibular' (walls of nasal capsule); 8, 'infrapharyngomandibular' (trabecles); 9, 'infrapharyngohyal' (otic shelf); 10, 'suprapharyngohyal' (lateral commissure). (A, based on Jarvik (1981*a*); B, from Bjerring 1979).)

terminal opening (Fig. 6.1, 1), surrounded by the terminal arch. The terminal opening would be the hypophyseal tube in lampreys, the nasopharyngeal duct in hagfishes, and the buccohypophysial canal in the gnathostomes. Some have criticized Bjerring's ideas, considering them to be mere idealistic morphology. Bjerring's search for a *Bauplan* for the craniates is not, however, fundamentally different from the Hennigian reconstruction of the hypothetical ancestral morphotype. The difference is that Bjerring's aim is to demonstrate that any organ of any vertebrate is potentially present in any segment of this ideal vertebrate, whereas Hennigian cladists consider that new structures or new organs can be added to the morphotype at each node of the cladogram. In Bjerring's view, any character is general by essence; in the Hennigian view, characters are always particular to a taxon. In other words, the consequence of Bjerring's model is that any character can occur independently in separate species or taxa, since its primordium is already there. This easily leads to an evolutionary tree that resembles S. J. Gould's (1991) much-praised 'bamboo field' model of arthropod phylogeny. In contrast, cladists assume that the hierarchy of character distribution can tell us something about an order of appearance in time, and that an organ does not necessarily develop, even if the embryonic tissues from which it should arise is present. Another example of theoretical reconstruction of the ancestral vertebrate or craniate is that of Gutmann and Bonik (1981*a,b*), which is based essentially on the laws of biomechanics. It will not be discussed here, since it rests on theories that are unrefutable for palaeontologists.

What is surprising about Bjerring's ideal craniate (Fig. 6.1B) is the lack of a tail. Most other reconstructions of an ideal craniate, inspired by the structure of the cephalochordate or tunicate larvae, postulate a long tail and a smaller head (Romer 1970). Bjerring's explanation is that all the segments of the ideal vertebrates must have possessed the same components along the entire body, and that the tail has no visceral component. His ideal 'myomerozoan' is reconstructed on the basis of two assumptions: first that ontogeny reflects phylogeny, and secondly that homogeny (common embryonic origin) is enough to assign both strict and serial homology. Bjerring also assumes that we known nothing of the relationships of the major craniate groups, and that any organ can develop independently in one particular metamere. Data can therefore be taken from any taxon to support the ideal model. For example, Bjerring's hypothesis that nasal sacs are modified gill slits is based on the condition in urodeles, where nasal sacs open towards the exterior by the nostrils, and towards the pharynx by the choanae, which are partly endodermal in origin. The fact that other tetrapods have entirely ectodermal choanae is regarded by him as secondary. In this way, the nasal sacs can be regarded as comparable to gill slits, which form when an endodermal process and an ectodermal invagination meet between two adjacent mesodermal segments. He then concludes that the olfactory placode is serially homologous to an epibranchial placode (related to taste buds), the folded olfactory epithelium is serially homologous to gills, and the dermal bones lining the nasal capsules are serially homologous to the branchial plates or gill cover of the premandibular arch (Bjerring 1979). In Bjerring's model, the anteriormost gill opening is a median passage surrounded by the terminal arch, which is supposed to arise by the junction between the Rathke's pouch and the anterior gut process of the endoderm (Sessel's pouch), and which originally had an inhalent function. This primordial inhalent opening would thus remain only in hagfishes among extant craniates.

Bjerring's radical 'segmentalism' is largely based on embryological inferences; and, to some extent, the series

of transformations leading from his morphotype to any known vertebrate structure reflects embryonic development. Now we meet with a dilemma: we know that extant adult craniates do not provide clear evidence for a segmented snout nor do fossil craniates. Extant hagfishes and lampreys, for example, or early osteostracans and heterostracans do not look particularly 'segmented' or 'metameric', apart from serial homologies in the branchial apparatus. In contrast, we know that the development of the embryo is the only transformation process of organisms that is directly visible and, as Agassiz (1874) once put it, is the only thing that seriously suggests that some evolution has existed. Why, then, should we suggest that reconstructions of evolution based on this source of data are less reliable than stratigraphically arranged series of fossils? Following the classical embryological practice, Bjerring's theory is based essentially on series of discrete developmental stages and on a number of experimental results mainly based on the extirpation method (i.e. the removal of parts of the embryonic tissues to see what will be missing in the adult). Embryology has, however, made considerable technical progress during the 1980s, e.g. by the use of radioactive tracers to follow the fate of the embryonic cells during the course of development; more recently, developmental genetics has yielded results that throw doubt on the immutability of the classical 'laws' of development (i.e. the endoderm–ectoderm barrier, the immutability of segments, etc.). In sum, development now appears as mix of very stable processes and 'tinkerings' that may occur here and there, and then disappear. These particular developmental pathways—let us call them developmental homologies—may be restricted to one particular taxon and have no equivalent in others (as examplified by the partly endodermal choanae of urodeles or the endodermal adenohypophysis of hagfishes, if this is not a general craniate feature). In consequence, many, but not all, of the inferences for homologies or transformations based on classical 'fate maps' produced by developmental morphology may turn out to be wrong. The construction of the vertebrate head may thus be the result of complex cell migrations and interchanges, in particular of neural crest cells, rather than of an almost geometrical anamorphosis of independent segments. The statement that this or that organ 'belongs to the mandibular metamere', which often occurs in segmentalist writings, may lose much of its explanatory power in the future.

Nevertheless, the fact that the pattern of at least a large part of the craniate head is metameric can hardly be denied, even if asymmetry occurs here and there in primitive craniates (hagfishes), other chordates (cephalochordates larvae), or even in presumed fossil craniate ancestors ('calcichordates', see p. 282). The question is mainly whether the metamery of the somitic mesoderm (somites) matches that of the parasomitic mesoderm and, consequently, whether the metamery of the neurocranium, cranial nerves (neuromery), and axial skeleton matches that of the visceral arches (branchiomery). The fact that, in lampreys and cephalochordates, the parasomitic mesoderm does not consist of a single lateral plate, but is segmented in the same way as the somitic mesoderm would rather support the view that there is a single metameric pattern in the craniate head, as suggested in Bjerring's model.

Further reading

Allis (1914); Bjerring (1967, 1968, 1977, 1979, 1984, 1989*a*); Gans (1989, 1993); Goodrich (1930); Hall (1992); Hanken and Hall (1993); Jarvik (1981*a*); Jefferies (1986); Jollie (1968, 1971, 1981, 1984); Northcutt and Gans (1983); Presley (1993); Thorogood (1993).

6.2 FOSSILS AND SEGMENTALISM

6.2.1 The earliest heads

As we have seen above, the fossil vertebrates contribute very little to the question of the formation of the skull, and even less to the question of the premandibular part of the skull. Osteostracans, which were once thought to possess a functional premandibular arch, have turned out to be quite similar to larval lampreys in their branchial organization. There is perhaps one detail that may possibly suggest that the premandibular part of the skull was initially very short: the peculiar morphology of the earliest known craniates, the arandaspids (Fig. 4.2). If the anterior position of their eyes is regarded as general for craniates, then the ethmoid region must have been initially very short.

6.2.2 Visceral contributions to the braincase

The theory of the segmented head implies that complete visceral arches, possibly with a respiratory function, once existed in each of the cranial segments of early craniates. As for the skeleton, there is no evidence that a complete gill arch skeleton (that is, with basi-, hypo-, cerato-, epi-, and pharyngobranchial elements for the gnathostomes) exists anteriorly to the first gill arch (with the possible exception of holocephalans, which were once considered as possessing pharyngohyals; see however Chapter 3, p. 64). The pharyngeal elements are usually missing from the hyoid and mandibular arch. Since the segmental theory says that all segments must have been initially similar, there are only three explanations: the missing elements have either disappeared, fused with the rest of the

arch, or fused with the overlying braincase. The third solution has been preferred by strict segmentalists, such as Jarvik (1954, 1981*a*) and Bjerring (1977), because there is some embryological evidence that parts of the braincase floor, namely the trabecles, are neural-crest derivatives, as is the visceral skeleton. These authors have thus suggested that, if these incorporated elements are no longer visible, they may be traced by means of the dermal tooth-bearing plates that are associated with them in gill arches. These are easy to find in osteichthyans: the parasphenoid and the vomers, for example, may thus be the very plates that once underlaid the pharyngeal components of the mandibular, premandibular, and terminal 'gill' arches. The endoskeletal elements of these arches may be represented in the braincase by the trabecles, trabecular commissure, and the walls of the nasal capsules (Fig. 6.2A). This theoretical model has been constructed essentially on the basis of the osteolepiform *Eusthenopteron*, and it may sound surprising that a rather advanced osteichthyan provides information on processes that are supposed to have taken place in the earliest gnathostomes at the latest. In the same way, the pharyngeal components of the hyoid arch are said to be incorporated into the braincase in the form of the lateral commissure (suprapharyngohyal) and otical shelf (infrapharyngohyal). Here again, the lateral commissure, which straddles the jugular vein and trigeminofacial chamber, is usually covered externally with dermal platelets lining the spiracular canal (Fig. 4.90, 10). There is also some experimental evidence that the otical shelf and part of the lateral commissure are neural crest derivatives (Langille and Hall 1993). Recent investigations on the development of the tetrapod braincase tend to show that nearly all the anterior part of the skull actually derives from neural crest cells (Couly *et al.* 1993), but this does not necessarily prove that these neural crest cells once contributed to the formation of true branchial arches.

These examples of presumed incorporation of visceral arch elements to the braincase are only a small part of the much detailed account given by Bjerring (1977). It is probably partly true, but I doubt whether it is entirely refutable as a theory, unless by the discovery of fossils showing both a pharyngohyal and a lateral commissure, for example. We ignore whether the neural crest derivation of the foremost part of the skull is evidence for once-existing gill arches or merely the result of a late neural-crest contribution to the braincase. In order to be fully consistent with his theory, Bjerring (1979) admits that even the external dermal bone covering of *Eusthenopteron* can be interpreted in terms of terminal, premandibular, and mandibular gill covers (Fig. 6.2B). The pattern of the various dermal plates of osteichthyans may, however, simply be linked with mechanical constraints, such as the movements of the palatoquadrate and gill arches, as well as with the pattern of the sensory-line neuromasts (Thomson 1993).

Further Reading

Bjerring (1977); Jarvik (1954, 1981*a*); Stensiö (1927, 1968); Thomson (1993)

6.2.3 The origin of jaws

The classical explanation of the origin of the jaws of the gnathostomes is that they are modified gill arches (Fig. 6.3A). This belief stems from segmentalist models, and also from the fact that jawless vertebrates have gill arches but no jaws, whereas the gnathostomes have both gill arches and jaws. Since the two structures derive from the same source of embryonic cells, the neural crest, it was logical to consider the jaws as serial homologues of the gill arches. Moreover, the jaws have the same gross morphology as the subsequent visceral arches, hyoid and branchial. This hypothesis of transformation also rests on the postulate that gill arches in jawless vertebrates and the gnathostomes are homologous, a postulate that may be supported by their embryonic origin (the neural crest cells) but certainly not by their morphology or relationships to other organs (see Chapter 3, p. 56). The hypothetical pre-gnathostome, with a gill-bearing mandibular arch, is a widespread scheme in textbooks (Fig. 6.3A1). There is, however, no instance of a craniate with either a gill-bearing mandibular arch or even a true spiracular gill slit. So far as can be seen from fossil and extant craniates, the foremost functional gill is the posterior hemibranch of the hyoid arch. (For the pseudobranch of some gnathostomes, see Chapter 3, p. 56). Nor does the embryonic development of the gnathostomes show any evidence of a mere rudiment of a mandibular gill.

Considering these facts, one may wonder whether this theory still holds, and whether a more parsimonious explanation could not be that jaws have always been jaws. In lampreys, the ectomesenchyme of the mandibular segment, which in gnathostomes gives rise to the endoskeletal component of the jaws (the palatoquadrate and Meckel's cartilage), develops into a complex cartilaginous element supporting the velum, a pumping and anti-reflux device of the larva (Fig. 6.3B, 1) that conveys food particles towards the 'mucous trap' produced by the endostyle (Fig. 6.3, 2, 3). This velar skeleton is among the first cranial elements to chondrify during development and, unlike the skeleton of the hyoid and branchial bars, it lies medially (Fig. 6.3, 8) against the wall of the pharynx, like the visceral arches of the gnathostomes. This may mean either that there is a homology between the velar skeleton of lampreys and the jaws of the gnathostomes, or

that the latter originate from a velum-like structure, which was almost certainly present also in osteostracans, as well as in most other jawless vertebrates (Fig. 4.15, 32). The velar skeleton of larval lampreys lies close to the hyoid bar, and the spiracular 'pouch' disappears early in development. With some imagination, one may then speculate that the velum, which plays a role in both respiration and feeding of the larval lamprey, could have been modified into a prehensile device in a hypothetical ancestral gnathostome, and have given rise to jaws (Fig. 6.3D). The medial wall of the pharynx would then have become strengthened by more medial endoskeletal bars: the internal hyoid and branchial arches (Fig. 6.3D2). These arches would have been organized in nearly the same way as the jaws for mechanical reasons, and would have helped to convey large food particles towards the oesophagus. The external gill arches would simultaneously have disappeared, because they were no longer necessary as gill supports (Fig. 6.3D3). This is a radically different story, in which gill arches do not become jaws but, instead, jaws trigger the rise of new gill arches.

Further Reading

Denison (1961); Forey and Janvier (1993, 1994); Janvier (1993); Jollie (1968, 1984); Lessertisseur and Robineau (1969–70); Maisey (1989*a*); Mallatt (1984*a,b*); Northcutt and Gans (1983); Wahlert (1970); Zangerl and Williams (1975).

6.2.4 The intracranial joint

Yet another consequence of the segmentalist theories is the attempt to find traces of a former segmentation, not only in the visceral skeleton of the vertebrate head, but also in the braincase itself. If we leave aside the possible incorporation of visceral components in it, the neurocranium forms essentially like the axial skeleton, from the sclerotomic part of the paraxial mesoderm (somites). It thus seems logical that the braincase is the anterior continuation of the vertebral column and, in consequence, that it may have looked more or less like a series of cranial vertebrae in some hypothetical ancestral craniate (the 'Goethe-Oken' theory). There is no evidence of such a neurocranial segmentation in any of the jawless craniates, fossil and extant, nor in chondrichthyans and placoderms. In contrast, some osteichthyans show that the ventral part of the neurocranium surrounding the notochord, i.e. the otico-occipital region, has a segmental arrangement. Evidence of this kind is found, for example, in such sarcopterygians as actinistians and osteolepiforms, and even in the embryo of some tetrapods, in which a series of basicranial muscles may occur. But a major argument in favour of neurocranial segmention is the intracranial joint in some sarcopterygians, where the

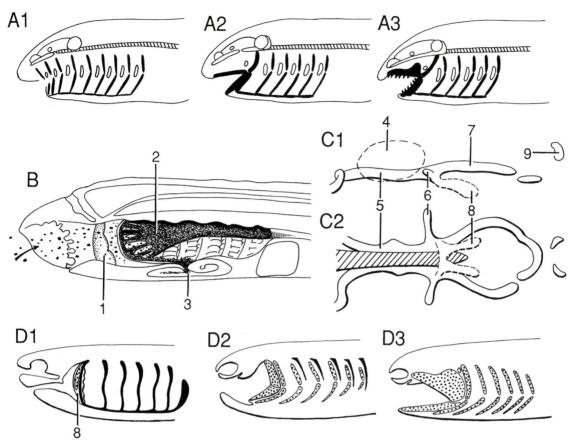

Fig. 6.3. The origin of jaws. A, the classical transformational scenario (A1–3; jaws, hyoid and gill arches in black) assuming that the endoskeletal jaws and hyoid arch (A3) derive from functional gill arches in a jawless ancestor (A1); B, sagittal section through a larval lamprey, showing the function of the velum and mucous trap; C, embryonic head skeleton of a lamprey, showing the medial velar skeleton in lateral (C1) and ventral (C2) view; D, transformational scenario (D1–3) assuming that the endoskeletal jaws formed from the velar skeleton before the subsequent hyoid and gill arches (medial visceral skeleton dotted, lateral visceral skeleton black). 1, velum; 2, mucous trap; 3, endostyle; 4, otic capsule; 5, parachordals; 6, pedicel; 7, trabecles; 8, medial velar skeleton; 9, nasal capsule. (A, from Reif (1982*a*); B, from Mallatt (1981); C, from Johnels (1948).)

braincase is divided into two halves just behind the hypophysial region (Fig. 4.27E). As already mentioned (Chapter 4, p. 129), this intracranial joint corresponds to a zone where, in gnathostome embryos, the two halves of the braincase are united only by the polar cartilage and the lateral commissure (Fig. 4.28). Despite the fact that the embryonic development of the coelacanth is still unknown, Bjerring (1967, 1977) has suggested that the large basicranial muscle (Fig. 3.21G, 23) that straddles the intracranial joint in actinistians (and presumably in many fossil sarcopterygians) is the homologue of the polar cartilage and is the remnant of an intersegmental muscle. This would be supported by the fact that the polar

cartilage is derived in ontogeny from the same source of mesodermal tissues that produces the epaxial musculature (myotomes). Consequently, the intracranial joint could be the remnant of the joint between two former 'cranial vertebrae'. This interesting explanation of the intracranial articulation has raised much discussion and remains unrefuted, unless by virtue of craniate phylogeny, which would imply many parallel disappearances of this joint (in jawless craniates and elasmobranchiomorphs). The intracranial joint is therefore, currently regarded as a secondary specialization of sarcopterygians, which has developed from a pre-existing ventral fissure, and then disappeared in some sarcopterygians taxa, such as lungfishes and tetrapods. The basicranial muscle is innervated by the abducent nerve and is regarded by Northcutt and Bemis (1993) as homologous to the retractor bulbi muscle of the tetrapod eyeball. Seen in this way, it would be a sarcopterygian character, and not a craniate character. For Bjerring (1973, 1978), however, the intracranial joint has nothing to do with the ventral fissure and belongs to a more anteriorly placed intersegmental joint. He also claims that the intracranial joint is not strictly homologous in the

various sarcopterygian groups, since it does not always have the same position in relation to the foramina for the cranial nerves (Fig. 6.4). He thus suggests that this joint passed between different elements of the primordial 'cranial vertebrae'. However, despite some difference between actinistians, dipnomorphs, and osteolepiforms (Fig. 6.4), this interpretation is no longer considered.

The intracranial joint of the living coelacanth, *Latimeria*, is paradoxical in being only slightly movable

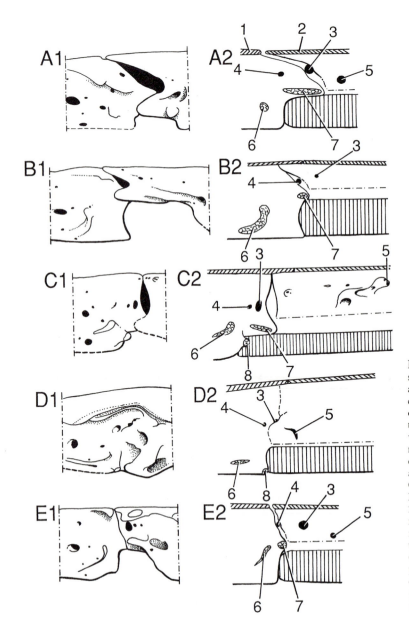

Fig. 6.4. The intracranial joint in sarcopterygians. A1–E1, as it appears in a lateral view of the braincase; A2–E2, corresponding schematic interpretation of the exits of some cranial nerves, with the notochord (vertically hatched; the ventral part of the otoccipital is removed to show the notochord) and dermal skull roof (obliquely hatched) added. A, Actinistia; B, Porolepiformes; C, *Powichthys*; D, *Youngolepis*; E, Osteolepiformes. 1, parietal; 2, postparietal; 3, maxillar and mandibular branches of trigeminus nerve; 4, profundus branch of trigeminus nerve; 5, facial nerve; 6, basipterygoid process; 7, connecting process; 8, descending process of ethmosphenoid. (A1–E1, from Ahlberg (1991a); A2, C2–E2, from Schultze (1987).)

and associated with one of the largest muscles of the head. The same condition may have occurred in some fossil sarcopterygians where, in addition, there is no functional dermal joint at this level of the skull-roof and, consequently, no possibility of movement (Fig. 6.4B2, C2, D2). It is more likely that the intracranial joint originally played a role in cranial mobility during biting movements, but may be secondarily involved in the process of swallowing prey.

As with most questions of this kind about cranial homologies, we have once again the choice between geometrical laws of Bjerring, where every part of the braincase has to be assigned to a particular segment, or metamere, and the common-sense approach to general resemblance, with some flexibility about the relative position of bones. Here we encounter the question of how much of the theories based on these different approaches can actually be refuted. It seems that Bjerring's model is able to absorb any contradiction, since new patterns can always be shoehorned into the geometrical metameric scheme by invoking examples from the adult or the embryo of any craniate. The other method, which *a priori* equates resemblance with homology until this is contradicted by the other, incongruent characteristics, is not immune from mistakes but seems to be more open to refutation.

Further Reading

Ahlberg (1991*a*); Bjerring (1967, 1973, 1977, 1978, 1993); Jarvik (1981*a*); Rocek (1986); Schultze (1987).

6.3 THE PATHS TO DERMAL BONE HOMOLOGIES

As we have seen in the preceding chapter, virtually no homology can be envisaged between the patterns of the dermal bones of the major jawless vertebrate groups, apart perhaps from the large median dorsal and ventral shields of arandaspids, astraspids, and heterostracans (Fig. 5.1, 3). Only in osteichthyans and, to some extent, placoderms do the dermal bone patterns show some resemblance from group to group. The dermal bones of osteichthyans were originally named after the condition in mammals (in particular Man), and then extended to bony fishes (in particular teleosts), with some additional names for bones found only in fishes, such as opercular, preopercular, gulars, etc. The question of the homology of these bones has been further complicated by the discovery of fossil fishes (e.g. osteolepiforms) that were believed to be close relatives of tetrapods but did not display a particularly tetrapod-like dermal bone pattern in the head. There have been lively debate about the possible

paths toward finding dermal bone homologies in osteichthyans and, as we have seen (Chapter 3, p. 67), the sensory lines are still regarded as useful guidelines for identifying some of them. But what, in turn, is the reason for the pattern of the sensory lines? Bjerring (1984) has assumed that the generalized pattern of the sensory lines in a hypothetical ancestral craniate consists of longitudinal and transverse lines (Fig. 6.1, 8). This seems to be supported by the pattern observed in some fossil jawless craniates (e.g. arandaspids, cyathaspids; Figs 4.2B, 4.7A), but as yet never observed in the anterior part of the head. Following the logic of segmentalism, one may assume that a craniate with a fully segmented head displayed this pattern even in the snout. Consequently, the dermal bones that formed along these lines must also have displayed a segmental pattern. In this way, Bjerring has formulated the hypothesis that all the bones of the cheek and snout of osteichthyans could be regarded as former gill covers of the mandibular, premandibular, and terminal arches respectively (Fig. 6.2B).

The homologization of dermal bones by reference to the underlying endoskeletal structures has also been used to a wide extent for those of the palate and skull-roof, but it has never proved to be very useful. Dermal bones form and develop according to the pattern of the sensory lines, but also according to other factors, such as mechanical constraints or simply chance mutation. Their homology is thus best estimated on the basis of resemblance and congruence with other characters.

The best-known problem in dermal bone homology is that of the fish and tetrapod skull-roof (Fig. 6.5). In mammals, the dorsal surface of the skull displays three median pairs of dermal bones: the parietals, frontals, and nasals (Fig. 6.5A, 1–3). In early tetrapods and some extant amniotes, there is, behind the parietals, an additional pair, the postparietals (Fig. 6.5B, 4,). In fishes, two or three pairs of dermal bones may occur. Their homology with the pairs in tetrapods has been determined on the basis either of their relationships to the underlying regions of the braincase, or of their relationships to the pineal foramen. Jarvik (1967) considers that the frontals invariably cover the orbital region and the parietals the otic region, whereas Westoll (1943*a*) considers that the parietals are invariably the bones that flank the pineal foramen. According to Jarvik's theory, the pineal foramen could be shifted backwards in tetrapods and lie between the parietals as a consequence of modifications in braincase morphology. Unfortunately, the bony fishes that were long regarded as the closest relatives to the tetrapods, the osteolepiforms (Fig. 6.5D, 3, 4) have only two median pairs of bones, the pineal foramen lying between the foremost pair, which are thus either the frontals or the parietals. What, then, could be, the homo-

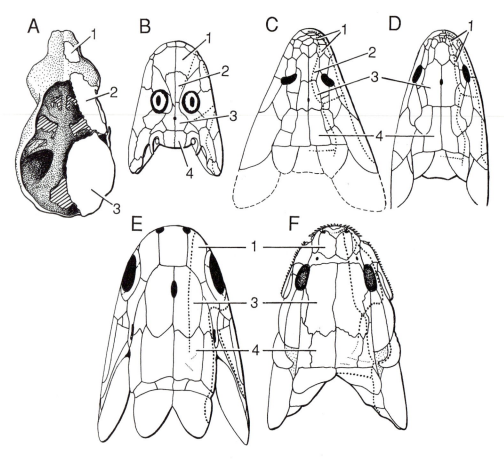

Fig. 6.5. Homology of the frontal, parietal, and postparietal in tetrapods and fishes. A–F, Skulls in dorsal view (sensory-line canals dotted on the right-hand side). A, mammal embryo (cartilage stippled); B, *Acanthostega*; C, *Panderichthys*; D, *Eusthenopteron*; E, early actinopterygian (*Moythomasia*); F, Recent actinopterygian (*Amia*). 1, nasal or nasal series; 2, frontal; 3, parietal; 4, postparietal. (A, from Starck (1967); B, from Clack and Coates (1993); C, from Vorobyeva and Schultze (1991); D, F, from Jarvik (1980); E, from Jessen (1968).)

logue of the tetrapod postparietal in these fishes? It has been claimed that it arose from a transverse series of post-cranial dermal bones, the extrascapulars, which bear the supratemporal commissure, a sensory line also found on the postparietal of some early tetrapods (Fig. 4.95D1, 37). The extrascapulars, however, lie mainly behind the posterior limit of the braincase, and the arguments about relationships that prevail for other bones cannot apply there. This dilemma may have been solved by the analysis of the skull-roof pattern in panderichthyids (Schultze and Arsenault 1985; Vorobyeva and Schultze 1991). This group of sarcopterygians is probably more closely related to the tetrapods by several other characters, than the osteolepiforms. In panderichthyids, there are three pairs

of bones in the skull-roof (Fig. 6.5C, 2–4), as in early tetrapods, and the pattern of the sensory lines clearly shows that the two rearmost pairs correspond to the two pairs in osteolepiforms (Fig. 6.5D, 3, 4). In front of them, there is one other pair that occupies the same position as the frontals in tetrapods (Fig. 6.5A, B, 2). A comparison between a panderichthyid and an early tetrapod like *Acanthostega* (Fig. 6.5B) shows a 'reasonable' agreement and leads to the conclusion that the rearmost pair in all osteichthyans is the postparietal and that the pineal foramen is between the parietals (except in dipnomorphs). The presence of a supratemporal commissure on the postparietals of early tetrapods may be explained by the *ad hoc* argument of the fusion of the extrascapulars to the postparietals. Here again, we see the weight given to transformation series and 'intermediate forms' in the construction of such evolutionary scenarios.

This story of the dermal roofing bones of sarcopterygians sounds consistent, although one must not forget that there are some facts that do not agree with it. One of these is the presence of a single median postparietal in one of the earliest tetrapods, *Ichthyostega* (Fig. 4.95, 14; this is currently interpreted as a specialization or an individual variation); others are the presence of paired 'frontals' in other sarcopterygians, such as some actinistians (Fig. 4.76) and lungfishes (Fig. 4.80), which suggest that paired frontals are not unique to panderichthyids and tetrapods. These 'frontals', however, occur only in advanced actinistians and lungfishes and are not traversed by the supraorbital sensory line. Are these contradictions to be neglected as homoplasies, or do they tell us that the entire story is wrong? What makes the panderichthyids have true frontals—and thereby makes a nice story—is that they share other unique characters with tetrapods. If we admit this interpretation, then the same names have to

be applied to the skull-roof bones of all osteichthyans, and the two rearmost bone pairs of actinopterygians, traditionally called frontals and parietals, have to be referred to as parietals and postparietals (Fig. 6.5E, F, 3, 4).

Further Reading

Ahlberg (1991*a*); Bjerring (1977); Borgen (1983); Jarvik (1967, 1980, 1981*a*); Panchen and Smithson (1987); Schultze (1991, 1993); Schultze and Arsenault (1985); Vorobyeva and Schultze (1991); Westoll (1943*a*).

6.4 MONORHINY VERSUS DIPLORHINY

In Chapter 3, we saw that the openings of the olfactory organ in extant craniates fall into two types: a single, median apparent nostril in jawless craniates (Fig. 6.6, 1) and two separate nostrils, or pairs of nostrils in gnathostomes (Fig. 6.6, 10); hence the major division into 'monorhinous' and 'diplorhinous' craniates used by early zoologists. The 'nostril' of jawless craniates (hagfishes

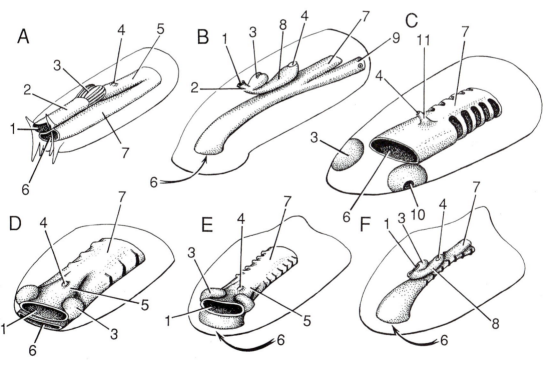

Fig. 6.6. The nasohypophysial complex in the Craniata. Schematic reconstruction of the relationships between the olfactory organ, hypophysis, and pharynx in some extant and fossil craniates. A, Hypetrotreti (hagfishes); B, Hyperoratia (lampreys); C, Gnathostomata (jawed vertebrates); D, Thelodonti and Heterostraci (hypothetical); E, Galeaspida; F, Osteostraci.
1, nasohypophysial opening; 2, prenasal sinus; 3, olfactory organ; 4, adenohypophysis; 5, nasopharyngeal duct; 6, mouth; 7, pharynx; 8, hypophysial tube; 9, pharyngobranchial duct (adult lampreys only); 10, external nostril; 11, buccohypophysial canal.

and lampreys) is not, however, really a nostril; it includes a median duct of endodermal or ectodermal origin (Fig. 6.6, 2, 5), which is involved in the formation of the adenohypophysis (Fig. 6.6, 4), and into which the actual nostrils open. In gnathostomes, the homologue of this duct is the embryonic Rathke's pouch, which becomes disconnected from the olfactory organ and migrates in the palate to form the buccohypophysial canal (Fig. 6.6, 11). If jawless craniates are paraphyletic, then the gnathostome condition is likely to be unique, and the condition in jawless craniates general for all craniates. The olfactory organ develops from a single median placode in extant jawless craniates (Fig. 3.7C2, 3), and from a paired placode in the gnathostomes, but the adult nasal sac is a single cavity in hagfishes (Fig. 6.6A, 3), whereas it is divided by a median septum into two cavities in lampreys (Fig. 6.6B, 3). This suggest a transformation series from a single median nasal sac to two separate nasal sacs (Fig. 6.6 A–C, 3). The pattern becomes more complicated when fossil jawless craniates are considered. There is some evidence for nasal or nasohypophysial structures in the heterostracans, osteostracans, and galeaspids (Fig. 6.6D–F). The supposedly paired nostrils of arandaspids are still too uncertain to be considered here, and the supposedly nasohypophysial opening of anaspids gives little information about the internal structures. Galeaspids are the only jawless craniates in which the nasal cavities are clearly separate and open into the median prenasal sinus by means of separate nostrils (Fig. 6.6E, 3). This condition has been regarded by N. Z. Wang (1991) as a precursor to the gnathostome condition (Fig. 6.6C, 3). The paired nasal impressions of heterostracans (Fig. 4.5, 1) are so similar to the nasal cavities in galeaspids that, despite the lack of perichondral ossification, one may assume that the condition was the same (Fig. 6.6D). In heterostracans, however, there is still no definite evidence of a separate prenasal sinus, and some authors claim that the nasal cavities opened ventrally into the mouth (Fig. 4.5E). In osteostracans, the shape of the ethmoid cavity fits perfectly the condition in lampreys, but the small, pear-shaped recess for the olfactory organ does not provide evidence for a median septum (Fig. 6.6F, 3). Finally some thelodonts show some evidence for paired olfactory organs, and a shagreen of forward-pointing denticles suggest the presence of a separate, inhalent prenasal sinus (Fig. 6.6D, 1, 3). Assuming that the condition in hagfishes (a single large nasal sac opening into a prenasal sinus) is general for craniates, one may consider that the division of the olfactory organ into two separate sacs is an advanced condition that is unique to the vertebrates (craniates minus hagfishes). The secondary reduction of the olfactory organ to a small median, but divided, nasal capsule may have occurred

once (if the cephalaspidomorphs are regarded as monophyletic) or twice (if osteostracans are regarded as the sister-group of the gnathostomes) and may be linked with the loss of the main inhalent function of the prenasal sinus. (In fact, the prenasal sinus of lampreys remains inhalent, because of the pumping action of the hypophysial tube, but it serves only for olfaction, and not for the intake of respiratory water. The rigidity of the endoskeleton, however, makes this mechanism impossible in osteostracans; there, the water movements into and out of the olfactory organ must have been effected by other means, possibly by ciliae.) The total disjunction between the Rathke's pouch and olfactory placodes in the gnathostomes (Fig. 3.11, 1, 2) makes prenasal sinus disappear and the olfactory organs open directly to the exterior by means of separate nostrils, which subsequently become divided into anterior and posterior nostrils (Fig. 6.6C). This scenario of transformation implies that the gnathostome condition arose from a condition with two separate olfactory organs (as in galeaspids, thelodonts or heterostracans, for example), and a prenasal sinus that progressively lost its inhalent function (as in lampreys and osteostracans).

Further Reading

Bjerring (1989b); Forey and Janvier (1994); Gorbman and Tamarin (1985); Halstead (1973a,b, 1982); Heintz (1962); Janvier (1974, 1975d, 1981a, 1993); Janvier and Blieck (1993); Jarvik (1980); Novitskaya (1983); Van der Brugghen and Janvier (1993); N. Z. Wang (1991).

6.5 NOSTRILS, TEAR DUCTS, AND CHOANAE

In gnathostomes, the primitive condition of the nasal cavities is assumed to be a single external opening, divided into two nostrils, anterior (incurrent) and posterior (excurrent), by a flap, as in chondrichthyans (Fig. 6.7A1, 2, 3). In osteichthyans, these two nostrils become completely separated (Fig. 6.7A2). In three extant taxa, holocephalans, lungfishes and possibly tetrapods, only the anterior nostril opens to the exterior; the posterior one opens inside the mouth in holocephalans and lungfishes. In tetrapods, there is also an intrabuccal opening, termed the choana (Fig. 6.7, 4) but whether it is the posterior nostril or a neomorph is a debated question. Also, in tetrapods alone, there is an additional duct, the tear duct or nasolacrymal duct, which connects the nasal cavity to the margin of the orbit and generally allows tears to keep the olfactory epithelium moist (Fig. 6.7A3, 5). Since the nineteenth century, there have been lively debates about the homology of the tetrapod choanae,

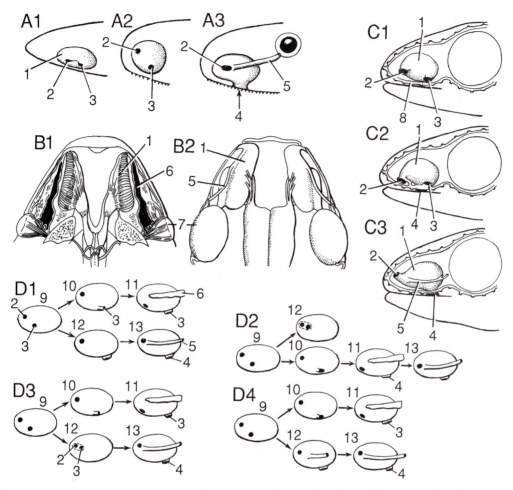

Fig. 6.7. The nasal apertures in sarcopterygians (nostrils, choanae, tear duct). A, the major nasal apertures in the gnathostomes: chondrichthyan condition (A1), generalized osteichthyan condition (A2), and tetrapod condition (A3). B, dissected snout of a lungfish (B1) and a urodele (B2), showing the comparable position of the upper labial cavity and tear duct. C, the theory of Schmalhausen (1968) on the origin of the tear duct and choana: C1, generalized osteichthyan condition; C2, hypothetical intermediate condition (the posterior nostril partly coalescent with the infraorbital sensory-line canal, and neomorphic choana); C3, tetrapod condition. D, four transformational scenarios of the origin of the choanae and tear duct: D1, scenario assuming that the choana is the posterior nostril and the tear duct a neomorph in tetrapods (e.g. Yvroud 1971); D2, scenario assuming that the posterior nostril of lungfishes and the choana are homologous, and that the upper labial cavity and the tear duct are homologous (e.g. Rosen *et al.* 1981); D3, scenario assuming that osteolepiforms have two external nostrils (the posterior one becoming the tear duct in tetrapods) and that the choana is a neomorph in tetrapods (e.g. Schmalhausen 1968); D4, scenario assuming that osteolepiforms have both a choana (neomorphic) and a tear duct (e.g. Jarvik 1942). 1, nasal sac; 2, anterior (incurrent nostril); 3, posterior (excurrent) nostril; 4, choana; 5, tear duct; 6, upper labial cavity; 7, eye; 8, infraorbital sensory-line canal; 9, generalized osteichthyan or sarcopterygian condition; 10, porolepiforms; 11, lungfishes; 12, osteolepiforms (or panderichthyid); 13, tetrapods. (B, after Jarvik (1942); C, based on Schmalhausen (1968).)

although there is now a consensus on the theory that the choanae are the posterior nostrils (or part of them), which have migrated to the palate. The embryonic development of extant lungfishes clearly shows that the internal nostrils (Fig. 3.24, 11) are in fact the posterior nostrils and

form, like the anterior nostrils, from an ectodermal invagination of the stomodeum. They form in virtually the same way as the choana of tetrapods, except in urodeles, where a process of the endoderm meets the nasal sac to form the choana. This partly endodermal origin of the choanae in urodeles (now regarded as a unique urodele feature), allied to the idea that lungfishes could not be the ancestors of the tetrapods, has led some anatomists (Allis 1919, Jarvik 1942, Bjerring 1977) to the conclusion that the tetrapod choana was not the posterior nostril but was either a neomorph or the remnant of a once existing gill slit. If this is accepted, one then has to determine the fate of the posterior nostril in tetrapods. It has been assumed by some that it was transformed into the tear duct, or took part to its formation (Jarvik 1942, Vorobyeva and Nitkin 1991; Fig. 6.7D3, D4). The condition in tetrapods (one pair of external nostrils and one pair of choanae) matched perfectly that in the fossil osteolepiforms and panderichthyids (Figs 4.88, 4.89), in which a gap occupies exactly the same position as the choanae, in relation to the surrounding dermal bones of the palate. The problem, however, was to find a tear duct in these fishes which, being essentially aquatic, did not really need tears. Jarvik (1942) assumed that osteolepiforms possessed a canal for the tear duct, whereas other authors interpreted this canal as for the maxillaris nerve branch. Rosen *et al.* (1981) even denied that osteolepiforms possessed any choana; they suggested that their two external nostrils were housed in the single opening classically referred to as the anterior nostril (Fig. 5.4A; Fig. 6.7D2, 12). The goal of Rosen *et al.* was to make lungfishes and tetrapods the only choanate osteichthyans, and they suggested that the homologue of the tear duct in lungfishes was the upper labial cavity, a peculiar elongated space inside the upper lip (Fig. 6.7B1, 6). This homology was also accepted by Bjerring (1989*b*). Others retained the old explanation, that the tear duct is either a neomorph in tetrapods (Fig. 6.7D1, Yvroud 1971), or derives from the infraorbital sensory-line canal and the posterior nostril (Fig. 6.7C; Schmalhausen 1968).

When the Early Devonian primitive lungfish *Diabolepis* was shown to have had two pairs of external nostrils, the posterior one being situated just on the margin of the upper jaw, but not inside the mouth (Fig. 4.84C1), it thus became clear that the internal migration of the posterior nostrils in lungfishes had occurred independently from the formation of the choanae in tetrapods, as suggested by Allis, whatever the origin of the tetrapod choanae. A detailed re-examination of osteolepiforms and panderichthyids apparently confirmed that both taxa show good evidence for choanae, in particular panderichthyids where the akinetic skull made it possible to reject the claim made by Rosen *et al.* that the osteolepiform 'choana' was a mere hinge between the snout and the palate (Fig. 4.89A2). For those who consider the choanae as posterior nostrils, the current consensus is that internal nostrils have appeared twice in sarcopterygians, in virtually the same way (parallelism): once in lungfishes, and

once in the common ancestor to osteolepiforms, panderichthyids and tetrapods (Fig. 6.7.D1). It is, however, probable that the general condition to both was quite similar to that in some porolepiforms (*Porolepis*) or *Youngolepis*: i.e. with a posterior nostril situated near the upper mouth margin, and a large endoskeletal ventrolateral fenestra that straddled the dermal bones of the upper jaw and the premaxillary-maxillary junction (Fig. 4.77C1, 36; Fig. 4.78, 14; Fig. 6.7, 10). The passage of the posterior nostril, or part of it, into the mouth was thus a minor evolutionary leap. The fact that the infraorbital sensory-line canal is interrupted by the intrabuccal migration of the posterior nostril in lungfishes, but not in choanates (osteolepiforms, panderichthyids and tetrapods) remains, however, a problem. It is one of the major arguments put forward by Jarvik (1980, 1981*b*) against the choana-posterior nostril theory.

There is no evidence for an upper labial cavity in osteolepiforms and panderichthyids, and there is no clear evidence for a tear duct (unless it passed medially to the maxilla (Bjerring 1989*b*). If the single external nostril in these groups is only the anterior nostril, then the fate of the posterior nostril may be either the tear duct (Fig. 6.7D3, D4), the choana (Fig. 6.7D1), or simply the loss. The earliest known tetrapods (e.g. *Ichthyostega*, *Acanthostega*) provide no decisive information as to the tear duct, except that they have a continuous infraorbital sensory-line canal and display virtually the same condition as in osteolepiforms and panderichthyids. The only theory that one may rule out at present, with regards to the choana, is that of Rosen *et al.* (Fig. 6.7.D2), if one admits that the common ancestor to all lungfishes had no internal nostrils. All other theories remain, however, ambiguous or rest on the interpretations of the 'choanal gap' in fossil forms. The test for these theories would perhaps be to find osteolepiforms (as defined here) with a continuous endoskeletal floor of the nasal capsule, or with two distinct external nostrils.

Further reading

Ahlberg 1991*a*), Bertmar (1969); Bjerring (1977, 1989*a*, 1991); Jarvik (1981*b*); Panchen (1967); Panchen and Smithson (1987); Rage and Janvier (1982); Rosen *et al.* (1981); Schmalhausen (1968); Schultze (1991); Vorobyeva and Nitkin (1991); Yvroud (1971)

6.6 THE ORIGIN OF THE TETRAPOD LIMBS

As soon as comparative biologists began to think in terms of homology and evolutionary relationships, the question of the origin of the limbs of tetrapods was raised, through consideration of the structure of the paired fins in their presumed fish relatives or ancestors. The first serious attempt at finding fin–limb homologies was made by Etienne Geoffroy-Saint-Hilaire in 1807 on the basis of the extant actinopterygian *Polypterus*. Later, the discovery of the extant lungfishes and, by the end of the nineteenth century, the increasing number of Palaeozoic lobe-finned fishes recorded from Britain and elsewhere, led anatomists and palaeontologists to leave aside *Polypterus* and choose some better anatomical model from which an archaetypic limb structure could be derived. After much discussion on the use of the lungfish in this purpose (see the historical review in Rosen *et al.* 1981; and Forey

1987*b*), there has been a consensus on the osteolepiform *Eusthenopteron*, whose pectoral fin was first regarded by the American anatomist W. K. Gregory (1911) as the ideal precursor of the tetrapod limb (Fig. 6.8A). The reason for this was that the endoskeleton of this pectoral fin has a proximal mesomere that recalls the shape of the early tetrapod humerus (Fig. 6.8A, 1), e.g. in having a large entepicondyle, and that this element articulates distally with two elements (the second mesomere and the first radial; Fig. 6.8A, 2, 3), which are suggestive of the ulna and radius of tetrapods. Since 1911, very little has changed, and the presence of choanae and folded teeth in *Eusthenopteron* were regarded as further evidence for its

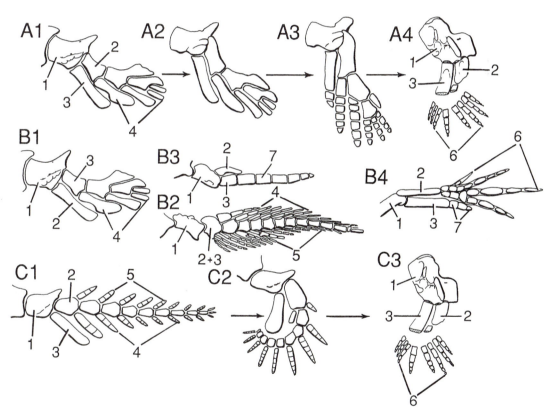

Fig. 6.8. Transformational scenarios and homologies of the pectoral fin and fore limb skeleton, all assuming that the phalanges are radials. A, the 'current consensus' (e.g. Jarvik 1981*a*) on the transformation of the pectoral fin of *Eusthenopteron* (A1) into that of a primitive tetrapod (A4) through successive hypothetical stages, assuming that phalanges are derived from radials. B, The theory of Rosen *et al.* (1981), assuming that the *Eusthenopteron* fin skeleton (B1) is general for sarcopterygians, and that the lungfish fin skeleton is rotated to bring the preaxial radials into a postaxial position (B2, B3), as are the digits of the tetrapods (B4, here a larval urodele); note the presence of separate ulna and radius in the juvenile extant lungfish (B2). C, the 'scrolling' theory (e.g. Coates and Clack 1990), in which a generalized biserial fin (C1) became recurved to bring postaxial radials to the shape of a digital arch (C2), and according to which the large number of digits in early tetrapods (C3) would accord with a large number of postaxial radials. 1, humerus; 2, ulna; 3, radius; 4, preaxial radials; 5, postaxial radials; 6, digits; 7, metapterygial axis (radius and prepollex in tetrapods). (A1–A3, from Jarvik 1964); B2–B4, from Rosen *et al.* (1981); A4, C3, from Coates and Clack (1990).)

close relationship to tetrapods (see Chapter 5, p. 246). Detailed investigations on the structure of the paired fin skeleton of this fish and other osteolepiforms, in particular the structure of the humerus and its relation to the endoskeletal shoulder girdle, all pointed toward their resemblance with the tetrapod limbs. This victory of evolutionary palaeontology has, however, been downgraded by criticisms made by Rosen *et al.* (1981). These authors have tried to demonstrate that most of the resemblances between *Eusthenopteron* and the tetrapods are in fact symplesiomorphies, i.e. general characters of osteichthyans or sarcopterygians, and that, in contrast, lungfishes were the closest relatives to tetrapods. Few papers in early vertebrate palaeontology have raised as much anger and violent or ironical responses as this one, probably because despite many inaccuracies, it touched the core of the palaeontological tradition, (see responses by Jarvik 1981*b*; Holmes 1985; Panchen and Smithson 1987). As far as the fin–limb transition is concerned, Rosen *et al.* argued that the dichotomous pattern of the paired fin skeleton in *Eusthenopteron* is merely the structure of the metapterygium of other gnathostome fishes (Figs 3.20, 4.29B6), and thus is general for sarcopterygians. In consequence, it could not be regarded as closer to the structure of the tetrapod limb in this respect. Instead, they tried to demonstrate that the biserial fin of lungfishes (Fig. 6.8B2, B3) shared some unique features with the tetrapod limb skeleton, namely the closely associated and similar-sized radius and ulna (Fig. 6.8B3, 2, 3). The fact that the larger preaxial radials of lungfishes (Fig. 6.8, 4) are upturned in a secondary postaxial position would accord with the postaxial origin of the digits in urodeles (Fig. 6.8B4, 6), where only the prepollex (Fig. 6.8B4, 7), which prolongs the radius, would be the remnant of the metapterygial axis (Fig. 6.8B3, 7). Since the publication of the paper by Rosen *et al.*, new data, in both palaeontology and biology, have shown that these authors had raised good questions, even if their answer was partly wrong. Among these new data are the strong support for the close relationship between lungfishes and porolepiforms, the confirmation of the homoplasy of the lungfish and tetrapod 'choanae' (as to their intrabuccal position), a better knowledge of the anatomy of the panderichthyids and earliest known tetrapod, and a better understanding of the genetic control of the development of the tetrapod limbs.

Most anatomists now agree that the three proximal bones of the tetrapod limbs (Fig. 6.8, 1, 2, 3) are homologous to the two or three proximal elements of the paired fin skeleton of other sarcopterygians, that is the humerus–femur, radius–tibia, and ulna–fibula (Schultze 1991*b*). Panchen and Smithson (1990) have suggested, however, that the first mesomere of the pectoral fin in os-

teolepiforms (and panderichtyids) is serially homologous to the pelvic girdle, and not to the first mesomere of the pelvic fin. This theory, based on the lack of a dorsal process in the first pelvic mesomere, implies that the humerus would not be serially homologous to the femur in tetrapods, but to the pelvic girdle. Panchen and Smithson regard this as corroborated by differences in the position and nature of the joints in the pectoral and pelvic fins of osteolepiforms, as pointed out by Rackoff (1980) in the osteolepidid *Sterropterygion*. Rackoff suggested that the paired fins in this (and other) osteolepiforms had one rotary articulation and one hinge-like articulation. The rotary articulation is between the fibula and the 'ankle' in the pelvic fin, and between the humerus and the ulna (elbow) in the pectoral fin. Conversely, the 'knee' and the 'wrist' of *Sterropterygion* are hinges. One may wonder, however, whether such a conclusion would have been reached without knowing in advance of the condition in tetrapods. The joints between the mesomeres of the osteolepiform paired fins are made up by 'unfinished' bone, and it is extremely hard to tell, even in remarkably well-preserved specimens, what may have been the actual shape and function of the articulation when it was covered with joint cartilage. Only the 'elbow', that is the articulation between the humerus, ulna, and radius, provides tenuous evidence of a rotary joint. Panchen and Smithson (1990) have also suggested that the ischium of tetrapods is of dermal origin and derives from the axillary scute of osteolepiforms. They arrived at this conclusion because of the presence of ornamentation on the ischium of the Early Carboniferous tetrapod *Crassigyrinus*. Whatever may be the serial homology between the fore and hind limb elements, it has no impact on their respective homologies with the elements of the pectoral and pelvic fins of piscine sarcopterygians.

The question of the origin of the tetrapod limbs rests mainly on the homology of the carpals–tarsals, metacarpal–metatarsals, and phalanges in fishes. All these elements are supposed to derive essentially from radials and from some of the distal mesomeres of the sarcopterygian fins (Fig. 6.8, 4, 6; Schultze 1991), unless they are neomorphs (the 'neopodium theory'; Fig. 6.9D). In piscine sarcopterygians, the paired fin skeleton possesses either a single, preaxial series of radials (rhizodontiforms, osteolepiforms, panderichtyids), or two series, pre- and postaxial (lungfishes, porolepiforms, and possibly actinistians; Ahlberg 1989). The major problem is to know whether the digits derive from pre- or postaxial radials, or both. It has long been admitted that the digits are of preaxial origin (as are the radials in *Eusthenopteron*). It has now been shown that the developmental processes that control the segmentation of the limb skeleton (into humerus, radius, ulna, etc.) and those that control the bi-

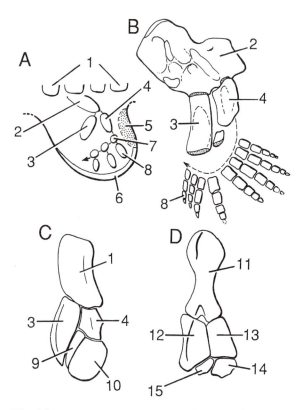

Fig. 6.9. Origin and development of the digits. A, early development of the limb skeleton in a tetrapod embryo (arrow indicates the direction of the further development of the digital arch; B, skeleton of the fore limb of *Acanthostega* (arrow indicates the presumed direction of the development of the digital arch). C, D, homologies of the paired fin and limb skeleton, seen in the light of the neopodium theory (Westoll 1943*b*; Holmgren 1949), i.e. assuming that phalanges are neoformations; C, skeleton of the pectoral fin of *Panderichthys*; D, skeleton of the hind limb of *Ichthyostega* (digits and some tarsals removed to emphasize the resemblance with the pectoral fin skeleton of *Panderichthys*. 1, somites; 2, humerus; 3, radius; 4, ulna; 5, ZPA (zone of polarizing activity); 6, AER (apical ectodermal ridge); 7, carpals; 8, digits; 9, second preaxial radial; 10, distal plate; 11, femur; 12, tibia; 13, fibula; 14, fibular; 15, intermedium. (A, from Hinchliffe (1991); B, from Coates and Clack (1990); C, from Vorobyeva (1992); D, from Jarvik (1980).)

ridge throughout the entire series of the tetrapods (Hinchliffe 1991; Hinchliffe and Johnson 1980, Vorobyeva and Hinchliffe 1991). It thus seems that it will be extremely difficult to trace a point-for-point homology between the paired fin and limb skeleton. Research in developmental biology has shown that the carpals, tarsals, and digits form from a digital arch that bends anteroventrally, in a crook-like pattern (Shubin and Alberch 1986; Fig. 6.9A). The root of this crook seems to be prolongation of the ulna (or fibula; Fig. 6.9B, 4), rather than of the radius (or tibia); the digits would thus be in the position of postaxial radials. The digits thus appear in the order 5, 4, 3, 2, 1; when digital reduction occurs, it follows the reverse order, i.e. the thumb disappears first (Raynaud and Clergue-Gazeau 1986). Two further discoveries have raised once more the question of the basic number of the digits in tetrapods. First, it has been shown that the number of digits (and phalanges in each digit) is controlled by five homeobox genes (Hox–4.4 to 4.8), which are successively activated during ontogeny (Tabin 1992); this condition implies that there cannot be more than five morphogenetic types of digits (the 'forbidden morphology'). The presence of more than five digits in *Acanthostega*, for example, would be explained by the fact that some digits are doubled (digits I, III, and IV), each pair of 'twin' digits having the same number of phalanges. Second, Raynaud and Brabet (1994) have shown that the number of somites involved in the limb bud matches the number of digits in the adult limb. They suggest that the ontogeny of the limb in the earliest tetrapods may have involved more than five somites. This theory is to some extent convergent with Jarvik's (1981*a*) model of limb evolution, by concentration of a number of radials and radial muscles. However interesting these mechanisms may be, they will remain frustrating as long as they are known only in the mouse, chick, and lizard. Similar investigations into the ontogeny of the lungfish fin, for example, are urgently needed. All that is known is that there seems to be no zone of polarizing activity in fishes, a fact that would seem to point towards the 'neopodium' theory, i.e. the carpals, tarsals, and phalanges are unique to tetrapods.

As we have seen above, the digits of the earliest known tetrapods (*Acanthostega*, *Ichthyostega*, *Tulerpeton*) are variable in number, ranging from eight to five, and the smallest ones are the most anterior or medial ones (Fig. 6.9B, 8). The number of digits is, after all, not far from that of the radials of lungfishes (Fig. 6.8B3, 4, 5) which, in addition, consist of segments, like the phalanges of the digits. As pointed out by Coates and Clack (1990), the tetrapod limb skeleton could be conceived as an anteriorly curled biserial fin skeleton, in which the carpals and tarsals would be preaxial and the digits postaxial (Figs

furcation of the digits are different. The growth of the skeletal elements is controlled by the apical ectoderm of the limb bud (the apical ectoderm ridge; Fig. 6.9, 6), and the branching, or anteroposterior multiplication of these elements is controlled by a posterior zone of polarizing activity (Fig. 6.9, 5). The posterior zone of polarizing activity is more stable and constant than the apical ectoderm

6.8C, 6.9B). Differences in limb structure and development between the urodeles and all other tetrapods (the urodele skeletal elements being developed from continuous strings of mesenchyme) gave rise to the theory of the diphyletism of the tetrapods (Holmgren 1933). These differences have since been shown by Saint-Aubain (1981) to be better regarded as a specialization of the urodeles.

One would have expected that the presumed closest relatives to tetrapods, the panderichthyids, would have had a paired fin skeleton that foreshadowed the tetrapod limbs, and in particular the position and number of radials. This is not the case; the panderichthyid skeleton shows the same basic pattern as that of *Eusthenopteron*, but most of the distal radials and mesomeres are fused into a large bony plate (Fig. 6.9, 10). This trend is thus opposite to that in tetrapods, which have numerous and segmented digits. If, however, we compare the structure of the panderichthyid pectoral fin skeleton (Fig. 6.9C) with that of the hind limb skeleton of *Ichthyostega* (Fig. 6.9D; the fore limb is incompletely known), we notice a striking resemblance between the second radial and terminal plate of the former (Fig. 6.9C, 9, 10), and the intermedium and fibular of the latter (Fig. 6.9D, 14, 15). A limb like this, without phalanges, would thus resemble more closely the pattern in the panderichthyid fin than that of any other fish. If we trust this resemblance and assume that the condition in panderichthyids is not a unique specialization of this taxon, then this may be another argument in favour of the 'neopodium' theory.

The dilemma is thus that the tetrapod limb skeleton could be more readily derived from a long, biserial skeleton of dipnomorph type than from that of osteolepiforms and panderichthyids, which, from other characters, appear to be more closely related to the tetrapods than the dipnomorphs. As far as palaeontology is concerned, the clue may come from better-preserved Devonian tetrapod material, which could more clearly show the pattern of the digital arch, and in particular the organization of the carpals and tarsals. At the moment, the question of the origin of the tetrapod limb is merely the question of whether the hand and foot are unique to tetrapods or have homologues in fishes, and it is unanswered.

Further reading

Ahlberg and Milner (1994); Alberch (1989); Coates (1993b); Coates and Clack (1990); Hinchliffe (1991, 1994); Hinchliffe and Johnson (1980); Holmes (1985); Holmgren (1933, 1949); Jarvik (1981a,b); Panchen and Smithson (1987, 1990); Raynaud and Brabet (1994); Rosen *et al.* (1981); Schmalhausen (1969); Schultze (1991); Shubin and Alberch (1986); Tabin (1992); Tabin and Laufer (1993); Vorobyeva (1992); Vorobyeva and Hinchliffe (1991); Vorobyeva and Kuznetzov (1992); Westoll (1943b).

6.7 HOMOLOGIES IN THE LABYRINTH

The labyrinth is assumed to be a derivative of one particular placode of the primitive lateral-line system (Jarvik 1981a). Notwithstanding the fact that the lateralis and acoustic nerve have the same origin, this theory is contradicted by the absence of a lateral line system in hagfishes, which, however, possess a labyrinth (Fig. 6.10A1). The labyrinth arises in ontogeny as a placodal thickening of the ectoderm, which becomes invaginated. Early in development a recess forms in the dorsal part of this invagination, and subsequently turns into the endolymphatic duct (Fig. 6.10, 4). Inside the labyrinth are a number of 'maculae', or sensory organs, whose function is to analyse the movements and the pressure of the internal liquid, or endolymph (Fig. 6.10, 15–18). The gnathostomes have three semicircular canals (two vertical ones and a horizontal one; Fig. 6.10A6, 5–7); lampreys have only two (the vertical ones; Fig. 6.10A2, 5, 6); and hagfishes one (Fig. 6.10A, 1, 1). The hagfish canal poses a problem that hitherto remains unresolved. This unique canal meets anteriorly and posteriorly two ampullae containing a crista (Fig. 6.10A1, 2, 3). The simplest interpretation would thus be that it is homologous to both the anterior and posterior vertical canals of other craniates. This condition could be either primitive or derived (due to 'degeneracy' for most authors). Some have suggested, however, that the single canal is homologous only to the posterior canal. Following Retzius (1881), Jarvik (1980) considered that this canal is not a semicircular canal but the utriculus, and that hagfishes once had two vertical semicircular canals, which have disappeared and remain in the hagfish embryo only in the form of two slight ridges on the dorsal surface of the saccus communis. This hypothesis was proposed to reconcile the hagfish labyrinth with the two vertical semicircular canal impressions in fossil heterostracans (Fig. 6.10A3, 5, 6), which Stensiö (1927) regarded as ancestors of hagfishes. Considering the primitive state of a large number of characters met with in hagfishes, it is certainly more parsimonious to regard the hagfish labyrinth as primitive, but a transition toward the lamprey or gnathostome condition, i.e. with two distinct vertical semicircular canals, remains a matter of imagination. The same applies to the rise of the horizontal semicircular canal in gnathostomes (Fig. 6.10A6, 7), for which no 'transitory' or 'precursor' state is known.

The labyrinth in all fossil jawless craniates in which it is recorded (i.e. heterostracans, galeaspids, and osteostracans) is basically comparable to that of lampreys in the position and number of the semicircular canals (Fig. 6.10A3, A4, A5, 5, 6). One should, however, note that the semicircular canals in lampreys do not make distinct

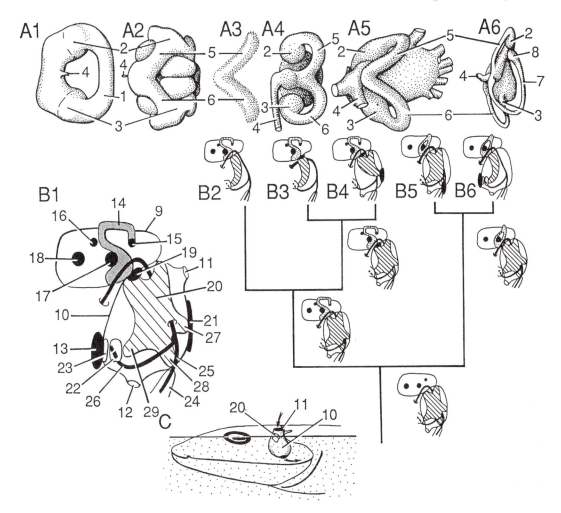

Fig. 6.10. Homologies and polarities in the labyrinth and the tetrapod middle ear. A, right membranous labyrinth or internal cast of the right labyrinth cavity in extant and fossil Craniata: Hyperotreti (A1), Hyperoartia (A2), Heterostraci (A3), Galeaspida (A4), Osteostraci (A5), and Gnathostomata (A6). B, Interpretation of the character states in the ear, according to Lombard and Bolt (1979): B1, key to characters; B2–B6, diagrammatic representation of the ear in extant tetrapods, and inferred ancestral states: Apoda (B2), Caudata (B3), Anura (B4), Sauropsida (B5), and Mammalia (B6). C, Clack's (1993) interpretation of the otic notch (here in *Acanthostega*) as a respiratory spiracle when the external nostrils are situated below the surface of the water (arrows indicate the direction of the air flow). 1, single semicircular canal; 2, anterior ampulla; 3, posterior ampulla; 4, endolymphatic duct; 5, anterior semicircular canal; 6, posterior semicircular canal; 7, horizontal semicircular canal; 8, ampulla of horizontal semicircular canal; 9, otic capsule; 10, hyoid (spiracular) pouch; 11, external hyoid opening (spiracle); 12, internal hyoid opening (Eustachian tube); 13, tympanic membrane; 14, special periotic system; 15, papilla basillaris; 16, papilla amphibiorum; 17, macula lagena; 18, macula neglecta; 19, otic opercular bone; 20, hyomandibula; 21, dermal opercular bone; 22, quadrate (incus); 23, articular (maleus); 24, ceratohyal; 25, hyoid branch of facial nerve; 26, internal branch of facial nerve (chorda tympani); 27, opercular process; 28, hyoid process; 29, quadrate process. (A1, A2, A6, from Marinelli and Strenger (1954, 1956, 1959); A4, modified from N. Z. Wang (1991); A5, from Stensiö (1927); B, from Lombard and Bolt (1979).)

loops and lie against the vestibular division, which is enlarged by vast ciliated chambers (regarded by Jarvik (1965a, 1980) as homologous to the 'sel' canals of os-

teostracans; see Chapter 4, p. 110). In contrast, the two vertical semicircular canals of galeaspids, osteostracans, and probably also heterostracans (the latter known only by impressions in the exoskeleton) form distinct loops that join medially into a crus commune, and thereby resemble more closely the vertical canals of the gnathostomes (Fig. 6.10A6, 5, 6). It is still undecided whether this represents a more derived condition, linked with a perfected balance function, or a generalized condition. (The canals in lampreys have a reduced function, owing to the prominent role of the ciliated sacs in balance.)

The endolymphatic duct opens to the exterior in some primitive gnathostomes, as well as in some galeaspids and all osteostracans (Fig. 6.10A4–6, 4). The fact that it is blind in hagfishes and lampreys has been regarded as a derived condition, because it is also blind in all 'higher' gnathostomes. Again, regarding this condition as primitive in hagfishes and lampreys, and thus as general for craniates, is more congruent with the distribution of other character states.

Further reading

Hagelin (1974); Lowenstein and Thornhill (1970).

6.8 THE ORIGIN OF THE MIDDLE EAR

One of the major tetrapod characteristics is the presence of a sound-conducting bone, the stapes, extending from the fenestra ovalis of the braincase either to the quadrate or to the tympanic membrane (Fig. 3.23E). It is now widely accepted that the stapes is the homologue of the hyomandibula of fishes, to which is fused an element of the wall of the otic capsule, the footplate. However, and contrary to what would have been expected, the stapes in the earliest known tetrapods (*Acanthostega*; Fig. 4.96C) is a small, massive bone, which differs from the long and slender epihyal of osteolepiforms and panderichthyids (Fig. 4.90, 11; Clack 1989, 1993, 1994; Vorobyeva and Schultze 1991), but is more like that of the actinistians and Devonian lungfishes (Fig. 4.75, 14; Fig. 4.81, 11). Here again, although the hyomandibula–stapes homology is hardly questionable, transitional stages in fossils are still sought that will bridge this morphological gap.

Another celebrated victory of evolutionary palaeontology is the origin of the maleus (hammer) and incus (anvil) in the mammalian middle ear. Fossils are said to have confirmed the embryology-based Reichert–Gaupp theory. This theory assumes that the maleus is the articular bone of the lower jaw and the incus the quadrate, which have both been incorporated into the middle ear, and that a new jaw articulation arose between the dentary and the squamosal. Jarvik (1981*a*) tried to refute this theory by showing that the ventral part of the hyoid arch in *Eusthenopteron* contained potentially all the elements that might have given rise to the mammalian middle ear, and

that it was not necessary to invoke a migration of the quadrate–incus and articular–maleus into the middle ear. Jarvik's comments have, however, never been taken seriously into consideration, perhaps because the Reichert–Gaupp theory is regarded as too well corroborated by fossils, and deep-rooted in the evolutionary paradigm.

The steps of the transformation of the tetrapod ear have been thoroughly analysed by Lombard and Bolt (1979). They proposed a theory that is different from the standard view, i.e., that a tympanic membrane was present in the hypothetical ancestor to all tetrapods (Goodrich 1915). Among extant tetrapods, only the urodeles and apods primitively lack a tympanic membrane. This condition was generally regarded as a secondary loss, as in snakes and some other burrowing reptiles. Lombard and Bolt's analysis leads to the conclusion that the tympanic membrane appeared three times, in anurans, mammals, and sauropsids (reptiles and birds), and is connected to the stapes in different ways (Fig. 6.10B4–6). The tympanic membrane is connected to the opercular process of the stapes (Fig. 6.10B1, 27) in anurans, to the hyoid process in sauropsids (Fig. 6.10B1, 28), and to the quadrate process (via the incus and maleus) in mammals (Fig. 4.10B1, 29). They also pointed out a major difference between amphibians and amniotes in the course of the periotic canal (Fig. 6.10B1, 14; passing laterally to the lagena in amphibians, and medially in amniotes). Which of these two conditions is general for tetrapods is, however, undecided, and this structure is not preserved in fossils. This periotic system is supposed to have preceded the rise of the tympanic membrane, and may have been an important sound-conducting device in early tetrapods. Lombard and Bolt's theory seems to be contradicted by the presence of an 'otic notch'—which is assumed to have housed a tympanic membrane—in a wide range of early tetrapods, such as *Ichthyostega*, *Acanthostega*, most temnospondyls, and anthracosaurs (Fig. 4.95, 59; Fig. 4.96, 4; Fig. 4.100, 6; Fig. 4.101, 1; Figs 4.102, 4.103, 5), even if colosteids and, presumably, generalized amniotes (captorhinomorphs) lack such a notch. There is, however, no clear evidence that what is called the 'otic notch' in these early forms was closed by a tympanic membrane. Panchen (1985) and Clack (1993), for example, have suggested that the 'otic notch' was in reality a respiratory device derived from the piscine spiracle, and had no auditory function (Fig. 6.10C). This interpretation was based on the fact that the stapes of the earliest tetrapods (*Acanthostega*, and also early anthracosauroids such as *Pholiderpeton* and amphibians such as the colosteid *Greererpeton*) is a massive rod that links the braincase either to the occipital arch or to the quadrate, but is not oriented towards the otic notch. In fact all these early tetrapods retain a relatively mobile palatoquadrate, and the stapes retains the suspensory function of the piscine hyomandibula. The air-filled spiracular cavity (Fig. 6.10C, 10) of these early forms could have played a role in sound conduction, but the stapes acquired a real sound-conducting function only when the palatoquadrate became firmly fused to the braincase. This was achieved, independently, only in the amniotes and temnospondyls. A tympanic membrane could then appear in the otic notch. The assumption that a tympanic membrane was present in stem-group temnospondyls would, however, imply that its absence in urodeles and apods is secondary, according to current phylogenies in which lissamphibiens are classified as temnospondyls.

Further reading

Clack (1993, 1994); Bjerring (1977); Bolt and Lombard (1985); Carroll (1980); Forey *et al.* (1991); Jarvik (1981*a*); Lombard and Bolt (1979); Panchen (1985); Rosen *et al.* (1981).

6.9 TRENDS IN BRAIN MORPHOLOGY

The brains of lampreys (Fig. 6.11B) and generalized gnathostomes (e.g. sharks, Fig. 3.8H) are somewhat similar in shape, the major differences being the lack of long olfactory tracts (the olfactory bulb is closely associ-

ated with the telencephalon) and the very small cerebellum in the former. The brain of hagfishes also lacks olfactory tracts and cerebellum, but in addition its shape is odd (Fig. 6.11A), because of an anteroposterior compression that takes place during ontogenetic development. It is therefore regarded as a uniquely derived feature of this taxon.

When considering fossils, and in particular extinct forms, such as several fossil jawless craniates and the placoderms among gnathostomes, the morphology of the brain can often be deduced from the shape of the brain cavity or, in the case of heterostracans, from the impression of the brain on the internal surface of the dorsal shield (Fig. 6.11C). How much the outline of this cavity mirrors the actual brain is not known, but in many cases the distinction between the divisions of the brain is so clear that it can reasonably be regarded as a fairly accurate image of the brain (Fig. 6.11D, E).

The morphology of the brain cavity in osteostracans (Fig. 6.11E) and galeaspids (Fig. 6.11D) is strikingly similar, and matches very well the impression that can also be observed in heterostracans (Fig. 6.11C). In turn, it looks quite similar to the aspect of the brain of lampreys (Fig. 6.11B), in particular larval lampreys. There remains, however, one important difference, which has led to ambiguous interpretations. In osteostracans, galeaspids, and heterostracans, there is a distinct, paired swelling of the

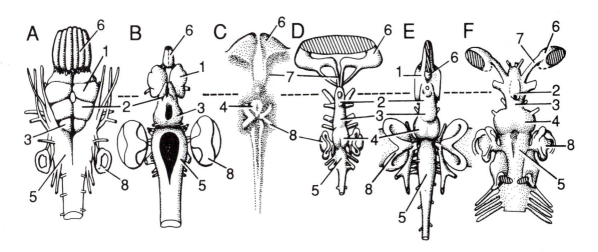

Fig. 6.11. Comparative morphology of the brain, cranial nerves, and associated olfactory and labyrinth cavities in the Craniata (C, impression on internal surface of exoskeleton, D–F, internal cast of cranial cavity; dashed line indicates the position of the pineal region as a point of reference. A, Hyperotreti (hagfishes); B, Hyperoartia (lampreys); C, Heterostraci ; D, Galeaspida; E, Osteostraci; F, Gnathostomata (acanthothoracid placoderm). 1, telencephalon and olfactory bulb; 2, diecephalon; 3, mesencephalon (optic lobes); 4, metencephalon (and cerebellum); 5, rhombencephalon and medulla oblongata; 6, olfactory capsule; 7, olfactory tract; 8, membranous labyrinth or labyrinth cavity. (A, B, modified from Marinelli and Strenger (1954, 1956); C, based on Novistkaya (1983); D, from N. Z. Wang (1991); E, from Janvier (1985*a*); F, based on Young (1980).)

roof of the brain cavity, which lies between the anterior semicircular canals (Fig. 6.11C–E, 4). This swelling was first interpreted by Stensiö (1927) as having housed the metencephalon (cerebellum), and then by Halstead Tarlo and Whiting (1965) as corresponding to the position of the optic lobes of the mesencephalon (Fig. 6.11B, 3). Janvier (1975a) followed the latter interpretation, but he later rejected it (Janvier 1985a) because it would imply a much too posterior position for the optic lobe in some osteostracans with a very elongated midpart of the brain cavity (Fig. 6.11E, 3). Despite the striking resemblance between the optic lobes of lampreys (Fig. 6.11B, 3) and this paired swelling in fossil jawless vertebrates (Fig. 6.11C–E, 4), one must admit that the problem remains unresolved. The position of the cranial nerve canals is of little help there, because we ignore the length of the intracranial part of these nerves, and it is likely that the exit of these canals from the brain cavity does not mirror the position of the exit of the nerves from the brain itself.

In Chapter 4, I used the 'cerebellum' interpretation for heterostracans, osteostracans, and galeaspids. In fact, all these jawless vertebrates had a well-developed lateral-line system, whose nervous component (the acoustico-lateral nerves) are connected to the cerebellum. It is thus probable that, like all piscine gnathostomes, these fossil forms had a well-developed cerebellum.

The brain cavity of placoderms has been described in a number of forms, in particular arthrodires and petalichthyids, and interpreted in the light of the brain morphology of modern sharks. In most instances, the brain cavity does not provide an accurate cast of the brain itself. In the acanthothoracid *Brindabellaspis* (Fig. 6.11F), the morphology of the brain cavity is, however, remarkably distinct, and displays the same, large dorsal paired swelling as that of osteostracans (Fig. 6.11F, 4). Young (1980) interpreted this paired swelling as having housed the cerebellum. The striking resemblance between the brain cavity of *Brindabellaspis* and that of osteostracans, galeaspids, and heterostracans thus suggests that this morphology represents the general condition, at least for the group that includes galeaspids, osteostracans, and gnathostomes (Fig. 5.1, 7).

Further reading

Janvier (1975a, 1985a); Halstead (1973b); Novitskaya (1983); N. Z. Wang (1991).

6.10 EVOLUTION OF THE HARD TISSUES: POLARITIES AND TRANSFORMATIONS IN HISTOLOGICAL STRUCTURES

The great diversity of the hard tissues in early vertebrates was noticed as early as the mid-nineteenth century, and knowledge of these tissues has increased as the techniques of observation have evolved. The search for trends in hard tissue evolution did not, however, begin before the mid-twentieth century. It was prompted mainly by the need for characters in analysing the relationships of fossil jawless vertebrates, which also display a wider diversity in exoskeletal structure. In the 1960s, the rise of scanning electron microscopy opened new paths into the ultrastructural level, i.e. the organization of the apatite crystals, in particular in hypermineralized tissues, such as enamel, enameloid, or acrodine. The results of these century-long investigations are, to say the least, quite chaotic, as far as the dermal skeleton is concerned. In contrast, the endoskeleton seems to provide a more simple pattern of transformation, which we shall deal with at first.

6.10.1 The endoskeleton

The histological structures in the endoskeleton in craniates can be grouped into five major types: hyaline cartilage (Fig. 6.12A1, 1), globular calcified cartilage (Fig. 6.12A2, 2), prismatic calcified cartilage (Fig. 6.12A3, 4), perichondral bone (Fig. 6.12A2, 3), and endochondral bone (Fig. 6.12A4, 5). Some of these are associated in various ways (e.g. prismatic calcified and normal cartilage; normal cartilage and perichondral bone; endochondral and perichondral bone), but the distribution of these tissues in a phylogeny of the craniates based on other characters (Fig. 5.1) roughly reflects their order of appearance in the ontogeny of osteichthyans; i.e. cartilage, perichondral bone, and endochondral bone in succession. The prismatic calcified cartilage, which lines the cartilaginous elements of the endoskeleton (Fig. 3.12B2, 5), is unique to chondrichthyans, and the apparent absence of bone in the chondrichthyan endoskeleton was long regarded as a primitive feature, until the discovery of perichondral bone lining the cartilage in several fossil jawless (osteostracans, galeaspids) and jawed vertebrates (placoderms, acanthodians). It thus became probable that chondrichthyans had lost perichondral bone. This was later confirmed by the discovery of thin layers of perichondral bone in some extant sharks (Fig. 3.12C, 6; Fig. 6.12A3, 3). As we see it now, the common ancestor to all craniates had an exclusively cartilaginous endoskeleton (Fig. 6.12A1), which, later in the history of the group, could be strengthened either by internal globular calcifications or by perichondral bone, or both (Fig. 6.12A2). The common ancestor to the gnathostomes also had cartilage lined with perichondral bone, again with some occasional globular calcifications. This condition remains in placoderms and acanthodians. Within the gnathostomes, two specializations appeared: on the one hand the loss of perichondral bone and gain of prismatic

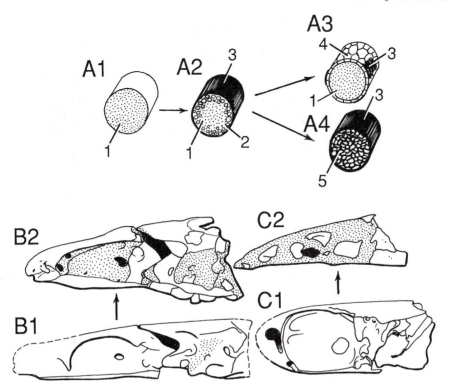

Fig. 6.12. Polarities and trends in endoskeleton structure. A, polarity of endoskeleton composition, according to both ontogeny and phylogeny: A1, cartilage (hagfishes, lampreys, ?heterostracans); A2, cartilage (with globular calcifications) and perichondral bone (galeaspids, pituriaspids, osteostracans, placoderms, acanthodians); A3, prismatic calcified cartilage, with remnants of perichondral bone (chondrichthyans); A4, perichondral and endochondral bone (osteichthyans). B, C, trends in endochondral bone reduction in the braincase of Devonian to Recent actinistians (B1, *Diplocercides*; B2, *Latimeria*) and actinopterygians (C1, *Mimia*; C2, *Amia*) (cartilage stippled). 1, normal cartilage; 2, globular calcified cartilage; 3, perichondral bone; 4, prismatic calcified cartilage; 5, endochondral bone (B1, C2, based on Jarvik (1980); B2, from Millot and Anthony (1958); C1, from Gardiner (1984*a*).)

calcified cartilage (Chondrichthyans, Fig. 6.12A3); and on the other hand the gain of endochondral bone (oste-ichthyans; Fig. 6.12A4).

Within each major gnathostome group, trends toward a reduction of the endoskeletal ossification have been observed and soon became regarded as a rule, if not a 'law' of vertebrate evolution. This is particularly clear in lungfishes, actinistians (Fig. 6.12B) and actinopterygians (Fig. 6.12C), for example; the endoskeleton of most post-Devonian forms (including Recent ones, Fig. 6.12B2, C2) is less ossified than that of the Devonian forms (Fig. 6.12B1, C1). In Devonian osteichthyans, in general, the braincase appears as a single mass of endochondral and perichondral bone, except at the level of the cranial fissures or joints, whereas in later forms of the same group, it consists of several units of membrane bone linked by cartilage. This trend is sometimes explained by heterochrony (see Chapter 7, p. 287).

Further reading

Halstead (1974); Maisey (1988); Ørvig (1951); Patterson (1977*a*); Ruben and Bennett (1987).

6.10.2 The exoskeleton

The evolutionary scenarios for the structure of the exoskeleton may be much more complex, since they comprise the histological history proper, i.e. the transformation and evolution of the types of hard tissue, and the macrostructural history, i.e. the growth, fusion, or breaking-down of the dermal units. The macrostructure will be dealt with in the next section (6.11).

The exoskeleton of the vertebrates generally consists of a superficial layer made up of some papillary dentinous tissue (mesodentine (Fig. 6.13A), semidentine (Fig. 6.13B) metadentine (Fig. 6.13C), and orthodentine (Fig. 6.13D), and a deeper layer of cellular bone ('true bone'; Fig. 6.14, 5) or acellular bone (aspidine; Fig. 6.14, 6). Both hard tissues are produced by ectomesenchymatous cells (derived from the neural crest). The dentinous tissues are generally covered with a hypemineralized layer, which may either be of epidermal origin (monotypic enamel, or 'true enamel'; Fig. 14, 4), or of ectomes-

enchymatous origin (acrodine), or both (bitypic enamel, or 'enameloid'; Fig. 6.14, 1). The dentinous tissue is generally metadentine or orthodentine, i.e. the tubules that house the processes of the dentine-producing cells (odontoblasts) all arise from a single pulp cavity (Fig. 6.14, 3). In some instances, however, these cells can be 'trapped' inside the dentine, either inside small, interconnected cavities (mesodentine; Fig. 6.13A) or in isolated drop-shaped cell spaces (semidentine, unique to placoderms; Fig. 6.13B). A special type of papillary tissue, the cement (Fig. 6.14, 10), links the superficial dentinous tissue to the underlying bone.

This diversity of dentine and dermal structure has been interpreted in various ways. Ørvig (1965, 1967), for example considers that orthodentine and acellular bone represent derived conditions, in particular because acellular bone occurs in advanced Recent fishes (e.g. teleosts) and because in such extinct groups as osteostracans acellular bone seems to occur only in the latest forms *Escuminaspis*. Since he considers cellular bone to be primitive, Ørvig regards dentinous tissues with enclosed cell spaces (i.e. mesodentine and semidentine) as closer to bone and, consequently, more primitive than orthodentine (Fig. 6.14A1). In contrast, other authors, such as Halstead (1974) and, more recently, Smith and Hall (1990) consider that the acellular dermal bone of early vertebrates (heterostracans, anaspids) was formed by cells arranged in a single front, which retreated as the layers of aspidine were deposited. This mode of formation is said to be comparable (but not similar) to that of orthodentine, in which the odontoblasts in the pulp cavity retreat as the dentinous matter is deposited. The comparison is supported by the fact that the aspidine of *Astraspis* is regarded as very close to dentine ('astraspidine'). This would suggest that acellular bone associated to metadentine or orthodentine (Fig. 6.14A4, B3) is the general condition for the vertebrates. There have been heated debates on this subject, with detailed arguments for and against each theory (for review see Ørvig 1989). As stressed by Ørvig, the mode of formation of the acellular bone in extant teleosts does not match this interpretation, since cells are trapped in the bone matter, and then become pycnotic and disappear. But was the process similar in the aspidine of heterostracans, for example? Until now, no true cell space has ever been observed in their exoskeleton, even in the basal layer, where it is supposed to grow. The stratigraphical order of appearance of dermal skeleton types does not help much in resolving this dilemma, since both cellular and acellular bone and both orthodentine and mesodentine occur by Middle Ordovician times (Fig. 4.3C, D). Only semidentine seems to be restricted to the placoderms and appears in the Silurian.

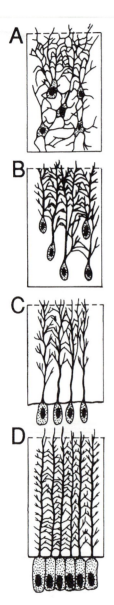

Fig. 6.13. Distribution of the cells (odontocytes or odontoblasts) in the four types of dentinous tissue. A, mesodentine; B, semidentine; C, D, dentine: metadentine (C), orthodentine (D). (From Smith and Hall (1993).)

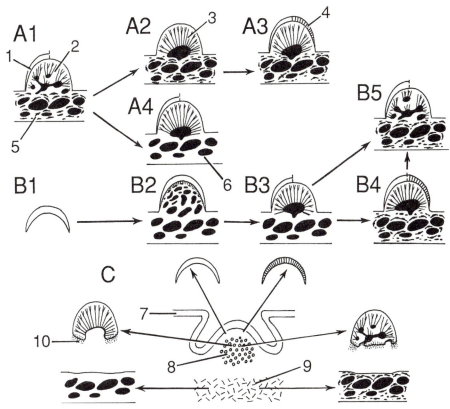

Fig. 6.14. Polarities in the histological structure of the odontodes and dermal bone. A, theory assuming that mesodentine and cellular bone is primitive: A1, odontode of mesodentine (with or without enameloid) attached to cellular dermal bone (e.g. osteostracans, some acanthodians); A2, odontode of metadentine or orthodentine attached to cellular dermal bone; A3, the same with a cap of either enameloid or monotypic enamel (e.g. osteichthyans); A4, the same (with or without enameloid) attached to acellular dermal bone, or aspidine (e.g. astraspids, heterostracans, chondrichthyans, some acanthodians). B, theory assuming that the odontodes were first made up of enameloid (B1), then filled with aspidine from the underlying acellular dermal bone, with an intervening layer of mesodentine or dentine (B2, e.g. astraspids), and then odontodes of meso-, meta-, or orthodentine (B3–5) attached on either acellular (B3) or cellular bone (B4, B5). C, theory assuming that the dermal papilla had the potentiality to produce any kind of dentinous tissue, which was covered by any kind of epidermal tissue, and was attached by means of cement on any kind of basal dermal bone. 1, enameloid; 2, mesodentine; 3, metadentine or orthodentine; 4, monotypic enamel; 5, cellular bone; 6, acellular bone (aspidine); 7, epidermis; 8, papillary cells; 9, dermis; 10, cement.

At present, considering acellular bone as plesiomorphous and cellular bone as apomorphous within the vertebrates is more congruent with other characters than the reverse polarity. So far as dentinous tissues are concerned, no satisfactory clue can be given. The Ordovician vertebrates have virtually no dentine at all, their tubercles being made up either of aspidine or of a thick layer of enameloid (Ørvig 1989; Gagnier 1993a). Only *Eriptychius* and *Pycnaspis* have some dentine-like tissue, and this was Ørvig's major argument for considering *Eriptychius* to be more closely related to heterostracans than *Astraspis*. Neither galeaspids nor anaspids have dentine, and osteostracans have mesodentine. So,

metadentine and orthodentine ('true dentine') are restricted to heterostracans, some thelodonts and most gnathostomes.

Maisey (1988; Fig. 6.14B) considers that the first odontodes to appear in vertebrates were possibly made up of enameloid (Fig. 6.14B1), and later filled with some aspidine-like tissue (Fig. 6.14B2). The dentinous (metadentine, orthodentine or mesodentine) tissue proper developed subsequently between these two layers, and the enameloid cap eventually disappeared or was replaced by enamel (Fig. 6.14B3–5). For the supporting dermal bone, Maisey accepts the theory that acellular bone is the generalized condition.

Smith and Hall (1993) suggested that the primordial type of dermal unit was an odontode formed in a dermal papilla that had the potential to produce any kind of dentinous tissue, and was covered with a layer of either monotypic or bitypic enamel (Fig. 6.14C). It later became attached basally to a deeper hard tissue of ectomesodermal derivation, either cellular or acellular, by means of some cement-like tissue (Fig. 6.14, 10).

The difference between these two opinions may be due to the fact that Smith and Hall use essentially embryological data, whereas Maisey relies on the stratigraphical order of occurrence of the various types of hard tissues, and seems to have been particularly impressed by the thick, glassy enameloid cap of the tubercles of *Astraspis* and *Pycnaspis*. It is also possible that Smith and Hall's theory, which rejects any polarity in the character states of the papillary tissues, is aimed at justifying the fact that all these states can be found in conodonts, which some regard as the sister-group of either craniates or vertebrates. Whatever may be the primitive types of hard tissue in the earliest vertebrate exoskeleton, most authors now agree on the hypothesis that there was initially a separation between the dentinous odontodes (the papillary tissue) and the underlying dermal bone.

The actual distribution of the major dermal hard tissues in early vertebrates may be as follows:

Monotypic enamel: (enamel, ganoine): ?*Eriptychius*, ?Galeaspida, Osteichthyes.

Bitypic enamel (enameloid, acrodine): ?*Astraspis* thyestiid Osteostraci, Chondrichthyes, Actinopterygii.

Metadentine and Orthodentine (including osteodentine, trabecular dentine, etc.): Heterostraci, thelodontid Thelodonti, most Gnathostomata.

Mesodentine: ?*Pycnaspis*, *Eriptychius*, ? katoporid and loganiid Thelodonti, Osteostraci, climatiid Acanthodii.

Semidentine: Placodermi.

Aspidine (or acellular bone in general): Astraspis, Pycnaspis, Eriptychius, ?Arandaspida, Heterostraci, Galeaspida, Anaspida, ?Thelodonti, Late Devonian Osteostraci, acanthodid Acanthodii, some Chondrichthyes, Teleostei.

Cellular bone: some undetermined Ordovician forms, most Osteostraci, most Gnathostomata.

Further reading

Halstead (1969, 1974, 1987*b*); Maisey (1988); Ørvig (1951, 1958, 1965, 1967*a*, 1989); Schaeffer (1977); M. M. Smith (1992), Smith and Hall (1990, 1993).

6.11 ORGANIZATION OF THE EXOSKELETON

Apart from the question of the histological structure of the exoskeleton, the trends in its organization are yet another matter of debate. Everyone agrees that the primary dermal skeletal unit is the odontode, i.e. a dentinous element, possibly covered with enamel or enameloid, which forms from a single dermal papilla. How these odontodes contribute to form larger dermal elements (scales, dermal plates, teeth) is, however the subject of two major current theories: the 'lepidomorial theory' (Stensiö 1962) and the 'odontode regulation theory' (Reif 1982). The lepidomorial theory assumes that large dermal units may arise by accretion of several, younger and younger odontodes (cyclomorial scales), and also that several odotondes may fuse at the papillary level to form a much larger element, such as placoid scales or teeth (synchronomorial scales). The odontode regulation theory assumes that any non-growing dentinous element always remains a single odontode, which differentiates into tooth or scale at the papillary level. The formation of larger, growing scales or dermal plates occurs in two ways: by accretion of individual odontodes, or their attachment to the underlying dermal bone. It depends on regulation by means of a more or less efficient inhibition zone around the developing odontode. The lepidomorial theory was initially put forward to explain the transition between the growing scales of most Palaeozoic chondrichthyans (Figs 4.30A, 4.35A2) and the non-growing, placoid scales of euselachians (Figs 3.13H, 4.40). Reif (1982) considers that there is no evidence of such a fusion of odontodes at the papillary level. There are, however, instances where the papillary tissue produces a continuous layer of hard tissue (e.g. cosmine). Reif's odontode regulation scenario (Fig. 6.15) explains virtually all possible exoskeletal structures. The odontodes of a given generation may be shed and replaced by a new generation of larger odontodes (Fig. 6.15A) as soon as the inhibition ceases, or may be covered by the new generations (Fig. 6.15B). New odontodes can be added peripherally to a non-inhibating core (primordium), thereby producing an onion-like structure (Fig. 6.15C), or they can become attached to a growing basal bony plate (Fig. 6.15D), and be covered by a new generation of adontodes. Finally, the odontodes may not be replaced by new generations; when they become abraded, the dermis then produces pleromic dentine, which fills and strengthens the lacunae of the basal bony plate (Fig. 6.15E). The odontode regulation theory also explains the rise of the dental lamina in the gnathostomes. It is assumed that the teeth arose from single, enlarged odontodes, as in sharks. When shed, teeth need to be replaced immediately; otherwise the animal cannot feed. This implies that the new teeth must be ready to serve as the old ones are shed. An ordinary skin odontode usually forms when the preceding one is shed, i.e. when the inhibition zone around the shed odontode disappears (Fig. 6.15A). But the development of a dental lamina (Fig. 6.16, 2) allows the teeth to form while the

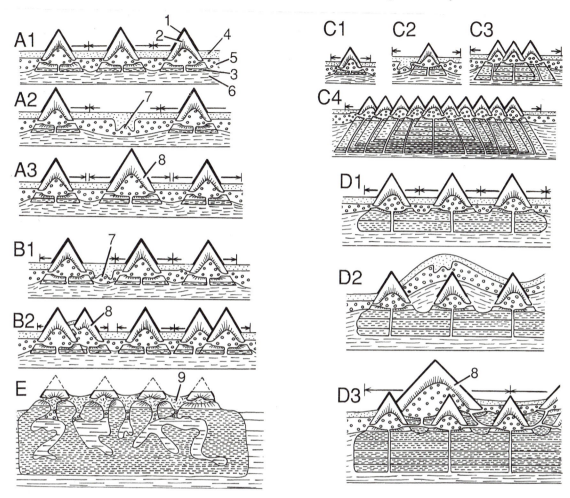

Fig. 6.15. Organization of the odontodes in the exoskeleton (according to the odontode regulation theory of Reif (1982*a*). Arrows indicate the extent of the zone of inhibition around the odontodes. A, successive stages in the replacement of the free odontodes (e.g. microsquamose placoid scales); B, successive generations of odontodes, the older ones not being shed (e.g. 'tesserae', mesosquamose scales); C, accretion of successive odontode generations to form a larger unit (e.g. macrosquamose scale of acanthodians, osteichthyans, and some fossil chondrichthyans); D, successive generations of larger and larger odontodes attached to a common bony base (e.g. macromeric toothed plates, scales, and ornamented dermal bone); E, filling and consolidation of abraded odontodes by pleromic dentine (e.g. holocephalan, lungfish or ptyctodontid tooth plates). 1, enamel; 2, dentine; 3, basal bone or cement; 4, epidermis; 5, vascular dermis; 6, fibrous dermis; 7, papilla; 8, second generation odontode; 9, pleromic dentine. (From Reif (1982*a*).)

preceding generation is still in use, thereby escaping the inhibition zone. The growth of the skin toward the lips then brings the newly formed tooth into a functional position (Fig. 6.16C). The fact that the denticles on the gnathal plates of placoderms are not shed, and form by apposition, in the same way as the tubercles of the external dermal bones, suggests that a dental lamina was not yet developed (Reif 1982). Also, the very low rate of tooth replacement in early chondrichthyans indicates a dental lamina that is still rudimentary (Patterson 1992).

Further reading

Maisey (1988); Ørvig (1977); Reif (1982*a*); Smith and Hall (1993); Stensiö (1962).

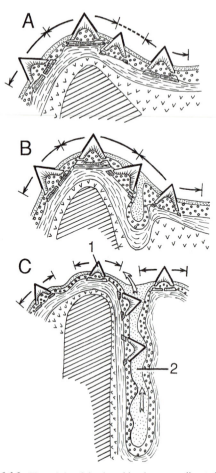

Fig. 6.16. The origin of the dental lamina, according to Reif
(1982*a*). Black arrows indicates the extent of the zone of
inhibition around the odontodes or teeth. White arrow indicates
the direction of tooth development. A–C, successive
hypothetical stages in the development of the dental lamina.
1, tooth; 2, dental lamina. (From Reif (1982*a*).)

7

Evolution, palaeobiogeography, and life history

One step further from reality, we enter here the realm of evolutionary processes and life history, i.e. all the flesh that one may put on fossils, homologies, and phylogenetic patterns: whence, how, and why early vertebrates evolved, how they lived, on what they preyed, etc. These stories are often amusing, sometimes plausible, but generally untestable as theories. Some are long-lived, even if unlikely; others are bound to geological or biological fashions. Nevertheless, this book would be incomplete if it did not include stories of this kind relating to early vertebrates.

7.1 ORIGINS

The search for the origin—rather than the relationships—of taxa has been the chief occupation of evolutionary palaeontologists for over a century. It is often said to stem from Darwin, but the idea that the origins of taxa can be imagined or actually discovered is deep-rooted in all the transformist views that have been expressed since antiquity. Nowadays, it stems largely from the belief that taxa are individuals, have a life and descendants, and are defined by their common ascent. Early vertebrate palaeontology does not escape this tradition and, since it deals with the earliest representatives of major taxa, it has been even more concerned with problems of this kind than any other field of vertebrate palaeontology. At best, the origin of a taxon is viewed as the origin of one character that defines it; this has been dealt with in the preceding chapter. This is the case for the origin of jaws for the gnathostomes, or the origin of limbs for the tetrapods. These anatomical questions, basically of homology and shared derived character distribution, are however, accompanied by a body of transformational theories on processes and on the role of the environment. The purpose of these additional theories is to provide an explanation, a justification, or worse, a number of *ad hoc* hypotheses to support character distributions—as if a taxon would not be real, enough if defined only by a character. We all have a propensity for such scenarios,

because we all (or nearly all) have been taught that there are ancestors and descendants, and that adaptation is a response to changes in the environment.

I have selected here three major questions on 'origins' that gave rise to various scenarios based on both palaeontology and neontology: the origins of the craniates, the gnathostomes, and the tetrapods.

7.1.1 Origin of the Craniata

For more than a century the question of the origin of the craniates (or of the vertebrates if no distinction is made between the two taxa) has rested almost exclusively on the study of extant chordates, i.e. the tunicates, cephalochordates, and craniates. Attempts have been made since the 1970s to include some exclusively fossil non-craniate forms in the construction of new scenarios: the 'calcichordates' (Jefferies 1968; and for an extensive discussion see Jefferies 1986) and more recently the conodonts (Briggs *et al.* 1983, Aldridge *et al.* 1993).

The 'calcichordate theory', foreshadowed by Gislén (1930) but developed extensively by Jefferies (1986), involves a number of Cambrian to Devonian animals classically referred to as stylophoran carpoids, and generally regarded as strongly specialized echinoderms. These forms have a 'head' and a 'tail' made up of calcite plates (hence the collective name given by Jefferies) that are identical in structure with those of typical echinoderms (Fig. 7.1A). But Jefferies suggested that this character is merely a general character of the dexiothetes, which has been lost independently in the various chordate groups, and subsequently replaced by a calcium phosphate skeleton in vertebrates. Some internal structures are visible inside or on the internal surface of the calcite plates, and were classically interpreted as canals and cavities for a water-vascular system comparable to that of Recent echinoderms. In contrast, Jefferies tried to interpret them in the light of chordate anatomy (Fig. 7.1A4). In spite of his efforts, he did not manage to convince many vertebrate paleontologists. The result of Jefferies's interpretation and comparisons with Recent

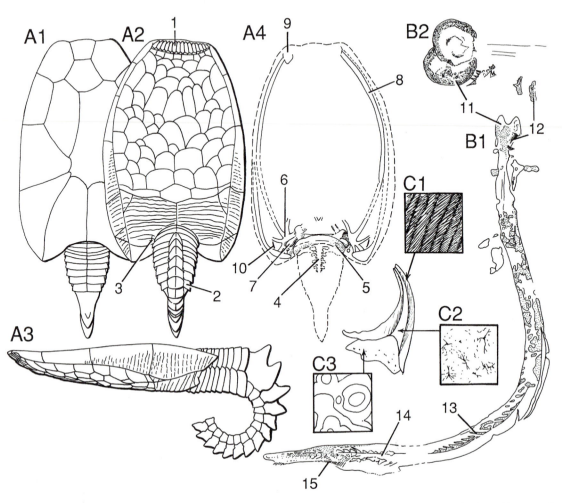

Fig. 7.1. Fossils and the ancestry of the Craniata. A, *Mitrocystella*, an Ordovician mitrate 'calcichordate', as interpreted in terms of craniate anatomy, in dorsal (A1, ×2), ventral (A2) and lateral (A3) view, and outline of the internal cavities in dorsal view (A4); B, outline of the body (B1, ×5) and detail of the eyes (B2, ×12) of the 'conodont animal' *Clydagnathus*, from the Carboniferous of Scotland, as interpreted in terms of craniate anatomy; C, histology of the conodonts, as interpreted in terms of craniate histology, with details of the structure of the 'enamel' (C1, about ×3000), 'cellular bone' (C2, about ×1000) and 'globular calcified cartilage' (C3, about ×1000). 1, oral plates; 2, fore tail; 3, lateral-line groove; 4, dorsal nerve cord; 5, brain; 6, palmar nerves (?branches of trigeminal nerve); 7, pyriform body (?trigeminus ganglion); 8, olfactory fibres; 9, olfactory organ; 10, atrial opening of gill chamber; 11, eye; 12, conodont organs; 13, myomeres; 14, notochord; 15, caudal fin radials. (A, from Jefferies (1986); B, from Briggs *et al.* (1983) and Aldridge *et al.* (1993); C, from Sansom *et al.* (1992).)

chordates is that chordates (including calcichordates) are basically asymmetrical animals whose common ancestor can be seen as resembling the larva of the hemichordate *Cephalodiscus*, in which one half of the head would have become reduced as a consequence of adaptation to benthic life. Apparent bilateral symmetry would thus have been gained independently in adult cephalochordates and craniates. Another consequence of Jefferies's scenario is that the tunicates would be more closely related to the

craniates than are the cephalochordates (Fig. 7.2). The 'calcichordates' would thus be an ensemble of stem-chordates, stem-cephalochordates, stem-tunicates, and stem-craniates, and thus paraphyletic (Fig. 7.2B, D, F). The closest relatives to the craniates among calci-chordates would be the mitrates (Figs 7.1A, 7.2F). Other specialists have interpreted these peculiar fossils quite differently. Ubaghs (1981), for example, regarded the 'tail' as an aulacophore, that is, a sort of 'arm' bearing a

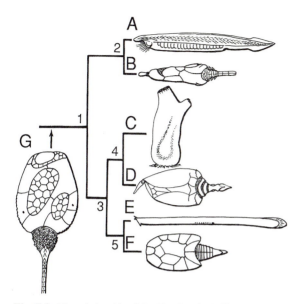

Fig. 7.2. The relationship of the Chordata, including 'calcichordates', as interpreted by Jefferies (1986). *Terminal taxa*: A, crown-group Cephalochordata; B, *Lagynocystis*; C, crown-group Tunicata; D, *Anatifopsis*; E, crown-group Craniata; F, *Mitrocystella*; G, hypothetical chordate ancestor. *Nested taxa and selected synapomorphies*: 1, Chordata (right gill slits and pharynx, left and right branchial atria); 2, Cephalochordata (median atrium with tertiary gill slits and mid-ventral atriopore); 3, tripartition of hind tail; 4, Tunicata (elongation and dorsal position of atrial openings); 5, Craniata (lateral-line pit or groove, trigeminus nerve complex). *Age*: A, ?Cambrian to Recent; B, Ordovician; C, ?Cambrian to Recent; D, Ordovician; E, Ordovician to Recent; F, Ordovician. (B, D, F, G, from Jefferies (1986).)

groove network of true vertebrates. One must, however, admit that, apart from the arthropods, no other Early Palaeozoic animal shows such a distinct head and a tail. I would say that a mitrate–craniate relationship is still unrefuted, even if difficult to accept, but the question is to define what could refute it. Perhaps the discovery of more unique characters shared by all calcichordates and one particular echinoderm taxon, which could not be interpreted as general for a higher taxon? Let us remember that the currently accepted bird–dinosaur relationship and the tetrapod–lungfish relationship were also doubted and regarded as 'impossible' in the past. The calcichordate theory has at least the merit of proposing one palaeontology-based theory for the origin of the chordates.

The idea that arthropods may have been ancestral to the craniates once had some success; the American zoologist W. Patten (1890, 1903, 1912) tried to reconcile it with the palaeontological data by suggesting that osteostracans were intermediate between xiphosurans (e.g. *Limulus*) and craniates. He was convinced that the gill openings of osteostracans were in fact the areas of attachment for the appendages. In order to support his theory, he collected hundreds of specimens of *Tremataspis* shields from the Silurian of Saaremaa (Estonia), in the hope of discovering the legs in situ. Patten failed to convince vertebrate palaeontologists, but his remarkable collection of tremataspids is now deposited in the collection of the American Museum of Natural History. Nevertheless, phosphatic arthropod carapace fragments are still often referred to heterostracan armours by palaeontologists, as in the case of *Anatolepis*, for example (see Chapter 4, p. 84). This may mean that they have features in common.

More recently, the discovery of a conodont-bearing animal from the Carboniferous of Scotland (Briggs *et al.* 1983, Aldridge *et al.* 1986, Briggs 1992; Aldridge *et al.* 1993) has generated fresh debates on the possible relationships between this hitherto enigmatic group and the craniates. The body of the 'conodont animal' admittedly displays V-shaped impressions that look like myomeres (Fig. 7.1B1, 13), and its tail seems to have possessed fin rays (Fig. 7.1B1, 15). There is also a longitudinal imprint that may be the notochord (Fig. 7.1B1, 14), and there are two large dark spots that are probably the eyes (Fig. 7.1B1, 11). New studies of the microstructure of the conodonts proper are claimed to show evidence of typical vertebrate hard tissues such as cellular bone (Fig. 7.1C2), globular calcified cartilage (Fig. 7.1C3), enamel-like matter (Fig. 7.1C1), and even dentine (Sansom *et al.* 1992, 1994). In consequence, the taxon Conodonta is now regarded by Briggs (1992) as the sister-group of the Vertebrata, i.e. as more advanced than hagfishes, because it has mineralized hard tissues and

ciliated groove to convey food particles to the mouth; he consequently oriented them the other way up. Philip (1979) considered the 'tail' to be a stele, i.e. a kind of stem serving as an anchoring device comparable to that of the sea lilies, and he denied that it could have contained a notochord.

Apart from a fairly striking superficial resemblance between the mitrates, for example, and some heterostracans (compare Figs 7.1A2 and 4.4.B2), which is due to the pattern of endoskeletal calcitic plates in the mitrates and exoskeletal phosphatic plates and scales in the heterostracans, the homologies proposed by Jefferies sound—to say the least—somewhat far-fetched, and his evolutionary scenario implies a large number of assumptions. The small groove on the right-hand side of the ventral head-skeleton of *Mitrocystella* (Fig. 7.1, 3), for example, which he regards as a sensory-line groove, is hard to reconcile with the complex and symmetrical large eyes. It is risky to draw a definitive conclusion about the affinities of the conodonts at the moment, because there is still no evidence of clear-cut craniate or vertebrate characters, apart from the large eyes. None of the histological arguments is fully satisfactory (e.g., the 'osteocyte spaces' are five times smaller than those in true vertebrate bone, and the enamel fibres are oblique); moreover, there is no evidence of anything that looks like a branchial apparatus, which is usually well preserved in

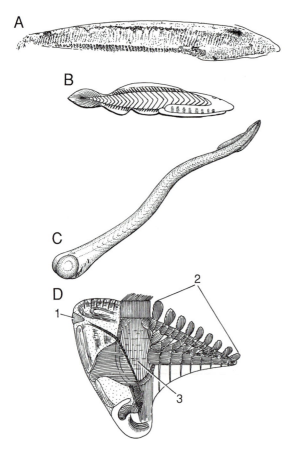

Fig. 7.3. A, *Pikaia*, Middle Cambrian of the Burgess Shale, British Columbia, a presumed cephalochordate (the head is classically regarded as being situated at the left-hand end) (×1); B, the hypothetical cephalochordate ancestor, as reconstructed by Holmgren and Stensiö in 1936 to suggest that cephalochodates derive from a 'cyclostome'-like lineage; C, *Clydagnathus*, Early Carboniferous of Scotland, attempted reconstruction (×2). D, *Ainiktozoon*, Early Silurian of Scotland, reconstruction in lateral view (×0.5). 1, presumed compound eye; 2, 'tail'; 3, musculous 'capsule'. (A, redrawn from Jefferies (1986); B, from Holmgren and Stensiö (1936); C, from Aldridge *et al.* (1993), D, from Ritchie (1985).)

similar vertebrate imprints. Nevertheless, one must admit that the tail of the conodonts is strikingly fish-like, and the clearly paired eyes are exactly similar to those found in some fossil lampreys and anaspids. Also, one cannot rule out the possibility that the conodonts are larval animals (as suggested by the very large eyes) and that the conodont organs are transitory organs of some kind used for larval feeding. In any case, if the conodonts are craniates or vertebrates, they probably represent a clade and do not answer the question of their 'origin'.

The Silurian enigmatic fossil *Ainiktozoon* (Fig. 7.3D) has also been considered as a possible chordate, despite the fact that it may have possessed compound eyes, like arthropods (Ritchie 1987). *Ainiktozoon* is more likely to have arthropod affinities, despite its aberrant morphology. The fact that the cephalochordates are generally believed to be the sister-group of the craniates implied that they must occur earlier than the craniates in the fossil record. This prediction was fulfilled with the discovery of *Pikaia* (Fig. 7.3A), from the Middle Cambrian Burgess Shale (Conway-Morris and Whittington 1979), a peculiar fossil with a metameric structure and preserved in imprint. *Pikaia* is regarded as either a cephalochordate or a pre-craniate, since it seems to have possessed eyes, if the interpretation is correct. When interpreted in another way (i.e., the 'tail' being regarded as the 'head'), it has an amazing resemblance to the hypothetical '*Amphioxus*-Stammform' (stem-form) illustrated in 1936 by Holmgren and Stensiö, which is supposed to depict an ancestral, cyclostome-like cephalochordate (Fig. 7.3B).

How far these interpretations stem from our subconscious is hard to tell. The 'calcichordates' are perhaps rejected by most palaeontologists because they do not fit our craniate archetype based of cephalochordates, i.e. a small, soft-bodied, metameric, swimming creature. In contrast, *Pikaia* (Fig. 7.3A) or the conodonts (Fig. 7.3C) would be in better agreement with this ideal ancestral craniate.

Further reading

Aldridge *et al.* (1993); Briggs (1992); Conway-Morris (1989); Denison (1971*a*); Gans (1989); Jarvik (1988); Jefferies (1986); Jollie (1973); Krejsa *et al.* (1990); Maisey (1986); Patten (1890, 1912); Ritchie (1987).

7.1.2 Origin of the Gnathostomata

The other questions on origins within the craniates are no better answered. We have seen in the preceding chapter that the origin of the jaws, and thus the origin of gnathostomes, is just as obscure as the origin of the skull or neural crest for craniates, despite the availability of much more abundant fossil and embryological data. We have also seen in the preceding chapter (p. 258) that it is plausible that the endoskeletal component of jaws, the palatoquadrate and Meckelian cartilage, derives from some medially placed visceral arch of the mandibular segment, comparable, if not homologous, to the medial velar skeleton of the larval lamprey (Fig. 6.3D). We may thus infer that the presence of a velum is a prerequisite for the appearance of jaws. The presence of a larval lamprey-like velum is highly probable in osteostracans, if not in all fossil jawless craniates. But the factual support for the story ends there. How craniates passed from a suspension-feeder diet to a microphagous diet, and then to a macrophagous diet is a matter of imagination. Mallatt (1984*a*,*b*) proposed a scenario that sounds satisfactory and explains how this transition may have been effected, in particular by the loss of the 'mucous trap' in gnathostome ancestors.

A microphagous diet or filter-feeding is far from being a handicap, as is shown by whales and basking sharks, and one may wonder why (and when) vertebrates 'suddenly' fed on larger prey. There is no sensible answer to this question, unless one links it with a decrease in the number of micro-organisms or organic debris in the sea water, in turn resulting from changes in the climate, sealevel, or the quantity of terrigenous sediment brought by rivers into the sea. Even the classical theory, which links predatory habits with the rise of perfected paired and unpaired fins (to chase the prey), does not hold, since osteostracans have virtually the same fins (except for the pelvic fins) as the gnathostomes and were probably microphagous. Another theory was proposed by von Wahlert (1970) and developed by Starck (1979) and Reif (1982*a*). It assumes that the origin of jaws is linked with suction-feeding in gnathostome ancestors. The need to create a depression in the pharynx, prior to the opening of the mouth, led to an enlargement of the foremost gill arches, in particular the mandibular arch. According to this theory, there would be a transition between the suspension-feeding of jawless vertebrates (with the help of a velum and mucous trap), and the suction-feeding of pre-gnathostomes. Teeth would then have developed to retain the prey in the oral cavity. This theory is, however, based on the traditional view that the gnathostome and agnathan visceral arches are homologous. If, as assumed here, the only endoskeletal visceral element that is common to lampreys and gnathostomes is the medial velar skeleton and mandibular arch, then one has to assume that it was once the only biting apparatus available to pre-gnathostomes before the internal gill arches appeared.

Further reading

Gutmann and Bonik (1981*b*); Lauder (1985); Lessertisseur and Robineau (1969, 1970); Maisey (1989*a*); Mallatt (1984a,*b*); Reif (1982); Romer (1966); Schaeffer (1975); Wahlert (1970).

7.1.3 Origin of the Tetrapoda

The origin of the tetrapods was also an ideal subject for ecological scenarios. There again, the ecological 'pressure' was first said to be the drying of the climate, forcing the tetrapod precursors (osteolepiforms), which had both lungs and choanae, to crawl from pool to pool in order to survive. The earliest known tetrapods (*Ichthyostega, Acanthostega, Tulerpeton*) were then shown to have lived together with fishes that occur in marine environments and commonly have a world-wide distribution (antiarchs, long-snouted lungfishes, groenlandaspid arthrodires, etc.). It thus became clear that these tetrapods could have lived in the sea, perhaps in lagoons or deltas, some of them possibly entering rivers. This ecological independence of the earliest tetrapods rather suggests that they diversified first in a aquatic—and probably marine—environment, and that their ability to crawl or walk on land was a considerable advantage in avoiding the concurrence of the other predators, such as other large sarcopterygian fishes (e.g. *Eusthenodon*). One may thus imagine that *Ichthyostega*, for example, used to prowl about on beaches or deltas, preying on fishes trapped in pools at low tide, and pulling them out of the water (Fig. 1.13). In contrast, Clack and Coates (1993) view these animals as virtually unable to walk on land, like seals and sea-lions. Whatever may have been the mode of life of the earliest tetrapods, it is now clear that their limbs were not initially an adaptation to walking on land, but rather a special type of 'fin', perhaps adapted to swimming in deltas, against the stream or through areas encumbered by decaying plants. This theory has also been used to explain the paddle-shaped paired fins of the Carboniferous rhizodontid *Strepsodus* (Andrews 1985; see Chapter 4, p. 216). The reduction or loss of the dermal fin web is in fact generally linked with this kind of mode of life. In sum, the 'conquest of land' by the vertebrates began first as the conquest of one particular type of aquatic environment.

Further reading

Ahlberg and Milner (1994); Clack and Coates (1993); Lebedev (1992); Romer (1956); Spjeldnaes (1982); Vorobyeva (1993).

7.2 EARLY VERTEBRATES AND EVOLUTIONARY PROCESSES

There is a tradition in palaeontology that only certain invertebrates and mammals can tell us something about the evolutionary processes. This is because invertebrates are abundant in the fossil record, and mammals have specific 'identity cards', the teeth. These two conditions are supposed to show what are species and populations,

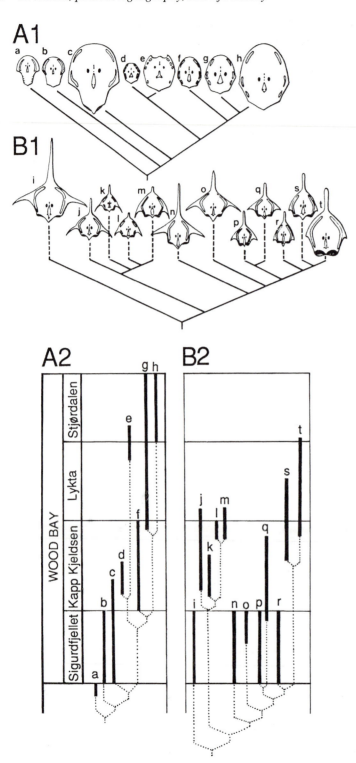

Fig. 7.4. Phylogeny (A1, B1) and stratigraphical distribution (A2, B2) of two osteostracan taxa, the Kiaeraspidida (A) and Boreaspididae (B), in the Wood Bay Formation of Spitsbergen. Note the relative congruence with stratigraphy in A, and the incongruence in B. (From Goujet *et al.* (1982).)

the classical evolutionary 'units'. Evolutionary processes inferred from palaeontology to match the models of population genetics, such as anagenesis or allopatric speciation, are thus rarely exemplified by early vertebrates, because these are traditionally regarded as rare and their specific status as uncertain. As a matter of fact, most early vertebrates are too rare to provide a continuous fossil record. There are, however, examples, such as some osteostracan groups of the Early Devonian of Spitsbergen (boreaspidids, kiaeraspidids) that could support the theories of the kind proposed by many invertebrate palaeontologists (Fig. 7.4). Boreaspidids seem to have had a definite growth and were fully ossified only when adult (more or less like mammal teeth); and they are just as abundant in the fossil record (throughout the whole of the Wood Bay Formation) as Jurassic brachiopods or Miocene rodents, for example. The boreaspidid species could be arranged in an evolutionary continuum according to their stratigraphical occurrence, but the cladistic analysis of their characteristics does not match such a pattern (Fig. 7.4B1): some of the most derived forms occur as early as the least derived ones (Fig. 7.4B2). In contrast, cladistic analysis of the kiaeraspidids reveals a pattern that is quite consistent with their stratigraphical order of occurrence (Fig. 7.4A; Goujet *et al.* 1982). This shows that, even with early vertebrates, evolutionary processes inferred from the fossil record (good—or bad—as it may be) are in fact limited by the chance of preservation; no one can tell whether the stratigraphical distribution actually reflects the phylogeny or is totally disconnected from it.

Since the 1980s, the classical models of evolutionary processes inherited from the Synthetic Theory, which are essentially based on population genetics, micromutations, and natural selection, have been criticized on various grounds, sometimes with good reason. The major objection against these half a century-old views was that they had served in building up an unrefutable—and thus unscientific—theory (see e.g. Riddiford and Penny 1984). This 'neo-Darwinian' evolutionary paradigm has been challenged by new ones that involve the genetics of development or even macromutation theories, and in which natural selection is merely seen as having a stabilizing effect, rather than being the main cause of evolution (see e.g. Løvtrup 1985, Alberch 1989). In sum, chance and the absence of selection would play a major role in the rise of new structures.

How long these paradigms will hold and how justified they are is not the subject of this book, but it is interesting to note that, here again, the world of early vertebrates has barely been touched by this revolution in evolutionary thought, although it could have been a favourite field for investigations on the rise of major structures (skull, jaws, limbs, etc.).

Heterochrony, i.e. changes in the timing of embryonic development, is now the pet subject of many palaeontologists. The various types of heterochronies (paedomorphosis, peramorphosis, etc.) can be combined in such a way that any pattern of transformation can be explained by this process (Dommergues *et al.* 1986, McNamara 1986). The ultimate consequence is that heterochrony, like anagenesis before it, can now be used to justify a transformation series which that is not consistent with a phylogenetic pattern, i.e. general character distribution. In this way, presumed reversions in character states can receive an *ad hoc* explanation. Heterochrony has not yet invaded the field of early vertebrate palaeontology, despite attempts by Long (1990) to explain the origin of the tetrapod skull morphology by means of paedomorphosis and the origin of the tetrapod limb by peramorphosis. In fact, the orbits are larger and the postorbital bones shorter in juvenile osteolepiforms (Fig. 4.88C2) and adult tetrapods than in adult osteolepiforms. The appearance of phalanges in the tetrapods is interpreted as the result of peramorphosis, i.e. the addition of new developmental stages to ontogeny (e.g. segmentation of radials or neoformation of phalanges). Paedomorphosis sounds to me particularly ambiguous because it serves to justify evolutionary relationships (based on relative geological age) that are contradicted by a cladogram. The embryonic or juvenile condition of a character, which would be the most general, or plesiomorphous, according to von Baer's law (see Nelson 1978), is thus turned into a more particular, or apomorphous, condition to fit a pre-existing theory of relationships. In spite of the numerous contradictions of 'heterochronism' (Janvier 1989), it will no doubt be increasingly invoked by early vertebrate palaeontologists to construct evolutionary scenarios. The genetic basis of heterochrony, i.e. drastic changes in the genetic control of ontogeny leading to 'macromutations' in the phenotype, is probably a major cause in the rise of new structures, if not lethal. As we have seen above, it is possible that such 'revolutions' in the construction of individuals may account for the origin of jaws, semicircular canals, tetrapod digits, etc. Such 'macromutations' of either genetic or epigenetic nature, are usually invoked when wide morphological gaps are not filled by palaeontological data.

Further reading

Long (1990); McNamara (1986).

7.3 SELECTION AMONG HIGHER GROUPS

Natural selection was originally regarded as having an effect at the level of individuals or populations. It was then extended to the species level and even to competition between higher taxa. This has led to widespread stories about one group 'eliminating' or 'outcompeting' another group. The history of the early vertebrates has been an important source of stories of this kind, put forward to account for the extinction of major taxa. Vertebrates have been said to have survived because they gained a dermal skeleton to repel the largest predators of their time, the 'sea scorpions' (eurypterids; Romer 1933). Conversely, Palaeozoic jawless vertebrates are sometimes said to have been eliminated by the jawed vertebrates that outcompeted them. Jawless vertebrates were, however, abundant and diversified while coexisting with gnathostomes for more than 40 Ma. They fed on different types of food but lived virtually in the same kind of environment, as is observed in the Early Devonian of Spitsbergen or the Late Devonian of the Baltic area, for example. Some traces of bites, with subsequent healing, have been observed on the branchial plates of Late Devonian psammosteid heterostracans (Mark-Kurik 1966; see below), and are thought to have been made by large sarcopterygians. Nevertheless, there is no other evidence that the gnathostomes caused massive destruction among jawless vertebrate populations (for example, no remains of a jawless vertebrate have yet been found in the stomach content of a gnathostome). After all, extant hagfishes and lampreys tend to prey on gnathostomes, but this is perhaps why they survived. The 'fall', i.e., the considerable decrease in number and diversity, of the jawless vertebrates, at the beginning of the Middle Devonian is more likely to be due to the reduction of their favourite types of environment, such as sandy marginal marine environments, lagoons, or large deltas (Janvier 1985c). The only Late Devonian jawless vertebrates, apart from a few thelodonts, which may have been more pelagic, are confined to sandy or muddy lagoonal and deltaic facies (the Miguasha 'anaspids' and osteostracans, the large psammosteids in the Baltic area, and a galeaspid in the Ningxia red sandstone in China). It is, however, possible that some placoderm groups, such as the phyllolepids or antiarchs, which were abundant in this type of facies, may have been competitors of such benthic jawless vertebrates as osteostracans or galeaspids, but this remains merely speculative. It is inconsistent with the fact that the Late Devonian localities where benthic jawless vertebrates (osteostacans and galeaspids) are most abundant are also well known for the abundance of antiarchs (*Bothriolepis*, *Remigolepis*).

There are many other patterns of stratigraphical distribution of early vertebrates that could be explained by such scenarios of competition, such as the placoderm–chondrichthyan or the tetrapod–osteolepiform concurrence, but they are no more valid than the story of the early mammals eating dinosaur eggs.

7.4 BIODIVERSITY AND MASS EXTINCTIONS

Changes in biodiversity, another pet subject of some palaeontologists, are now beginning to gain fame in early vertebrate studies. The concept of biodiversity stems from evolutionary or phenetic approaches to morphology, since it implies that 'strongly divergent' morphologies account for more taxa, and thus a high diversity, regardless of the phylogenetic relationships. Organisms that look quite similar but bear a wide variety of discrete characters shared with several major subsequent taxa would thus be said to have a 'low diversity'. The thelodonts (Fig. 5.1K), for example, display a comparatively low diversity but may include stem pteraspidomorphs, stem galeaspids, stem osteostracans, and stem gnathostomes. By virtue of their descent and relationships, they are thus much more diverse than mammals or birds. The same can apply to 'leptolepids' among Mesozoic teleosts (Fig. 4.73). If, however, we see the world through the eyes of evolutionary biologists and palaeontologists, i.e. by considering that taxa are defined by their common ascent, and not by characters, then it is true that some periods seem to display a higher taxonomic diversity than others, if only in the numbers of fossil species recorded. So far as early vertebrates are concerned, this impression may be due to mere sampling or conditions of preservations. In the 1970s, the Carboniferous marine vertebrate faunas were said to display a low diversity, by comparison with the rich Late Devonian placoderm and sarcopterygian faunas. The discovery of two major localities, Mecca Quarry (Indiana) and Bear Gulch (Montana), has, however, increased the evidence of marine Carboniferous biodiversity to such a large extent, in particular for chondrichthyans, actinopterygians, and coelacanths, that it largely equals the richest Devonian ecosystems, e.g. the vertebrate faunas of Spitsbergen or Gogo, Australia. Ordovician vertebrates admittedly, seem to display a low diversity, both in morphology and number of recorded taxa, yet they show a greater diversity in the histology of the dermal skeleton than all placoderms. In contrast, a look at vertebrate diversity from the Silurian to the Early Permian reveals only short periods of fall, which often correspond to the extinction of some taxa or may simply reflect gaps in the fossil record. Such 'falls' in biodiversity occurs in the Late Silurian, Early Eifelian, Late Frasnian, Late Famennian, Late Carboniferous, and Late Permian. Those who like 'nomothetic' palaeontology, i.e. the search for great laws and periodicity in life history, may count numbers of species, genera, and families from treatises and draw curves of extinction and radiations, as did Raup and Sepkoski (1982, 1988, see also Sepkoski 1982, 1986), but while they are doing this, the number of

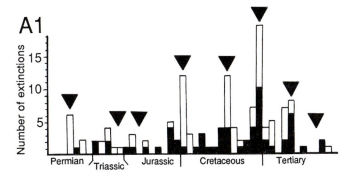

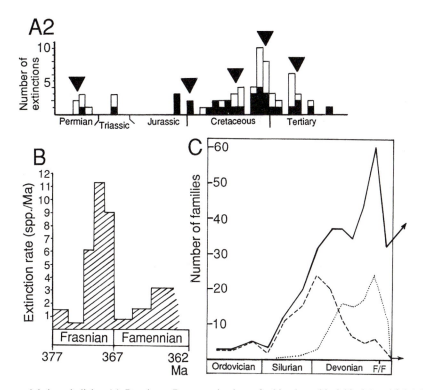

Fig. 7.5. Extinctions and their periodicity. A1, Permian to Recent extinctions of echinoderm (black blocks) and fish (white blocks) families according to Sepkoski's (1982) data; black triangles indicate the major extinctions, with a periodicity of about 25 Ma. A2, The same, when 'noise' (paraphyletic taxa, erroneous datings and monospecific families) is removed; black triangles indicate the only major extinctions that remain conspicuous. B, extinction rates (species per million years) of Appalachian marine invertebrates for the Late Devonian. C, diversity (number of families) of placoderms (dotted line), jawless vertebrates (dashed line), and all fish families (continuous line) from the Ordovician to Late Devonian. (A, B, modified from Patterson and Smith (1987); C, modified from McGhee (1988); D, modified from Long (1993*b*).)

new taxa discovered or old taxa synonymized will increase to bias their results. Patterson and Smith (1987) have shown that 75 per cent of the data used in such studies generate only noise resulting from the use of non-

monophyletic groups, mistaken dating, monospecific families, etc. A curve that accounts for periodic extinctions when based on the number of families often shows no major peak when based on the number of species, whatever the group under consideration. Extinction curves based on families of marine fishes and echinoderms have shown data-to-noise ratios of 27/73 and 23/77 respectively (Fig. 7.5A). 'Diversity curves' based on the number of families or rates of origination and extinction of families have been drawn by Long (1993*b*; Fig. 7.5C) for Ordovician to Carboniferous fishes, showing two periods of decreasing diversity, at the beginning of the Middle Devonian, and at the end of the late Devonian. This probably means that Emsian–Eifelian and Famennian vertebrate localities are comparatively scarce. In contrast, the peaks of high diversity in the Early Devonian and Frasnian may be linked with the large number of taxa in a few famous localities (Spitsbergen, Yunnan, Gogo, Wildungen, Miguasha).

Let us admit that the results of this kind of study mean that periods of fall in biodiversity, or of simultaneous extinctions of major taxa, are a reality. The best-known examples of mass-extinction periods are the Permian–Triassic and the Cretaceous–Tertiary boundaries. During the time of the early vertebrates, there were two major periods of extinction or reduction in diversity, which are said to bear on all animal groups. One is at the end of the Ordovician, and seems to coincide with a rather widespread glaciation; the other is at the end of the Frasnian, during the Late Devonian. A number of other, minor 'extinction events' are now recognized, mostly by invertebrate palaeontologists. For the Silurian and Devonian alone, no less than eight such 'events' have been suggested, essentially on the basis of particular invertebrate groups (e.g. brachiopods, conodonts): The Ireviken event (Llandovery–Wenlock), the Pentamerid event (end-Ludlow), the end-*pesavis* event (end-Lochkovian), the mid-Emsian event, the late-Eifelian event, the Stringocephalid event (mid-Givetian), two Kellwasser events (end-Frasnian), and the Hangenberg event (end-Famennian) (Talent *et al.* 1993).

The Late Ordovician extinctions are classically linked with the glaciation that took place at that time and may have caused a fall in sea-level and an oceanic overturn (Brenchley 1990). The subsequent melting of the ice caps also caused flooding of the continental shelves with equally important consequences for the benthic faunas. So far as vertebrates are concerned, their diversity was apparently so low in the Ordovician that one cannot speak of mass extinction. However, few of the Ordovician vertebrate taxa, the arandaspids and astraspids, seem to straddle the Ordovician– Silurian boundary. Only astraspid-like tesserae are known from the Early Silurian of Siberia.

The improverishment of the invertebrate faunas at the Frasnian–Famennian boundary was noticed long ago (Newell 1967) and was studied in more detail, using

'quantitative' analyses, during the 1980s in connection with increasing interest in catastrophic global events (McLaren 1982, Raup and Sepkoski 1982; McGhee 1988, 1990). As for heterochronies (see above, p. 287), the quantitative analysis of past biodiversity now rests on a combination of factors that permit reconciliation of contradictory data sets: rates of extinction, rates of origination, etc. A study of the Frasnian–Famennian extinction event (McGhee 1988) concluded that the drop in species diversity took place during the last 4 Ma of the Frasnian period (Fig. 7.5B), but this could not be attributed to a mass extinction. Instead, the rates of species origination dropped suddenly and became lower than the rates of extinction. As McGhee puts it: 'the most important question is not "What triggered the elevated extinction rates?", but rather "What was the inhibiting factor that caused the cessation of new species originations?".' Of course, this analysis 'does not rule out a catastrophic component (asteroidal impact) as a cause of extinctions' (McGhee 1988), which is now a dogma of palaeobiology. Opportunely, evidence for a Frasnian–Famennian asteroid impact has turned up in the Montagne Noire (France) and in Australia, in the form of an iridium concentration, and in Belgium in the form of microtectites, i.e. droplets of glass originating from meteorite impact. The Australian iridium is now said to have been accumulated by bacteria or algae, and this probably applies to many other cases of platinoid anomalies (Sawlowicz 1993). Another supposedly dramatic event took place in the Late Frasnian: the 'Kellwasser event' (Schindler 1993). In several parts of the world (Poland, Germany, Southern France, Morocco, the Middle East), the Frasnian–Famennian boundary is marked by two supposedly synchronous layers of black limestone, which is full of invertebrate debris, the 'Kellwasserkalk'. These layers are supposed to be the result of considerable turmoil in the sea during anoxic or dysaerobic conditions (due either to climatic cooling, a 'tsunami' or, indirectly, an asteroid impact). The pecularity of this deposit is that it must have been formed under anoxic conditions in open, shallow seas. Above this level, the faunal diversity is considerably impoverished. Copper (1986) has suggested that this 'Kellwasser event' might also be due to drastic changes in the course of oceanic streams, in turn due to global tectonic changes.

Although the search for the congruence between abnormal platinoid concentrations and major Middle Palaeozoic extinction 'events' has been rather disappointing for catastrophists, the study of the carbon and oxygen isotopes (^{13}C and ^{18}O) in the rocks seems to give more interesting results. The amount of these isotopes of organic origin is supposed to reflect the biomass at a given time. An increase in the sea temperature reduces the concentration of these isotopes, which pass into the atmosphere in the form of carbon dioxide. In the Silurian and Devonian, reductions in ^{13}C and ^{18}O seem to coincide with previously recognized extinction events, and thus may mean something (Talent *et al.* 1993).

How did vertebrates cope with the Frasnian–Famennian extinction event? Most of them did well. The only noticeable extinction is that of the armoured jawless vertebrates. The latest representatives of the osteostracans, anaspids, galeaspids, heterostracans, and thelodonts are all Frasnian in age, but they consist of only about eight species. They are so rare, and so restricted to specific types of marginal environments, that this pattern of distribution may well be due to sampling chance. It can also be suggested that the considerable Frasnian marine transgression rapidly reduced the extent of the shallow platforms and deltas in which they used to survive. Young (1986*a*, 1990) and Long (1993*b*) also pointed out that many of the extinctions that occur during the Givetian and Frasnian are probably due to the fact that there were many endemic groups in the Early and Middle Devonian, which had been eliminated by the great global faunal interchange of the Givetian–Frasnian. This interchange brought many widespread taxa into hitherto highly 'protected' areas, such as China or Western Gondwana, and may have caused the extinction of endemic taxa. (It is interesting that this explanation, which is acceptable, has been put forward by Australian palaeontologists, who, by the virtue of their habitat, are much concerned with ecological problems of this kind!)

All the gnathostomes pass through the Frasnian–Famennian boundary without any great change in diversity. In contrast, the diversity of actinopterygians increases in Famennian times. Huge arthrodires, like *Dunkleosteus* or *Titanichthys*, lived in Famennian seas until the very end of this stage. A sudden extinction of the major placoderm taxa occurs, however, at the Famennian–Carboniferous boundary. Although represented by a variety of taxa (antiarchs, arthrodires, ptyctodonts, and phyllolepids) up to the latest levels of the Famennian, no placoderm survives into the Carboniferous. The same happens to the cosmine-covered lungfishes, such as the 'dipterids', which are fairly abundant until the very end of the Famennian, but never occur in the Carboniferous. Other sarcopterygian taxa, such as onychodontiforms, porolepiforms, rhizodontiforms, osteolepiforms, and panderichthyids, either disappear or survive until the Late Carboniferous or Early Permian in the form of a few species (see below, section 7.6). Acanthodians also show a drop in diversity at the end of the Devonian. Only acanthodids survive until the Early Permian, and gyracanthids (possibly climatiids) until the Late Carboniferous. In contrast, chondri-

chthyans, actinopterygians, actinistians (coelacanths), and tetrapods pass into the Carboniferous with a seemingly increased diversity. Placoderms are thus the only vertebrates that show a pattern of extinction comparable to that of the 'dinosaurs' by the end of the Cretaceous. The cause of this extinction may possibly be a drop in sea-level (marine regression) that took place at the end of the Famennian, which may have reduced their favourite environments, such as shallow platforms with reefs and abundant benthic life. The comparatively small number of Famennian placoderm species is probably due to the paucity of Famennian vertebrate localities.

The end-Permian extinction event, which is said to have caused the disappearance of 83 per cent of marine genera (Sepkoski 1986; Erwin 1990), and is regarded as the major extinction in the history of life, had little impact on the fish faunas (Fig. 7.5A2). As far as fishes are concerned, most of the relics of Devonian times (osteolepiforms, acanthodians) became extinct by the Early Permian or somewhat earlier; those that survived the Early Permian continued into the Triassic and later.

Again, it must be stressed that the 'science' of global events would gain credibility by considering taxonomy with as much care as isotopes and sediments. As long as many taxa are defined by privative characters (absence of characters), the basis for bio-events will remain spurious. The only extinction that exists is the extinction of a character.

Further reading

Donovan (1989); Geldsetzer and Nowlan (1993); Halstead (1987*a*); House (1989); Long (1993*b*); McGhee (1990); Patterson and Smith (1987); Raup and Sepkoski (1982); Schindler (1993); Sepkoski (1986); Talent *et al.* (1993); Thomson (1976*b*).

7.5 ENVIRONMENTAL CONSTRAINTS AND EVOLUTIONARY RADIATIONS

The sudden appearances of major and highly diversified taxa in the fossil record are frequently referred to as evolutionary 'radiations' or 'explosions', and their causal interpretation is also a major concern to many palaeontologists (who are generally those dealing also with mass extinctions!). The early history of the vertebrates, as seen by these palaeontologists, displays some of these 'radiations', now sometimes called 'recovery events', when it deals with taxa that survived an extinction event. The first 'radiation' was in the Early Silurian (Llandovery–Wenlock), with the sudden rise of the major vertebrate taxa (heterostracans, thelodonts, osteostracans, galeaspids, anaspids, chondrichthyans, acanthodians, and placoderms). The second was in the Early Devonian, with the diversification of the major gnathostome taxa (in particular placoderms and dipmomorphs). A third was in the Early Carboniferous, with the radiation of the chondrichthyans (in particular holocephalans), actinopterygians, and tetrapods. Given that these 'radiations' are not mere illusions due to gaps in the fossil record, the search for their possible causes is limited by the unavailability—or at best uncertainty—of many physical parameters of the environment, such as temperature, salinity, etc. The Early Silurian radiation may be linked with the marine transgression that followed the end-Ordovician glaciation, but it may also be an illusion resulting from the fact that the Llandovery is a relatively long period (about 10 Ma) with a very sparse vertebrate record. In contrast, the Early Devonian radiation seems to be due to the wide extension of flood-plains and deltas, which favoured the diversity and preservation of many jawless vertebrate groups (Janvier 1985*c*). The Early Carboniferous radiation is perhaps more clearly linked with the extinction of the placoderms, the scarcity of the various sarcopterygian fishes in marine environments, and the conquest of land by the tetrapods. Here again, the duration of the Tournaisian and Early Viséan (about 20 Ma) may explain the diversity of Late Viséan tetrapods (Ahlberg and Milner 1994).

One must, however, be extremely cautious when considering these patterns of radiation, which are often interpreted in the light of evolutionary systematics, and rooted in the idea of competition between taxa. A detailed cladistic analysis of the major Silurian vertebrate clades, in particular heterostracans, osteostracans, galeaspids, anaspids, and gnathostomes, shows that the most generalized and the most derived forms already occur together in their earliest record, except perhaps in heterostracans. This means that their diversification occurred long before this time, unless evolution was in some way accelerated. There is so far no clear evidence for higher rates of evolution or diversification; and, as palaeontological findings progress, the roots of evolutionary radiations are found in an increasingly remote past, in the form of generalized taxa that were not classified with their younger relatives. The placoderms, for example, which in the 1980s were seen as a typical early Devonian radiation, are now found with a fairly wide range of diversity (arthrodires and antiarchs) as early as the Lower Silurian.

Further reading

Blieck and Janvier (1991); Janvier (1985*c*).

7.6 SURVIVAL AND LIVING FOSSILS

Why did some species survive after the extinction of most of the taxon to which they belong? This is a question that

I regard as at least as interesting as the cause of mass extinctions, because it deals with positive data. In early vertebrate history, there seems to be a rather constant characteristic of such survivors: they lived in restricted, and often marginal, marine environments, such as lagoons, tidal flats, or coastal marshes, or even passed from marine to freshwater habits. This feature is particularly striking where the Carboniferous survivors of the major Devonian taxa are concerned. Among Sarcopterygians, for example, rhizodontiforms survive until the Late Carboniferous; osteolepiforms until the Early Permian; and coelacanths, lungfishes, and tetrapods until the present. The post-Devonian lungfishes commonly occur in coal deposits, corresponding to lagoonal to freshwater environments, and become freshwater by the Late Carboniferous. The two post-Devonian osteolepiform groups, megalichthyids and rhizodopsids, are very large fishes, but are restricted to a Coal Measure type of environment or to red beds that were supposedly deposited in brackish waters (the Early Permian of Texas and Germany). The same applies to the rhizodontiforms, whose latest occurrence is in Carboniferous coal deposits. Only Carboniferous actinistians seem to have lived in both truly marine and brackish or even freshwater deposits (Forey 1991*b*). The same applies to the acanthodid acanthodians and the xenacanthiform sharks, whose Carboniferous and Early Permian representatives are bound to marginal marine environments, often associated with coal deposits. To some extent, this can be tested by comparison with an epipelagic marine fauna trapped in a lagoon, such as the Early Carboniferous fauna of Bear Gulch (Montana), which contains mainly chondrichthyans (except xenacanthiforms), actinopterygians, and actinistians, but no—or very few—acanthodians, osteolepiforms, rhizodontiforms, or lungfishes. This protective effect of the marginal marine environments, which favours the survival of a few species of a taxon that became extinct elsewhere, has also been noticed in the invertebrate fauna of the Triassic red sandstone of Europe.

All extant 'survivors' of Palaeozoic fish groups, i.e. taxa that retain a rather generalized state, do not, however, live in marshy environments. The extant coelacanth, *Latimeria chalumnae*, lives in a deep-sea environment and may come up near the surface as cool water is pulled up. Nevertheless, *Latimeria* is very restricted geographically (off the Comoro Islands) and is bound to a particular environment. The extant hagfishes are all marine, although they generally live in shallow, muddy bays. The extant lampreys are either anadromous or catadromous, but in the latter case they are pelagic forms. Only lungfishes are bound to rivers and muddy ponds, and they do not tolerate salt water. Survival may, after all, be due to mere chance.

In this connection, one may question the concept of a 'living fossil', which is widespread in popular science (Eldredge and Stanley 1984). Living fossils are seen as animals or plants that are regarded as almost unmodified since, say, the Palaeozoic (or, more restrictively, before the Mid-Tertiary where mammals are concerned). This lack of modification is rated by reference to the fossils, and may be due either to a large number of general characters of the taxon they belong to, or to the early appearance of a conspicuous specialization (Janvier 1983). Lampreys are 'living fossils', because we now know that Carboniferous lampreys (*Mayomyzon*) already had a piston cartilage and a sucker. Before *Mayomyzon* was discovered, lampreys were often regarded as a recently specialized group, admittedly derived from some Early Palaeozoic armoured jawless vertebrate. In contrast, the coelacanth is regarded as a 'living fossil' not so much because it is an actinistian, but because it retains 'crossopterygian' features, such as the intracranial articulation, which are general sarcopterygian characters, lost in other extant members of this taxon (lungfishes and tetrapods). In sum, the coelacanth is a living fossil because it resembles the presumed ancestors of the tetrapods in shared primitive characters. The concept of 'living fossils' or 'survivors' is thus deeply rooted in evolutionary systematics, and loses most of its significance when used in a phylogenetic, cladistic, context.

Further reading

Eldredge and Stanley (1984); Janvier (1983).

7.7 MARINE VERSUS FRESH WATER

The early discoveries of Palaeozoic vertebrates were essentially in the Carboniferous and Devonian (see Chapter 8). Most Carboniferous fishes were regarded as freshwater fishes because they occurred in Coal Measures strata, together with abundant plant remains, and also in association with amphibians, which, by analogy with extant lissamphibians, were thought to be freshwater dwellers. Most of the Devonian vertebrates from these early discoveries were found in the 'Old Red Sandstone' of Great Britain, but some were from the red sands and clays of Estonia and Latvia. In these facies, they were virtually the only fossils, except for some plant, bivalve, or crustacean remains. There was thus a widespread belief that these early vertebrates lived in fresh water. This belief may also have been reinforced by the early misinterpretation of the antiarch plates and sarcopterygian teeth from the Devonian of Estonia and Latvia as plates and teeth of freshwater turtles and crocodiles respectively (see Chapter 8). In addition, the fact that Agassiz (1835) referred most of the Old Red Sandstone fishes either to

the catfishes or to the sturgeons (the 'ganoids'), most of which are freshwater fishes, may have also played a role in the survival of this belief. As a consequence, there have been numerous publications on the freshwater origin of the vertebrates (e.g. Romer 1955; Denison 1956; White 1958; Halstead 1969), or about the possibly anadromous habits of their early representatives (Halstead 1973b). Detailed investigations of a large number of early vertebrate localities, ranging from the Ordovician to the Late Devonian (Gross 1951, Denison 1956, 1967; Spjeldnaes 1979, 1982; Märss and Einasto 1978; Goujet 1984a; Blieck 1985), have shown that there is generally some evidence of a marine, if usually marginal, environment. The authors arrived at this conclusion either by direct evidence, such as minute fragments of marine invertebrates (usually lingulid brachiopods) in the sediment, or by indirect evidence, such as the occurrence of the same vertebrate genera or species elsewhere in a typically marine environment. Märss (1992) suggested that alternating hyaline and semiopaque zones in the crown of Silurian thelodonts may be interpreted as a sign of changes in the environment, possibly indicating migrations from marine to fresh waters.

There is now an even more direct and reliable means of determining the environment in which these early vertebrates lived: the measurement of the ratio of two strontium isotopes ($^{87}Sr/^{86}Sr$) and the amount of the sodium (Na), fluorine (F), strontium (Sr), and lanthanum (La) preserved in the apatite of the bones and teeth. Most present-day rivers and lakes have an $^{87}Sr/^{86}Sr$ ratio of 0.704 to 0.703, whereas the sea-water ratio is 0.709 and is quite uniform all over the world. Variations in the sea-water ratio during Phanerozoic times have been evaluated on the basis of a large number of marine sediments. They appear to be very small (0.7079 for the Late Devonian, and 0.708 for the Late Silurian, for example), and the ratio remains very close to the present-day value. Measurements made on Recent fish bone mirror these differences. For example, bones of a cod from the North Sea yield an $^{87}Sr/^{86}Sr$ ratio of 0.709, and those of a pike from a lake give a ratio of 0.703. This direct chemical measurement has the advantage of being made on the fossil, and not on the surrounding rocks. Thus, the 'signature' it bears, in the form of an Sr isotope ratio, for example, comes from the environment where it lived, fed, and grew, and not from that in which it eventually died or to which was subsequently transported. The Sr isotope ratio varies according to the salinity of the water. It is assumed that these isotopes were little affected by migration due to percolation after fossilization, particularly in hypermineralized hard tissues such as dentine or enamel. Analyses of this kind were first carried out by Dasch and Campbell (1970) on lungfish bones from the Late Devonian locality of Gogo, Australia, which is clearly a marine, inter-reef environment. The isotope 'signature' there was also clearly marine, and showed that the Gogo lungfishes had not been washed out from fresh water to the sea. Schmitz *et al.* (1991) later made an extensive survey of Sr isotope ratios on a large number of samples from various vertebrate-bearing localities. Most of the $^{87}Sr/^{86}Sr$ ratios agree with results of previous palaeoecological or sedimentological studies. Permian coelacanths, for example, are found to be either marine or freshwater. Carboniferous Coal Measure fishes (rhizodontiforms) are found to be freshwater, whereas those from the Carboniferous limestone (bradyodonts) are definitely marine. Marine ratios were found in Devonian and Silurian fishes from reputedly marine localities (Cleveland Shale, the Ludlow bone-bed, Gotland), and also from localities that had long been believed to be freshwater. For example, the antiarch *Bothriolepis*, from Miguasha, appears to be marine, with a ratio of 0.708, and thus in agreement with the conclusions of Schultze and Arsenault (1985), based on vertebrate distribution, and of Vézina (1991), based on the geochemistry of the rock itself. Interestingly, the ratios of fishes from Old Red Sandstone deposits (reputedly freshwater) give contradictory results. A pteraspid from Podolia appears as clearly marine (0.708), whereas another pteraspid from Wales and a psammosteid from Latvia are both clearly freshwater (0.701). This discrepancy leaves the question of the environment of the Old Red Sandstone deposits frustratingly unresolved. If there is no error or bias in the isotope ratio analysis, it may be explained by the fact that some of these Old Red Sandstone fishes could be either marine or freshwater, or even anadromous.

Further reading

Blieck (1985); Chaloner and Lawson (1985); Darby (1982); Denison (1956); Goujet (1984*a*); Halstead (1973*b*); Janvier (1985*c*); Graffin (1992); Romer (1955); Schmitz *et al.* (1991); Spjeldnaes (1979); Vézina (1991).

7.8 HISTORICAL PALAEOBIOGEOGRAPHY

Historical biogeography has long been largely narrative, invoking processes of casual dispersal to explain the distribution of living beings. It tends now to be replaced by an analytical approach, referred to as vicariance biogeography, which is linked with the development of cladistics (Nelson and Platnick 1981; Humphries and Parenti 1986). The inclusion of fossil taxa in historical biogeography was seen as an essential support to narrative biogeography, whereas it is regarded as secondary in analytical biogeography, because of the incompleteness of the fossil record (Patterson 1981*c*). In sum, fossil taxa

are rated in the same way as in cladistic analysis, and for the same reasons. The problem is, however, somewhat different when one is dealing with very ancient (e.g. Palaeozoic) biogeographical patterns, involving essentially extinct taxa. Here, one has to cope with the incompleteness of the fossil record and sometimes weakly supported theories of relationships. The basic work is thus to reconstruct theories of phylogenetic relationships between the endemic clades that are available. Comparisons with palaeogeographical maps based only on geological and geophysical data may then show either congruences or inconsistencies with the phylogeny-based relationships between the areas of endemism. Inconsistencies have to be explained either by erroneous phylogenetic analysis (e.g. due to missing data or a wrong choice of out-group), erroneous geophysical and geological inferences, or random dispersal.

Marine fishes are often said to be of poor significance in historical biogeography, because they are supposed to be able to disperse easily through the oceans, and they are more bound to climatic zones and thus relevant to ecological biogeography. However, extant marine fishes show biogeographical patterns that are not exclusively due to streams or climatic zones, but are largely bound to the geography and history of the continents. Most extant fishes live on the continental shelf or around islands and never disperse through wide oceanic areas, as is evidenced by patterns of endemism. They can thus almost be considered as land animals. The same applies to Palaeozoic fishes, many of which were bound to marginal, shallow-water environments and are endemic to particular areas, whereas others were probably epipelagic and widespread.

The Devonian is probably the best period for the analysis of early vertebrate biogeography and for making comparisons with geophysical models. Devonian vertebrate localities are abundant all over the world, with a variety of types of environments ranging from lagoonal through, deltaic to truly marine.

The first task in analytical historical biogeography is the search for endemic clades. Even if endemism is theoretically unrefutable in palaeontology, the consideration of roughly synchronous (say, Early Devonian, for example) localities provides reasonable support for hypotheses of endemism. For instance, in Early Devonian times, the jawless vertebrates show patterns of endemism (Young 1981, 1986*a*, 1990, 1993*b*) that may have a geographical significance. Galeaspids are unique to China, Tibet, and northern Vietnam (Fig. 7.6, 3). Osteostracans are known only in Euramerica (North America, northern Europe), northern Siberia, and Tuva (Fig. 7.6, 2). Heterostracans have nearly the same distribution as osteostracans (Fig. 7.6, 1), with a possible, yet unconfirmed, occurrence in

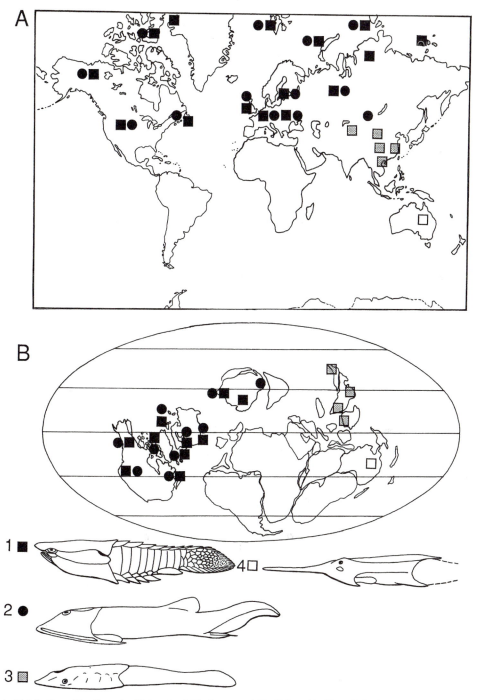

Fig. 7.6. A, Distribution of the Devonian Heterostraci (1), Osteostraci (2), Galeaspida (3), and Pituriaspida (4) on a Recent world map (the Algerian and Spanish heterostracan occurrences need confirmation and, therefore, have been omitted here); B, the same data on an Early Devonian palaeogeographical reconstruction based on geophysical data. (Reconstruction in B based on Li *et al.* (1993).)

Spain and Algeria (two areas which are supposed to belong to the Gondwana). Pituriaspids are hitherto unknown outside Australia (Fig. 7.6, 4). Moreover, within osteostracans and heterostracans, there are also several endemic taxa, such as the amphiaspidid heterostracans in north-western Siberia (Fig. 4.8, 6), the boreaspid osteostracans in Spitsbergen and Severnaya Zemlya (Fig. 4.18F2), the thyestidian osteostracans in Europe (from Britain to the Urals; Fig. 4.18H), and the tannuaspid osteostracans in Tuva. Blieck (1982*b*, 1984) suggested that the rise of the Protopteraspididae and Pteraspididae in North America and Europe respectively may have been a vicariance event resulting from the opening of the Iapetus ocean. Only in very rare instances, however, can these endemic taxa serve in resolving biogeographical relationships between separate areas by using the method of vicariance biogeography: their relationships within their respective higher taxa are too uncertain. In some instances the areas of endemism of Early Devonian jawless vertebrates are congruent with areas of endemism of some gnathostomes. The best example is that of the 'galeaspid province' (South and North China, and northern Vietnam), where the yunnanolepiform and procondylolepiform antiarchs (Fig. 4.54A, B) also occur, together with particular dipnomorphs, such as *Youngolepis* or *Diabolepis* (Fig. 4.84B, C). This unique 'Chinese' vertebrate fauna (Fig. 1.7) is generally attributed to the isolation of the Chinese terranes from all other continents in Devonian times (Young 1990; Fig. 7.6B, 3), but it may also be interpreted as the effect of a closed sea. Recent palaeogeographical reconstructions of this area for Silurian–Devonian times show that this part of Asia was occupied by a large epicontinental gulf surrounded by land, which was intermittently open to the south. So far as galeaspids are concerned, their endemism lasted from the Early Silurian to the Late Devonian, i.e. for more than 60 Ma. It shows at least that the wide separation between the South and North China blocks advocated by some geologists for the Middle Palaeozoic is unlikely, since galeaspid-bearing faunas with no major differences between them occur on both blocks (Fig. 7.7A).

Young (1981) defined five major areas of endemism, or 'provinces' for Devonian vertebrates: (1) the Euramerican 'cephalaspid' province, characterized by osteostracans (except tannuaspids); (2) the Siberian 'amphiaspid' province, characterized by amphiaspidid heterostracans; (3) the Tuvan 'tannuaspid' province, characterized by tannuaspid osteostracans; (4) the South Chinese 'yunnanolepid-galeaspid' province mentioned above, characterized by the galeaspids and yunnanolepiform antiarchs; and (5) the East Gondwanan 'wuttagoonaspid–phyllolepid' province, characterized by the actinolepid-related arthrodire *Wuttagoonaspis* and

phyllolepids (Fig. 4.44A, C). Some taxa are widespread, but the distribution of other taxa straddles several of these provinces and indicates area relationships. Heterostracans, for example, occur in both Euramerica and Siberia (Fig. 7.6, 1), and osteostracans in Euramerica, Siberia, and Tuva (tannuaspids are osteostracans; Fig. 7.6, 2). The sinolepid antiarchs are now known to occur in China, Vietnam and Australia (Fig. 7.7B). Young (1993*b*) discussed the biogeographical relationships of other areas that have yielded an increasingly large number of Devonian vertebrates. Kazakhstan, which is regarded as having been a separate landmass in Devonian times, has yielded euantiarchs (e.g. *Tenizolepis*, *Stegolepis*) that seem to be most closely related to the Chinese genus *Hunanolepis*. The North China block and western Tibet have yielded galeaspids, which extend this area of endemism westward (Fig. 7.7A). West Gondwana (South America, Africa, the Middle East, and probably also Armorica) has yielded vertebrate faunas that indicate bio-

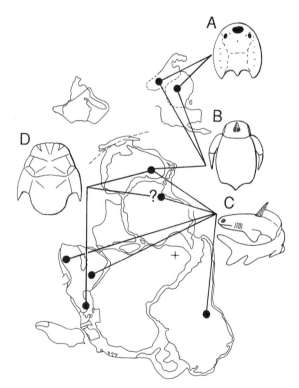

Fig. 7.7. Distribution of some Devonian vertebrate taxa on a palaeogeographical reconstruction of the Gondwana and adjacent terranes. A, Galeaspida; B, Sinolepidae (antiarchs); C, *Antarctilamna* (chondrichthyan); D, Phyllolepida (the Antarctic occurrence is based on the disputed genus *Antarctaspis*). (Based on Young (1981).)

geographical affinities either with East Gondwana or with Euramerica. In particular, the Devonian vertebrate faunas of Bolivia displays the closest resemblances to Australian and Antarctic faunas, whereas the Turkish, Iranian, and Saudi vertebrate faunas are a mix of taxa that also occur in the East Gondwanan (Australian), Euramerican, and possibly also the Chinese provinces. Some Devonian vertebrates seem to be more abundant in Gondwana, yet not endemic. The spines of the acanthodian *Gyracanthides* and of the Chondrichthyan *Antarctilamna* (Fig. 7.7C) for example, are among the most common fish remains in the Early and Middle Devonian of South America, South Africa, Iran, Saudi Arabia, Australia, and Antarctica.

The general features of the Devonian vertebrate biogeography are thus a well-marked endemism in the Early Devonian (Fig. 7.6) and a striking homogeneity in the Late Devonian. The breakdown of the Early Devonian endemism may be due to various causes. One is geographical, and would imply that all landmasses, and in particular Euramerica and West Gondwana, as well as the China blocks and East Gondwana, came in closer contact towards the end of the Devonian; but this is quite contrary to the current palaeogeographical models produced by geophysicists (Kent and Van der Voo 1990). Other possible causes are climatic or eustatic; the Frasnian marine transgression may have destroyed the marginal environments in which the Early Devonian endemic groups thrived (Young 1993b). Yet another possibility is that there was always a Palaeozoic Pangea (Boucot and Gray 1979), and that the patterns of endemism observed in the Early Devonian are due to other kinds of barriers than oceanic spaces. The possible presence of heterostracans in the Early Devonian of Spain and Algeria (Lelièvre et al. 1993), for example, contradicts models that assume a complete isolation of Euramerica from Gondwana.

The distribution of the Ordovician vertebrates also displays some peculiar features. Astraspids seem to be unique to North America and Siberia (Fig. 7.8, 1), two areas that current geophysical reconstructions put near the Ordovician equator (Fig. 7.8B). In contrast, the distribution of arandaspids (Fig. 7.8, 2), unique to Australia and Bolivia, is not satisfactorily consistent with these reconstructions, unless one assumes that they were widespread along the east coast of Gondwana. They are perhaps more consistent with heretical expanding-Earth reconstructions (Shields 1979), now unanimously rejected by geophysicists, or with the Pacifica theory (Nur and Ben-Avraham 1981).

Further reading

Blieck (1982b, 1984); Long (1986a, 1993a); McKerrow and Scotese (1990); Milner (1993a); Patterson (1981c); Young (1981, 1982b, 1986a, 1990, 1993b).

7.9 DIET

The diet of early vertebrates can be inferred from their presumed mode of life (e.g. benthic *versus* pelagic), which is in turn inferred from their morphology. This ecological and biomechanical approach is, however, flawed with strong contradictions, as is shown by radically divergent interpretations of the same morphology (see section 7.11, e.g. anaspids or heterostracans). The most reliable information on diet is provided by stomach or intestinal contents, which are rarely preserved in Palaeozoic vertebrates.

Stomach contents are known in some jawless craniates, such as the anaspid-like *Endeiolepis* and thelodonts (Stensiö 1939; Wilson and Caldwell 1993; Fig. 7.9A). In either form, these contents consist of an amorphous mass of fine-grained sediment, which merely indicates that they were active microphagous suspension-feeders. Among gnathostomes, large prey have been found inside placoderms, acanthodians, and various osteichthyans, indicating that for most of these fishes the mechanism of swallowing involved suction generated by the entire branchial apparatus. Only durophagous forms, such as some placoderms, lungfishes, or bradyodonts, may have torn or chewed their prey. In the Late Devonian locality of Miguasha, specimens of the osteolepiform *Eusthenopteron* contain various large osteichthyans, in particular the actinopterygian *Cheirolepis*, and acanthodians (Fig. 7.9C). Also, specimens of the actinopterygian *Cheirolepis* from Miguasha contain either acanthodians or young of *Eusthenopteron*. In contrast, the stomach contents of these acanthodians, which are toothless acanthodids (*Mesacanthus*, *Triazeugacanthus*), consist of minute invertebrates, in particular ostracods, which are also found in lungfishes at the same locality. In some instances, the predator seems to have died while swallowing the prey, which suggests that the latter was either too large or was unusual. The prey found in these Devonian fishes had always been swallowed head first. This is shown, for example, by the orientation of the scales or spines (Fig. 7.9, 2), of acanthodians swallowed by the osteolepiform *Eusthenopteron* (Arsenault 1982) or by the porolepiform *Glyptolepis* from the Devonian of Scotland (Ahlberg 1992). In the Late Devonian of Gogo, remains of young arthrodires have been found inside the abdominal cavities of larger individuals, and pebbles occur inside the thoracic armour of *Holonema* (Fig. 7.9, 3). The pebbles are not gastroliths, since they show no trace of wear; they were probably ingested together with food (Miles 1971).

The Late Devonian Cleveland Shale has provided much information on the diet and behaviour of the large arthrodires (dinichthyids) and chondrichthyans

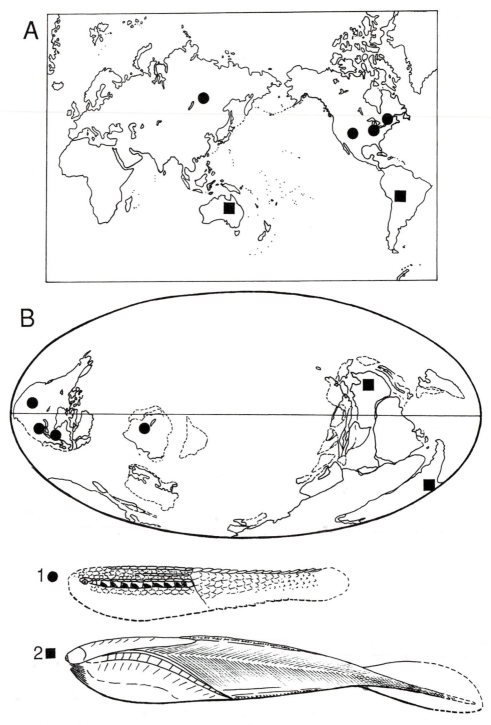

Fig. 7.8. Distribution of the Ordovician Astraspida (1) and Arandaspida (2) on a Recent world map (A) and a current palaeomagnetism-based Ordovician map (B). (B, from Li *et al.* (1993).)

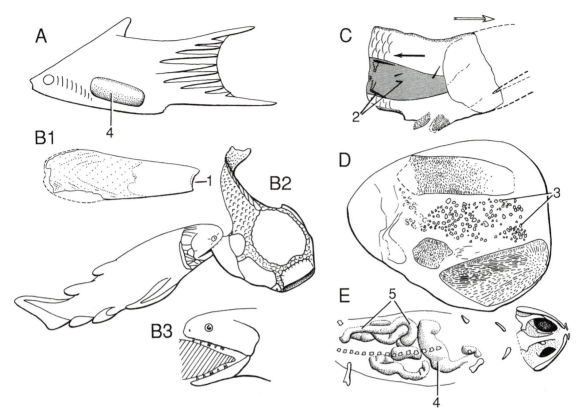

Fig. 7.9. Predation, diet, and stomach contents in early vertebrates. A, thelodont, Early Devonian of Canada, outline of the body in lateral view, showing the internal cast of the barrel-shaped stomach (×1). B, predation of a sarcopterygian on a large psammosteid heterostracan; B1, broken and subsequently healed psammosteid branchial plate (*Tartuosteus*), Late Devonian of Estonia (×0.3); B2, interpretation of the way in which *Glyptolepis* could have preyed on a large psammosteid, by biting the branchial plate; B3, comparison of the size of a large psammosteid with the gape of *Glyptolepis*; C, specimen of the acanthodian *Homalacanthus* (stippled) inside a large specimen of the osteolepiform *Eusthenopteron*, from the Late Devonian of Quebec (black arrow points towards the front of the acanthodian, white arrow towards the front of the osteolepiform) (×0.5); D, pebbles inside the trunk-shield of an articulated specimen of the arthrodire *Holonema*, Late Devonian of Australia (×0.5); E, stomach and intestine cast in a larval dissorophoid temnospondyl (*Amphibamus*) from the Pennsylvanian of the USA, in ventral view (×3). 1, bitten off and subsequently healed extremity of the branchial plate; 2, pectoral fin spines of the acanthodian; 3, pebbles; 4, stomach; 5, intestine. (A, from Wilson and Caldwell (1993); B, from Mark-Kurik (1966); C, from Arsenault (1982); D, from Miles (1971); D, from Milner (1982).)

(*Cladoselache*, *Symmorium*). Traces of bites by *Symmorium* and the petalodontid *Petalodus* have been observed on various nautiloids; and a large arthrodire (*Holdenius*) has been found with a huge spine of ctenacanthid shark that pierced its palate and probably caused its death (Hansen and Mapes 1991; Williams 1991). Because their spiral intestine does not permit the passage of large undigested food particles, sharks regurgitate pellets of compact food material (spines, plates). These pellets occur quite frequently in the Cleveland Shale. The nature of the stomach contents of the Cleveland Shale fishes does not seem to follow other rules than that of size, i.e. bigger fishes ate smaller ones.

Sharks seem to have fed chiefly on actinopterygians, crustaceans, and conodont animals. In contrast, a large actinopterygian (*Tegeolepis*) is shown to have eaten a small arthrodire and a small shark. There is no instance of jawless vertebrate remains in the stomach content of a gnathostome: Gnathostomes apparently preferred to eat other gnathostomes.

The coprolites of early vertebrates commonly have a spiral structure, owing to the generalized distribution of the spiral intestine in early gnathostomes. Some spirally coiled coprolites are in fact entospirae, that is, infillings of the spiral intestine of fossilized fishes, which were later reworked and displaced from their original position

(McAllister 1985). The spiral structure may, however, remain even when the faecal mass is expelled, in particular in sharks, and there are true spirally coiled coprolites. The difference between entospirae (or cololites) and coprolites rests on the presence, in the former, of delicate impressions of the folds of the intestinal mucosa. The earliest known spirally coiled vertebrate coprolites have been recorded from the Late Silurian of Ireland (Gilmore 1992) and are associated with scales of anaspids (*Birkenia*) and thelodonts (*Loganellia*). Their fine-grained structure has been interpreted as evidence for particulate feeding, and they have consequently been attributed to these jawless fishes. However, a fully coiled spiral intestine is so far known only in gnathostomes. Coprolites of the Carboniferous chondrichthyan *Cobelodus* contain possible parasitic helminth eggs (perhaps cestodes) (Zangerl and Case 1976). Traces of the stomach and intestine are also found in early tetrapods (Fig. 7.9E).

Traces of bites may indicate predator–prey relationships, but subsequent healing of the wound may also indicate that the predator failed. This is often the case for the large branchial plates of the psammosteid heterostracans in the Late Devonian of Estonia (Mark-Kurik 1966), where traces of bites, probably due to large sarcopterygians (*Laccognathus*, *Panderichthys*), are overgrown by blisters of dentine tubercles, thus indicating that the predators did not manage to kill these large harmless fishes: their gape was not large enough (Fig. 7.9B3). In some instances they could snap off only the tip of the branchial plate (Fig. 7.9B1, B2, 1).

Further reading

Ahlberg (1992); Arsenault (1982); Hansen and Mapes (1991); Mark-Kurik (1966); McAllister (1985); Williams (1991).

7.10 ORGANIC MATTER IN EARLY VERTEBRATES

Finding phylogenetically significant macromolecules in fossils is believed by some to be a new path in palaeontology (Runnegar 1986). As yet, however, it has not proved to be a more fruitful approach than a thorough comparative analysis of morphological characters. Where early vertebrates are concerned, L. B. Halstead (Armstrong and Halstead Tarlo 1966; Halstead Tarlo 1967a) was a pioneer in discovering various macromolecules associated with the process of calcification (e.g. collagen, mucopolysaccharides) in the dermal bone of Devonian heterostracans. Investigations of this kind have not been continued further on early vertebrates, apart from some consideration of the biomineralization

process in the dermal plates of placoderms (Ivanov *et al.* 1992), which is linked with the organization of the collagen fibrils. The main problem with research of this kind on early vertebrates is that most of the material, however well preserved it may be, has been more or less heated up at one time or another. This heating may result from the depth of burial of the sediment (geothermy) or from the proximity of some volcanic activity. Specimens in which the bone is dark or opaque white have been submitted to a high temperature, and thus offer no hope of finding macromolecules. Only in very rare cases (the Baltic, Gogo) are the conditions of fossilization and subsequent diagenesis likely to be favourable to the preservation of such molecules.

Further reading

Armstrong and Halstead Tarlo (1966); Halstead Tarlo (1967a); Halstead Tarlo and Mercer (1966); Runnegar (1986).

7.11 BIOMECHANICS OR FUNCTIONAL MORPHOLOGY

How early vertebrates swam and moved has been the subject of various biomechanical studies, most of which have depended on analogies with extant fishes that display comparable morphologies (see Braun and Reif 1985). Abel (1907) and Dollo (1910), were the first to try and interpret the morphology of various Silurian and Devonian jawless fishes in terms of function, by analogy with the rays, trilobites, and extant xiphosurans. It was in fact quite obvious that the shape of the head-shield of most osteostracans and the dorsal position of their eyes and 'nostril' indicated a benthic mode of life. However, this interpretation becomes less clear when less generalized osteostracans, such as boreaspidids, are concerned. The long and thin cornual and rostral processes that occur on the shields of some osteostracans, galeaspids, pituriaspids, and even heterostracans can receive a variety of interpretations (Fig. 7.10G). They may have served in digging, have housed special cutaneous sense-organs, have helped in swimming, or may simply have given the animal an apparently larger size, as seen by a predator.

In some instances, the morphology of early vertebrates has no close analogue among extant fishes. Their mode of swimming has then been inferred on the basis of hydrodynamic studies, sometimes performed on models (Mark-Kurik 1992; Belles-Isles 1987, 1992). The results of such studies can often, however, be surprisingly contradictory. Janvier (1987), for example, pointed out the diversity of the conclusions reached by several authors about the function of the hypocercal tail of anaspids. It

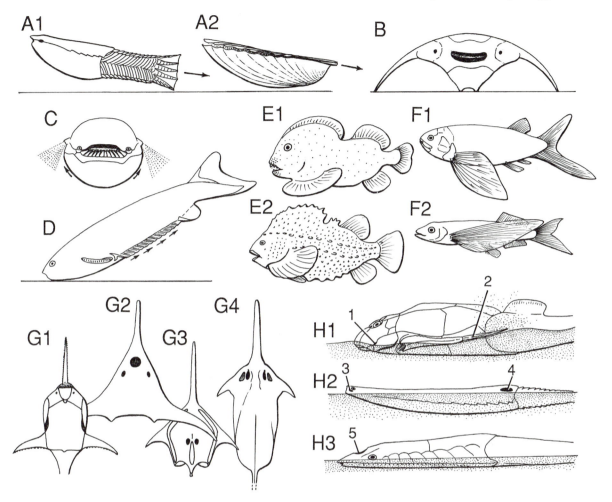

Fig. 7.10. Functional morphology of early vertebrates. A, comparison between a heterostracan (A1) and a pectinid bivalve (A2) in lateral view (arrow indicates the main propulsion force); B, presumed sustaining function of the downwardly curved branchial plates in psammosteid heterostracans; C, cyathaspidiform heterostracan in anterior view, illustrating the hypothesis of the stabilizing function of the water jets from the common external branchial openings (stipple); D, interpretation of the paired fins of the anaspid *Pharyngolepis* as a slow displacement device (arrows indicate the direction of the undulation of the fin); E, F, similarities in habitat, as inferred from gross morphology of fossil and extant fishes, in bottom-dwellers (the Carboniferous petalodontids, E1, and the Recent lumpfish, E2) and flying fishes (the Triassic actinopterygian *Thoracopterus*, F1, and the Recent exocet, F2); G, convergent morphologies of the head-shield in heterostracans (*Doryaspis*, G1), galeaspids (*Asiaspis*, G2) osteostracans (*Boreapis*, G3), and pituriaspids (*Pituriaspis*, G4); H, adaptation to partly burrowing habits in the antiarchan placoderm *Bothriolepis* (H1), the amphiaspid heterostracan *Kureykaspis* (H2), and the galeaspid *Polybranchiaspis* (H3). 1, spiracle; 2, pectoral fin; 3, adorbital opening; 4, common external branchial opening; 5, median dorsal inhalent opening. (A1, from Dineley (1976); B, from Mark (1961); C, From Bendix-Almgreen (1986); D, from Janvier (1987); E1, from Lund (1989); F1, from Lehman (1979); G1, from Heintz (1968), modified; G2, from P'an *et al.* (1975); G3, from Janvier (1985a); G4, from Young (1991a).)

has been said either to generate an anterior pitch of the fish (useful to a bottom-feeder) or an anterior lift, or to function in the same way as a diphycercal tail. The elongated paired fins of the anaspid *Pharyngolepis* have been interpreted by analogy with the anal fin of the gymnotes, and have been supposed to undulate and generate a slow displacement of the fish (Fig. 7.10D). The strongly convex ventral shield of heterostracans has been interpreted as generating upward lift, much in the same way as the convex valve of the bivalve *Pecten* (Fig. 7.10A).

Further analogy with *Pecten* was proposed by Bendix-Almgren (1986), who suggested that the common external branchial opening of heterostracans may have been a jet-propulsion or stabilizing device (making them resemble Cousteau's diving saucer!) that served mainly for balance, and compensated for the lack of paired fins (Fig. 7.10C). Conversely, Belles-Isles (1987) regarded pteraspids as fast swimmers, comparable to tunas, and provided with a powerful tail musculature. The traces of wear on the ventral shield of many heterostracan would, however, rather support their having benthic habits. The downward-bent branchial plates of the large psammosteid heterostracans (Fig. 7.10B) also show traces of wear, which suggests that they served as supports for these heavy animals when resting on the bottom (Mark 1961).

Some benthic or burrowing jawless vertebrates display adaptations for the intake of the respiratory water. In amphiaspid heterostracans, the adorbital openings and dorsally placed external branchial openings are possibly linked with partial burrowing habits (Fig. 7.10H2, 3, 4). The median dorsal opening of galeaspids (Fig. 7.10H3, 5) is also linked with the benthic mode of life of these animals. Recent hagfishes, too, use their nasopharyngeal duct for the intake of respiratory water when buried in mud.

The function of the thoracic armour of placoderms has also been the subject of diverse interpretations. The main question is how mobile the cranio-thoracic joint could be, and how far it served in opening the mouth. In some arthrodires, such as *Holonema*, the presence of large 'extrascapular' plates in the nuchal gap rendered almost impossible any movement of the cranio-thoracic joint, unless these plates could be lifted above the level of the skull-roof and thoracic armour. In contrast, some dinichthyids have a very large nuchal gap and no extrascapular, thereby allowing some dorsal flexure (Fig. 4.48B1). Heintz (1931) and Miles (1969) considered that in these arthrodires the skull could be lifted up by means of powerful muscles attached to the ventral side of the median dorsal plate. It is, however, surprising that those arthrodires that had almost no cranio-thoracic mobility (e.g. *Holonema*) retained a hinge with a complex ball-and-socket device.

One of the most peculiar structures of the dermal armour of placoderms is certainly the appendage-like fin of antiarchs, which Patten (1890) considered as further evidence for the direct descent of vertebrates from arthropods. Wells and Dorr (1985), among others, have interpreted these appendages either as an anchoring device to counteract strong currents, or as an adaptation for walking on land over short distances. Detailed reconstruction of the articular device shows, however, that these appendages were oriented slightly posterodorsally, which suggests that they probably served in covering the back of

the fish with sand, in the way that some tropical crabs use their posterior appendages (Fig. 7.10H, 12). The dorsal position of the eyes and nostrils in antiarchs would also accord with some kind of burrowing habit, but the spiracle must have remained above the level of the sand (Fig. 7.10, 1).

Because of their supposedly similar basic structure, the biomechanics of the gnathostomes that have close extant relatives, such as sharks and osteichthyans, is certainly easier to interpret. The interpretation of the bizarre stethacanthid 'sharks' (Fig. 4.33D–F) and bradyodonts (Fig. 4.36) is, however, open to fancy. Williams (1985), for example, suggested that the dorsal fin 'brush' of stethacantids and the large spiny scutes on their head could serve in the display behaviour of males, whereas Zangerl (1984) suggested that the 'brush' was used for threatening predators (see Section 7.14). It is, however, more probable that this 'brush' had to do with mating behaviour. What bradyodonts did with their frontal spines, combs, and whips is left open to the unbridled fancy of the reader. The stout-bodied petalodontids (Fig. 7.10E1) have no morphological analogue among extant chondrichthyans, and are more like extant lumps (Fig. 7.10E2) or scorpaeniid teleosts. The most peculiar symphysial tooth spirals of edestoids also have no extant analogue, except perhaps the large upper jaw tusks of barracudas (Fig. 4.31, 1). Although it certainly served in catching prey, there is no explanation of the fact that these teeth were not shed, and were scrolled into a huge spiral, unless it is a general condition for the gnathostomes (Patterson 1992).

The functional morphology of Palaeozoic actinopterygians is often considered in terms of analogy with the wide diversity of extant teleosts. The deep-bodied platysomids or bobasatraniids (Fig. 4.70P, Q), for example, are directly compared to various extant reef-dwellers (*Platax*, *Chaetodon*, etc.), whereas the long-snouted and slender-bodied saurichtyids (Fig. 4.71C) are compared to the extant green bone (*Belone*). Some Palaeozoic and Triassic actinopterygians that have very large pectoral fins (Fig. 7.10F1) are also compared to extant exocets (Fig. 7.10F2), although there are many examples of benthic teleosts that also possess such enlarged pectorals.

With the sarcopterygians, we meet a particular case. Many studies of functional anatomy are made in the light of the theory that some sarcopterygians are closely related to the tetrapods, and can therefore tell us something about the origin of walking. The function of the paired fins of the osteolepiform *Eusthenopteron* has often been interpreted in terms of a walking limb (rotation of the elbow, hinge-like knee, etc. see e.g. Andrews and Westoll 1970*a*; Rackoff 1980), whereas its general morphology is more comparable to that of a pike and does not suggest any adaptation for walking on land (Belles-Isles 1992). The limb structures of the early tetrapods *Ichthyostega* and *Acanthostega* are also easier to interpret in terms of a swimming function, rather than walking (Coates and Clack 1990).

Further reading

Alexander (1981); Belles-Isles (1987, 1992); Bonde (1983); Braun and Reif (1985); Janvier (1985a, 1987); Lund (1967); Mark-Kurik (1992); Miles (1969); Novitskaya (1992); Pridmore and Barwick (1993); Rackoff (1980); Thomson (1976a); Webb and Weihs (1985); Young and Zhang (1992).

7.12   INDIVIDUAL VARIATION

Like all other animals, early vertebrates show some individual variation and 'abnormalities'. The study of individual variation has admittedly been limited, by comparison with that of, say, mammals, because there are only few localities where there are enough specimens of the same species. Excellent work on individual variations in the sensory line and dermal bone pattern has, however, been done, for example, by Graham-Smith (1978) on large collections of the antiarch *Bothriolepis canadensis* from the Late Devonian of Miguasha, as well as on various osteolepiforms and lungfishes from the Middle Devonian of Scotland. Such studies are certainly of value to avoid using as specific a character that is not general, or that is polymorphous in the species under consideration. One may, however, wonder what is the rationale of such analyses in palaeontology. A palaeontological species is defined on a type-specimen, or holotype, and the 'population' that is referred to it is merely a collection of more or less similar specimens from the same locality and horizon. There is thus no guarantee that one is dealing with the same biological species; only a strong probability. Making a character analysis and a cladogram from such a collection will generally result in a cluster with a basal multiple branching; and one or two specimens will perhaps be outside this cluster, because they lack one or two defining characteristics of the species (see, for example, McCune 1987). What, then, is one to do with these aberrant specimens? Is each of them a member of a different species, or are they 'stem-individuals' of the species? And, if some specimens of the collection cluster together within the main cluster, are these a population, a subspecies, or does this mean that part the species is paraphyletic? This fundamental debate, which relates to the species concept, is far from closed, and it certainly marks a major gulf between cladists and evolutionary palaeontologists.

There are occasions when individual variation may throw some light on possible evolutionary processes. When some variations are linked with growth processes (e.g. as for the lateral-line canals or grooves), they may serve in providing an explanation for their generalization in later species. Even 'abnormalities', that is, very conspicuous and unique variations may simulate (or foreshadow) widespread characteristics in other groups. Gross (1951), for example, recorded an unjointed pectoral fin in a specimen of the asteolepidoid antiarch *Asterolepis*, which normally has jointed appendages, like most other antiarchs. This variation may be of interest when considering that one asterolepidoid genus, *Remigolepis*, has unjointed appendages. It suggests at least that the loss of this joint may be due to a single mutation. Gross (1942) also recorded various other abnormal or pathological features in antiarchs, including a double median ventral plate and an atrophied brachial process.

Further reading

Graham Smith (1978); Gross (1942)

7.13   GROWTH

Like variation, the growth of early vertebrates can be studied only when large collections are available, or when adults retain a 'memory' of their growth in the form of growth lines. The latter is the case, for example, in the dermal plates of heterostracans or various placoderms (Fig. 7.11A). The cosmine of lungfishes also shows concentric grooves, referred to as Westoll lines, which were formerly considered to be simple growth lines, but are now regarded as the result of a complex process of resorbtion and redeposition of the dentinous tissue, which is independent of growth itself and is to some extent linked with a seasonal rhythm. Growth series of complete individuals have been studied, for example, in the antiarch *Bothriolepis* (Fig. 7.11C), the acanthodian *Acanthodes* (Fig. 7.11D), the osteolepiform *Eusthenopteron* (Fig. 7.11E), and in various Palaeozoic actinopterygians (Fig. 7.11F). All these studies show some allometry. In nearly all gnathostomes, the eyes of the young are larger than in the adult, and the chordal lobe of the tail is more elongated (Fig. 7.11, 2). In young antiarchs, the pectoral appendages are much more elongated and slender than in adults (Fig. 7.11C, 3; Fig. 4.53F). The squamation may also vary during growth, as in the acanthodian *Acanthodes*, in which the squamation is restricted to the caudal region in smaller individuals, and progressed forward as the animal grew, with a more rapid growth along the main lateral line (Fig. 7.11D). A reverse trend is observed in early

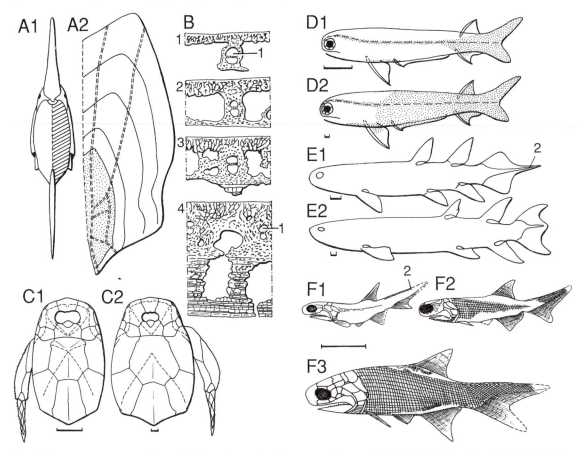

Fig. 7.11. Growth in early vertebrates. A, allometric growth of the sensory-line canals (dashed) in the dorsal disc (A1, hatched) of the pteraspidid heterostracan *Rhinopteraspis* (primordium of dorsal disc stippled in A2); B, four successive stages in the development of the exoskeleton of the thyestiid osteostracan *Tremataspis*, showing the centripetal growth (×40); C, juvenile (C1) and adult (C2) armour of the antiarch *Bothriolepis* in dorsal view, brought to the same size to show allometry in the growth of the orbitonasal fenestra and pectoral fins; D, juvenile (D1) and adult (D2) specimens of the acanthodian *Acanthodes* brought to the same size to show the progessive forward development of the body squamation (stippled) along the main lateral line; E, juvenile (E1) and adult (E2) specimens of the osteolepiform *Eusthenopteron* brought to the same size to show allometry in the size of the unpaired fins and shape of the chordal lobe; F, three successive growth stages in the Carboniferous actinopterygian '*Elonichthys*', showing the shortening of the chordal lobe and the backward development of the body squamation along the main lateral line. Scale bars: 10 mm. 1, sensory-line canal; 2, chordal lobe. (A, from E. I. White (1958); B, from Denison (1947); C, from Long and Werdelin (1986); D, from Zidek (1985, 1988); E, from Thomson and Hahn (1968), modified; F, from Schultze and Bardack (1987).)

actinopterygians (Fig. 7.11F). In some instances, such as the tremataspid and boreaspidid osteostracans (Fig. 4.18F2, H4), no growth series have ever been observed, even in large populations. This suggests that ossification occurred only when the animal was fully grown. In *Tremataspis*, various degrees of ossification of the exoskeleton have been observed in individuals of exactly the same size (Fig. 7.11B), showing that mineralization started with the superficial cosmine and sensory-line canals (Fig. 7.11B, 1) and proceeded centripetally.

The study of growth or, at least, of juvenile individuals, is important because it can provide ontogenetic arguments to determine which state of a character should be regarded as plesiomorphous when out-group comparison provides no clue. This is particularly true for the major fossil jawless vertebrate taxa, whose sister-group is not clearly established. A good example is that of the psammosteid heterostracans. Their partly platelet-covered armour was regarded as primitive by reference to astraspids and stem-group heterostracans, but their young are devoid of tesserae and show the typical pteraspidiform plate pattern (Fig. 2.7A).

Further reading

Blieck (1984); Denison (1947; 1973); Long and Werdelin (1986); Schultze (1972, 1984*a*); Schultze and Bardack (1987); Thomson and Hahn (1968); Upeniece and Upenieks (1992); White (1973); White and Toombs (1983); Zidek (1985).

7.14 SEXUAL DIMORPHISM AND BEHAVIOUR

Sexual dimorphism can only rarely be identified in early vertebrates: it requires not only excellent preservation of numerous specimens of the same species, but also some comparable extant models. The only clear evidence for sexual dimorphism in early vertebrates is in fact met with in fossil chondrichthyans and some placoderms, where the sex is indicated by the presence of pelvic claspers in males (Figs 4.51A5, B, 7.12A). Stethacanthid and holocephalan males are adorned with a variety of peculiar organs, whether modified fins or dermal scutes and spines on the head, which are probably related to display or mating behaviour. Some male and female specimens of the stethacanthid *Falcatus* (Fig. 7.12D) have been found side by side in the locality of Bear Gulch, and it has been suggested that they died during courtship or mating. Lund (1985*a*, 1990) pointed out that males of this genus largely outnumber the females, and that the latter are generally larger than the former, although there is evidence that males can be very large. He interpreted this as an indication that *Falcatus* had a display-courtship behaviour that entailed high male mortality rates in earlier growth stages and few, sexually dominant males. The dorsal 'brush' in the males of other stethacanthids (*Stethacanthus*) may also have played a role in courtship or mating. Zangerl (1984) has suggested, however, that it was inflatable and may more probably have served as a threat device. Together with the head denticles (Fig. 7.13, 2), it may have simulated a very large mouth (Fig. 7.13B) and threatened predators coming from above. A possible case of viviparity and intrauterine feeding (oophagy) has been suggested by Lund (1980) for the Carboniferous holocephalan *Delphyodontos*, but this remains speculative.

Fossil egg capsules of various chondrichthyans have been recorded (Fig. 7.12C); these imply internal fertilization. One of the most intriguing of these fossil egg capsules is a specimen from the Late Devonian of South Africa (Fig. 7.12B1; Chaloner *et al.* 1980), which was found in association with abundant placoderm (arthrodire)

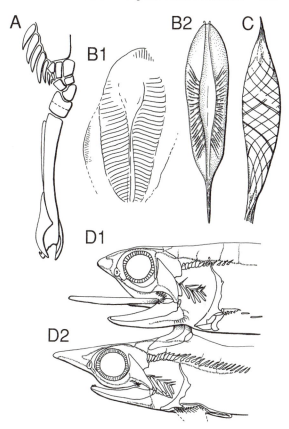

Fig. 7.12. Sexual dimorphism and reproduction in early chondrichthyans. A, mixipterygial (pelvic) clasper of the symmoriid *Cobelodus*, in ventral view (×0.3); B, chondrichthyan or placoderm egg capsule from the Late Devonian of South Africa (B1, ×1.5), compared with that of a Recent holocephalan (B2); C, chondrichthyan (possibly hybodontiform) egg capsule from the Late Carboniferous of the USA, showing spiral twisting, as in the egg capsule of the extant shark *Heterodontus*; D, heads of female (D1) and male (D2) specimens of the stethacanthid *Falcatus*, lying side by side and presumably dead during courtship (×1). (A, from Zangerl and Case (1976); B1, from Chaloner *et al.* (1980); C, from Zidek (1976); D, from Lund (1985*b*).)

remains, but has a striking resemblance to a modern holocephalan egg capsule (Fig. 7.12B2). Although there are records of Devonian holocephalans (bradyodonts), one may wonder whether this capsule might not be that of a placoderm.

Further reading

Lund (1990); Schultze (1985); Zangerl (1984).

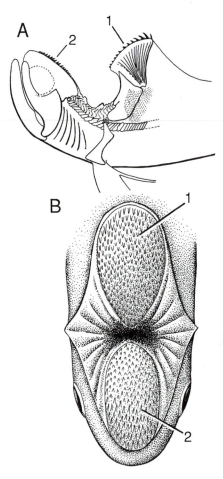

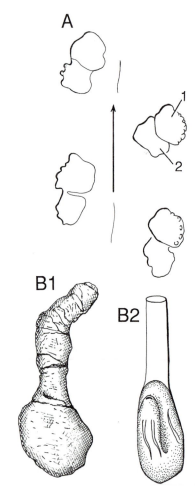

Fig. 7.13. Interpretation of the dorsal 'brush' and head denticles of the stethacanthid chondrichthyan *Stethacanthus* as an inflatable threat device. A, lateral view; B, reconstruction in dorsal view when the 'brush' is inflated. 1, dorsal 'brush'; 2, head denticles. (From Zangerl (1984).)

7.15 TRACE FOSSILS

Most of the early vertebrates are fishes, and can therefore hardly leave any trace fossils, apart from traces of grazing or burrowing. The record of trace fossils due to early vertebrates is therefore very poor. Some peculiar Ordovician trace fossils referred to as *Agnathichnis* have been regarded as traces of grazing made by some jawless fish (Fischer 1978). One can in fact imagine that the oral plates of arandaspids or heterostracans could have left some traces in the mud, if they actually served in grabbing.

Various types of fossil burrows, from Devonian to Recent, have been interpreted as lungfish aestivation burrows, but only the burrows that contain the remains of the lungfish *Gnathorhiza* (Fig. 7.14B), from the Permian and Triassic of the USA, can at present be reliably regarded as lungfish burrows. All other records are doubtful (McAllister 1992).

The earliest known tetrapod track has been recorded from the Late Devonian (possibly Frasnian) of Australia (Warren and Wakefield 1972). Although it differs in many respects from most post-Devonian tracks, in particular in the lateral orientation of the digits, there is little doubt that it is a tetrapod track (Fig. 7.14A). It is sur-

Fig. 7.14. Early vertebrate trace fossils. A, the earliest known undisputed tetrapod trackway, from the Late Devonian (Frasnian) of Australia (arrow points forward) (×0.3); B, internal cast (B1) and reconstruction (B2) of the aestivation burrow of the Permian lungfish *Gnathorhiza* (×0.7). 1, pes; 2, manus. (B, from Warren and Wakefield (1972); C, from McAllister (1992).)

prisingly narrow, given the size of the hand and foot, and the hands and feet overlap in an unusual way (Fig. 7.14A, 1, 2). It also shows a median, sinuous, groove left by the tail. Another Devonian track has been recorded from Brazil (Leonardi 1983); it is represented only by a single, doubtful, hand or foot impression. Yet another possible tetrapod trackway has been recorded from Australia, but it was found in a paving-stone of an old courtyard, and is therefore of uncertain origin. It is thought to have come from a neighbouring quarry in what is presumed to be Early Devonian red sandstone. If this turns out to be the case, then the earliest occurrence of tetrapods would be brought back by about 35 Ma (Warren *et al.* 1986).

Tetrapod trackways are extremely common in the Late Palaeozoic and provide a large amount of information on the mode of life. Particularly interesting are the impressions of the scales of the belly and limbs that are associated with the tracks of various amphibians.

Further reading
McAllister (1992); Warren and Wakefield (1972).

8

A short history of research on early vertebrates

The reader may be surprised to see this chapter on the history of early vertebrate research at the end of the book. I think, however, that it is more fruitful to consider the history after the present-day data, theories, and beliefs have been presented and discussed. The diversity of opinion on the structure and relationships of early vertebrates stems largely from the beginning of this century, and has been initiated on the one hand by the work of the anatomists E. S. Goodrich and E. P. Allis and on the other by that of the palaeontologists E. Stensiö and A. S. Romer. Opinions on extrinsic data, such as the mode of life, stratigraphy, and biogeography, are more deeply rooted into the history of geology and can be traced back to the nineteenth century. I have divided this brief historical account into two sections: first, the history of the discovery of the structure and relationships of early vertebrates, which is closely linked with the progress of comparative biology; second, the history of exploration and field discoveries, which has more to do with the history of geology.

8.1 EARLY VERTEBRATES AS ORGANISMS

Fossil fishes, and in particular teleosts, which have a familiar aspect, have been noticed, and sometimes used for decoration, since Roman times (pachycormid teleosts on limestone slabs adorned Roman swimming-pools in Fontaines Salées, Burgundy, France). Strangely, they were not much considered in the early creationist versus transformist debates, but some questions arose when the Devonian osteostracans and placoderms of Scotland and Russia began to turn up and, later, when the distribution of fossil fishes in time became known to a wider extent (see Patterson 1977b, 1981b). Were these to be put in modern groups, or did they represent entirely extinct forms?

8.1.1 Early vertebrates and Scotsmen

The first discoveries of Early Palaeozoic fishes took place independently in two places: Scotland and the Baltic area.

In Scotland interest in these peculiar fossil creatures arose first among societies of amateur naturalists and geologists, whose members were generally learned clerks or teachers (e.g. John Flemming, the Revd John Anderson of Newburgh, Robert Jameson of Newcastle, Samuel Hibbert, the Revd George Gordon of Birnie, Patrick Duff, and John Martin of Elgin) but also included self-made scientists (Hugh Miller; Fig. 8.1C) and members of the gentry (Dr John Grant Malcolmson of Madras, Lady Eliza Maria Gordon Cumming of Altyre; Fig. 8.1A,B)). Andrews (1982) has provided a masterly record of the activities and debates on early vertebrates among these Scottish scientists during the first half of the nineteenth century, and here I shall highlight only some of the events that had an impact on subsequent theories.

The first early vertebrate remains recorded from Scotland were of Carboniferous age and are now referred to the sarcopterygians *Rhizodus* and *Megalichthys* and the petalodontid *Petalodus*, but most inhabitants of various localities (e.g. Caithness) knew of the 'ichthyoliths' (fish remains) that were frequently discovered in the surrounding quarries. Interest in these fossils began when the famous British geologist Roderick I. Murchison started a long-term project to investigate the geology of Scotland with the collaboration of various colleagues. The fish remains were then expected to provide an age for several sandstone formations that were poor in other kinds of fossils. As early as 1828, Sedgwick and Murchison described the first Devonian fishes, *Dipterus* and *Osteolepis*, but considered them to be of Permian age, by comparison with palaeoniscids from the Late Permian of Germany, which had similar shiny, diamond-shaped scales. [Later, the 'Old Red Sandstone' was first regarded as part of the Carboniferous; then, after a long dispute between Murchison and Henry De La Beche, the name 'Devonian' was erected by Sedgwick and Murchison (1839) for the equivalent marine rocks lying below the Carboniferous and above the Silurian (for details see Rudwick 1985)]. Notwithstanding the merits of these English geologists, the activity and originality of the Scottish community seems to have been crucial in trig-

Fig. 8.1. Four famous Scots who triggered the interest in early vertebrates: A, Lady Eliza Maria Gordon Cumming of Altyre (1798–1842); B, Dr John Grant Malcolmson of Madras (1802–44); C, Hugh Miller of Cromarty (1802–56); D, Professor R. Heathley Traquair (1840–1913). (A, from Gordon Cumming (1904), after a painting by George Sanders, reproduced by permission of Sir William Gordon Cumming, her great grandson. B, Moray District Council, Museums Division; C, photo Patrick, Edinburgh; D, from Anonymous (1909).)

gering the debates on the affinities of these fishes. One of the most important characters in this story is certainly Professor Robert Jameson, of Edinburgh University, who taught and influenced most of the others. His knowledge of theoretical geology was highly stimulating to many of the Scottish naturalists of his time. Controversies, such as the 1835 battle between Jameson, Flemming, and Hibbert on the reptilian or piscine nature of a large Carboniferous tooth from Burdiehouse, also excited the interest of learned people in Scotland.

Andrews (1982) recognized two centres of interest in the Scottish fossil fishes, besides Edinburgh, between 1834 and 1840: the Midland Valley and the Moray Firth region. In the former, John Anderson discovered between 1836 and 1838 a number of fish remains in the Old Red Sandstone, among which was the porolepiform *Holoptychius*. Later, it was the discovery of a scale of the same fish in the marine Devonian of Belgium that convinced Murchison that the Old Red Sandstone was not part of the Carboniferous, but an equivalent of his marine 'Devonian' (Rudwick 1985).

Further north, the Moray Firth region also yielded a large quantity of fish remains, which were actively collected by George Gordon, Patrick Duff, and John Martin around Elgin. It is remarkable that, despite their isolation, these people came together and formed the 'Elgin and Moray Scientific Association' (now the Moray Society) in order to circulate books and publications among themselves. One of them, Alexander Robertson, a wealthy farmer of Inverugie, even subscribed to Agassiz's *Recherches sur les poissons fossiles* (see below) as early as 1838, thereby showing a level of culture and interest rarely met with outside major cities in Europe at this time. In Cromarty, on the opposite shore of the Moray Firth, lived Hugh Miller (Fig. 8.1C), who was originally a stonemason but also a talented writer. Despite his isolation (he defined himself as a 'Robinson Crusoe in geology'), and thanks to his friends in the Elgin and Moray Scientific Association, he managed to give brief descriptions of his palaeontological discoveries in his book *Scenes and legends of the North of Scotland* (Miller 1835). These writings aroused the interest of an army surgeon, John Grant Malcolmson (Fig. 8.1B), who had just come back from a long stay in India. Malcolmson had a great enthusiasm for fossils and a good geological background. His social status and his wealth allowed him to travel easily to London and Paris, and he acted as a link between the scientists in the Moray Firth and the famous geologists and palaeontologists of the time, such as Murchison, Lyell, Cuvier, and Agassiz. As is pointed out by Andrews (1982), Malcolmson's activity was crucial to the success of Agassiz's endeavours. In particular, he and Miller resolved the riddle of the placoderms. Plates of an-

tiarchs (*Bothriolepis, Pterichthyodes*; Figs 4.52, 4.53E) and arthrodires (*Coccosteus*; Fig. 4.47A) had been recorded from the Old Red Sandstone as early as 1828 by Sedgwick and Murchison, who referred them to as plates of a 'tortoise allied to *Trionyx*' (as a consequence of Cuvier's early opinion on fragmentary placoderm material). Miller had made reconstructions of these strange armoured fishes, combining antiarchs and arthrodires into a peculiar 'winged creature' (according to Agassiz's terms), which his friends of the Moray Firth rather regarded as some arthropod (a 'beetle'!). After seeing Miller's drawings, however, Agassiz was more and more convinced of the piscine nature of the creature, for which he coined the name *Coccosteus*.

In about 1839, Malcolmson made the acquaintance of Lady Gordon Cumming (Fig. 8.1A) of Altyre House, who became passionately interested in collecting the well-preserved fossil fishes from the neighbouring quarry of Lethen Bar. Lady Gordon Cummings and her daughter, Lady Seymour, then spent years making drawings and watercolour paintings of these fishes, which later were largely used by Agassiz, with no special acknowledgement. Stiven, of Elgin, was another artist who contributed largely to the information of scientists at a time when photographs did not exist. In 1840, Malcolmson went back to India. He died there in 1844, leaving a large amount of geological and palaeontological data unpublished. Lady Gordon Cumming continued to provide drawings and specimens to Agassiz until she died in 1842. It is now, however, clear evident from the archives that Malcolmson had clearly seen the difference in structure between the arthrodire *Coccosteus* and the antiarch '*Pterichthys*' (now *Pterichthyodes*), and that he understood their anatomy very much better than Agassiz (Andrews 1982).

In 1840, Hugh Miller moved to Edinburgh to become the editor of a religious newspaper, *The Witness*, which he used to publish extensively on the Old Red Sandstone and its fossils, thus arousing the interest of many local people in fossil collecting. His three books *The Old Red Sandstone, or new walks in an old field* (1841), *Footprints of the Creator* (1850), and *Testimony of the rocks or geology in its bearing on the two theologies, natural and revealed* (1857), in which he tried to reconcile geology and the Bible, had a considerable success. Miller committed suicide in 1856.

In addition to these famous Scots of the Moray Firth, a number of other, humble Scottish collectors greatly contributed to the knowledge of Devonian fishes at that time. Robert Dick, for example, was a shy, retiring baker at Thurso in northern Scotland who tirelessly collected fishes in his area and wrote poems about his findings:

Hammers an' chisels an' a',
  Chisels an' fossils an' a';
Resurrection's our trade, by raising the dead
  We've grandeur an' honour an' a';
It's good to be breaking a stone,
  The work now is lucky an' braw;
It's grand to be finding a bone—
  A fish-bone the grandest of a'.

Although Agassiz's works on fossil fishes in general prompted the investigation of early vertebrates in other regions, Scotland has remained a centre of interest in this field until now. Ramsay Heatley Traquair (1840–1913; Fig. 8.1D) is one of the famous characters in the history of early vertebrate research: we owe to him extensive studies on the Carboniferous actinopterygians, as well as the discovery of the Silurian jawless fishes in Scotland. He first studied medicine, although his interest had always been in natural history, and as soon as he entered the University of Edinburgh in 1857 he rushed to the library to read Agassiz's works in order to identify the Carboniferous fishes he had found during his excursions on the Early Carboniferous rocks of Wardie, near Edinburgh. Traquair is linked with the early history by his contacts with two famous collectors, Sir Philip Egerton and the Earl of Enniskillen, who had been Agassiz's 'fossil hunters'. During his long career as the keeper of the Royal Scottish Museum, Traquair published a large quantity of notes (see Anonymous 1909), which show a special care in the description and restoration of the dermal bones of the head in fishes, probably as a reaction against the weight given by Agassiz to the shape of the scales. Traquair discovered the anaspids and the thelodonts and was the first to suggest that the microsquamose condition of the exoskeleton in the latter may have been ancestral to that in armoured ostracoderms. He spent many holidays in Germany, and was the first to describe in detail the Devonian fishes (e.g. *Drepanaspis*; Fig. 4.9E) of the Hunsrück Slates. We also owe to him a number of discoveries on actinopterygians, e.g. that some Palaeozoic actinopterygians are more closely related to sturgeons than to gars, despite their well-developed ganoid scales, or that *Cheirolepis* is an actinopterygian, and not an acanthodian, as was previously believed on account of its minute scales.

In modern times, the tradition of early vertebrate studies in Scotland has been particularly maintained by T. S. Westoll, a professor in the University of Newcastle upon Tyne, and by his disciples S. M. Andrews and A. Panchen. Westoll's works, which bear chiefly on sarcopterygians, are characterized by a particular interest in the evolution of major structures and the construction of transformational theories (e.g. origin of paired limbs, homologies in the dermal bones of the skull). We owe to

him the current idea that lungfishes are merely derived sarcopterygians and are closely allied to porolepiforms. The Scottish amateur palaeontologists have kept up this tradition with many amazing discoveries made by a 'new Hugh Miller': the fossil collector Stanley Wood (see e.g. Wood 1982).

Further reading
Andrews (1982); Anonymous (1909).

8.1.2 Louis Agassiz

Louis Agassiz (1807–1873; Fig. 8.2A) is regarded as the founder of early vertebrate palaeontology, in particular because his monographs on fossil fishes were the basis for all research in this field until the 1860s. He was born in Neuchâtel, Switzerland, and showed an early interest in natural history. He first studied medicine in Zürich, and later went to Heidelberg and Munich to acquire a doctorate in natural history. During his stay in München, around 1827, he became interested in freshwater fishes, and was given by von Martius a collection of fishes from the Amazon, which he described in 1829. This book on Brazilian fishes (dedicated to Cuvier) gave great weight to the morphology of the scales, which he later used to a wide extent in his studies of fossil fishes. Then, after the publication of some works on the freshwater fishes of Europe, Agassiz decided to write a series of monographs describing all the fossil fishes available at that time, about 1700 species. Back in Neuchâtel in 1832, and with the support of von Humboldt, he became a professor of the College there, a post that he held until he left for a new career in America in 1846. Agassiz is described as a brilliant, ambitious, enthusiastic, hard-working, and charming person. In 1831, he made a trip to Paris, where he first met Alexander von Humboldt, who recommended him to Cuvier. It was certainly the latter who persuaded him to study fossil fishes and passed on to him the first Palaeozoic fishes that he had received from Scotland and the Baltic. Agassiz had been impressed by Cuvier's method of comparing Recent and fossil animals (outlined in the *Recherches sur les ossements fossiles*; Cuvier 1812–22), and he used it as a guideline in his own work. Agassiz, like Cuvier, allied French and German cultures, which certainly influenced his way of thinking (he is sometimes depicted as a 'romantic naturalist'), but during his four trips to Great Britain (including two to Scotland) he discovered with great enthusiasm the Anglo-Saxon culture.

Dupuis (1988) praised Agassiz for being a 'prophet of the memory of communication in zoology'. In fact, he was obsessed by the idea that progress in science would be more rapid if scientists could easily obtain data from

Fig. 8.2. A, Louis Agassiz (1807–73), the founder of early vertebrate palaeontology; B, the Estonian palaeontologist Hermann Asmuss (1812–59), who demonstrated independently of Hugh Miller, that placoderms were fishes. (A, from Lurie (1960); B, from Mark-Kurik (1991).)

their colleagues. This was the goal of his monographs on fossil fishes and also of his *Nomenclator zoologicus*, which records all generic names of animals known at that time. Agassiz also considered that widely distributed casts of fossils or Recent animals could be regarded as a means of scientific communication comparable to publications. His '*Recherches sur les poissons fossiles*' was published in the form of '*livraisons*' to which anyone could subscribe. He planned twelve such '*livraisons*' gathered into five volumes (Agassiz 1833–44), but as this monograph series appeared, new fossils were sent to him, in particular from Scotland and Russia (*via* Murchison). He felt thus obliged to add some supplements, of which only one was actually published, on the Old Red Sandstone fishes (Agassiz 1844–5). However, despite substantial grants from the British Association for the Advancement of Science and from Sir Philip Egerton, Agassiz's private resources were under serious strain. In 1846, he went on an expedition to America and was teaching at Harvard when the Swiss revolution against the Prussians broke out in 1848. The College of Neuchâtel was then dismantled, and he had to stay for the rest of his life as a professor in Harvard. When Alcide d'Orbigny died, in 1857, The French government proposed Agassiz the chair of palaeontology in the Museum National d'Histoire Naturelle, in Paris. Agassiz hesitated until 1861 and finally the chair went to a former cavalry officer, A. d'Archiac Desmier de Saint-Simon, who had some knowledge of geology.

After his departure to America, Agassiz stopped work on fossil fishes to concentrate on extant fishes and glacial geology, his other pet subject. (For a detailed biography of Agassiz, see Lurie (1960)).

Agassiz's *Recherches sur les poissons fossiles* had a considerable impact on the discovery of the diversity of fossil fishes, but, as could be expected from a disciple of Cuvier, it contains little theory. His discovery of the 'three-fold parallelism' (Agassiz 1842), i.e. the congurence between the taxonomic series, the geological succession, and the embryonic stages, is largely based on observations initially made by his friend Carl Vogt on the development of the trout (Dupuis 1988). Agassiz (1859) nevertheless later criticized Vogt for his transformist ideas, which were partly derived from his own works. Agassiz did not consider this parallelism to be a consequence of a gradual evolutionary progress, but rather, like von Baer (1828)—although he rejected von Baer's law— as being oriented from the most general to the most particular; the most general being, in his sense, the most 'perfect'. He insisted many times on the fact that the earliest fishes (i.e. osteostracans, placoderms, sharks, etc) were the most 'perfect', but this was not, to him, in conflict with the embryonic development, for he regarded the embryo as more 'perfect' than the adult. The embryo of a fish, for example, is more perfect than an adult fish because it comprises the essential structures to make also a tetrapod. In his *Monographie des poissons fossiles du Vieux Grès Rouge*, he went so far to suggest that Devonian fishes 'represent, by the virtue of their particular structure, the embryonic age of fishes'. Nowadays, we are used to seeing the three-fold parallelism as oriented the other way, as did Haeckel (1866), but Agassiz's way of considering it was widespread in pre-evolutionary times and is rooted in Buffon's notion of 'degeneracy'. Dupuis (1988) pointed out that Agassiz often used terms, such as *progrès, génétique, développement, marche*, which may suggest that he believed in some kind of evolution that mirrors the development of the individuals (Agassiz 1874). As suggested by Patterson (1981*b*), Agassiz and Vogt were certainly, on this matter, the precursors of Haeckel.

A look at Agassiz's (1844) chart of the classification and distribution of fish taxa through time shows how close it is to the traditional evolutionary trees (Patterson 1981*b*) and how it already shows 'evolutionary radiations', which Agassiz regarded as 'creations'. The best example is certainly the 'sudden' appearance of the teleosts in the Lower Cretaceous (see discussion in Patterson 1981*b*), which is a consequence of classification (early fossil relatives of the major teleost taxa not being recognized as teleosts). In this chart, the 'bradyodonts' were correctly put in the chondrichthyans (*Placoïdes*), but osteostracans (*Céphalaspides*) were put among the *Ganoïdes*, together with lungfishes, gars, and other actinopterygians, because of their thick, diamond-shaped scales.

Further reading

Agassiz (1833–44, 1839, 1844–45, 1857, 1859, 1874); Dupuis (1988); Gaudant (1980); Lurie (1960); Patterson (1981*b*).

8.1.3 Early vertebrates, the Russians, and the Baltics

After the death of Lady Gordon Cumming in 1842, the Altyre estate was visited by a member of the old Prussian Baltic aristocracy, Dr. von Hamel, of the Russian Academy of Science, who served as a fossil collector for a Russian palaeontologist in St Petersburg, Christian Heinrich Pander (1794–1865; see biography in Siegfried and Gross 1971). He obtained from Lady Seymour and from many other private collectors in Great Britain (including Hugh Miller) a large number of Devonian fishes, which were immediately shipped to St Petersburg. From 1832 onwards many fish remains were collected from the Silurian and Devonian marls and sandstones of the Baltic provinces of Russia (now Estonia, Latvia, and

Lithuania). Unlike to those from Scotland, they were generally in the form of isolated bones, though much better preserved than the latter. They also had the advantage of being in comparatively softer sediments and could be completely freed from the matrix. The Marquis P. E. Poulletier de Verneuil, a wealthy Frenchman and learned fossil amateur, as well as Murchison, had had the opportunity of bringing back from Russia a number of these fish remains, which Agassiz included in his *Monographie des Poissons Fossiles des Vieux Grès Rouges*, but Pander needed comparative material in order to give a more detailed description of his specimens. The result was a remarkably illustrated series of monographs (Pander 1856, 1857, 1858, 1860). Because of the nature of his material, Pander could not only reconstruct, plate by plate, the armour of the placoderms (in particular *Coccosteus* and *Asterolepis*) but was also able to describe for the first time their histological structure. Pander's works are certainly the first step towards the modern way of describing the exoskeleton of early vertebrates. However, before Pander and before Agassiz, in the 1830s and early 1840s, there had been lively discussions on the Devonian fishes in Estonia among scientists at Tartu (Dorpat) University. Kutorga had collected large bony plates and conical teeth (now known to belong to homostiid arthrodires, antiarchs, and sarcopterygians) from the Late Devonian sands of Estonia, and he referred to them as turtle and crocodile remains, possibly on the advice of Cuvier. At the same time, Hermann Asmuss (1812–59; Fig. 8.2B) collected a large quantity of this material, which he correctly referred to fishes (Asmuss *in litteris* 1840). He sent plaster casts of his specimens to Agassiz, who illustrated them in his monograph.

Beside Asmuss and Pander, other scientists and fossil collectors developed early vertebrate palaeontology in the Baltics and the Russian Platform during the mid-nineteenth century, in particular F. Schmidt, E. Eichwald, and J. V. Rohon, who, in addition to Devonian fishes, described the first Silurian heterostracans, thelodonts, and osteostracans. This tradition was continued into the twentieth century with the works of the Latvian-born German palaeontologist Walter Gross (1903–74, Fig. 8.3), who began to describe the Baltic material he had collected himself, and later founded modern palaeohistology. Early vertebrate research in the Baltic countries was rather quiet during the first half of this century (except for some work on Silurian fishes by the Estonian A. Luha). Scientists from Sweden and the USA collected a large amount of material, in particular from the Silurian of Saaremaa island in Estonia. Noteworthy are the activities of the American zoologist William Patten (1861–1932), who carried out extensive excavations in the quarries of Saaremaa in 1930–2 with the aim of supporting his theory

Fig. 8.3. Walter Gross (1903–74) was born in Latvia and developed early vertebrate palaeontology in Germany, with particular reference to the palaeohistology of the dermal skeleton. (From Schultze (1974).)

that osteostracans and antiarchs were the 'missing link' between the xiphosuran arthropods and vertebrates. Most of his material was later described by his student G. M. Robertson and by R. H. Denison.

During the Second World War, the Baltic countries became included into the USSR, and a new generation of Baltic early vertebrate palaeontologists (e.g. Elga Mark-Kurik, Valentina Karatayute-Talimaa) was trained by the Russian biologist and palaeontologist Dimitri Vladimirovitch Obruchev (1900–70), who already had wide experience of the comparable Devonian material from the Russian Platform (St Petersburg area). The competence and activity of these Baltic palaeontologists was so outstanding that a large part of the Ordovician to Devonian vertebrate remains collected in the entire USSR were brought to them for identification. Their interest was mainly oriented toward biostratigraphy and palaeohistology, in the tradition of Pander and Eichwald. Since the Baltic countries obtained their independence in the early 1990s, these scientists have kept up their activities, despite drastic cuts, and remain leading specialists in this field.

Further reading

Mark-Kurik (1991); Märss (1982, 1986*b*); Pander (1857, 1858, 1860); Siegfried and Gross (1971).

8.1.4 Thomas H. Huxley and the time of evolution

After Agassiz's emigration to America, interest in early vertebrates declines somewhat in Britain, but with the rise of the Darwinian theory a new period of activity then began. This was initiated by Thomas H. Huxley, who in 1854 was employed by the Geological Survey of Great Britain to identify fossils. He became interested in the variety of Palaeozoic fishes that geologists and amateurs (e.g. McCoy, Egerton, Salter) continued to collect, not only from Scotland but also from the Welsh Borderland. As early as 1858, Huxley began to publish notes on the distinction between *Cephalaspis* and *Pteraspis* (which Agassiz had treated as the same genus). His most famous note on early vertebrates is probably his 1861 review of the systematics of Devonian fishes, in which he erected the taxon 'Crossopterygidae' for extant cladistians (*Polypterus*) as well as for various Palaeozoic fishes with lobed fins that are now known as osteolepiforms, lungfishes, actinistians, and even some fossil actinopterygians. (A. S. Woodward later said that this note 'completely revolutionised existing knowledge of the subject'.) Huxley failed to recognize that Devonian lungfishes were related to extant ones, but he regarded the latter as 'next of kin' to his 'Crossopterygidae', and later (Huxley 1880) considered crossopterygians to be descendants of lungfishes. By 'creating' the crossopterygians, Huxley paved the way for the theory that the tetrapods are descendants of some of these fossil lobe-finned fishes. This became the received view when the American palaeontologist E. D. Cope (1892) published a short but famous *Phylogeny of the Vertebrata*, in which the crossopterygians were associated for the first time with the tetrapods, and armoured jawless vertebrates were put with lampreys and hagfishes in the 'Agnatha'.

Another important character in early vertebrate research in Britain is E. Ray Lankester (1847–1928). His father, a prominent physician, was a close friend of Huxley, and he soon became interested in fossils. Following Huxley's indications, E. R. Lankester collected fossil fishes in the Silurian and Devonian of South Wales. At the age of 16 he discovered the first heterostracan scales in 1863, thereby supporting Huxley's claim that *Pteraspis* was a fish, and not a squid as Kner had suggested in 1847. This was the subject of his first paper (Lankester 1864). Later, at the age of 21, he published a masterly review of the 'Cephalaspidae' of the Old Red Sandstone of Britain (Lankester 1868), in which he erected the taxa Osteostraci and Heterostraci that we still use. Lankester then became a famous anatomist (he published on the embryology of molluscs, malaria parasites, blood cells, the fish heart, the brain of monkey, the homology concept, etc.) and finally became Director of the British Museum (Natural History) in London. There, he had a crucial influence on A. Smith Woodward's research on fossil fishes. (The *Catalogue of the fossil fishes in the British Museum* (Woodward 1889–1901) has long served as a basis for the classification of fossil fishes, and still remains one of the most cited works on this subject.)

In 1909, Lankester (now Sir Ray) was the editor of a famous *Treatise on zoology* in which the chapter on vertebrates was written by an outstanding anatomist, Edwin S. Goodrich (Oxford), who described Recent and fossil fishes together, and proposed phylogenies that later strongly influenced the growing generation of early vertebrate palaeontologists. Following Woodward, Goodrich put lungfishes with placoderms. The position he gave to cladistians is interesting and reflects the hesitations of his time: he placed them in a multifurcation with coelacanths, actinopterygians, and other fossil sarcopterygians. Left open was the question of the affinities of the tetrapods, since they branch off from a basal trifurcation of the osteichthyans. Goodrich evaded the question of the affinities of ostracoderms. In his third phylogenetic tree (Goodrich 1909, p. 229), there are, however, two short, unspecified dashed lines arising from the basal teleostomes, which may represent the osteostracans and antiarchs (regarded as 'ostracoderms') respectively, following Lankester's ideas. In the same chapter, however, Goodrich mentions in a footnote that osteostracans may have a median nostril and that 'some authors believe the Cephalaspids to be monorhinal, and allied to the Cyclostomes'. Strangely, no previous author, to my knowledge had written clearly about this, except perhaps the 'Baltic' Russian palaeontologist F. Schmidt when he described the first osteostracans from Saaremaa. I presume that this idea was spreading orally at the time, before it became firmly established by Kiaer (1924) and Stensiö (1927).

With Huxley and Lankester, Palaeozoic fishes began to be treated in a zoological manner, i.e. with regard to their relationships to extant fishes. This tendency was later developed in the United States by Bashford Dean (1895) and his disciple W. K. Gregory, in the context of the growing interest in evolutionary processes. Bashford Dean (1867–1928) was an interesting character in the history of the early vertebrate studies in North America. He was a student of J. S. Newberry (see below) and a remarkable ichthyologist, and he had prophetic insights into the relationships of ostracoderms. He worked chiefly on the structure of the dermal armour of the large arthrodires from the Late Devonian of North America, but was also a world-famous specialist in medieval armour (Fig. 8.5A). After his retirement he became an expert in arms and armour at the Metropolitan Museum of Art and published 'phylogenies' of battle-axes and picks.

Further reading

Goodrich (1909); Huxley (1861, 1880); Lankester (1868–70, 1873); Patterson (1981*b*).

8.1.5 Erik Stensiö and the time of the endoskeleton

After Agassiz, and until early twentieth Century, Palaeozoic fishes were essentially described and classified by means of their exoskeletal characters. Agassiz gave much weight to the aspect of the scales, Traquair to the pattern of the dermal bones of the head, but very few works mentioned internal, endoskeletal structures, even in material that clearly displayed a well-preserved endoskeleton, such as the osteostracans from the Welsh Borderland or Estonia. The Swedish palaeontologist Erik (Anderson) Stensiö (1891–1984; Fig. 8.4) understood as early as the 1910s that the relationship of fossil fishes would be better elucidated through the study of their internal anatomy, and that this goal could be reached only by the techniques of anatomy, i.e. dissections and serial sectioning. First trained as a biologist, he also followed Carl Wiman's lectures in geology and palaeontology at Uppsala University. In 1912, thanks to Wiman's advice and help, he organized with some other students his first expedition to Spitsbergen and there collected the Triassic fishes that later became the basis of his thesis. Stensiö made six visits to Spitsbergen to collect Triassic vertebrates between 1912 and 1918, and travelled also to England (where he met Arthur Smith Woodward, already a world-famous specialist in fossil fishes, and the anatomist E. S. Goodrich), Switzerland, Italy, and Austria (where he met Othenio Abel). His thesis was published in 1921 (with a supplement in 1925). His supervisor, C. Wiman, was amazed by the preparation and interpretation of the internal anatomy of the braincase of these Triassic fishes, which Stensiö here presented for the first time. This amount of information on fossils made him suspicious, and he asked the famous anatomist Nils Holmgren to assess Stensiö's work. Holmgren's enthusiasm for the thesis meant that Stensiö not only got his degree with the highest marks, but also immediately gained international fame. Despite rumours from palaeontologists who accused him of destroying fossils (by dissecting or sectioning them), he received material from various sources. Johannes Kiaer, from Oslo, sent him osteostracans from the Devonian of Spitsbergen (collected by Norwegian expeditions between 1917 and 1920), and the famous German palaeontologist Otto Jaekel Devonian coelacanths from Germany. In 1922, he travelled to North America, where he discovered in museum collections a number of early vertebrates in which the internal cranial anatomy was preserved (e.g. the placoderm *Macropetalichthys*). In Quebec, he visited the site of Miguasha and there purchased one of the best-preserved specimens of the osteolepiform *Eusthenopteron*. In 1923, he became the head of the department of Palaeozoology of the Swedish Museum of Natural History. From this time on, Stensiö was highly active, not only in describing and interpreting the anatomy of early vertebrates, but also in training a large number of Swedish and foreign students. He continued to organize collecting expeditions, e.g. to Podolia and Spitsbergen (in particular the famous English–Norwegian–Swedish expedition of 1939 (Fig. 8.4B), which provided material for the major research centres in Europe) and sent his disciples (e.g. Gunnar Säve-Söderbergh, Erik Jarvik (Fig. 8.4B, 2), and Eigil Nielsen) to Greenland as members of the numerous expeditions organized by the Dane Lauge Koch. From the 1920s to the 1950s, Stensiö's scientific production was characterized by care for anatomical details and a constant reference to the structure of extant fishes. Osteostracan and placoderm anatomy was compared to that of lampreys and extant chondrichthyans respectively. His monograph on the Spitsbergen cephalaspids is certainly a milestone in the history of early vertebrate research, since it contains not only a detailed description of the cranial cavities and canals, but also a well-argued analysis of the anatomy of all fossil jawless fishes known at that time and a first attempt at relating fossil to extant jawless vertebrate taxa on the basis of a number of anatomical characteristics. However, in this work there appeared for the first time a notion that later became the leitmotiv in Stensiö's writings, as well as in those of his disciples: polyphyletism, i.e. the idea that some of the derived characters shared by extant taxa have appeared separately several times. In the 1920s, the extant cyclostomes (hagfishes and lampreys) were unanimously regarded as a natural group, characterized by e.g. a 'rasping tongue'. In contrast, Stensiö (1927) suggested that hagfishes and lampreys were derived independently from (or allied to) fossil taxa (heterostracans and osteostracans respectively) that did not possess these defining characters. Later, Stensiö (1969) also proposed that extant sharks, rays, and chimeras were derived independently from three distinct groups of placoderms: the arthrodires, rhenanids, and ptyctodonts. Following Holmgren (1933, 1949, see also Säve Söderbergh, 1934) he accepted the independent origin of the urodeles and all other tetrapods from two distinct fish groups, an idea that was later largely worked out by his disciple Erik Jarvik, as well as Jarvik's disciple Hans Bjerring.

The success of Stensiö's investigations is largely due to his use of the technique of grinding sections (however destructive it may be) initiated by the British palaeontologist E. Sollas (Oxford), and which allowed him to reconstruct minute and otherwise unavailable structures. But part of his success is also due to the fact that he considered fossil fishes independently of their geological age, and thus with no regard to the evolutionary process.

Although there had been some earlier descriptions of endoskeletal structures of lower vertebrates (e.g. the endoskeleton of *Acanthodes* by Reis (1896) or the brain-case of *Eusthenopteron* by Bryant (1919)), it was Stensiö who pioneered our modern way of looking at these fossils as extant animals, even if we may differ in the way of assessing the characteristics. For this reason, the memorial medal coined in his honour by the Swedish State bears the sentence '*Quae Stensioe perscrutatus est fossilia prope revixerunt*' ('The fossils studied by Stensiö became almost alive'). Outside Sweden, his new methods and aims were soon taken up by some famous palaeontologists of the time; e.g. D. M. S. Watson (1886–1973) in Britain and A. S. Romer (1894–1973) in the USA.

Notwithstanding the great admiration and respect that all early vertebrate palaeontologists owe to Stensiö, many, including some of his own disciples, have followed different paths in the interpretation of the anatomical structures. First of all, Gunnar Säve-Söderbergh (Uppsala), who died prematurely in 1947 but to whom we owe the discovery of the first Devonian tetrapods in Greenland in 1932, developed a somewhat different way of looking at and ordering characters. Like Stensiö, Säve-Söderbergh was an outstanding anatomist and his views on vertebrate phylogeny, expressed as early as 1934, were very close to modern ones in the search for a hierarchic distribution and a classification that reflects sister-group relationships rather than ancestor–descendant relationships. Others, like Gustav Wängsjö (Uppsala), have strongly disagreed with Stensiö on questions of anatomical interpretation. Stensiö's strong personality has certainly played a major role in the promotion of early vertebrate palaeontology in Europe and elsewhere between the 1920s and 1950s, but it may have later become a handicap, for it has to some extent hindered theoretical progress. However friendly and cheerful he may have been, Stensiö in common with many supervisors liked his students to follow his theoretical framework. Interestingly, Stensiö and his disciples in Stockholm remained completely outside the development of 'evolutionary palaeontology' that arose from the Modern Synthesis in the 1940s–60s while these transformational approaches (now rejected by cladists) were the major preoccupation of most other palaeontologists. This originality of the 'Stockholm school' has, paradoxically, been one of the reasons for the leading role played by specialists in fossil fishes in the rise of the cladistic method (see next paragraph).

We should not leave the 'time of the endoskeleton' without mentioning the considerable role played by the introduction of a new technique, acid preparation, in the field of early vertebrate research. Although acetic or formic acid preparation had been used on invertebrates since the nineteenth century, it was developed considerably in the 1960s, in particular at the British Museum (Natural History) (Toombs 1948). The impulse may have been due to the discovery of the Devonian locality of Gogo, in Australia, where fossil fishes are preserved in calcareous concretions and have been extensively prepared by this method in London. The result is that since the 1970s a large number of papers and monographs have been published on acid-prepared Gogo specimens, including two *magna opera*, one by R. S. Miles (1977) on lungfishes and the other by B. G. Gardiner (1984*a*) on actinopterygians. The outstanding quality of the Gogo material has provided a quantity of information that certainly equals Stensiö's early works. The acid preparation technique has rapidly been taken up by most workers on early vertebrate, except in Stockholm, where Stensiö and Jarvik have always been somewhat suspicious about its reliability (Ørvig, however, used this method for preparing isolated scales and even placoderm skulls). Grinding sections do in fact provide information (on internal structures of on the organization of discrete, articulated elements) that is complementary to that provided by acid preparation. Acid preparation has even been used with success on Stensiö's favourite material, the Spitsbergen cephalaspids, thereby corroborating most of his early descriptions based on grinding sections (Janvier 1985*a*).

Special mention should also be made on the development of palaeohistology, i.e. the study of the structure of hard tissues, as an important part of this trend towards considering early vertebrates as extant animals. Although founded by C. Pander (see above), modern palaeohistology was developed by the German palaeontologist Walter Gross (1903–74, Fig. 8.3) who in turn trained Hans-Peter Schultze. Tor Ørvig (1917–94), a student of Stensiö, has also provided a considerable body of data and theory on the evolution of hard tissues, having benefitted in the 1970s from the development of scanning electron microscopy. Ørvig's palaeohistological studies are very much biologically oriented and provide a theoretical framework for the evolution of hard tissues, whereas Gross's work is more descriptive and aimed at defining histological characters.

Further reading
Jarvik (1992); Stensiö (1921, 1925, 1927); Schultze (1974).

8.1.6 Gareth Nelson and the time of cladistics

In the 1960s, the Swedish Museum of Natural History had a pleasant cafeteria called 'Kemikum' where staff members and visitors used to meet for lunch. Among the *habitués* of this cafeteria were Stensiö, Jarvik, a vertebrate zoologist, Alf Johnels, and an entomologist Lars Brundin. There, the young visitors who were doing postdoctoral work in the palaeozoological department were often stimulated by Lars Brundin's comments (with

touches of irony) on the methods and concepts used by palaeontologists for the reconstruction of evolutionary trees. He used to mention the name of one of his friends, Willi Hennig, who was virtually unknown at that time outside entomological circles. In 1966, a young American, Gareth Nelson met Brundin in Stockholm and became acquainted with the methods of phylogenetic reconstruction and classification that Hennig called 'phylogenetic systematics', which is now known as cladistics (see Chapter 2, p. 37). Nelson immediately applied Hennig's principles to his study of the gill arches (Nelson 1969, 1970) and developed further this theory of systematics. The same happened to several other young visitors, such as the Briton Roger Miles, the Frenchman Daniel Goujet and the German Hans-Peter Schultze. In turn, these people have spread in their own countries what they regarded as a revolutionary method of assessing character resemblances and relationships. A common feature of these early cladists was that they were more interested in anatomical structures and systematics than in evolutionary processes, hence their ties with the 'Stockholm school'. A prophet is without honour in his own country! Brundin's irony had virtually no effect on the Stockholm palaeozoologists, although Jarvik (1968) has published a paper in which he admits that 'the sister-group concept recently introduced by Hennig... is certainly a useful tool for this purpose [phylogeny reconstruction], because it enforces a phylogenetic way of thinking and careful consideration of the evidence, but as far as the vertebrates are concerned it is at present not so easy to handle'. However, he rightly pointed out that Säve-Söderbergh's (1934) vertebrate classification was perfectly concordant with Hennig's principles, i.e., that classifications must reflect phylogeny, and had already overturned many of the classical evolutionary taxa. But the Paris meeting on 'vertebrate evolution' in 1973 dealt the *coup de grace* to Hennig's method for the 'Stockholmers' (Jarvik, Bjerring, and also, at that time, the present writer). At the meeting, two of the young researchers 'contaminated' by Brundin, namely Roger Miles and Niels Bonde (Copenhagen), presented communications that, on the basis of Hennig's method, scoured not only the theoretical basis of the celebrated results of Stensiö and his disciples, but also that of the current evolutionary palaeontology. Miles's communication was on the position of lungfishes (which he put with 'crossopterygians'); Bonde's was on the gnathostome interrelationships and tetrapod monophyly, but also more generally on methods, tests, and schools in science (Miles 1975; Bonde 1975). After this meeting, Jarvik used to refer to cladistics as a 'shameful desease' (because of its rapid spread!) or more poetically as the 'King's new dress' (i.e. the Emperor's New Clothes)

(Jarvik 1981*b*). One can easily understand Jarvik's attitude, because Stensiö, Säve-Söderbergh, and he were already familiar with some of Hennig's principles, i.e. that resemblance may be due to shared primitiveness (this is how Jarvik assesses the resemblance between porolepiforms and osteolepiforms, for example) and that a classification must reflect phylogeny (which is why he classifies porolepiforms and urodeles as 'urodelomorphs', for example). The problem rests, however, on the use of the principle of parsimony, and this difficulty is not unique to the 'Stockholmers', but is widespread among evolutionary palaeontologists. The idea that one can 'play' with the theories of relationships, and overturn them as new contradictory data appear is for many difficult to accept. This results in a coagulation of the phylogenies, and an *ad absurdum* appeal to convergencies in order to cope with the contradictory data. This does not mean that cladistic theories of relationships are necessarily true (many are certainly based on fanciful data), but they are open and, as such, transitory. The constant game of 'conjecture and refutation' on every single characteristic (the saga of the choanae is a good example; see Chapter 6, p. 264) paves the way for better and better corroborated phylogenies. Most comparative biologists have, however, some difficulty in understanding that a good theory is to some extent a refuted theory, because it has proved to be refutable.

Another famous opponent of cladistics among early vertebrate specialists was Berverley Halstead (1979). His arguments were different from Jarvik's, and concerned the rejection of paraphyletic taxa (grades), a kind of grouping that he regarded as essential for making palaeontology a support for evolutionary theories. Panchen (1992) rejects cladistics on other grounds, namely abuse of parsimony and the empirical observation of mosaic evolution. He criticizes cladograms for often being asymmetrical (the 'Hennigian comb'), and thus in no way different from a 'Tree of Porphyry', a chain of beings, a *'Stufenreihe'* or orthogenetic series, and thus contradictory to what we know of 'mosaic evolution'. Cladograms are in fact Hennigian combs only when, for practical reasons, terminal taxa are not detailed.

Following the pioneering works of G. Nelson and R. Miles, major contributions to the use of cladistics in early vertebrate studies came from the Department of Palaeontology of the British Museum (Natural History). There was in particular a very important paper by Colin Patterson (1975) on the significance of fossil taxa in teleost phylogeny (see also Patterson 1977*b*, Patterson and Rosen 1977). Here, for the first time, the question of how to rate fossil data was considered in detail.

Cladistics is now widely used among early vertebrate specialists, and I doubt whether it can any longer be regarded as a mere 'mode'; its power of generating discoveries in both biology and palaeontology has been amply demonstrated. Nor is it a question of generation (the 'young' against the 'elderly'), since many of the earliest users of cladistics in palaeontology, such as Bobb Schaeffer in the USA or Robert Hoffstetter in France, were not particularly young when they grasped this method. Even if some palaeontologists are still reluctant to use it, biologists know that they must do so, because of the large amount of data they have to cope with.

Further reading

Farris and Platnick (1989); Hull (1988, 1989); Jarvik (1968, 1981*b*); Nelson (1969); Bonde (1975).

8.1.7 Present-day centres and perspectives in early vertebrate studies

The 'scientific radiation' from Stockholm between the 1920s and 1960s was considerable, and many researchers who had spent a few months or a few years in the department of palaeozoology at the Swedish Museum of Natural History later developed centres, or at least, teams of early vertebrate or 'lower' vertebrate specialists. Among them are Eigil Nielsen in Copenhagen, Jean-Pierre Lehman in Paris, Mee-Mann Chang in Beijing, Gareth Nelson in New York, Roger Miles in London, Hans-Peter Schultze in Lawrence, Kansas, (and now Berlin) and Elga Mark-Kurik in Tallinn. Britain and the USA are special cases, because work on early vertebrates had been in progress in these countries since the nineteenth century. The universities of Newcastle and Oxford and the British Museum (Natural History) were already autonomous centres of research, with such famous scientists as T. S. Westoll, J. A. Moy-Thomas (Fig. 8.4B, 1) and E. I. White respectively, when Stensiö was at the summit of his career. In the USA, Bashford Dean (Fig. 8.5A) in New York had initiated the study of Devonian fishes and has been followed by a series of outstanding early vertebrate palaeontologists, such as his student W. K. Gregory and Gregory's student B. Schaeffer (Fig. 8.5B). A. S. Romer (Harvard) was to some extent a rival of Stensiö in his early career, but then worked mainly on early amniotes. Noteworthy also is the role of immigrants from various European countries, such as R. Zangerl, J. Zidek, J. Maisey, D. Elliott, and H. P. Schultze, who all contributed greatly to the development of early vertebrate palaeontology in the USA.

From Britain, some early vertebrate students migrated to Australia in the 1970s, and Australian palaeontologists were later trained by Roger Miles in London. This resulted in the formation of yet another community in the antipodes (Ken Campbell, Alex Ritchie, Susan Turner, Gavin Young, and John Long, to date). These brilliant scientists in a country that was virtually unexplored (so far as early verte-

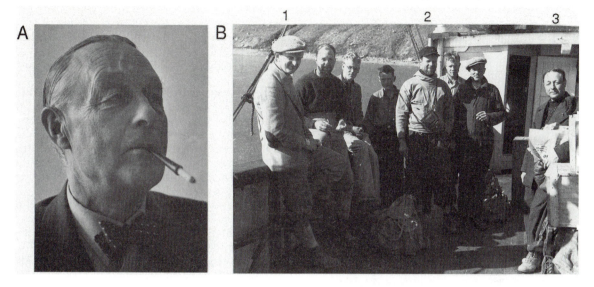

Fig. 8.4. Erik Stensiö (1891–1984), the founder of modern early vertebrate palaeontology. A, with his favourite working tool, the cigarette; B, on board the *Heimen* during the 1939 expedition to Spitsbergen (3, together with J. A. Moy-Thomas (1) and E. Jarvik (2).) (A, courtesy of the Naturhistoriska Riksmuséet, Stockholm; B, photo E. I. White).

brates are concerned) have produced an amazing body of results, both factual and theoretical. When considering the equally rich results produced by the group of Chinese palaeontologists trained by Mee-Mann Chang in Beijing, one realizes that, like the business world, early vertebrate palaeontology is now developing along the Pacific coast. Only the Californians missed the opportunity of following this trend when the rich Silurian and Devonian fish localities of Nevada were discovered in the 1970s.

Present trends indicate that the most promising developments in early vertebrate palaeontology will remain in the field of palaeoanatomy and phylogeny, i.e. in the study of early vertebrates as organisms, although progress in this field ultimately depends on field investigations to provide more new material. Closer and closer links with neontology, and in particular with modern developmental biology and ultrastructural histology, will also be particularly beneficial to early vertebrate palaeontology. On the other hand, I do not think that, with present-day techniques, fossil molecules extracted from early vertebrates will provide anything more than anecdotal data. Isotopic studies on diet and environment may possibly be an exception here.

8.2 EXPLORATIONS, GEOLOGY AND EARLY VERTEBRATES

What we know now of early vertebrates is also the ultimate result of field explorations in Palaeozoic rocks

all over the world (Fig. 8.6). We have seen above how the early discoveries of Devonian fishes in Scotland and the Baltic were triggered by the works of Agassiz. In the case of Scotland and Wales, one may also invoke the role of the rising wealth of the industrial society. Many quarries were opened to build houses and railway cuttings were made, which yielded a large number of fossils (Rudwick 1985). It was probably also rising prosperity that accounted for the discovery of the rich placoderm faunas of the Cleveland Shale in North America. Long-distance explorations were initiated by de Verneuil, Sedgwick, and Murchison in the Baltic and Russia in the 1830s, and the remains of Palaeozoic vertebrates were eventually collected by geologists during expeditions at the end of the nineteenth century, such as A. E. Nordenskjöld in Spitsbergen (1868), Nathorst in Greenland (1897, during the expedition aimed at rescuing Andrée), and Katzer in Brazil (1894). Major expeditions aimed essentially at finding early vertebrates did not take place before Stensiö's first expedition to Spitsbergen in 1912, with perhaps the exception of J. F. Whiteave's collecting expeditions to Miguasha in Quebec (1879). Most of the new localities have been discovered since the Second World War, in the wake of large-scale oil exploration and mapping programmes (Fig. 8.6) which, unfortunately, have since slowed down.

The use of early vertebrates in biostratigraphy was long neglected because they were regarded as too rare and had been worked out, mainly by anatomists. Gross (1951*a*)

was among the first to point out the fact that Palaeozoic fishes can provide rather accurate stratigraphical correlations. This has been largely confirmed by a number of works published since the 1970s (for review, see Long 1993a).

The biogeographical significance of early vertebrates was also neglected or, at best, seen in the light of Buffon's Law, i.e. that the northern hemisphere forms from the 'Old Red Sandstone continent' were supposed to have dispersed toward the southern hemisphere. In the 1970s, the rebirth of the continental drift theory, in the form of plate tectonics, aroused some interest among Palaeozoic fish students. The then British palaeontologist Susan Turner was the first to interpret Siluro–Devonian fish distribution in the light of the new mobilistic palaeogeographical reconstructions. She was probably influenced by the geophysicist Don Tarling (Turner 1970; Turner and Tarling 1982). Here again, the interpretation was essentially based on migrations from one landmass to another, along marine routes. G. C. Young (1981, 1982b, 1986a, 1990, 1993b) is among the first who considered early vertebrate biogeography in the light of vicariance biogeography (see Chapter 2, p. 40).

8.2.1 The Arctic

During the first half of the twentieth century, a number of important collecting expeditions were made to the Arctic by the Norwegians, Swedes, and Danes. The results of these expeditions have been crucial for modern early vertebrate palaeontology. They provided the basic material that was worked on by Stensiö, Jarvik, Wängsjö, and Säve-Söderbergh in Sweden and by J. Kiaer and A. Heintz in Norway. One can say that, after Britain, Spitsbergen and Greenland furnished the major factual basis for the early theories on structure and interrelationships of Devonian fishes, with some contributions of material from the USA (e.g. *Macropetalichthys*), Canada (e.g. *Eusthenopteron*), Germany (the Wildungen coelacanths and placoderms), and Podolia (e.g. *Kujda-*

Fig. 8.5. A, Bashford Dean (1867–1928) was a famous specialist in Devonian armoured fishes and chondrichthyans, as well as a world-famous expert in medieval armour, which he liked to wear himself; B, Bobb Schaeffer (here seen in the American Museum of Natural History in 1955) is a leading figure in North America in the field of early vertebrates. His contributions have been mainly on early elasmobranchs, coelacanths, and actinopterygians, but he has also worked on the interrelationships of the major vertebrate and chordate taxa. (A, reproduced with the permission of the Metropolitan Museum of Art, New York; B, from Colbert and Gaffney (1984).)

nowiaspis and heterostracans). Why was the Arctic so important to early vertebrate palaeontologists? Here again, we find a conjunction of circumstances: the need for the various Scandinavian countries to establish their sovereignty over Spitsbergen and Greenland (the Norwegians and Swedes in Spitsbergen, and the Danes in Greenland), the fact that Palaeozoic rocks, in particular the Old Red Sandstone facies of the Devonian, are widely exposed in these regions, because of the Quaternary glaciation, and the fact that Stensiö was Swedish and thus could either become a member of these expeditions himself or make his disciples participate in them. Had Stensiö been French, for example, he would no doubt have made collecting expeditions in the Devonian of North Africa or Indochina. One can imagine that he would have been excited by the first description of Devonian fishes from Tonkin by Mansuy (1915) and that he would have unravelled the anatomy of galeaspids or *Youngolepis* before that of osteostracans or *Eusthenopteron*, with perhaps a totally different outcome for theoretical vertebrate phylogeny. Another important circumstance was that many of the best lower vertebrate anatomists and embryologists at this time worked in Sweden (e.g. Nils Holmgren) and stimulated Stensiö's interest in theoretical issues on fossil fish anatomy and relationships.

The Arctic remains a favourite source of early vertebrate material, as is exemplified by the results of the international expeditions to Spitsbergen in 1939 and 1969 and, more recently, of the Canadian expeditions to the North-West Territories of Canada (Ellesmere Land, Mackenzie, etc.) as well as the later Anglo-Danish collecting expeditions (1987) to the Devonian tetrapod localities of Greenland.

Further reading

Jarvik (1969, 1992); introduction in Dineley and Loeffler (1976); introduction in Stensiö (1927).

8.2.2 Russia and Siberia

Western Russia has, like Great Britain, been a cradle of early vertebrate palaeontology, but it was not before the 1920s that substantial Palaeozoic fish material was recorded from Siberia, although some sporadic discoveries had been made in the Minusinsk Basin as early as 1876 (see review by Obruchev 1941). Most of these discoveries were made by field geologists. When D. V. Obruchev began his work for the Geological Committee of the USSR in 1926, he had to identify a large number of these scraps, and he generally did this by reference to the fauna he knew best, that of the Main Devonian Field of the Russian Platform. His brother, the geologist Sergei Obruchev, who studied the Siberian

Platform in the 1920s and 1930s, certainly played a role in this record of Palaeozoic fishes. Because of the difficulty of gaining access to the Siberian localities, most of this material was represented by small collections, but it was clearly quite different from what was known from Europe and the western Arctic (e.g. the tannuaspid osteostracans from Tuva, or the amphiaspid heterostracans from north-western Siberia). Later, when the field conditions became more comfortable, D. V. Obruchev and his Russian students (e.g. E. I. Vorobjeva and L. Novitskaya) were able to make extensive collections, which resulted, for example, in Novitskaya's (1971) monograph on the Siberian amphiaspids. In the 1970s and subsequently, some of Obruchev's former students (e.g. E. Mark-Kurik and V. Karatayute-Talimaa, now Estonian and Lithuanian respectively), participated in a number of geological expeditions to the Siberian Arctic (e.g. Severnaya Zemlya) and Central Asia (Tuva, Mongolia), and collected there an amazing variety of Ordovician to Devonian vertebrates, including the earliest known gnathostomes (Early Silurian chondrichthyans).

Further reading

Blieck and Janvier (1993); Obruchev (1941).

8.2.3 China and Indochina

When exploring the Devonian of South China and Tonkin, the French geologist Jacques Deprat discovered some fish remains, which Mansuy (1915) correctly referred to as 'ostracoderms' and placoderms. We now know that these specimens are the first evidence of galeaspids and of yunnanolepiform and procondylolepiform antiarchs respectively (see Chapter 4, p. 166). These remains went virtually unnoticed (owing to the fact that Deprat was later accused of fraud—wrongly so—and that suspicion was thrown on his findings). The remarkable and largely endemic Silurian and Devonian fish faunas of China consequently remained unknown until the 1960s, when Chinese palaeontologists (Liu Yühai, Pan Jiang, and Chang Mee-Mann) revealed the structure of these new taxa. The discovery of complete galeaspids, in particular, puzzled scientists for some time, and they tried, according to the classical practice, to 'shoehorn' them into previously known jawless fish taxa (Liu 1965, 1975; Janvier 1975*b*). Since then, under the leading influence of Mrs Chang Mee-Mann, early vertebrate palaeontology has been flourishing in China, and has produced key discoveries, such as the primitive lungfish *Diabolepis* and the lungfish-relative *Youngolepis*, which have settled a century-long debate on lungfish affinities. In the 1970s, Deprat's findings were confirmed by the discovery of these 'Chinese' taxa in northern Vietnam, but this material was first submitted to D. V. Obruchev , in accordance with the close ties between Vietnam and the USSR. Obruchev was not aware of the Chinese faunas, because at that time, relations between China and the USSR were cool. He therefore referred this fragmentary Vietnamese material to classical European antiarch and porolepiform taxa (*Asterolepis, Porolepis*), without any reference to galeaspids. (This illustrates how we are prone to assess new taxa in the light of those that are familiar to us.) In consequence, the Vietnamese palaeontologists, who had had no contact with China since the 1970s, discovered the 'Chinese' nature of their Devonian vertebrate faunas only through a collaboration with French vertebrate palaeontologists in the mid-1980s.

Further reading

Liu and P'an (1958); Mansuy (1915); Pan (1992); Pan and Dineley (1988); Tong Dzuy and Janvier (1990).

8.2.4 North America

Until the late 1800s the history of early vertebrate research in North America is more chaotic than in Europe, because of the lack of an early scientific 'leader' comparable to Agassiz, Huxley, or Stensiö. (Although Agassiz lived in the USA from 1848 until his death in 1873, he was at that time no longer interested in Palaeozoic fishes. He planned a review of the fossil fishes of the USA in collaboration with William C. Redfield, but it never materialized.) Only E. D. Cope and B. Dean raised some theoretical questions about material of this kind. Most of the findings were made by field geologists of the geological surveys of the USA and Canada, in particular during the exploration of the Far West. I have already mentioned the early and excellent work by Whiteaves (1881) on the Devonian fishes of Miguasha (Quebec), but the discovery of the earliest known vertebrates by C. D. Walcott (1892) in the Ordovician of Colorado was equally important to theorists of vertebrate palaeontology, since it showed, for the first time, that there were vertebrates before the Silurian. The first Devonian fish findings in North America date back to the 1830s, i.e. to nearly the same time as in Europe. In 1838, J. Sullivan of Columbus, Ohio, found a skull of *Macropetalichthys* in the 'Corniferous limestone' (Late Devonian), and this probably represents the earliest finding in North America. Sporadic findings were subsequently made by amateur collectors and geologists in the north-eastern United States, with the result that in the late 1850s more than fifty species of fossil fishes had been recorded from the Devonian formations alone. Newberry's (1889) famous monograph *The Paleozoic*

fishes of North America sums up the discoveries of Silurian to Carboniferous fishes made in North America during the nineteenth century.

Interestingly, the earliest discoveries of large arthrodires in the Late Devonian Cleveland Shale started in very much the same manner as the early discoveries of Devonian fishes in Scotland, although thirty years later, about 1870. Many specimens from Ohio were discovered by private collectors, such as the Revd Herman Herzer, Jay Terrell (a horticulturalist), Dr William Clark (a physician), C. Fyler, etc. Many of them were quite fanatic about their pet localities and findings. The abundant material they collected along creeks (e.g. Big Creek and Vermilion River), essentially around Cleveland, was then deposited either in the American Museum of Natural History or in the Cleveland Museum. It has been worked on by J. Newberry, his student B. Dean, and Hussakof and Hyde, two students of Dean. (In this century, these collections have been considerably enlarged by the material collected and studied by David H. Dunkle, in whose honour the giant arthrodire *Dunkleosteus* was named.) The difference between these Cleveland Shale collectors and the Scotsmen of the Moray Firth is that, with the exception of A. Wright and B. Branson, his successor in Oberlin College (Ohio), both professional geologists, the former do not seem to have been motivated by scientific or philosophical questions. Most of the other Ohio collectors were motivated by the pleasure of finding the largest arthrodire, or the most complete cladoselachian in the rare concretions that crop out there along creeks. By this time, the work of describing and interpreting fossils was already a matter for the specialist, and there was no place in science for a Malcolmson or a Lady Gordon Cumming.

Large-scale collecting expeditions in the west began in the 1920s. Particularly noteworthy are the discovery of the Devonian osteostracan and heterostracan faunas of Utah and Wyoming (e.g. in field expeditions by W. L. Bryant, E. B. Branson, and R. H. Denison) and, more recently, that of the Carboniferous faunas of Indiana and Illinois (D. Bardack and R. Zangerl) and Bear Gulch (R. Lund) in Montana. The Bear Gulch and Indiana specimens have probably resulted in the greatest advances in our knowledge of Palaeozoic chondrichthyans since Dean's (1909) studies of the cladoselachians. The latest important discoveries of early vertebrate localities in North America have been in the Red Hills in Nevada (Silurian and Devonian fishes) and the North-West Territories of Canada.

Further reading

Denison (1953); Gregory (1930, 1931); Hyde (1926); Ilyes and Elliott (1994); Newberry (1889); Nitecki (1979); Scheele (1965); Yochelson (1983).

8.2.5 Africa

After geological explorations in the 1950s, numerous Devonian fishes, mainly placoderms, have been recorded from North Africa, most of them from Morocco, but others from Algeria and Lybia (Lehman 1956; see also review in Lelièvre *et al.* 1993). The first findings were enormous bone fragments of the giant arthrodire *Titanichthys* (Fig. 1.14) discovered in the Late Devonian of Tafilalt (Morocco) by the geologist Henri Termier, who had great difficulty in convincing his colleagues that these were Devonian fish and not dinosaur bones. Later, Early to Late Devonian arthrodires, petalichthyids, chondrichthyans, onychodonts, osteolepiforms, and actinistians were recorded from these regions, providing excellent acid-prepared material. The war between the former Spanish Sahara and Morocco has long been an obstacle to further field-work in the Tafilalt, but there is now evidence for rich outcrops in southern Lybia as well. Some Devonian fish remains and a rich Early Carboniferous actinopterygian fauna have also been discovered in South Africa (Gardiner 1969; Chaloner *et al.* 1980).

Further reading

Lehman (1956); Lelièvre *et al.* (1993); Anderson *et al.* (1994).

8.2.6 Australia and Antarctica

There were sporadic findings of Palaeozoic fishes in Australia in the nineteenth and early twentieth century (see, e.g. the description of the lungfish *Dipnorhynchus* from Taemas by Etheridge in 1906). Systematic early vertebrate studies were initiated there by the geologist Edwin Sherbon Hills (1906–86), of Melbourne University. During field-work in Victoria, he discovered his first Late Devonian fishes, and became interested in this field. He came to London in 1929 and was trained by D. M. S. Watson and A. Smith Woodward on fossil fishes. Back in Australia, in 1932, Hills began to collect more Devonian fishes all over the continent and showed that they could be fruitfully used in stratigraphy. In the late 1950s, however, he decided to give up his activities in this field and concentrate on geology. He had been sensitive to Stensiö's—ill-founded—criticisms of his interpretation of the braincase of the primitive brachythoracid arthrodire *Buchanosteus*, but he received the *coup de grâce* when he heard that E. I. White of the British Museum (Natural History) had sent his assistant H. Toombs to make extensive collections of Devonian fishes in various Australian localities (Taemas, Buchan, and above all Gogo) (Turner and Long 1989). Consequently, most of the important results provided by

this outstanding material were, until the late 1980s, published by English scientists. This sad start for early vertebrate palaeontology in Australia has since been compensated for by the skill and activity of a number of young scientists, some of them immigrants from Britain, who have collected more material in Australia and Antarctica than was ever obtained by the famous Arctic expeditions led by Europeans. Among their discoveries, noteworthy is Ritchie's and Gilbert-Tomlinson's (1977) finding of Ordovician vertebrates, the arandaspids, in the Northern Territory of Australia, and G. C. Young's (1991*a*) discovery of a hitherto unknown group of Devonian jawless fishes, the pituriaspids.

Devonian fishes from Antarctica were first discovered by Scott's ill-fated Terra Nova expedition in 1910–13 and were described by A. Smith Woodward in 1921. Much later, in 1968–9, more material was found in southern Victoria Land during the New Zealand Transantarctic Expedition (VUWAE 13). The first early vertebrate palaeontologist to 'conquer' this realm was Alex Ritchie (Sydney) in 1970–1. A number of expeditions have since been organized to the Devonian localities of the Transantarctic Mountains and Victoria Land, and Australian palaeontologists (G. C. Young, J. Long) have collected a considerable amount of material, essentially of placoderms and sarcopterygians that are quite similar to the southern Australian forms.

Further reading

Grande and Eastman (1986); Ritchie (1971); Turner and Long (1989); Young (1991*b*, 1993*a*).

8.2.7 The Middle East, India

Devonian fish remains had been recorded by geologists from Turkey and Iran since the nineteenth century, but collecting expeditions were first organized in the 1970s in collaboration with the Turkish and Iranian geological surveys and also in Afghanistan in the frame of the French–Afghan mapping programme. Although no spectacular locality comparable to the Australian ones was found, the material collected was interesting in providing a mix of Euramerican and Gondwanan taxa (see reviews in Lelièvre *et al.* 1993 and Young 1990, 1993*b*). Particularly amazing was the discovery of the Old Red Sandstone (Late Devonian) fish fauna of Antalya, Turkey, on the shore of the Mediterranean, among Greek ruins, with much the same taxa as in Greenland. The Early Devonian placoderm faunas recently discovered in Saudi Arabia are more promising and consist of remarkably preserved acid-prepared material.

Further east, in northern India, some Devonian fish remains were recorded in the 1970s and early 1980s in

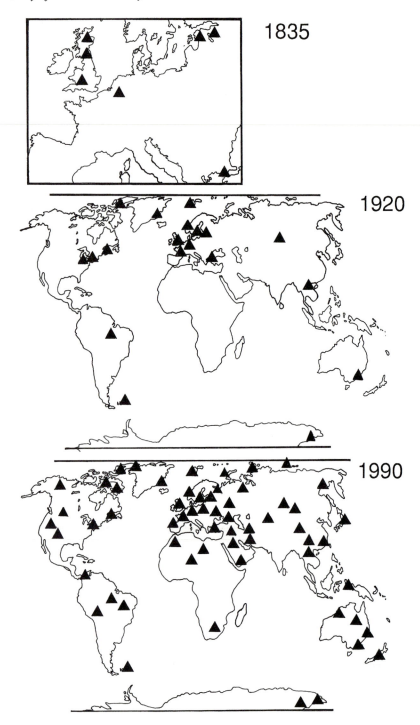

Fig. 8.6. The major Devonian vertebrate localities and occasional occurrences in the world that were known in about 1835, 1920 and 1990. Note the role of the Arctic and Antarctic expeditions in the early twentieth century in the discovery of new localities.

various notes by early vertebrate palaeontologists, in collaboration with the Indian geologist V. J. Gupta, who provided the material. There is now a high probability that this material is apocryphal; some specimens (e.g. *Youngolepis*) are probably from China, and others from the USA (for details see Talent *et al.* 1988). Before doubt was thrown on their origin, these now infamous fish remains aroused some interest among specialists in this field, because they appeared to approach the bounds of the area of endemism of the Chinese Devonian vertebrate faunas, somewhere in the Himalayas.

Further reading

Janvier (1980*a*); Lelièvre *et al.* (1993); Talent *et al.* (1988).

8.2.8 South America

Strangely, some Devonian fish remains were recorded from Brazil and the Falklands as early as the late nineteenth century, but virtually nothing more was found until the 1970s. Field-work in Bolivia (Janvier and Suarez-Riglos 1986) and the Amazon Basin in Brazil (Janvier and Melo 1986) has yielded some peculiar Devonian chondrichthyan remains, associated with acanthodian, actinopterygian, and thelodont scales and a rhenanid placoderm. This material is poor, but it shows special affinities with the Antarctic and Australian material. In contrast, the Ordovician of Bolivia has yielded impressive material of articulated arandaspids (Gagnier 1993*a,b*), yet another resemblance with Australia. The recent discovery of Late Devonian antiarchs and sarcopterygians in the Maracaibo area of Venezuela (Young, personal communication) may also yield new biogeographic information.

Further reading

Gagnier (1993*a*); Janvier and Suarez-Riglos (1986); Lelièvre *et al.* (1993).

8.2.9 New discoveries in Europe

Early vertebrate palaeontologists have long had a tendency to neglect Europe, partly because they considered that nothing more could be discovered there, and partly because it is easier to obtain funding for expeditions to South America than to Belgium or Wales. Since the Second World War, a number of famous Palaeozoic vertebrate localities have, however, been discovered in Germany (Bergisch-Gladbach, Zweifall), France (Montceau-les-Mines), and Scotland (Cowdenbeath, Dora), many of them through amateur fossil collectors (e.g. S. P. Wood in Scotland and P. Bardenheuer in Germany). Sparse vertebrate remains have also turned up in Brittany and Spain. Investigations into old collections or new excavations in century-old localities have also yielded unsuspected taxa, such as a presumed Late Devonian tetrapod from Scat Craig, Scotland (Ahlberg 1991*b*), or the Early Carboniferous amniote from East Kirkton, Scotland (Smithson 1989). Large-scale excavations made by L. Lyarskaya and E. Mark-Kurik (1972) in the Lode quarry, in Latvia, have yielded the first articulated panderichthyids and have thereby played a crucial role in the understanding of tetrapod origin. Ahlberg *et al.* (1994) have pointed out that some of the sarcopterygian remains formerly described as 'fish' from the Famennian of Latvia turn out to belong to *Ventastega*, probably the most primitive tetrapod known at present.

Further reading

Ahlberg (1991*b*); Ahlberg *et al.* (1994); Heyler and Poplin (1994); Lyarskaya and Mark-Kurik (1972); Milner *et al.* (1986); Ørvig (1960); Wood (1982).

9

Epilogue

Writing a book, in which statements should be as unambiguous as possible, always imposes some self-censorship. I have pointed out, here and there, various facts that may contradict the apparently harmonious hierarchy of characters supporting phylogenies and classifications. These are assessed as homoplasies or simply overlooked, but some of them may be the threads that lead us to challenge current phylogenies. The aim of this epilogue is to point out briefly these inconsistencies and suggest new paths. I am aware that many authors in the past tried to orient research in what they believed to be promising directions, but most of them failed to predict major theoretical changes. Novelties generally come from fields in which they are not expected. In the 1960s, palaeontologists believed that the major theoretical advances in palaeontology would come from evolutionary models based on ammonites or mammals, for example, but it was systematic studies of insects and teleost fishes that prompted the rise of cladistics, and thereby deeply modified the way in which palaeontologists as a whole look at fossils. In some instances, however, it is possible to point out unanswered questions and contradictions that may generate profound changes, at least at the level of phylogeny and systematics. It is also possible simply to indicate where data are lacking and how or where they might be sought.

9.1 ON THE LARGE AMOUNT OF DATA

In the introduction to Chapter 5, I mentioned that the now widespread use of cladistic analysis (although not to blame) sometimes makes discussions difficult at the more elementary level. Evolutionary trees, based on the stratigraphical history of a few selected characters (usually from the skeleton), were formerly rather simple, yet often ambiguous, and just as easy to read as old history books in which societies were assessed as 'primitive' when not industrial. Today, most cladistic analyses take into consideration a very large number of characters; even more so when molecular sequence data are added (the so-called

'total evidence'). Some of these characters have several states (multi-state characters) and there may be no means of ordering them. Some of them may be ordered, but display conflicting polarities that imply more steps in the resulting trees. The search for the most parsimonious trees in such cases is beyond the human brain in a reasonable amount of time, and it entails the use of cladistic computer programs. (The performance of these programs will no doubt improve in years to come and they will provide more extensive and more rapid search for the shortest trees.) Most discussions about phylogenies, other than in highly specialized papers, now therefore bear on the comparative length, consistency, and retention indexes or robustness of the trees, rather than on the 'history' of each character throughout the trees (see, for example, Forey *et al.* 1991; Schultze and Marshall 1993; Schultze 1994).

In the field of vertebrate phylogeny, for example, the volume of anatomical data is potentially immense and is certainly under-exploited. The enormous body of data produced by anatomists and histologists over the past hundred years is generally reliable but needs to be reassessed. Many structures were in fact described with a certain evolutionary aim (e.g., the glottis and 'epiglottis' of lungfishes, described by some as a precursor to the glottis and epiglottis of tetrapods, and by others as an homoplasy) and, in this respect, they are suspect. Checking and re-checking characters is a necessity but it is extremely time-consuming and poorly rewarding in a scientific career. There is nevertheless no instance where such work did not provide either much new data or completely new views about character-state polarities. However detailed and reliable may be the description of an anatomical structure, there will always be some ambiguous interpretations, and most phylogeneticists then rely on the power of statistics. Given a data matrix of, say, 150 characters, they bet that most of the characters are correctly observed (i.e. they are actually there). Molecular phylogeneticists even have methods of evaluating the percentage of mistaken data, which, they say, is generally negligible. Nelson (1994), for example, considers that 'there might exist a correlation between the care in collecting and organizing data and either index (consistency or retention), but there is no evidence to support such a claim, however intuitively reasonable it might appear, or however hopefully one may wish that it were true' and that 'recent progress in systematics stems not from better data but from better methods and more data of roughly the same quality.' When dealing with fossils, however, the number of available data is significantly smaller than for extant organisms, and missing data in a data matrix are optimized according to the most parsimonious distribution of the available data. This process of optimization is sometimes very sensitive to the assessment or to slight mistakes in the record of the available data; this implies that increased care is needed in the construction of the data matrix. However numerous are the data, one should never be afraid of or impressed by them. There are always means of checking them point for point. In contrast, the theoretical basis of their assessment (arguments of polarity, for example) should be constantly questioned.

Future textbooks on comparative biology will perhaps leave no place for the description of characters, but will instead contain series of data matrices, taken from various authors, for the reader to assess with his portable computer.

9.2 ON SUCCESS IN PHYLOGENETICS

The words 'advance' or 'progress' are commonly used to describe new steps in the history of a discipline. These terms, however, are generally understood as synonyms of 'success'. In the particular field of phylogenetic reconstruction, the notion of success can be viewed in many different ways. For cladists in general, a fully resolved tree or a shorter tree is viewed as a success. For palaeontologists, a success is when fossils meet their expectations or fit the stratigraphical order. In many cases, corroboration of theories of relationships based on different data sets (morphological, physiological, or molecular character) is also viewed as a success. But remember that success proves nothing; only refutation generates some progress in the form of eliminated solutions. In a sense, a good theory is a dead theory, because it has been refuted and thus has proved to have been accessible to refutation. The fact that phylogenies based on morphological data are roughly consistent with some pre-evolutionary classification, and, in turn, that molecular phylogenies are roughly consistent with morphology may show that, since the eighteenth century (and even perhaps since Aristotle), comparative biologists have come quite close to the 'true phylogeny' or the true history of life, but it may also

mean that we all assess characters in the same way, and thus make the same mistakes (in particular by using privative characters to define taxa). This 'success' in phylogeny reconstruction sometimes has the effect of congealing theories or orienting future investigations. If we consider only the example of craniate phylogeny, there have been some attempts at overturning a number of widely accepted relationships, often rooted in the early evolutionary systematics. Each time this was done, it was regarded as a disturbance, rather than a 'success'. The debates about the paraphyly of the cyclostomes, the plesiomorphy of *Eusthenopteron*, and bird–mammal relationships (Løvtrup 1977, 1985; Rosen *et al.* 1981; Gardiner 1982, 1993) illustrate this difficulty in the endeavour of refuting the consensus, whether it is cladistics-based or not. This stems perhaps from the confusion between incongruence and ignorance. Before a theory is clearly refuted, there is a period of time when the challenging theory is not robust enough, and the scientists supporting the former theory fear that this equal-weight contradiction will lead to ignorance and chaos. They therefore prefer—sometimes unconsciously—to stick to a theory that has been supported by time and tradition (in particular when palaeontological arguments are challenged). In addition, many biologists and palaeontologists consider that taxa are real and are defined by their ascent rather than by their characters. To them, it is harder to abandon a theory about ascent (meaning that one can make mistakes about evolution and ancestry) than a theory about homology. This attitude may be due to the fact that scientists consider themselves, in relation to the public, as the guardians of the history of life. Changing their mind too often would weaken their credibility.

This is yet another thing one should not be afraid of in comparative biology. Most of the branchings of the currently accepted consensus tree of the craniates (Figs 3.25, 4.1) hold only by the virtue of a few homologies. If this were regarded as a success, then there would be little reason to continue with research in this field. The phylogeneticists ought not to be afraid of revising phylogenetic trees, perhaps drastically.

9.3 ON 'ODD PHYLOGENIES'

During the past hundred years, only a few comparative biologists and palaeontologists have proposed phylogenies that are strongly at odds with the majority, i.e. what people call the consensus (essentially the tree derived from Haeckel's). First came the Swedish anatomists and palaeontologists, chiefly Holmgren, Säve-Söderbergh, Stensiö, Jarvik, and Bjerring (see Chapter 8), and more recently the biologist Løvtrup (1977). The first five highlighted a number of characters that were inconsistent with the currently accepted theories, and from this derived a theory of interrelationships of craniates that emphasized homoplasies on what were considered to be the most widely accepted taxon-defining characters (gills, limbs, etc.). Løvtrup emphasized biochemical characters (not necessarily molecular sequences), on the ground that morphological characters were unreliable, and too much involved in epigenesis. Needless to say, these phylogenies were criticized at their time of publication. Strangely, when odd (in the sense of unconventional or unexpected) molecular sequence-based phylogenies arrived on the scene, in the late 1980s, the criticisms from morphologist were less severe, as if molecules had more power than morphology or, conversely, as if morphologist considered that molecular phylogenies were so odd that they had no future (see, for example, Marshall and Schultze 1992). Whatever the phylogenies they provide, molecular sequence data represent new kinds of characters and are *a priori* just as good or as bad as dermal bone or sensory-line patterns, tooth histology, or nerve foramina. The fact that the molecular phylogeny of the gnathostomes proposed by Lê *et al.* (1993) shows the tetrapods as polyphyletic will probably satisfy some palaeontologists; others will attribute this odd result to biases in the method (as admitted by these authors), but the result exists and cannot be flatly ignored. Another example of 'odd phylogeny', at the level of tetrapod interrelationships, is Gardiner's (1982, 1993) resurrection, essentially on the basis of soft anatomy, of the theory that birds and mammals are sister-groups (the Haemothermia). There, the criticisms came mainly from palaeontologists (e.g. Gauthier *et al.* 1988), since Gardiner's theory would have forced them to question seriously a number of century-old interpretations and inferences that are now regarded as facts, i.e. archosaurian characters (see Gauthier *et al.* 1988). The phylogeny of sarcopterygians put forward by Rosen *et al.* (1981) was 'odd' in the sense that it assessed fossils quite differently from the way they were assessed by palaeontologists of this period, in particular by rejecting some osteolepiform-tetrapod characters as osteichthyan symplesiomorphies. We have seen that this phylogeny was fruitful in forcing palaeontologists to become conscious of their own idiosyncrasies. Such 'odd phylogenies', as long as they are supported by homologies (often by new kinds of characters), wake up our minds and, in this sense, are really worth putting forward. The evolutionary trees (put in the form of cladograms by Schultze and Trueb (1981) produced by the 'Stockholm school' of early vertebrate palaeontologists were largely derived from Holmgren's and Stensiö's views, which account for a widespread polyphyletism. Many of these proposed instances of polyphyletism were inferred from

the morphological divergence between a generalized taxon, on the one hand, and a much derived one on the other (e.g. sharks and rays or holocephalans, urodeles, and other tetrapods, etc.). However, the enormous volume of data, both anatomical and embryological, used to protect these theories against refutation has revealed details that would otherwise have gone unnoticed. It is important to note again that these theories were not phylogenies in the present-day sense, but rather inferences based on the transformation of omnipotent archetypes. Løvtrup's phylogeny of metazoans, and vertebrates in particular, is constructed cladistically, but its originality is that it takes into consideration—probably for the first time—a number of physiological and biochemical characters that had been ignored by morphologists, and are useless to palaeontologists. It nevertheless showed that such neontological characters alone can help in resolving relationships that morphology and fossils could not resolve. For example, it had the merit of raising the question of the paraphyly of the cyclostomes, and by doing so, prompted further anatomical research which has largely corroborated this theory.

In Chapters 4 and 5 of this book, I have alluded to a number of characters that make one suspicious of some segments of the currently accepted phylogeny of the vertebrates. Let us now return briefly to these questions.

As mentioned in Chapter 6, many anatomists have been struck by the fact that jawed vertebrates share a clearer segmental pattern of the head than hagfishes and lampreys. This is why, for example, Jarvik (1981*a*) has always used *Squalus*, *Amia*, or the fossil *Eusthenopteron* as a basis for what I here call 'anatomical philosophy', i.e. imaginary constructs relating to transformations of characters. All attempts at finding a more extensively segmented pattern in the head of jawless craniates (see, e.g. Stensiö 1927 on the premandibular arch) have failed, with perhaps the exception of two trigeminal ganglia in hagfishes and lampreys, and the fully segmented parasomitic mesoderm (branchiomery), which roughly matches the somitic segmentation in the lamprey embryo. Even more puzzling is the fact that the braincase of osteichthyans displays more evidence of segmentation (in particular in the otico-occipital region) than that of chondrichthyans and placoderms. In other words, and if we believe that head segmentation is a general craniate feature (see discussion in Chapter 6), this aspect of vertebrate anatomy runs counter to the currently admitted phylogenies; hence Jarvik's and Bjerring's rejection of the latter. (Jarvik's 1980 book on the major vertebrate taxa deals first with osteichthyans, then chondrichthyans, and finally jawless vertebrates, which he clearly considers as much derived.) Does this mean that jawed vertebrates or, at any rate, osteichthyans cannot be proved to be monophyletic? Considering the currently accepted defining characters, this would be most unparsimonious. I doubt whether one can find any evidence for the paraphyly of the gnathostomes, i.e. that hagfishes or lampreys can be derived from jawed vertebrates (as was believed until the mid-nineteenth century). I similarly doubt whether all jawless craniates can be more closely related to cephalochordates than to jawed vertebrates, as suggested by Bjerring (1984). In contrast, it is not ruled out that some major gnathostome groups, when they include fossils, may turn out to be paraphyletic.

These impressions are based essentially on data from various fossil taxa, perhaps only because of their incompleteness. For example, I feel that there is something strange with the so-called Silurian 'chondrichthyans' (see Chapter 4, section 13), which are now known by many scales but no teeth at all. Are they really chondrichthyans, or stem-group gnathostomes? Another question is the status of placoderms. In this book (Chapter 4, section 14) I have followed the consensus that they are a clade and the sister-group either of chondrichthyans or all other gnathostomes (Fig. 5.2). But is their monophyly well supported? The bipartite dermal armour with a median dorsal plate and the presence of semidentine in teeth and tubercles (not found in all of them) are meagre evidence for monophyly. Suppose now that placoderms are an ensemble of generalized, toothless gnathostomes (as they were sometimes placed in evolutionary trees, on the grounds of their early geological age). Their osteichthyan-like characters (large dermal plates, large macrosquamose scales) would then turn out to be generalized gnathostome characters, lost in chondrichthyans and acanthodians. Yet another questionable taxon is the acanthodians, also regarded here as a clade on the basis of the spines in front of the paired fins and their unique scale structure. There are, however, some peculiar acanthodian taxa which may lead us to reconsider the monophyly of this group. One of these is *Machairacanthus*, whose scales do not have the typical onion-like structure and thereby recall the mesosquamose scales of some Palaeozoic sharks. Moreover, no symmetrical spine of *Machairacanthus* has ever been found; it may therefore have possessed spines in front of the paired fins only.

Questions can also be raised about the supposedly actinopterygian remains from the Silurian of the Baltic and Sweden, *Andreolepis* and *Lophosteus*. The latter displays an ornamentation of the dermal bones which recalls that of some acanthothoracid placoderms and it seems to possess fin spines, yet it is uncertain whether they are paired or unpaired fin spines (Otto 1993). Moreover, it has groove-like sensory-line canals, like some placoderms, and its teeth, if correctly identified, recall those of onychodontiform sarcopterygians (Gross 1971b). This

may well be explained by the fact that we are there dealing with scraps from different animals, but there remains a slight possibility that *Lophosteus* displays an assemblage of characters unknown in post-Silurian taxa, which might overturn theories about the monophyly of some of them, in particular placoderms and acanthodians.

Finally, onychodontiforms, currently regarded as a clade of rather generalized sarcopterygians, may well not be a clade. They are known by two distinct types, the large *Onychodus* and *Grossius* type on the one hand, and the small *Strunius* type on the other. Both are put in the same taxon on the basis of their large sigmoid and striated parasymphysial teeth and large premaxilla. Outside sarcopterygians, large parasymphysial teeth or tooth spirals are also known in some acanthodians. They may thus be a generalized osteichthyan character. We know very little of the internal anatomy of onychodontiforms, but the paired fins of *Onychodus*, at least, are said to be monobasal, the condition in *Strunius* being unknown. One should thus not rule out the possibility that onychodontiformes are in fact an assemblage of generalized sarcopterygians, but not a clade.

On a smaller scale, there are several other gnathostome taxa that can be considered as poorly supported or probably misplaced. The relationships of all non-euselachian chondrichthyan taxa deserve a thorough research programme, despite the attempts made by Zangerl (1981), Maisey (1984a), and Gaudin (1991). The same applies to the relationships of all non-neopterygian fossil actinopterygian taxa (see Gardiner and Schaeffer 1989). The position of the rhizodontiform sarcopterygians remains poorly supported as long as the question of the structure of their nostrils is not clearly settled. The characters they share only with tristichopterid osteolepiforms (e.g. *Eusthenopteron*) have to be more critically assessed, and should not be regarded as homoplasies on the ground that they possibly posses two external nostrils (and thus no choanae). The few characters that support the monophyly of the osteolepiforms, choanates (osteolepiforms, panderichthyids, and tetrapods), and even tetrapodomorphs (rhizodontiforms and choanates) may turn out to be general rhipidistian characters when more is known of forms like *Youngolepis* and *Kenichthys*. *Youngolepis*, for example, displays a puzzling assemblage of characters found elsewhere only in choanates (dermintermedial process, polyplocodont teeth) or dipnomorph (cosmine histology, rostral tubules). Osteolepiforms may have the same fate as 'leptolepids' among teleosts.

As to fossil jawless craniates, the phylogeny is even more unstable. Gagnier's recent proposal (Gagnier 1993b; see also Forey and Janvier 1994) that 'ostracoderms' (i.e. all bony jawless craniates) are more closely related to the gnathostomes than to any of the two extant groups (lampreys and hagfishes) sounds, after all, more parsimonious than the phylogeny proposed here (Fig. 5.1) and avoids a number of *ad hoc* assumptions such as reversals in lampreys (exoskeleton, sensory-line canals, paired fins, etc.). I am still confident about the monophyly of arandaspids, heterostracans, osteostracans, galeaspids and probably also pituriaspids. All other higher taxa (i.e. astraspids, anaspids in the broad sense, and thelodonts) may be non-monophyletic.

Figure 9.1 shows an 'odd' tree of the craniates which may be constructed from basically the same data that have been used for the other trees in this book, with only a slight loss of parsimony, by considering some characters (annular cartilage, fin spines, dermintermedial process, etc.) as having a higher degree of generality (with a different assessment of missing data or with a different interpretation of controversial characters, such as the external nostrils and choanae of osteolepiforms). This tree, which should not be taken too seriously, is not aimed at contradicting most of what I have written in this book, but just to show that there may be radically different ways of assessing the few, sometimes ambiguous characters provided by fossils.

9.4. ON NEW KINDS OF DATA, NEW TECHNIQUES, NEW METHODS

As we have seen in Chapter 8, the gain of new data in early vertebrate palaeontology has occurred stepwise, and has been advanced by new techniques of preparation and observation, as well as by the discovery of better material. Both conditions have always led towards better-resolved phylogenies. So far as techniques are concerned, an increasingly large volume of data has been obtained by means of grinding and slicing, acid preparation, and the scanning electron microscope. I see the use of computer tomography as a promising means of collecting data from a certain type of fossil, i.e. three-dimensionally preserved specimens with calcified endoskeleton, perhaps not so much because of its accuracy (CT images are, at present, far less clear than real sections through the fossil), but because it is non-destructive and can be used rapidly on a large number of specimens of the same species, and can thereby provide information on individual variations in the endoskeletal structure that are used as key characters in phylogenies (e.g. the choanae of osteolepiforms). In addition, the search for exceptional preservation of soft parts is also a potential source of new kinds of anatomical characters. The use of histological characters remains difficult, partly because the definitions of the types of hard tissues are fluctuating (there should be a strict code of nomenclature for them), and partly because there is a great problem of homology between histological structures (homology versus homonymy).

On a theoretical level, progress should be made in the way in which characters are described and included in a data matrix, in particular when one is dealing with transformation series. How to define the shape, or the different states of the shape, of a dermal bone is often a problem.

Fig. 9.1. An 'odd phylogeny' of the Recent and fossil Craniata, based on the assumption that some characters defining the currently recognized fossil groups have a higher degree of generality, and that some characters currently assessed with doubt as homoplasies are in fact synapomorphies. Rhizodontiforms have not been considered here, but are interchangeable with 'some osteolepiforms' above the node 20. 'Some Silurian chondrichthyans' refers to the taxa and scale types described by Karatayute-Talimaa (1992). Placoderms are here still regarded as a clade. The choanae and tear duct are here regarded as unique to tetrapods. *Selected synapomorphies and nested taxa*: 1, Craniata (same characters as in Fig. 5.1); 2, Vertebrata (arcualia, extrinsic eye muscles, hypocercal tail, sensory-line system, non-muscularized paired fin fold (this implies a different interpretation of the paired fins of anaspids)); 3, Gnathostomata (exoskeleton); 4, strongly mineralized exoskeleton made up of aspidine; 5, sensory-line system extending on the body, well-separated olfactory organs (with a reversion in osteostracans); 6, open endolymphatic ducts, perichondral bone; 7, paired fins strongly muscularized and concentrated in the postbranchial region, endoskeletal shoulder girdle, sclerotic ring, cellular bone, epicercal tail, two dorsal fins (homoplasic in extant lampreys); ?gill covers; 8, jaws, paired fin spines, olfactory organs opening to the exterior (loss of the prenasal sinus and hypophysial tube); 9, Crown-group Gnathostomata (adductor jaw musculature lateral to mandibular arch, superior oblique eye muscle inserted in the anterior part of the orbital cavity, median fin spines, ?intermediate fin spines, mesosquamose scales (the placoid scales of neoselachians being a reversion)); 10, pectinate or nodose ornamentation on fin spines (this would imply that all non-elasmobranch chondrichthyans have lost this character); 11, Osteichthyes, or Teleostomi (three otoliths, separate anterior and posterior nostrils); 12, diamond-shaped scales, teeth attached to large dermal bones, large dermal bones on head and shoulder girdle; 13, Crown-group Osteichthyes (loss of paired and median fin spines, endochondral bone, posteriorly expanded maxilla, lepidotrichiae); 14, Sarcopterygii (more than four sclerotic plates); 15, monobasal (metapterygial) paired fin skeleton, large basal lobe in fins; 16, crown-group Sarcopterygii (maxilla reduced or lost); 17, cosmine, dermintermedial process, closely set external nostrils passing through a common opening in exoskeleton, ventrolateral fenestra in nasal cavity, weakly folded polyplocodont teeth, axillary (basal) scute in fins; 18, Rhipidistia (strongly folded polyplocodont teeth); 19, rostral tubules, posterior nostril near oral margin, special features of cosmine structure, elongated biserial pectoral fin skeleton; 20, large entepicondylar process on humerus; 21, loss of the cosmine, narrow and anteriorly pointed parasphenoid, long posterior process of vomers, reduction of adsymphysial toothed plates; 22, frontals, elongated humerus, jugal contacting quadratojugal; 23, reduced coronoid fangs, enlarged teeth on dentary, maxilla, and premaxilla.

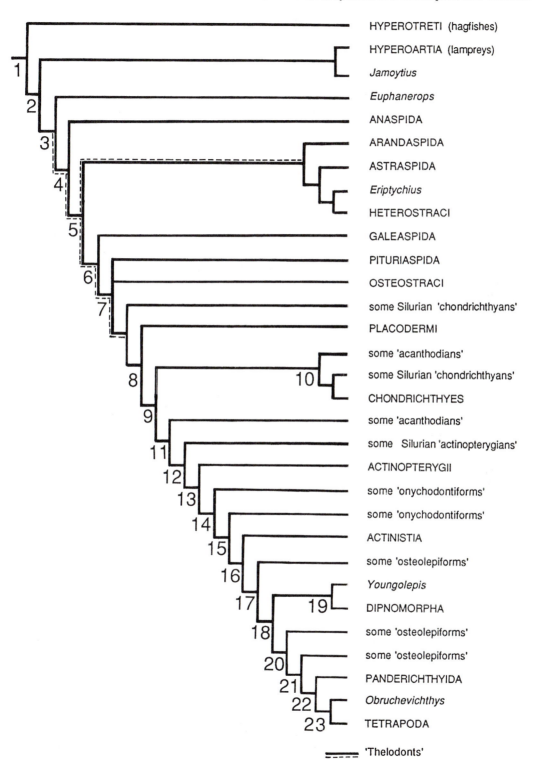

HYPEROTRETI (hagfishes)
HYPEROARTIA (lampreys)
Jamoytius
Euphanerops
ANASPIDA
ARANDASPIDA
ASTRASPIDA
Eriptychius
HETEROSTRACI
GALEASPIDA
PITURIASPIDA
OSTEOSTRACI
some Silurian 'chondrichthyans'
PLACODERMI
some 'acanthodians'
some Silurian 'chondrichthyans'
CHONDRICHTHYES
some 'acanthodians'
some Silurian 'actinopterygians'
ACTINOPTERYGII
some 'onychodontiforms'
some 'onychodontiforms'
ACTINISTIA
some 'osteolepiforms'
Youngolepis
DIPNOMORPHA
some 'osteolepiforms'
some 'osteolepiforms'
PANDERICHTHYIDA
Obruchevichthys
TETRAPODA

'Thelodonts'

Measurements alone are poorly informative, and transformation series based on measurements are always arbitrary (e.g. 'less than 1 mm' or 'more than 1 mm'). Clearly, biometrics has to become adapted to binary coding in some way.

Whatever the source of data and the technique used for yielding characters from fossils, the major tools in future will be methodological and theoretical: i.e. they will be concerned with the way we assess characters and the way we look for the congruence of homologies. This advance will certainly come from more efficient cladistic computer programs, which are able to provide the most extensive set of equally parsimonious trees, and the shortest trees, but also from the improvement of the cladistic analysis itself. The 'three-taxon statement' of Nelson and Platnick (1991), for example, is one of these developments of cladistics, and will have important bearings on the optimization of missing data—a crucial issue for palaeontologists. The question of the weight of characters in cladistic analyses has constantly been addressed since the birth of this method, and will be even more important when morphological and molecular sequence data are combined in a 'total evidence' data matrix. I have always been reluctant to use character weighting, not because characters do not have different weights, but because we generally have no means of assessing these weights (except perhaps in molecular sequences, where transition substitutions are more frequent than transversion substitutions). How comparable are a mutation of a base on the site of a molecular sequence on the one hand, and the appearance of paired fins on the other? If the former can easily be assessed as one step, does the latter also represent one step, or many more steps at the genetic level? Instead of relying on assumptions based on vague comparisons, case studies, etc., it seems wiser to count both as one step for the sake of parsimony. The most important thing is to be aware that there is a potential bias.

9.5 ON RELATIONSHIPS WITH OTHER DISCIPLINES

Studies on early vertebrates have bearings on, or use information from, a number of other disciplines, some of them being totally unrelated. In the Earth sciences, relationships are mostly with biostratigraphy, palaeoecology, and global tectonics. In the life sciences, they are with systematics, comparative anatomy, and developmental biology. In all these interconnections, early vertebrate palaeontology has some specific data to provide. In biostratigraphy, it provides unique information on correlations between marine and non-marine environments, and provides dating for geological formations that are almost devoid of other fossils. In palaeoecology, it provides information on the still poorly studied marine marginal environments. This nearly continental habitat of many early vertebrate groups, allied to the fact that their phylogeny is perhaps better resolved than that of many invertebrate groups, gives them more weight in assessing the biogeographical relationships of Palaeozoic landmasses. In the realm of biology, early vertebrates provide associations of characters that no longer exist, but are crucial to our knowledge of the order of appearance of major vertebrate structures. Such associations of characters may be used as a test for models in developmental biology and genetics. They also provide a minimum age for characters and taxa, which are useful data for some molecular phylogeneticists. Conversely, developmental biology provides insights into character-state polarity and possible cases of heterochrony. How to classify these interconnections, from the more general to the most particular? I would put relationships with biology first, as it is the basis for the definition of characters and taxa, and thus for systematics. Then, systematics is the basis for biogeography and palaeobiogeography. In turn, palaeobiogeography is a test for geophysical models (endemic taxa) and a support for biostratigraphy (widespread taxa). Systematics is also a test for nomothetic palaeontology. I therefore consider that early vertebrates will be more informative if we look at them as organisms rather than as geological objects.

References

Abel, O. (1907). Die Lebensweise der altpaläozoischen Fische. *Verhandlungen der Kaiserlichen Zoologisch-Botanischen Gesellschaft in Wien*, **57**, 158–68.

Adrain, J. M., and Wilson, M. V. H. (1994). Early Devonian cephalaspids (Vertebrata: Osteostraci: Cornuata) from the southern Mackenzie Mountains, N.W.T., Canada. *Journal of Vertebrate Paleontology*, **14**, 301–19.

Afanassieva, O. (1991). [*The cephalaspids of the Soviet Union.*] Trudy Paleontologicheskogo Instituta Akademia Nauk, **248**, Moscow. (In Russian.)

Afanessieva, O., and Janvier, P. (1985). *Tannuaspis, Tuvaspis* and *Ilemoraspis*, endemic osteostracan genera from the Silurian and Devonian of Tuva and Khakassia (USSR). *Géobios*, **18**, 493–505.

Agassiz, J. L. R. (1833–44). *Recherches sur les poissons fossiles*, (5 vols and atlas). Imprimerie Petitpierre, Neuchâtel.

Agassiz, J. L. R. (1839). Fishes of the Upper Ludlow Rock. In *The Silurian System*, (ed. R. I. Murchison), Vol. 2, pp. 605–7, London.

Agassiz, J. L. R. (1844–5). *Monographie des poissons fossiles des Vieux Grès Rouges ou Système Devonien (Old Red Sandstone) des Iles Britanniques et de Russie*. Jent and Gassmann, Neuchâtel.

Agassiz, J. L. R. (1857). Classification of fishes. *Proceedings of the American Academy of Arts and Science*, **4**, 8–9.

Agassiz, J. L. R. (1859). *Essay on classification*. London.

Agassiz, J. L. R. (1874). Evolution and permanence of type. *Atlantic Monthly*, **33**, 92–101.

Ahlberg, P. E. (1989). Paired fin skeletons and relationships of the fossil group Porolepiformes (Osteichthyes: Sarcopterygii). *Zoological Journal of the Linnean Society*, **96**, 119–66.

Ahlberg, P. E. (1991a). A re-examination of sarcopterygian interrelationships, with special reference to the Porolepiformes. *Zoological Journal of the Linnean Society*, **103**, 241–87.

Ahlberg, P. E. (1991b). Tetrapod or near-tetrapod fossils from the Upper Devonian of Scotland. *Nature*, **354**, 298–301.

Ahlberg, P. E. (1992). The palaeoecology and evolutionary history of the porolepiform sarcopterygians. In *Fossil fishes as living animals*, (ed. E. Mark-Kurik). *Akademia*, **1**, 71–90.

Ahlberg, P. E., and Milner, A. R. (1994). The origin and early diversification of tetrapods. *Nature*, **368**, 507–14.

Ahlberg, P. E., Luksevics, E., and Lebedev, O. (1994). The first tetrapod finds from the Devonian (Upper Famennian) of Latvia. *Philosophical Transactions of the Royal Society of London*, **B 343**, 303–28.

Alberch, P. (1989). The logic of monsters: evidence for internal constraint in development and evolution. In *Ontogenese et evolution*, (ed. B. David, J. L. Dommergues, J. Chaline, and B. Laurin). *Géobios, Mémoire Spécial*, **12**, 21–57.

Aldridge, R. J., Briggs, D. E. G., Clarkson, E. N. K., and Smith, M. P. (1986). The affinities of conodonts—new evidence from the Carboniferous of Edinburgh, Scotland. *Lethaia*, **19**, 279–91.

Aldridge, R. J., Briggs, D. E. G., Smith, M. P., Clarkson, E. N. K., and Clark, N. D. L. (1993). The anatomy of conodonts. *Philosophical Transactions of the Royal Society of London*, **B 340**, 405–21.

Alexander, R. M. (1981). *Functional design in fishes*, (3rd edn). Cambridge University Press.

Allis, E. P. (1914). The pituitary fossa and trigemino-facialis chamber in selachians. *Anatomische Anzeiger*, **45**, 22–253.

Allis, E. P. (1919). The lips and nasal apertures in the gnathostome fishes. *Journal of Morphology*, **32**, 145–205.

Allis, E. P. (1931). Concerning the mouth opening and certain features of the visceral endoskeleton of *Cephalaspis*. *Journal of Anatomy*, **65**, 509–27.

Anderson, M. E., Hiller, N., and Gess, R. W. (1994). The first *Bothriolepis*-associated fish fauna from Africa. *Syid-Afrikaanse Tydskrift vir Wetenskap*, **90**, 397–403.

Andrews, S. M. (1973). Interrelationships of crossopterygians. In *Interrelationships of fishes*, (ed. P. H. Greenwood, R. S. Miles, and C. Patterson), pp. 137–77. *Zoological Journal of the Linnean Society, Supplement* **1**. Academic Press, London.

Andrews, S. M. (1982). *The discovery of fossil fishes in Scotland up to 1845*. Royal Scottish Museum, Edinburgh.

Andrews, S. M. (1985). Rhizodont crossopterygian fish from the Dinantian of Foulden, Berwickshire, Scotland, with a re-evaluation of this group. *Transactions of the Royal Society of Edinburgh*, **76**, 67–95.

Andrews, S. M., and Westoll, T. S. (1970a). The postcranial skeleton of *Eusthenopteron foordi* Whiteaves. *Transactions of the Royal Society of Edinburgh*, **68**, 207–39.

Andrews, S. M., and Westoll, T. S. (1970b). The postcranial skeleton of rhipidistian fishes, excluding *Eusthenopteron*. *Transactions of the Royal Society of Edinburgh*, **68**, 391–489.

Anonymous (1909). Ramsay Heatley Traquair. *Geological Magazine*, **6**, 242–50. (With a list of R. H. Traquair's works.)

Aquesbi, N. (1988). Étude d'un Onychodontiforme (Osteichthyes, Sarcopterygii) du Dévonien moyen (Eifélien) du Maroc. *Bulletin du Muséum national d'Histoire Naturelle, Paris*, **10**, 181–96.

Arambourg, C., and Guibé, J. (1958). Sous-classe des Dipneustes. In *Traité de zoologie*, (ed. P. P. Grassé), Vol. 13 (3), pp. 2522–52, Masson, Paris.

Armstrong, W. G., and Halstead-Tarlo, L. B. (1966). Amino-acid components in fossil calcified tissues. *Nature*, **210**, 481–2.

Arratia, G. (1991). The caudal skeleton of Jurassic teleosts: a phylogenetic analysis. In *Early vertebrates and related problems of evolutionary biology*, (ed. M. M. Chang, Y. H. Liu, and G. R. Zhang), pp. 249–340. Science Press, Beijing.

Arratia, G., and Schultze, H. P. (1991). Palatoquadrate and its ossifications: development and homology within osteichthyans. *Journal of Morphology*, **208**, 1–81.

Arsenault, M. (1982). *Eusthenopteron foordi*, a predator on *Homalacanthus concinus* from the Escuminac Formation, Miguasha, Quebec. *Canadian Journal of Earth Sciences*, **19**, 2214–17.

Arsenault, M., and Janvier, P. (1991). The anaspid-like craniates of the Escuminac Formation (Upper Devonian) from Miguasha (Québec, Canada), with remarks on anaspid–petromyzontid relationships. In *Early vertebrates and related problems of evolutionary biology*, (ed. M. M. Chang, Y. H. Liu, and G. R. Zhang), pp. 19–40. Science Press, Beijing.

Baer, K. E. von (1828). *Über entwickelungsgeschichte der Thiere. Beobachtung und Reflexion*. Borntrager, Königsberg.

Balfour, F. M. (1878). *A monograph on the development of elasmobranch fishes*. Macmillan, London.

Bardack, D. (1991). First fossil hagfish (Myxinoidea): a record from the Pennsylvanian of Illinois. *Science*, **254**, 701–3.

Bardack, D., and Richardson, E. S., Jr. (1977). New agnathous fishes from the Pennsylvanian of Illinois. *Fieldiana: Geology*, **33**, 489–510.

Bardack, D., and Zangerl, R. (1971). Lampreys in the fossil record. In *The biology of lampreys*, (ed. M. W. Hardisty and I. C. Potter), Vol. 1, pp. 67–84. Academic Press, London.

Beaumont, E. H. (1977). Cranial morphology of the Loxommatidae (Amphibian: Labyrinthodontia). *Philosophical Transactions of the Royal Society of London*, **B 280**, 29–101.

Belles-Isles, M. (1987). La nage et l'hydrodynamique de deux Agnathes du Paléozoïque: *Alaspis macrotuberculata* et *Pteraspis rostrata*. *Neues Jahrbuch für Geologie und Paläontologie, Abhandlungen*, **175**, 347–76.

Belles-Isles, M. (1992). The modes of swimming of sarcopterygians. In *Fossil fishes as living animals*, (ed. E. Mark-Kurik), *Academia*, **1**, 117–40.

Bemis, W. E., and Grande, L. (1992). Early development of the actinopterygian head. 1. External development and staging of the paddlefish *Polyodon spatula*. *Journal of Morphology*, **213**, 47–83.

Bemis, W. E., and Northcutt, G. (1992). Skin and blood vessels of the snout of the Australian lungfish, *Neoceratodus forsteri*, and their significance for interpreting the cosmine of Devonian lungfishes. *Acta Zoologica, Stockholm*, **73**, 115–39.

Bemis, W. E., Burggren, W. W., and Kemp, N. E. (ed.) (1987). *The biology and evolution of lungfishes*. Liss, New York.

Bendix-Almgreen, S. E. (1971). The anatomy of *Menaspis armata* and the phyletic affinities of the menaspid bradyodonts. *Lethaia*, **4**, 21–49.

Bendix-Almgreen, S. E. (1975). The paired fins and shoulder girdle in *Chadoselache*, their morphology and phyletic significance. In *Problèmes actuels de paléontologie: evolution des Vertébrés*. Colloques internationaux du Centre national de la Recherche scientifique, No. 218, (ed. J. P. Lehman), pp. 111–23. Paris.

Bendix-Almgreen, S. E. (1983). *Carcharodon megalodon* from the Upper Miocene of Denmark, with comments on elasmobranch tooth enameloid: coronoin. *Bulletin of the Geological Society of Denmark*, **32**, 1–32.

Bendix-Almgreen, S. E. (1986). Silurian ostracoderms from Washington Land (North Greenland), with comments on cyathaspid structure, systematics and phyletic position. *Rapporter fra Grønlands geologiske Undersøgerlser*. **132**, 89–123.

Benton, M. J. (ed.) (1988). *The phylogeny and classification of the tetrapods*. (2 vols.) Systematics Association Special Volume No. 35. Clarendon Press, Oxford.

Benton, M. J. (ed.) (1993). *The fossil record*, (2nd edn). Chapman and Hall, London.

Bertran, G. (1969). The vertebrate nose: remarks on its structural and functional adaptation and evolution. *Evolution*, **23**, 131–52.

Bischoff, T. L. W. von (1840). *Lepidosiren paradoxa anatomisch untersucht und beschrieben*. Voss, Leipzig.

Bjerring, H. (1967). Does a homology exist between the basicranial muscle and the polar cartilage? In *Problèmes actuels de paléontologie: evolution des Vertébrés*. Colloques Internationaux du Centre national de la Recherche scientifique, No. 163, (ed. J. P. Lehman), pp. 223–67. Paris.

Bjerring, H. (1968). The second somite with special reference to the evolution of its myotomic derivatives. In *Current problems of lower vertebrate phylogeny*, (ed. T. Ørvig), pp. 341–57. Almqvist and Wiksell, Stockholm.

Bjerring, H. (1973). Relationships of coelacanthiformes. In *Interrelationships of fishes*, (ed. P. H. Greenwood, R. S. Miles, and C. Patterson), pp. 179–205. *Zoological Journal of the Linnean Society, Supplement* **1**. Academic Press, London

Bjerring, H. (1977). A contribution to structural analysis of the head of craniate animals. *Zoologica Scripta*, **6**, 127–83.

Bjerring, H. (1978). The 'intracranial joint' versus the 'ventral otic fissure'. *Acta Zoologica, Stockholm*, **59**, 203–14.

Bjerring, H. (1979). Quondam gill-covers. *Zoologica Scripta*, **8**, 235–40.

Bjerring, H. (1984). Major anatomical steps toward craniotedness: a heterodox view based largely on embryological data. *Journal of Vertebrate Palaeontology*, **4**, 17–29.

Bjerring, H. (1985). Facts and thoughts on piscine phylogeny. In *Evolutionary biology of primitive fishes*, (ed. R. E. Foreman, A. Gorbman, J. M. Dodd, and R. Olsson), pp. 31–57. Plenum, New York.

Bjerring, H. (1986*a*). The question of a dermohyal in brachiopterygian fishes. *Acta Zoologica, Stockholm*, **67**, 1–4.

Bjerring, H. (1986*b*). Tofsstjärtfiskarnas elsinnesorgan—'ett sjätte sinne'. *Fauna och Flora*, **81**, 215–22.

Bjerring, H. (1987). Notes on some annexa oculi in the osteolepiform freshwater fish *Eusthenopteron foordi* from the Upper

Devonian Escuminac Formation of Miguasha, eastern Canada. *Acta Zoologica, Stockholm*, **68**, 173–78.

Bjerring, H. (1989*a*). Evolution of the tear duct. *Plzensky lékarsky Sbornik, Supplement* **59**, 99–108.

Bjerring, H. (1989*b*). Apertures of craniate olfactory organs. *Acta zoologica, Stockholm*, **70**, 71–85.

Bjerring, H. (1991). Some features of the olfactory organ in a Middle Devonian porolepiform, *Glyptolepis groenlandica*. *Palaeontographica*, **A 219**, 89–95.

Bjerring, H. (1993). Yet another interpretation of the coelacanthiform basicranial muscle and its innervation. *Acta Zoologica, Stockholm*, **74**, 289–99.

Blieck, A. (1982*a*). *Les Hétérostracés de l'Horizon* Vogti, *Dévonien inferieur du Spitsberg*. Cahiers de Paléontologie, Centre national de la Recherche scientifique, Paris.

Blieck, A. (1982*b*). Les grandes lignes de la biogéographie des Hétérostracés du Silurien supérieur–Dévonien inférieur dans le domaine Nord-Atlantique. *Palaeogeography, Palaeoclimatology, Palaeoecology*, **38**, 283–316.

Blieck, A. (1984). *Les Hétérostracés Ptéraspidiformes. Systématique, phylogénie, biostratigraphie, biogéographie*. Cahiers de Paléontologie, Centre national de la Recherche scientifique, Paris.

Blieck, A. (1985). Paléoenvironnements des Hétérostracés, Vertébrés agnathes ordoviciens à dévoniens. *Bulletin du Muséum national d'Histoire naturelle, Paris*, **7**, 143–55.

Blieck, A. (1992). At the origin of chordates. *Géobios*, **25**, 101–13.

Blieck, A., and Goujet, D. (1983). *Zascinaspis laticephala* nov. sp. (Agnatha, Heterostraci) du Dévonien inférieur du Spitsberg. *Annales de Paléontologie*, **69**, 43–56.

Blieck, A., and Heintz, N. (1983). The cyathaspids of the Red Bay Group (Lower Devonian) of Spitsbergen. *Polar Research*, **1**, 49–74.

Blieck, A., and Janvier, P. (1991). Silurian vertebrates. In *The Murchison Symposium*, (ed. M. G. Bassett, P. D. Lane, and D. Edwards), *Special Papers in Palaeontology*, **44**, 345–89.

Blieck, A., and Janvier, P. (1993). Silurian–Devonian vertebrate biostratigraphy of Siberia and neighbouring terranes. In *Palaeozoic vertebrate biostratigraphy and biogeography*, (ed. J. A. Long), pp. 87–138.

Blieck, A., Elliott, D. K., and Gagnier, P. Y. (1991). Some questions concerning the phylogenetic relationships of heterostracans, Ordovician to Devonian jawless vertebrates. In *Early vertebrates and related problems of evolutionary biology*, (ed. M. M. Chang, Y. H. Liu, and G. R. Zhang), pp. 1–17. Science Press, Beijing.

Bockelie, T. G., and Fortey, R. (1976). An Early Ordovician vertebrate. *Nature*, **260**, 36–8.

Bolk, L., Göppert, E., Kallius, E., and Lubosch, W. (ed.) (1931–9). *Handbuch der vergleichenden Anatomie der Wirbeltiere*. Urban and Schwarzenberg, Berlin. (Reprinted in 1967 by Ascher and Co., Amsterdam.)

Bolt, J. R. (1991). Lissamphibian origins. In *Origins of the higher groups of tetrapods*, (ed. H. P. Schultze and L. Trueb), pp. 194–22. Cormstock, Ithaca.

Bolt, J. R., and Lombard, R. E. (1985). Evolution of the amphibian tympanic ear and the origin of frogs. *Biological Journal of the Linnean Society*, **24**, 83–99.

Bonde, N. (1975). Origin of 'higher groups': viewpoints of phylogenetic systematics. In *Problèmes actuels de paléontologie: Evolution des Vertébrés*. Colloques internationaux du Centre national de la Recherche scientifique, No. 218, (ed. J. P. Lehman), pp. 293–324. Paris.

Bonde, N. (1977). Cladistic classification as applied to vertebrates. In *Major patterns in vertebrate evolution*, (ed. M. K. Hecht, P. C. Goody, and B. M. Hecht), pp. 741–804. Plenum, New York.

Bonde, N. (1983). Functional anatomy and reconstruction of phylogeny. In *Actes du Symposium paléontologique G. Cuvier*, (ed. E. Buffetaut, J. M. Mazin, and E. Salmon), pp. 11–26. Montbéliard.

Bonde, N. (1985). Inledning-isaer om palaeoøkologisk rekonstruktion. *Dansk geologiske Foreningen*, **1984**, 117–22.

Borgen, U. (1983). Homologization of skull roofing bones between tetrapods and osteolepiform fishes. *Palaeontology*, **26**, 735–53.

Borgen, U. (1992). The function of the cosmine pore canal system. In *Fossil fishes as living animals*, (ed. E. Mark-Kurik), *Academia*, **1**, 141–50.

Boucot, A. J., and Gray, J. (1979). Epilogue. A Palaeozoic Pangaea? In *Historical biogeography, plate tectonics and the changing environment*, (ed. A. J. Boucot and J. Gray), pp. 465–82. Oregon State University Press, Corvallis.

Braun, J., and Reif, W. E. (1985). A survey of aquatic locomotion in fishes and tetrapods. *Neues Jahrbuch für Geologie und Paläontologie, Abhandlungen*, **169**, 307–32.

Brenchley, P. J. (1990). End Ordovician. In *Palaeobiology. A synthesis*, (ed. D. E. Briggs and P. R. Crowther), pp. 181–4. Blackwell Scientific Publications, Oxford.

Briggs, D. E. G. (1992). Conodonts: A major extinct group added to the vertebrates. *Science*, **256**, 1285–6.

Briggs, D. E. G., and Crowther, P. R. (ed.) (1990). *Palaeobiology. A synthesis*. Blackwell Scientific Publications, Oxford.

Briggs, D. E. G., and Fortey, R. A. (1982). The cuticle of the aglaspidid arthropods, a red-herring in the early history of the vertebrates. *Lethaia*, **15**, 25–9.

Briggs, D. E. G., Clarkson, E. N. K., and Aldridge, R. J. (1983). The conodont animal. *Lethaia*, **20**, 1–14.

Broad, D. S., and Dineley, D. L. (1973). *Torpedaspis*, a new Upper Silurian and Lower Devonian genus of Cyathaspididae (Ostracodermi) from Arctic Canada. Contributions to Canadian palaeontology. *Bulletin of the Geological Survey of Canada*, **222**, 53–91.

Brodal, A., and Fänge, R. (ed.) (1963). *The biology of Myxine*. Universitetsforlaget, Oslo.

Bryant, W. L. (1919) On the structure of *Eusthenopteron*. *Bulletin of the Buffalo Society of Natural Sciences*, **13**, 1–59.

Campbell, K. S. W., and Barwick, R. E. (1982). The neurocranium of the primitive dipnoan *Dipnorhynchus sussmilchi* (Etheridge). *Journal of Vertebrate Paleontology*, **2**, 286–327.

Campbell, K. S. W., and Barwick, R. E. (1984*a*). The choanae, maxillae, premaxillae and anterior palatal bones of early dipnoans. *Proceedings of the Linnean Society of New South Wales*, **107**, 147–70.

Campbell, K. S. W., and Barwick, R. E. (1984*b*). *Speonesydrion*, an Early Devonian dipnoan with primitive toothplates. *Palaeo Ichthyologica*, **2**, 1–48.

Campbell, K. S. W., and Barwick, R. E. (1987). Paleozoic lungfishes—a review. In *Biology and evolution of lungfishes*, (ed. W. E. Bemis, W. W. Burggren, and N. E. Kemp). *Journal of Morphology, Supplement*, **1**, 93–131, Liss, New York.

Campbell, K. S. W., and Barwick, R. E. (1988*a*). *Uranolophus*: a reappraisal of a primitive dipnoan. In *Devonian and Carboniferous fish studies*, (ed. P. A. Jell), *Memoirs of the Association of Australasian Palaeontologists*, **7**, 87–144.

Campbell, K. S. W., and Barwick, R. E. (1988*b*). Geological and palaeontological information and phylogenetic hypotheses. *Geological Magazine*, **125**, 207–27.

Campbell, K. S. W., and Barwick, R. E. (1990). Palaeozoic dipnoan phylogeny: functional complexes and evolution without parsimony. *Paleobiology*, **16**, 143–69.

Campbell, K. S. W., and Bell, M. W. (1977). A primitive amphibian from the Late Devonian of New South Wales. *Alcheringa*, **1**, 369–81.

Campbell, K. S. W., and Le Duy Phuoc (1983). A Late Permian actinopterygian fish from Australia. *Palaeontology*, **26**, 33–70.

Cappetta, H. (1987). Chondrichthyes II. In *Handbook of paleoichthyology*, Vol. 3B, (ed. H. P. Schultze). Gustav Fischer, Stuttgart.

Carr, R. K. (1991). Reanalysis of *Heintzichthys gouldii* (Newberry), an aspinothoracid arthrodire (Placodermi) from the Famennian of northern Ohio, with a review of brachythoracid systematics. *Zoological Journal of the Linnean Society*, **103**, 349–90.

Carroll, R. L. (1967). Labyrinthodonts from the Joggins Formation. *Journal of Paleontology*, **41**, 111–42.

Carroll, R. L. (1980). The hyomandibular as a supporting element in the skull of primitive tetrapods. In *The terrestrial environment and the origin of land vertebrates*. Systematics Association Special Volume No. 15, (ed. A. L. Panchen), pp. 293–315. Academic Press, London.

Carroll, R. L. (1987). *Vertebrate paleontology and evolution*. Freeman, New York.

Carroll, R. L. (1990). A tiny microsaur from the Lower Permian of Texas: size constraints in Palaeozoic tetrapods. *Palaeontology*, **33**, 893–909.

Carroll, R. L., and Currie, P. J. (1975). Microsaurs as possible apodan ancestors. *Zoological Journal of the Linnean Society*, **57**, 229–47.

Carroll, R. L., and Gaskill, P. (1978). The order microsaurs. *Memoirs of the American Philosophical Society*, **126**, 1–211.

Casey, J., and Lawson, R. (1981). A histological and scanning electron microscope study of the teeth of caecilian amphibians. *Archives of Oral Biology*, **26**, 49–58.

Castanet, J., Meunier, F., Bergot, C, François, Y., Francillon, H., Meunier, F., *et al.* (1975). Données préliminaires sur les structures histologiques du squelette de *Latimeria chalumnae*. In *Problèmes actuels de paléontologie: evolution des vertébrés*. Colloques internationaux du Centre national de la Recherche scientifique, No. 218, (ed. J. P. Lehman), pp. 159–73. Paris.

Chaloner, W. G., and Lawson, J. D. (ed.) (1985). Evolution and environment in the Late Silurian and Early Devonian. *Philosophical Transactions of the Royal Society of London*, **B 309**.

Chaloner, W. G., Forey, P. L., Gardiner, B. G., Hill, A. J., and Young, V. T. (1980). Devonian fish and plants from the Bokkeveld Series of South Africa. *Annals of the South African Museum*, **81**, 127–57.

Chang, M. M. (1982). The braincase of *Youngolepis*, a Lower Devonian crossopterygian from Yunnan, south-western China. Doctoral dissertation, Department of Geology, Stockholm University.

Chang, M. M. (1991*a*). Head exoskeleton and shoulder girdle of *Youngolepis*. In *Early vertebrates and related problems of evolutionary biology*, (ed. M. M. Chang, Y. H. Liu, and G. R. Zhang), pp. 355–78. Science Press, Beijing.

Chang, M. M. (1991*b*). 'Rhipidistians', dipnoans and tetrapods. In *Origins of the higher groups of tetrapods. Controversy and consensus*, (ed. H. P. Schultze and L. Trueb), pp. 4–28. Cormstock, Ithaca.

Chang, M. M., and Smith, M. M. (1992) Is *Youngolepis* a porolepiform? *Journal of Vertebrate Paleontology*, **12**, 294–312.

Chang, M. M., and Yu, X. B. (1984). Structure and phylogenetic significance of *Diabolichthys speratus* gen. et sp. nov., a new dipnoan-like form from the Lower Devonian of eastern Yunnan, China. *Proceedings of the Linnean Society of New South Wales*, **107**, 171–84.

Chang, M. M., and Zhu, M. (1993). A new Middle Devonian osteolepidid from Qujing, Yunnan. *Memoirs of the Association of Australasian Palaeontologists*, **15**, 183–98.

Cheng, H. (1992) [On the meshwork of bony tubules in the rostrum of *Youngolepis*] *Vertebrata PalAsiatica*, **30**, 102–10. (In Chinese with English abstract.)

Clack, J. A. (1988). New material of the early tetrapod *Acanthostega* from the Upper Devonian of Greenland. *Palaeontology*, **31**, 699–724.

Clack, J. A. (1989). Discovery of the earliest-known tetrapod stapes. *Nature*, **342**, 425–27.

Clack, J. A. (1993). The stapes of *Acanthostega gunnari* and the role of the stapes in early tetrapods. In *The evolutionary biology of hearing*, (ed. D. B. Webster, R. R. Fay, and A. N. Popper), pp. 405–20. Springer, Berlin.

Clack, J. A. (1994). Earliest known tetrapod braincase and the evolution of the stapes and fenestra ovalis. *Nature*, **369**, 392–4.

Clack, J. A., and Coates, M. I. (1993). *Acanthostega gunnari*: our present connection. In *Deciphering the natural world and the role of collections and museums*, pp. 39–43. Geological Museum, Copenhagen University.

Clack, J. A., and Holmes, R. (1988). The braincase of the anthracosaur *Archeria crassidisca*, with comments in the interrelationships of primitive tetrapods. *Palaeontology*, **31**, 85–107.

Clark, R. B. (1964). *Dynamics in metazoan evolution. The origin of the coelom and segments*. Clarendon Press, Oxford.

Cloutier, R. (1991*a*). Interrelationships of Palaeozoic actinistians: patterns and trends. In *Early vertebrates and related problems of evolutionary biology*, (ed. M. M. Chang, Y. H. Liu, and G. R. Zhang), pp. 379–428. Science Press, Beijing.

Cloutier, R. (1991*b*). Patterns, trends, and rate of evolution within the Actinistia. *Environmental Biology of Fishes*, **32**, 23–58.

Coates, M. I. (1993*a*). New actinopterygian fish from the Namurian Manse Burn Formation of Bearsden, Scotland. *Palaeontology*, **36**, 123–46.

Coates, M. I. (1993*b*). *Hox* genes, fin folds and symmetry. *Nature*, **364**, 195–6.

Coates, M. I., and Clack, J. A. (1990). Polydactyly in the earliest known tetrapod limbs. *Nature*, **347**, 66–9.

Colbert, E. H., and Gaffney, E. S. (1984). Bobb Schaeffer, a biographic sketch. *Journal of Vertebrate Paleontology*, **4**, 285–91.

Compagno, L. J. V. (1977). Phyletic relationships of living sharks and rays. In *Recent advances in the biology of sharks*. *American Zoologist*, **17**, 303–22.

Conway Morris, S. (1989). Conodont palaeobiology: recent progress and unresolved problems. *Terra Nova*, **1**, 135–50.

Conway Morris, S., and Whittington, H. B. (1979). The animals of the Burgess Shale. *Scientific American*, **241**, 122–33.

Cope, E. D. (1889). Synopsis of the families of Vertebrata. *American Naturalist*, **23** (2), 1–29.

Cope, E. D. (1892). On the phylogeny of the Vertebrata. *Proceedings of the American Philosophical Society*, **30**, 278–81.

Cope, E. D. (1898). *Syllabus of the lectures on the Vertebrata*. University of Pennsylvania, Philadelphia.

Copper, P. (1986). Frasnian–Famennian mass extinction and cold-water oceans. *Geology*, **14**, 835–39.

Couly, G. F., Coltey, P. M., and Le Douairin, N. M. (1993). The triple origin of the skull in higher vertebrates: a study in quail-chick chimaeras. *Development*, **117**, 409–29.

Crowther, P. R., and Collins, C. J. (ed.) (1987). The conservation of geological material. *Geological Curator*, **4**, 375–474.

Cuvier, G. (1812–22). *Recherches sur les ossements fossiles des quadrupèdes*. Dufour et d'Ocagne, Paris.

Damas, H. (1944). Recherches sur le développement de *Lampetra fluviatilis* L. Contribution à l'étude de la céphalogenèse des Vertébrés. *Archives de Biologie*, **55**, 1–285.

Damas, H. (1954). La branchie préspiraculaire des Céphalaspides. *Annales de la Société zoologique de Belgique*, **85**, 89–102.

Daniel, J. F. (1937). *The elasmobranch fishes*, (3rd edn). University of California Press, Berkeley.

Darby, D., 1982. The early vertebrate *Astraspis*, habitat based on a lithologic association. *Journal of Paleontology*, **56**, 1187–96.

Darlu, P. and Tassy, P., (1993). *Reconstruction phylogénétique: concepts et méthodes*. Masson, Paris.

Dasch, E. J. and Campbell, K. S. W. (1970). Strontium-isotope evidence for marine or freshwater origin of fossil dipnoans and arthrodires. *Nature*, **227**, 1159.

Dean, B. (1895). *Fishes, living and fossil*. MacMillan, New York.

Dean, B. (1909). Studies on fossil fishes (sharks, chimaeroids and arthrodires). *Memoirs of the American Museum of Natural History*, **9**, 209–87.

De Beer, G. R. (1937). *The development of the vertebrate skull*. Clarendon Press, Oxford.

Denison, R. H. (1941). The soft anatomy of *Bothriolepis*. *Journal of Paleontology*, **15**, 553–61.

Denison, R. H. (1947). The exoskeleton of *Tremataspis*. *American Journal of Science*, **245**, 337–65.

Denison, R. H. (1951). The exoskeleton of early Osteostraci. *Fieldiana: Geology*, **11**, 198–218.

Denison, R. H. (1953). Early Devonian fishes from Utah, part II: Heterostraci. *Fieldiana: Geology*, **11**, 291–355.

Denison, R. H. (1956). A review of the habitat of the earliest vertebrates. *Fieldiana: Geology*, **11**, 359–407.

Denison, R. H. (1961). Feeding mechanisms of agnatha and early gnathostomes. *American Zoologist*, **1**, 177–81.

Denison, R. H. (1963). The early history of the vertebrate calcified skeleton. *Clinical Orthopaedics*, **31**, 141–52.

Denison, R. H. (1964). The Cyathaspididae, a family of Silurian and Devonian jawless vertebrates. *Fieldiana: Geology*, **13**, 311–473.

Denison, R. H. (1966). *Cardipeltis*, and Early Devonian agnathan of the order Heterostraci. *Fieldiana: Geology*, **16**, 89–116.

Denison, R. H. (1967). Ordovician vertebrates from Western United States. *Fieldiana: Geology*, **16**, 269–88.

Denison, R. H. (1968). The evolutionary significance of *Uranolophus*. In *Current problems of lower vertebrate phylogeny*, (ed. T. Ørvig). pp. 247–57, Almqvist and Wiksell, Stockholm.

Denison, R. H. (1970). Revised classification of Pteraspididae, with description of new forms from Wyoming. *Fieldiana: Geology*, **20**, 1–41.

Denison, R. H. (1971*a*). The origin of the vertebrates: a critical evaluation of current theories. *Proceedings of the North American Paleontological Convention*, **H**, **1969**, 1132–46.

Denison, R. H. (1971*b*). On the tail of Heterostraci (Agnatha). *Forma et Functio*, **4**, 87–99.

Denison, R. H. (1973). Growth and wear of the shield in Pteraspididae (Agnatha). *Palaeontographica*, **A 143**, 1–10.

Denison, R. H. (1978). Placodermi. In *Handbook of paleoichthyology*, Vol. 2, (ed. H. P. Schultze). Gustav Fisher, Stuttgart.

Denison, R. H. (1979). Acanthodii. In *Handbook of paleoichthyology*, Vol. 5, (ed. H. P. Schultze). Gustav Fisher, Stuttgart.

Denison, R. H. (1983). Further consideration of placoderm evolution. *Journal of Vertebrate Paleontology*, **3**, 69–83.

Dennis, K. D., and Miles, R. S. (1979). Eubrachythoracid arthrodire with tubular rostral plates from Gogo, Western Australia. *Zoological Journal of the Linnean Society*, **67**, 297–328.

Dennis, K. D., and Miles, R. S. (1981). A pachyosteomorph arthrodire from Gogo, Western Australia. *Zoological Journal of the Linnean Society*, **73**, 213–58.

Dennis-Bryan, K. D. (1987). A new species of eastmanosteid arthrodire (Pisces, Placodermi) from Gogo, Western Australia. *Zoological Journal of the Linnean Society*, **90**, 1–64.

De Pinna, M. C. C. (1991). Concepts and tests of homology in the cladistic paradigm. *Cladistics*, **7**, 367–94.

Derycke, C. (1990). Application de la cathodoluminescence à la paléohistologie. *Comptes-Rendus de l'Académie des Sciences, Paris*, **310**, 1561–5.

Devillers, C. (1958). Le crâne des poissons. In *Traité de zoologie* (ed. P. P. Grassé), Vol. 13(1), pp. 1303–41. Masson, Paris.

Dick, J. R. F., and Maisey, J. (1980). The Scottish Lower Carboniferous shark *Onychoselache traquairi*. *Palaeontology*, **23**, 363–74.

Didier, D. A., Stahl, B. J., and Zangerl, R. (1994). Development and growth of compound tooth plates in *Callorhinchus milii* (Chondichthyes, Holocephali). *Journal of Morphology*, **222**, 73–89.

Dineley, D. L. (1976). New species of *Ctenaspis* (Ostracodermi) from the Devonian of Arctic Canada. In *Essays in palaeontology in honour of Loris Shano Russell*, (ed. C. S. Churcher), pp. 26–44. Royal Ontario Museum, Toronto.

Dineley, D. L., and Loeffler, E. J. (1976). Ostracoderm faunas of the Delorme and associated Siluro–Devonian formations, North West Territories, Canada. *Palaeontology, Special Papers*, **18**, 1–214.

Dollo, L. (1910). La paléontologie écologique. *Bulletin de la Société belge de Géologie, de Paléontologie et d'Hydrologie*, **23**, 377–421.

Dommergues, J. L., David, B., and Marchand, D. (1986). Les relations ontogenèse–phylogenèse: applications paléontologiques. *Géobios*, **19**, 335–56.

Donovan, S. K. (ed.) (1989). *Mass extinctions. Processes and evidence*. Belhaven Press, London.

Duffin, C. J., and Ward, D. J. (1983). Neoselachian shark's teeth from the Lower Carboniferous of Britain and the Lower Permian of the USA. *Palaeontology*, **26**, 93–110.

Duméril, A. M. C. (1806). *Zoologie analytique, ou méthode naturelle de classification des animaux*. Didot, Paris.

Duméril, A. M. C. (1807). *Dissertation sur la famille des poissons Cyclostomes pour démontrer leurs rapports avec les animaux sans vertèbres*. Didot, Paris.

Dunkle, D. H., and Bungart, P. A. (1946). The antero-supragnathal of *Gorgonichthys*. *American Museum Novitates*, **1316**, 1–10.

Dupuis, C. (1988). Savoirs et idées à Neuchâtel, terroir séculaire fertile pour l'histoire naturelle. *Bulletin de la Société neuchâteloise des Sciences Naturelles*, **3**, 105–24.

Eastman, C. R. (1917). Fossil fishes in the collection of the United States National Museum. *Proceedings of the U.S. National Museum*, **52**, 235–304.

Eldredge, N., and Stanley, S. M. (ed.) (1984). *Living fossils*. Springer, Berlin.

Elliott, D. K. (1983). New Pteraspididae from the Lower Devonian of the Northwest Territories, Canada. *Journal of Vertebrate Paleontology*, **6**, 389–406.

Elliott, D. K. (1984). A new subfamily of Pteraspididae (Agnatha, Heterostraci) from the Upper Silurian and Lower Devonian of Arctic Canada. *Palaeontology*, **27**, 169–97.

Elliott, D. K. (1987). A reassessment of *Astraspis desiderata*, the oldest North American vertebrate. *Science*, **237**, 190–2.

Elliott, D. K., and Loeffler, E. J. (1988). A new agnathan from the Lower Devonian of Arctic Canada, and a review of the tessellated heterostracans. *Palaeontology*, **32** (4), 883–91.

Elliott, D. K., Blieck, A., and Gagnier, P. Y. (1991). Ordovician vertebrates. In *Advances in Ordovician geology*, (ed. C. R. Barnes and S. H. Williams). *Geological Survey of Canada Papers*, **90**, 93–106.

Erwin, D. H. (1990). End-Permian. In *Palaeobiology. A synthesis*, (ed. D. E. Briggs and P. R. Crowther), pp. 187–94. Blackwell Scientific Publications, Oxford.

Fänge, R. (ed.) (1973). *Myxine glutinosa. Biochemistry, physiology and structure*. Kungliga Vetenskaps-och Vitterhets-Samhället, Göteborg.

Farris, J. S. (1988). HENNIG86. Version 1.5. Program and user's manual. Published by the author. Port Jefferson Station, New York.

Farris, J. S., and Platnick, N. I. (1989). Lord of the Flies; the systematist as study animal. (Review of Hull 1988.) *Cladistics*, **5**, 295–310.

Fernholm, B. (1985). The lateral line system of cyclostomes. In *Evolutionary biology of primitive fishes*, (ed. R. E. Foreman, A. Gorbman, J. M. Dodd, and R. Olsson), pp. 113–22. Plenum, New York.

Fernholm, B., and Holmberg, K. (1975). The eyes in three genera of hagfish (*Eptatretus, Paramyxine* and *Myxine*)—a case of degenerative evolution. *Vision Research*, **15**, 253–9.

Fernholm, B., Bremer, K. and Jörnvall (ed.) (1989). *The hierarchy of life: Molecules and morphology in phylogenetic analysis*. Nobel Symposium **70**. Excerpta Medica, Amsterdam.

Fischer, W. A. (1978). The habitat of the early vertebrates: trace and body fossil evidence from the Harding Formation (Middle Ordovician), Colorado. *The Mountain Geologist*, **15**, 1–26.

Foreman, R. E., Gorbman, A., Dodd, J. M., and R. Olsson (ed.) (1985). *Evolutionary biology of primitive fishes*, Plenum, New York.

Forey, P. L. (1973). A revision of the elopiform fishes, fossil and Recent. *Bulletin of the British Museum (Natural History), Geology, Supplement*, **10**, 1–222.

Forey, P. L. (1980). *Latimeria*: a paradoxical fish. *Proceedings of the Royal Society of London*, **B 208**, 369–84.

Forey, P. L. (1981). The coelacanth *Rhabdoderma* in the Carboniferous of the British Isles. *Palaeontology*, **24**, 203–29.

Forey, P. L. (1984a). Yet more reflections on agnathan–gnathostome relationships. *Journal of Vertebrate Paleontology*, **4**, 330–43.

Forey, P. L. (1984b). L'origine des Tétrapodes. *La Recherche*, **15**, 476–87.

Forey, P. L. (1987a). The Downtonian ostracoderm *Sclerodus* Agassiz (Osteostraci: Tremataspididae). *Bulletin of the British Museum (Natural History), Geology*, **41**, 1–30.

Forey, P. L. (1987b). Relationships of lungfishes. In *Biology and evolution of lungfishes*, (ed. W. E. Bemis, W. W. Burggren, and N. E. Kemp). *Journal of Morphology, Supplement*, **1**, 75–91.

Forey, P. L. (1988). Golden jubilee for the coelacanth *Latimeria chalumnae*, *Nature*, **336**, 727–32.

Forey, P. L. (1991a). Blood lines of the Coelacanth. *Nature*, **351**, 347–48.

Forey, P. L. (1991b). *Latimeria chalumnae* and its pedigree. *Environmental Biology of Fishes*, **32**, 75–97.

Forey, P. L., and Cloutier, R. (1991). Literature relating to fossil coelacanths.*Environmental Biology of Fishes*, **32**, 391–401.

Forey, P. L., and Gardiner, B. G. (1981). J. A. Moy-Thomas and his association with the British Museum (Natural History). *Bulletin of the British Museum (Natural History), Geology*, **35**, 131–44.

Forey, P. L., and Gardiner, B. G. (1986). Observations on *Ctenurella* (Ptyctodontida) and the classification of placoderm fishes. *Zoological Journal of the Linnean Society*, **86** 43–74.

Forey, P. L., and Janvier, P. (1993). Agnathans and the origin of jawed vertebrates. *Nature*, **361**, 129–34.

Forey, P. L., and Janvier, P. (1994). Evolution of the early vertebrates. *American Scientist*, **82**, 554–65.

Forey, P. L., and Young, V. T. (1983). Upper Stephanian fishes from the Puertollano Basin, Ciudad Real, Spain. *Anais da Facultad de Ciências, Porto, Supplement*, **64**, 233–44.

Forey, P. L., Gardiner, B. G., and Patterson, C. (1991). The lungfish, the coelacanth and the cow revisited. In *Origins of the higher groups of tetrapods, controversy and consensus*, (ed. H. P. Schultze and L. Trueb), pp. 145–72. Cormstock, Ithaca.

Forey, P. L., Humphries, C. J., Kitching, I. L., Scotland, R. W., Siebert, D. J., and Williams, D. M. (ed.) (1992). *Cladistics, a practical course in systematics*. Clarendon Press, Oxford.

Forster-Cooper, C. (1937). The Middle Devonian fish fauna of Achanarras. *Transactions of the Royal Society of Edinburgh*, **59**, 223–39.

Franz, V. (1934). Vergleichende Anatomie des Wirbeltierauges. In *Handbuch der Vergleichenden Anatomie der Wirbeltiere*, (ed. L. Bolk, E. Göppert, E. Kallius, and W. Lubosch), Vol. 2 (2), pp. 989–1292. Urban and Schwarzenberg, Berlin.

Fritzsch, B. (1987). Inner ear of coelacanth fish *Latimeria* has tetrapod affinities. *Nature*, **327**, 153–54.

Gaffney, E. S. (1979). Tetrapod monophyly: a phylogenetic analysis. *Bulletin of the Carnegie Museum of Natural History*, **13**, 92–105.

Gagnier, P. Y. (1989). The oldest vertebrate: a 470-million-year-old jawless fish, *Sacabambaspis janvieri*, from the Ordovician of South America. *National Geographic Research*, **5**, 250–3.

Gagnier, P. Y. (1992). Ordovician vertebrates from Bolivia. In *Fosiles y facies de Bolivia. I. Vertebrados*, (ed. R. Suarez-Soruco). *Revista Tecnica de YPFB*, Santa Cruz, **12**, 371–9.

Gagnier, P. Y. (1993a). *Sacabambaspis janvieri*, Vertébré ordovicien de Bolivie. 1, Analyse morphologique. *Annales de Paléontologie*, **79**, 19–69.

Gagnier, P. Y. (1993b). *Sacabambaspis janvieri*, Vertébré ordovicien de Bolivie. 2. Analyse phylogénétique. *Annales de Paléontologie*, **79**, 119–66.

Gagnier, P. Y., and Blieck, A. (1992). On *Sacabambaspis janvieri* and the vertebrate diversity in Ordovician seas. In *Fossil fishes as living animals*, (ed. E. Mark-Kurik). *Akademia*, **1**, 9–20.

Gagnier, P. Y., Blieck, A., and Rodrigo, G. (1986). First Ordovician vertebrate from South America. *Géobios*, **19**, 629–34.

Gans, C. (1989). Stages in the origin of vertebrates: analysis by means of scenarios. *Biological Reviews*, **64**, 221–68.

Gans, C. (1993). Evolutionary origin of the vertebrate skull. In *The skull*, (ed. J. Hanken and B. K. Hall), Vol. 2, pp. 1–35. University of Chicago Press.

Gans, C., and Northcutt, R. G. (1983). Neural crest and the origin of vertebrates: A new head. *Science*, **220**, 268–74.

Gardiner, B. G. (1967). Further notes on palaeoniscoid fishes, with a classification of the Chondrostei. *Bulletin of the British Museum (Natural History), Geology*, **14**, 143–206.

Gardiner, B. G. (1969). New palaeoniscoid fish from the Witteberg Series of South Africa. *Zoological Journal of the Linnean Society*, **48**, 423–52.

Gardiner, B. G. (1973). Interrelationships of teleostomes. In *Interrelationships of fishes*, (ed. P. H. Greenwood, R. S. Miles,

and C. Patterson), pp. 105–35. *Zoological Journal of the Linnean Society, supplement* **1**, Academic Press, London.

Gardiner, B. G. (1982). Tetrapod classification. *Zoological Journal of the Linnean Society*, **74**, 207–32.

Gardiner, B. G. (1983). Gnathostome vertebrae and the classification of the Amphibia. *Zoological Journal of the Linnean Society*, **79**, 1–59.

Gardiner, B. G. (1984a). The relationships of the palaeoniscid fishes, a review based on new specimens of *Mimia* and *Moythomasia* from the Upper Devonian of Western Australia. *Bulletin of the British Museum (Natural History), Geology*, **37**, 173–428.

Gardiner, B. G. (1984b). The relationships of placoderms. *Journal of Vertebrate Paleontology*, **4**, 379–95.

Gardiner, B. G. (1993). Haematothermia: warm blooded amniotes. *Cladistics*, **9**, 369–95.

Gardiner, B. G., and Miles, R. S. (1990). A new genus of eubrachythoracid arthrodire from Gogo, Australia. *Zoological Journal of the Linnean Society*, **99**, 159–204.

Gardiner, B. G., and Schaeffer, B. (1989). Interrelationships of lower actinopterygian fishes. *Zoological Journal of the Linnean Society*, **97**, 135–87.

Garman, S. (1904). The chimaeroids (Chismopnea Raf. 1815; Holocephala Müll. 1834), especially *Rhinochimaera* and its allies. *Bulletin of the Museum of Comparative Zoology, Harvard*, **41**, 234–72.

Garman, S. (1913). The Plagiostomia (sharks, skates, and rays). *Memoirs of the Museum of Comparative Zoology, Harvard*, **36**, 1–528.

Gaskell, W. H. (1908). *The origin of vertebrates*. Longmans, Green, and Co., London.

Gaudant, J. (1968). Contribution à une révision des *Anaethalion* de Cérin (Ain). *Bulletin du Bureau de Recherches Géologiques et Minières*, **2**, 95–115.

Gaudant, J. (1980). Louis Agassiz (1807–1873), fondateur de la paléoichthyologie. *Revue d'Histoire des Sciences*, **33**, 151–62.

Gaudin, T. J. (1991). A re-examination of elasmobranch monophyly and chondrichthyan phylogeny. *Neues Jahrbuch für Geologie und Paläontologie, Abhandlungen*, **182**, 133–60.

Gauthier, J. A. (1989). Tetrapod phylogeny. In *The hierarchy of life*, (ed. B. Fernholm, K. Bremer, and H. Jörnvall), pp. 337–53. Excerpta Medica, Amsterdam.

Gauthier, J. A., Kluge, A. G., and Rowe, T. (1988). Amniote phylogeny and the importance of fossils. *Cladistics*, **4**, 105–209.

Gayet, M., and Meunier, F. J. (1992). Polyptériformes du Maastrichien et du Paléocène de Bolivie. *Géobios, Mémoire Special*, **14**, 159–168.

Gegenbaur, C. (1872). Ueber die Kopfnerven von *Hexanchus* und ihr Verhältnis zur 'Wirbeltheorie' des Schädels. *Jenaische Zeitschrift für Medizin und Naturwissenschaften*, **6**, 497–560.

Geldsetzer, H. H. J., and Nowlan, G. S. (ed.) (1993). *Event markers in Earth history*. Palaeogeography, Palaeoclimatology, Palaeoecology, Special Issue, **104**, (1–4).

Geoffroy Saint-Hilaire (1807). Premier mémoire sur les poissons, où l'on compare les piéces osseuses de leurs nagoires pectorales avec les os de l'extrémité antérieure des autres animaux à vertèbres. *Annales du Muséum*, **9**, 357–70.

Geoffroy, Saint-Hilaire, E. (1818). *Philosophie anatomique. Des organes respiratoires sous le rapport de la détermination et de l'identité de leurs pièces osseuses*. Paris.

Geoffroy Saint-Hilaire, E. (1822). *Philosophie anatomique. Des monstruosités humaines*. Paris.

Gilmore, B. (1992). Scroll coprolites from the Silurian of Ireland and the feeding of early vertebrates. *Palaeontology*, **35**, 319–33.

Gislén, T. (1930). Affinities between the Echinodermata, Enteropneusta and Chordonia. *Zoologiska Bidrag, Uppsala*, **12**, 199–304.

Goethe, J.W. von (1890–1906). Naturwissenschaftlische Schriften. *Weimarer Sophienausgabe*, **2** (1–13).

Goette, A. (1901). Ueber die kiemen der Fische. *Zeitschrift für wissenschaftlichen Zoologie*, **69**, 533–77.

Goodrich, E. S. (1909). Vertebrata Craniata. First fascicle: Cyclostomes and fishes. In *A treatise on zoology*, (ed. E. R. Lankester), Vol. 9. Black, London.

Goodrich, E. S. (1915). The chorda tympani and middle ear in reptiles, birds and mammals. *Quarterly Journal of the Microscopical Society*, **61**, 137–60.

Goodrich, E. S. (1930). *Studies on the structure and development of vertebrates*. Macmillan, London. (Reprinted in 1958 by Dover Publications, New York, and Constable, London.)

Gorbman, A., and Tamarin, A. (1985). Early development of oral, olfactory and adenohypophyseal structures of agnathans and its evolutionary implications. In *Evolutionary biology of primitive fishes*, (ed. R. E. Foreman, A. Gorbman, J. M. Dodd, and R. Olsson), pp. 165–85. Plenum, New York.

Gordon Cumming, C. F. (1904). *Memories*. Blackwood, Edinburgh.

Gorr, T., Kleinschmidt, T., and Fricke, H. (1991). Close tetrapod relationships of the coelacanth *Latimeria* indicated by haemoglobin sequences. *Nature*, **351**, 394–7.

Goujet, D. (1973). *Sigaspis*, un nouvel Arthrodire du Dévonien inférieur du Spitsberg. *Palaeontographica*, **A 143**, 73–88.

Goujet, D. (1975). *Dicksonosteus*, un nouvel arthrodire du Dévonien du Spitsberg; remarques sur le squelette viscéral des Dolichothoraci. In *Problèmes actuels de paléontologie: evolution des Vertébrés*. Colloques internationaux du Centre national de la Recherche Scientifique, No. 218, (ed. J. P. Lehman), pp. 81–99. Paris.

Goujet, D. (1976). Les Poissons. In *Les schistes et calcaires de l'Armorique (Dévonien inférieur, Massif Armoricain, France)*. *Mémoires de la Société Géologique et Minéralogique de Bretagne*, **24**, 305–8.

Goujet, D. (1982). Les affinités des Placodermes, une revue des hypothèses actuelles. *Géobios, Mémoire Spécial*, **6**, 27–38.

Goujet, D. (1984*a*). *Les poissons Placodermes du Spitsberg. Arthrodires Dolichothoraci de la Formation de Wood Bay (Dévonien inferieur)*. Cahiers de Paléontologie, Centre national de la Recherche scientifique, Paris.

Goujet, D. (1984*b*). Placoderm interrelationships: a new interpretation, with a short review of placoderm classification. *Proceedings of the Linnean Society of New South Wales*. **107**, 211–41.

Goujet, D. (1993). Vertebrate microremains from northern Spain, their importance in correlation between marine and 'non-marine' Lower Devonian sediments. Abstracts of the Gross Symposium, Göttingen, August 1993, p. 28.

Goujet, D., Janvier, P., Rage, J. C., and Tassy, P. (1982). Structure ou modalités de l'évolution: point de vue sur l'apport de la paléontologie. In *Modalités, rythmes et mécanismes de l'évolution biologique*. Colloques internationaux du Centre national de la Recherche scientifique, No. 330, (ed. J. Chaline), pp. 137–43. Paris.

Goujet, D., Janvier, P., and Suarez-Riglos, M. (1985). Un nouveau Rhénanide (Vertebrata, Placodermi) de la Formation de Belen (Dévonien moyen), Bolivie. *Annales de Paléontologie*, **71**, 35–53.

Gould, S. J. (1991). The disparity of the Burgess Shale arthropod fauna and the limits of cladistic analysis: why we must strive to quantify morphospace. *Palaeobiology*, **17**, 411–23.

Graffin, G. (1992). A new locality of fossiliferous Harding sandstone: Evidence for freshwater Ordovician vertebrates. *Journal of Vertebrate Paleontology*, **12**, 1–10.

Graham Smith, W. (1978). On some variation in the laterosensory lines of the placoderm fish *Bothriolepis*. *Philosophical Transactions of the Royal Society of London*, **B 282**, 1–39.

Grande, L. (1989). The Eocene Green River lake system, fossil lake, and the history of the North American fish fauna. In *Mesozoic/Cenozoic vertebrate paleontology: classic localities, contemporary approaches*, (ed. J. Flynn), pp. 18–28. American Geophysical Union.

Grande, L., and Bemis, W. E. (1991). Osteology and phylogenetic relationships of fossil and Recent paddlefishes (Polyodontidae), with comments on the interrelationships of Acipenseriformes. *Society of Vertebrate Palaeontology, Memoir* **1**, *Journal of Vertebrate Paleontology*, **11** (Supplement to No. 1), 1–121.

Grande, L. and Eastman, J. T. (1986). A review of Antarctic ichthyofaunas in the light of new fossil discoveries. *Palaeontology*, **29**, 113–37.

Greenwood, P. H., Miles, R. S., and Patterson, C. (ed.) (1973). *Interrelationships of fishes. Zoological Journal of the Linnean Society, Supplement* **1**. Academic Press, London.

Gregory, W. K. (1911). The limbs of *Eryops* and the origin of paired limbs from fins. *Science*, **33**, 55–69.

Gregory, W. K. (1915). Present status of the problem of the origin of the Tetrapoda, with special reference to the skull and paired limbs. *Annals of the New York Academy of Sciences*, **26**, 317–83.

Gregory, W. K. (1930). Memorial of Bashford Dean. In *Bashford Dean memorial volume*, **1**, (ed. E. W. Gudger), pp. 1–40. American Museum of Natural History, New York.

Gregory, W. K. (1931). Cope's contribution to ichthyology. In *Cope: master naturalist*, (ed. H. F. Osborn), pp. 30–42. Princeton University Press.

Gregory, J., Morgan, T., and Reed, J. (1977). Devonian fishes in central Nevada. *University of California Riverside, Campus Museum Contributions*, **4**, 112–21.

Griffith, J. (1977). The Upper Triassic fishes from Polzberg bei Lunz, Austria. *Zoological Journal of the Linnean Society*, **60**, 1–93.

Gross, W. (1932). Die Arthrodira Wildungens. *Geologische und paläontologische Abhandlungen, Neue Folge*, **19**, 1–61.

Gross, W. (1942). Ueber Knochen-Missbildungen bei Asterolepiden. *Paläontologische Zeitschift*, **23**, 206–18.

Gross, W. (1947). Die Agnathen und Acanthodier des obersilurischen Beyrichienkalks. *Palaeontographica*, **A, 96**, 91–161.

Gross, W. (1951). Die paläontologische und stratigraphische Bedeutung der Wirbeltierfaunen des Old Reds und der marinen altpaläozoischen Schichten. *Abhandlungen der deutschen Akademie der Wissenschaften, Berlin, matematische und naturwissenschaftlische Klasse*, **1949**, 1–130.

Gross, W. (1954). Zur Conodonten-Frage. *Senckenbergiana Lethaea*, **35**, 73–85.

Gross, W. (1956). Über Crossopterygier und Dipnoer aus dem baltischen Oberdevon im Zusammenhang einiger vergleichenden Untersuchung des Porenkanalslystem paläozoischer Agnathen und Fische. *Kungliga Svenska Vetenskapsakademiens Handlingar*, **5** (6), 1–140.

Gross, W. (1958). Anaspiden-Schuppen aus dem Ludlow des Ostseegebiets. *Paläontologische Zeitschrift*, **32**, 24–37.

Gross, W. (1959). Arthrodiren aus dem Obersilur der Prager Mulde. *Palaeontographica*, **A 113**, 1–35.

Gross, W. (1961a). Aufbau des Panzers obersilurischer Heterostraci un Osteostraci Norddeutschlands (Geschiebe) und Oesels. *Acta Zoologica, Stockholm*, **42**, 73–150.

Gross, W. (1961b). *Lunaspis broilii* und *Lunaspis heroldi* aus dem Hunsrückschiefer (Unterdevon, Rheinland). *Notizblatt des hessisches Landesanstalt für Bodenforschung*, **89**, 17–43.

Gross, W. (1962a). Neuuntersuchung der Stensiöellida (Arthrodira, Unterdevon). *Notizblatt des hessisches Landesanstalt für Bodenforschung*, **90**, 48–86.

Gross, W. (1962b). Peut-on homologuer les os des Arthrodires et des Téléostomes? In *Evolution des Vertébrés*. Colloques internationaux du Centre national de la Recherche scientifique, No. 104, (ed. J. P. Lehman), pp. 69–74. Paris.

Gross, W. (1963a). *Drepanaspis gemuendenensis* Schlüter. Neuuntersuchung. *Palaeontographica*, A 121, 133–55.

Gross, W. (1963b). *Gemuendina stuerzi* Traquair. Neuuntersuchung. *Notizblatt des hessisches Landesanstalt für Bodenforschung*, **91**, 36–73.

Gross, W. (1964). Polyphyletische Stämme im System der Wirbeltiere? *Zoologische Anzeiger*, **173**, 1–22.

Gross, W. (1967). Über Thelodontier-Schuppen. *Palaeontographica*, A 127, 1–67.

Gross, W. (1968a). Porenschuppen und Sinneslinien des Thelodontiers *Phlebolepis elegans* Pander. *Paläontologische Zeitschrift*, **42**, 131–46.

Gross, W. (1968b). Fragliche Actinopterygier-Schuppen aus den Silur Gotlands, *Lethaia*, **1**, 184–218.

Gross, W. (1968c). Beobachtung mit dem Elektronenraster-Auflichtmikoskop an den Siebplatten un dem Isopedin von *Dartmuthia* (Osteostraci). *Paläontologische Zeitschrift*, **42**, 74–82.

Gross, W. (1969). *Lophosteus superbus* Pander, ein Teleostome aus dem Silur Oesels. *Lethaia*, **2**, 15–47.

Gross, W. (1971a). Downtonische und dittonische Acanthodier-Reste des Ostseegebietes. *Palaeontographica* A, **136**, 131–52.

Gross, W. (1971b). *Lophosteus superbus* Pander: Zähne, Zahnknochen und besondere Schuppenformen. *Lethaia*, **4**, 131–52.

Günther, A. C. (1871). Description of *Ceratodus*, a genus of ganoid fishes recently discovered in rivers of Queensland, Australia. *Philosophical Transactions of the Royal Society of London*, **161**, 511–71.

Gutmann, W. F., and Bonik, K. (1981a). *Kritische Evolutionstheorie*. Gerstenberg, Hildesheim.

Gutmann, W. F., and Bonik, K. (1981b). Hennigs Theorem und die Strategie des stammegeschichtlichen Rekonstruierens. Die Agnathen–Gnathostomen-Beziehung als Beispiel. *Paläontologische Zeitschrift*, **55**, 51–70.

Haeckel, E. (1866). *Generelle Morphologie der Organismen*. Reimer, Berlin.

Hagelin, L. O. (1974). Development of the membranous labyrinth in lampreys. *Acta Zoologica, Stockholm*, **55** (supplement), 1–218.

Hall, B. K. (1992). *Evolutionary developmental biology*. Chapman and Hall, London.

Hall, B. K. (ed.) (1994). *Homology: the hierarchical basis of comparative biology*. Academic Press, London.

Halstead, L. B. (1969). The origin and early evolution of calcified tissues in the vertebrates. *Proceedings of the Malacological Society of London*, **38**, 552–3.

Halstead, L. B. (1971). The presence of a spiracle in the Heterostraci (Agnatha). *Zoological Journal of the Linnean Society*, **50**, 195–7.

Halstead, L. B. (1973a). Affinities of the Heterostraci (Agnatha). *Biological Journal of the Linnean Society*, **5**, 339–49.

Halstead, L. B. (1973b). The heterostracan fishes. *Biological Reviews*, **48**, 279–332.

Halstead, L. B. (1974). *Vertebrate hard tissues*. Wykeham Publication, London.

Halstead, L. B. (1978). The cladistic revolution—can it make the grade? *Nature*, **276**, 759–60.

Halstead, L. B. (1979). Internal anatomy of the polybranchiaspids (Agnatha, Galeaspida). *Nature*, **282**, 833–6.

Halstead, L. B. (1982). Evolutionary trends and the phylogeny of the Agnatha. In *Problems of phylogenetic reconstruction*. Systematics Association Special Volume No. 21, (ed. K. A. Joysey and A. E. Friday), pp. 159–96. Academic Press, London.

Halstead, L. B. (1987a). Agnathan extinctions in the Devonian. *Mémoires de la Société Géologique de France*, **150**, 7–11.

Halstead, L. B. (1987b). Evolutionary aspects of neural crest-derived skeletogenic cells in the earliest vertebrates. In *Developmental and evolutionary aspects of the neural crest*, (ed. P. Maderson), pp. 339–58. Wiley, New York.

Halstead, L. B., Liu, Y. H., and P'an, K. (1979). Agnathans from the Devonian of China. *Nature*, **282**, 831–3.

Halstead Tarlo, L. B. (1967a). Biochemical evolution and the fossil record. In *The fossil record*, (ed. W. B. Harland, C. H. Holland, M. R. House, *et al.*) pp. 119–32. Geological Society Special Publication No. 2.

Halstead Tarlo, L. B. (1967b). The tessellated pattern of dermal armour in the Heterostraci. *Journal of the Linnean Society (Zoology)*, **47**, 45–54.

Halstead Tarlo, L. B., and Mercer, J. R. (1966). Decalcified fossil dentine. *Journal of the Royal Microscopical Society*, **86**, 137–40.

Halstead Tarlo, L. B., and Whiting, H. P. (1965). A new interpretation of the internal anatomy of the Heterostraci (Agnatha). *Nature*, **206**, 148–50.

Hanken, J., and Hall, B. K. (ed.) (1993). *The skull*, (3 vols). University of Chicago Press.

Hansen, M. C. (1978). A presumed lower dentition and a spine of a Permian petalodontiform chondrichthyan, *Megactenopetalus kaibabanus*. *Journal of Paleontology*, **52**, 55–60.

Hansen, M. C., and Mapes, R. H. (1991). A predator–prey relationship between sharks and cephalopods in the Late Paleozoic. In *Evolutionary paleobiology of behavior and coevolution*, (ed. A. J. Boucot), pp. 189–95. Elsevier, Amsterdam.

Hardisty, M. W. (1979). *The biology of the cyclostomes*. Chapman and Hall, London.

Hardisty, M. W. (1982). Lampreys and hagfishes: analysis of cyclostome relationships. In *The Biology of Lampreys*, (ed. M. W. Hardisty and I. C. Potter), Vol. 4B, pp. 165–259. Academic Press, London.

Hardisty, M. W., and Potter, I. C. (ed.) (1974–82). *The biology of lampreys*, (4 vols). Academic Press, London.

Harland, W. B., Armstrong, R. L., Cox, A. V., Craig, L. E., Smith, A. G., and Smith, D. D. G. (1990). *A geologic time scale 1989*. Cambridge University Press.

Heaton, M. J. (1980). The Cotylosauria: a reconsideration of a group of archaic tetrapods. In *The terrestrial environement and the origin of land vertebrates*, (ed. A. L. Panchen). *Systematics Association Special Volume* **15**, pp. 497–551. Academic Press, London.

Heintz, A. (1931). The structure of *Dinichthys*: A contribution to the knowledge of the Arthrodira. In *Archaic fishes. Bashford Dean memorial volume*, **4**, 115–224. American Museum of Natural History, New York.

Heintz, A. (1934). A revision of the Estonian Arthrodira. Part 1, Family Homosteidae. *Arkhiv für die Naturkunde Estlands*, Tartu, **10** (4), 1–114.

Heintz, A. (1939). Cephalaspida from the Downtonian of Norway. *Skrifter utgitt av det Norske Videnskaps-Akademi (Matematiske-naturvidenskapslige Klasse)*, **1939** (5), 1–119.

Heintz, A. (1958). The head of the anaspid *Birkenia elegans* Traq. In *Studies on fossil vertebrates*, (ed. T. S. Westoll), pp. 71–85. Athlone Press, London.

Heintz, A. (1962). Les organes olfactifs des Hétérostracés. In *Evolution des Vertébrés*. Colloques internationaux du Centre national de la Recherche scientifique, No. 104, (ed. J. P. Lehman), pp. 13–29. Paris.

Heintz, A. (1967). Some remarks about the structure of the tail in cephalaspids. In *Evolution des Vertébrés*. Colloques internationaux du Centre national de la Recherche scientifique, No. 163, (ed. J. P. Lehman), pp. 21–35. Paris.

Heintz, N. (1968). The pteraspid *Lyktaspis* n.g. from the Devonian of Vestspitsbergen. In *Current problems of lower vertebrate phylogeny*, (ed. T. Ørvig), pp. 73–88. Almqvist and Wiksell, Stockholm.

Hemmings, S. K. (1978). The Old Red Sandstone antiarchs of Scotland: *Pterichthyodes* and *Microbrachius*. *Palaeontographical Society Monograph* **131**, 1–64.

Hennig, W. (1950). *Grundzüge einer Theorie der phylogenetischen Systematik*. Deutscher Zentralverlag, Berlin.

Hennig, W. (1966). *Phylogenetic systematics*, (transl. D. D. Davis and R. Zangerl). University of Illinois Press, Urbana.

Hennig, W. (1983). Stammgeschichte der Chordaten. *Fortschritten in zoologischen Systematik und Evolutionsforschung*, **2**, 1–208.

Heyler, D. (1969). *Vertébrés de l'Autunien de France*. Cahiers de Paléontologie, Editions du Centre national de la Recherche scientifique, Paris.

Heyler, D., and Poplin, C. (ed.) (1994). *Quand le Massif Central était sous l'équateur: un écosystème carbonifère à Montceau-les-Mines*. Edition du Comité des Travaux Historiques et Scientifiques, Paris.

Hilliard, R. W., Bird, D. J., and Potter, I. C. (1983). Metamorphic changes in the intestine of three species of lampreys. *Journal of Morphology*, **176**, 181–96.

Hillis, D. M., and Dixon, M. T. (1989). Vertebrate phylogeny: evidence from 28S ribosomal DNA sequences. In *The hierarchy of life* (ed. B. Fernholm, K. Bremer, and H. Jörnvall), pp. 355–67. Excerpta Medica, Amsterdam.

Hinchliffe, J. R. (1991). Developmental approaches to the problem of transformation of limb structure in evolution. In *Developmental patterning of the vertebrate limb*, (ed. R. Hinchliffe, J. M. Hurle, and D. Summerbell). Plenum, New York.

Hinchliffe, J. R. (1994). Evolutionary developmental biology of the tetrapod limb. *Development*, Supplement, **1994**, 163–8.

Hinchliffe, J. R., and Johnson, D. R. (1980). *The development of the vertebrate limb*. Clarendon Press, Oxford.

Holmes, E. B. (1985). Are lungfishes the sister-group of tetrapods? *Biological Journal of the Linnean Society*, **25**, 379–94.

Holmgren, N. (1933). On the origin of the tetrapod limb. *Acta Zoologica, Stockholm*, **14**, 184–295.

Holmgren, N. (1946). On two embryos of *Myxine glutinosa*. *Acta Zoologica, Stockholm*, **27**, 1–90.

Holmgren, N. (1949). On the tetrapod limb again. *Acata Zoologica, Stockholm*, **30**, 485–508.

Holmgren, N., and Stensiö, E. A. (1936). Kranium un Visceralskelett des Akranier, Cyclostomen und Fische. In *Handbuch der Vergleichenden Anatomie der Wirbeltiere*, (ed. L. Bolk, E. Göppert, E. Kallius, and W. Luboscch), Vol. 4, 233–500.

House, M. R. (1989). Analysis of mid-Palaeozoic extinctions. *Bulletin de la Société belge de Géologie*, **98**, 99–107.

Hull, D. (1988). *Science as a process: an evolutionary account of the social and conceptual development of science*. University of Chicago Press.

Hull, D. (1989). The evolution of phylogenetic systematics. In *The hierarchy of life*, (ed. B. Fernholm, K. Bremer, and H. Jörnvall), pp. 3–15. Excerpta Medica, Amsterdam.

Humphries, C. J., and Parenti, L. (1986). *Cladistic biogeography*, Clarendon Press, Oxford.

Huxley, T. H. (1858*a*). On *Cephalaspis* and *Pteraspis*. *Quarterly Journal of the Geological Society of London*, **14**, 267–80.

Huxley, T. H. (1858*b*). The Croonian Lecture. On the theory of the vertebrate skull. *Proceedings of the Royal Society of London*, **9**, 381–457.

Huxley, T. H. (1861). Preliminary essay upon the systematic arrangement of the fishes of the Devonian epoch. *Memoirs of the Geological Survey of the United Kingdom*, **10**, 1–40.

Huxley, T. H. (1880). On the application of the laws of evolution to the arrangement of the Vertebrata and more particularly of the Mammalia. *Proceedings of the Zoological Society of London*, **1880**, 649–62.

Hyde, J. E. (1926). Collecting fossil fishes from the Cleveland Shales. *Natural History*, **26**, 497–504.

Ilyes, R. R., and Elliott, D. K. (1994). New Early Devonian pteraspidids (Agnatha, Heterostraci) from east-central Nevada. *Journal of Paleontology*, **68**, 878–92.

Ivachnenko, M. F. (1978). [Urodeles from the Triassic and Jurassic of Soviet Central Asia.] *Paleontologicheskii Zhurnal*, **12**, 362–68. (In Russian.)

Ivanov A., Pavlov, D., and Cherepanov, G. (1992). Some aspects of biomineralization of vertebrate skeleton. In *Fossil fishes as living animals*, (ed. E. Mark-Kurik), *Academia*, **1**, 159–64.

Jaekel, O. (1911). *Die Wirbeltiere*. Borntraeger, Berlin.

Janvier, P. (1974). The structure of the naso-hypophysial complex and the mouth in fossil and extant cyclostomes, with remarks on amphiaspiforms. *Zoologica Scripta*, **3**, 193–200.

Janvier, P. (1975a). Spécialisations précoces et caractères primitifs du système circulatoire des Ostéostracés. In *Problèmes actuels de paléontologie. Evolution des Vertébrés*. Colloques internationaux du Centre national de la Recherche scientifique, No. 218, (ed. J. P. Lehman), pp. 15–31. Paris.

Janvier, P. (1975b). Anatomie et position systématique des Galéaspides (Vertebrata, Cyclostomata), Céphalaspidomorphes du Dévonien inférieur du Yunnan (Chine). *Bulletin du Muséum national d'Histoire naturelle, Paris*, **278**, 1–16.

Janvier, P. (1975c). Les yeux des Cyclostomes fossiles et le problème de l'origine des Myxinoïdes. *Acta Zoologica, Stockholm*, **56**, 1–9.

Janvier, P. (1975d). Remarques sur l'orifice naso-hypophysaire des Céphalaspidomorphes. *Annales de Paléontologie (Vertébrés)*, **61**, 3–16.

Janvier, P. (1978a). Les nageoires paires des Ostéostracés et la position systématique des Céphalaspidomorphes. *Annales de Paléontologie (Vertébrés)*, **64**, 113–42.

Janvier, P. (1978b). On the oldest known teleostome fish *Andreolepis hedei* Gross (Ludlow of Gotland), and the systematic position of the lophosteids. *Eesti NSV Teaduste Akadeemia Toimetised, Geoloogia*, **27**, 86–95.

Janvier, P. (1980a). Osteolepid remains from the Devonian of the Middle East, with particular reference to the endoskeletal shoulder girdle. In *The terrestrial environment and the origin of land vertebrates*. Systematics Association Special Volume No. 15, (ed. A. L. Panchen), pp. 223–54. Academic Press, London.

Janvier, P. (1980b). L'origine des Tétrapodes: nouveaux aspects du problème. *Comptes-Rendus du 105e Congrès national des Sociétés Savantes, Caen*, **3**, 35–46. Bibliothèque Nationale, Paris.

Janvier, P. (1981a). The phylogeny of the Craniata, with particular reference to the significance of fossil 'agnathans'. *Journal of Vertebrate Paleontology*, **1**, 121–59.

Janvier, P. (1981b). *Norselaspis glacialis* n.g., n.sp. et les relations phylogénétiques entre les Kiaeraspidiens (Osteostraci) du Dévonien inférieur du Spitsberg. *Palaeovertebrata*, **11** (2–3), 19–131.

Janvier, P. (1983). Groupes panchroniques, 'fossiles vivants' et systématique: l'exemple des 'Crossopterygii' et des Petromyzontida. *Bulletin de la Société Zoologique de France*, **108**, 609–16.

Janvier, P. (1984). The relationships of the Osteostraci and Galeaspida. *Journal of Vertebrate Paleontology*, **4**, 344–58.

Janvier, P. (1985a). *Les Céphalaspides du Spitsberg. Anatomie, phylogénie et systématique des Ostéostracés siluro–dévoniens. Révision des Ostéostracés de la Formation de Wood Bay (Dévonien inférieur du Spitsberg)*. Cahiers de Paléontologie. Centre national de la Recherche scientifique, Paris.

Janvier, P. (1985b). Les Thyestidiens (Osteostraci) du Silurien de Saaremaa (Estonie). *Annales de Paléontologie*. Première partie: Morphologie et anatomie, **71** (2), pp. 83–147. Deuxième partie: Analyse phylogénétique, répartition stratigraphique, remarques sur les genres *Auchenaspis, Timanaspis, Tyriaspis, Didymaspis, Sclerodus* et *Tannuaspis*. **71** (3), pp. 187–216.

Janvier, P. (1985c). Environmental framework of the diversification of the Osteostraci during the Silurian and Devonian. *Philosophical Transactions of the Royal Society of London*, **B 309**, 259–72.

Janvier, P. (1986). Les nouvelles conceptions de la phylogénie et de la classification des 'Agnathes' et des Sarcoptérygiens. *Océanis*, **12**, 123–38.

Janvier, P. (1987). The paired fins of anaspids: one more hypothesis about their function. *Journal of Paleontology*, **61**, 850–53.

Janvier, P. (1989). Ontogénie, phylogénie et homologie: les tests de l'hétérochronie. *Géobios, Mémoire Spécial* **12**, 245–55.

Janvier, P. (1990). La structure de l'exosquelette des Galéaspides (Vertebrata). *Comptes-Rendus de l'Académie des Sciences, Paris*, **310**, 655–9.

Janvier, P. (1991). The Permian and Triassic vertebrates of Bolivia. In *Fosiles y facies de Bolivia*, (ed. R. Suarez-Soruco). *Revista tecnica de YPFB, Santa Cruz*, **12**, 389–91.

Janvier, P. (1992). Les écailles des Trématosaures (Tetrapoda, Temnospondyli): nouvelles données sur les Trématosaures du Trias inférieur de Madagascar. *Bulletin du Muséum national d'Histoire naturelle, Paris*, **14**, 3–13.

Janvier, P. (1993). Patterns of diversity in the skull of jawless fishes. In *The skull*, (ed. J. Hanken and B. K. Hall), Vol. 2, pp. 131–88. University of Chicago Press.

Janvier, P., and Blieck, A. (1979). New data on the internal anatomy of the Heterostraci (Agnatha), with general remarks on the phylogeny of the Craniota. *Zoologica Scripta*, **8**, 287–96.

Janvier, P., and Blieck, A. (1993). L. B. Halstead and the heterostracan controversy. *Modern Geology*, **18**, 89–105.

Janvier, P., and Dingerkus, G. (1991). Le synarcual de *Pucapampella* Janvier et Suarez-Riglos: une preuve de l'existence d'Holocéphales (Vertebrata, Chondrichthyes) dès le Dévonien moyen. *Contres Rendus de l'Académie des Sciences, Paris*, **312**, 549–52.

Janvier, P., and Lund, R. (1983). *Hardistiella montanensis* n. gen. et sp. (Petromyzontid*a*) from the Lower Carboniferous of Montana, with remarks on the affinities of the lampreys. *Journal of Vertebrate Paleontology*, **2**, 407–13.

Janvier, P., and Lund, R. (1985). Ces étranges bêtes du Montana. *La Recherche*, **162**, 98–100.

Janvier, P., and Melo, J. H. G. 1986. Acanthodian fish remains from the Upper Silurian or Lower Devonian of the Amazon Basin, Brazil. *Palaeontology*, **31**, 771–77.

Janvier, P., and Pan, J. (1982). *Hyrcanaspis bliecki* n.g., n.sp., a new primitive euantiarch (Antiarcha, Placodermi) from the Middle Devonian of northeastern Iran, with a discussion on antiarch phylogeny. *Neues Jahrbuch für Geologie und Paläontologie, Abhandlungen*, **164**, 384–92.

Janvier, P., and Suarez-Riglos, M. 1986. The Silurian and Devonian vertebrates of Bolivia. *Bulletin de l'Institut français d'Etudes andines*, Lima, **15**, 73–114.

Janvier, P., Lethiers, F., Monod, O., and Balkas, Ö. (1984). Discovery of a vertebrate fauna at the Devonian–Carboniferous boundary in S. E. Turkey (Hakkari Province). *Journal of Petroleum Geology*, **7**, 147–68.

Janvier, P., Percy, Lord R., and Potter, I. C. (1991). The arrangement of the heart chambers and associated blood vessels in the Devonian osteostracan *Norselaspis glacialis*. A reinterpretation based on recent studies of the circulatory system in lampreys. *Journal of the Zoological Society of London*, **223**, 567–76.

Janvier, P., Tong-Dzuy, T., and Ta-Hoa, P. (1993). A new Early Devonian galeaspid from Bac Thai Province, Vietnam. *Palaeontology*, **36**, 297–309.

Jarvik, E. (1942). On the structure of the snout of crossopterygians and lower gnathostomes in general. *Zoologiska Bidrag, Uppsala*, **21**, 235–675.

Jarvik, E. (1944). On the exoskeletal shoulder girdle of teleostomian fishes, with special reference to *Eusthenopteron foordi* Whiteaves. *Kungliga Svenska Vetenskapsakademiens Handlingar*, **21**, 1–32.

Jarvik, E. (1948*a*). On the morphology and taxonomy of the Middle Devonian osteolepid fishes of Scotland. *Kungliga Svenska Vetenskapsakademiens Handlingar*, **25**, 1–301.

Jarvik, E. (1948*b*). Note on the Upper Devonian vertebrate fauna of East Greenland and on the age of the ichthyostegid stegocephalians. *Arkiv för Zoologi*, **13**, 1–8.

Jarvik, E. (1952). On the fish-like tail in the ichthyostegid stegocephalians with descriptions of a new stegocephalian and a new crossopterygian from the Upper Devonian of East Greenland. *Meddelelser om Grønland*, **114**, 1–90.

Jarvik, E. (1954). On the visceral skeleton in *Eusthenopteron*, with a discussion of the parasphenoid and palatoquadrate in fishes. *Kungliga Svenska Vetenskapsakademiens Handlingar*, **5**, 1–104.

Jarvik, E. (1959). Dermal fin-rays and Holmgren's principle of delamination? *Kungliga Svenska Vetenskapsakademiens Handlingar*, **6**, 1–51.

Jarvik, E. (1964). Specializations in early vertebrates. *Annales de la Société Royale Zoologique de Belgique*, **94** (1), 11–95.

Jarvik, E. (1965*a*). Die Raspelzunge der Cyclostomen und die pentadactyle Extremität der Tetrapoden als beweise für monophyletische Herkunft. *Zoologische Anzeiger*, **175**, 101–43.

Jarvik, E. (1965*b*). On the origin of girdles and paired fins. *Israel Journal of Zoology*, **14**, 141–72.

Jarvik, E. (1966). Remarks on the structure of the snout in *Megalichthys* and certain other rhipidistid crossopterygians. *Arkiv för Zoologi*, **19**, 41–98.

Jarvik, E. (1967). The homologies of the frontal and parietal bones in fishes and tetrapods. In *Evolution des Vertébrés*. Colloques internationaux du Centre national de la Recherche scientifique, Vol. 163, (ed. J. P. Lehman), pp. 181–213. Paris.

Jarvik, E. (1968). Aspects of vertebrate phylogeny. In *Current problems of lower vertebrate phylogeny*, (ed. T. Ørvig), pp. 497–527. Almqvist and Wiksell, Stockholm.

Jarvik, E. (1969). På jakt efter den fyrbenta fisken och vår egen anfader. [Hunting the four-legged fish and our own ancestors.] *Fauna och Flora*, **5-6**, 225–67. (In Swedish.)

Jarvik, E. (1972). Middle and Upper Devonian Porolepiformes from East Greenland with special reference to *Glyptolepis groenlandica* n.sp., and a discussion on the structure of the head in the Porolepiformes. *Meddelelser om Grønland*, **187**, 1–307.

Jarvik, E. (1977). The systematic position of acanthodian fishes. In *Problems in vertebrate evolution*, (ed. S. M. Andrews, R. S. Miles, and A. D. Walker), pp. 199–225. Academic Press, London.

Jarvik, E. (1980). *Basic structure and evolution of vertebrates*, Vol. 1. Academic Press, London.

Jarvik, E. (1981*a*). *Basic structure and evolution of vertebrates*, Vol. 2. Academic Press, London.

Jarvik, E. (1981*b*). Review of D. E. Rosen *et al.*: Lungfishes, tetrapods, paleontology and plesiomorphy. *Systematic Zoology*, **30**, 378–84.

Jarvik, E. (1986). The origin of the Amphibia. In *Studies in herpetology*, (ed. Z. Rocek), pp. 1–21. Charles University, Prague.

Jarvik, E. (1988). The early vertebrates and their forerunners. In *L'évolution dans sa réalité et ses diverses modalités*, pp. 36–64. Fondation Singer-Polignac and Masson, Paris.

Jarvik, E. (1992). Erik Stensiö. *Levnadsteckningar över Kungliga Vetenskapsakademiens ledamöter*, **190**, 1–36.

Jefferies, R. P. S. (1968). The subphylum Calcichordata (Jefferies 1967)—Primitive fossil chordates with echinoderm affinities. *Bulletin of the British Museum (Natural History), Geology*, **16**, 243–339.

Jefferies, R. P. S. (1986). *The ancestry of the vertebrates*. British Museum (Natural History), London.

Jefferies, R. P. S. and Lewis, N. D. (1978). The English fossil *Placocystites forbesianus* and the ancestry of the vertebrates. *Philosophical Transactions of the Royal Society of London*, **B 282**, 205–323.

Jenkins, F. A., and Walsh, D. M. (1993). An Early Jurassic caecilian with limbs. *Nature*, **365**, 246–50.

Jessen, H. (1966). Die Crossopterygier des Oberen Plattenkalkes (Devon) des Bergisch-Gladbach-Paffrather Mulde (Rheinisches Schiefergebirge) unter Berücksichtigung von amerikanischem und europäischem *Onychodus*-Material. *Arkiv för Zoologi*, **18**, 305–89.

Jessen, H. (1968). *Moythomasia nitida* Gross und *M.* cf. *striata* Gross, devonische Palaeonisciden aus dem Oberen Plattenkalk der Bergisch-Gladbach-Paffrather Mulde (Rheinisches Schiefergebirge). *Palaeontographica*, **A 128**, 87–114.

Jessen, H. (1972). Schultergürtel und Pectoralflosse bei Actinopterygiern. *Fossils and Strata*, **1**, 1–110.

Jessen, H. (1973). Weitere Fischreste aus dem Oberen Plattenkalk der Bergisch-Gladbach-Paffrather Mulde (Oberdevon, Rheinisches Schiefergebirge).*Palaeontographica*, **143**, 159–87.

Jessen, H. (1980). Lower Devonian porolepiformes from the Canadian Arctic with special reference to *Powichthys thorsteinssoni* Jessen. *Palaeontographica*, **A 167**, 180–214.

Johnels, A. (1948). On the development and morphology of the skeleton of the head of *Petromyzon*. *Acta Zoologica, Stockholm*, **29**, 140–278.

Johnels, A. (1950). On the dermal connective tissue of lampreys. *Acta Zoologica, Stockholm*, **31**, 177–85.

Johnson, G. D., and Anderson, W. D. Jr. (ed.) (1993). Proceedings of the Symposium on the Phylogeny of Percomorpha, Charleston. *Bulletin of Marine Science*, **52** (1).

Johnson, G. D., and Patterson, C. (1993). Percomorph phylogeny. In *Proceedings of the Symposium of Phylogeny of Percomorpha, Charleston*, (ed. G. D. Johnson and W. R. Anderson Jr.). *Bulletin of Marine Science*, **52**, 515–623.

Jollie, M. (1962). *Chordate morphology*. Reinhold, New York.

Jollie, M. (1968). Some implications of the acceptance of a delamination principle. In *Current problems of lower vertebrate phylogeny*, (ed. T. Ørvig), pp. 89–107. Almqvist and Wiksell, Stockholm.

Jollie, M. (1971). A theory concerning the early evolution of the visceral arches. *Acta Zoologica, Stockholm*, **52**, 85–96.

Jollie, M. (1973). The origin of the chordates. *Acta Zoologica, Stockholm*, **54**, 81–100.

Jollie, M. (1981). Segment theory and the homologizing of cranial bones. *American Naturalist*, **118**, 783–802.

Jollie, M. (1984). The vertebrate head—segmented or a single morphogenetic structure? *Journal of Vertebrate Paleontology*, **4**, 320–9.

Karatayute-Talimaa, V. N. (1973). *Elegestolepis grossi* gen. et sp. nov., ein neuer Typ der Placoidschuppe aus dem Obersilur der Tuwa. *Palaeontographica*, **A 143**, 35–50.

Karatayute-Talimaa, V. N. (1978). [*Silurian and Devonian thelodonts of the USSR and Spitsbergen.*] Mosklas, Vilnius. (In Russian.)

Karatayute-Talimaa, V. N. (1992). The early stages of the dermal skeleton formation in chondrichthyans. In *Fossil fishes as living animals*, (ed. E. Mark-Kurik). *Akademia* **1**, 223–31.

Karatayute-Talimaa, V. N., Novitskaya, L. I., Rosman, K. C., and Sodov, J. (1990). [*Mongolepis*, a new elasmobranch genus from the Lower Silurian of Mongolia.] *Paleontologicheskii Zhurnal*, **1990**, 7–86. (In Russian.)

Kent, D. V., and Van der Voo, R. (1990). Palaeozoic palaeogeography from palaeomagnetism of the Atlantic-bordering continents. In *Palaeozoic palaeogeography and biogeography*, (ed. W. S. McKerrow and C. R. Scotese). *Geological Society Memoir* **12**, 49–56.

Kesteven, H. L. (1950). The origin of the tetrapods. *Proceedings of the Royal Society of Victoria*, **59**, 93–138.

Kiaer, J. (1924). The Downtonian fauna of Norway. 1. Anaspida. *Skrifter utgitt af det Norske Videnskapsakademien, 1, matematisk-naturvidenskaplige Klasse*, **6**, 1–139.

Kiaer, J. (1928). The structure of the mouth of the oldest known vertebrates, pteraspids and cephalaspids. *Palaeobiologica*, **1**, 117–34.

Kiaer, J. (1932). The Downtonian and Devonian vertebrates of Spitsbergen. IV: Suborder Cyathaspidida. *Skifter om Svalbard og Ishavet*, **40**, 1–138.

Kiaer, J., and Heintz, A. (1935). The Downtonian and Dittonian vertebrates of Spitsbergen. IV. Suborder Cyathaspida. *Skifter om Svalbard og Ishavet*, **40**, 1–138.

Kingsbury, B. F. (1926). Branchiomerism and the theory of head segmentation. *Journal of morphology*, **42**, 83–109.

Koenen, A. von (1895). Ueber einige Fischreste des norddeutschen und böhmischen Devons. *Abhandlungen der Königlischen Gesellschaft für Wissenschaften, Göttingen*, **40**, 1–37.

Krejsa, R. J., Bringas, P., and Slavkin, H. (1990). A neontological interpretation of conodont elements based on agnathan cyclostome tooth structure, function, and development. *Lethaia*, **23**, 359–78.

Kulczycki, J. (1960). *Porolepis* (Crossopterygii) from the Lower Devonian of the Holy Cross Mountains. *Acta Palaeontologica Polonica*, **5**, 65–104.

Kupffer, C. von (1892). Entwicklungsgeschichte des Kopfes. Anatomische Hefte, Abt. 2. *Ergebnisse der Anatomie und Entwicklungsgeschichte*, **2**, 501–64.

Lake, J. A. (1990). Origin of the Metazoa. *Proceedings of the National Academy of Sciences of the USA*, **87**, 763–6.

Lambers, P. (1992). *On the ichthyofauna of the Solnhofen lithographic limestone (Upper Jurassic, Germany)*. Rijksuniversiteit Groningen.

Langille, R. M., and Hall, B. K. (1993). Pattern formation and the neural crest. In *The skull*, (ed. J. Hanken and B. K. Hall), Vol. 1, pp. 77–111. University of Chicago Press.

Lankester, E. R. (1864). Scales of *Pteraspis*. *Quarterly Journal of the Geological Society*, **20**, 194.

Lankester, E. R. (1868–70). *A monograph of the fishes of the Old Red Sandstone of Britain. 1. The Caphalaspidae.* Palaeontographical Society, London.

Lankester, E. R. (1873). On *Holaspis sericeus*, and on the relationships of the fish-genera *Pteraspis*, *Cyathaspis* and *Scaphaspis*. *Geological Magazine*, **10**, 241–5.

Lauder, G. V. (1980*a*). On the evolution of the jaw adductor musculature in primitive gnathostome fishes. *Breviora*, **460**, 1–10.

Lauder, G. V. (1980*b*). Evolution of the feeding mechanism in primitive actinopterygian fishes: a functional anatomical analysis of *Polypterus*, *Lepisosteus*, and *Amia*. *Journal of Morphology*, **163**, 283–317.

Lauder, G. V. (1985). Aquatic feeding in lower vertebrates. In *Functional vertebrate morphology*, (ed. M. Hildebrand, D. Bramble, K. F. Liem, and D. B. Wake), pp. 210–29. Harvard University Press, Cambridge, Mass.

Lauder, G. V., and Liem, K. F. (1983). The evolution and interrelationships of the actinopterygian fishes. *Bulletin of the Museum of Comparative Zoology, Harvard*, **150**, 95–197.

Laurent, R. F. (1985). Sous-classe des Lissamphibiens (Lissamphibia). In *Traité de zoologie*, (ed. P. P. Grassé), Vol 14 (1b), pp. 594–797. Masson, Paris.

Lê, H. L. V., Lecointre, G., and Perasso, R. (1993). A 28S rRNA-based phylogeny of the gnathostomes: first steps in the analysis of conflict and congruence with morphology-based cladograms. *Molecular Phylogenetics and Evolution*, **2**, 31–51.

Lebedev, O. A. (1984). [First discovery of a Devonian tetrapod vertebrate in the USSR.] *Dokhlady Akademia Nauk SSSR*, **278**, 1470–3. (In Russian.)

Lebedev, O. A. (1985). [Old and new data on the origin of tetrapods.] *Priroda*, **11**, 26–36. (In Russian.)

Lebedev, O. A. (1990). *Tulerpeton*, l'animal à six doigts. *La Recherche*, **225**, 1274–5.

Lebedev, O. A. (1992). The latest Devonian, Khovanian vertebrate assemblage of Andreyevka-2 locality, Tula region Russia. In *Fossil fishes as living animals*, (ed. E. Mark-Kurik). *Akademia*, **1**, 265–72.

Lebedev, O. A., and Clack, J. A. (1993). Upper Devonian tetrapods from Andreyevka, Russia. *Palaeontology*, **36**, 721–34.

Lecointre, G. (1994). Aspects historiques et heuristiques de l'ichtyologie systématique. *Cybium*, **18**, 339–430.

Lehman, J. P. (1956). Les Arthrodires du Dévonien supérieur du Tafilalt (Sud Marocain). *Notes et Mémoires du Service Géologique du Maroc*, **129**, 1–70.

Lehman, J. P. (1966). Actinopterygii, Dipnoi, Crossopterygii, Brachiopterygii. In *Traité de paléontologie*, (ed. J. Piveteau), Vol. 4 (3), pp. 1–387, 398–420. Masson, Paris.

Lehman, J. P. (1968). Remarques concernant la phylogénie des Amphibiens. In *Current problems of lower vertebrate phylogeny*, (ed. T. Ørvig), pp. 307–15. Almqvist and Wiksell, Stockholm.

Lehman, J. P. (1979). Notes sur les Poissons du Trias de Lunz. 1. *Thoracopterus* Bronn. *Annalen des Naturhistorischen Museum, Wien*, **82**, 56–66.

Lelièvre, H. (1984). *Antineosteus lehmani, n.g., n.sp.*, nouveau Brachythoraci du Dévonien inférieur du Maroc présaharien. *Annales de Paléontologie*, **70**, 115–58.

Lelièvre, H. (1991). New information on the structure and the systematic position of *Tafilalichthys lavocati* (placoderm, arthrodire) from the Late Devonian of Tafilalt, Morocco. In *Early vertebrates and related problems of evolutionary biology*, (ed. M. M. Chang, Y. H. Liu, and G. R. Zhang), pp. 121–30. Science Press, Beijing.

Lelièvre, H., Goujet, D., and Henn, A. (1990). Un nouveau spécimen d'*Holonema radiatum* (Placodermi, Arthrodira) du Dévonien moyen de la région d'Oviedo, Espagne. *Bulletin du Muséum national d'Histoire naturelle, Paris* **12**, 53–83.

Lelièvre, H., Janvier, P., and Blieck, A. (1993). Silurian-Devonian vertebrate biostratigraphy of western Gondwana and related terranes (South America, Africa, Armorica–Bohemia, Middle East). In *Palaeozoic vertebrate biostratigraphy and biogeography*, (ed. J. A. Long), pp. 139–73. Belhaven Press, London.

Leonardi, G. (1983). *Notopus petri* nov. gen., nov. sp.: une empreinte d'amphibien du Dévonien du Parana (Brésil). *Géobios*, **16**, 233–9.

Lessertisseur, J., and Robineau, D. (1969–70). Le mode d'alimentation des premiers Vertébrés et l'origine des mâchoires. I. Les faits et les théories. *Bulletin du Muséum national d'Histoire naturelle, Paris*, **41** (6), 13–23; II. Les cor-

rélations et les conséquences. *Bulletin du Muséum national d'Histoire naturelle, Paris*, **42** (1), 102–21.

Li, Z. X., McA. Powell, C., and Trench, A. (1993). Palaeozoic global reconstructions. In *Palaeozoic vertebrate biostratigraphy and biogeography*, (ed. J. A. Long), pp. 25–53. Belhaven Press, London.

Linnaeus, C. (1758). *Systema naturae per regna tria naturae*, (10th edn). Laurentii Salvii, Stockholm.

Liu, Y. H. (1965). [New Devonian agnathans of Yunnan.] *Vertebrata PalAsiatica*, **9**, 125–34. (In Chinese with English summary.)

Liu, Y. H. (1973). [On the new forms of Polybranchiaspiformes and Petalichthyida from Devonian of Southwest China.] *Vertebrata PalAsiatica*, **11**, 132–43. (In Chinese.)

Liu, Y. H. (1975). [Lower Devonian agnathans of Yunnan and Sichuan.] *Vertebrata PalAsiatica*, **13**, 215–23. (In Chinese with English summary.)

Liu, Y. H. (1986). [On the sensory-line system of Galeaspida (Agnatha).] *Vertebrata PalAsiatica*, **24**, 245–59. (In Chinese with English summary.)

Liu, Y. H. (1991). On a new petalichthyid, *Eurycaraspis incilis* gen. et sp. nov. (Placodermi, Pisces) from the Middle Devonian of Zhanyi, Yunnan. In *Early vertebrates and related problems of evolutionary biology*, (ed. M. M. Chang, Y. H. Liu, and G. R. Zhang), pp. 137–77. Science Press, Beijing.

Liu, T. S., and P'an, K. (1958). Devonian fishes from the Wutung Series near Nanking, China. *Palaeontologica Sinica*, **141**, 1–41.

Lombard, R. E., and Bolt, J. R. (1979). Evolution of the tetrapod ear: an analysis and reinterpretation. *Biological Journal of the Linnean Society*, **11**, 19–76.

Long, J. A. (1983). A new diplacanthoid acanthodian from the Late Devonian of Victoria. *Memoirs of the Association of the Australasian Palaeontologists*, **1**, 51–65.

Long, J. A. (1984). New phyllolepids from Victoria and the relationships of the group. *Proceedings of the Linnean Society of New South Wales*, **107**, 263–308.

Long, J. A. (1985*a*). New information on the head and shoulder girdle or *Canowindra grossi* Thomson, from the Late Devonian Mandagery Sandstone, New South Wales. *Records of the Australian Museum*, **37**, 91–9.

Long, J. A. (1985*b*). The structure and relationships of a new osteolepiform fish from the Late Devonian of Victoria, Australia. *Alcheringa*, **9**, 1–22.

Long, J. A. (1985*c*). A new osteolepid fish from the Upper Devonian Gogo Formation, western Australia. *Records of the Western Australian Museum*, **12**, 361–7.

Long, J. A. (1986*a*). A new Late Devonian acanthodian fish from Mt Howitt, Victoria, Australia, with remarks on acanthodian biogeography. *Proceedings of the Royal Society of Victoria*, **98**, 1–17.

Long, J. A. (1986*b*). New ischnacanthid acanthodians from the Early Devonian of Australia, with comments on acanthodian interrelationships. *Zoological Journal of the Linnean Society*, **87**, 321–39.

Long, J. A. (1987). An unusual osteolepiform fish from the Late Devonian of Victoria, Australia. *Palaeontology*, **30**, 839–52.

Long, J. A. (1988*a*). Late Devonian fishes from Gogo, Western Australia. *National Geographic Research*, **4**, 436–50.

Long, J. A. (1988b). New palaeoniscoid fishes from the Late Devonian and Early Carboniferous of Victoria. In *Devonian and Carboniferous fish studies*, (ed. P. A. Jell). *Memoirs of the Association of the Australasian Palaeontologists*, **7**, 1–64

Long, J. A. (1988c). A new camuropiscid arthrodire (Pisces: Placodermi) from Gogo, Western Australia. *Zoological Journal of the Linnean Society*, **94**, 233–58.

Long, J. A. (1989). A new rhizodontiform fish from the Early Carboniferous of Victoria, Australia, with remarks on the phylogenetic position of the group. *Journal of Vertebrate Paleontology*, **9**, 1–17.

Long, J. A. (1990). Heterochrony and the origin of tetrapods. *Lethaia*, **23**, 157–66.

Long, J. A. (ed.) (1993a). *Palaeozoic vertebrate biostratigraphy and biogeography*. Belhaven Press, London.

Long, J. A. (1993b). Early–Middle Palaeozoic vertebrate extinction events. In *Palaeozoic vertebrate biostratigraphy andbiogeography*, pp. 54–63. Belhaven Press, London.

Long, J. A. (1995). *Rise of fishes*. University of New South Wales Press and Johns Hopkins University Press, Sydney.

Long, J. A., and Werderlin, L. (1986). Allometry in the placoderm *Bothiolepis canadensis* and its significance to antiarch evolution. *Lethaia*, **19**, 161–9.

Løvtrup, S. (1974). *Epigenetics—A treatise on theoretical biology*. University of Chicago Press.

Løvtrup, S. (1977). *The phylogeny of Vertebrata*. Wiley, New York.

Løvtrup, S. (1985). On the classification of the taxon Tetrapoda. *Systematic Zoology*, **34**, 463–70.

Lowenstein, O. and Thornhill, R. A. (1970). The labyrinth of *Myxine*: anatomy, ultrastructure and electrophysiology. *Proceedings of the Royal Society of London*, **B 176**, 21–42.

Lund, R. (1967). An analysis of the propulsive mechanism of fishes, with reference to some fossil actinopterygians. *Annals of the Carnegie Museum*, **39**, 195–218.

Lund, R. (1977a). A new petalodont (Chondrichthyes, Bradyodonti), from the Upper Mississippian of Montana. *Annals of the Carnegie Museum*, **46**, 129–55.

Lund, R. (1977b). New information on the evolution of the bradyodont Chondrichthyes. *Fieldiana: Geology*, **33**, 521–39.

Lund, R. (1977c). *Echinochimaera meltoni*, new genus and species (Chimaeriformes), from the Mississippian of Montana. *Annals of the Carnegie Museum*, **46**, 195–221.

Lund, R. (1980). Viviparity and intrauterine feeding in a new holocephalan fish from the Lower Carboniferous of Montana. *Science*, **209**, 697–9.

Lund, R. (1982). *Harpagofututor volsellorhinus* new genus and species (Chondrichthyes, Chondrenchelyformes) from the Namurian Bear Gulch Limestone, *Chondrenchelys problematica* Traquair (Visean), and their sexual dimorphism. *Journal of Paleontology*, **56**, 938–58.

Lund, R. (1983). The dentition of *Polyrhizodus* (Chondrichthyes, Petalodontiformes) from the Namurian Bear Gulch Limestone of Montana. *Journal of Vertebrate Paleontology*, **3**, 1–6.

Lund, R. (1984). On the spines of the Stethacanthidae (Chondrichthyes) with a description of a new genus from the Mississippian Bear Gulch Limestone. *Géobios*, **17**, 281–95.

Lund, R. (1985a). Stethacanthid elasmobranch remains from the Bear Gulch Limestone (Namurian E2b) of Montana. *American Museum Novitates*, **2828**, 1–24.

Lund, R. (1985b). The morphology of *Falcatus falcatus* (St. John and Worthen), a Mississippian stethacanthid chondrichthyan from the Bear Gulch Limestone of Montana. *Journal of Vertebrate Paleontology*, **5**, 1–19.

Lund, R. (1986a). On *Damocles serratus*, nov. gen. et sp. (Elasmobranchii: Cladodontida) from the Upper Mississippian Bear Gulch Limestone of Montana. *Journal of Vertebrate Paleontology*, **6**, 12–19.

Lund, R. (1986b). The diversity and relationships of the holocephali. In *Indo-Pacific Fish Biology*, (ed. T. Uyeno, R. Arai, T. Taniuchi, and K. Matsuura), pp. 97–106. Ichthyological Society of Japan, Tokyo.

Lund, R. (1986c). New Mississippian Holocephali (Chondrichthyes) and the evolution of the Holocephali. In *Teeth revisited: Proceedings of the VIIe International Symposium on Dental Morphology, Paris 1986*, (ed. D. E. Russell, J. P. Santoro, and D. Sigogneau-Russell). Mémoires du Muséum national d'Histoire naturelle, Paris, **C 53**, 195–205.

Lund, R. (1989). New petalodonts (Chondrichthyes) from the Upper Mississippian Bear Gulch Limestone (Namurian E2b) of Montana. *Journal of Vertebrate Paleontology*, **9**, 350–68.

Lund, R. (1990). Chondrichthyan life history styles as revealed by the 320 million years old Mississippian of Montana. *Environmental Biology of Fishes*, **27**, 1–19.

Lund, R., and Janvier, P. (1986). A second lamprey from the Lower Carboniferous (Namurian) of Bear Gulch, Montana (U.S.A.). *Géobios*, **19**, 647–52.

Lund, R., and Lund, W. (1984). New genera and species of coelacanths from the Bear Gulch limestone (Lower Carboniferous) of Montana (U.S.A.). *Géobios*, **17**, 237–44.

Lund, R., and Zangerl, R. (1974). *Squatinactis caudispinatus*, a new elasmobranch from the Upper Mississippian of Montana. *Annals of the Carnegie Museum*, **45** (4), 43–55.

Lurie, E. (1960). *Louis Agassiz: a life in science*. University of Chicago Press.

Lyarskaya, L. A. (1981). *Baltic Devonian Placodermi. Asterolepididae*. Znatne, Riga.

Lyarskaya, L. A., and Mark-Kurik, E. (1972). Eine neue Fundstelle oberdevonischer Fische im Baltikum. *Neues Jahrbuch für Geologie und Paläontologie, Monatshefte*, **1972**, 407–14.

McAllister, J. A. (1985). Reevaluation of the formation of spiral coprolites. *University of Kansas Paleontological Contributions*, **114**, 1–11.

McAllister, J. A. (1992). *Gnathorhiza* (Dipnoi): life aspects, and lungfish burrows. In *Fossil fishes as living animals*, (ed. E. Mark-Kurik). *Akademia*, **1**, 91–105.

McCoy, F. (1848). On some new fossil fish of the Carboniferous period. *Annals and Magazine of Natural History*, **2**, 1–10.

McCune, A. (1987). Toward the phylogeny of a fossil species flock: semionotid fishes from a lake deposit in the Early Jurassic Towaco Formation, Newark Basin. *Bulletin of the Peabody Museum of Natural History*, **43**, 1–108.

McGhee, G. R. (1988). The Late Devonian extinction event: evidence for abrupt ecosystem collapse. *Palaeobiology*, **14**, 250–7.

McGhee, G. R. (1990). The Frasnian–Famennian extinction event. In *Mass extinctions: processes and evidence*, (ed. S. K. Donovan), pp. 217–34. Belhaven Press, London

McGhee, G. R., Gilmore, J. S., Orth, C. J., and Olsen, E. (1984). No geochemical evidence for an asteroidal impact at late Devonian mass extinction horizon. *Nature*, **308**, 629–31.

McKerrow, W. S., and Scotese, C. R., (ed.) (1990). *Palaeozoic palaeogeography and biogeography. Geological Society Memoir* 12.

McLaren, D. J. (1982). Frasnian–Famennian extinctions. *Geological Society of America Special Papers*, **190**, 447–84,

McNamara, K. J. (1986). A guide to the nomenclature of heterochronies. *Journal of Paleontology*, **60**, 4–13.

Maddison, W. P., and Maddison, D. R. (1987). MACCLADE, version 2.1. A computer program distributed by the authors. Cambridge, Massachusetts.

Mader, H. (1986). Schuppen und Zähne von Acanthodiern und Elasmobranchiern aus dem Unter-Devon Spaniens (Pisces). *Göttinger Arbeiten zür Geologie und Paläontologie*, **28**, 1–59.

Maisey, J. G. (1975). The interrelationships of phalacanthous sharks. *Neues Jahrbuch für Geologie und Paläontologie, Monatshefte*, **1975**, 553–67.

Maisey, J. G. (1976). The Jurassic selachian fish *Protospinax* Woodward. *Palaeontology*, **21**, 733–47.

Maisey, J. G. (1980). An evaluation of jaw suspension in sharks. *American Museum Novitates*, **2706**, 1–17.

Maisey, J. G. (1982). The anatomy and interrelationships of Mesozoic hybodont sharks. *American Museum Novitates*, **2724**, 1–48.

Maisey, J. G. (1983). Cranial anatomy of *Hybodus basanus* Egerton, from the Lower Cretaceous of England. *American Museum Novitates*, **2758**, 1–64.

Maisey, J. G. (1984a). Chondrichthyan phylogeny: a look at the evidence. *Journal of Vertebrate Paleontology*, **4**, 359–71.

Maisey, J. G. (1984b). Higher elasmobranch phylogeny and biostratigraphy. *Zoological Journal of the Linnean Society*, **82**, 33–54.

Maisey, J. G. (1985). Cranial morpholgy of the fossil elasmobranch *Synechodus dubrisiensis*. *American Museum Novitates*, **2804**, 1–28.

Maisey, J. G. (1986). Heads and tails: a chordate phylogeny. *Cladistics*, **2**, 201–56.

Maisey, J. G. (1987). Notes on the structure and phylogeny of vertebrate otoliths. *Copeia*, **1987**, 495–9.

Maisey, J. G. (1988). Phylogeny of early vertebrate skeletal induction and ossification pattern. *Evolutionary Biology*, **22**, 1–36.

Maisey, J. G. (1989a). Visceral skeleton and musculature of a Late Devonian shark. *Journal of Vertebrate Paleontology*, **9**, 174–90.

Maisey, J. G. (1989b). *Hamiltonichthys mapesi* g. and sp. nov. (Chondrichthyes; Elasmobranchii), from the Upper Pennsylvanian of Kansas. *American Museum Novitates*, **2931**, 1–42.

Mallatt, J. (1981). The suspension feeding mechanism of the larval lamprey *Petromyzon marinus*. *Journal of Zoology*, **194**, 103–42.

Mallatt, J. (1984a). Feeding ecology of the earliest vertebrates. *Zoological Journal of the Linnean Society*, **82**, 261–72.

Mallatt, J. (1984b). Early vertebrate evolution: pharyngeal structure and the origin of gnathostomes. *Journal of Zoology*, **204**, 169–83.

Mallatt, J., and Paulsen, C. (1986). Gill ultrastructure of the Pacific hagfish *Eptatretus stouti*. *American Journal of Anatomy*, **177**, 243–69.

Mansuy, H. (1915). Contribution à l'étude des faunes de l'Ordovicien et du Gothlandien du Tonkin. *Mémoires du Service géologique d'Indochine, Hanoi*, **4**, 1–7.

Marcus, H. (1937). Lungen. In *Handbuch der vergleichenden Anatomie der Wirbeltiere*, (ed. L. Bolk, E. Göppert, E. Kallius, and W. Lubosch), Vol. 3, pp. 909–88. Urban and Schwarzenberg, Berlin.

Marinelli, W., and Strenger, A. (1954–73). *Vergleichende Anatomie und Morphologie der Wirberltiere*. 1. *Lampetra fluviatilis* (1954); 2. *Myxine glutinosa* (1956); 3, *Squalus acanthias* (1959); 4, *Lucioperca sandra* (1973). Franz Deuticke, Vienna.

Mark, E. (1961). [Moningatest kohastumisnähtustest psammosetiididel.] (Certain adaptation phenomena in the psammosteids). In *Geological Notes* 1, pp. 30–4. Estonian Academy of Sciences, Tallinn. (In Estonian.)

Mark-Kurik, E. (1966). [On some injuries on the exoskeleton of the psammosteids (Agnatha).] In [*Organisms and environments in the geological past*], pp. 55–60, Nauka, Moscow. (In Russian.)

Mark-Kurik, E. (1977). [The structure of the shoulder girdle in early ptyctodonts.] In [*Essays on phylogeny and systematics of fossil agnathans and fishes*], (ed. V. V. Menner), pp. 61–70, Nauka, Moscow. (In Russian.)

Mark-Kurik, E. (1991). [The study of the Devonian fishes in the first half of the 19th century in Tartu.] *Daba un muzejs*, **3**, 28–31. (In Russian.)

Mark-Kurik, E. (1992). Functional aspects of the armour in the early vertebrates. In *Fossil fishes as living animals*, (ed. E. Mark-Kurik). *Akademia*, **1**, 107–15.

Mark-Kurik, E., and Janvier, P. (1995). New osteostracans from the Lower Devonian of Severnaya Zemlya. *Journal of Vertebrate Paleontology*, **15**, 449–62.

Mark-Kurik, E., Ivanov, A., and Obrucheva, O. (1991). The endoskeleton of the shoulder girdle in ptyctodonts (Placodermi). *Eesti Teaduste Akadeemia Toimetised, Geoloogia*, **40**, 160–4.

Marshall, C. R. (1987). Lungfish: phylogeny and parsimony. In *Biology and evolution of lungfishes*, (ed. W. E. Bemis, W. W. Burggren, and N. E. Kemp), *Journal of Morphology, Supplement*, **1**, 151–62.

Marshall, C. R., and Schultze, H. P. (1992). Relative importance of molecular, neontological, and paleontological data in understanding the biology of the vertebrate invasion of land. *Journal of Molecular Evolution*, **35**, 93–101.

Märss, T. (1979). [Lateral line sensory system of the Ludlovian thelodont *Phlebolepis elegans* Pander.] *Eesti NSV Teaduste Akadeemia Toimetised, Geoloogia*, **28**, 108–11. (In Russian with English summary.)

Märss, T. (1982). Vanimate selgrootsete leiud Saaremaal [The oldest vertebrate finds in Saaremaa.] *Eesti Loodus (Estonian Nature)*, **1982**, 646–54. (In Estonian with English abstract.)

Märss, T. (1986a). Squamation of the thelodont agnathan *Phlebolepis*. *Journal of Vertebrate Paleontology*, **6**, 1–11.

Märss, T. (1986*b*). [Silurian vertebrates of Estonia and West Latvia.] *Fossilia Baltica*, **1**, 1–103. (In Russian.)

Märss, T. (1992). The structure of growth layers of Silurian fish scales as a potential evidence of the environmental changes. In *Fossil fishes as living animals*, (ed. E. Mark-Kurik). *Akademia*, **1**, 41–48.

Märss, T., and Einasto, R. (1978). [Distribution of the vertebrates in deposits of various facies in the North Baltic Silurian.] *Esti NSV Teaduste Akadeemia Toimetised Geoloogia*, **27**, 16–22. (In Russian with English summary.)

Martill, D. M. (1988). Preservation of fishes in the Cretaceous of the Santana Formation of Brazil. *Palaeontology*, **31**, 1–18.

Meinke, D. (1984). A review of cosmine: its structure, development and relationships to other forms of the dermal skeleton in osteichthyans. *Journal of Vertebrate Palaeontology*, **4**, 457–70.

Meinke, D., and Thomson, K. S. (1983). The distribution and significance of enamel and enameloid in the dermal skeleton of osteolepiform rhipidistian fishes. *Palaeobiology*, **9**, 138–49.

Meyer, A., and Wilson, A. C. (1992). Origin of tetrapods inferred from their mitochondrial DNA affiliation to lungfish. *Journal of Molecular Evolution*, **31**, 359–64.

Miles, R. S. (1962). *Gemuendenaspis* n. gen., an arthrodiran fish from the Lower Devonian Hunsrückschiefer of Germany. *Transactions of the Royal Society of Edinburgh*, **65**, 59–77.

Miles, R. S. (1966). The acanthodian fishes of the Devonian Plattenkalk of the Paffrath Trough in the Rhineland. *Arkiv för Zoologi*, **18**, 147–94.

Miles, R. S. (1967). Observations on the ptyctodont fish *Rhamphodopsis* Watson. *Zoological Journal of the Linnean Society*, **47**, 99–120.

Miles, R. S. (1969). Features of placoderm diversification and the evolution of the Arthrodire feeding mechanism. *Transactions of the Royal Society of Edinburgh*, **68**, 123–70.

Miles, R. S. (1971). The Holonematidae (placoderm fishes), a review based on new specimens of *Holonema* from the Upper Devonian of Western Australia. *Philosophical Transactions of the Royal Society of London*, **263**, 191–234.

Miles, R. S. (1973*a*). Articulated acanthodian fishes from the Old Red Sandstone of England, with a review of the structure and evolution of the acanthodian shoulder-girdle. *Bulletin of the British Museum (Natural History), Geology*, **24**, 113–213.

Miles, R. S. (1973*b*). Relationships of acanthodians. In *Interrelationships of fishes*, (ed. P. H. Greenwood, R. S. Miles, and C. Patterson). *Zoological Journal of the Linnean Society, Supplement 1*, **53**, 63–103.

Miles, R. S. (1975). The relationships of the Dipnoi. In *Problèmes actuels de paléontologie: evolution des Vertébrés*. Colloques internationaux du Centre national de la Recherche scientifique, No. 218, (ed. J. P. Lehman), pp. 138–48. Paris.

Miles, R. S. (1977). Dipnoan (lungfish) skulls and the relationships of the group: a study based on new species from the Devonian of Australia. *Zoological Journal of the Linnean Society*, **61**, 1–328.

Miles, R. S., and Dennis, K. D. (1979). A primitive eubrachythoracid arthrodire from Gogo, Western Australia. *Zoological Journal of the Linnean Society*, **65**, 31–62.

Miles, R. S., and Westoll, T. S. (1968). The placoderm fish *Coccosteus cuspidatus* Miller *ex* Agassiz, from the Middle Old Red Sandstone of Scotland. Part 1. Descriptive morphology. *Transactions of the Royal Society of Edinburgh*, **67**, 373–476.

Miles, R. S., and Young, G. C. (1977). Placoderm interrelationships reconsidered in the light of new ptyctodontids from Gogo, Western Australia. In *Problems in vertebrate evolution*, (ed. S. M. Andrews, R. S. Miles, and A. D. Walker), pp. 123–98. Academic Press, London.

Miller, H. (1835). *Scenes and legends of the north of Scotland*, Black, Edinburgh.

Miller, H. (1841). *The Old Red Sandstone or new walks in an old field*. Johnstone, Edinburgh.

Miller, H. (1850). *Footprints of the Creator*. Shepherd and Elliot, Edinburgh.

Miller, H. (1857). *Testimony of the rocks or geology in its bearing on the two theologies, natural and revealed*. Shepherd and Elliot, Edinburgh.

Millot, J., and Anthony, J. (1958). *Anatomie de* Latimeria chalumnae. *I. Squelette et muscles*. Centre national de la Recherche scientifique, Paris.

Millot, J., and Anthony, J. (1965). *Anatomie de* Latimeria chalumnae. *II. Système nerveux et organes des sens*. Centre national de la Recherche scientifique, Paris.

Millot, J., Anthony, J., and Robineau, D. (1978). *Anatomie de* Latimeria chalumnae. *III. Appareil digestif, appareil respiratoire, appareil uro-génital, glandes endocrines, appareil circulatoire, téguments, écailles, conclusions générales*. Centre National de la Recherche scientifique, Paris.

Milne-Edwards, H. (1844). Considérations sur quelques principes relatifs à la classification naturelle des animaux. *Annales des Sciences Naturelles*, **1**, 65–99.

Milner, A. R. (1982). Small temnospondyl amphibians from the Middle Pennsylvanian of Illinois. *Palaeontology*, **25**, 635–64.

Milner, A. R. (1988). The relationships and origin of living amphibians. In *The phylogeny and classification of the tetrapods*, (ed. M. J. Benton), Vol. 1, pp. 59–102. Systematics Association Special Volume No. 35a. Clarendon Press, Oxford.

Milner, A. R. (1993*a*). Biogeography of Palaeozoic tetrapods. In *Palaeozoic vertebrate biostratigraphy and biogeography*, (ed. J. A. Long), pp. 324–53, Belhaven Press, London.

Milner, A. R. (1993*b*). The Paleozoic relatives of lissamphibians. *Herpetological Monographs*, **7**, 8–27.

Milner, A. R., Smithson, T. R., Milner, A. C., Coates, M. I., and Rolfe, W. D. I. (1986). The search for early tetrapods. *Modern Geology*, **10**, 1–28.

Moskalenko, T. A. (1973). [Distribution of the conodonts in the Ordovician of the Siberian platform). in [*The news in palaeontology of Siberia and Middle Asia*], ed. A. B. Ivanovski, pp. 87–95. Nauka (Siberian Branch), Novosibirsk. (In Russian.)

Moy-Thomas, J. A. (1936*a*). On the structure and affinity of the Carboniferous cochliodont *Helodus simplex*. *Geological Magazine*, **73**, 488–503.

Moy-Thomas, J. A. (1936*b*). The structure and affinities of the fossil elasmobranch fishes from the Lower Carbonierous of Glencartholm, Eskdale. *Proceedings of the Zoological Society of London*, **B, 1936**, 762–88.

Moy-Thomas, J. A. (1939). *Palaeozoic fishes*. Methuen and Co., London.

Moy-Thomas, J. A., and Miles, R. S. (1971). *Palaeozoic fishes* (2nd edn). Revised by R. S. Miles. Chapman and Hall, London.

Müller, J. (1846). On the structure and characters of the ganoidei, and the natural classification of fish. *Scientific Memoirs*, **4**, 499–552.

Murchison, R. I. (1839). *The Silurian System*. J. Murray, London.

Musick, A., Bruton, M. N., and Balon, E. K. (ed.) (1991). The *biology of* Latimeria chalumnae *and evolution of coelacanths*. *Environmental Biology of Fishes*, **32**, 1–446.

Neal, H. V., and Rand, H. W. (1936). *Comparative anatomy*. Blakiston, Philadelphia.

Nelson, G. J. (1968). Gill-arch structure in *Acanthodes*. In *Current problems of lower vertebrate phylogeny*, (ed. T. Ørvig), pp. 129–43. Almqvist and Wiksell, Stockholm.

Nelson, G. J. (1969). Gill arches and the phylogeny of fishes, with notes on the classification of vertebrates. *Bulletin of the American Museum of Natural History*, **141**, 475–552.

Nelson, G. J. (1970). Pharyngeal (placoid) scales of sharks, with notes on the dermal skeleton of vertebrates. *American Museum Novitates*, **2415** 1–26.

Nelson, G. J. (1974). Classification as an expression of phylogenetic relationships. *Systematic Zoology*, **22**, 344–59.

Nelson, G. J. (1978). Ontogeny, phylogeny, paleontology, and the biogenetic law. *Systematic Zoology*, **21**, 364–74.

Nelson, G. J. (1989). Phylogeny of major fish groups. In *The hierarchy of life*, (ed. B. Fernholm, K. Bremer, and H. Jörnvall), pp. 325–36. Excerpta Medica, Amsterdam.

Nelson, G. J. (1994). Homology and systematics. In *Homology: the hierarchical basis of comparative biology*, (ed. B. K. Hall), pp. 101–49. Academic Press, London.

Nelson, G. J., and Ladiges, P. I. (1991). TAX: MS–DOS program for cladistic systematics. Distributed by the authors, New York and Melbourne.

Nelson, G. J. and Platnick, N. I, (1981). *Systematics and biogeography. Cladistics and vicariance*. Columbia University Press, New York.

Nelson, G. J., and Platnick, N. I. (1991). Three-taxon statements: a more precise use of parsimony. *Cladistics*, **7**, 351–66.

Nelson, J. S. (1994). *Fishes of the world*, (3rd edn). Wiley, New York.

Newberry, J. S. (1885). Descriptions of some gigantic placoderm fishes recently discovered in the Devonian of Ohio. *Transactions of the New York Academy of Sciences*, **5**, 25–8.

Newberry, J. S. (1889). The Paleozoic fishes of North America. *Monographs of the U.S. Geological Survey*, **16**, 1–228.

Newell, N. D. (1967). Revolutions in the history of life. *Geological Society of America, Special Paper*, **89**, 63–91.

Nielsen, E. (1942). Studies on Triassic fishes from East-Greenland, I. *Glaucolepis* and *Boreosomus*. *Meddelelser om Grønland*, **138**, 1–403.

Nielsen, E. (1952). No new and little known Edestidae from the Permian and Triassic of East Greenland. *Meddelelser om Grønland*, **144**, 1–55.

Nitecki, M. H. (ed.) (1979). *Mazon Creek fossils*. Academic Press, New York.

Nolf, D. (1985). *Otolithi piscium*. In *Handbook of paleoichthyology*, (ed. H. P. Schultze), Vol. 10. Gustav Fischer, Stuttgart.

Northcutt, R. G. (1987). Lungfish neural characters and their bearing on sarcopterygian phylogeny. In *The biology and evolution of lungfishes*, (ed. W. E. Bemis, W. W. Burggren, and N. E. Kemp), pp. 277–97. Liss, New York.

Northcutt, R. G., and Bemis, W. E. (1993). *Cranial nerves of the coelacanth* Latimeria chalumnae *(Osteichthyes: Sarcopterygii: Actinistia) and comparison with other Craniata*. Karger, Basel.

Northcutt, R. G., and Grans, C. (1983). The genesis of neural crest and epidermal placodes: a reinterpretation of vertebrate origins. *Quarterly Review of Biology*, **58**, 1–28.

Novitskaya, L. I. (1971). *Les Amphiaspides (Heterostraci) du Dévonien de la Sibérie*. Cahiers de Paléontologie, Centre national de la Recherche scientifique, Paris.

Novitskaya, L. I. (1983). [*Morphology of ancient agnathans. Heterostracans and the problem of relationships of agnathans and gnathostome vertebrates*.] *Trudi Paleontologicheskogo Instituta*, No. 196, Akademia Nauk SSSR, Moscow. (In Russian.)

Novitskaya, L. I. (1992). Heterostracans: their ecology, internal structure and ontogeny. In *Fossils fishes as living animals*, (ed. E. Mark-Kurik). *Akademia*, **1**, 51–9.

Novitskaya, L. I., and Karatayute-Talimaa, V. N. (1988). [The consequence of the phylogenetic method: are tremataspids (*Tremataspis*) archaic or specialized osteostracans?] *Paleontologicheskii Zhurnal*, **567**, 3–13. (In Russian.)

Novitskaya, L. I., and Karatayute-Talimaa, V. N. (1989). [Type of ontogeny as a criterion in the reconstruction of phylogenetic relationships of lower vertebrates.] *Paleontologicheskii Zhurnal*, **1**, 3–16. (In Russian.)

Nur, A., and Ben-Avraham, Z. (1981). Lost Pacifica continent: a mobilistic speculation. In *Vicariance biogeography: a critique*, (ed. G. Nelson and D. E. Rosen), pp. 341–58. Columbia University Press, New York.

Obruchev, D. V. (1941). [Devonian fishes from the Minusinsk Basin.] *Trudy paleontologicheskogo Instituta*, **8** (4), 23–48. (in Russian.)

Obruchev, D. V. (1964). [Agnathans and fishes.] In [*Fundamentals of Palaeontology*], (ed. Y. O. Orlov), Vol. 11. Nauka, Moscow. (In Russian; transl. Israel Program for Scientific Translations, 1967.)

Obruchev, D. V., and Mark-Kurik, E. (1965). [Devonian psammosteids (Agnatha, Psammosteidae) of the USSR.] *Eesti NSV Teaduste Akadeemia Toimetised*, **16**, 1–304. (In Russian.)

Obruchev, D. V., and Mark-Kurik, E. (1968). On the evolution of the psammosteids (Heterostraci). *Eesti NSV Teaduste Akadeemia Toimetised, Geoloogia*, **17**, 279–84.

Oken, L. (1807). *Über die Bedeutung der Schädelknochen. Ein Programm beim Antritt der Professur und der Gesammt Universität zu Jena*. Bamberg bei Godhardt, Jena.

Olsen, P. O., and McCune, A. R. (1991). Morphology of the *Semionotus elegans* species group from the Early Jurassic part of the Newark Supergroup of eastern North America, with comments on the family Semionotidae (Neopterygii). *Journal of Vertebrate Paleontology*, **11**, 269–92.

Ørvig, T. (1951). Histologic studies of placoderm and fossil elasmobranchs. 1. The endoskeleton, with remarks on the hard tissues of lower vertebrates in general. *Arkiv för Zoologi*, **2**, 321–454.

Ørvig, T. (1957). Remarks on the vertebrate fauna of the Lower Devonian of Escuminac Bay, P. Q., Canada, with special reference to the porolepiform crossopterygians. *Arkiv för Zoologi*, **10**, 367–426.

Ørvig, T. (1958). *Pycnaspis splendens* new genus, new species, a new ostracoderm from the Upper Ordovician of North America. *Proceedings of the United States National Museum*, **108**, 1–23.

Ørvig, T. (1960). New finds of acanthodians, arthrodires, crossopterygians, ganoids and dipnoans in the Upper Middle Devonian calcareous flags (Oberer Plattenkalk) of the Bergisch Gladbach–Paffrath Trough, Part 1. *Paläontologische Zeitschrift*, **35**, 10–27.

Ørvig, T. (1962). Y a-t-il une relation directe entre les Arthrodires ptyctodontides et les Holocéphales? In *Problèmes actuels de paléontologie: evolution des Vertébrés*. Colloques internationaux du Centre national de la Recherche scientifique, No. 104, (ed. J. P. Lehman), pp. 49–61, Paris.

Ørvig, T. (1965). Palaeohistological notes. 2. Certain comments on the phyletic significance of acellular bone tissue in early lower vertebrates. *Arkiv för Zoologi*, **16**, 551–6.

Ørvig, T. (1967*a*). Phylogeny of tooth tissues: evolution of some calcified tissues in early vertebrates. In *Structural and chemical organization of teeth*, (ed. A. E. W. Miles), Vol. 1, pp. 45–110. Academic Press, New York.

Ørvig, T. (1967*b*). Some new acanthodian remains from the Lower Devonian of Europe. *Journal of the Linnean Society (Zoology)*, **47**, 131–53.

Ørvig, T. (1968*a*). The dermal skeleton: general consideration. In *Current problems of lower vertebrate phylogeny*, (ed. T. Ørvig), pp. 373–97. Almqvist and Wiksell, Stockholm.

Ørvig, T. (1968*b*). Tanden i kultur, fantasi och verklighet. *Naturhistoriska Riksmuséets Småskrifter*, **3**, 1–23.

Ørvig, T. (1969). Cosmine and cosmine growth. *Lethaia*, **2**, 261–71.

Ørvig, T. (1972). The latero-sensory component of the dermal skeleton in lower vertebrates and its phyletic significance. *Zoologica Scripta*, **1**, 139–55.

Ørvig, T. (1973). Acanthodian dentition and its bearing on the relationships of the group. *Palaeontographica*, **A 143**, 119–50.

Ørvig, T. (1975). Description, with special reference to the dermal skeleton, of a new radotinid arthrodire from the Gedinnian of Arctic Canada. In *Problèmes actuels de paléontologie: évolution des Vertébrés*. Colloques internationaux du Centre National de la Recherche scientifique, No. 218, (ed. J. P. Lehman), pp. 41–71, Paris.

Ørvig, T. (1977). A survey of odontodes ('dermal teeth') from developmental, structural, functional, and phyletic points of view. In *Problems in vertebrate evolution*, (ed. S. M. Andrews, R. S. Miles, and A. D. Walker). *Linnean Society Symposium Series*, **4**, 53–75.

Ørvig, T. (1980). Histologic studies of ostracoderms, placoderms and fossil elasmobranchs. 4. Ptyctodontid tooth plates and their bearing on holocephalan ancestry: the condition of *Ctenurella* and *Ptyctodus*. *Zoologica Scripta*, **9**, 219–39.

Ørvig, T. (1985). Ptyctodontid tooth plates and their bearing on holocephalan ancestry: the condition of chimaerids. *Zoologica Scripta*, **14**, 55–79.

Ørvig, T. (1989). Histologic studies of ostracoderms, placoderms and fossil elasmobranchs. 6. Hard tissues of Ordovician vertebrates. *Zoologica Scripta*, **18**, 427–46.

Otto, M. (1993). Zur systematischen Stellung der Lophosteiden (Obersilur, Pisces inc. sedis). *Paläontologische Zeitschrift*, **65**, 345–50.

Owen, R. (1861). *Palaeontology, or a systematic summary of extinct animals and their geological relations*. Black, Edinburgh.

Pan, J. [P'an, K.] (1984). The phylogenetic position of the Eugaleaspida in China. *Proceedings of the Linnean Society of New South Wales*, **107**, 309–19.

Pan, J. (1988). Note on Silurian vertebrates of China. *Bulletin of the Chinese Academy of Geological Sciences*, **1988**, 228–49.

Pan, J. (1992). *New Galeaspids (Agnatha) from the Silurian and Devonian of China*. Geological Publishing House, Beijing.

Pan, J., and Chen, L. (1993). [Geraspididae, a new family of polybranchiaspidida (Agnatha), from Silurian of Northern Anhui.] *Vertebrata PalAsiatica*, **31**, 225–30. (In Chinese with English summary.)

Pan, J., and Dineley, D. (1988). A review of early (Silurian and Devonian) vertebrate biogeography and biostratigraphy of China. *Proceedings of the Royal Society of London*, **B 235**, 29–61.

P'an, K. [Pan, J.], and Wang S. T. (1978). [Devonian Agnatha and Pisces of South China.] In [*Symposium on the Devonian System of South China*], pp. 299–333. Geological Publishing House, Beijing. (In Chinese.)

P'an, K., and Wang, S. T. (1980). [New findings of galeaspidiforms in South China.] *Acta Palaeontologica Sinica*, **19**, 1–7. (In Chinese with English summary.)

P'an, K., and Wang, S. T. (1981). [New discoveries of polybranchiaspids from Yunnan Province.] *Vertebrata PalAsiatica*, **19**, 113–21. (In Chinese with English summary.)

P'an, K., and Wang, S. T. (1983). [Xiushuiaspidae, a new family of polybranchiaspiformes from Xiushui of Jianxi Province.] *Acta Palaeontologica Sinica*, **22**, 505–9. (In Chinese with English summary.)

P'an, K., Wang, S. T., and Liu, Y. H. (1975). [The Lower Devonian Agnatha and Pisces from South China.] In *Professional Papers on Stratigraphy and and Paleontology*, **1**, 135–69. Geological Publishing House, Beijing. (In Chinese.)

Panchen, A. L. (1967). The nostrils of choanate fishes and early tetrapods. *Biological Reviews*, **42**, 374–420.

Panchen, A. L. (1970). Anthracosauria. In *Handbuch der Paläoherpetologie*, Vol. 5a. Fischer, Stuttgart.

Panchen, A. L. (ed.) (1980*a*). *The terrestrial environment and the origin of land vertebrates*. Systematics Association Special Volume No. 15. Academic Press, London.

Panchen, A. L. (1980*b*). The origin and relationships of the anthracosaur Amphibia from the Late Palaeozoic. In *The terrestrial environment and the origin of land vertebrates*. Systematics Association Special Volume No. 15, (ed. A. L. Panchen), pp. 321–50. Academic Press, London.

Panchen, A. L. (1981). A jaw ramus of the Coal Measure amphibian *Anthracosaurus* from Northumberland. *Palaeontology*, **24** (1), 85–92.

Panchen, A. L. (1985). On the amphibian *Crassigyrinus scoticus* Watson from the Carboniferous of Scotland. *Philosophical Transactions of the Royal Society of London*, **B 309**, 505–68.

Panchen, A. L. (1992). *Classification, evolution, and the nature of biology*. Cambridge University Press.

Panchen, A. L., and Smithson, T. R. (1987). Character diagnosis, fossils and the origin of tetrapods. *Biological Reviews*, **62**, 341–438.

Panchen, A. L., and Smithson, T. R. (1988). The relationships of the earliest tetrapods. In *The phylogeny and classification of the tetrapods, Volume 1: Amphibians, reptiles, birds*. Systematics Association Special Volume No. 35a, (ed. M. J. Benton), pp. 1–32. Clarendon Press, Oxford.

Panchen, A. L. and Smithson, T. R. (1990). The pelvic girdle and hind lib of *Crassigyrinus scoticus* (Lydekker) from the Scottish Carboniferous and the origin of the tetrapod pelvic skeleton. *Transactions of the Royal Society of Edinburgh*, **81**, 31–44.

Pander, C. H. (1856). *Monographie der fossilen Fische des Silurischen Systems der Russich–Baltischen Gouvernements*. Kaiserliche Akademie der Wissenschaften, St Petersburg.

Pander, C. H. (1857). *Über die Placodermen des devonischen Systems*. Kaiserlichen Akademie des Wissenschaften, St Petersburg.

Pander, C. H. (1858). *Über die Ctenodipteriden des devonischen Systems*. Kaiserlichen Akademie des Wissenschaften, St Petersburg.

Pander, C. H. (1860). *Über die Saurodipterinen, Dendrodonten, Glyptolepiden und Cheirolepiden des devonischen Systems*. Kaiserlichen Akademie des Wissenschaften, St Petersburg.

Panteleyev, N. (1992). New remigolepids and high armoured antiarchs of Kirgizia. In *Fossil fishes as living animals*, (ed. E. Mark-Kurik), *Akademia*, **1**, 185–91.

Parrington, F. R. (1958). On the nature of the Anaspida. In: *Studies on fossil vertebrates* (ed. T. S. Westoll), pp. 108–28, Athlone Press, London.

Parrington, F. R., and Westoll, T. S. (1974). David Meredith Seares Watson. *Biographical Memoirs of Fellows of the Royal Society*, **20**, 483–504. (With a list of D. M. S. Watson's works.)

Patten, W. (1890). On the origin of vertebrates from arachnids. *Quarterly Journal of Microscopical Sciences*, **31**, 359–65.

Patten, W. (1903). On the structure and classification of the Tremataspididae. *Mémoires de l'Académie Impériale des Sciences, St Petersbourg. Classe de Physique et Mathématique*, **13**, 1–33.

Patten, W. (1912). *The evolution of the vertebrates and their kin*. Blakiston, Philadelphia.

Patterson, C. (1965). The phylogeny of the chimaeroids. *Philosophical Transactions of the Royal Society of London*, **B 249**, 101–219.

Patterson, C. (1968). *Menaspis* and the bradyodonts. In *Current problems of lower vertebrate phylogeny*, (ed. T. Ørvig), pp. 171–205. Almqvist and Wiksell, Stockholm.

Patterson, C. (1973). Interrelationships of holosteans. In *Interrelationships of fishes*, (ed. P. H. Greenwood, R. S. Miles, and C. Patterson), pp. 235–306. *Zoological Journal of the Linnean Society, supplement* **1**, Academic Press, London.

Patterson, C. (1975). The braincase of pholidophorid and leptolepid fishes, with a review of the actinopterygian brain-case. *Philosophical Transactions of the Royal Society of London*, **B 269**, 275–579.

Patterson, C. (1977a). Cartilage bones, dermal bones and membrane bones, or the exoskeleton versus the endoskeleton. In *Problems in vertebrate evolution*, (ed. S. M. Andrews, R. S. Miles, and A. D. Walker), pp. 77–121. Linnean Society Symposium Series, No. 4.

Patterson, C. (1977b). The contribution of palaeontology to teleostean phylogeny. In *Major patterns in vertebrate evolution*, (ed. M. K. Hecht, P. C. Goody, and B. M. Hecht), pp. 579–643. Plenum Press, New York.

Patterson, C. (1980). Origin of tetrapods: historical introduction to the problem. In *The terrestrial environment and the origin of land vertebrates*. Systematics Association Special Volume No. 15, (ed. A. L. Panchen), pp. 159–175. Academic Press, London.

Patterson, C. (1981a). Significance of fossils in determining evolutionary relationships. *Annual Reviews of Ecology and Systematics*, **12**, 195–223.

Patterson, C. (1981b). Agassiz, Darwin, Huxley, and the fossil record of teleost fishes. *Bulletin of the British Museum (Natural History), Geology*, **35**, 213–24.

Patterson, C. (1981c). Methods of paleobiogeography. In *Vicariance biogeography: a critique*, (ed. G. Nelson and D. E. Rosen), pp. 446–500. Columbia University Press, New York.

Patterson, C. (1982a). Morphology and interrelationships of primitive actinopterygian fishes. *American Zoologist*, **22**, 241–59.

Patterson, C. (1982b). Morphological characters and homology. In *Problems of phylogenetic reconstruction*. Systematics Association Special Volume No. 21, (ed. K. A. Joysey and A. E. Friday), pp. 21–74. Academic Press, London.

Patterson, C. (1992). Interpretation of the toothplates of Chimaeroid fishes. *Zoological Journal of the Linnean Society*, **106**, 33–61.

Patterson, C., and Rosen, D. E. (1977). A review of ichthyodectiform and other Mesozoic teleost fishes and the theory and practice of classifying fossils. *Bulletin of the American Museum of Natural History*, **158**, 81–172.

Patterson, C., and Smith, A. (1987). Is the periodicity of extinctions a taxonomic artefact? *Nature*, **330**, 248–52.

Pearson, D. M. (1981). Functional aspects of the integument in polypterid fishes. *Zoological Journal of the Linnean Society, London*, **72**, 93–106.

Pearson, D. M. (1982). Primitive bony fishes, with especial reference to *Cheirolepis* and palaeonisciform actinopterygians. *Zoological Journal of the Linnean Society*, **74**, 35–67.

Pearson, D. M., and Westoll, T. S. (1979). The Devonian actinopterygian *Cheirolepis* Agassiz. *Transactions of the Royal Society of Edinburgh*, **70**, 337–99.

Peel, J. S., and Higgins, A. C. (1977). *Anatolepis*—a problematic Ordovician vertebrate re-interpreted as an arthropod. *Rapport fra Grønland Geologiske Undersøgelser*. **85**, 108–9.

Pehrson, T. (1940). The development of the dermal bones in the skull of *Amia calva*. *Acta Zoologica, Stockholm*, **21**, 1–50.

Peignoux-Deville, J., Lallier, F., and Vidal, B. (1982). Evidence for the presence of osseous tissue in dogfish vertebrae. *Cell and Tissue Research*, **222**, 605–14.

Peyer, B. (1968). *Comparative odontology*. University of Chicago Press, Chicago.

Philip, G. M. (1979). Carpoids—echinoderms or chordates? *Biological Reviews*, **54**, 439–71.

Poplin, C. (1974). *Études de quelques Paléoniscidés pennsylvaniens du Kansas*. Cahiers de Paléontologie, Centre national de la Recherche scientifique, Paris.

Poplin, C., and Heyler, D. (1989). Evolution et phylogénie des Xénacanthiformes (=Pleuracanthiformes, Pisces, Chondrichthyes). *Annales de Paléontologie*, **75**, 187–222.

Poplin, C., and Ricqlès, A. de (1970). A technique of serial sectioning for the study of undecalcified fossils. *Curator*, **13**, 7–20.

Potter, I. C., and Hilliard, R. W. (1987). A proposal for the functional and phylogenetic significance of differences in the dentition of lampreys (Agnatha: Petromyzontiformes). *Journal of Zoology*, **212**, 713–37.

Presley, R. (1993). Preconception of adult structural pattern in the analysis of the developing skull. In *The skull*, (ed. J. Hanken and B. K. Hall), Vol. 1, pp. 347–77. University of Chicago Press.

Pridmore, P. A., and Barwick, R. E. (1993). Post-cranial morphology of the Late Devonian dipnoans *Griphognathus* and *Chirodipterus* and locomotor implications. *Memoirs of the Association of Australasian Palaeontologists*, **15**, 161–82.

Quenstedt, A. (1838). Ueber die fossilen knochen im roten Sandsteine Livlands und Esthlands. *Neues Jahrbuch für Mineralogie, Geologie und Paläontologie*. **1838**, 12–6.

Quenstedt, A. (1851). *Handbuch der Petrefacten*. Laupp, Tübingen.

Rackoff, J. S. (1980). The origin of the tetrapod limb and the ancestry of tetrapods. In *The terrestrial environment and the origin of land vertebrates*. Systematics Association Special Volume No. 15, (ed. A. L. Panchen), pp. 255–92. Academic Press, London.

Rage, J. C., and Janvier, P. (1982). Le problème de la monophylie des Amphibiens actuels, à la lumière des nouvelles données sur les affinités des Tétrapodes. *Géobios, Mémoire Spécial* No. 6, pp. 65–83.

Rage, J. C., and Rocek, Z. (1989). Redescription of *Triadobatrachus massinoti* (Piveteau 1936), an anuran amphibian from the Early Triassic. *Palaeontographica*, **A 206**, 1–16.

Raup, D. M., and Sepkoski, J. J. (1982). Periodicity of extinctions in the geologic past. *Proceedings of the National Academy of Sciences of the U.S.A.*, **81**, 801–5.

Raup, D. M., and Sepkoski, J. J. (1988). Testing for periodicity of extinctions. *Science*, **241**, 94–6.

Raynaud, A. (1990). Developmental mechanisms involved in the embryonic reduction of limbs in reptiles. *International Journal of Developmental Biology*, **34**, 233–43.

Raynaud, A., and Brabet, J. (1994). Nouvelles observations sur le développement des membres chez l'Orvet (*Anguis fragilis* Linné, 1758). *Annales des Sciences Naturelles, Zoologie*, **15**, 97–113.

Raynaud, A., and Clergue-Gazeau, M. (1986). Identification des doigts réduits ou manquants dans les pattes des embryons de lézard vert (*Lacerta viridis*) traités par la cytosine-arabinofuranoside. Comparison avec les réductions digitales naturelles des espèces de reptiles serpentiformes. *Archives de Biologie*, **97**, 279–99.

Rayner, D. H. (1951). On the cranial structure of an early palaeoniscid, *Kentuckia* gen. nov. *Transactions of the Royal Society of Edinburgh*, **62**, 58–83.

Reese, A. M. (1910). The lateral line of *Chimaera colliei*. *Journal of Experimental Zoology*, **9**, 349–70.

Reif, W. E. (1973). Morphologie und ultrastruktur des Hai 'Schmelzes'. *Zoologica Scripta*, **2**, 231–50.

Reif, W. E. (1982a). Evolution of dermal skeleton and dentition in vertebrates. The odontode-regulation theory. *Evolutionary Biology*, **15**, 287–368.

Reif, W. E. (1982b). Morphogenesis and function of the squamation in sharks. *Neues Jahrbuch für Geologie und Paläontologie, Abhandlungen*, **164**, 171–83.

Reis, O. M. (1896). Über *Acanthodes bronni* Agassiz. *Morphologischen Arbeiten*, **6**, 143–220.

Repetski, J. (1978). A fish from the Upper Cambrian of North America. *Science*, **200**, 529–31.

Retzius, G. (1881–4). *Das Gehörorgan der Wirbelthiere*, Vol. 1 (1883), Vol. 2 (1884). Samson and Wallin, Stockholm.

Riddiford, A., and Penny, D. (1984). The scientific status of modern evolutionary theory. In *Evolutionary theory, Paths into the future*, (ed. J. W. Pollard), pp. 1–38. Wiley, Chichester.

Rieppel, O. (1988). *Fundamentals of comparative biology*. Birkhäuser, Basel.

Ritchie, A. (1964). New light on the morphology of the Norwegian Anaspida. *Skrifter utgitt av det Norske Videnskaps-Akademi, 1, Matematisk-Naturvidenskapslige Klasse*, **14**, 1–35.

Ritchie, A. (1967). *Ateleaspis tessellata* Traquair, a non-cornuate cephalaspid from the Upper Silurian of Scotland. *Zoological Journal of the Linnean Society*, **47**, 69–81.

Ritchie, A. (1968). New evidence on *Jamoytius kerwoodi* White, an important ostracoderm from the Silurian of Lanarkshire, Scotland. *Palaeontology*, **11**, 21–39.

Ritchie, A. (1971). Fossil fish discoveries in Antarctica. *Australian Natural History*, **17**, 65–71.

Ritchie, A. (1973). *Wuttagoonaspis* gen. nov., an unusual arthrodire from the Devonian of western New South Wales, Australia. *Palaeontographica*, **A 143**, 58–72.

Ritchie, A. (1980). The Late Silurian anaspid genus *Rhyncholepis* from Oesel, Estonia, and Ringerike, Norway. *American Museum Novitates*, **2699**, 1–18.

Ritchie, A. (1984a). Conflicting interpretations of the Silurian agnathan, *Jamoytius*. *Scottish Journal of Geology*, **20**, 249–56.

Ritchie, A. (1984b). A new Placoderm, *Placolepis* (Phyllolepidae), from the Late Devonian of New South Wales, Australia. *Proceedings of the Linnean Society of New South Wales*, **107**, 321–53.

Ritchie, A. (1985). *Ainiktozoon loganense* Scourfield, a protochordate? from the Silurian of Scotland. *Alcheringa*, **9**, 117–42.

Ritchie, A., and Gilbert-Tomlinson, J. (1977). First Ordovician vertebrates from the southern hemisphere. *Alcheringa*, **1**, 351–68.

Ritchie, A., Wang, S. T., Young, G. C., and Zhang, G. R. (1992). The Sinolepidae, a family of antiarchs (Placoderm fishes) from the Devonian of South China and Eastern Australia). *Records of the Australian Museum*, **44**, 319–70.

Rixon, A. E. (1976). *Fossil animal remains: their preparation and conversation*. Athlone Press, London.

Robineau, D. (1987). Sur la signification phylogénétique de quelques caractères anatomiques remarquables du Coelacanthe *Latimeria chalumnae* Smith, 1939. *Annales des Sciences Naturelles, Zoologie*, **8**, 43–60.

Rocek, Z. (1986). An 'intracranial joint' in frogs. In *Studies in Herpetology* (ed. Z. Rocek), pp. 49–54. Charles University, Prague.

Rohon, J. V. (1893). Die Obersilurischen Fische von Oesel. *Mémoires de l'Académie des Sciences de St Pétersbourg*, **41**, 1–124.

Romer, A. S. (1933). Eurypterid influence on vertebrate history. *Science*, **78**, 114–17.

Romer, A. S. (1937). The braincase of the Carboniferous crossopterygian *Megalichthys nitidus*. *Bulletin of the Museum of Comparative Zoology, Harvard*, **82**, 1–73.

Romer, A. S. (1955). Fish origins—Fresh or salt water? In *Papers in marine biology and oceanography*, pp. 262–80. Pergamon Press, London.

Romer, A. S. (1956). The early evolution of land vertebrates. *Proceedings of the American Philosophical Society*, **100**, 157–67.

Romer, A. S. (1958). The Texas Permian redbeds and their vertebrate fauna. In *Studies on fossil vertebrates*, (ed. T. S. Westoll), pp. 157–79. Athlone Press, London.

Romer, A. S. (1966). *Vertebrate paleontology*, (2nd edn). University of Chicago Press.

Romer, A. S. (1970). *The vertebrate body*, (4th edn). Saunders, Philadelphia.

Rosen, D. E., Forey, P. L., Gardiner, B. G., and Patterson, C. (1981). Lungfishes, tetrapods, paleontology and plesiomorphy. *Bulletin of the American Museum of Natural History*, **167**, 159–276.

Rowe, T., Carlson, W., and Bottorff, W. W. (1992). *Thrinaxodon: digital atlas of the skull, a compact disc*. University of Texas Press, Austin.

Ruben, J. A., and Bennett, A. A. (1987). The evolution of bone. *Evolution*, **41**, 1187–97.

Rudwick, M. J. S. (1985). *The great Devonian controversy*. Chicago University Press.

Rudwick, M. J. S. (1992). *Scenes from deep time. Early pictorial representations of the prehistoric world*. University of Chicago Press.

Runnegar, B. (1986). Molecular palaeontology. *Palaeontology*, **29**, 1–24.

Saint-Aubain, M. L. de (1981). Amphibian limb ontogeny and its bearing on the phylogeny of the group. *Zeitschift für Zoologie und Evolutionsforschung*, **19**, 175–94.

Sansom, I. J., Smith, M. P., Armstrong, H. A., and Smith, M. M. (1992). Presence of the earliest vertebrate hard tissues in conodonts. *Science*, **256**, 1308–11.

Sansom, I. J., Smith, M. P., and Smith, M. M. (1994). Dentine in conodonts. *Nature*, **368**, 591.

Säve-Söderbergh, G. (1932). Preliminary note on Devonian stegocephalians from East Greenland. *Meddelelser om Grønland*, **94**, 1–107.

Säve-Söderbergh, G. (1934). Some points of view concerning the evolution of the vertebrates and the classification of this group. *Arkiv för Zoologi*, **26A**, 1–20.

Säve-Söderbergh, G. (1941). On the dermal bones of the head in *Osteolepis macrolepidotus* Ag. and the interpretation of the lateral-line system in certain primitive vertebrates. *Zoologiska Bidrag, Uppsala*, **20**, 523–41.

Sawlowicz, Z. (1993). Iridium and other platinum-group elements as geochemical markers in sedimentary environments. *Palaeogeography, Palaeoclimatology, Palaeoecology*, **104**, 253–70.

Schaeffer, B. (1941). The morphological and functional evolution of the tarsus in amphibians and reptiles. *Bulletin of the American Museum of Natural History*, **78**, 395–472.

Schaeffer, B. (1973). Interrelationships of chondrosteans. In *Interrelationships of fishes*, (ed. H. P. Greenwood, R. S. Miles, and C. Patterson), pp. 207–26. *Zoological Journal of the Linnean Society, supplement* **1**, Academic Press, London.

Schaeffer, B. (1975). Comments on the origin and basic radiation of the gnathostome fishes with particular reference to the feeding mechanism. In *Problèmes actuels de paléontologie: evolution des Vertébrés*. Colloques internationaux du Centre national de la Recherche scientifique, No. 218, (ed. J. P. Lehman), pp. 101–9. Paris.

Schaeffer, B. (1977). The dermal skeleton in fishes. In *Problems in vertebrate evolution*. Linnean Society Symposium Series No. 4, (ed. S. M. Andrews, R. S. Miles, and A. D. Walker), pp. 25–52.

Schaeffer, B. (1981). The xenacanth shark neurocranium, with comments on elasmobranch monophyly. *Bulletin of the American Museum of Natural History*, **69**, 3–66.

Schaeffer, B. (1987). Deuterostome monophyly and phylogeny. *Evolutionary Biology*, **21**, 179–235.

Schaeffer, B., and Dalquest, W. W. (1978). A palaeonisciform braincase from the Permian of Texas, with comments on cranial fissures and posterior myodome. *American Museum Novitates*, **2658**, 1–15.

Schaeffer, B., and Thomson, K. S. (1980). Reflections on agnathan–gnathostome relationships. In *Aspects of vertebrate life*, (ed. L. L. Jacobs), pp. 19–33. Museum of Northern Arizona Press, Flagstaff.

Schaeffer, B., and Williams, M. E. (1977). Relationships of fossil and living elasmobranchs. *American Zoologist*, **17**, 293–302.

Scheele, W. E. (1965). Fossil dig. *The Explorer*, **7** (3), 5–8.

Schindler, E. (1993). Event-stratigraphic markers within the Kellwasser Crisis near the Frasnian/Famennian boundary (Upper Devonian) in Germany. *Palaeogeography, Palaeoclimatology, Palaeoecology*, **104**, 115–25.

Schmalhausen, I. I. (1968). *The origin of terrestrial vertebrates*. Academic Press, London. (Translated by K. S. Thomson.)

Schmitz, B., Aberg, G., Werdelin, L., Forey, P., and Bendix-Almgren, S. E. (1991). $^{87}Sr/^{86}Sr$, Na, F, Sr and La in skeletal fish debris as a measure of the paleosalinity of fossil-fish habitats. *Bulletin of the Geological Society of America*, **103**, 786–94.

Schoch, R. M. (1986). *Phylogeny reconstruction in paleontology.* Van Nostrand Reinhold, New York.

Schultze, H. P. (1968). Palaeoniscoidea-Schuppen aus dem Unterdevon Australiens und Kanadas und aus dem Mitteldevon Spitzbergens. *Bulletin of the British Museum (Natural History), Geology,* **16**, 343–68.

Schultze, H. P. (1969). Die Faltenzähne der rhipidistiden Crossopterygier, der Tetrapoden un der Actinopterygier-Gattung *Lepisosteus,* nebst einer Beschreibung der Zahnstruktur von *Onychodus* (struniiformer Crossopterygier). *Palaeontographica Italiana,* **65**, 63–136.

Schultze, H. P. (1970). Folded teeth and the monophyletic origin of tetrapods. *American Museum Novitates,* **2408**, 1–10.

Schultze, H. P. (1972). Early growth stages in Coelacanth fishes. *Nature,* **236**, 90–1.

Schultze, H. P. (1973). Crossopterygier mit heterozerker Schwanzenflosse aus dem Oberdevon Kanadas, nebst einer Beschreibung von Onychondontida-Resten aus dem Mittel devon Spaniens und aus dem Karbon der USA. *Palaeontographica,* A **143**, 188–208.

Schultze, H. P. (1974). Walter Robert Gross (1903–1974). *Paläontologische Zeitschift,* **48**, 143–8.

Schultze, H. P. (1977). Ausgangform und Entwicklung der rhombischen Schuppen der Osteichthyes (Pisces). *Palaeontologische Zeitschrift,* **51**, 152–68.

Schultze, H. P. (1980). Eier legende und lebend gebärende Quastenflosser. *Natur und Museum,* **110**, 101–8.

Schultze, H. P. (1984*a*). Juvenile specimens of *Eusthenopteron foordi* Whiteaves, 1881 (osteolepiform rhipidistian, Pisces) from the Late Devonian of Miguasha, Quebec, Canada. *Journal of Vertebrate Paleontology,* **4**, 1–16.

Schultze, H. P. (1984*b*). The head shield of *Tiaraspis subtilis* (Gross), Pisces, Arthrodira. *Proceedings of the Linnean Society of New South Wales,* **107**, 355–65.

Schultze, H. P. (1985). Reproduction and spawning sites of *Rhabdoderma* (Pisces, Osteichthtes, Actinistia) in Pennsylvanian deposits of Illinois, USA. Proceedings of the 9th International Symposium on Carboniferous Stratigraphy and Geology, Washington and Champaign-Urbana, **5**, 326–30.

Schultze, H. P. (1987). Dipnoans as sarcopterygians. In *Biology and evolution of lungfishes,* (ed. W. E. Bemis, W. W. Burggren, and N. E. Kemp). *Journal of Morphology, Supplement* **1**, 39–74. Liss, New York.

Schultze, H. P. (1988). Notes on the structure and phylogeny of vertebrate otoliths. *Copeia,* **1988**, 257–59.

Schultze, H. P. (1989). Three-dimensional muscle preservation in Jurassic fishes of Chile. *Revista Geologica de Chile,* **16**, 183–215.

Schultze, H. P. (1990). A new acanthodian from the Pennsylvanian of Utah, U.S.A., and the distribution of otoliths in gnathostomes. *Journal of Vertebrate Paleontology,* **10**, 49–58.

Schultze, H. P. (1991). A comparison of controversial hypotheses on the origin of tetrapods. In *Origins of the higher groups of tetrapods. Controversy and consensus,* (ed. H. P. Schultze and L. Trueb), pp. 29–67. Comstock, Ithaca.

Schultze, H. P. (1992). Early Devonian actinopterygians (Osteichthyes, Pisces) from Siberia. In *Fossil fishes as living animals,* (ed. E. Mark-Kurik), *Akademia* **1**, 233–42.

Schultze, H. P. (1993). Patterns of diversity in the skull of jawed fishes. In *The skull,* (ed. J. Hanken and B. K. Hall), Vol. 2, pp. 189–254. University of Chicago Press.

Schultze, H. P. (1994). Comparison of hypotheses on the relationships of sarcopterygians. *Systematic Zoology,* **43**, 155–73.

Schultze, H. P., and Arsenault, M. (1985). The panderichthyid fish *Elpistostege*: a close relative of tetrapods? *Palaeontology,* **28**, 293–309.

Schultze, H. P., and Bardack, D. (1987). Diversity and size changes in palaeonisciform fishes (Actinopterygii, Pisces) from the Pennsylvanian Mazon Creek fauna, Illinois, USA. *Journal of Vertebrate Paleontology,* **7**, 1–23.

Schultze, H. P., and Campbell, K. S. W. (1986). Characterization of the Dipnoi, a monophyletic group. In *Biology and evolution of lungfishes,* (ed. W. E. Bemis, W. W. Burggren, and N. E. Kemp). *Journal of Morphology, supplement* **1**, 25–37.

Schultze, H. P., and Heidtke, U. (1986). Rhizodopside Rhipidistia (Pisces) aus dem Perm der Pfalz (W-Deutschland). *Neues Jahrbuch für Geologie und Paläontologie, Monatshefte,* **1986**, 165–70.

Schultze, H. P., and Marshall, C. R. (1993). Contrasting the use of functional complexes and isolated characters in lungfish evolution. *Memoirs of the Association of the Australasian Palaeontologists,* **15**, 211–24.

Schultze, H. P. and Trueb, L. (1981). Review of E. Jarvik: *Basic structure and evolution of vertebrates. Journal of Vertebrate Paleontology,* **1**, 389–97.

Schultze, H. P., and Trueb, L. (ed.) (1991). *Origins of the higher groups of tetrapods: controvery and consensus.* Cormstock, Ithaca.

Schultze, H. P., and Zidek, J. (1982). Ein primitiver Acanthodier (Pisces) aus dem Unterdevon Lettlands. *Paläontologische Zeitschrift,* **56**, 95–105.

Sedgwick, A., and Murchison, R. I. (1828). On the structure and relations of the deposits contained between the Primary Rocks and the Oolitic Series in the North of Scotland. *Transactions of the Geological Society of London,* **3**, 125–60.

Sedgwick, A., and Murchison, R. I. (1839). On the classification of the older rocks of Devonshire and Cornwall. *Proceedings of the Geological Society of London,* **3**, 121–3.

Sepkoski, J. J. (1982). Compendium of fossil marine families. *Contributions in Biology and Geology of the Milwaukee Public Museum,* **51**, 1–125.

Sepkoski, J. J. (1986). Phanerozoic overview of mass extinction. In *Pattern and processes in the history of life,* (ed. D. M. Raup and D. Jablonski), pp. 277–95. Springer, Berlin.

Shergold, J. H. (1991). Late Proterozoic and Early Palaeozoic palaeontology and biostratigraphy of the Amadeus Basin. In *Geological and geophysical studies in the Amadeus basin, Central Australia,* (ed. R. J. Korsch and J. M. Kennard). *Bulletin of the Bureau of Mineral Resources,* **236**, 97–111.

Shields, O. (1979). Evidence for initial opening of the Pacific Ocean in the Jurassic. *Palaeogeography, Palaeoclimatology, Palaeoecology,* **26**, 181–220.

Shubin, N. H., and Alberch, P. (1986). A morphogenetic approach to the origin and basic organization of the tetrapod limb. *Evolutionary Biology,* **20**, 319–87.

Siegfried, P., and Gross, W. (1971). Christian Heinrich Pander 1794–1865 und seine Bedeutung für Paläontologie. *Münsterische Forschung in Geologie und Paläontologie*, **9**, 101–83.

Sire, J. Y., Geraudie, J., Meunier, F. J., and Zylberberg, L. (1987). On the origin of ganoine: histological and ultrastructural data on the experimental regeneration of the scales of *Calamoichthys calabricus* (Osteichthyes, Brachyopterygii, Polypteridae). *American Journal of Anatomy*, **180**, 391–402.

Smith, C. L., Rand, C. S., Schaeffer, B., and Atz, J. W. (1975). *Latimeria*, the living coelacanth, is ovoviviparous. *Science*, **190**, 1105–6.

Smith, I. C. (1957). New restorations of the heads of *Pharyngolepis oblongus* Kiaer and *Pharyngolepis kiaeri* sp. nov., with a note on their lateral-line systems. *Norsk Geologisk Tidsskrift*, **33**, 373–402.

Smith, M. M. (1978). Enamel in the oral teeth of *Latimeria chalumnae* (Pisces: Actinistia): a scanning electron microscope study. *Journal of Zoology*, **185**, 355–69.

Smith, M. M. (1991). Putative skeletal neural crest cells in early Late Ordovician vertebrates from Colorado. *Science*, **251**, 301–3.

Smith, M. M. (1992). Microstructure and evolution of enamel amongst osteichthyan fishes and early tetrapods. In *Structure, function and evolution of teeth*, (ed. P. Smith and E. Tchernov), pp. 72–101. Freund, Tel Aviv.

Smith, M. M., and Campbell, K. S. W. (1987). Comparative morphology, histology and growth of the dental plates of the Devonian dipnoan *Chirodipterus*. *Philosophical Transactions of the Royal Society of London*, **B 317**, 329–63.

Smith, M. M., and Chang, M. M. (1990). The dentition of *Diabolepis speratus* Chang and Yu, with further consideration of its relationships and the primitive dipnoan dentition. *Journal of Vertebrate Paleontology*, **10** (4), 420–33.

Smith, M. M., and Hall, B. K. (1990). Development and evolutionary origins of vertebrate skeletogenic and odontogenic tissues. *Biological Reviews*, **65**, 277–373.

Smith, M. M., and Hall, B. K. (1993). A developmental model for evolution of the vertebrate exoskeleton and teeth. The role of cranial and trunk neural crest. *Evolutionary Biology*, **27**, 387–448.

Smith, M. M., Boyde, A., and Reid, A. (1984). Cathodoluminescence as an indicator of growth increments in the dentine in tooth plates of Triassic lungfish. *Neues Jahrbuch für Geologie und Paläontologie, Monatshefte*, **1984**, 39–45.

Smithson, T. S. (1989). The earliest known reptile. *Nature*, **342**, 676–8.

Soehn, K. L., and Wilson, M. V. H. (1990). A complete, articulated heterostracan from Wenlockian (Silurian) beds of the Delorme Group, Mackenzie Mountains, Northwest territories, Canada. *Journal of Vertebrate Paleontology*, **10**, 405–19.

Sollas, W. J. (1904). A method for the investigation of fossils by serial sections. *Transactions of the Royal Society of London*, **B 196**, 259–65.

Spjeldnaes, N. (1979). The paleoecology of the Ordovician Harding Sandstone (Colorado, USA). *Palaeogeography, Palaeoclimatology, Palaeoecology*, **26**, 317–47.

Spjeldnaes, N. (1982). Palaeoecology of *Ichthyostega* and the origin of the terrestrial vertebrates. In *Paleontologia come scienza geostorica*, (ed. E. Montanaro-Gallitelli), pp. 323–43. Mocchi, Modena.

Stahl, B. (1967). Morphology and relationships of the Holocephali, with special reference to the venous system. *Bulletin of the Museum of Comparative Zoology, Harvard*, **135**, 141–213.

Stahl, B. (1980). Non-autostylic Pennsylvanian iniopterygian fishes. *Palaeontology*, **23**, 315–24.

Starck, D. (1967). Le crâne des Mammifères. In *Traité de zoologie*, (ed. P. P. Grassé), Vol. 16, pp. 495–549. Masson, Paris.

Starck, D. (1979). *Vergleichenden Anatomie der Wirbeltiere*, Vol. 2. Springer, Berlin.

Stensiö, E. A. (1921). *Triassic fishes from Spitzbergen*, Part 1, A. Holzhausen, Vienna.

Stensiö, E. A. (1925). On the head of the macropetalichthyids, with certain remarks on the head of the other arthrodires. *Publications of the Field Museum of Natural History (Geology)*, **4**, 87–197.

Stensiö, E. A. (1927). The Devonian and Downtonian vertebrates of Spitsbergen. 1. Family Cephalaspidae. *Skrifter om Svalbard og Ishavet*, **12**, 1–391.

Stensiö, E. A. (1931). Upper Devonian vertebrates from East Greenland. *Meddelelser om Grønland*, **86**, 1–212.

Stensiö, E. A. (1932). *The cephalaspids of Great Britain*. British Museum (Natural History), London.

Stensiö, E. A. (1937). On the Devonian coelacanthids of Germany, with special reference to the dermal skeleton. *Kungliga Svenska Vetenskapsakademiens Handlingar*, **16**, 1–56.

Stensiö, E. A. (1939). A new anaspid from the Upper Devonian of Scaumenac Bay, in Canada, with remarks on other anaspids. *Kungliga Svenska Vetenskapsakademiens handlingar*, **18**, 3–25.

Stensiö, E. A. (1948). On the Placodermi of the Upper Devonian of East Greenland. II. Antiarchi: subfamily Bothriolepidae. *Meddelelser om Grønland*, **139**, 1–622.

Stensiö, E. A. (1959). On the pectoral fin and shoulder girdle of the arthrodires. *Kungliga Svenska Vetenskapsakademiens Handlingar*, **8**, 1–229.

Stensiö, E. A. (1962). Origine et nature des écailles placoïdes et des dents. In *Problèmes actuels de paléontologie: evolution des Vertébrés*. Colloques internationaux du Centre national de la Recherche scientifique, No. 104, (ed. J. P. Lehman), pp. 75–85. Paris.

Stensiö, E. A. (1963). Anatomical studies on the arthrodiran head. Part 1. *Kungliga Svenska Vetenskapsakademiens Hanlingar*, **9**, 1–419.

Stensiö, E. A. (1964). Les Cyclostomes fossiles ou Ostracodermes. In: *Traité de paléontologie*, (ed. J. Piveteau), Vol. 4 (1), pp. 96–383. Masson, Paris.

Stensiö, E. A. (1968). The cyclostomes with special reference to the diphyletic origin of the Petromyzontida and Myxinoidea. In *Current problems of lower vertebrate phylogeny*, (ed. T. Ørvig), pp. 14–80. Alqvist and Wiksell, Stockholm.

Stensiö, E. A. (1969). Elasmobranchiomorphi, Placodermata, Arthrodires. In *Traité de paléontologie*, (ed. J. Piveteau), Vol. 4 (2), pp. 71–692. Masson, Paris.

Stetson, H. C. (1928). A restoration of the anaspid *Birkenia elegans*. *Journal of Geology*, **34**, 458–70.

Stock, D. W., and Swofford, D. L. (1991). Coelacanth's relationships. *Nature*, **353**, 217–8.

Stock, D. W., and Whitt, G. S. (1992). Evidence from 18S ribosomal RNA that lampreys and hagfishes form a natural group. *Science*, **257**, 787–9.

Swofford, D. L. (1990). *PAUP, version 3.0. User's manual.* Illinois Natural History Survey, Champaign.

Szarski, H. (1977). Sarcopterygii and the origin of tetrapods. In *Major patterns in vertebrate evolution*, (ed. M. K. Hecht, P. C. Goody, and B. M. Hecht), pp. 517–40. Plenum Press, New York.

Tabin, C. J. (1992). Why we have (only) five fingers per hand: Hox genes and the evolution of paired limbs. *Development*, **116**, 289–96.

Tabin, C. J., and Laufer, E. (1993). Hox genes and serial homology. *Nature*, **361**, 692–3.

Talent, J. A., Goel, R. K., Jain, A. K., and Pickett, J. W. (1988). Silurian and Devonian of India, Nepal and Bhutan: biostratigraphic and palaeobiogeographic anomalies: *Courier Forschungsinstitut Senckenberg*, **106**, 1–57.

Talent, J. A., Mawson, R., Andrew, A. S., Hamilton, P. J., and Whitford, D. J. (1993). Middle Palaeozoic extinction events: faunal and isotopic data. *Palaeogeography, Palaeoclimatology, Palaeoecology*, **104**, 139–52.

Tarlo, L. B. H. (1964). Psammosteiformes (Agnatha). 1. General part. *Palaeontologia Polonica*, **13**, 1–135.

Tarlo, L. B. H. (1965). Psammosteiformes (Agnatha). 2. Systematic part. *Palaeontologia Polonica*, **15**, 1–164.

Tarrant, P. R. (1991). The ostracoderm *Phialaspis* from the Lower Devonian of the Welsh Borderland and South Wales. *Palaeontology*, **34**, 399–438.

Taverne, L. (1981). Ostéologie et affinités systématiques de *Leptolepides sprattiformis* (Pisces, Teleostei) du Jurassique supérieur de l'Europe. *Annales de la Société Royale Zoologique de Belgique*, **110**, 7–28.

Taylor, K., and Adamec, T. (1977). Tooth histology and ultrastructure of a Palaeozoic shark, *Edestus heinrichii*. *Fieldiana: Geology*, **33**, 29–39.

Thies, D. (1982). A neoselachian shark's tooth from the Lower Triassic of the Kocaeli (=Bithynian) Peninsula, W. Turkey. *Neues Jahrbuch für Geologie und Paläontologie, Monatshefte*, **1982**, 272–8.

Thies, D., and Reif, W. E. (1985). Phylogeny and evolutionary ecology of Mesozoic Neoselachii. *Neues Jahrbuch für Geologie und Paläontologie, Abhandlungen*, **169**, 333–61.

Thomson, K. S. (1964). The comparative anatomy of the snout in rhipidistian fishes. *Bulletin of the Museum of Comparative Zoology, Harvard*, **131**, 316–57.

Thomson, K. S. (1965). The endocranium and associated structures in the Middle Devonian rhipidistian fish *Osteolepis*. *Proceedings of the Linnean Society of London*, **176**, 181–95.

Thomson, K. S. (1973). Observations on a new rhipidistian fish from the Upper Devonian of Australia. *Palaeontographica*, **A 143**, 209–20.

Thomson, K. S. (1975). On the biology of cosmine. *Bulletin the Peabody Museum of Natural History*, **40**, 1–59.

Thomson, K. S. (1976a). On the heterocercal tail in sharks. *Paleobiology*, **2**, 19–38.

Thomson, K. S. (1976b). Explanation of large scale extinctions of lower vertebrates. *Nature*, **261**, 578–80.

Thomson, K. S. (1977). On the individual history of cosmine and a possible electroreceptive function of the pore-canal system in fossil fishes. In *Problems in vertebrate evolution*. Linnean Society Symposium Series No. 4, (ed. S. M. Andrews, R. S. Miles, and A. D. Walker), pp. 247–70. Academic Press, London.

Thomson, K. S. (1993). Segmentation, the adult skull, and the problem of homology. In *The skull*, (ed. J. Hanken and B. K. Hall), Vol. 2, pp. 36–68. University of Chicago Press.

Thomson, K. S., and Campbell, K. S. W. (1971). The structure and relationships of the primitive Dipnoan lungfish *Dipnorhynchus sussmilchi* (Etheridge). *Bulletin of the Peabody Museum of Natural History*, **38**, 1–109.

Thomson, K. S., and Hahn, K. V. (1968). Growth and form in fossil rhipidistian fishes (Crossopterygii). *Journal of Zoology*, **156**, 199–223.

Thorogood, P. (1993). Differentiation and morphogenesis of cranial skeletal tissues. In *The skull*, (ed. J. Hanken and B. K. Hall), Vol. 1, pp. 112–52. University of Chicago Press.

Thorsteinsson, R. (1967). Preliminary note on Silurian and Devonian ostracoderms from Cornwallis and Somerset Islands, Canadian Artic Archipelago. In *Problèmes actuels de paléontologie: evolution des Vertébrés*. Colloques Internationaux du Centre National de la Recherche Scientifique, No. 163, (ed. J. P. Lehman), pp. 45–7. Paris.

Thorsteinsson, R. (1973). Dermal elements of a new lower vertebrate from the Middle Silurian (Upper Wenlockian) rocks of the Canadian Arctic archipelago. *Palaeovertebrata*, **A 143**, 51–7.

Tong-Dzuy, T., and Janvier, P. (1990). Les Vertébrés du Dévonien inférieur du Bac Bo oriental (provinces de Bac Thaï et Lang Son, Viêt Nam). *Bulletin du Muséum national d'Histoire Naturelle, Paris*, **2**, 143–223.

Tong-Dzuy, T., Janvier, P., Ta Hoa, P., and Doan Nhat, T. (1995). Lower Devonian biostratigraphy and vertebrates of the Tong Vai Valley (Ha Giang Province, Vietnam). *Palaeontology*, **38**, 169–86.

Toombs, H. A. (1948). The use of acetic acid in the development of vertebrate fossils. *Museums Journal*, **48**, 54–5.

Traquair, R. H. (1877–1914). The ganoid fishes of the British Carboniferous formations. *Palaeontographical Society Monographs, 1877–1914*, 1–186.

Traquair, R. H. (1899). Report of fossil fishes collected by the Geological Survey of Scotland in the Silurian rocks of South Scotland. *Transactions of the Royal Society of Edinburgh*, **39**, 827–64.

Traquair, R. H. (1900). The bearings of fossil ichthyology on the problem of evolution. *Geological Magazine*, **7**, 463–70.

Trueb, L., and Cloutier, R. (1991). A phylogenetic investigation of the inter- and intrarelationships of the Lissamphibia (Amphibia: Temnospondyli). In *Origins of the higher groups of tetrapods: controversy and consensus*, (ed. H. P. Schultze and L. Trueb), pp. 175–88. Cormstock, Ithaca.

Turner, S. (1970). Fish help to trace continental movements. *Spectrum*, **79**, 8–10.

Turner, S. (1982). A new articulated thelodont (Agnatha) from the Early Devonian of Britain. *Palaeontology*, **25**, 879–89.

Turner, S. (1991). Monophyly and interrelationships of the Thelodonti. In *Early vertebrates and related problems of evolutionary biology*, (ed. M. M. Chang, Y. H. Liu, and G. R. Zhang), pp. 87–119. Science Press, Beijing.

Turner, S. (1992). Thelodont lifestyles. In *Fossil fishes as living animals*, (ed. E. Mark-Kurik). *Akademia*, **1**, 21–40.

Turner, S., and Long, J. A. (1989). Professor Edwin Sherbon Hills: his palaeontological works. In *Pathways in geology, essays in honour of Edwin Sherbon Hills*, (ed. R. W. Le Maître), pp. 41–51. Blackwell, Melbourne.

Turner, S., and Tarling, D. H. (1982). Thelodont and other agnathan distributions as tests of Lower Palaeozoic continental reconstructions. *Palaeogeography, Palaeoclimatology, Palaeoecology*, **39**, 295–311.

Turner, S., and Van der Brugghen, W. (1993). The Thelodonti, an important but enigmatic group of Palaeozoic fishes. *Modern Geology*, **18**, 125–40.

Turner, S., and Young, G. C. (1987). Shark teeth from the Early–Middle Devonian Cravens Peak Beds, Georgina Basin, Queensland. *Alcheringa*, **11**, 233–44.

Ubaghs, G. 1981. Réflexions sur la nature et la fonction de l'appendice articulé des carpoïdes Stylophora (Echinodermata). *Annales de Paléontologie (Invertébrés)*, **67**, 33–48.

Upeniece, I., and Upenieks, J. (1992). Young Upper Devonian antiarch (*Asterolepis*) individuals from the Lode Quarry, Latvia. In *Fossil fishes as living animals*, (ed. E. Mark-Kurik). *Academia*, **1**, 167–74.

Vaillant, L. (1902). Sur la présence de tissu osseux chez certains poissons des terrains paléozoïques de Canyon City (Colorado). *Comptes Rendus de l'Académie des Sciences, Paris*, **134**, 1321–2.

Valiukevicius, J. (1992). First articulated *Poracanthodes* from the Lower Devonian of Severnaya Zemlya. In *Fossil fishes as living animals*, (ed. E. Mark-Kurik). *Akademia* **1**, 193–213.

Van der Brugghen, W., and Janvier, P. (1993). Denticles in thelodonts. *Nature*, **364**, 107.

Van Wijhe, J. W. (1882). Ueber die Mesodermsegmente und die Entwicklung der Nerven des Selachierkopfes. *Koninklijke Akademie van Wetenschappen, Natuurkundige Verhandlingen*, **22**, 1–50.

Véran, M. (1988). Les éléments accessoires de l'arc hyoïdien des poissons téléostomes (Acanthodiens et Ostéichthyens) fossiles et actuels. *Mémoires du Muséum national d'Histoire naturelle*, **54**, 1–98.

Vézina, D. (1990). Les Plourdosteidae fam. nov. (Placodermi, Arthrodira) et leurs relations phylétiques au sein des Brachythoraci. *Canadian Journal of Earth Sciences*, **27**, 677–83.

Vézina, D. (1991). Nouvelles observations sur l'environnement sédimentaire de la Formation d'Escuminac (Dévonien supérieur, Frasnien), Québec, Canada. *Canadian Journal of Earth Sciences*, **28**, 225–30.

Vieth, J. (1980). Thelodontier- Acanthodier- ond Elasmobranchier-Schuppen aus dem Unter-Devon der Kanadischen Arktis (Agnatha, Pisces). *Göttinger Arbeiten in Geologie und Paläontologie*, **23**, 1–69.

Vogt, C. (1842). *Untersuchungen über die Entwicklungsgeschichte der Geburtshelferkroete (Alytes obstetricans)*. Jent and Gassmann, Solothurn.

Vogt, C. (1845). Quelques observations sur les caractères qui servent à la classification des poissons ganoïdes. *Annales des Sciences Naturelles (Zoologie)*. **4**, 53–68.

Vorobyeva, E. I. (1977). [*Morphology and Evolution of Sarcopterygian Fishes*]. Trudy, Paleontologischeskogo Instituta, No. 163, Akademia Nauk SSSR, Moscow. (In Russia.)

Vorobyeva, E. I. (1992). The role of development and function in formation of 'tetrapod-like' pectoral fins. *Zhurnal Obshechei Biologii*, **53**, 149–58.

Vorobyeva, E. I. (1993). [*The problem of the terrestrial vertebrate origin*.] Nauka, Moscow. (In Russian.)

Vorobyeva, E. I., and Hinchliffe, J. R. (1991). [The problem of fish fin transformations to tetrapod limbs.] *Zhurnal Obshechei Biologii*, **52**, 192–204. (In Russian.)

Vorobyeva, E. I., and Kuznetzov, A. (1992). The locomotor apparatus of *Panderichthys rhombolepis* (Gross), a supplement to the problem of fish–tetrapod transition. In *Fossil fishes as living animals*, (ed. E. Mark-Kurik). *Akademia*, **1**, 131–40.

Vorobyeva, E. I., and Nitkin, V. B. (1991). On the homology of nasal structure in ancient fishes and lower tetrapods. In *Early vertebrates and related problems of evolutionary biology*, (ed. M. M. Chang, Y. H. Liu, and G. R. Zhang), pp. 477–85. Science Press, Beijing.

Vorobyeva, E. I., and Obruchev, D. V. (1964). [Subclass Sarcopterygii.] In (*Oznovii Paleontologii*), (ed. Y. A. Orlov), Vol. 2, pp. 268–322. Nauka, Moscow. (In Russian.)

Vorobyeva, E. I., and Schultze, H. P. (1991). Description and systematics of panderichthyid fishes with comments on their relationship to tetrapods. In *Origins of the higher groups of tetrapods. Controversy and consensus*, (ed. H. P. Schultze and L. Trueb), pp. 68–109. Cormstock, Ithaca.

Wahlert, G. von (1970). Die Entstehung des Kieferapparates der Gnathostomen. *Verhandlungen der deutschen Zoologischen Gesellschaft*, **64**, 344–7.

Wake, M. H. (ed.) (1979). *Hyman's comparative vertebrate anatomy*, (3rd edn). University of Chicago Press.

Walcott, C. (1892). Notes on the discovery of a vertebrate fauna in the Silurian (Ordovician) strata. *Bulletin of the Geological Society of America*, **3**, 153–72.

Wang, J. Q. (1989). [The Antiarchi from Early Silurian of Hunan.] *Vertebrata PalAsiatica*, **29**, 240–4. (In Chinese with English summary.)

Wang, J. Q., and Wang, N. Z. (1992). [Early Devonian galeaspid agnatha from Southeast of Yunnan, China.] *Vertebrata PalAsiatica*, **30**, 185–94. (In Chinese with English summary.)

Wang, N. Z. (1986). [Note on two Middle Silurian agnathans (*Hanyangaspis* and *Latirostraspis*) of China.] In *Collected papers of the the 13th annual conference of the Paleontological Society of China*, pp. 49–57. Beijing. (In Chinese with English summary.)

Wang, N. Z. (1991). Two new Silurian galeaspids (jawless craniates) from Zhejiang province, China, with a discussion of galeaspid–gnathostome relationships. In *Early vertebrates and related problems of evolutionary biology*, (ed. M. M. Chang, Y. H. Liu, and G. R. Zhang), pp. 41–65. Science Press, Beijing.

Wang, N. Z., and Dong Z. H. (1989). [Discovery of Late Silurian microfossils of Agnatha and fishes from Yunnan, China.] *Acta Palaeontologica Sinica*, **28**, 192–206. (In Chinese with English summary.)

Wang, S. T. (1991). Lower Devonian vertebrate paleocommunities from South China. In *Early vertebrates and related problems of evolutionary biology*, (ed. M. M. Chang, Y. H. Liu, and G. R. Zhang), pp. 487–97. Science Press, Beijing.

Wängsjö, G. (1952). The Downtonian and Devonian vertebrates of Spitsbergen. 9. Morphologic and systematic studies of the Spitsbergen cephalaspids. Results of Th. Vogt's Expedition 1928 and the English–Norwegian–Swedish Expedition in 1939. *Norsk Polarinstitutt Skrifter*, **97**, 1–611.

Warren, J. W., and Wakefield, N. A. (1972). Trackway of tetrapod from the Upper Devonian of Victoria, Australia. *Nature*, **238**, 469–70.

Warren, A. A., Jupp, R., and Bolton, R. (1986). Earliest tetrapod trackway. *Alcheringa*, **10**, 183–6.

Waterman, A. J. (ed.) (1971). *Chordate structure and function*. Macmillan, New York.

Watson, D. M. S. (1927). The reproduction of the coelacanth fish *Undina*. *Proceedings of the Zoological Society, London*, **1927**, 453–7.

Watson, D. M. S. (1928). On some points in the structure of palaeoniscid fishes. *Proceedings of the Zoological Society of London*, **1928**, 49–70.

Watson, D. M. S. (1937). The acanthodian fishes. *Philosophical Transactions of the Royal Society of London*, **B 228**, 49–146.

Watson, D. M. S. (1954). A consideration of ostracoderms. *Proceedings of the Royal Society of London*, **B 238**, 1–25.

Webb, P. H., and Weihs, D. (ed.) (1983). *Fish biomechanics*. Praeger, New York.

Wells, N. A., and Dorr, J. A. (1985). Form and function of the fish *Bothriolepis* (Devonian; Placodermi, Antiarchi): the first terrestrial vertebrate? *Michigan Academian*, **17**, 157–73.

Westoll, T. S. (1943a). The origin of the tetrapods. *Biological Reviews*, **18**, 78–98.

Westoll, T. S. (1943b). The origin of the primitive tetrapod limb. *Proceedings of the Royal Society of London*, **B 131**, 373–93.

White, E. I. (1935). The ostracoderm *Pteraspis* Kner and the relationships of the agnathous vertebrates. *Philosophical Transactions of the Royal Society of London*, **B 225**, 381–457.

White, E. I. (1946). *Jamoytius kerwoodi*, a new chordate from the Silurian of Lanarkshire. *Geological Magazine*, **83**, 89–97.

White, E. I. (1958). Original environment of the craniates. In *Studies of fossil vertebrates*, (ed. T. S. Westoll), pp. 213–33. Athlone Press, London.

White, E. I. (1973). Form and growth of *Belgicaspis* (Heterostraci). *Palaeontographica*, **A 143**, 11–24.

White, E. I. (1968). Devonian fishes from the Mawson–Mulock area, Victoria Land, Antarctica. *Scientific Report of the Transantarctic Expeditions*, **16**, 1–26.

White, E. I., and Toombs, H. A. (1983). The cephalaspids from the Dittonian at Cwm Mill, near Abergavenny, Gwent. *Bulletin of the British Museum (Natural History), Geology*, **37**, 149–71.

White, T. E. (1939). Osteology of *Seymouria baylorensis* Broili. *Bulletin of the Museum of Comparative Zoology, Harvard*, **85**, 325–409.

Whiteaves, J. F. (1881). On some remarkable fossil fishes from the Devonian rocks of Scaumenac Bay, Province of Québec, with descriptions of a new genus and three new species. *Canadian Naturalist*, **10**, 27–35.

Whiting, H. P. (1977). Cranial nerves in lampreys and cephalaspids. In *Problems in vertebrate evolution*. Linnean Society Symposium Series No. 4, (ed. S. M. Andrews, R. S. Miles, and A. D. Walker), pp. 1–23. Academic Press, London.

Whybrow, P., and Lindsay, W. (1990). Preparation of macrofossils. In *Palaeobiology. A synthesis*, (ed. D. E. Briggs and P. R. Crowther), pp. 499–502. Blackwell Scientific Publications, Oxford.

Wickstead, J. M. (1969). Some further comments on *Jamoytius kerwoodi* White. *Zoological Journal of the Linnean Society*, **48**, 421–2.

Wiley, E. O. (1979). Ventral gill arch muscles and the phylogenetic relationships of gnathostomes, with a new classification of the Vertebrata. *Journal of the Linnean Society (Zoology)*, **67**, 149–79.

Wiley, E. O. (1981). *Phylogenetics: the theory and practice of phylogenetic systematics*. Wiley, New York.

Wiley, E. O., Siegel-Causey, D., Brooks, D. R., and Fink, V. A. (1991). *The compleat cladist*. Special publication of the University of Kansas Natural History Museum, Lawrence.

Williams, M. E. (1985). The 'cladodont' level sharks of the Pennsylvanian black shales of Central North America. *Palaeontographica*, **A 190**, 83–158.

Williams, M. E. (1991). Feeding behavior in Cleveland Shale fishes. In *Evolutionary paleobiology of behavior and coevolution*, (ed. A. J. Boucot), pp. 273–87. Elsevier, Amsterdam.

Wilson, M. V. H., and Caldwell, M. W. (1993). New Silurian and Devonien fork-tailed 'thelodonts' are jawless vertebrates with stomachs and deep bodies. *Nature*, **361**, 442–4.

Wood, S. P. (1982). New basal Namurian (Upper Carboniferous) fishes and crustaceans found near Glasgow. *Nature*, **297**, 574–7.

Woodward, A. S. (1889–1901). *Catalogue of the fossil fishes in the British Museum (Natural History)*, (3 vols). British Museum (Natural History), London.

Woodward, A. S. (1935). On the affinities of the acanthodian and arthrodiran fishes. *Annals and Magazine of Natural History*, **15**, 392–5.

Yalden, D. W. (1985). Feeding mechanisms as evidence for cyclostome monophyly. *Zoological Journal of the Linnean Society*, **84**, 291–300.

Yochelson, E. L. (1983). Walcott's discovery of Middle Ordovician vertebrates. *History of Geology*, **2**, 66–75.

Young, G. C. (1978). A new Early Devonian petalichthyid fish from the Taemas/Wee Jasper region of New South Wales. *Alcheringa*, **2**, 103–16.

Young, G. C. (1979). New information on the structure and relationships of *Buchanosteus* (Placodermi, Euarthrodira) from the Early Devonian of New South Wales. *Zoological Journal of the Linnean Society*, **66**, 309–52.

Young, G. C. (1980). A new Early Devonian placoderm from New South Wales, Australia, with a discussion of placoderm phylogeny. *Palaeontographica*, **A 167**, 10–76.

Young, G. C. (1981). Biogeography of Devonian vertebrates. *Alcheringa*, **5**, 225–43.

Young, G. C. (1982*a*). Devonian sharks from south-eastern Australia and Antarctica. *Palaeontology*, **25**, 87–9.

Young, G. C. (1982*b*). Devonian palaeontological data and the Armorica problem. *Palaeogeography, Palaeoclimatology, Palaeoecology*, **60**, 283–304.

Young, G. C. (1984). Reconstruction of the jaws and braincase in the Devonian placoderm fish *Bothriolepis*. *Palaeontology*, **27**, 625–61.

Young, G. C. (1986*a*). Cladistic methods in Paleozoic continental reconstruction. *Journal of Geology*, **94**, 523–37.

Young, G. C. (1986*b*). The relationships of placoderm fishes. *Zoological Journal of the Linnean Society*, **88**, 1–57.

Young, G. C. (1988). Antiarchs (placoderm fishes) from the Devonian Aztec sandstone, southern Victoria Land, Antarctica. *Palaeontographica*, **A 202**, 1–125.

Young, G. C. (1990). Devonian vertebrate distribution patterns and cladistic analysis of palaeogeographical hypotheses. In *Palaeozoic palaeogeography and biogeography*, (ed. W. S. McKerrow and C. R. Scotese). *Geological Society Memoir* **12**, 243–55.

Young, G. C. (1991*a*). The first armoured agnathan vertebrates from the Devonian of Australia. In *Early vertebrates and related problems of evolutionary biology*, (ed. M. M. Chang, Y. H. Liu, and G. R. Zhang), pp. 67–85. Science Press, Beijing.

Young, G. C. (1991*b*). Fossil fishes from Antarctica. In *The geology of Antarctica*, (ed. R. J. Tingey), pp. 538–67. Oxford Monographs in Geology and Geophysics No. 17. Clarendon Press, Oxford.

Young, G. C. (1993*a*). Middle Palaeozoic macrovertebrate biostratigraphy of eastern Gondwana. In *Palaeozoic vertebrate biostratigraphy and biogeography*, (ed. J. A. Long), pp. 208–51. Belhaven Press, London.

Young, G. C. (1993*b*). Vertebrate faunal provinces in the Middle Palaeozoic. In *Palaeozoic vertebrate biostratigraphy and biogeography*, (ed. J. A. Long), pp. 293–323. Belhaven Press, London.

Young, G. C., and Zhang, G. R. (1992). Structure and function of the pectoral joint and operculum in antiarchs. Devonian placoderm fishes. *Palaeontology*, **35**, 443–64.

Young, G. C., Barwick, R. E., and Campbell, K. S. W. (1989). Pelvic girdles of lungfishes (Dipnoi). In *Pathways in geology. Essays in honour of Edwin Sherbon Hills*, (ed. R. W. Le Maître), pp. 59–75. Blackwell, Melbourne.

Young, G. C., Long, J. A., and Ritchie, A. (1992). Crossopterygian fishes from the Devonian of Antarctica: systematics, relationships and biogeographic significance. *Records of the Australian Museum, Supplement* **14**, 1–77.

Yvroud, M. (1971). Le canal nasolacrymal et son intérêt évolutif. *Annales de Biologie*, **10**, 465–509.

Zangerl, R. (1966). A new shark of the family Edestidae, *Ornithoprion hertwigi*, from the Pennsylvanian Mecca and Logan Quarry Shales of Indiana. *Fieldiana: Geology*, **16**, 1–43.

Zangerl, R. (1973). Interrelationships of early chondrichthyans. In *Interrelationships of fishes*, (ed. P. H. Greenwood, R. S. Miles, and C. Patterson). *Zoological Journal of the Linnean Society, Supplement 1*, **53**, 1–14.

Zangerl, R. (1981). Chondrichthyes I. Palaeozoic Elasmobranchii. In *Handbook of paleoichthyology*, Vol. 3A, (ed. H. P. Schultze). Gustav Fisher, Stuttgart.

Zangerl, R. (1984). On the microscopic anatomy and possible function of the spine-'brush' complex of *Stethacanthus* (Elasmobranchii: Symmoriida). *Journal of Vertebrate Paleontology*, **4**, 372–8.

Zangerl, R., and Case G. R. (1973). Iniopterygia, a new order of chondrichthyan fishes from the Pennsylvanian of North America. *Fieldiana: Geology*, **6**, 1- 67.

Zangerl, R., and Case, G. R. (1976). *Cobelodus aculeatus* (Cope), an anacanthous shark from Pennsylvanian black shales of North America. *Palaeontographica*, **A 154**, 107–57.

Zangerl, R., and Williams, M. (1975). New evidence on the nature of the jaw suspension in Palaeozoic anacanthous sharks. *Palaeontology*, **18**, 333–41.

Zangerl, R., Winter, H. F., and Hansen, M. C. (1993). Comparative microscopic dental anatomy in the Petalodontida (Chondrichthyes, Elasmobranchii). *Fieldiana: Geology*, **26**, 1–43.

Zhang, G. R. (1978). [The antiarchs from the Early Devonian of China.] *Vertebrata PalAsiatica*, **16**, 147–86. (In Chinese with English abstract.)

Zhang, M. M. (Chang M. M.) (1980). Preliminary note on a Lower Devonian antiarch from Yunnan, China. *Vertebrata PalAsiatica*, **18**, 179–90.

Zhu, M. (1992). [Two new eugaleaspids, with a discussion on eugaleaspid phylogeny.] *Vertebrata PalAsiatica*, **30**, 169–84. (In Chinese with English summary.)

Zhu, M., and Janvier, P. (1994). Un Onychodontide (Vertebrata, Sarcopterygii) du Dévonien inférieur de Chine. *Comptes-Rendus de l'Académie des Sciences, Paris*, **319**, 951–6.

Zidek, J. (1976). A new shark egg-capsule from the Pennsylvanian of Oklahoma, and remarks on the chondrichthyan egg capsules in general. *Journal of Paleontology*, **50**, 907–15.

Zidek, J. (1985). Growth in *Acanthodes* (Acanthodii; Pisces). Data and implications. *Paläontologische Zeitschrift*, **59**, 147–66.

Zidek, J. (1988). Hamilton Quarry *Acanthodes* (Acanthodii; Kansas, Late Pennsylvanian). *Kansas Geological Survey Guidebook Series*, **6**, 155–9.

Zidek, J. (1992). Late Pennsylvanian Chondrichthyes, Acanthodii, and deep-bodied Actinopterygii from Kinney Quarry, Manzanita Mountains, New Mexico. *New Mexico Bureau of Mines and Mineral Resources Bulletin*, **138**, 145–82.

Appendix

While this book was in press, a number of new and important data have been published, some of which modify part of its contents. (Page numbers refer to the sections which are completed or modified by these new references.)

pp. 1–23. A large number of reconstructions of Palaeozoic fishes are provided by Long (1995). This book also contains numerous colour photographs of specimens, in particular the first articulated skull of *Onychodus* (see p. 198).

pp. 42, 281–3. Peterson (1995) proposes a phylogenetic test of the calcichordate scenario and shows that Jefferies' chordate phylogeny is less parsimonious than the trees in which the 'calcichordates' appear as the sister-group of echinoderms.

He also shows that Jefferies' theory of deuterostome interrelationships is less parsimonious than the current consensus, where enteropneusts are the sister-group of chordates, and cephalochordates the sister-group of craniates. In addition, it seems now quite clear that the hemichordates are not a clade and that only enteropneusts are the sister-group of the chordates. Christoffersen and Araujo-de-Almeida (1994) propose a morphology-based phylogeny of the Enterocoela (which they regard as the sister-group of the tubicolous annelids Oweniidae). They accept Jefferies' theory that tunicates are closer to craniates than cephalochordates, but propose that chaetognaths are the sister-group of craniates, a theory which may raise new insights as to the conodont controversy. They also suggest that, among 'hemichordates', pterobranchs are paraphyletic, cephalodiscids being the closest relatives of echinoderms, enteropneusts, and chordates.

p. 45. Wicht and Northcutt (1995) provide evidence for lateralis nerve fibres in hagfishes, although they fail to find neuromasts in the grooves on the hagfish head. They describe eye-lens and lateral-line placodes in an embryonic hagfish, but admit that the lack of neuromasts and lens in hagfishes is the primitive craniate condition.

pp. 84, 283. M. P. Smith and Sansom (1995) provide new histological support for the vertebrate affinity of *Anatolepis*, whose tubercles seem to have been made up by a kind of dentine with large tubules.

p. 84. Karatayute-Talimaa and Predtschenskyj (1995) describe various Late Ordovician and Early Silurian vertebrate remains from Siberia, including peculiar growing elements of *Tesakoviaspis*, which they doubtfully refer to the astraspids.

pp. 84, 284. The new conodont animal described by Gabbott *et al.* (1995) from the Ordovician of South Africa provides additional evidence for the craniate affinities of conodonts. It shows remarkably well preserved optic capsules and muscle fibres. The doubts expressed here as to conodont affinities fade away as new material turns up (Janvier 1995). Moreover, the very large size of the South African conodont animal rules out the suggestion that conodonts may be larval forms.

pp. 87, 276. M. M. Smith *et al.* (1995) provide evidence for dentine with fine tubules in the core of the tubercles of *Astraspis*. Consequently, *Astraspis* is in no way different from *Pycnaspis*. These authors also confirm the histological resemblance between *Eriptychius* and heterostracans.

pp. 95, 99. The caudal fin of *Nahanniaspis* is now known to be fork-shaped, and made up by five large scale-covered digitations with no intervening web, as in *Athenaegis* (Pellerin and Wilson 1995). This probably applies to all cyathaspidiform heterostracans.

p. 139. The soft tissues preserved in the Bear Gulch chondrichthyans provide support for the holocephalans affinities of *Heteropetalus* (Grogan and Lund 1995).

p. 149. Karatayute-Talimaa (1995) provides detailed information on the structure and diversity of the scales of the Silurian mongolepids, which she still refers to as chondrichthyans.

pp. 158–9. Lelièvre (1995) proposes a new phylogeny of the Arthrodira, where *Holonema* appears as an Eubrachythoraci. He also proposes two characters for the Migmatocephalia: the elongate nuchal and paranuchal plates, and the median dorsal plate shorter than broad. He no longer regards the sensory-line pattern on the sub-orbital plate as a migmatocephalan character, but this may still be retained as a general brachythoracid character. He now excludes buchanosteids from the Migmato cephalia.

pp. 154, 171–2. Goujet and Young (1995) make a new analysis of the interrelationships of the placoderms. Their strict consensus tree of six cladograms differs from the cladogram in Fig. 4.57 as to the position of the Acanthothoraci and Rhenanida. The Acanthothoraci appear as the sister-clade of all other placoderms, and the Rhenanida are the sister-group of the clade that includes the antiarchs, petalichthyids, ptyctodonts, and arthrodires. The relationships of the last three taxa is the same as in Fig. 4.57.

pp. 172, 291. Carr (1995) reviews placoderm diversity throughout the Devonian and provides an updated list of all the placoderm species (his mention of Carboniferous placoderms is, however, based on doubtful stratigraphical record). He also discusses various aspects of placoderm palaeobiology and functional anatomy, and proposes an improved cladogram of the Eubrachythoraci, where brachydeirids are placed among the aspinothoracid pachyosteomorphs, and not in the coccosteomorphs, as shown in Fig. 4.49.

pp. 184, 186, 329. Burrow (1995*a*) describes a new Lower Devonian actinopterygian from Australia, *Terenolepis*, whose scales lack the dorsal peg and have a single ganoine layer. It can be inserted in Fig. 4.67 between *Naxilepis* and *Orvikuina*. Burrow (1995*b*) also describes a new species of *Lophosteus* from the Silurian of Australia and provides further arguments in support of the possible placoderm affinities of this enigmatic fish, doubtfully referred to here as an actinopterygian.

p. 200. The lack of radials in the anterior dorsal fin is not unique to porolepiforms, as it is now know to exist in lungfishes (Ahlberg 1994). It is thus a character of the Dipnomorpha.

p. 211. Chang (1995) provides new information on the lungfish *Diabolepis*, with a reconstruction of the lower jaw (somewhat different from that in Fig. 4.84C2). She also shows that *Diabolepis* retains an independent palato-quadrate.

p. 216. Lebedev (1995) makes a detailed description of the acid-prepared skull of the osteolepiform *Medoevia*, probably the best preserved osteolepiform specimen known to date. This form is suggestive of canowindrids.

pp. 225–36, 285 Clack and Coates (1995) and Coates and Clack (1995) review the tetrapod characters on the basis of the earliest fossil forms and suggest a process of early tetrapod evolution toward terrestralization. They provide the first full reconstruction of the skeleton of *Acanthostega* and a new reconstruction of *Ichthyostega*. They also suggest that both were exclusively aquatic animals (contrary to the reconstructions in Fig. 1.13). They suggest a link between the Frasnian/Famennian extinction event and the rise of the tetrapods. Carroll (1995) proposes a theory of Palaeozoic tetrapod interrelationships based on the analysis of 184 characters. His strict consensus tree of the three shortest trees mainly differs from that in Fig. 4.105 in the position of the nectrideans, aistopods, and microsaurs, which appear as more closely related to the amniotes than to either the temnospondyls, loxommatids, seymouriamorphs, or anthracosaurs. Ahlberg (1995) describes the Late Frasnian tetrapod *Elginerpeton*, closely related to *Obruchevichthys*, both being the sister-group of all other tetrapods, including *Ventastega* and *Acanthostega*. Daeschler *et al.* (1994) describe yet another Devonian tetrapod, *Hynerpeton* from North America, and Lebedev and Coates (1995) describe the postcranial skeleton of *Tulerpeton* and place this taxon among the Reptiliomorpha (i.e. above node 5 in Fig. 4.105). This position of *Tulerpeton* is supported by the reptiliomorph L-shaped intermedium in tarsus, large and lateral supraglenoid foramen, and loss of the post-branchial lamina (closure of the gill slit). This, however, imposes a number of reversions in reptiliomorphs and *Tulerpeton*. It also implies that pentadactyly is no longer a character of the Neotetrapoda (node 4 in Fig. 4.105), which are now best defined by the structure of wrist and ankle joints (hinged and rotary, respectively) and the scapulocoracoid separated from the cleithrum.

pp. 211, 221–5, 247–50. Cloutier and Ahlberg (1995) make a thorough analysis of sarcopterygian interrelationships based on 140 characters, and they test the robustness of several clades. Their tree differs from that in Fig. 5.3 (left side) in the position of the Onychodontiformes, which are more closely related to the Rhipidistia than are the Actinistia (although, the authors concede that this is the most parsimonious tree by only three steps). The interrelationships of the major sarcopterygian clades also differ somewhat from those proposed here. The tree of the Dipnomorpha show porolepiforms as a clade, but that of lungfishes is the same as in Fig. 4.85, except for

Uranolophus, which is placed more crownward than *Dipnorhynchus*. The tree of the osteolepiformes differs from that in Fig. 4.93 in suggesting the monophyly of osteolepids plus the Canowindridae, the latter being the sister-group of *Osteolepis*. This would be more consistent with the late stratigraphical occurrence of the Canowindridae. The authors also point out that the name Elpistotegalia should replace Panderichthyida, by priority. They allude to the possible paraphyly of the Osteolepiformes and Elpistostegalia. They also place *Crassigyrinus* as the sister-group of all other tetrapods (see Fig. 4.105).

p. 225. Vorobyeva (1995) gives a new reconstruction of the shoulder girdle of *Panderichthys*.

pp. 268–9. Vorobyeva and Hinchliffe (1995) consider that the distal part of the tetrapod limb skeleton, in particular phalanges and carpals/tarsals, is an evolutionary novelty and, therefore, cannot be homologized with the fish radials. However, both develop under the influence of much the same, highly instable developmental process.

p. 284. Insom *et al.* (1995) describe a new fossil from the Middle Cambrian Burgess Shale, which they consider as a craniate. They also discuss chordate and craniate phylogeny, in the framework of a strong attack against cladistics. They conclude that protochordates (tunicates and cephalochordates) are a clade (including *Pikaia*) and that all monorhinous jawless craniates (cyclostomes, osteostracans, anaspids) are the sister-clade of the heterostracans and gnathostomes.

p. 287. Trinajstic (1995) provides an example of heterochronic evolution in placoderms.

p. 296. Young (1995*a, b*) makes a new consideration of the palaeogeographic significance of early vertebrate biogeography. He also proposes a new analytical method for inferring terrane accretions from palaeontological data.

p. 297. Mark-Kurik (1995) provides a detailed account of the possible trophic relationships between Devonian fishes.

REFERENCES

Ahlberg, P. E. (1994). The postcranial skeleton of the Middle Devonian lungfish *Dipterus valenciennesi*. *Transactions of the Royal Society of Edinburgh, Earth Sciences*, **85**, 159–75.

Ahlberg, P. E. (1995). *Elginerpeton pancheni* and the earliest tetrapod clade. *Nature*, **373**, 420–5.

Burrow, C. J. (1995*a*). A new palaeoniscoid from the Lower Devonian Trundle beds of Australia.In *Premiers Vertébrés et Vertébrés inférieurs* (ed. H. Lelièvre, S. Wenz, A. Blieck, and R. Cloutier), pp. 319–25, *Géobios, Mémoire Spécial*, **19**.

Burrow, C. J. (1995*b*). A new lophosteiform (Osteichthyes) from the Lower Devonian of Australia. In *Premiers Vertébrés et Vertébrés inférieurs* (ed. H. Lelièvre, S. Wenz, A. Blieck, and R. Cloutier), pp. 327–33, *Géobios, Mémoire Spécial*, **19**.

Carr, R. K. (1995). Placoderm diversity and evolution. In *Etudes sur les Vertébrés inférieurs* (ed. M. Arsenault, H. Lelièvre, and P. Janvier), pp. 85–125. *Bulletin du Muséum national d'Histoire naturelle, Paris*, C, **1–4**.

Carroll, R. L. (1995). Problems of the phylogenetic analysis of Paleozoic choanates. In *Etudes sur les Vertébrés inférieurs* (ed. M. Arsenault, H. Lelièvre, and P. Janvier), pp. 389–445. *Bulletin du Muséum national d'Histoire naturelle, Paris*, C, **1–4**.

Chang, M. M. (1995). *Diabolepis* and its bearing on the relationships between porolepiforms and dipnoans. In *Etudes sur les Vertébrés inférieurs* (ed. M. Arsenault, H. Lelièvre, and P. Janvier), pp. 235–68. *Bulletin du Muséum national d'Histoire naturelle, Paris*, C, **1–4**.

Clack, J. A. and Coates, M. I. (1995). *Acanthostega gunnari*, a primitive, aquatic tetrapod? In *Etudes sur les Vertébrés inférieurs* (ed. M. Arsenault, H. Lelièvre, and P. Janvier), pp. 359–72. *Bulletin du Muséum national d'Histoire naturelle, Paris*, C, **1–4**.

Christoffersen, M. L. and Araujo-de-Almeida, E. (1994). A phylogenetic framework of the Enterocoela (Metameria: Coelomata). *Revista Nordestina de Biologia*, **9**, 173–208.

Cloutier, R. and Ahlberg, P. E. (1995). Sarcopterygian interrelationships: how far are we from a phylogenetic consensus? In *Premiers Vertébrés et Vertébrés inférieurs* (ed. H. Lelièvre, S. Wenz, A. Blieck, and R. Cloutier), pp. 241–8, *Géobios, Mémoire Spécial*, **19**.

Coates, M. I. and Clack, J. A. (1995). Romer's gap: tetrapod origins and terrestriality. In *Premiers Vertébrés et Vertébrés inférieurs* (ed. H. Lelièvre, S. Wenz, A. Blieck, and R. Cloutier), pp. 373–88, *Géobios, Mémoire Spécial*, **19**.

Daeschler, E. B. Shubin, N. H., Thomson, K. S., and Amaral W. W. (1994). A Devonian tetrapod from North America. *Science*, **265**, 639–642.

Gabbott, S. E., Aldridge, R. J., and Theron, J. N. (1995). A giant conodont with preserved muscle tissue from the Upper Ordovician of South Africa. *Nature*, **374**, 800–3.

Goujet, D. and Young, G. C. (1995). Interrelationships of placoderms revisited. In *Premiers Vertébrés et Vertébrés inférieurs* (ed. H. Lelièvre, S. Wenz, A. Blieck, and R. Cloutier), pp. 89–95, *Géobios, Mémoire Spécial* **19**.

Grogan, E. D. and Lund, R. (1995). Pigment patterns, soft anatomy and relationships of Bear Gulch chondrichthyes (Namurian E2b, Lower Carboniferous, Montana, USA). In *Premiers Vertébrés et Vertébrés inférieurs* (ed. H. Lelièvre, S. Wenz, A. Blieck, and R. Cloutier), pp. 145–6, *Géobios, Mémoire Spécial*, **19**.

Insom, E., Pucci, A., and Simonetta, A. M. (1995). Cambrian Protochordata, their origin and significance. *Bolletino di Zoologia*, **62**, 243–53.

Janvier, P. (1995). Conodonts join the club. *Nature*, **374**, 761–2.

Karatayute-Talimaa, V. (1995). The Mongolepidida: Scale structure and systematic position. In *Premiers Vertébrés et Vertébrés inférieurs* (ed. H. Lelièvre, S. Wenz, A. Blieck, and R. Cloutier), pp. 35–7, *Géobios, Mémoire Spécial*, **19**.

Karatayute-Talimaa, V. and Predtechenskyj, N. (1995). The distribution of the vertebrates in the Late Ordovician and Early Silurian palaeobasins of the Siberian Platform. In *Etudes sur les Vertébrés inférieurs* (ed. M. Arsenault, H. Lelièvre, and P. Janvier), pp. 39–55. *Bulletin du Muséum national d'Histoire naturelle, Paris*, C, **1–4**.

Lebedev, O. A. (1995). Morphology of a new osteolepidid fish from Russia. In *Etudes sur les Vertébrés inférieurs* (ed. M. Arsenault, H. Lelièvre, and P. Janvier), pp. 287–341. *Bulletin du Muséum national d'Histoire naturelle, Paris*, C, **1–4**.

Lebedev, O. A. and Coates, M. I. (1995). The postcranial skeleton of the Devonian tetrapod *Tulerpeton curtum* Lebedev. *Zoological Journal of the Linnean Society*, **114**, 307–48.

Lelièvre, H. (1995). Description of *Maideria falipoui* n. g., n. sp., a long snouted brachythoracid (Vertebrata, Placodermi, Arthrodira) from the Givetian of Maider (South Morocco), with a phylogenetic analysis of primitive brachythoracids. In *Etudes sur les Vertébrés inférieurs* (ed. M. Arsenault, H. Lelièvre, and P. Janvier), pp. 163–207. *Bulletin du Muséum national d'Histoire naturelle, Paris*, C, **1–4**.

Long, J. A. (1995). *The rise of fishes*. Johns Hopkins University Press, Baltimore

Mark-Kurik, E. (1995). Trophic relations of Devonian fishes. In *Premiers Vertébrés et Vertébrés inférieurs* (ed. H. Lelièvre, S. Wenz, A. Blieck, and R. Cloutier), pp. 121–3, *Géobios, Mémoire Spécial*, **19**.

Pellerin, N. M. and Wilson, M. V. H. (1995). New evidence for structure of irregular easpididae tails from Lochkovian beds of the Delorme Group, Mackenzie Mountains, Northwest Territories, Canada. In *Premiers Vertébrés et Vertébrés inférieurs* (ed. H. Lelièvre, S. Wenz, A. Blieck, and R. Cloutier), pp. 45–50, *Géobios, Mémoire Spécial*, **19**.

Peterson K. J. (1995). A phylogenetic test of the calcichordate scenario. *Lethaia*, **28**, 25–38.

Smith, M. M., Sansom, I. J., and Smith, M. P. (1995). Diversity of the dermal skeleton in Ordovician to Silurian vertebrate taxa from North America: Histology, skeletogenesis and relationships. In *Premiers Vertébrés et Vertébrés inférieurs* (ed. H. Lelièvre, S. Wenz, A. Blieck, and R. Cloutier), pp. 65–70, *Géobios, Mémoire Spécial*, **19**.

Smith, M. P. and Sansom, I. J. (1995). The affinity of *Anatolepis* Bockelie and Fortey. In *Premiers Vertébrés et Vertébrés inférieurs* (ed. H. Lelièvre, S. Wenz, A. Blieck and R. Cloutier), pp. 61–63, *Géobios, Mémoire Spécial*, **19**.

Trinajstic, K. (1995). The role of heterochrony in the evolution of eubrachythoracid arthrodires with special reference to *Compagopiscis croucheri* and *Incisoscutum ritchiei* from the Late Devonian Gogo Formation, Western Australia. In *Premiers Vertébrés et Vertébrés inférieurs* (ed. H. Lelièvre, S. Wenz, A. Blieck, and R. Cloutier), pp. 125–8, *Géobios, Mémoire Spécial*, **19**.

Vorobyeva, E. I. and Hinchliffe, R. J. (1995). The fin-limb transition and sarcopterygian relationships. In *Premiers Vertébrés et Vertébrés inférieurs* (ed. H. Lelièvre, S. Wenz, A. Blieck, and R. Cloutier), pp. 289–92, *Géobios, Mémoire Spécial*, **19**.

Wicht, H. and Northcutt, R. G. (1995). Ontogeny of the head of the Pacific hagfish (*Eptatretus stouti*, Myxinoidea): Development of the lateral line system. *Philosophical Transactions of the Royal Society of London*, B, **349**, 119–34.

Young, G. C. (1995a). Early vertebrates and paleogeographic models. In *Premiers Vertébrés et Vertébrés inférieurs* (ed. H. Lelièvre, S. Wenz, A. Blieck, and R. Cloutier), pp. 129–34, *Géobios, Mémoire Spécial*, **19**.

Young, G. C. (1995b). Application of cladistics to terrane history—parsimony analysis of qualitative geological data. *Journal of Southeast Asian Earth Sciences*, **11**, 167–76.

Author index

(references to portraits in italics)

Abel, O. 300, 315
Adamec, T. 139
Adrain, J. M. 115
Afanassieva, O. 115
Agassiz, J. L. R. 7, 244, 256, 293,
310–*12*, 313–15, 319, 322
Ahlberg, P. 81–2, 198, 200, 204–5, 214,
227, 229, 231, 234–5, 249–50,
260–1, 263, 266, 270, 285, 292,
297, 300, 325
Alberch, P. E. 27, 269–70, 287
Aldridge, R. J. 85, 281–4
Alexander, R. M. 303
Allis, E. P. 115, 254, 257, 266, 308
Anderson, J. 308, 310
Anderson, M. E. 323
Anderson, W. D. 82
Andrews, S. M. 17, 196, 199, 215–16,
219, 223, 249, 251, 302, 308,
310–11
Anthony, J. 74, 82, 175
Aquesbi, N. 196, 199
Arambourg, C. 75
Archiac, A. d.' 312
Aristotle 37
Armstrong, W. G. 300
Arratia, G. 135, 193–4
Arsenault, M. 10–11, 104, 225, 262–3,
294, 297, 299–300
Asmuss, H. *312*–13

Baer, K. E. von 287, 312
Balfour, F. M. 254
Bardack, D. 85, 104–5, 304–5, 322
Bardenheuer, P. 325
Barwick, R. E. 206–7, 209, 210,
213–14, 251, 303
Beaumont, E. H. 231
Bell, M. W. 233, 235
Belles–Isles, M. 300, 302–3
Bemis, W. E. 60, 70, 82, 205, 207, 214,
260
Ben-Avraham, Z. 297
Bendix-Almgreen, S. E. 62, 100, 132,
136, 138, 145–6, 150, 301–2
Bennett, A. A. 75
Benton, M. J. 24, 79, 82, 235
Bertmar, G. 266

Bjerring, H. 42–3, 53, 56, 74, 82, 129,
135, 194, 198, 200, 203–4, 214,
218–19, 223, 226, 232, 247, 250,
253–4, 266, 270, 273, 316–17,
319
Blieck, A. 6, 39, 84–5, 87, 89, 93, 96–7,
100, 239, 251, 264, 292–4,
296–7, 305, 321, 327–9
Bloch, J. 44
Bockelie, T. G. 84–5
Bolk, L. 42, 52, 77, 82
Bolt, J. R. 234–5, 271–3
Bonde, N. 41, 251, 303, 317–18
Bonik, K. 256, 285
Borgen, U. 67, 214, 263
Boucot, A. J. 297
Brabet, J. 269–70
Branson, E. B. 322
Braun, J. 300, 303
Brenchley, P. J. 290
Briggs, D. E. G. 41, 84–5, 281–4
Broad, D. S. 92, 96
Brodal, A. 82
Brundin, L. 317
Bryant, W. L. 316, 322
Buffon, G. M. 313, 320
Bungart, P. A. 161
Bystrov, A. P. 202

Caldwell, M. W. 124, 127, 297, 299
Campbell, K. S. W. 191, 205–7
209–14, 233, 235, 251, 293,
318
Cappetta, H. 82, 144, 150–1
Carr, R. K. 158, 173
Carroll, R. L. 233–5, 273
Case, G. R. 41, 140, 146, 150, 300,
305
Casey, R. L. 78
Castanet, J. 74
Chaloner, W. G. 294, 305, 323
Chang, M. M. 204, 207, 210–14,
221–3, 246–7, 249–51, 318–19,
321;
see also Zhang, M. M.
Chen, L. 122
Cheng, H. 211
Clack, J. 14, 27, 226–8, 230, 232, 249,

262, 267, 269–70, 272–3, 285,
303
Clark, R. B. 253
Clark, W. 322
Clergue-Gazeau, M. 269
Cloutier, R. 10, 200, 231, 233, 235
Coates, M. 14, 133, 177, 187, 190, 192,
194, 226–7, 230, 235–6, 246,
262, 267, 269–70, 285, 303
Colbert, E. H. 320
Collins, C. J. 27, 41
Compagno, L. J. V. 82
Conway-Morris, S. 284
Cope, E. D. 44, 166, 188, 246, 314, 322
Copper, P. 290
Couly, G. F. 257
Crowther, P. R. 27, 41
Currie, P. J. 233
Cuvier, G. 252–3, 310–11, 313

Dalquest, W. W. 189
Damas, H. 115
Daniel, J. F. 82
Darby, D. 85, 294
Darlu, P. 40–1
Dasch, E. J. 293
De la Beche, H. 308
De Beer, G. 42, 82, 254
De Pinna, M. 41
Dean, B. 53, 65, 82, 150–1, 242, 315,
318, *320*, 322
Denison, R. H. 84–5, 88–9, 97, 100,
112, 115, 166, 169, 172–3, 181,
183, 210, 214, 240, 251, 258,
284, 293–4, 304–5, 314, 322
Dennis, K. D. 13, 160, 173
Dennis-Bryan, K. D. 13, 161, 173
Deprat, J. 321
Derycke, C. 34
Devillers, C. 52, 62
Dick, J. R. F. 143
Dick, R. 310
Didier, D. A. 145, 150
Dineley, D. L. 39, 92, 96, 99–100, 122,
301, 321–2
Dingerkus, G. 147
Dixon, M. T. 81
Dollo, L. 300

Dommergues, J. L. 287
Dong, Z. H. 186
Donovan, S. K. 291
Dorr, J. A. 302
Duff, P. 308, 310
Duffin, C. J. 150–1
Duméril, A. M. C. 43
Dunkle, D. H. 161, 322
Dupuis, C. 311–13

Eastman, C. R. 83
Eastman, J. T. 323
Egerton, P. 311–14
Eichwald, E. 313, 314
Einasto, R. 293
Eldredge, N. 293
Elliott, E. K. 85, 88–9, 93, 97, 100,
        239, 318, 322
Erwin, D. H. 291
Etheridge, R. 323

Fänge, R. 82
Farris, J. S. 40–1, 241, 318
Fernholm, B. 46, 82
Fischer, W. A. 306
Flemming J. 308, 310
Foreman, R. E. 82
Forey, P. L. 40–1, 76, 81–2, 84, 104,
        115, 173, 175, 183, 195,
        199–201, 214, 238–41, 244–6,
        249–51, 258, 264, 267, 273, 292,
        326, 329
Forster-Cooper, C. 207, 214
Fortey, R. 84–5
Franz, V. 48
Fritzch, B. 81–2
Fyler, C. 322

Gaffney, E. S. 229, 235, 320
Gagnier, P. Y. 1–3, 84–7, 239, 241,
        250, 277, 325, 329
Gans, C. 255, 257–8, 284
Gardiner, B. 13, 35, 79–80, 82, 104,
        128, 135, 158, 172–3, 175, 177,
        182–3, 185, 188–94, 198, 229,
        231, 235, 238–40, 243–5, 251,
        275, 317, 323, 327–9
Garman, S. 62, 82
Gaudant, J. 195, 313
Gaudin, T. J. 135, 141, 147, 329
Gauthier, J. 79, 82, 231, 235, 328
Gayet, M. 70
Gegenbaur, C. 253–4
Geldsetzer, H. H. J. 291
Geoffroy Saint–Hilaire, E. 69, 252–3,
        267
Gesner, A. 10
Gilbert-Tomlinson, J. 27, 84–7, 323
Gilmore, B. 300
Gislén, T. 81
Goethe, J. W. von 252–4, 258
Goette, A. 44, 254

Goodrich, E. S. 42, 52, 61–2, 65, 75,
        77–9, 82, 188, 239–41, 257, 272,
        308, 315
Gorbman, A. 47, 82, 264
Gordon, G. 308, 310
Gordon Cumming, M. E. 308–9, 310,
        313, 322
Gordon Cumming, Sir W. 309
Gorr, T. 81
Goujet, D. 28, 128, 135, 152, 154, 166,
        170–3, 179, 181, 183, 243–5,
        251, 286–7, 293–4, 317
Gould, S. J. 6, 256
Graffin, G. 84, 294
Graham-Smith, W. 303
Grande, L. 28, 31, 41, 60, 70, 323
Gray, J. 297
Greenwood, P. H. 82
Gregory, W. K. 234, 267, 315, 318, 322
Griffith, J. 192
Gross, W. 31, 39 ,97, 100–1, 103–4,
        115, 126–7, 160, 163, 169, 171,
        173, 179, 182–3, 186, 194, 202,
        206, 214, 217, 224–5, 240, 244,
        251, 293, 303, 313–*14*, 317, 319,
        329
Guibé, J. 75
Günther, A. C. 75, 82
Gupta, V. J. 325
Gutmann, W. F. 256, 285

Haeckel, E. 44, 253, 313, 327
Hagelin, L. O. 272
Hahn, K. V. 304–5
Hall, B. K. 41–3, 82, 254, 257, 276–9
Halstead, L. B. 87–9, 91, 95, 100, 122,
        240, 250, 264, 274–5, 278, 291,
        293–4, 300, 318; *see also*
        Halstead-Tarlo, L. B.; Tarlo,
        L. B. H.
Halstead-Tarlo, L. B. 89, 100, 274, 300
        see also Halstead, L. B.; Tarlo,
        L. B. H.
Hamel, von 313
Hanken, J. 42, 82, 254, 257
Hansen, M. C. 82, 139, 299–300
Hardisty M. W. 82
Harland, W. B. 24
Heaton, M. J. 232
Heidtke, U. 223
Heintz, N. 6, 89, 91, 96–7
Heintz, A. 39, 100, 104, 114–15, 158,
        161, 173, 239, 264, 301–2, 320
Hemmings, S. K. 9, 167–8, 173
Hennig, W. 37, 40–1, 82, 317–18
Herzer, H. 322
Heyler, D. 39, 150, 183, 191, 240, 325
Hibbert, S. 308, 310
Higgins, A. C. 84–5
Hill, A. J. 323
Hilliard, R. W. 48, 82
Hillis, D. M. 81

Hills, E. S. 323
Hinchliffe, J. R. 269–70
Hoffstetter, R. 318
Holmberg, K. 46
Holmes, E. B. 76, 81, 214, 218, 232,
        246, 249, 251, 268, 270
Holmgren, N. 46, 65, 82, 246, 254,
        269–70, 284, 315–16, 321,
        327–8
House, M. R. 291
Hull, D. 318
Humboldt, A. von 311
Humphries, C. J. 294
Hussakof, L. 322
Huxley, T. H. 70, 73, 89, 188, 244–5,
        253–4, 314–15, 322
Hyde, J. E. 22

Ilyes, R. R. 322
Ivachnenko, M. F. 233–4
Ivanov, A. 300

Jaekel, O. 315, 244
Jameson, R. 308, 310
Janvier, P. 4–6, 8, 11, 17, 22, 53, 75,
        82, 85, 91, 100, 104–6, 108–9,
        114–16, 119, 122, 125, 127, 147,
        167, 173, 182, 194, 199, 214,
        220, 223, 234, 239–41, 250–1,
        258, 264, 273–4, 287–8, 292–4,
        300–3, 317, 321–2, 325, 329
Jarvik, E. 6, 9, 11, 14, 32, 41–2, 54, 56,
        66–7, 69–70, 75–6, 81–2, 115,
        128, 132, 135, 175, 177, 183,
        197–8, 200, 202–4, 210, 213,
        214, 216–20, 223–6, 235–6, 240,
        242, 245–7, 249–51, 254–5,
        257–8, 261–73, 275, 284, 316,
        319–21, 327–8
Jefferies, R. P. S. 32, 41–3, 82, 254,
        257, 281–4
Jenkins, F. A. 233, 235
Jessen, H. 182, 186, 189, 191, 194,
        196–7, 199–200, 211–12, 214,
        223, 249, 262
Johnels, A. 48, 82, 113, 259, 317
Johnson, D. R. 269–70
Johnson, G. D. 71–2, 82
Jollie, M. 42, 82, 251, 253–4, 257–8,
        284

Karatayute-Talimaa, V. N. 115, 126–7,
        136, 148–50, 166, 240, 251, 314,
        321, 330
Katzer, F. 319
Kent, D. V. 297
Kesteven, H. L. 82
Kiaer, J. 89, 100, 103–4, 115, 239, 251,
        315, 320
Kingsbury, B. F. 254
Kner, R. 314
Koenen, A. von 171

Koch, L. 316
Kreijsa, R. J. 284
Kulczycki, J. 214
Kupffer, C. von 254
Kutorga, A. 313
Kuznetzov, A. 225, 270

Ladiges, P. 41
Lake, J. A. 42, 82
Lamarck, J. B. 252
Lambers, P. 194–5
Langille, R. M. 43, 257
Lankester, E. R. 89, 314–15
Lauder, G. V. 53, 68–9, 71–2, 82, 135,
    183, 189, 194, 285
Laufer, E. 270
Laurent, R. F. 78
Lawson, J. D. 294
Lawson, R. 78
Le Duy Phuoc 191
Lê, H. L. V. 70, 328
Lebedev, O. 18, 229–30, 235–6, 285
Lecointre, G. 72
Lehman, J. P. 15–16, 78, 191–5, 301,
    318, 323
Lelièvre, H. 16, 158, 161, 173, 297,
    323, 325
Leonardi, G. 233, 235, 307
Lessertisseur, J. 258, 285
Lewis, N. D. 32, 41
Li, Z. X. 295, 298
Liem, K. F. 68–9, 71–2, 82, 183, 189,
    194
Lindsay, W. 41
Linnaeus, C. 43, 44
Liu, Y. H. 116–17, 120, 122, 163, 173,
    321–2
Liu, T. S. 167
Loeffler, E. J. 96, 99–100, 321
Lombard, R. E. 271–3
Long, J. A. 13, 29, 127, 157, 164,
    173–4, 176, 181–3, 190–1, 194,
    199, 215–18, 220, 221–4, 244,
    246, 287, 289–91, 297, 304–5,
    318, 320, 323, 327–8
Løvtrup, S. 44, 82, 287, 327–8
Lowenstein, O. 272
Luha, A. 313
Lund, R. 19, 20, 104–5, 137, 139–41,
    144–7, 150, 200–1, 301, 303,
    305, 322
Lund, W. 20
Lurie, E. 312–13
Lyarskaya, L. 12, 41, 167–8, 173, 325
Lyell, R. 310

McAllister, J. A. 211, 300, 306–7
McCoy, F. 244, 314
McCune, A. 71, 182, 193–4, 303
McGhee, G. R. 289–91
McKerrow, W. S. 297
McLaren, D. J. 290

McNamara, J. J. 287
Maddison, W. P. 40–1
Maddison, D. R. 40–1
Mader, H. 135, 141, 147, 149–51, 179
Maisey, J. G. 28, 41, 54, 56–7, 61–2, 64,
    82, 103, 125, 135–6, 141, 143–4,
    147, 149–51, 181, 183, 251, 258,
    275, 277–9, 284–5, 318, 329
Malcolmson, J. G. 166, 308–9, 310,
    322
Mallatt, J. 56–7, 82, 258–9, 285
Mansuy, H. 321–2
Mapes, R. H. 299–300
Marcus, H. 53, 75
Marinelli, W. 45–6, 48, 52–3, 62, 82,
    271, 273
Mark-Kurik (Mark), E. 41, 97, 100,
    114–15, 173, 245, 288, 299,
    300–3, 312, 314, 318, 321, 325
Marshall, C. R. 210, 214, 326, 328
Märss, T. 4, 103–4, 124, 127, 184, 293,
    314
Martill, D. M. 27, 30, 41
Martin, J. 308, 310
Martius, C. von 311
Meinke, D. 205, 214
Melo, J. H. 325
Mercer, J. R. 300
Meunier, F. 70
Meyer, A. 81
Miles, R. S. 9, 13, 67, 128, 158, 160,
    164, 171–8, 183, 191, 206, 208,
    214, 243–5, 251, 297, 299,
    302–3, 317–18
Miller, H. 166, 308–9, 310–11, 313
Millot, J. 74, 82, 275
Milne Edwards, H. 44
Milner, A. R. 21, 23, 230, 233–6, 270,
    285, 292, 297, 299, 325
Moskalenko, T. A. 84, 88
Moy-Thomas, J. A. 9, 67, 143, 145–6,
    150, 191, 251, 318–*19*
Müller, J. 188
Murchison, R. I. 308, 310, 312–13, 318
Musick, A. 82

Nathorst, A. 319
Neal, H. V. 253
Nelson, J. 82
Nelson, G. J. 38, 40–1, 71, 82, 135,
    175, 177, 183, 244, 251–2, 287,
    294, 317–18, 326, 332
Newberry, J. S. 315, 322
Newell, N. D. 290
Nielsen, E. 138, 150, 194, 316, 318
Nitecki, M. H. 322
Nitkin, V. B. 266
Noden, D. M. 43
Nolf, D. 187
Nordenskjöld, A. E. 319
Northcutt, G. 82, 205, 207, 214, 255,
    257–8, 260

Novitskaya, L. 89, 91–2, 94, 96, 100,
    115, 166, 240, 251, 264, 273–4,
    303, 321
Nowlan, G. S. 291
Nur, A. 297
Obruchev, S. 321
Obruchev, D. V. 12, 97, 100, 224, 314,
    321–2
Oken, L. 252–4, 258
Olsen, P. O. 71, 182, 193–4
Orbigny, A. d' 312
Ørvig, T. 11, 62, 83–5, 87–9, 112, 115,
    135, 156, 163–4, 170–1, 173,
    178–9, 183, 202, 209, 214, 243,
    275–9, 317, 325
Otto, M. 184, 194, 329
Owen, R. 254

P'an, K. 116, 119–20, 167, 301, 321–2;
    see also Pan, J.
Pan, J. 116, 118, 120, 122, 172–3, 322
    see also P'an, K.
Panchen, A. L. 67, 76, 81–2, 122, 214,
    221, 230–3, 235–6, 249, 251,
    263, 266, 268, 270–3, 311,
    318
Pander, C. 224–5, 313–14, 317
Panteleyev, N. 12, 167
Parenti, L. 294
Parrington, F. R. 101–2, 104
Patten, W. 267, 283–4, 302, 313
Patterson, C. 36, 38, 41, 64, 68, 70–2,
    76, 82, 130, 144–6, 149–50,
    188–9, 193–5, 244, 251, 275,
    279, 289, 291, 294, 297, 302,
    308, 313, 315, 318
Paulsen, C. 56, 82
Pearson, D. M. 9, 11, 69, 182, 186, 189,
    191, 194
Peel, J. S. 84–5
Pehrson, T. 66
Peignoux-Deville, J. 61
Penny, D. 287
Peyer, B. 214
Philip, G. M. 283
Platnick, N. I. 40–1, 294, 318, 332
Poplin, C. 32–3, 41, 150, 185, 188, 194,
    325
Potter, I. C. 82
Presley, R. 36, 257
Pridmore, P. A. 303

Quenstedt, A. 245

Rackoff, J. S. 223, 268, 302–3
Rage, J. C. 233–4, 251, 266
Rand, H. W. 253
Raup, D. M. 288, 290–1
Raynaud, A. 269–70
Rayner, D. H. 189, 194
Redfield, C. 322
Reese, A. M. 65

Reif, W. E. 62, 82, 131, 135, 144, 150, 259, 278–80, 285, 300, 303
Reis, O. 183, 316
Repteski, J. 85
Retzius, G. 52, 270
Richardson, E. S. 85, 104–5
Ricqlès, A. de 32–3, 41
Riddiford, A. 287
Rieppel, O. 41
Ritchie, A. 17, 27, 39, 84–7, 101–2, 104, 106, 114–15, 132, 157, 167–8, 173, 239, 284, 318, 323
Rixon, A. E. 27, 41
Robertson, A. 310
Robertson, G. M. 314
Robineau, D. 74, 82, 258, 285
Rocek, Z. 233–4, 261
Rohon, J. V. 313
Romer, A. S. 23, 42, 82, 220, 223, 256, 285, 288, 293–4, 308, 316, 318
Rosen, D. E. 38, 45, 62, 65, 73, 76, 81–2, 135, 183, 193–5, 204, 214, 218, 243, 245, 246, 249–51, 265–8, 270, 273, 318, 327–8
Rowe, T. 32–3, 41
Ruben, J. A. 275
Rudwick 1, 308, 310, 319
Runnegar, B. 300

Saint-Aubain, M. L. 270
Salter, J. W. 314
Sanders, G. 309
Sansom, I. J. 84, 282
Säve–Söderbergh, G. 135, 251, 316–18, 320, 327
Sawlowicz, Z. 290
Schaeffer, B. 43, 77, 82, 128, 130, 135–6, 140, 142, 147, 150, 188–92, 194, 239, 243, 245, 251, 255, 278, 285, 318, *320*, 329
Scheele, W. 322
Schindler, E. 291
Schmalhausen, I. I. 265–6, 270
Schmidt, F. 313, 315
Schmitz, B. 294
Schoch, R. M. 40–1
Schultze, H. P. 11–12, 27–8, 41, 67–8, 76, 81–2, 135, 174–8, 183–4, 186–7, 189, 194, 196–7, 199–202, 205, 210, 213–15, 217–18, 223–5, 230, 235, 237, 246, 249–51, 260–3, 266, 268, 270, 272, 294, 304–5, 314, 317–18, 326, 328
Scotese, C. R. 297
Sedgwick, A. 308, 310, 319
Sepkoski, J. J. 288–91
Seymour, Lady 310, 313
Shergold, J. H. 84–5
Shields, O. 297
Shubin, N. H. 269–70
Siegfried, P. 313–14

Sire, J. Y. 194
Smith, A. 289, 291
Smith, C. L. 82, 197, 199,
Smith, I. C. 101
Smith, M. M. 34, 43, 84, 85, 88–9, 187, 207, 211, 214, 276–9
Smithson, T. R. 67, 76, 81–2, 214, 221, 230–1, 233, 235–6, 249, 251, 263, 266, 268, 270, 325
Soehn, K. L. 90, 96, 99–100, 240
Sollas, W. J. 41, 316
Spjeldnaes, N. 84, 285, 293
Stahl, B. J. 82, 146, 150
Stanley, S. M. 27, 293
Starck, D. 262, 285
Stensiö, E. A. 11, 32, 41, 65, 82, 85, 89, 94, 100–1, 103–4, 109, 111, 113–15, 127, 131, 135, 156, 161–2, 165, 168–9, 171, 173, 188, 197, 200, 239–41, 243–4, 251, 254–5, 258, 270–1, 274, 278–9, 284, 297, 308, 316–*19*, 320–3, 327–8
Stetson, H. C. 102, 104
Stiven, J. 310
Stock, D. W. 81, 237
Strenger, A. 45–6, 48, 52–3, 62, 82, 271, 273
Suarez-Riglos, M. 325
Sullivan, J. 322
Swofford, D. L. 40–1, 81
Szarski, H. 251

Tabin, C. J. 269–70
Talent, J. A. 290–1, 325
Tamarin, A. 47, 51, 82, 264
Tarling, D. H. 297, 320
Tarlo, L. B. 97, 100; *see also* Halstead, L. B., Halstead Tarlo, L. B.
Tarrant, P. 96, 99–100
Tassy, P. 40–1
Taverne, L. 195
Taylor, K. 139
Termier, H. 323
Terrell, J. 322
Thies, D. 150
Thomson, K. S. 82, 205–6, 214, 219–20, 223–4, 239, 251, 258, 291, 303–5
Thornhill, R. A. 272
Thorogood, P. 254, 257
Thorsteinsson, R. 136, 149
Tong–Dzuy, T. 7, 8, 116, 122, 167–8, 173, 322
Toombs, H. A. 41, 106, 113, 115, 305, 316, 323
Traquair, R. H. 104, 127, 188, 245, *309*, 311, 315
Trueb, L 82, 231, 233, 235, 328
Turner, S. 124, 126–7, 149–51, 239–40, 251, 297, 318, 320, 323

Ubaghs, G. 282
Upeniece, I. 167, 305
Upenieks, J. 167, 305
Vaillant, L. 84
Valiukevicius, J. 178, 183
Van Wijhe, J. W. 254
Van der Voo, R. 297
Van der Brugghen, W. 125, 127, 264
Vèran, M. 203, 214
Verneuil (Poulletier de), P. E. 313, 319
Vèzina, D. 11, 294
Vieth, J. 151
Vogt, C. 188, 254, 312–13
Vorobyeva, E.I. 12, 220–1, 223–5, 249–51, 262–3, 266, 269–70, 272, 285, 321

Wahlert, G. von 258, 285
Wake, M. H. 42, 82
Wakefield, N. A. 233, 235, 306–7
Walcott, C. 83, 322
Walsh, D. M. 233, 235
Wang, S. T. 116, 119–22, 240
Wang, N. Z. 116, 119–22, 186, 240, 251, 264, 271, 273–4
Wang, J. Q. 120–2, 173
Wängsjö, G. 108, 112, 114–15, 251, 316, 320
Ward, D. J. 150–1
Warren, A. A. 234, 307
Warren, J. W. 233–5, 306–7
Waterman, A. J. 42, 52, 79, 82
Watson, D. M. S. 115, 176, 183, 188, 194, 199, 242, 245, 251, 316, 323
Webb, P. H. 303
Weihs, D. 303
Wells, N. A. 302
Werdelin, L. 304–5
Westoll, T. S. 9, 11, 67, 160, 171, 173, 182, 186, 189, 191, 194, 216, 219, 223, 242, 251, 261, 263, 269–70, 302, 311, 318
White, T. E. 100, 106, 113, 115, 157, 233, 293, 304–5, 319, 323
White, E. I. 89, 91
Whiteaves, J. F. 319, 322
Whiting, H. P. 100, 115, 274
Whitt, G. S. 237
Whittington, H. B. 284
Whybrow, P. 41
Wickstead, J. M. 104
Wiley, E. O. 40–1, 68
Williams, M. E. 136, 139–40, 147, 150, 245, 299, 300, 302
Wilson, M. V. H. 90, 96, 99–100, 114, 124, 127, 142, 240, 297, 299
Wilson, A. C. 81
Wiman, C. 315
Wood, S. P. 311, 325
Woodward, A. S. 188, 244–5, 314–15, 323

Wright, A. 322

Yalden, D. W. 44, 82
Yochelson, E. 83, 322
Young, V. T. 245
Young, G. C. 13, 122, 129–30, 135,
    141–2, 147, 149–51, 158–9,
    164–6, 169–70, 172–3, 175,

215–17, 220–4, 239–40, 243–6,
    251, 273, 291, 294, 296–7, 301,
    303, 318, 320, 323, 325
Yu, X. B. 210–14
Yvroud, M. 266

Zangerl, R. 41, 104–5, 132, 135–7,
    139–40, 145–7, 150, 245, 258,

300, 302, 305–6, 318, 322,
    329
Zhang, G. R. 173, 178, 303
Zhang, M.M. 167–8, *see also*
    Chang, M. M.
Zhu, M. 122, 199, 221–3, 250
Zidek, J. 175, 183, 304, 305, 318

Index of genera

(references to figures in italics)

Acanthodes 15, *17*, *128*, 130, 173–5, 176–7, *179–80*, 245, 303–4, 316
Acanthostega 11, 13, 225, 227, 228–35, 262, 266–7, *269*, *271*, *272*, 285, 303
Acipenser 69, 71, *192*
Acrolepis 191
Acronemus 144
Aeduella 190–1
Aetheretmon 190–1
Agnathichnis 306
Ainiktozoon 284
Allenypterus 14, *20*, 199, 201
Amaltheolepis 126
Amblypterus 190, *191*, 192
Amia 66, 67–71, *129*, *193*, 262, 275, 328
Amiopsis 193
Amphibamus 299
Amphicentrum 191
Amphioxus 284
Anachronistes 144, 150, *151*
Anaethalion 193, 195
Anatifopsis 283
Anatolepis 84, 283
Anchipteraspis 97
Andinaspis 84
Andreievichthys 18
Andreolepis 4, *182–4*, *186–8*, 329
Anglaspis 89, 90–2, 95–96
Antarctaspis 157
Antarctilamna 141–2, 150, *296–7*
Anthracosaurus 230, *232*
Antineosteus 158–9
Apalolepis 126
Arandaspis 1, 83–6
Arauzia 150–1
Archaegonaspis 4
Archeria 232
Arctolepis 6, 28, 157
Aserotaspis 98
Asiaspis 120, 301
Asterolepis 9, *12*, 13, *18*, 166–8, 303, 313
Asterosteus 170
Astraspis 83–4, 87–88, 98, 120, 276–8
Ateleaspis 106–7, 113–5, 127, 133
Athenaegis 90, 92, 95–6, 99–100
Atractosteus 71

Australichthys 190, 191–2
Australosomus 190–1
Austrophyllolepis 157, 244

Bannhuanaspis 7, 8, *116*, 120
Barameda 215–6
Beelarongia 221
Belantsea 14, *20*, 137, *139*
Belichthys 191–2
Belonaspis 5, 6, 113–14
Belone 302
Benneviaspis 113–14
Besania 191
Birgeria 191–2
Birkenia 100–4
Bobasatrania 191–2
Bobbodus 137–8
Bolivosteus 170
Boothiaspis 90
Boreaspis 5–6, *301*
Boreosomus 190–1
Bothriolepis 8, *10–11*, 13, 165–8, 288, 294, *301*, 303–4
Brachyacanthus 176–7
Brachydeirus 159–60
Branchiosaurus 234
Brindabellaspis 170, 274
Brookvalia 191
Bryantolepis 156
Buchanosteus 155, *159*, 323

Callorhinchus 64, 145
Campbellodus 10, *13*, *164*
Camuropiscis 159
Canobius 191
Canonia 126
Canowindra 217, 221
Cardipeltis 90, 98
Caridosuctor 14, *20*, 199, *201*
Caseodus 137–8
Caturus 193
Cephalaspis 106, *114*, 314
Cephalodiscus 282
Ceratodus 210
Chagrinia 200
Changxingaspis 120
Cheiracanthus 7, *179–80*
Cheirodopsis 191

Cheirolepis 7, *9*, *11*, *182–9*, *190*, 194, 297, 311
Chimaera 64–5, 144
Chirodipterus 10, 12, *17*, *206–10*
Chirodus 191
Chlamydoselachus 61, *63–4*, 135, 177
Chondrosteus 191–2
Chrysolepis 18
Chuchinolepis 8, 166–8
Cladodus 53
Cladoselache 28, 131–3, 135–7, 244, 299
Cleithrolepis 191–2
Climatius 176, 177, 180
Clydagnathus 284
Cobelodus 139, *140*, 300
Coccosteus 7, 9, 155, 159–60, 310, 313
Cochliodus 144–5
Coelacanthus 199, *201*
Colobodus 191
Colosteus 230
Commentrya 191
Cornuboniscus 191
Corvaspis 95, 98
Cosmolepis 191
Cosmoptychius 191
Crassigyrinus 15, *21*, 225, 230–1, 235, 268
Ctenacanthus 12, *16–17*, 141, *143*
Ctenaspis 92, 95–6, 100
Ctenodus 210
Ctenurella 10, 163–4
Culmacanthus 178, *180*, 183
Cycloptychius 191
Cyrtaspidichthys 97

Damocles 14
Dartmuthia 4, 114–15
Davelaspis 95
Dayaoshania 168
Dayongaspis 121
Decazella 191
Delphyodontos 305
Deltodus 144–5
Deltoptychius 171
Denaea 139–40
Dendrerpeton 233–4
Desmiodus 139

Diabolepis 7, 197, 200, 205, 207,
    210–14, 247, 249, 266, 296, 321
Diademaspis 113–14
Dialipina 184, *186*, 189
Dicellopyge 191
Dicksonosteus 6, *128*, 152, *154*–5, 157
Didymaspis 115
Diplacanthus 7, *9*, *11*, *176*, *180*
Diplocercides 196–7, 199–200
Diplomystus *31*
Diploselache 141–2
Dipnorhynchus 197, 207–10, *212*, 323
Dipteronotus 191
Dipterus 7, *9*, *206*–7, 209–10, 308
Doleserpeton 233–4
Dollopterus 191
Dongfangaspis *120*–1
Doragnathus 15
Doryaspis 97
Dorypterus *190*–1
Drepanaspis 97
Dunkleosteus 12, *16*, 155, 160–*1*, 291,
    322
Duyunolepis 117–18, *119*–21

Eastmanosteus 10, *13*, 160–1
Ebenaqua *190*–1
Echinochimaera 14, *20*, 144–6
Echinorhinus 151
Ectosteorhachis 15, *23*, 216–*17*,
    219–21, 223
Ecrinosomus 191
Eglonaspis 92, *96*
Elegestolepis *136*, 144, 148–50
Ellesmereia 150–*1*
Elonichthys 191
Elops 69–70, *195*
Elpistostege *9*, *11*, 197, 224–5, 246
Empedaspis 96
Endeiolepis 8, *11*, *102*–5, 297
Eocoecilia 233, *235*
Eogyrinus 230, 232
Eoherpeton 15, *21*, 230
Eptatretus 44–6
Eriptychius 83–4, 87–8, 277–8
Erpetoichthys 69, 190
Errivaspis 89, 90–2, 95, *97*
Eryops 234
Escuminaspis 8, *11*, 112, 115
Eugaleaspis *116*–17, *119*–20
Euphanerops 8, *11*, *102*–4, 238, 241
Euporosteus 199–200
Eurycaraspis 162–*3*
Eusthenodon 11, 13, *18*, 221, *223*, 285
Eusthenopteron *9*, *11*–12, 128, 196,
    216–*1*, *223*, 225, 247, 249, *250*,
    255, 257, *262*, 267–8, 272, 297,
    299, 302, *304*, 315–6, 321, 327–9
Euthacanthus 177
Evencodus 84
Expleuracanthus 132, 141–2, 150
Fadenia 137–8

Falcatus 14, *19*, *20*, 139–*40*, 305

Gabreyaspis *91*, 96
Gadus 53
Gantarostrataspis 120
Gemuendenaspis *158*, *162*
Gemuendina 153, *169*–70
Gephyrostegus 232
Gigantaspis 5
Gilpichthys 85
Glyptolepis 196, 200, *202*–5, 247, 297,
    299
Glyptopomus 221, *223*
Gnathorhiza 210–11, *306*
Gogonasus 10, *13*, 218
Gomphonchus 4, *179*
Gonatodus 191
Goodradigbeeon 159
Gorgonichthys 160
Greererpeton 230, 272
Grenfellaspis 166–7
Griphognathus 10, 13, 197, *206*,
    208–10, *212*
Groenlandaspis 12, *17*, 156–7, 205
Grossius 196, 198, 329
Gustavaspis 5, *6*, 113–14
Gyracanthides *176*, *181*, 183, 297
Gyracanthus 15, *21*
Gyrolepidotus 191
Gyroptychius 216–*17*, 221–*3*
Gyrosteus 191

Habroichthys 191
Hadronector 201
Hamiltonichthys *125*, *143*
Hamodus 202
Hanyangaspis *116*–17, 119–21, 239
Haplolepis *190*–1
Hardistiella 14, *20*, *104*–5, 238
Harpagofututor 14, *20*, 144–5
Harriotta 6
Harrytoombsia 10, *13*
Heimenia 205
Heintzichthys 160
Helenolepis 126
Helicampodus 150
Helichthys 191
Helicoprion 137–8
Helodus 144–6
Hemicyclaspis 113–*14*
Heptranchias 61
Herasmius 160
Heterodontus 53, 135
Heteropetalus 139
Heterosteus 160
Hexanchus 61
Hirella 113–15
Hoelaspis 113–*14*
Holdenius 299
Holodipterus 10, *13*, 197
Holonema 10, *158*, *162*, 299, 302
Holophagus 199

Holoptychius 8, *11*, 200, 202, 205, 310
Homalacanthus *176*, 179–80
Homo 73
Homostius 158–9
Hopleacanthus 144
Howittacanthus 174
Howqualepis 182, 184, *190*–1
Hunanolepis 296
Hybodus 143
Hydrolagus 64–5
Hydropessum 191

Ichthyemidion 195
Ichthyomyzon 104–5
Ichthyostega 11, 13–*14*, 196, 225–7,
    228–32, 235–6, 246, 263, 266,
    269–70, 272, 285, 303
Igornella 191
Ilemoraspis 115
Incisoscutum 10, *13*
Iniopteryx 146
Irregulareaspis 95
Ischnacanthus 178, *180*

Jagorina 169–70
Jamoytius 101, *102*–5, 238, 241
Janassa 137
Jarvikina 221

Kallostracon 98
Kansasiella *33*, *185*, 187
Karaurus 233–4
Katoporodus 126
Kawalepis 126
Kendrickichthys 10
Kenichthys 222, 329
Kentuckia *190*–1
Kimaspis 170
Kimberleyichthys 10
Koharalepis 217, 221–*3*
Kossoraspis 170
Kujdanowiaspis 156, 320
Kureikaspis 92, *96*, *301*

Laccognathus *9*, *12*, 200
Lagynocystis 283
Lampetra 45, 47–8
Lamprotolepis 221
Lanarkia 123–4
Lasanius 100–4
Latimeria 71, *73*–4, 129, 187, *197*–9,
    260, 292
Latocamurus 13
Latviacanthus 178
Legendrelepis 101–3, 105
Lepidaspis 98–100, 127
Lepidosiren 75
Lepisosteus 69–71, *193*
Leptolepides *193*, *195*
Leptolepis *193*, *195*
Leptosteus 159
Ligulalepis 184, *186*

Limulus 283
Listraspis 95
Livosteus 9
Loganellia 123–6, 240, 300
Lophosteus 4, *182–4*, *186–8*, 329
Loxomma 230–1
Luganoia 192
Lunaspis 153, *162–3*
Lungmenshanaspis *120–1*

Machairacanthus 179, 181, 329
Machairaspis *6*, 113
Macroaethes 191
Macropetalichthys 162, 315, 320
Macropoma 199–201
Mahalalepis 221
Manlietta 191
Marsdenichthys 221–3
Mawsonia 199, *201*
Mayomyzon 104–5, 293
Mcmurdodus 150–1
Megactenopetalus 22, 138–9
Megalichthys 216–17, 220–1, 223, 308
Megalocephalus 230–*1*
Megapleuron 210–11
Megistolepis 221
Meidiichthys 191
Menaspis *146*, 171
Mendocinichthys 191
Mentzichthys 191
Meridensia 191
Mesacanthus *6*, *176*, 179–80
Mesonichthys 191
Mesopoma 191
Metaxygnathus 233
Microbrachius 166, *168*
Microhoplonaspis 117–18
Miguashaia 9, *11*, 134, *197*, 199–201
Mimia 10, *13*, *35*, *47*, 184–5, 187–8,
190–1, 194, 214, 218
Mitrocystella *282–3*
Mongolepis *136*, 148–9
Moythomasia *182*, 184, 186, 188,
190–1, 194
Mylostoma 173
Myxine 44–6, 90
Myxinikela 85

Nahanniaspis 95–6, 98–100
Namaichthys 191
Naxilepis 184, *186*
Neeyambaspis *122*
Nematoptychius 191
Neoceratodus 73, 75–6, 207, 209,
211–*12*, 214
Neomyxine 44
Netsepoye 138
Niualepis 148
Norselaspis *5*, *6*, *108–9*, 113–14
Nostolepis *4*, *179*
Notelops 30
Notorhizodon 215–16

Obruchevia 98
Obruchevichthys 224, *331*
Oeselaspis 4
Olbiaspis 96
Oniscolepis 98
Onychodus 13, *18*, *196*, 198, 329
Onychoselache 143
Orestiacanthus *139–40*
Ornithoprion 137–8
Orvikuina 184, *186*
Osorioichthys 184
Osteolepis 7, 9, 216–*17*, *219–21*, 223,
308
Oxyosteus 159
Oxypteriscus 191

Palaeacanthaspis 170
Palaeoniscum 188, 191–2
Palaeospinax 143, *151*
Palaeospondylus 240
Panderichthys 9, *12*, 197, 224–5,
269–300
Paraduyunaspis 118
Parahelicampodus 150
Parahelicoprion 22
Paramblypterus 191
Paramesolepis 191
Parameteoraspis *5*, *6*, *28*, 113–14
Paramyxine 44
Paraplesiobatis 171
Paratarrasius 14, *20*
Peltopleurus 191–2
Pelurgaspis 96
Pentathyraspis 117–18
Perleidus 191–2
Persacanthus 178
Petalodus 139
Petromyzon 45
Phaneropleuron 210, *212*
Phanerorhynchus 191
Phanerosteon 191
Pharyngolepis 100–4, 127, *301*
Phialaspis 99
Phlebolepis 4, 123–4, *126*
Pholiderpeton 272
Pholidogaster 230
Pholidophorus 193, *195*
Pholidopleuron 191
Physonemus 140
Pikaia 284
Pilolepis *136*, 149
Pipiscius 104–5
Pituriaspis 122–3
Platysiagum 191
Platysomus *190*, 191–2
Plectrolepis 191
Plourdosteus 8–9, *11–12*
Polybranchiaspis 7–8, *116*–7, *120*–1,
125
Polyodon 60, 69–71
Polypterus 69–70, 189–90, 267, 314
Polysentor 147

Poracanthodes 177–8
Poraspis 95
Porolepis *6*, 7, 202–*3*, 205, 207, 218,
266, 322
Porophoraspis 83–5, *86*
Powichthys 197, *211*–14, *260*
Procephalaspis 3–4, *39*, *112–14*
Procheirichthys 191
Proleptolepis 195
Prosarctaspis 96
Proterogyrinus 15, *21*
Proteurynotus 191
Protopteraspis *91*, 97
Protopterus 75, 209
Psammolepis *12*, 97–8
Psephurus 71
Pseudacanthodes 179
Pseudobeaconia 191
Pseudogonatodus 191
Pseudopetalichthys 171
Pterichthyodes 7, 9, 153, 166–8, 310
Pterichthys 310
Pteronisculus *190*–1
Pterygolepis 103–4
Ptomacanthus *178*
Ptycholepis 191
Pucapampella 147
Pycnaspis 84, 87–8, *277–8*
Pycnosteus 97–8
Pyritocephalus 191

Quebecius 200

Radotina 154, 170–1
Redfieldius *190*–1
Remigolepis *18*, 121, 166–8, 288, 303
Rhabdoderma 197, 199, *201*
Rhabdolepis 191
Rhachiaspis 97
Rhadinichthys 191
Rhamphodopsis 7, *163–4*
Rhinochimaera 64
Rhinodipterus 206
Rhinopteraspis *91*, 97, *304*
Rhinosteus *161*
Rhizodus 215–*16*, 308
Rhyncholepis 4, 100–4
Rolfosteus 159–*60*
Romerodus 137
Romundina 154, *170–2*

Sacabambaspis *1–3*, 83–7, 241
Sagenodus 210
Sanchaspis 116, 120–1
Sanqiaspis 116, 120
Sarcoprion 137–8
Saurichthys 191–2
Sauripterus 215–16
Scanilepis 191
Scaphirhynchus 71
Scaumenacia 8, *10*–11, 210, *212*
Scolenaspis 108

Semionotus 182
Seymouria 233
Shirolepis 250
Sibyrhynchus 146
Sigaspis 6, 153, 156–7
Simosteus 10
Sinogaleaspis 120–1
Sinolepis 166–7
Sodolepis 148
Spathicephalus 15, 21
Spathobatis 144, 151
Speonesydrion 197, 208, 210
Squalus 43, 45, 53, 62, 142, 253, 328
Squatina 61, 63
Squatinactis 147
Stegotrachelus 184, 188, 191
Stegolepis 296
Stensioella 171
Sterropterygion 216, 268
Stethacanthus 14, 20, 139–40, 305–6
Strepheoschema 190–1
Strepsodus 13, 17, 215, 285
Strunius 196, 198, 329
Styracopterus 191
Symmorium 139, 299
Synauchenia 151, 159–60
Synechodus 143, 151

Taemasosteus 159
Tamiobaris 141
Tannuaspis 115
Tarseodus 191
Tartuosteus 98, 299

Tauraspis 114
Tegeolepis 188, 191, 299
Tenizolepis 296
Tesseraspis 98
Tharsis 193, 195
Thelodus 126
Thrinaxodon 32–3
Thursius 221, 223, 250
Thyestes 4, 39, 113–14
Thysanolepis 221
Tiaraspis 157–8
Titanichthys 12, 15–16, 155, 161, 291,
    323
Tollodus 163
Tolypelepis 4, 95, 98
Toombsaspis 99
Torosteus 13
Torpedaspis 92, 95–6
Traquairaspis 95–6, 99–100
Traquairichthys 179–80
Tremataspis 3, 4, 39, 106–7, 109–10,
    112, 114–15, 283, *304*
Trematosaurus 234
Triadobatrachus 233–4
Triazeugacanthus 297
Tristychius 143
Tristichopterus 7
Tubonasus 10, 13, 159
Tulerpeton 13, 18, 225, 229–1, 235–6,
    269, 285
Turinia 6, 124

Udalepis 148

Ulutitaspis 97
Unarkaspis 97
Undina 199
Uraniacanthus 178
Uranolophus 209–10, 212, 214
Uronemus 210
Utahacanthus 174, 177

Vanchienolepis 8
Ventastega 227, 229, 231
Viliuichthys 221

Waengsjoeaspis 115
Wangolepis 172
Watsonichthys 191
Willomorichthys 191
Witaaspis 4
Wumengshanaspis 121
Wuttagoonaspis 123, 156–7, 296

Xenacanthus 128, 141–2
Xiushuiaspis 117–21
Xylacanthus 178

Youngolepis 7, 8, 195, 197, 200, 204,
    211–14, 218, 220, 222–3, 247,
    249, 250, 266, 296, 321, 325, 329
Yunnanogaleaspis 120–1
Yunnanolepis 8, 166–8

Zamponiopteron 147, 150
Zascinaspis 5, 6
Zenaspis 106–7, 114

Subject index

(Including suprageneric taxa, geographical names, and stratigraphical units)

(References to figures in italics. The names of the taxa are given in Latin, followed by the corresponding vernacular or Anglicized names. A cross-reference is provided when the vernacular names are too different from the Latin name, i. e. birds, *see* Aves)

abdominal cavity *108*, 111, *122*, 199
abnormalities 303
Acanthodida, acanthodids *180*–1
Acanthodidae 174, 179, *180*–1, 297
Acanthodiformes, acanthodiforms 173,
    178–81
Acanthodii, acanthodians
    anatomy 174–7
    characters 173
    classification 181
    diet 183, 297
    diversity 177–81
    growth 303–*4*
    habitat, mode of life 4–15, 181–3
    histology 177–*9*
    monophyly questioned 181, 329
    phylogeny 177–81
    relationships 245–6
    stratigraphical distribution *24*, 181–3
Acanthodinae *180*–1
Acanthomorpha *72*
Acanthopterygii *72*
Acanthothoraci, acanthothoracids
    170–2, 184, 245–6, 273, 329
accommodation of eye
    in gnathostomes 50
    in lampreys 49
    see also eye (muscle)
acellular bone
    in acanthodians 177
    in anaspids 103
    in arandaspids 87
    in astraspids 84, 87
    formation 54, 275–8
    in galeaspids 120
    in heterostracans 95
    primitive versus derived 275–8
    in teleosts 276
    see also aspidine
acetabular fossa 229
achanolepid scale, *see* scale (in
    thelodonts)
acid preparation
    chemicals 27, 31
    techniques applied to early
        vertebrates 27–9, *197*, 316–17
    see also technique (of preparation)
Acipenseridae *69*

acipenseroids 188
acrodine 34, 68–*9*, 70, *182*–4, 187, 191,
    276–8
Actinistia, actinistians, coelacanths
    anatomy 71–*4*, *197*, 199
    characters 71, 199
    fossil *197*, 199, *201*
    habitat, mode of life 11, *20*, 200
    phylogeny *201*
    Recent 71–*4*
    relationships 247–8
    stratigraphical distribution *24*, 200
    young 75, *197*, 199
Actinolepida, actinolepids 156, *158*,
    162, 172
Actinolepidoidei, actinolepidoids 156,
    162, *171*
Actinopteri, actinopterans *69*–70,
    190–1, *192*–4
Actinopterygii, actinopterygians
    anatomy 66, 68–71, 183–7
    characters 68, 183–4
    classification 66, *69*, *72*, 194
    diversity 68–*72*, 188–95
    fossil 183–95
    functional morphology 302
    growth *304*
    habitat, mode of life 8–23, 193
    histology 187
    phylogeny 71–*2*, 188–95
    Recent 68–*72*
    relationships 68, 80–1, *242*
    stratigraphical distribution *24*, 191–4
adenohypophysis *46*–8, 59, *263*–4; *see*
    also hypophysis
adipose fin, *see* fin
adorbital depression *122*
adorbital opening *91*–2, 94, 96, *301*
adsymphysial toothed plate, *see* tooth
    (parasymphysial)
aestivation burrows 75–6, 211, *306*
aglaspidids 84
Agnatha, agnathans 43–44, 237, 241;
    see also jawless fishes,
    vertebrates
air bladder 53, 59; *see also*
    swimbladder
Aistopoda, aistopods 230–1, 235–6

Algeria 296, 323
allometric growth, *see* growth
Altyre 310, 313
Amadeus Basin 84
Amiidae *193*–4
ammocoete *48*
amnion, *79*, 232; *see also* egg
    (amniotic)
Amniota, amniotes
    characters 79, 232
    fossil 232–3
    phylogeny 79–*80*, *236*
    Recent 79–81
    relationships 80–1, 232, *236*
amphiaspid province 296
Amphiaspidida, amphiaspids *91*, *92*,
    96, 286, *301*–2
Amphibia, amphibians
    anatomy *78*–9, 231–*4*
    characters 78, 231
    fossil 231–*4*
    phylogeny *78*, *236*
    Recent 78–9
    relationships 231, *see also*
        diphyletism (of tetrapods)
    see also Lissamphibia,
        Temnospondyli
ampulla(e) 44, *46*, 48–*9*, *109*–10,
    118–*19*, *270*–*1*
anal fin, *see* fin (anal)
analogy 252, 302
anamestic bone, *see* bone (anamestic)
Anaspida, anaspids
    anatomy *100*–3
    characters 100
    classification 104
    functional morphology *301*
    habitat, mode of life *4*, 104
    histology *101*, 103
    phylogeny *102*–3
    relationships 238–40
    stratigraphical distribution *24*, 104
Anchipteraspididae, anchipteraspidids
    97
Andreyevka 13, *18*
Anguilloidea *72*
angular 66–7
ankle 76, 229, 269

annular cartilage *48*, 49, 102, 104, *105*,
    330
anocleithrum *66–7*, *195*, 204, 207, 218,
    227, 229
Antarctica 296, 323
anterior orbital notch *231*
Anthracosauroidea, anthracosaurs *21*,
    232, 236
Antiarcha, antiarchs
    abnormalities in 303
    anatomy *165–6*
    biogeography 296
    characters *164–5*
    classification 172
    early discoveries 310
    functional morphology *301–2*
    habitat, mode of life *7–11*, 166–7,
        288, 293–14
    individual variation 303
    phylogeny 166, *168*
    stratigraphical distribution 168, 171
antorbital *66–7*
antorbital cartilage *61–3*
antorbital fossa *80–1*
Anura, anurans *78, 82; see also*
    Salientia
Anzaldo Formation 84
aorta, dorsal, ventral *108*, 110, 118–*19*
apalolepid scale, *see* scale (in
    thelodonts)
aphetohyoidians 245
apical ectoderm ridge *269*
apical fossa *218*
Apoda, apodans, *78; see also*
    Gymnophiona
Arandaspida, arandaspids
    anatomy 85, *86–7*
    biogeography *297–8*
    characters 85
    habitat, mode of life *1–3*
    histology 87
    relationships *238–41*
    stratigraphical distribution *24*, 83–4
Archosauria, archosaurs 79, *80–1*
Arctic 320–1
arcualia 34, *45*, 49, 55, 57, 61, 73, *91*,
    94, 111, 156, 220, 240; *see also*
    basidorsal, basiventral, inter-
    dorsal, interventral, vertebra
Arenig *24*, 83, *77*
Armorica 296
arterial cone *75–6*
artery
    brachial *108–10*
    branchial
        afferent *57, 108*, 101
        efferent *57, 108*, 110, 118–*19*
    carotid *109*–10, 153
    extrabranchial *108*, 110
    marginal *109*, 110
    occipital *109*–10
    pulmonary 71, *75*–6, *77*

subclavian 110, 118
    see also aorta
Arthrodira, arthrodires
    anatomy 152, *154*, 155–6
    characters 155
    classification 172
    functional anatomy 302
    habitat, mode of life 5–13, 173,
        297–9
    histology 156
    phylogeny 156–62
    relationships *171–2*
    stratigraphical distribution *162, 171*
arthropods
    fragments referred to vertebrates 84,
        283
    origin of vertebrates from 283
ascending process
    of palatoquadrate 195, *203–4*
    of parasphenoid *190–1*
Ashgill *24*, 84
Asia 7, 15, 115, 136, 149, 296, 321
aspidine 84–5, *87–9*, 95, 98, *101, 103*,
    116, 120, 276–7, 279; *see also*
    acellular bone
Aspinothoraci, aspinothoracids *160–1*
asteroidal impact 290
Asterolepididae, asterolepids *166–8*
Asterolepidoidei 166, *168*, 172, 303
astragalus 79
Astraspida, astraspids
    anatomy *87–8*
    biogeography *297–8*
    characters 87
    habitat 84
    histology 87–8, *276–8*
    history of discovery 83, 322
    relationships *238–41*
    stratigraphical distribution *24*, 84, 88
astraspidine 87, 276
Atherinidae 72
Atherinomorpha 72
atrial opening *283*
atriopore *283*
atrium, *see* heart
aulacophore 282
Australia 10–*13*, 296–8, 323
Austria 315
Aves, birds 79, *80–1*
axial skeleton, *see* skeleton (axial)
axillary foramen *165–167*
axillary scute 216, *217*, 223, 246, 268

B-bone 75, *206–7, 210*, 213
Bac Bo 121; *see also* Tonkin
Bac Bun Formation 8
Balistidae 72
Baltic countries, Baltics
    role in history of early vertebrate
        research 313–14
    Silurian and Devonian vertebrate
        localities, 3–4, 9, *12*

basal articulation 56, *66, 128*, 130; *see
    also* basipterygoid articulation
basal plate (element) of fins 58, 60, 70,
    73, 133, 138, 141, 146–7, 153,
    163, 177, 187, 199, *204*, 220; *see
    also* basipterygium, fin (support)
basal process, *see* basipterygoid
    articulation
Bashkirian *24*
basibranchial copula 45, *62*, 65, 131,
    147
basibranchial 56, 60, *175*, 177, *185*
    187, *203*, 208, *219*, 220
basibranchial tooth plate 72, 195
basicranial muscle *73–74, 201, 219*,
    259
basidorsal *45*, 57, 70, 134, 153, *219*,
    220, *226, 232–4*
basihyal 56, 60, 72, 149, 208
basioccipital 232
basipterygium 58, 134, 153, *164; see
    also*, basal plate, fin (support)
basipterygoid articulation, process 56,
    128, 130, *174*, 176, *185*, 187,
    203, 219, 227, 232; *see also*
    basal articulation, dermal
    basipterygoid process,
    palatoquadrate (suspension)
basisphenoid *174*, 176
basiventral *45*, 57, 70, 134, 153, 220,
    226
Batomorphii, batomorphs
    characters 61
    derived from rhenanids 169, 242, 316
    earliest known 144, *148*
    relationships 61, *63*
Batrachia 78, 235
Batrachomorpha 231; *see also*
    Amphibia
Bear Gulch 14, *20*, 137, 144, 288, 292,
    322
Beijing 318–19
Belgium 325
Beloniformes 72
Benneviaspidida, benneviaspidids
    113–14
Bergisch–Gladbach 325
Beryciformes 72
Big Creek 322
biodiversity 288–90
bio-event, *see* event
biogeography 40–1, *294–7, see also*
    palaeobiogeography
biomechanics *see* morphology
    (functional)
birds, *see* Aves
Birkeniidae *104*, 104
Birnie 308
bite (trace of) 288, *297–9*
bitypic enamel, *see* enamel
blastopore *43*
Bolivia 1–2, 15, 83–4, 147, 297, 325

bone
  acellular, *see* acellular bone, aspidine
  anamestic 207, 210, *192, 208*
  cellular, *see* cellular bone
  embryonic origin of 43, 53–4,
      274–80; *see also* neural crest
  endochondral 54, 65–6, 112, 133,
      183, 244, 246, 274–5
  endoskeletal 53
  evolution of 133, 274–8
  exoskeletal 54
  membrane 54
  perichondral 27, 33, 54, *61, 55*, 104,
      107, 109–10, 112, *116*–17, 120,
      123, 129, 133–4, 153–*4*, 155,
      185, 207, *238*–40, 244, 246, 264,
      274–5
  regression 275
  see also endoskeleton, exoskeleton
bony fishes, *see* Osteichthyes
Boreaspididae, boreaspidids 113–*14*,
      286–7, 300–*1*, 304
Bothriolepididae, bothriolepids *168*
Bothriolepidoidei 166, *168*
bottle-shaped cavity, *see* cosmine
      (flask-shaped cavities of)
brachial process *165*, 166, *167*–8
Brachydeiroidea, brachydeiroids 159
Brachythoraci, brachythoracids 158–60,
      172, *158*–62
Bradyodonti, bradyodonts
  diversity 144–*6*
  habitat, mode of life 14, *18, 20*
  relationships to chimaeroids 144–*6*
  sexual dimorphism 144, 302
  tooth structure and organization
      144–5
  stratigraphical distribution 146
  see also Holocephali
brain cavity *108*, 109–10, 118–19, 122,
      142, 154, 185, 187, *203*, 204,
      208, 273–4
brain 46–7, 48–9, *52*, 55, 58–9, 68, 94,
      109, 255, *273*–4
braincase, *see* neurocranium
branchial arch, *see* gill (arch)
branchial atrium, *see* atrium
branchial basket 49, *102*
branchial cover, *see* gill (cover)
branchial duct *46, 108*
branchial fossa(e) *108*, 111, 118,
      119–20
branchial nerve, *see* nerve
branchial opening, *see* gill (opening, slit)
branchial platelet *86*
branchial pouch, *see* gill (pouch)
branchial ray, *see* gill (ray)
branchial unit, *see* gill (unit)
branchiostegal, branchiostegal ray 67,
      174–5, *176*, 181, 191, 199, *202*,
      207
Brazil 233, 325

Britain 5, 7, 115, 267, 293, 296,
      308–18, 325
Brittany 325
'brush' (of stethacanthids) 139–40, 302,
      305–6
buccohypophysial canal, duct, foramen
      59, *74*, 141–2, 143, 153–*4, 174*,
      176, 107–8, 264
Buchan 323
Buchanosteidae, buchanosteids, 159,
      162
Buffon's law 320
Burdiehouse 310
Burgundy 308
buttress of scapulocoracoid *185*, 187,
      209

Caithness 308
calcaneum 79
Calcichordates 281–3
calcified cartilage, *see* cartilage
calcified rings in sensory-line canals,
      see sensory line (calcified rings
      in)
calcium phosphate 26, 54
Callorhinchidae 64, 65
Cambrian *24*, 84, 281, 284
camera lucida 32, 34
Camuropiscidae, camuropiscids 159
cancellar structure (of exoskeleton)
      86–7, *89*, 95
canibalism 199
Canowindridae, canowindrids 221, *223*
Capitosauroidea, capitosauroids 232,
      249
caput humeri 215–*16*
Caradoc *24*, 83–4
Carboniferous
  radiation of chondrichthyans and
      actinopterygians 13, 292
  stratigraphy *24*
  survivors of Devonian fish faunas
      292
Carcharhinidae *63*
Cardipeltida, cardipeltids 90, 98
carinal process, *see* keel
carotid, *see* artery
cartilage
  adventitious 54
  calcified (globular, prismatic,
      spherulitic) 54, 59, *61*, 88, 112,
      133, 135, 274–*5*
  as a primitive tissue in endoskeleton
      53–4, 274–5
  secondary cartilaginous state 54, 275
cathodoluminescence 34
Caturidae 193
caudal heart, *see* heart (accessory)
caudal fin, *see* fin (caudal), tail
Caudata 78, 233, 246–7, 270; *see also*
      Urodela
cell space, lacunae 87, *88, 112, 156*,

276; *see also* osteocyte,
      odontocyte
cellular bone
  in gnathostomes 54, 133, 275–8
  in Ordovician vertebrates 84, *88*
  in osteostracans *112*–113
  primitive versus derived 275–8
cement 277–8
centrum(a)
  in actinopterygians 57, *70*, 193, 195
  in lungfishes 134, 208
  in neoselachians 62, 64, 143
  in osteolepiforms 134, 220
  in tetrapods 57, 231–3
cephalaspid province 296
Cephalaspidida, cephalaspidids 113,
      114
Cephalaspidomorphi,
      cephalaspidomorphs *238*–9
cephalic fields in osteostracans 105,
      106, 108
cephalic heart, *see* heart (accessory)
Cephalochordata, cephalochordates
      42–*3*, 102, 282–*3*, 284
ceratobranchial *45*, 55–6, 62, 65,
      130–*1, 175, 185, 208*
ceratohyal *45*, 56, *175, 185, 203*–4, *219*
ceratotrichia(e) *66*, 68, *175*
cerebellar auricle *52*
cerebellum 49, *52*, 58, *91*, 94, 109,
      273–4
cestode 300
Chaleur Bay 8
character, characteristic
  adaptive, non-adaptive 37, 239
  analysis 36–40, 326–7
  apomorphous 38
  definition 34
  histological 34–5, 330
  homologous 36–7
  homoplasic 37
  kinds of 34–5
  ordered versus non-ordered 38
  plesiomorphous 38
  polarity criteria 38–9
  'privative' 37
  state 38
  taxa defined by 39–40, 252, 281, 288
  transformation series 38
  unique 39
  weight 330
Cheiracanthidae 179–81
Chelonii 79–81
chemical preparation of fossils 27–30;
      see also acid preparation
Chiarugi's vesicle *253*–4
chimaeras, *see* Chimaeriformes,
      Holocephali
Chimaeridae 64–5
Chimaeriformes 65
Chimaeroidei *65*, 144–*146*; *see also*
      Holocephali

Chlamydoselachidae 62–*3*
choana(e)
  derived from a gill slit 256
  derived from posterior nostril 264–6
  neomorphic 264–6
  questioned in lungfishes 76, 249,
    264–6
  questioned in osteolepiforms *218*,
    266
  questioned in porolepiforms 204
  in tetrapods 77, *226–7*, 228, 247,
    264–*5*, *330*–1
  in urodeles 247, 266
Choanata, choanates
  characters *248–9*
  classification 251
  phylogeny *248*–50
Chondrenchelyidae, chondrenchelyids
  144–*6*
Chondrichthyes, chondrichthyans
  characters 59, *61*
  classification 149
  derived from placoderms 241–2, 316
  earliest known *136*, 148–9
  fossil 135–50
  mode of life 14–15, *20*–2, 293–4,
    297–9, 305
  phylogeny *147*–8
  questionable Silurian remains 148,
    329
  Recent 59–65
  relationships 241–6
  stratigraphical distribution *24*,
    148–50
Chondrostei, chondrosteans
  characters 71
  including palaeonisciforms 188
  loss of posterior myodome in 189
  phylogeny *69*, *192*
chorda tympani 271
chordal lobe (in caudal fin) 53, *74*, 86,
    100–*1*, *124*, 137, 141–2, *148–9*,
    174, *182*, 189, *192*, *196*–9,
    216–*17*, 227–8, 303–*4*; *see also*
    fin (caudal)
circum-areal canal 112
clade 37–9
cladistic computer programs 40
cladistics
  oponents and pioneers 317–8
  principles 37–9
  role of ichthyology in the
    development of 40, 68, 317–8
cladodont tooth, *see* tooth
cladogram 38–40
Cladoselachidae, cladoselachids *136*,
    137, 147, *148*
clasper
  frontal (tenaculum) 64–5, 144
  morphology
  pelvic (mixipterygium)
    absence in cladoselachids and

eugeneodontids 135–7, 147
  in chondrichthyans 58–*61*, 134–5,
    139–*40*, *146*–7, 149, *305*
  in holocephalans 64–5, 144–5
  homoplasy among chondrichthyans
    64–5, *144–5*
  questioned in placoderms 153,
    157, 163–*4*, 242–4
  prepelvic
    in holocephalans 64–5, 144–*5, 146*
    questioned in ptyctodonts 163–*4*,
      242–*4*
clasper scutes *164*
classification
  of acanthodians 181
  of actinopterygians 82, 194
  of anaspids 104
  of chondrichthyans 82, 149
  of craniates 82, 251
  of dipnomorphs 214
  of galeaspids 121
  of heterostracans 100
  of lampreys (Hyperoartia) 105
  of osteolepiformes 223
  of osteostracans 115
  of placoderms 172
  principles of 38
  of sarcopterygians 82, 251
  of tetrapods 235
clavicle 67, 214, *216*, *220*, *226*, *228*,
    229
cleidoic egg, *see* egg (amniotic)
cleithrum 66–7, *185*, *204*, *209*, 214,
    216, *220*, *226*–7, 228–9, 236
Cleveland Shale 28, 297–9, 322
Cleveland 322
Climatiidae *180*–1
Climatiiformes, climatiiforms 177–81
Clupeidae *72*
Clupeocephala, clupeocephalans *72*, *195*
Clupeomorpha, clupeomorphs *72*, *195*
Coal Measures 214, 235, 292–3
Coccosteidae, coccosteids *162*
Coccosteomorphi, coccosteomorphs
    159, *160*, 162
Cochliodontidae, cochliodonts 144–5,
    146
coelacanths, *see* Actinistia
coelom, coelomic cavity 43, *53*, 59
collagen 54, 300
Colorado 83, 322
Colosteidae, colosteids 230, 233–5, *236*
Columbus 322
commissure
  ethmoid *66*, 191, *202*
  lateral *185*, 187, *220*, 255, 257
  supratemporal (occipital,
    extrascapular) *66*, *202*, *226*,
    262–3, *296*
  trabecular 257
Comoro Islands 71, 272
competition between higher groups 288

computerized tomography, *see*
    tomography
concretion 27
connecting process of ethmosphenoid
    260
Conodonta, conodonts 26, 84, *282*–4
Copacabana Formation 15, *22*
Copenhagen 317
coprolite 299–300
coracoid 58, 79, 134, *185*, 228, 232
corneal muscle *48*–9, 103
Corniferous limestone 322
cornual process *106*–7, *113*–*14*, *116*,
    120, *122*
Cornuata, cornuate osteostracans
    113–*14*
coronoid *66*–7, 184, *203*, 207, *215*, *218*,
    226, 228, 249
coronoid fang, tusk 203, 218, 330
coronoid process 69
Corvaspidida, corvaspids 90, 98
cosmine
  in actinistians 199
  flask (bottle)-shaped cavities of *112*,
    202, 205–6, *217*, 221
  function 112, 205
  in lungfishes *206*, 209
  in onychodontiforms 199
  in osteolepiforms *217*, 221
  in osteostracans 107, *112*–13
  in porolepiforms *202*, 205
  see also pore-canal system
courtship behaviour 305
Cowdenbeath, *see* Dora
cranial vertebra, *see* vertebra
Craniata
  characters 42–4
  classification 82, 251
  diphyletic 42, 329
  monophyletic 42
  origin of 281–5
  phylogeny *80*–1, 237–50
  relationships 42–3, 281–5
  versus Vertebrata 42–4
cranio-ethmoidian fissure, *see* fissure
cranio-thoracic articulation 155–6,
    160–162, 165, 302
craniospinal process 192
Cretaceous 27, 30, 70, 74, 141, 200,
    210, 290–1, 313
crista(e) (in labyrinth) 270
crista rostrocaudalis 200
Crocodylia, crocodilians 79, *80*–1
Cromarty 310
Crossopterygidae 314
crossopterygians 198
crus commune 110, 272
crystalline; *see* eye (lens)
Ctenacanthidae, ctenacanthids 135, 299
Ctenacanthiformes, ctenacanthiforms
    141, *143*
Cukurca 12, *17*

Culmacanthidae, culmacanthids *180*–1
Cyathaspidida, cyathaspidids 95–*6*, 100
Cyathaspidiformes, cyathaspidiforms
    89–*92*, 95–*6*, 98, 100, 301
cycloid scale, *see* scale (shape)
Cyclostomi, cyclostomes 44–*5*, 238
Cypriniformes *72*

dactylic limb, *see* limb
dark dentine, *see* dentine
delamination (Holmgren's principle of)
    54
deltoid process 249
dendrodont teeth, *see* tooth
dental lamina 130, 242–3, 278–*80*; *see
    also* tooth (origin of)
dentary *66*–7, 73–5, *77*, 184, *197*, 199,
    201–3, *206*–7, *211*, 214–*15*, *218*,
    226, 228, 272
denteon 132, *146*; *see also* dentine
    (tubate)
denticle
    backward-pointing *125*
    forward-pointing *125*
    whorls *125*
dentigerous cartilage *45*, *46*
dentigerous jaw-bone *178*, 181
dentine (see also metadentine,
    orthodentine)
    dark 229–*30*, 232
    folded, *see* tooth (folded), plicidentine
    pleromic 64, 132, 144, 163, 242,
    278–*9*
    as a primitive tissue 276–7
    structure 54, *276*
    trabecular 132, 138, 177, 278
    tubate 132, 139, 144, *146*, 147, 163;
    see also denteon
    tubules, canalicules 34, 87, 95, 126,
    149, 187, 229, 276–*17*
    see also mesodentine, metadentine,
    orthodentine, osteodentine,
    plicidentine, semidentine.
dermal basipterygoid process *190*–1
dermal bone, *see* bone, plate
dermal papilla 275–9
dermal skeleton, *see* exoskeleton
dermatome 55
dermintermedial process 213, *217*–8,
    249–*50*
dermohyal *185*, 191, 244
dermometapterygoid *66*
dermopalatine *66*–7, 79, 184, *203*–4,
    208, *211*, *218*, *224*, *226*, 228,
    249; *see also* palatine
dermopterotic 67, *190*–2
dermosphenotic *66*–7, 184, *189*, 191
descending process of sphenethmoid
    185, 187, *211*, 214
Deuterostomi, deuterostomes 42–*3*
developmental genetics 254, 256
Devonian

biogeography 294–7
    diversity of vertebrate faunas 5–13
    extinction events 288–91
    stratigraphy 24
Dexiothetes *43*, 281
Diadectomorpha, diadectomorphs 235
Diapsida, diapsids 80–1
diencephalon 47, *52*, *109*–10, *273*
diet of early vertebrates 297–300
digestive tract, *see* tract
digits 76–7, 225–6, 227, 229, 232, 236,
    266–70
Dinantian *24*
Dinichthyidae, dinichthyids 160–*1*, 297
Diodontidae *72*
diphycercal tail, *see* tail
diphyletism
    of craniates 42, 329
    of cyclostomes 44, 239–40, 316
    of tetrapods 246, 264–6, 270, 316
    see also taxon (polyphyletic)
Diplacanthida, dilacanthids 178, *180*–1
Diplacanthidae *180*–1
Diplorhina 240
diplorhiny, diplorhinous 240, 263–6;
    see also nostril (paired)
Dipnoi, dipnoans (lungfishes)
    aestivation 75–6, 211, *306*
    anatomy 75–6, 206–9
    characters 75, 205, *212*
    choanae questioned in 76, 213,
    265–6
    classification 75–6, 214
    fossil 205–11
    habitat, mode of life 7–25, 75–6,
    210–11, 292, *306*
    phylogeny 75, 209–10, *212*–13
    Recent 75–6
    relationships 76, 81, 247–8
    stratigraphical distribution *24*, 210–11
Dipnoiformes *212*–13
Dipnomorpha, dipnomorphs
    characters 200
    classification 214
    phylogeny *212*–13
    relationships 247–8
Dipteridae, dipterids *17*, 210, 214, 291
disc (dorsal, ventral, in heterostracans)
    89, 93–4, *238*–9
dispersal 40, 294
display behaviour 305
Dissorophidae, dissorophids 233, 235
Dissorophoidea, dissorophoids 230,
    233, *234*
Dora 14, *21*, 325
dorsal fin, *see* fin (dorsal)
durophagous diet 7, 14–15, 64, 130,
    144, 160, 173, 194, 207, 297

East Kirkton 27
Echinochimaeridae 146
Echinodermata, echinoderms 42–*3*,

281–3
ectethmoid process *128*, 141, 155
ectomesenchyme *43*, 50, 53, 54, 258;
    see also neural crest
ectopterygoid *66*–7, 184, 207, *211*, *224*,
    226, 228
Edestida, edestids, edestoids 137, 302;
    see also Eugeneodontida
Edinburgh 310–11
Edopoidea 235
egg
    of actinistians 73
    amniotic (cleidoic) *79*, 232
    capsule, case 243, *305*
    fossil *31*
    of hagfishes *46*–7
Eifelian *24*, 290
Elasmobranchii
    anatomy 60–2
    characters 60, 141
    fossil 141–4
    phylogeny 61, *63*
    Recent 60–4
    relationships 80, 147–8
    stratigraphical distribution *24*,
    149–50
    see also Euselachii, Neoselachii
Elasmobranchiomorphi,
    elasmobranchiomorphs *242*, 259
elbow 76, 229, 268
electric organ 105, 110
electroreceptor (electrosensory organ)
    73, 107, 205
Elgin 308–10
Ellesmere Land 321
Elopocephala, elopocephalans *72*
Elopomorpha, elopomorphs *72*, *195*
embryonic development
    as a criterion of polarity in character
    analysis 38
    in gnathostomes 59–60
    in hagfishes 47
    in lampreys 50–*1*
    as a support to segmentalist theories
    253–7
    of the tetrapod limb 268
    and three-fold parallelism 312
Emsian *24*, 290
enamel
    bitypic 68, 187, 276, 278; *see also*
    enameloid
    collar 68–*9*, *182*, 184, 195
    in cosmine, *see* cosmine
    monotypic (true enamel) 54, 68, 71,
    133, 187, 195, 205, 209, 221,
    276–8; *see also* ganoine
enameloid
    haphazardly fibered 61, 62–*3*, 143,
    150
    parallel fibered 61–2
    origin 54, 276–8
    shiny 61, 62–*3*

enameloid (*cont.*)
structure, distribution 54, 87–8, 107, *112*, *116*, 120, 143–4, 276–8
see also enamel (bitypic)
endemism 40, 294–7
endochoanal opening 219
endochondral bone, *see* bone
endoderm (–al) *43*–4, *47*, 54, 256, 265
endolymph 110
endolymphatic fossa 60, *62*, *142*–3, *151*
endolymphatic duct 52, 59, *106*, *109*–10, 117–*19*, *152*–3, 162–3, 166, 270–2
endoskeleton
basic structure in gnathostomes *52*–8, 128–34
calcification, ossification of 53–4, 274–5
in *Eriptychius* 88
in galeaspids 117–20
in hagfishes *45*–6
in lampreys *48*–9
in osteostracans 107–12
in pituriaspids 122–3
see also bone (endochondral, perichondral), cartilage, fin, neurocranium, skeleton, splanchnocranium
endostyle 94, 258–9
Engraulidae 72
entepicondylar process 215–*16*, 220–1, *227*, *229*, 249
entopterygoid *66*–7, 184, *197*, *203*, 207–8, *210*–*11*, *218*, 266–8, *234*
entopterygoid tooth plate 75, 205, *208*, *211*, 213
entospirae 299, 300
environment
of early vertebrates 1–23, 291–4
marginal 291–3, 297
marine versus fresh water 84, 292–4
see also habitat
epibranchial placode, *see* placode
epibranchial *32*, *62*, *65*, *130*
epicercal tail, *see* tail
epichordal lobe, *see* fin (epichordal lobe of caudal), tail
epihyal *45*, 56, *64*–5, 175, 185; *see also* hyomandibula
epiphyse 81
epipleural intermuscular bone 195
epitegum(a) 90–1, 97
epurals *70*–1
Escuminac Formation 8, *10*, 194
Esocidae 72
Estonia 3, 313
ethmoid articulation 56, *128*, 130
ethmoid cavity 110–11, 117
ethmoid commissure, *see* commissure
ethmosphenoid 73–*4*, *203*, *211*, 214, *219*
Euacanthomorpha 72

Euacanthopterygii 72
Euantiarcha, euantiarchs 166, *168*, 172, 296
Eubrachythoraci, eubrachythoracids 159, *160*–2, 172
Eugaleaspidiformes, eugaleaspidiforms *120*–1
Eugeneodontida, eugeneodontids 15, 22, 137, *138*, 147–8, 150, 302, *see also* Edestida
Euramerica 183, 294–7, 323
Europe 294–7, 325
eurybasal fin, *see* fin
eurypterids 3, 8, 288
Euselachii, euselachians *149*, *151*, 278, 329
Eustachian tube *271*
eustatism, eustatic movements 297
eusthenodont teeth, *see* tooth (eusthenodont)
Euteleostei, euteleosteans 72, 193
event 288–91
evolutionary process 40, 281, 285–7, 303, 315–16
evolutionary radiations 291–2
excretory system, apparatus 59
exoccipital 232
exoskeleton
basic structure in gnathostomes 54–5, 131–3, 275–80
basic structure in osteichthyans 66–8, 183, 261–3
in fossil jawless craniates 83–127, 239–41, 275–80
macromeric, -y 55, 131, *279*
macrosquamose 55, 131, 278–9
mesomeric, -y 131–2
mesosquamose 55, 131–2, 278–9
micromeric, -y 55, 131, *279*
microsquamose 55, 131, 278–9
neural crest and 43, 54, 276, 278
tessellate 38–9, 97, 98, 171; *see also* tessera(e)
see also acellular bone, cellular bone, dentine, enamel, enameloid, histology, ornamentation
expanding earth theory 297
extinction 288–93
extrabranchial *45*, 57
extrascapular *66*, 67, 184, 191, *202*–3, *206*–7, 214–*15*, 218, 222, *224*–5, 228, 269
extratemporal *222*
eye
lens, 44, *48*, 49, 255
muscle
extrinsic 44, 47–50, 80–1, 110, 155, *170*, 189, *238*–9, 242–3, 254
intrinsic 50, *52*, 81
retractor 231, *236*
see also myodome

spot *46*
-stalk 61–2, *153*–*4*, *159*, 243–4, 255
eyelid 77
eye-stalk, *see* eye

facial nerve, *see* nerve
Fahrenholz organs 207
Falklands 325
Famennian 10, *24*, 288–91
fate map 43, 256
fecundation (internal) 243, 305
feeding
intrauterine 305
particulate 300
succion- 285
suspension- 50, 115, 174, 183, 285, 297
filter- 285
see also diet, jaw, lingual apparatus, stomach, bite (trace of)
femur 77, 220, 226, 229, 268–*269*
fenestra ovalis 77, 81, 228, 236, 271–2
fenestration
in galeaspids 117–8
in sinolepid antiarchs 165–7
fibula 77, 220–1, 226, 229, 233, 268–9
fibular 226, 229, *269*–70
field (cephalic) 3, 105–6, 110, 117, 118
field research 26–7, 309–25
filter-feeding, *see* feeding
fin
adipose 72
anal 53, 58, *101*, *105*, 134, 141, 149, 153, 174, *182*, 199, 210, 213, 215, 238, 249
biserial paired 268–70
caudal 47, 48, 50, 58, *70*–1, *132*, 134, 139, 153, 178, 183, 200; *see also* tail
dorsal 48, 50, *53*, 58, 68, 71, 101, *105*, 107, 123, 134, 139, 153, 178, 183, 200
epichordal lobe of caudal *101*, 103, 123–*4*
eurybasal *132*–3
-fold theory 113
horizontal lobe of caudal 105–6
hypochordal lobe of caudal 141–2, 147, 177
-limb homology 52, 76, 266–79
loss of 39–9, 113, 137, 224
monobasal 71, *73*, *132*, 134, 144–5, 194, 198, 247, 329
paired (pectoral, pelvic) 50, *52*–3, 58, *73*, 81, 100–*101*, 103, 105–6, 122, *124*, *132*–4, 195, 329, 266–70, 302
radials, *see* radials
rays (dermal) *34*–5, *66*, 68, *69*, 200–*1*, 214–*15*, 204, 226, 228, 242; *see also* lepidotrichia(e)
spine, *see* spine (fin)

stenobasal *132–4*, 156
support *34–5*, 58, *69*, 71, 139
uniserial paired 248–9
see also spine (fin)
fissure
cranio-ethmoidian (ethmo-otic) *128–129*, 155, *159*
neurocranial 55, *128–9*
otico-occipital *128–9*, 141–2, *174*, *185–7*, *188*, *195*, *242*, 246
ventral *128–9*, *174*, 177, *185–6*, *188–9*, 198, 205, 213, *226–7*, 228, 243, 246, 259–60
flask-shaped cavity, *see* cosmine
folded tooth, *see* tooth, plicidentine
fontanelle
anterior (precerebral) 60, *62*, *142*, *151*
dorsal *185*, *203*
footplate (of stapes) *227*, 272
fossils
in cladograms and classifications 38
unique properties of 41
fossa Bridgei *190–1*
France 290, 318
Frasnian *24*, 8–10, *24*, 90, 98, 99, 115, 127, 172, 178, 184, 194, 200, 234, 288, 290, 291, 306
Frasnian-Famennian extinction event 288–91
freshwater versus marine environement, *see* environment
fringing fulcrum(a), *see* fulcrum(a)
frontal
morphology *224–5*, *226*, 249, 261–3
homology 68, 261–3
frontoparietal 77
fulcrum(a) *69–70*, *185*, *190–1*
functional anatomy, morphology *see* morphology (functional)
funnel pit 166, *167–8*

Gadiformes 72
Galeaspida, galeaspids
anatomy 116–20, *125*
biogeography 294, *295–6*
characters 115–16
classification 121
forward–pointing tubercles in 118, *125*
functional morphology *301–2*
habitat, mode of life 7, *8*, 121, *301–2*
histology *116*, 120
phylogeny 120–1
relationships *238–41*
stratigraphical distribution *24*, 121
Galeomorphii, galeomorphs 61, *63*, 64
ganoid fishes 188, 244–5, 293
ganoid scale, *see* scale (structure)
ganoine 68, *182*, 184, *186–7*, 278
Gasterosteiformes *72*
gastroliths *297*

Gemuendenaspididae, gemuendenaspids 158
genital apparatus 53
Georgina Basin 122
Germany 290, 308, 311, 314–15
gill
arch 45, *48–9*, 56–7, *62*, *65*, *108*, 111, 175, 177, *185*, 187, 195, *203*, *208*, *219*, 227, *255*, 257–8, *259*
cover *52*, 64, 67, 173, *176*, 178, *255–6*, *330–1*
filaments *30*
lamellae *46*, *48*, 56–7, *91*, 94, *102*, *108*, 111, 131
opening, slit 44–5, 49, 51–2, 60–2, 65, 89, *101–2*, 103, *106*, *116*, *134*, *136*, *152*, *174*, *176*, 226
pouch *46*, *48–9*, *85*, 95, 104–5, 111, *124*, 239
prespiracular 111, 247, *253*
-raker *175* 177
rays 56–7, 131, *136–7*, *174*, 177
septum, *see* interbranchial septum
spiracular 111, *253*, 258
unit *57*, 88, 95, 102, *125*, 240
Ginglymodi *69–70*, 71, *193*, 247
girdle
endoskeletal
pectoral 58, *63*, *132*, 133–4; *see also* scapulocoracoid, scapula, coracoid
pelvic *53*, 58, 64, 68, *74*, *80*, 133–4, 144, 153, *160*, *185*, 187, *204*, 205, *209*, *220–1*, *226*, 229, 230, 268–9
exoskeletal
pectoral 66, 67, 134, *176*, 178, 228, 244; *see also* cleithrum, clavicle, interclavicle
pelvic 268
Givetian 7, *24*, 290
glenoid fossa
of girdle *216*, *220*, 228
of squamosal 79
glossopharyngeus nerve, *see* nerve
glottis 73–4, *75*, 77, 326
gnathal plate, *see* plate (gnathal)
Gnathostomata, gnathostomes
basic structure 50–9
characters 50–1, 127–35
classification 82, 251
fossil 127–236
origin 284–5
phylogeny *80–2*, 241–50
Recent 50–82
stratigraphical distribution *24*
Gogo 10, *13*, *29*, 172, 290, 300, 317, 323
Gondwana 127, 150, 173, 183, 223, 291, 296–7, 323
Gotland 294
Green River Formation 31

Greenland 10, *14*, 27, 316, 320–1, 323
grey matter *52*, 58
grinding section 32, 109, 196, 315–16
growth
allometric *167*, *197*, 199, 216–17, *303–4*
and character polarity 38–39
Guangxi 121
Guizhou 121
Gymnophiona, gymnophiones *78*, 233, *235*; *see also* Apoda
Gyracanthidae, gyracanthids 178, *180–1*, 183
Gzelian *24*

habenular ganglion 49, 110
habitat (of early vertebrates) 1–23, 84, 85, 99, 104, 115, 121, 123, 127, 150, 167, 173, 182–3, 193–4, 199, 200, 205, 211, 216, 223, 235, 285, 291–4; *see also* environment
haemal arch, spine *70–1*, 137, *219*, *226*
hagfishes, *see* Hyperotreti
Halecomorphi, halecomorphs *69*, 71, 193
Halecostomi, halecostomes 67, *69*, 71, 193
haphazardly fibred enameloid, *see* enameloid
Harding Sandstone Formation 83–4
head-shield, *see* shield
heart
accessory (caudal, cephalic, portal) *46–7*
in actinistians 73–4
atrium *46–7*, *48–9*, *52*, 59, *74*, *75*, *109*, 110
auricle 59, 76
in gnathostomes *52*, 59
in hagfishes *46–7*
innervation 44, 59, 239
interauricular septum in 76–7
in lampreys *48–9*
in lungfishes *75–6*, 81
main *46–7*
in osteostracans *109–10*
in tetrapods 77, 81
ventricle *46–7*, *48–9*, *52*, 59, *74–5*, *109–10*
Heidelberg 311
Helodontidae *146*
hemibranch 49, 56, 95, 111, 259
Hemichordata, hemichordates *42–3*
Hennigian comb 318
heterochrony 287
Heterodontiformes *62–3*
Heterostraci, heterostracans
anatomy 89–95
biogeography 294–6
characters 89–90
classification 100
early discoveries of 314

Heterostraci, heterostracans (*cont.*)
  functional morphology *301*–2
  growth 38–*9*, 303–*4*
  habitat, mode of life 4–6, 9, 99,
    301–2
  histology 95
  phylogeny 95–8
  problematical taxa 98–9
  relationships 90, *238*–9, 240–1
  stratigraphical distribution *24*, 99
Heterostracomorphi *238*–9
Hexanchidae 61–*3*
Hexanchiformes, hexanchiforms 150
Himalayas 324
histology (of hard tissues) 34, 53–4, 61,
    84, 87–8, *89*, 95, *101*, 103,
    112–*13*, *116*, 120, *126*–7, 131–3,
    136, 149–50, 152, *156*, 163–4,
    177, *179*, *182*, 184, 187, *202*,
    205, *206*, 209, *217*, 221, 225,
    229–*30*, 274–8; *see also*
    palaeohistology
Holacanthopterygii 72
Holocephali, holocephalans
  anatomy 64–*5*
  characters 64
  classification 64, 149
  fossil 144–*5*
  phylogeny *65*, *146*
  Recent 64–*5*
  relationships 59, *80*, *148*, 163, 242
  stratigraphical distribution *24*, *146*
  see also Bradyodonti, Chimaeroidei
Holonematidae, holonematids *158*–9,
    162
Holoptychiidae, holoptychiids 205
holosteans 188–*9*
holostyly, holostylic 56, 64, 144, 146
homeobox genes (Hox genes) 269
homology 36–9, 252
homoplasy 37
Homostiidae, homostiids *158*, *162*
horizontal caudal fin lobe of
    osteostracans, *see* fin (horizontal
    lobe of caudal)
horizontal septum (in body musculature)
    57–8, 78, 103
horny plates, teeth, *see* tooth (horny)
Huananaspidiformes,
    huananaspidiforms *120*–1
humerus *73*, *77*, 215–*16*, 220–1, *224*–5,
    226–7, 229, 249, 267–*9*
Hunsrück Shale 29, 90, 171, 311
hyale *77*
hybodont tooth, *see* tooth
Hybodontiformes, hybodontiforms *23*,
    141–3, *148*–9, 150
hyobranchial skeleton *227*, *234*
hyoid
  arch *48*–9, 55–6, 64, 108 111, 128,
    129–30, 153, 177, 207, 245,
    257–8, 272

plate *77*
  process *271*–2
  somite 253
  see also basihyal, ceratohyal, epihyal,
    hyomandibula, hypohyal
hyoidean gill cover *176*
hyomandibula
  diversity of morphology 56, *66*, *75*,
    140, 144, *154*, *175*, *185*, 187,
    189, 192, *197*, 199, *203*, *208*,
    219–20, 244–5,
  double-headed 195
  homologous to stapes 77, 225, 228,
    249, *271*–3
  in jaw suspension *128*, 130
  sound-conducting 77, *80*–1, *271*–3
  see also epihyal, stapes
hyostyly 56
hypermineralized tissues, *see* enamel,
    enameloid, acrodine
Hyperoartia, lampreys
  anatomy *45*, 47–50, *259*
  behaviour 47
  characters 47, 104
  classification 105
  embryonic development 50, *51*
  fossil 14, *20*, 104–*5*
  larval development *48*, 50
  phylogeny *102*
  Recent 47–50
  relationships 42–4, *45*, *102*, 237–9
  stratigraphical distribution *24*
Hyperotreti, hagfishes
  anatomy 44–*7*
  behaviour 47
  characters 44
  embryonic development *47*
  fossil 85
  Recent 44–*7*
  relationships 42–4, 237–9
  stratigraphical distribution *24*
hypobranchial *45*, 56, 60, *62*, 64–*5*,
    130–1, 141, 147, *149*, *175*, 185,
    203–4
hypocercal tail, *see* tail (hypocercal)
hypochordal lobe, *see* fin, hypochordal
    lobe of caudal
hypohyal 56, 60, 64, *203*, *208*, *219*
hypophyseal cavity, fossa, pit *74*, *142*,
    188; *see also* pituitary cavity
hypophysial tube *48*, 50–*1*, *109*–10,
    239, 269
hypophysis 44, *46*–7, *48*, 59, *81*, 256,
    263–4; *see also*
    adenohypophysis,
    neurohypophysis
hypural 70–1, *195*

ichthyolith 26, *136*, 147, 170, 308
ichthyosaurs 247
idealistic morphology, *see* morphology
iliac process 229

iliac portion of pelvic girdle 220–1, 229
ilium *77*, 134
Illinois 29, 85, 104, 288, 322
incertae sedis (taxon) 38
incus *271*–2
India 310, 323
Indiana 29, 288
individual variation 303
Indochina 321–2
inferognathal 153, *154*–5, *160*–1,
    162–3, *165*, *169*, 188
infradentary 73–4, *197*–9, *202*–3,
    206–7, *215*, 226, *see also*
    splenial
infraorbital 71, 184, *191*
infraorbital sensory line, *see* sensory
    line
infrapharyngohyal *255*, 257
infrapharyngomandibular *255*
infrapharyngopremandibular *255*
infundibular cavity, *see* hypophyseal
    cavity
Iniopterygia, iniopterygians
  anatomy 145–6
  characters 145
  relationships 147–8
  stratigraphical distribution *24*
interbranchial ridge *108*, 111, 118–*19*
interbranchial septum *52*, 56–7, 125,
    131, 137, 177, 241
intercentrum *219*, 231, *232*–4; *see also*
    ventral vertebral arch
interclavicle 67, 218, *226*, 228
interdorsal *45*, 49, 57, 134, 204,
    219–20, *226*, 228, 234, *see also*
    pleurocentrum
interhyal 56, 130, *175*, 177, *185*, 187,
    187, *219*
intermedial process 218
intermediate spine, *see* spine
intermedium *226*, 229, 269–70
internasal fossa, pit *203*, 214
interopercular 66–7, *193*
interpterygoid vacuity *232*, 233–*4*
intertemporal 67, 183–4, *189*, *226*–7
interventral *45*, 57, 134
intra-areal canal *112*
intracranial joint
  diversity in sarcopterygians 73–4,
    128–9, 195, 198, 204, *211*, 213,
    216, 219, 247, *260*
  as an intersegmental joint 129,
    258–61
  homologous to ventral fissure *128*–9,
    258–61
  relation to cranial nerves *260*
intramural cavity, *see* pericardic cavity
intrinsic eye muscle, *see* eye (muscle)
Iran 297, 323
iridium anomalies, *see* platinoid
    anomalies
iris 81

ischiadic process *220–1*
ischium 134
Ischnacanthiformes, ischnacanthiforms
    178, 180–1
isocercal tail, *see* tail
isotope ratio
    as evidence for bio-event 290–1
    as evidence for marine or freshwater
        environment 293–4

jaw
    basic structure *52–3*, 56, 58, *128*,
        129–30
    dermal bones of *66–7*
    musculature *53*, 58, 129–30, *154*,
        243
    origin of 258–9
    suspension 56, *128*–30
    see also mandibular arch
jawed fishes, vertebrates, *see*
        Gnathostomata
jawless fishes, vertebrates 44–50,
        85–127, 237–41; *see also*
        Agnatha
jet-propulsion device *301–2*
jugal sensory line, *see* sensory line (jugal)
Jurassic 24
juvenile stage
    in acanthodians 304
    in actinistians 73–4, *197*, 199
    in actinopterygians *31, 304*
    in antiarchs *167, 304*
    in heterostracans 38–9
    in lungfishes 211
    in osteolepiforms 216–17, 304
    see also growth

Khazakhstan 296
Kasimovian 24
Katoporida 127, *see also* scale (in
        thelodonts)
keel (of median dorsal plate) 158,
        160–2
Kiaeraspidida, kiaeraspidids *113–14,
        286–7*
kidney tubules *53*
kidneys 27, 59, 137
knee 76, 229, 268

labial cartilage *45*, 56
labial cavity *265–6*
labial pit *206–7*
labyrinthodont teeth, *see* tooth
labyrinth, labyrinth cavity 36, 44, *46,
        48*–9, 50, 52, 59, 77, 94, 110–11,
        118–*19*, *154*, 156, *174*, *208*, *219*,
        270–2; *see also* otic capsule
labyrinth cavity
Lacertilia *81–2*
lachrymal *66*, *202*, *215*, *224*, *226*
lagenar otolith, *see* otolith
Lamniformes 62–3

lamprey, *see* Hyperoartia
Lampridiformes *72*
larva(e)
    lamprey *48*, 50, 104–*5*, 113, 258
    lissamphibian 79
    lungfish 76
    Palaeospondylus as a 240
    thelodonts as 127
    tunicate *43*, 256
lateral commissure, *see* commissure
lateralis nerve 47, 49, 59, 135, 270
lateral-line, *see* sensory line
lateral plate, *see* mesoderm
Latvia 9, *12*, 30, *314*
Lebach 173
Lepidaspidida 90
lepidomorial theory 131, 278
Lepidosauria, lepidodaurs 79, *80*–1
Lepidosirenidae, lepidosirenids *75–6*,
        210, *213*
lepidotrichia(e), lepidotrichs *66*, 68, *69,
        71, 81*, 134, 175, 185, 199, *201,
        204*, 214–*15*, *224*, *226*, 228
leptolepids 193, 329
Lethen Bar 7, 9, 310
limb
    bud 269
    dactylic 235
    development 38, 246, 269–70
    homologies with sarcopterygian
        paired fin skeleton 50–2, *73,
        216*, 220–1, *224*, 266–70
    origin of 69, 266–70, 285, 287
    serial homology between fore and
        hind 268
    structure *73*, 76–7, 220, 223, 227,
        229
lingual apparatus 42, 44, *46*, 48–9, 87,
        105
Lissamphibia, lissamphibians
    characters 78
    classification 82, 235
    earliest known 233
    fossil 232–3, *234–5*
    phylogeny *78*, 233
    Recent 78–9
    relationships *80*
Lithuania 9
living fossil 292–3
Llandeilo 24
Llandovery *24*, 101, 103, 115, 121,
        290–2
Llanvirn *24*, 83–4
lobe-finned fishes, *see* Sarcopterygii
Lochkovian *24*, 290
Lode 325
Loganiida 127, *see also* scale (in
        thelodonts)
Lophiiformes *72*
Loxommatidae, loxommatids *231*, 235,
        236
Ludlow 4, *24*, 183, 290, 294

lungfish, *see* Dipnoi
lung
    in actinistians 73, *74*, 199
    alveolae; *see* pulmonary alveolae
    in antiarchs 166
    in cladistians *52*, 70
    fatty 73–*4*, 199
    in lungfishes *75–6*
    in tetrapods 76–7
    see also, artery (pulmonary), vein
        (pulmonary)
Lybia 323

Mackenzie 321
macromeric, -y, *see* exoskeleton
        (macromeric)
macromutation theory 287
macrosquamose, *see* exoskeleton
        (macrosquamose)
macula lagena *271*
macula neglecta *271*
macula(e) 270
Madras 309
Main Devonian Field 321
main lateral line, *see* sensory line
maleus *271–2*
Mammalia, mammals 38, 68, *79*, *80*–1,
        261, 272, 328
mandibular arch 52, *56*, *108*, 111,
        128–30, *140*, 254–8; *see also*
        jaw
mandibular dermal bone (of
        acanthodians) *174*–5
mandibular fenestra *80*–1
mandibular line, *see* sensory line
        (mandibular)
mandibular spine, *see* spine
mandibularis nerve, *see* nerve
        (mandibular branch of
        trigeminus)
marginal environment, *see* environment
marginal teeth, *see* tooth
marine regression 291
mass extinction 288–91
mating behaviour 302, 305
maxilla *66–7*, 68, 71, 73, 77, 198–9,
        202, 204, 207, 214, *218*, *224*,
        226–8, 247, 249, 266
maxillaris nerve, *see* nerve (maxillar
        branch of trigeminus)
Mazon Creek 104, 322
Mecca Quarry, Shale 29, 136, 288,
mechanoreceptors 207
Meckelian cartilage, bone 52, 56, *65*,
        128, *140*, *146*, 153–*4*, *160*, *161*,
        165, *169*, *174*–5, 177, *178*, 253,
        258; *see also* mandibular arch
medial ventral process *108*, 111
median dorsal opening, duct (in
        galeaspids) 115, *116–17*,
        118–19, *120–1*, *125*, 239, *263–4*
median dorsal ridge scales, *see* scale

medulla oblongata *91, 109, 273*
Megalichthyidae, megalichthyids 221,
    223
Melbourne 323
membrane bone, *see* bone
Menaspidae, menaspids 144, *146*
mentomandibular 153, *160*
Mesacanthidae, mesacanthids 179,
    180–1
mesencephalon *109, 119, 185, 273–4*
mesodentine 54, 87, *88, 112*, 126–7,
    131–2, 173, 177, *179, 275, 276,
    277–8*
mesoderm
    lateral plate 50, 53, 58, 257
    paraxial *43*, 258
    parasomitic 50, *253–7*, 328
    sclerotomic (sclerotome) *43*, 50, 53,
        55, 57, 254, 258
    segmentation of 254
    somitic *43*, 50, 53, 55, 57, *253–8,
        269*, 321
mesomere *73–4*, 76, 133, *204–5*, 209,
    220–1, 224, 267–8
mesomeric, -y, *see* exoskeleton
    (mesomeric)
mesopterygium *53*, 58, 134, 141, *143*
mesosquamose, *see* exoskeleton
    (mesomeric)
metacarpal *73*, 268
metadentine 54, *88*, 131–3, 275–8
metamery 252–8
metapod 76
metapterygial axis, rod 58, *73, 182*,
    134, *139–40*, 267–8
metapterygium, -al *53*, 58, 59, *61*, 68,
    71, *73*, 127, *132–7, 139–40*, 153,
    175, 187, 204, 221, 243–9,
    266–70
metapterygoid *175*, 177
metatarsal 268
metencephalon *52, 91, 109*, 118–*19*,
    170, *273–4*; *see also* cerebellum
micromeric, -y, *see* exoskeleton
    (micromeric)
microphagous diet 115, 285, 297
Microsauria, microsaurs 230, 233
microsquamose exoskeleton, *see*
    exoskeleton (microsquamose)
microtectites 290
microvertebrates, *see* ichthyoliths
middle ear *77, 271–3*
Middle East 296, 323
Midland Valley 310
Migmatocephalia, migmatocephalans
    158–9, 162
Miguasha 8, 10, *11*, 294, 315, 319, 322
Minusinsk Basin 321
mitochondria 27, *30*
Mitrata, mitrates *282–3*
mixipterygium, mixipterygial clasper,
    see clasper (pelvic)

Modern Synthesis 316
molecular sequence, data 38, 42, 70, 80,
    81, 83, 237–8, 247, 326–32
Molidae 72
Mongolia 148, 321
monobasal articulation 71, *73, 132*,
    134, 141, 144, 194, 247; *see also*
    fin (monobasal)
monophyletic, -y, *see* taxon
monorhiny, monorhinal, monorhinous
    263, 315; *see also* nostril
    (median)
monotypic enamel, *see* enamel
Montana 14, 29, *139, 145*, 288, 292,
    322
Montceau-les-Mines 325
Moray Firth 310
Mormyridae 72
Morocco 11, *15–16, 158*, 290, 323
morphology
    characters used in comparative *35–6*
    functional 300–3
    idealistic 252, 255
morphotype 127–35, 153, 183, 244,
    255–5
Moscovian *24*
muco-cartilage *48*, 50, 113
mucopolysaccharides 300
mucous trap 258–9, 285
mucous cocoon *75–6*
Munich 311
muscular tissues (fossil) 27, *30, 136–7*
Myliobatidae, myliobatids 61, 63
myodome
    in actinopterygians 71, *188–9*
    in galeaspids 118
    in osteostracans 110, 243
    in placoderms *154–5*, 243
    posterior 71, *114, 154*, 188–9
    for superior oblique muscle 100, *170,
        242–3*
    see also eye (muscle, extrinsic)
Myomerozoa, myomerozoans *43*
Myopterygii, myopterygians *239*
Myxinoidea, *see* Hyperotreti

Namurian 14, *24*, 77
nasal *66–7, 183, 189, 215, 223*, 226,
    262
nasal capsule *45*, 49, 55, *62, 105, 171,
    185*
nasal cavity *52, 109*, 115, *119, 142,
    154–5, 159, 165, 169, 170*, 204,
    207–8, 211, 218, 219, 225, *250*,
    264–6
nasal gland *80–1*
nasal placode 50, *51*, 256, 264
nasal sac 77, *253, 264–5*
nasobuccal shelf *91*, 94
nasohypophysial opening *48*, 51,
    106–7, 108, 111, *238–9, 263–4*
nasolacrymal duct *77*, 264–6

nasopharyngeal duct 44, *45–6, 91*, 94,
    118, *125, 253*, 240, *263–4*
neck canal *132, 136*, 149
Nectridea, nectridians 230, 233, 235–6
Neognathi *72*
neopodium theory 268–70
Neopterygii, neopterygians
    characters *69*, 71, *192–3*
    classification 194
    phylogeny, relationships *69, 192–3*
Neoselachii, neoselachians
    'anachronic' 144, 150–*1*
    characters 61, 143
    classification 62–*3*
    earliest known 143–4, *151*
    relationships 143–4, *148*
Neotetrapoda 231, 235–6
nephric opening *253*
nerve
    abducens 47, 49
    acoustic 47, 49, 59, 110, *154, 159*,
        270
    brachial 166
    branchial 56–7, 58, 110–11, 118
    facial 47, *108–9*, 110–11, *154, 159,
        260*
    glossopharyngeus 47, *108–9*,
        110–11, *154, 159, 170*
    lateralis 47, 135
    myelinated 50, 58
    oculomotor 47, 49
    olfactory, *see* tract (olfactory)
    optic, *see* tract (optic)
    root *45*, 49, *52*, 69
    spinal *43, 45*, 49, *52*, 55, 110, 135,
        159
    spino–occipital 110, *159, 170*
    trigeminus
        ganglion 58, 254
        mandibular branch (*ramus
            mandibularis*) 47, *108–11*, 135,
            154, 159, 260
        maxillar branch (*ramus maxillaris*)
            47, 49, 64, *108–11*, 135, *154,
            159, 260* 266
        profundus branch (*ramus
            profundus*) 47, *109–11*, 135,
            204, *260*
    trochlear 47, 49, 110
    vagus 47, *109*, 110, *119, 142, 154,
        159, 170*
nervous system 43, 47, 49, 55, 58–9,
    253
Neuchatel 311
neural arch, spine 70, 137, 148, 153,
    200, 208, 219, 225–6, 228,
    232–3, 234; *see also* basidorsal
neural crest *43–4*, 50, 53–7, 113, 254,
    256–8, 276, 284; *see also*
    ectomesenchyme
neurectoderm *43*
neurocranial fissures, *see* fissure

neurocranium *45, 49, 53*, 55, *62, 65*, *73–4*, 77, 113, 128, 147, 189, 219, 244, 246, 257, 258–60
neurohypophysis 47; *see also* hypophysis
neuromast, *see* sensory line
Nevada 319
New York 318
Newburgh 308
Newcastle 311
nictitating membrane *80–1*
nomothetic palaeontology 288
non–segmentalist views *254*
North Africa 323
North America 2, 11–15, 83, 88, 99, 104, 115, 150, 172, 200, 234, 294–5, 315, 319, 322
North China 296
Northern Territory 83, 323
North-West Territories 321
Norway 104, 320
nostril
    anterior (incurrent) 50, *62, 64–5, 66*, *75–6, 78–9*, 198, *200–3, 207–8*, *211, 215, 217–18, 224–5*, 250, 264–6
    choanae as 264–6
    internal *64–5, 75–6, 161*, 205
    median 44, *46, 47–8, 106–7*, 239, 315, 363–4,
    paired 50, *52*, 59–60, 127, 263–4
    posterior (excurrent) 50, *52*, 65–6, *74*, 78, *202*, 205, *208*, 211, 213, 215, 248–50, 264–6
    questioned in heterostracans 94, 264
    questioned in pituriaspids 122–3, 264
    see also diplorhiny, monorhiny
notochord *45–6*, 49, 57, *73–4*, *108*, 110, 134, *282–3*
nuchal gap *160–1, 302*
nuchal spine, *see* spine
numerical taxonomy, 37; *see also* phenetics

occipital condyles 60, *62*, 143, *151*, *232, 233, 236*
occipital process *162*
odondoblast 54, 112, 229, 276–8
odontocyte 152, *156*
odontocyte space, lacuna *156*
odontode 55, 131–3, *136–7*, 149, 173, 277–80
odontode regulation theory 278–80
oesophagocutaneous duct 44, *46*
oesophagus *108*, 110
Ohio 322
Old Red Sandstone 5, 7, 26, 83, 115, 167, 183, 233, 235, 293–4, 308–10, 314, 321, 323
olfactory bulb *52*, 58, *273*
olfactory capsule, *see* nasal capsule
olfactory organ 110, *125*
olfactory placode, *see* placode (olfactory)

olfactory tract, *see* tract (olfactory)
ontogeny, *see* embryonic development
Onychodontiformes, onychodontiforms
    anatomy *196*, 198–9
    characters 198
    habitat, mode of life *18*, 199
    histology *196*, 198
    monophyly questioned 329
    relationships 247–*8*
    stratigraphical distribution *24*, 199
oophagy, *see* feeding (intrauterine)
opercular *66–7*, 73, *197*, 199, *202*, *206–7*, 218, 228, 244, 261
opercular bone (of lissamphibians) *77–8*
opercular cartilage *64–5, 169*, 245
opercular (epihyal) rays *64–5, 146*
opercular process *271–2*
opisthotic *231–2*
optic lobes, tectum *48–9, 52*, 274
optic tract, *see* tract
oral hood *51*
oral notch *122*
oral plates, *see* plate (oral)
oral tube *92*
oralobranchial fenestra 114, 117, 122
oralobranchial cavity, chamber *106*, *108*, 110–11, 117, *119, 122*
orbit, orbital cavity *109–10, 116–17*, *119, 122–3, 128*
orbital process 61, *63*, 245
orbitostylic sharks 61, *62–3*, 245
Orcadian Basin 7
Ordovician
    extinction event 290
    glaciation 2–3, 290
    palaeobiogeography 297–8
    stratigraphy *24*
    vertebrates 1–2, 83–9, 182
Orectolobiformes *62–3*
organic matter (in fossils) 300
origin
    of craniates 281–4
    of gnathostomes 284–5
    of jaws 258–9
    of limbs 266–70
    of tetrapods 285
ornamentation of dermal bones and scales
    in acanthodian *176–9*
    in actinistians 72, *74, 197*, 199
    in actinopterygians *182*, 183–4, *186–7*
    in anaspids *101*, 103
    in arandaspids *86*
    in astraspids 87–8
    in chondrichthyan 141–3
    in dipnomorphs *202*, 205, 209
    in galeaspids *116–7*
    in heterostracans 89–90, 95–6, 98
    oak leaf–shaped tubercles in 85–6, *98–9*, 239, 241

in onychodontiforms *196*, 198–9
in osteolepiforms 216–*17*
in osteostracans 107, *112*
in panderichthyids 224–5
in placoderms 152, 155–6, 158, 163, 166, 171
in rhizodontiforms 214–15
as a source of characters 34, 54
in tetrapods 228, 232
see also exoskeleton, histology, scale
ornithuric acid *80–1*
orthodentine 54, 131–3, 275–8
orthotrabeculine 139, 144; *see also* tubate dentine
os basale *78*
Oslo 315
osmoregulation *45*
Ostariophysi, ostariophysans *72*
Osteichthyes, osteichthyans
    anatomy 65–8, 127–35
    characters 65, 183
    classification 82, 251
    fossil 183–236
    phylogeny *80–1*, *242*, 248, 329–31
    Recent 68–82
    relationships 59, 241–6
    stratigraphical distribution *24*
osteocyte 54, *112*, 284
osteodentine 132, *202*, 205, 278
Osteoglossidae *72*
Osteoglossomorpha, osteoglossomorphs *72*, 193
Osteolepididae, osteolepids 221, *222–3*
Osteolepiformes, osteolepiforms
    anatomy 216–21
    characters 216
    choanae in *218*, 264–6
    classification 223
    external nostrils in *250*, 264–6
    functional morphology 302–3
    growth 216–*17, 304*
    habitat, mode of life 7–13, 15, *16*, *18, 23, 302–3*
    histology *217–21*
    monophyletic versus paraphyletic 246, 329–31
    paired fin skeleton *220*, 266–8
    phylogeny *222–3*
    relationships 248–50
    tear duct in *225–6*
Osteostraci, osteostracans
    anatomy 106–112, 263–4, *273–4*
    characters 104–6
    classification *115*
    early discoveries 313–14
    functional morphology 300–*1*
    habitat, mode of life 4–6, 8, *11*, 115
    histology *112–13*, 275–8
    phylogeny 113–15
    and the premandibular arch theory 111
    relationships *238–40*, 241

Ostracodermi, ostracoderms 241, 311,
    315, 321, 329
otic articulation *128*, 130, *174*–5, 177
otic capsule 36, *45*, *46*, 49, 68, *109*,
    129, *232*, 241, *259*, 271–2
otic condyles *174*
otic notch 226, *227*–8, *231*, *232*–4,
    271–3
otico-occipital fissure, *see* fissure
otico-occipital region, 129, 141–2, 207,
    228, 258, 328
otic shelf 255–7
otoccipital 73–*4*, *197*, *203*–4, 214, *219*,
    221–2, *260*
otolith 65, *174*, 176, 187, *190*–1, *242*,
    246, 330–*1*; *see also* statolith
out-group comparison 38–*9*
oviduct 73, 199
ovoviviparity 73, 79, 199
Oxford 315–16, 318

Pachycormidae *195*
Pachyosteomorphi, pachyosteomorphs
    159–60, *161*–2
Pacifica theory 297
paedomorphosis 59, 275, 287
Palaeacanthaspida, palaeacanthaspids,
    see Acanthothoraci
palaeobiogeography
    of early vertebrates 294–7
    methods in 40, 294
    see also biogeography
palaeohistology
    characters obtained from 34
    techniques in 31, 34
    pioneers in 313, 317
    see also histology
Palaeonisciformes, palaeonisciforms,
    palaeoniscids 188–9, 208
palaeontology
    as a character polarity criterion
        38
    evolutionary process inferred from
        287
Palaeozoic
    environments 1–25, 291–4
    events 288–11
    palaeogeography 294–7
    stratigraphy *24*
palatal platelets 203–*4*
palatine 78, 207, 244; *see also*
    dermopalatine
palatobasal articulation, *see* basal
    articulation, basipterygoid
    articulation
palatoquadrate
    composite origin *235*, 257
    diversity of morphology *45*, *53*–3,
        56, 61, *63*, 71, *154*, 157, 159,
        163–*4*, *165*, *189*, *175*, 177, *178*,
        185, *197*, *203*–4, *219*–20, 243,
        245, 247

fusion to the braincase 56, 64–5,
    137–8, 144–6, 207; *see also*
    holostyly
multiple ossifications in 159, 163–4,
    175, 177, *197*
suspension *35*, 56, *128*–30, 137,
    140–1, 144, *146*, 153–4, *185*,
    187, *211*, 220, 226, 232, 272
palmar nerves *282*
pancreas *45*, 49–50, 73
Panderichthyida, panderichthyids
    anatomy 221, *224*–5, 228, *262*–3,
        272
    characters 224
    choanae in *224*, 266
    habitat, mode of life 9, *11*–*12*, 225,
        300
    histology *224*, 225, 229
    paired fin skeleton in *224*, 269–70
    relationships 68, 197, 221, 246–51
    stratigraphical distribution *24*, 225,
        291
Pangea 297
papilla
    amphibiorum 78, *271*
    basillaris *271*
    dermal 55, 131, 275–8
    neglecta *271*
papillary cells, tissue, *see* papilla
    (dermal)
Paracanthopterygii 72
parachordals *259*
parallel–fibered enameloid, *see*
    enameloid
paraphyletic, *see* taxon
paraphyly versus monophyly
    of acanthodians 181, 329–*31*
    of acanthothoracids 170–*1*
    of bradyodonts *148*
    of cyclostomes 44–5
    of holosteans 69, 188–9, *193*
    of leptolepids 193–5
    of onychodontiforms 320–*1*
    of osteolepiforms 246, 329–*31*
    of palaeonisciforms 188, *190*
    of sharks 61–*3*
    of thelodonts *127*
parapineal opening *86*
parapineal canal *203*–4, *207*–8
Paraselachiomorpha 137, 147
Parasemionotidae, parasemionotids *193*
parasomitic mesoderm, *see* mesoderm
parasphenoid *66*–7, 71, *74*–6, 78,
    153–4, 155, *157*, *159*, *161*,
    184–5, 192, 200, *203*, 207–8,
    210–11, 214, *215*–16, *218*,
    223–4, *226*–8, *232*–4, 244, 247,
    257
parasymphysial (adsymphysial) teeth,
    toothed plates, *see* tooth
paratemporal articulation *128*, 130,
    219–20

parietal
    homology 68, 261–3
    morphology *66*–67 *189*, *196*–8, 200,
        202–3, 207, *210*, 213, *215*, 218,
        222, *224*–6, 231–*4*, 244, 249,
parieto-ethmoidal shield, *see* shield
Paris 310, 312, 317–8
parsimony 38, 40, 44, 229, 237, 246,
    318, 330
particulate feeding, *see* feeding
    (particulate)
pectoral fin, *see* fin (paired)
pectoral fenestra, notch 152, 155, *158*,
    160–1
pedicel (in teeth of lissamphibians) 78,
    274
pedicel (in lamprey embryo) *259*
peg-and-socket articulation, *see* scale
    (peg-and-socket articulation
    of)
pelvic clasper, *see* clasper (pelvic)
pelvic fin, *see* fin (paired)
pelvic plate, *see* girdle, pelvic,
    endoskeletal
pelvic process, dorsal 144, 146
pentadactyly 226, 266–70
peramorphosis 287
Perciformes 72
Percomorpha 72
perforated lamina, *see* pore-canal
    system
pericardic cavity *108*–9, 110
pericardic cartilage *45*, 48
perichondral bone, *see* bone
periotic system *171*–2
Permian
    extinction 291
    stratigraphy *24*
    vertebrate faunas 15, *22*, 23
Petalichthyida, petalichthyids, *161*–3,
    171–2
Petalodontida, petalodontids 14, 15, *20*,
    22, 137–9, 147–8
phalange 76, *77*, 266–70, *286*; *see also*
    radials
phaner *80*–1
pharyngeal slit *43*
pharyngobranchial duct *48*–9, *263*
pharyngohyal 64–5, 130, *255*, 257
phenetics 37, *288*; *see also* numerical
    taxonomy, systematics (phenetic)
Phlyctaenii, phlyctaeniids *152*–3, 155,
    157, *158*, *162*, 172
Phlyctaenioidei, phlyctaenioids 161,
    162
photography (techniques of) 34
phyletic sequencing 38
Phyllolepida, phyllolepids 156–7, 162,
    171–3, 244, 288, 291, *296*
phylogenetic tree 40, 315
phylogeny
    of acanthodians 177–80, *181*

of actinistians (fossil and Recent)
199–200, *201*
of actinopterygians (fossil and
Recent) 188–93, *190*, *192–3*
of actinopterygians (Recent) *69–71*,
72
of amphibians (fossil and Recent)
236
of amphibians (Recent) *78*
of anaspids *102–3*
of chondrichthyans (fossil and
Recent) 147, *8*
of chondrichthyans (Recent) *61–3*
of chordates (fossil and Recent) *43*,
283
of craniates (fossil and Recent) *283*,
329–30
of craniates (Recent) *80–1*
of deuterostomes (Recent) *43*
of dipnomorphs (fossil and Recent)
212–14
of galeaspids *120–1*
of gnathostomes (fossil and Recent)
241, *242–50*
of gnathostomes (Recent) *80–1*
of heterostracans *96–7*, *98–100*
of holocephalans (fossil and Recent)
146
of holocephalans (Recent) 144–*146*
of lampreys (fossil and Recent) *102*
of lungfishes (fossil and Recent)
209–10, *212–14*
of lungfishes (Recent) *75*
methods of reconstruction of *36–40*
of osteolepiforms *221–3*
of osteostracans 113–*3*
of placoderms *271–2*
of sarcopterygians (fossil and Recent)
248–50
of sarcopterygians (Recent) *80–1*
of tetrapods (fossil and Recent)
229–33, *236*
of tetrapods (Recent) *80–1*
pila occipitalis 129
pineal canal 109, *203–4*, *207–8*, *219*
pineal foramen, opening 86, 90, 100–*1*,
106–7, *116–7*, 123, 153, *165*,
218, *221–3*, 261–*2*
pineal line, *see* sensory line
pineal organ 47, 49, 90, 94
pineal recess *91*, 94, 97, *154*, *165*
piston cartilage *48–9*, *104–5*, *293*
pit line
anterior (parietal) *66*, *202*, *210*
middle (postparietal) *66*, *152*, *202*,
210
posterior (postparietal) 162–*4*, 166,
172, *202*
and sensory-line system 34, 51, 68,
155
pituitary cavity *159*; *see also*
hypophyseal cavity

Pituriaspida, pituriaspids
anatomy, characters *122–3*
biogeography *295–6*
relationships *238*, 240, 331
stratigraphical distribution *24*, 123
placode
epibranchial 256
olfactory 50–*1*, 256, 264
sensory-line 270
Placodermi
anatomy 151–5
biogeography 172–3, 194–7
characters 151–3
classification 172
extinction 289–91
functional morphology 302
habitat, mode of life 5–*18*, 194–7
history of discovery 310, 313
monophyly questioned 329
phylogeny *171–2*
relationships *241–5*
stratigraphical distribution *24*, 194–7
placoid scale, *see* scale (structure)
planum viscerale 45
plate (dermal)
anterior dorsolateral *152*, 160, *163–5*,
169
anterior median dorsal *165*, *167*
anterior median ventral *152*
anterior ventrolateral *152*, *165*, 170
anterolateral *152*, *164*, *169*–70
branchial 88–*9*, 97
branchio-cornual 97
central (in pectoral fins of antiarchs)
168
central (in skull-roof) *152*, *159*, 162
circumorbital 67, *174*, 190
cornual 89, *93*, 97
epibranchial 86
extrascapular *160*, 302
gnathal 152–*169*, *154*, *157*, *160–1*,
164–5, *169*, *243–4*, *279*; *see also*
inferognathal, superognathal
gular *66–7*, *101*, 103, 207, 224, 228,
261
interolateral *152*, 172
lorical *176*
marginal *152*, *157*, *163–4*, *170*
marginal (in pectoral fin of antiarchs)
165
median dorsal *152–3*, *157*, *158*, *160*,
162–5, *167–9*, 302, 329
mixilateral *165*
nuchal *152–3*, 156, *157–8*, *160–1*,
163–5, *168*
oral *86*, *89*, *91*, *93*, 94, 95, 100, *101*,
106, 107, *116*, 117
orbital *89*, 90, *97–8*
paranuchal *152–3*, *155*, *157*, 160,
162–5, *170–2*, *244*
parasphenotic *211*
parotic *219–20*

pineal 89, 93, 97, 107, 131, *152–3*,
164–5, *189*, *190–1*
pinnal *176–7*
postanal 89
posterior dorsolateral *152*, 156, *163*,
165, *167*
posterior median dorsal *165*, *167*
posterior median ventral *152*
posterior paranuchal *152*, *165*, *167*
posterior ventrolateral *152*, *165*, *167*
posterolateral *152*, *165*, *167*
postmarginal *152*, *165*, *168*
postnasal *152–3*, *157*
postorbital *152*, *157*, *162–3*, *164*, *170*
postsuborbital *152*, *154*, *160*
prelateral *165*
premedian *165–6*, *167–8*, *170–1*
preorbital *152*, *157*, 159, *163–4*, *171*
rostral *89*, *93*, *152*, *160*, *163*, *165–6*
sclerotic 71, *86*, 100, *106–7*, *109*,
111, 131, *152–3*, *183*, *196*, 239,
240–1, *247–9*
semilunar *165*
spinal (in heterostracans) *89*, *93*, *97*
spinal (in placoderms) *152*, 155, *157*,
160, *163*, 170
submarginal *152–3*, *155*, *157*,
159–60, *162*, *164–5*, *169*–70
suborbital *89*, 90, 130, *152–3*, *154–5*,
156, *157*, 159, *160*, *165*, *170*,
243
see also disc
plate tectonics 320
platinoid anomalies, concentrations,
290
Platt's vesicle *253–4*
pleromin, *see* dentine (pleromic)
plesiomorphous, -y 38
pleural ribs, *see* rib
pleurocentrum *219–20*, *226*, 231, 234;
see also interdorsal
Pleuronectiformes *72*
Pleurotremata *62–3*
plicidentine *196*, *198*, *202*, 217, *224*,
230, 247; *see also* tooth
(folded)
Podolia *294*, 316; *see also* Ukraine
poisonous gland 147
polar cartilage *129*, *198*, *258*, *259*
polarity (criteria of) *38–9*
Polybranchiaspidida,
polybranchiaspidids *120–1*
Polymyxiiformes *72*
Polyodontidae *69*
polyphyletic, *see* taxon (polyphyletic)
polyphyly
of chondrichthyans 241, 316
of tetrapods *246–7*, 316
polyplocodont teeth, *see* tooth
(polyplocodont)
polypter, *see* Cladistia
population genetics 40, 287

pore canal system
  in acanthodians 177
  canals in *112*
  flask (bottle)-shaped cavities in 107,
    112, 205
  function 107, 205
  horizontal lamina in *112*, *217*, 221
  perforated lamina in 31, 107, *112*
  pores 107, *112*, 205
  in osteostracans 31, 107, *112*
  in sarcopterygians 202, 205, *206*,
    209, *217*, 221
  see also cosmine
Porolepidae, porolepids 214
Porolepiformes, porolepiforms
  anatomy 200–5, *202*, *203*, *204*
  characters 200
  choanae in 204, 264–6
  histology 202, *205*
  relationships *212*–13, 247–9
  stratigraphical distribution 24, 205
portal heart, *see* heart (accessory)
post-temporal 66, 67, 204, 207, 225
post-temporal fossa 231
postaxial radial, *see* radial
postbranchial spine, *see* spine
postbranchial wall 108, 110, 117
postcleithrum 67, 199, *190*–1; *see also*
  anocleithrum
posterior myodome, *see* myodome
posterior superognathal, *see*
  superognathal
posthyoid segment 255
postmarginal sensory line, *see* sensory
  line
postminimus 225
postorbital 66–7, 78, 202, 214–5,
  217–8, 221, 222, *223*, 287
postorbital articulation 56, *128*, 130,
  177–8, 243; *see also*
  palatoquadrate
postorbital process 62, *136*–7, *142*,
  154–5, *169*
postorbital sensory line, *see* sensory line
postparietal
  homology 68, 260–3
  morphology 66–7, *189*, *196*–8,
    200–3, 207, *210*, 213–15, 222,
    224, *227*–8, 231–4
postparietal shield, *see* shield
Pragian 5–6, *24*, 99, 149, 199
preanal ridge 89
prearticular 66–7, 73, 75, 184, 205,
  208, 210–11, *226*
prearticular tooth plate 75, 205, *208*,
  211–13
preaxial radials, *see* radial
prebranchial fossa *108*
prebranchial ridge *108*
precerebral (anterior) fontanelle, *see*
  fontanelle
prehallux 225

premandibular arch 56, 64, 110–11,
  254–7, 261, 328
premandibular ridge 111
premandibular segment *253*
premandibular somite 254–5
premaxilla 66–7, 71–2, 77, *192*, *195*,
  197–9, 204, 207–8, *211*, 213–14,
  215, *218*, 221, 224, *226*–7, 249
premaxillo–antorbital *190*–1
prenasal groove 89, 94
prenasal sinus 44–6, *85*, *91*, 93–4, 119,
  125, *263*–4
preopercular 66–7, *182*, 184
preopercular sensory line, *see* sensory
  line
preorbital depression 166–7, *168*,
  170–1
prepectoral corner *167*
prepectoral spine, *see* spine
  (prepectoral)
prepelvic clasper; *see* clasper
  (prepelvic)
prepollex *267*–8
pre-premandibular arch, segment, *see*
  terminal arch, segment
prespiracle, prespiracular gill opening
  94, *253*
prespiracular 67, *202*
pretrematic branch (of branchial nerves)
  58
Pridoli *24*
prismatic calcified cartilage, *see*
  cartilage (calcified)
Pristidae 61–*3*
Pristiophoriformes 61–*3*
'privative' characters, *see* character
processus ascendens, *see* ascending
  process
processus dermintermedius, *see*
  dermintermedial process
Procondylolepiformes,
  procondylolepiforms 166,
  167–8, 296
procoracoid *175*, 177
profundus nerve, *see* nerve
  (trigeminus)
pronephros 59, *78*
propterygium *53*, 58, 134, 141, *143*,
  148–9, *186*, 187, 62–*3*
Protaspididae 97, 100
Protaspidoidea 97, 100
Protopteraspididae, protopteraspids 97,
  100, 296
protractor muscle (in lingual apparatus)
  46
Psammosteidae, psammosteids
  functional anatomy *301*–2
  habitat, mode of life 9, *12*
  juvenile 38, *39*, *304*
  relationships 38, 97, 100
  traces of bite on 288, *299*–300
pseudobranch 56, 258

Pseudopetalichthyida 148, 171
Pteraspidida *97*, 100
Pteraspididae, pteraspids ,3, *39*, *97*, 100,
  294, 296, 302, *304*
Pteraspidiformes, pteraspidiforms 89,
  91–3, 95–100
Pteraspidina *97*, 100
Pteraspidomorphi, pteraspidomorphs
  arandaspids and astraspids as *238*–9,
    240–1
  classification 251
  hagfishes as 94, 240, 316
  Palaeospondylus as a 240
  phylogeny *238*–9
pterygoid flange 79, *80*–1, 232–3
pterygoid, *see* entopterygoid
pterygoquadrate 56, *197*, 255
Ptyctodontida, ptyctodonts
  anatomy 163–*4*
  characters 163
  claspers in 163–*4*, 243–4
  histology 163–*4*, 242
  as holocephalan ancestors 242, 316
  relationships *171*–2
  stratigraphical distribution 163,
    291
pubic process 220–1
pubo-ischiadic portion 229
pulmonary alveolae, *75*–77
pulp cavity 54, *62*, *74*, 112, 123, *126*,
  131–2, *136*–7, 149, 205, *217*,
  221, 276
Pycnodontidae, pycnodonts 193–*4*
pyriform body 282

quadrate process *271*–2
quadrate 67, 71, 79, 163, *175*, 177, 187,
  204, 207–*8*, 228, *271*–2
Quebec 8, *10*, 102, 104, 115, 225, 315,
  327
Queensland 122

radial muscle *48*, 50, 58, 71, 95, 100,
  240, 269
radials
  metapterygial *53*, 58, *132*–4
  paired fin *53*, 58, 71 100, *132*–4,
    136–8, *140*–1, 153, *175*, 177,
    185, 187, *215*–*16*, 220–1, *224*–5,
    266–70
  phalanges derived from 76, 266–70;
    see also phalanges
  postaxial 73, *76*, 141, *204*–5,
    266–70
  preaxial 73, *74*, *76*, *204*–5, 220–1,
    224–5, 266–70
  unpaired fin *46*, *48*, *53*, 57, *69*, 71,
    95, 103, 134, *136*–7, *138*, *204*,
    225–6, 240,
radiating vascular canals *112*
radius 73, *76*, 220–1, *224*, *226*–7, 229,
  237, *267*–9

Rajidae, rajids 61, 63–4, 77
rasping tongue, *see* lingual apparatus
Rathke's pouch 47, 59–*60*, *253*, 256, 264
ray-finned fishes, *see* Actinopterygii
recovery (after mass extinction) 291
Red Hills 322
Reptilia *82*, *see also* Sauropsida
Reptiliomorpha, reptiliomorphs
   characters 231
   classification 235
   phylogeny *236*
retina *46*, 49
retractor bulbi muscle 231, 260
retractor muscle (in lingual apparatus) *46*
retractor pit *232*
retroarticular *195*
retroarticular process *74*, *197*, 199
reversion 37
Rhenanida, rhenanids
   anatomy *169–70*
   as ancestors of batomorphs 169, 242, 316
   characters 169
   habitat, mode of life 170
   relationships *171–2*
Rhinobatidae, rhinobatids 61, 62–*3*, 144
rhinocapsular bone, ossification 129, 153–4, 155, *159*, 165, 243
Rhinochimaeridae 64–*5*
Rhipidistia, rhipidistians
   characters 198, 221–3, 247
   choanae in 198, 249, 264–6
   classification 251
   folded teeth in 198, 213, 247
   lungfishes as *80–2*
   phylogeny 247–8, 329–31
   relationships 247, *248–50*
Rhizodontiformes, rhizodontiforms
   anatomy 214, *215–16*
   characters 214–*15*
   habitat, mode of life 13, *17*, 216, 285
   relationships 248–9, 329
   stratigraphical distribution *24*, 216
Rhizodopsidae, rhizodopsids 223, 292
Rhyncholepidida *102*, 104
ribs
   calcified in hybodontiforms 134, *142–3*
   cranial *75*
   dorsal 57
   epipleural *195*
   in *Ichthyostega* *226–7*, 228
   in lungfishes 134, 209, *212–13*
   in osteolepiforms 134, *219–20*
   myoseptal 57
   in panderichthyids 225
   pleural (ventral) *45*, *53*, 57, *72*
   sacral 77, *80*–1, 229

rostral
   lateral *217*, 221, *224*, *225*, *226*, 228, 230, 249
   median 66–7, 184, *189*, *190*–1, *224*–5, *226*, 228
rostral articulation, *see* ethmoid articulation
rostral cartilage *72*
rostral organ *72*, *74*, *197*, 199, *201*
rostral process 113, 115–16, *120*, *122*–3, *301*–2
Rotliegendes *24*
Russia 5, 13, 15, *18*, 99, 115, 136, 183, 234–5, 308, 312, 313–14, 319, 321
Russian Platform 313–14

Saaremaa 3–*4*
Sacabamba 1
Saccopharyngoidea *72*
saccus communis 270
sacral rib, *see* ribs
sacrum 77
Salientia 78, 233–4, 235; *see also* Anura
Salmoniformes *72*
Santana Formation 30
Sarcopterygii, sarcopterygians
   anatomy 57–9, 67–8, 71–9, 194–8
   characters 71, 194–5
   classification 82, 251
   fossil 194–236
   habitat, mode of life 5–23
   phylogeny *80*–1, 241–2
   Recent 71–82
   relationships 80–2, 241–2
   stratigraphical distribution *24*
   see also Crossopterygidae, crossopterygians
Saudi Arabia 297, 323
Sauropsida, sauropsids 79, *80*–1, 82
scale
   in acanthodians 174, *179*
   in actinistians *72*, *74*, *197*, 199
   in actinopterygians 68–9, *182*, 184, *186*, 195
   in anaspids 100–*1*
   in arandaspids *86*–7
   in astraspids *87*–8
   in chondrichthyans *53*, 61, 64, 135–7, 141–*3*, 147–51
   circumorbital 123–*4*
   in early classifications of fishes 188, 311
   evolution of 54, 68, 131–3, 276–9
   fulcral 69–70, *182*, 184, *see also* fulcrum
   in galeaspids *116*
   growth 126, 131, 276–9, 303–4
   in heterostracans 89, 92–3, 95
   histology, *see* histology
   in lungfishes 75, *206*–7

median (dorsal, ventral) ridge 89, 92, *101*–2, 103, *106*–7, 113
   in onychodontiforms *196*, 199
ornamentation, *see* ornamentation of dermal bones and scales
   in osteolepiforms *217*
   in osteostracans *106*–7
   in panderichthyids *224*–5
   peg-and-socket articulation of 68–*9*, 184, *186*, *191*
   in porolepiforms 202, 205
   in placoderms 153, *158*, *163*, *167*
   in rhizodontiforms 214–*15*
   shape
      cycloid (rounded) 68, 72, *74*–5, *195*–7, 199, 202, 205, *206*–7, 214–*15*, *226*, 228, 229
      diamond-shaped 68–*9*, *87*–8, 93–3, *106*–7, 174, *179*, *182*, *186*, 187, 202, 205, *206*–7, *217*, *224*–5
      rod-shaped *86*, 100–*1*
      small, mesosquamose (growing) *136*–7, *143*, 174, *179*, 184, *186*
      small, microsquamose (non-growing) *53*, 61, 64, *116*, 123–6, 149–*51*
   structure
      cosmine-covered, *see* cosmine
      ganoid 68–*9*, *182*, *186*, 187
      placoid 54, *62*, 131–2, *136*, *143*, 150–*1*
   in tetrapods *226*, 228, 229, 234
   in thelodonts 123–6
   see also odontode, ornamentation
scapula, scapular ossification 58, 79, 232
scapular blade 77
scapulocoracoid *53*, 58, 60, 70, *73*–4, 77, 133–4, *136*, 146, 152, 153, 155–6, *161*, 163, 166, 169, *175*, 177, *186*, 187, 204, 209, 216, 220–1, *224*–5, *226*–7, 228, 229, 241, 244
Scat Craig 325
scenario 40
scleral ossification *109*–10, 153–4, 239, 241
sclerotic ring, *see* plate (sclerotic)
sclerotome, *see* mesoderm (sclerotomic)
Scorpaeniformes, scorpaenids *72*, 302
Scotland 7, *9*, 14, *21*, 26–7, 101, 106, 123, 234, 283, 297, 303, 308–12, 319, 325
Scyliorhinidae 62–*3*
sedis mutabilis (taxon) 38
segmental theory, segmentalism 252–61
'sel'canal *109*–11, 118, 271
semidentine 54, 131–2, 152, 155–6, 170–*1*, 184, 275–8, 329

Semionotidae, semionotids *182, 193*
sensory line
  calcified rings in 64–*5*, 140, 145–6
  canals 51–*2*, 71, 73, 90, *93*, 97–8,
    106–7, 113, *116*–7, 120–1,
    123–*4*, 162–3, 171, 225, 227,
    241
  commissural transverse *93, 116*–7,
    120; *see also* commissure
  generalized vertebrate pattern of
    134–5, *253*
  grooves *86*–8, 100–*1, 152*, 155, 157,
    184, 234
  as guideline to dermal bone
    homologies 66–8, 261–3
  innervation 49, 59, 135, *see also*
    nerve (lateralis)
  inter-scale 175, 245
  infraorbital 66, *106*–7, *116*–17, 135,
    152, 158, 164–5, 176, *202, 206*,
    207, *212*, 222, 226–8, 247,
    265–6
  jugal *202*, 222
  main lateral 66, *116, 152, 176, 224*,
    247
  mandibular 66, *176*, 184, 198, *202*–3,
    206–7, 226
  neuromasts 45, *48*–9, 51, *52*, 55, *66,
    80*–1, *253*, 259
  origin of *253*, 256
  pattern 134–5, *152, 176, 202, 226*,
    261
  pineal *93*, 97
  pores *52, 65, 117, 126*, 162–3, *206*,
    227
  postmarginal *152*
  postoral 135
  postorbital *152, 202, 224*, 226
  preopercular *66*–7, *182*, 184
  questioned in hagfishes 44, *46*–7
  relation to the pore canal system of
    cosmine 107, 112
  role in dermal bone formation 66–8
  supramaxillar *176*
  supraoral *152, 158*
  supraorbital 88, *92*–3, 95, 97, 107,
    116–17, 123, 135, *152*, 161,
    163–4, 172, *176, 192, 202*,
    206–7, *213*–14, *224*, 226, 247,
    263
  supratemporal, *see* commissure
  in tetrapods 52, *226, 227, 234*
septal rods 43
Serpentes *80*–1
Serpukovian 24
Sessel's pouch *252*, 256
Severnaya Zemlya 115, 296, 321
sexual behaviour 139, 302, 305–4
sexual dimorphism 64–5, 135, 163,
  305–6
Seymouriamorpha, seymouriamorphs
  232–3, 235

sharks
  monophyletic versus paraphyletic 61,
    62–3
  orbitostylic 61, 245
  see also Elasmobranchii, Neoselachii
Sharpey's fibres 36
shield
  dorsal 85–*6*, 88, 89, 239, 241
  endoskeletal 107–12, 117–19, 122–3
  ethmoidal *206*
  exoskeletal 85–*6, 87*–8, *89*–93,
    106–7, *116*–117, *118, 122*
  parieto-ethmoidal 201–2, *212*–13
  postparietal *201*, 221
  ventral *86*–7, *89*–93, 98, 241, 261,
    301–2
Siberia 88, 96, 99, 148, 163, 250, 290,
  294–7, 321
Siberian Platform 83–4, 321
Sichuan 121
Silurian
  palaeobiogeography 294–7
  radiation 291–2
  stratigraphy *24*
  vertebrate faunas 2–5, 89–127,
    148–9, 172, 181–4, 195
Siluriformes 72
Sinolepidae, sinolepids 166, *167*–8, 296
sister-group, -species, -taxa 37
skeleton
  appendicular, *see* fin (paired), girdle,
    limb
  axial 45, 55, 57, 95, 111, 153, 156,
    160, 164, 177, 187, 204, 208,
    228, 257; *see also* arculia,
    centrum, vertebra, fin (unpaired),
    tail
  branchial, *see* gill (arch)
  calcitic 281
  cartilaginous, *see* cartilage
  dermal, *see* exoskeleton,
    lepidotrichia, plate, scale
  neurocranial, *see* neurocranium
  phosphatic, *see* bone, cartilage
    (calcified)
  visceral, *see* splanchnocranium
skull
  in gnathostomes *45*, 55–7, *66*–8,
    128–34
  in hagfishes *45*–6
  in lampreys *45*, 49
  segmentation of 252–61
  see also exoskeleton, neurocranium,
    splanchnocranium
skull-roof
  in actinopterygians *66*–7, *189*
  homologies in 67, 261–3
  in osteichthyans *66*–7
  in placoderms *152*–3
  in sarcopterygians 196, *197*–8, *202*,
    206–7, *215*, 221–2, *224*
  in tetrapods 226–7, 228, 231–*4*

slicing technique 32–*3*
slime gland 44–*5*
sof tissues (preserved in fossils) 27–*30*,
  102, 136–7
somite, *see* mesoderm (somitic)
South China 121, 296
South Africa 297, 323
South America 150, 189, 296–7, 325
South Wales 314
Spain 99, 296
sphenethmoid, *see* ethmosphenoid
Sphenodontia *80*–1, 82
spinal cord 45, 49, *52*, 57–8, 94, *119*
spinal nerve, *see* nerve (spinal)
spinal process 113, 122–3
spinal roots, *see* nerve (root)
spine
  fin
    paired 173–4, *175*–81
    unpaired 54, 64–*5, 136, 140*–5,
      173–*4*, 184, 329–1
  frontal 139, 144–*5*
  haemal, *see* basiventral
  intermediate 133, 173, *176*–7,
    178–81
  mandibular *145*–6
  neural, *see* basidorsal
  nuchal 144, *145*–6
  postbranchial 100, *101*–2, 103
  prepectoral *176*, 178
  see also Acanthodii, Actinopterygii,
    Chondrichthyes
spiracle 45, 51–*2*, 56, *62*, 94, 96, 165,
  187, *271*–2, *301*–2
spiracular canal *190*–1, *195*, 220
spiracular cavity *271*–2
spiracular cleft, slit 49, 56, 225, *153*
spiracular gill, *see* gill
spiracular groove *174*, 176, *185, 190*–1,
  246
spiracular opening 49, 56, 61, *182, 193*,
  228
spiracular platelets *219*–20, 257
spiral intestine, valve 49, *53*, 59, 69,
  147, 166, 299–300; *see also*
  typhlosole
Spitsbergen 5–*6*, 26, 106, 115, 155–6,
  187, 288, 296, 315–16, *319*–21
splanchnocranium *53*, 55, *see also* gill
  (arch), hyoid (arch), mandibular
  arch
spleen *45*, 49
splenial 67, *74*, 79, *197*; *see also*
  infradentary
Squaliformes, squaliforms 61, *62*–3
Squalomorphii, squalomorphs 61, *62*–*3*,
  151
Squamata, squamates *80*–2
squamation, *see* scale
squamosal 67, *77*, 79, *190*–1, *202*–3,
  215, 222, 224, 226, 228, 272
Squatiniformes *62*–3

St Petersburg 313
Stairway Sandstone 84
stapedial foramen *227*
stapes 56, *77–8*, *227–8*, *234*, *236*, 249, 272–3; *see also* hyomandibula
statoconiae 187
statoliths 176, 187, *242*; *see also* otolith
stele 283
stenobasal fin, *see* fin
Stensioellida 148–171
Stephanian *24*
Stephanoberyciformes *72*
Stethacanthidae, stethacanthids *19–20*, 136, 139–41, *146–7*, *148–9*, 302, *305–6*
Stockholm 32, 316–18
stomach 103, 124, 127, 183, 240, 288, 297, *299*
stomodeum *47*, 77–9, 247, *253*
stratigraphy
    distribution of taxa in *24, 26*, 38, 83–5, 99, 115, 121, 123, 127, 148–50, 172–3, 181–3, 193–4, 199–200, 205, 210–11, 223, 225, 233–5, *286–8*
    use of early vertebrates in 127, 308, 314 319–20, 323, 326, 332
striated muscles *30*, 80–1
Struniiformes, *see* Onychodontiformes
Stufenreihe 318
subaponevrotic vascular canals *88*, *92*, 94, 112, *116*, 120
sublingual rod *208*, 219–20
submandibular *202*, 228, 247
subnasal cartilage *45*
subopercular *66–7*, *202–3*, 207, 218, *226*, 228
subsidiary gill covers, *see* gill (cover)
Subterbranchialia *146–7*, *148–9*
suction-feeding, *see* feeding
sulcus (in otoliths) 187
superognathal (anterior, posterior) *153–4*, *155–7*, *161–2*, *163–4*, 165
supinator process *248–9*
supracleithrum *66–7*, 77, 204, 207
supracoracoid canal (foramen) *185*, 187, 205, *209*, 213, *220*, *225*, 228, 244
supraglenoid canal (foramen) *185*, 187, 205, *209*, 213, *220*, *225*, 228
supramaxilla *66*
supramaxillar sensory line, *see* sensory line (supramaxillar)
supraneural 58, *75*, 187, 200, 204, *219–20*
supraoral sensory line, *see* sensory line (supraorbital)
supraoral field *108*, 111, 113
supraorbital 67, 200
supraorbital line, *see* sensory line (supraorbital)
supraorbito-tectal *217*, 221

suprapharyngohyal *255*, 257
suprapterygoid articulation, process *128*, 130, *219–20*
suprascapular ossification *175*, 177
supratemporal *66–7*, 183, 213, *222*, *224*, *226*
supratemporal fossa *192*
supratemporal commissure, *see* commissure
surangular *66–7*, *78–9*, *192*
survival (of taxa) 292–3
suspension-feeding, *see* feeding
suspensorium 189, *190–1*, 193
swimbladder *72*, *see also* air bladder
Symmoriida, symmoriids *24*, 137, 139–41, 144, 147–9
Symmoriidae 189–41, *148–9*
symphysial teeth, *see* tooth
symplectic 56, 71, *193*, *197*, 199, *203*
symplesiomorphy 37
synapomorphy 37
synapticulae *43*
synarcual 61–4, *144–7*, 153, 155–6, *169*, 171
Synthetic Theory 287
systematics
    as an artefact in extinction patterns 288–90
    and biogeography 40
    evolutionary 37–8, 288, 293
    phenetic 37, 288
    phylogenetic (cladistic) 38–40, 237, 252, 294, 317–18, 326, 332

tabular 67, *197*, *202*, 213, *215*, *222*, *224*, *226*, *227–8*, *231–4*
Taemas 323
Tafilalt *15*, 323
tail
    diphycercal 50, 58, 66, 70, 72, 76, 87, 92–3, *116*, 123, *132*, 134, 145, 198, *201*, 216
    epicercal 50, 53, 58, 66, 70, 107, 111, *116*, *132*, 134, 153, 173, 198–200, 216, *238–9*
    hind *43*, *283*
    hypocercal 92, *100–1*, *102–4*, *116*, 123, *238–301*
    isocercal *72*, *132*
    structure 38–9, 44, 46, 48, *70*, 95, *226*, 240, 256, 283
    see also fin (caudal)
Tallinn 318
Tannuaspidida, tannuaspids 115, 296, 321
tannuaspid province 296
tarsal *77*, *233–4*, 268–9
Tartu 313
taxon, taxa 37–8, 281, 288
    basal 82
    character versus common ascent in definition of 39–40, 252, 281, 288

as clades 38
    monophyletic 37; *see also* paraphyly versus monophyly
    paraphyletic 35, 37; *see also* paraphyly versus monophyly
    polyphyletic 37; *see also* diphyletism, polyphyly
tear duct, *see* nasolacrymal duct
techniques (used in fossil studies)
    of drawing 34
    of photography and radiography *28*, 34
    of preparation 27–34, 330–2
    special 32–34
    see also acid preparation, grinding section, slicing technique, tomography
tectal *217*, 230
tectal cartilage *45*
teeth, *see* tooth
telencephalon 52, *109*, 273
Teleostomi, teleostomes 130, *242*, 246, 251, 315
Temnospondyli, temnospondyls 231–3, *235*, *236*, 249, 272–3
temporal fossa(e) *80–1*
tenaculum, *see* clasper (frontal)
tentacle 44–5, 85, 94
terminal arch, segment, *253–7*, 261
terminal opening 253
terrane 296
Tertiary 200, 290, 293
tessellate pattern, *see* exoskeleton (tessellate)
tessera(e) 38–9, 84, *87–8*, *97–8*, *106–7*, 112, 120, 131, 133, 153, *171*, *279*, 290, 305
Tesseraspidida 90
Tetraodontiformes *72*
Tetrapoda, tetrapods
    anatomy *52*, 76–9
    characters 76, 225
    choanae in *77–8*, *226–8*, *264–6*
    classification 82, 235
    dermal bone homologies in 261–3
    Devonian 10–18, 225–30, 233–5
    earliest trackways 234, *306*
    fossil 225–36
    habitat of the earliest 10–11, *13–14*, *18*
    limb origin 266–70
    middle-ear *227–8*, 272–3
    origin of 285
    phylogeny 80–1, *236*
    Recent 76–81
    relationships *80–1*, 246–50
    tear duct in 264–6
Tetrapodomorpha, tetrapodomorphs 223, *248–9*, 251, 329
Texas *15*, *23*, *142*, 293
Thelodonti, thelodonts
    anatomy 123–6, *263–4*

Thelodonti, thelodonts (*cont.*)
  characters 127
  endemic to Gondwana in the Late
    Devonian 127, 325
  habitat, mode of life 3–6, 127, 288
  histology *126*, 277–8
  monophyly questioned 127, 237,
    238, 240, 288, 311
Thelodontida 127, *see also* scale (in
  thelodonts)
threat device 140, 302, 305–6
three-taxon statement 332
Thyestiida, thyestiids 3–4, 107, *112*,
  113–115, 278, 296
Tibet 121, 294
tibia 76–7, *220–1*, *226*, 229, 233, 268–9
Timan 115
Tolypelepidida, tolypeleids 90, 98
tomography 32–*33*, 330
Tonkin 121, 321; *see also* Bac Bo
tooth
  adsymphysial, *see* parasymphysial
  apical 48–9
  attachment bone in *202*, 205 *215*,
    217, 221, *224*, 225
  bicuspidate 78
  cladodont 135–7, 140, 147
  dendrodont 200, *202*, 205
  diplodont 141–2, 149
  eusthenodont *215*, 221
  family 61–2, 130, 144, *178*
  folded 200, *202*, 205, *211*, 213,
    214–*5*, *217*, 221, *224*, 225, 229,
    230, 247; *see also* plicidentine,
    Rhipidistia
  histology; *see also* acrodine, dentine,
    enamel, enameloid
  horny 44, *46*, 47–9, *85*, 104–5
  hybodont 142–3
  labyrinthodont 225, 229
  origin of, 130, 285, *see also* dental
    lamina
  parasymphysial 67, 138, *196*, 198,
    203, 214–15, *218*, 222, 329
  pedicellate 78, 233–4
  plate 54, 64–5, 68, 72, 75–6, *91*, 95,
    144–*5*, 163–4, 184, 195, 205,
    207–8, *211*–14, 249; *see also*
    entopterygoid, prearticular,
    vomerine tooth plates
  polyplocodont 213, 214–*5*, *217*, 221,
    224–5, 229–*30*, 250, 329
  spiral, whorl *125*, 137–8, *178*, 241,
    302
  symphysial 137–*9*, 147–9, *178*,
    180–1, 188
  see also dentine, enamel, enameloid,
    metadentine, orthodentine, pulp
    cavity,
Torpediniformes 62–3
total evidence 326
Tournaisian *24*, 205, 234

trabecles 50, 55, 104–*5*, *129*–30, *253*,
  255, 257, 259
trabecular dentine, *see* dentine
trabecular commissure, *see* commissure
trace fossils 233–4, *306*–7
Transantarctic Mountains 323
Traquairaspidiformes, traquairaspidiforms 90, *96*, 98–9
tract
  digestive 29, 49, 59, 70, 73–*4*, *125*
  olfactory 47, *52*, 58, *91*, 94, 118–*19*,
    122, *154*, *170*, *185*, *203*, 204,
    208, *219*, *273*
  optic 47, 94, 110, *154*, *159*, *170*, *203*,
    208
Tree of Porphyry 318
Tremadoc *24*, 83–4
Tremataspididae, tremataspids 31,
  38–*9*, *106*, 112–*14*, 115, 205,
  283, *304*
trematic rings 45
Trematosauria, trematosaurs 230,
  293–*4*
Triassic 15, *24*, 150, 183, 190–2,
  200–*1*, 233, 288–91, 315
trigeminal cavity *119*
trigeminofacial chamber 257
trigeminus, *see* nerve
Tristichopteridae 221, *223*
Tristichopterinae *223*
tubular dentine, *see* dentine
tubules
  of dentine, *see* dentine
  rostral 200, 207–*8*, *211*–14, 248–9,
    329
  symphysial 205, 207–*8*, *212*–13, 302;
    see also rostral tubules
Tula 13, *18*
Tunicata, tunicates 42–*3*, 237, 254, 256,
  281, *283*
Turkey 12, *17*, 143, 323
turtles, *see* Chelonii
Tuva 113, 149, 294–*6*, 321
tympanic membrane *77*, *80*–1, 228,
  271–3
tympanic ring 21
typhlosole 48–*9*, 59

Ukraine 99, 115; *see also* Podolia
ulna 76–7, *220–1*, *226*–9, 233,
  267–70
uniserial paired fin, *see* fin (uniserial)
unpaired fin, *see* fin (unpaired)
Uppsala 315–16
urinary bladder *80*–1
urinary duct *53*
Urodela, urodeles
  anatomy 57, *77*–8, 234
  choanae partly endodermal in 204,
    247, 256, 265–6
  and diphyletism of tetrapods 200,
    246–7, 316, 318, 328

fossil 233–*4*
limb ontogeny in 246, 270
middle ear *271*–12
Recent 78–9
relationships 78–9, 235
see also Caudata
urogenital apparatus *53*, 59
urohyal 195, *203*–4, *208*, *219*–20, 225,
  244–5
uroneural 70–1, *72*, 195
Utah 322
utriculus 270

vein
  brachial *108*–11
  cardinal
    anterior 111, 118
    posterior 71, 111
  extrabranchial *108*, 111
  hepatic *108*, 111
  jugular *109*, 111, 118–*9*, *154*, *159*,
    170, *174*, 187, *203*, 239
  marginal *109*, 110
  pituitary *159*
  pulmonary 71, 73, *75*–6, *77*
  see also vena cava
velar skeleton *45*–6, *48*, *108*, 130,
  258–*9*, 285
velum *46*, 48–*9*, 50, 108, 110, 111,
  258–*9*
vena capitis lateralis, *see* vein
  (jugular)
vena cava 71, 73, *80*–11
Venezuela 325
ventral fissure, *see* fissure
ventral myodome, *see* myodome
ventral vertebral arch *219*–20, *231*,
  234; *see also* intercentrum
ventricle, *see* heart
ventrolateral fenestra *203*–4, *211*, 213,
  266
Vermilion River 322
vertebra(e)
  arcualia in lampreys *45*, 49
  basic structure in gnathostomes *45*,
    53, 57, 134
  cranial 252–*3*, 258–60
  see also arcualia, basidorsal,
    basiventral, centrum,
    intercentrum, interdorsal,
    interventral, pleurocentrum,
    skeleton (axial)
Vertebrata, vertebrates
  characters 42–4, *45*, *89*–1
  classification 82, 25
  fossil 83–236
  origin of 281–4
  phylogeny 237–41
  Recent 42–82
  relationships 42–5
  stratigraphical distribution *24*
vestibular fontanelle 186, *219*

vicariance biogeography
 applied to early vertebrates 296, 320
 principles 40
Victoria (Australia) 323
Victoria Land 323
Vietnam 6–8, 121, 166, 172, 294–6,
 321
visceral skeleton *see* splanchnocranium
visceral contribution to braincase 55,
 257–8
Viséan 14, *24*, 234
vomer *66–7, 74–5, 78–9,* 184–*5, 203*–4,
 207–8, 210–11, 213–14, *218,*
 221, *224–5, 226, 227–8, 232,*
 234, 244, 249, 257, *331–2*
vomerine tooth plate *75, 208*
von Baer's law 287, 312

Wales 90, 294, 315
Wardie 311

wax model 32
water-vascular system 281
Welsh Borderland 314, 315
Wenlock *4, 24,* 84, 90, 115, 121, 151,
 290–1
Westoll lines *206,* 209, *210,* 303
Westphalian *24*
Wichita Group 15, *23*
Wildungen 320
Wood Bay Formation *5–6,* 286–7
wrist 76, 268
wuttagoonaspid-phyllolepid province
 296
Wuttagoonaspida, wuttagoonaspids
 156–7, *162,* 172–3, 296
Wyoming 84, 322

Xenacanthiformes, xenacanthiforms,
 'xenacanth sharks' *23, 128*–9,
 132–4, 135, 141–*2,* 147–9, 292

Xinjiang, *see* Tibet
xiphosurans 283

yolk sac 74, *197,* 199
Youngolepiformes *212*–3
Yunnan 121, 166, 172, 213
Yunnanolepiformes, yunnanolepiforms
 27, 166, *167–8,* 172–3, 296,
 321
yunnanolepid-galeaspid province 296

Zechstein *24*
Zeiformes *72*
Zenaspidida, zenaspids *106,* 113–*14*
zone of polarizing activity *269*
Zürich 311
Zweifall 325